ELECTRICAL PRINCIPLES
for the Electrical Trades

8TH EDITION

ELECTROTECHNOLOGY
SERIES

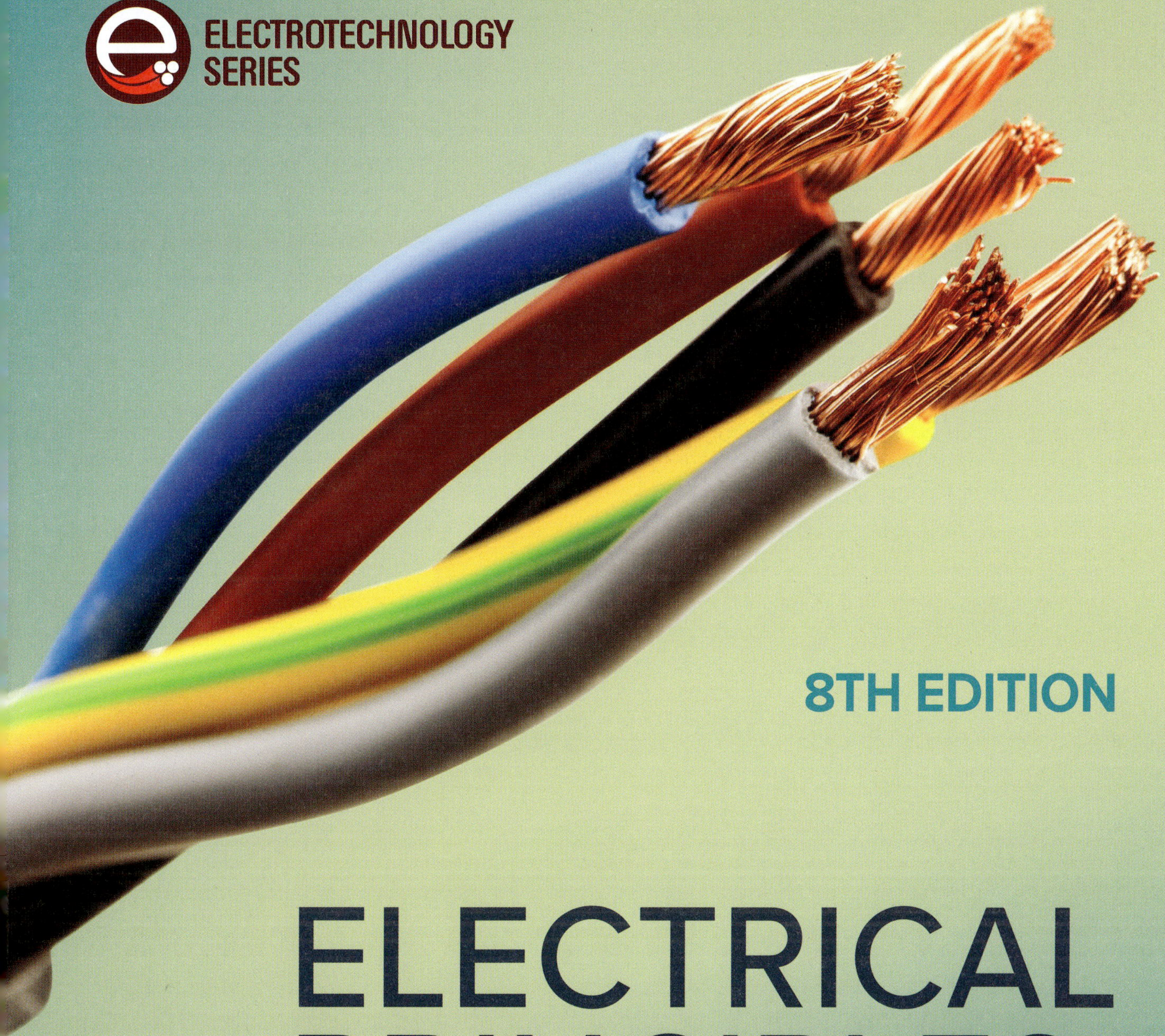

8TH EDITION

ELECTRICAL PRINCIPLES
for the Electrical Trades

Mc
Graw
Hill

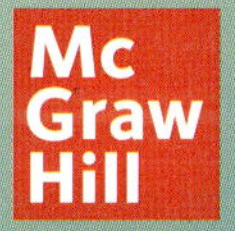

JIM JENNESON, BOB HARPER, BOB MOORE
SIMON DAND, TONY JONES, MICHAEL SCOTT

Reprinted 2022, 2023, 2024, 2025

Authors: Jim Jenneson, Bob Moore, Bob Harper, Simon Dand, Tony Jones, Michael Scott
Title: *Electrical Principles for the Electrical Trades,* eighth edition
ISBN: 9781743767801 (paperback)

NATIONAL LIBRARY OF AUSTRALIA

A catalogue record for this work is available from the National Library of Australia

Published in Australia by

McGraw-Hill Education (Australia) Pty Ltd,
Level 33, 680 George Street, Sydney NSW 2000

Publisher: Matthew Coxhill
Production editor: Elmandi du Toit
Copy editor: Paul Hines
Permissions editor: Debbie Gallagher, Legend Images
Technical editor: Steve Waye
Indexer: Straive, India
Interior design: Simon Rattray, Squirt Creative
Cover design: Simon Rattray, Squirt Creative
Cover image: Mariusz Szczygiel/Shutterstock
Typeset by Straive, India
Printed in Singapore by Markono Print Media Pte Ltd

CONTENTS IN BRIEF

CONTENTS

CHAPTER 14

PREFACE

WELCOME TO THE EIGHTH EDITION of *Electrical Principles for the Electrical Trades.* This well-established and much respected text has undergone a range of changes to bring all its chapters into alignment with the current National UEE—Electrotechnology Training Package. The chapters have been refreshed and updated to make them as relevant as ever to the modern electrical student and tradesperson. Along with new content and updated figures, the chapters' learning objectives have been streamlined to focus on the critical competencies and include short questions for each topic to help students understand new topics as they are introduced.

We have taken into account the many suggestions and comments from teachers, students and industry to produce a comprehensive publication to complement McGraw Hill's *Electrical Wiring Practice* publication. The content has been presented in an easier-to-read format and all revisions have been undertaken by teachers who are currently involved in the training and assessment of Electrical Trade students in Australia.

It has been a pleasure to work with Tony Jones, Mike Scott and Steve Waye, who have provided practical hands-on backgrounds combined with a sound technical knowledge of the subject matter. Their insights and experiences in the classroom, along with their combined talents, have shaped both the sequence and content of this new edition.

The content of this publication is designed to augment and assist in the classroom-based and blended delivery of Electrical Trade qualifications in Australia. The content is not exhaustive but represents comprehensive details of the fundamental principles and theories that will assist and inform an Electrical Trade student.

I would also like to thank the staff members of McGraw Hill who have contributed to this project. I also acknowledge the ongoing support of McGraw Hill in the production of leading resources in electrical training.

Simon Dand

ABOUT THE AUTHORS

Jim Jenneson, Bob Harper and **Bob Moore** provided the original content upon which this text is based. This eighth edition would not exist without the work of these authors, whose distinguished careers spanned many years.

Simon Dand has almost 30 years' experience in the electrotechnology industry and is currently a full-time Electrical Trades teacher for TAFE NSW. His career started as an electrical apprentice in the coal-fired power stations on the Central Coast of NSW and includes many years working as both a software engineer developing industrial process control software and a maintenance electrician in the food and beverage FMCG industry.

Simon's passion for imparting electrical knowledge led him to TAFE NSW where he has taught for almost 20 years both part-time while as a practicing electrician and now full-time as an acting Head Teacher. During his career at TAFE NSW, Simon has taught post-trade, diploma, and TVET students along with electrical apprentices on a wide range of subjects, including computers and computer networks, electrical theory, electrical practices and industrial automation.

Simon holds a qualified Electrical Contractors Licence, an Electrical Trade Certificate, a Certificate III in Electrical Fitting, an Advanced Diploma in Electrical Engineering, a Bachelor's Degree in Computer Science and a Certificate IV in Workplace Training and Assessment.

When he isn't working, Simon likes to spend his spare time with his wife and four boys.

Tony Jones has worked in the electrotechnology industry for many years. His electrical trade and industrial electronics experience includes working with PLC controls, OCR and detection systems. He has performed equipment maintenance, breakdown, diagnostics and repairs in servicing and production environments for more than 30 years.

He has worked in the VET sector for the past 16 years; 11 of those as an Electrical Trade teacher for ElectroGroup, Canberra Institute of Technology, and as Senior Technical Trainer with NECA Training ACT. Tony has served as a technical advisory committee member, NECA Excellence Awards Judge and industry mentor for apprentices.

Over the past two years Tony has continued to develop learning and assessment resources for the UEE20 Training Package.

Michael Scott is currently a Senior Academic staff member at the Toi Ohomai Institute of Technology (formerly the Bay of Plenty Polytechnic) in the Bay of Plenty, New Zealand. His responsibilities include the moderation and development of new qualifications and resources. Michael works closely with the Industry Training Organisation and is involved in the integration of all 16 national providers into one organisation known as 'Te Pukenga'. He is also a subject matter expert for a number of external bodies.

Michael began his working life as an electromechanical apprentice with the Royal Mail in England. He gained a Higher National Certificate in Electrical Engineering at Newham College of Further Education in East London, where he took up his first teaching position while studying. Michael taught at the college for 15 years, gaining a BA (Hons) in Adult Education.

Michael emigrated to New Zealand in 2005, where he has been teaching at the Toi Ohomai Institute of Technology for the past 16 years.

Michael is honoured to contribute his expertise to this book.

TEXT AT A GLANCE

SETTING A CLEAR AGENDA ▶

Each chapter begins with a list of objectives that the reader can aim to achieve, as well as a list of prerequisite knowledge required before commencing the chapter.

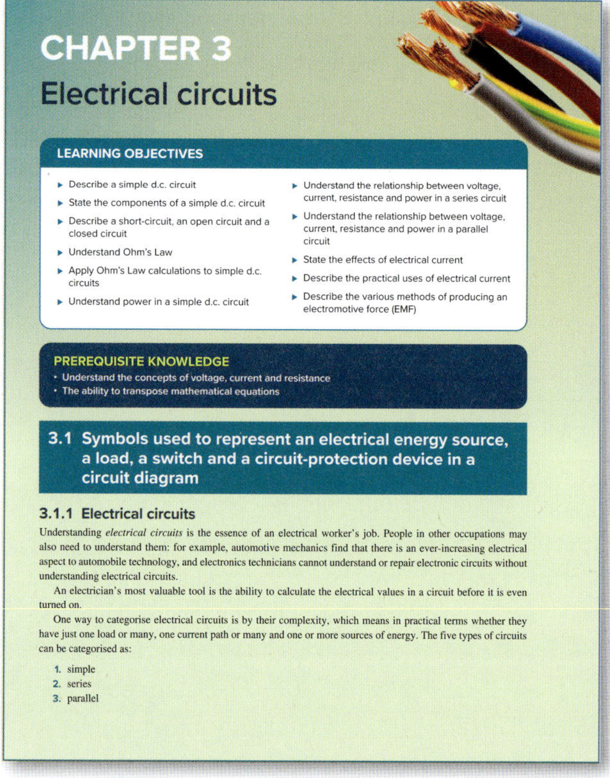

CHAPTER 3

Electrical circuits

LEARNING OBJECTIVES

- Describe a simple d.c. circuit
- State the components of a simple d.c. circuit
- Describe a short-circuit, an open circuit and a closed circuit
- Understand Ohm's Law
- Apply Ohm's Law calculations to simple d.c. circuits
- Understand power in a simple d.c. circuit
- Understand the relationship between voltage, current, resistance and power in a series circuit
- Understand the relationship between voltage, current, resistance and power in a parallel circuit
- State the effects of electrical current
- Describe the practical uses of electrical current
- Describe the various methods of producing an electromotive force (EMF)

PREREQUISITE KNOWLEDGE

- Understand the concepts of voltage, current and resistance
- The ability to transpose mathematical equations

3.1 Symbols used to represent an electrical energy source, a load, a switch and a circuit-protection device in a circuit diagram

3.1.1 Electrical circuits

Understanding *electrical circuits* is the essence of an electrical worker's job. People in other occupations may also need to understand them: for example, automotive mechanics find that there is an ever-increasing electrical aspect to automobile technology, and electronics technicians cannot understand or repair electronic circuits without understanding electrical circuits.

An electrician's most valuable tool is the ability to calculate the electrical values in a circuit before it is even turned on.

One way to categorise electrical circuits is by their complexity, which means in practical terms whether they have just one load or many, one current path or many and one or more sources of energy. The five types of circuits can be categorised as:

1. simple
2. series
3. parallel

WORKED EXAMPLES ▶

Each chapter supports the theoretical aspects by providing practical applications of the theory covered. The theory is illustrated by fully worked examples. These examples give students a template to use when completing similar exercises.

EXAMPLE 3.5

An electric heater rated at 2400 W is switched on for an average of 6 hours per day for the whole of winter, from 1 June to 31 August.

How much electricity is used in that period by the heater?

Step 1. Calculate the time in hours:

$$\text{Days} = 30 + 31 + 31 = 92$$
$$\therefore t = 92 \text{ days} \times 6 \text{ hrs}$$
$$\therefore t = 552 \text{ hrs}$$

Step 2. Calculate the power in kilowatts:

$$\text{Power} = \frac{2400 \text{ W}}{1000} = 2.4 \text{ kW}$$

Step 3. Calculate kilowatt hours:

$$W = P \times \text{hr}$$
$$= 2.4 \text{ kW} \times 552 \text{ hrs}$$
$$W = 1324 \text{ kWh (ans)}$$

CHECK YOUR UNDERSTANDING ▶

'Check your understanding' boxes within the chapters encourage understanding and reinforce learning of the chapter content.

CHECK YOUR UNDERSTANDING

3.26 What are the methods and units that are used to calculate electrical power?

3.27 If electrical current passing through a lamp is doubled, what will happen to the electrical power used by the lamp?

3.28 What is the commercial reason for using the kilowatt hour?

Open circuit characteristic (OCC) shows the relationship between generated EMF at no load and the field current at a given fixed speed. The OCC curve mimics the magnetisation curve and is similar for all types of generators. OCC curve data is obtained by operating the generator at no load and keeping a constant speed. It is also known as 'magnetic characteristic' or 'no-load saturation characteristic'.

The OCCs for a separately excited d.c. generator can be seen in Figure 7.13(a).

Shunt generators

In a *shunt generator*, the armature voltage is applied to the generator field that is connected in parallel with the armature. The connection is shown in Figure 7.14. The machine is said to be 'self-excited' because, when running, it does not rely on an external power source for field excitation. The shunt generator does, however, require a field from some source in order to start generating.

The open circuit characteristics for a shunt d.c. generator can be seen in Figure 7.15(a) which shows the relationship between field current and generated voltage. In the load characteristic curve shown in Figure 7.15(b), the decreasing voltage characteristic that is typical of the separately excited machine can still be seen, although in the shunt machine the voltage drop is more pronounced at full load. This is because the load is also in parallel with the field and a decreasing output voltage means a decreasing field current, which further decreases the output voltage.

◀ KEY TERMS

Key terms are highlighted in bold where they first appear in the text and are defined in the Glossary at the end of the book.

SUMMARY

- The essence of an electrical worker's job is to understand electric circuits.
- There are five basic types of electrical circuits: simple circuit, series circuit, parallel circuit, compound circuit and complex circuit.
- Electrical circuit diagrams use standard symbols and labelling techniques so that electrical workers find them easy to understand.
- There are many common electrical abbreviations used with electrical circuits. These include 'V', 'I', 'R' and 'P'.
- A basic circuit has three main parts: source, path and load.
- Simple circuits are basic circuits that may also contain control devices such as a switch and protective devices such as a fuse.
- Electrical workers use measuring devices to measure the values of electrical properties in circuits.
- An ammeter is an electrical measuring device that is used to measure current.
- A voltmeter is an electrical measuring device that is used to measure voltage.
- Ammeters are connected in series.
- Voltmeters are connected in parallel.
- Generally speaking, there are three conditions of electric circuits: open circuit, closed circuit and short-circuit.
- Open and closed circuits are normal conditions of circuits, whereas a short-circuit is normally a fault condition.
- When dealing with very large and very small values in electrical circuits, engineering notation and electrical prefixes are used to make those values easier to represent.

◀ SUMMARY

Each chapter ends with a comprehensive summary listing the core concepts covered, making it an excellent tool for revision and reference.

END-OF-CHAPTER QUESTIONS

3.1 State the five main components that are required to construct a basic electrical circuit.
3.2 Explain what is meant by 'open circuit', 'closed circuit', 'short-circuit' and 'fault'.
3.3 State Ohm's Law in your own words.
3.4 If the voltage in a basic d.c single path circuit is increased, will the current increase, decrease or stay the same?
3.5 Explain what is meant by the 'power rating' of a device and what the effects of exceeding this value would be
3.6 If the resistance of a basic d.c. single path circuit is reduced, will the output power of the circuit increase, decrease or remain the same?
3.7 Explain the terms 'basic protection' and 'fault protection', as described in AS/NZS 3000.
3.8 What are the two effects of electricity that are always present when an electrical current is flowing?
3.9 What is the difference between a primary and a secondary cell?
3.10 Is the output power of an electrical device equal to the input power plus the lost power or minus the lost power?
3.11 In a series circuit, what is the relationship between the voltage drop across all the components and the supply voltage?
3.12 Give an example of where a series circuit is used in the electrotechnology industry.
3.13 In a parallel circuit, what is the relationship between the current in each parallel branch and the current drawn from the supply?
3.14 Give an example of where a parallel circuit is used in the electrotechnology industry.
3.15 Give an example of where a series/parallel (compound) circuit is used in the electrotechnology industry.

◀ END-OF-CHAPTER QUESTIONS

End-of-chapter questions act as a revision and checkpoint for each chapter's skills, knowledge and understanding.

CHAPTER 1

Use routine equipment/plant/technologies in an energy sector environment

LEARNING OBJECTIVES

- Understand the structure of the electrotechnology industry and the different types of jobs within it
- Understand electrical supply and distribution
- Identify how circuits are arranged and the use of protective methods in meeting safety requirements
- State the difference between alternating current (a.c.) and direct current (d.c.)
- Understand the calculation and measurement of voltage, current, resistance and power in practical circuits
- Define magnetism and electromagnetic induction and explain their application
- State transformer operating principles and their application
- Identify and discuss hazards associated with electrical systems and apparatus
- Understand how to work safely with electricity
- Use workplace documentation
- Understand energy sector tools, equipment and technology
- Understand work health and safety (WHS) and occupational health and safety (OHS) and the related legislative requirements
- Understand the use of operating instructions for tools, equipment and technologies
- Identify sustainable work practices
- Discuss workplace policies, procedures and instructions

PREREQUISITE KNOWLEDGE

Understand how work health and safety regulations, codes and practices are applied in the workplace

INTRODUCTION

This chapter is an overview of the electrotechnology industry. As well as explaining how the industry is currently configured, it discusses the generation and distribution of energy and the maintenance of electrical and electronic equipment. It also covers the hazards involved in working with electricity, as well as various ways of minimising the associated risk. The chapter outlines the types of knowledge, skills and qualities that are required to work in the industry. The information presented here is suitable for electrical apprentices or others who are embarking on a career within electrotechnology.

The chapter is arranged in three sections:

1.1 *The electrotechnology industry*. This section provides a general description of the industry and identifies the various types of job within it, as well as the procedures that direct how those jobs are carried out.

1.2 *Electrical concepts*. This section introduces key concepts for electrical apprentices and core information for Certificate II in Electrotechnology (Career Start) students.

1.3 *Useful terms, definitions and industry knowledge*. This section introduces some common industry terms, codes of practice and tools and equipment. It also explains the purpose and application of workplace policies, procedures and documentation.

The information that is introduced here is explored and explained in greater detail in other chapters of this textbook and its companion volume *Electrical Wiring Practice*.

1.1 The electrotechnology industry

The electrotechnology industry deals with the generation, transmission and distribution of electricity and the design, installation, repair and maintenance of electrical and electronic equipment. The hundreds of thousands of people who are employed in the industry across Australia work in the following sectors:

- appliances
- business equipment
- computers
- data communications
- electrical
- electrical machines
- electronics
- fire protection
- instrumentation
- refrigeration and air-conditioning
- renewable/sustainable energy
- security technology.

These sectors can be grouped into the *electrical industry* and the *electronics industry*.

1.1.1 The electrical industry

The electrical industry is organised into the following four subsectors:

1. generation, transmission and distribution
2. industrial and mining
3. commercial construction, installation and maintenance
4. domestic construction, installation and maintenance.

Generation, transmission and distribution

This subsector is responsible for electrical power generation, as well as its transmission and distribution from power stations to end users. It employs a range of people, including electrical engineers and operators, electrical lineworkers and electricians. These professionals design, construct and maintain the *electrical network* (or *grid*) that supplies electricity across the country.

The generation, transmission and distribution subsector has become increasingly complex over the last 20 years or so. Its assets and infrastructure have transferred from public to private ownership, and the advent of *renewable technology* has brought many new entrants. These range from companies that provide large-scale wind or solar generation to individuals who supply the grid with solar power fed in from their roof.

Industrial and mining

The industrial and mining subsector employs electricians to install and maintain machinery and electrical wiring and equipment at industrial installations, factories and mining sites. Larger companies engage full-time electricians, while smaller ones use specialist electrical contractors.

This subsector also includes instrumentation, which involves the installation, maintenance and calibration of instruments used in processing and manufacturing. These instruments measure data such as current, flow rates, temperature and location. The functions of the industrial and mining subsector include working with control wiring and equipment to regulate motors, valves and other devices used in electrical equipment such as conveyor belts, smelters, lifts and escalators. Also employed in this subsector are refrigeration mechanics, who are responsible for the installation and servicing of air-conditioning and refrigeration equipment for both large systems and individual appliances. Other categories of industrial and mining are motor servicing and repair, and small appliance servicing.

Commercial and domestic construction, installation and maintenance

In terms of the types of work done by the people employed in them, the commercial (offices, large retail outlets, shops, restaurants and hotels) and domestic (residential dwellings) subsectors can be bracketed together. Both involve the construction of new buildings, the refurbishment of existing ones and servicing and maintenance. Both employ licensed electricians to install and maintain wiring as well as control and protection equipment, lighting, socket-outlets, motors, and appliances for cooking, heating and cooling.

1.1.2 The electronics industry

While the electrical industry generally works with higher power rated electrical equipment and apparatus, the electronics industry tends to focus on electronic components. However, it is not unusual for either sector to cross over and perform tasks associated with the other. For example, people who are employed in data and telecommunications are required to install associated wiring and cables to connect equipment as well as to test, commission and maintain the equipment itself. Electrical licence holders now often obtain specialised electronic qualifications (and vice versa with electronic licence holders).

The subsectors of the electronics industry are:

- data and telecommunications
- radio communications
- computer systems
- security systems
- industrial electronics
- commercial electronics
- consumer electronics.

Data and telecommunications involves the installation of copper cable and optical fibre across the country, and has seen many changes in recent years with the introduction of the NBN (National Broadband Network). It also includes the installation of associated equipment and apparatus such as telephone and data systems, networking equipment, and satellite and microwave equipment (including towers). Those employed in data and communications require appropriate telecommunications cabling registration if their work involves connecting customer cabling to the telecommunications network.

Radio communications involves the installation, maintenance and servicing of equipment associated with radio systems for radio and television stations, aviation and marine radio, emergency services and other public radio networks such as taxi systems and UHF/VHF CB repeaters.

Computer systems involves the installation, servicing and maintenance of IT equipment, including computers, data hubs and uninterruptable power supplies (UPSs).

Security systems involves the installation, servicing and maintenance of commercial and domestic security and monitoring equipment such as CCTV, infrared and motion detectors and X-ray scanners.

Industrial electronics involves the installation, servicing, maintenance and repair of electronic equipment associated with processes such as electronic motor controls and industrial computer process integration for factories.

Commercial electronics involves the servicing, maintenance and repair of photocopiers, cash registers and similar equipment used in offices and retail outlets.

Consumer electronics involves the servicing, maintenance and repair of consumer electronic equipment such as televisions and microwaves.

1.2 Electrical concepts

1.2.1 Electrical supply and distribution within a building or premises

Buildings and premises include factories, schools, shopping centres, cafés and residences. These structures typically contain electric lights, power points (socket-outlets) and a range of different appliances (e.g. water heaters, cooking appliances and air-conditioners).

The term 'electrical supply' refers to the energy source that supplies a network which distributes electricity to buildings and premises. 'Energy source' simply means where the energy comes from. Sources can be either renewable such as wind farms, solar farms and water turbines or non-renewable such as coal-fired power stations.

Figure 1.1 shows a generation, transmission and distribution network.

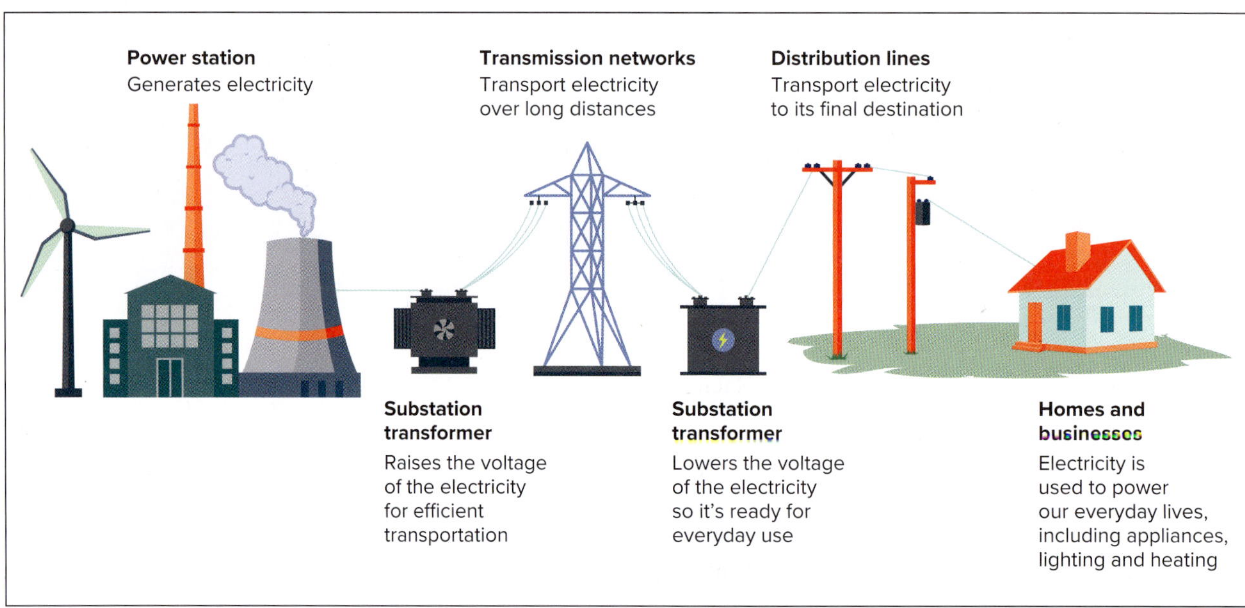

FIGURE 1.1 **A generation, transmission and distribution network**

1.2.2 Arrangement of circuits

Before supply can be provided to a new installation ('installation' means any electrical equipment installed from an electrical supply), an electrician needs to submit an application for supply or notice to the local energy distributor. The electrician should also check the manufacturer installation specification before installing new equipment or a cable circuit in direct sunlight. The relevant standard for determining and verifying the **voltage drop** of a selected cable is AS/NZS 3008.

When installing equipment, plant and technologies, electricians must refer to AS/NZS 3000 Wiring Rules to ensure their work is compliant. An example would be when trying to find out the correct height for installing a cable above a driveway.

In domestic installations, there will generally be two lighting circuits and at least two circuits for power, each with multiple points connected on a circuit. For larger appliances like hot water systems, stoves and hotplates, a dedicated circuit is usually installed. These circuits will have separate control and protection.

Figure 1.2 shows a residential floor plan layout and the position of lighting points, light switches and socket-outlets for power.

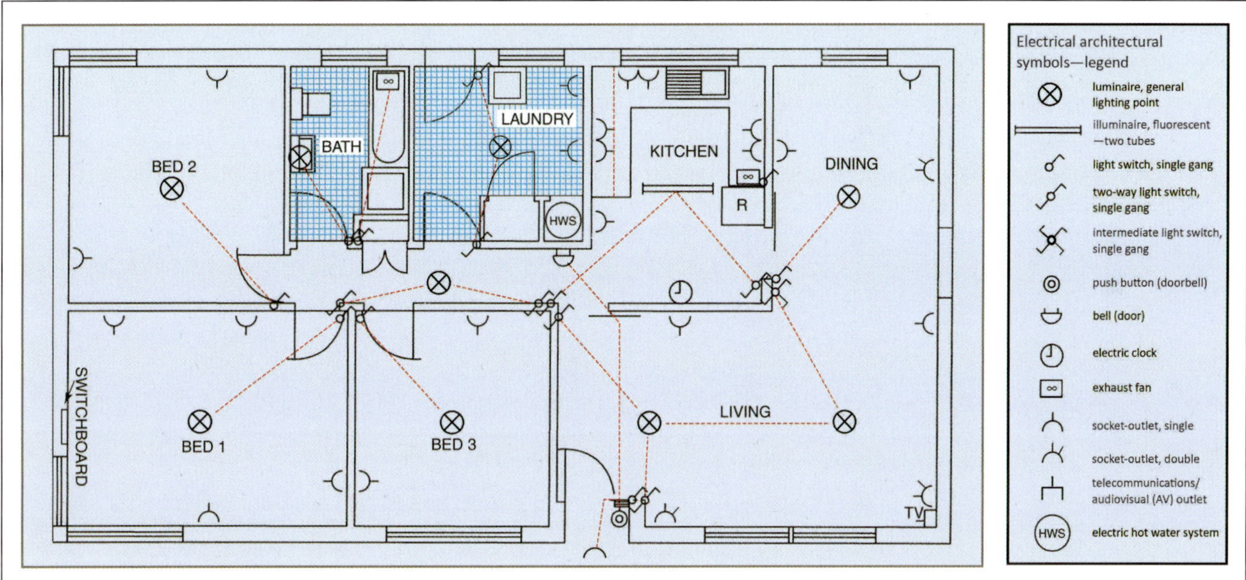

FIGURE 1.2 Floor plan layout for a three-bedroom residence

Cable installation and the coordination of current ratings with those cables and associated protective devices is critical so that, under **load** conditions, the wiring does not overheat, the insulation melt and fire result. A *maximum demand calculation* is carried out to ensure that the current demand of the property is not exceeded.

Electrical installations will have a switchboard (some will have sub-boards), where circuits originate. Switchboards are usually referred to as a 'main switchboard' (MSB) or a 'distribution board' (DB). Located within those boards will be protective devices connected to each circuit.

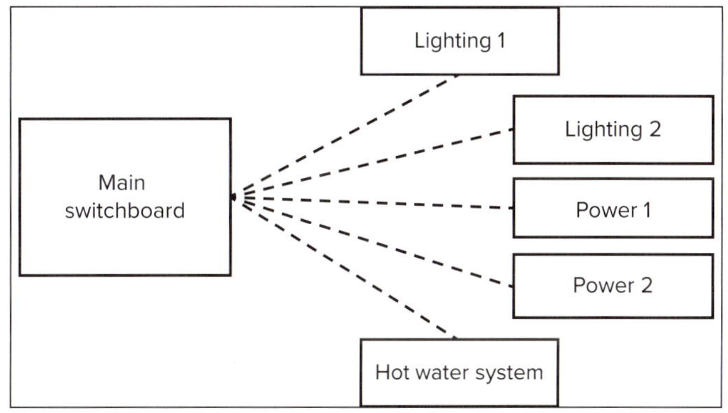

FIGURE 1.3 Line drawing of a main switchboard (MSB) with light and power final subcircuits

Circuits that connect one switchboard to another are called 'submains'. Circular multi-core cable is used for installation of consumer mains and submains in a commercial installation. Circuits supplying end-user equipment are called 'final subcircuits', as shown in Figure 1.3.

Load devices are the working parts that are connected to a circuit. Examples are lighting, a hot water system (as in Figure 1.3), an electric stove or plug-in devices like a toaster, kettle, television or refrigerator. In the case of power circuits, there is a maximum current (the circuit load) that can be sustained safely before a protective device operates.

It is important to appreciate the installation considerations necessary for protection, isolation and control of electrical final subcircuits and what happens if the cable size is too small. Consider Examples 1.1 and 1.2.

EXAMPLE 1.1

A final subcircuit load current is 20 A (amps) protected by a 10 A circuit-breaker. In this situation, the circuit-breaker will operate, disconnecting the supply. This disconnect operation is referred to as a 'trip', with the circuit-breaker tripping under overload conditions.

EXAMPLE 1.2

If the cable wiring is rated at 20 A and the protective device is rated higher at 25 A, under load conditions the cable could overheat and fail before the protective device can operate, resulting in a possible electrical installation fire.

Final subcircuits in an installation normally have five main components:

1. an energy source or supply
2. circuit protection
3. control devices, such as switches and isolators
4. a conducting path, such as cable or wires
5. the load, such as a heater, fan, motor or lamp.

The arrangement of final subcircuits in any installation is similar to the basic d.c. circuit configuration shown in the circuit diagram in Figure 1.4. Circuit diagrams are a pictorial method of recording the connections of a circuit quickly, simply and in a way that is easy to understand. The symbols generally represent a concept rather than simply an object.

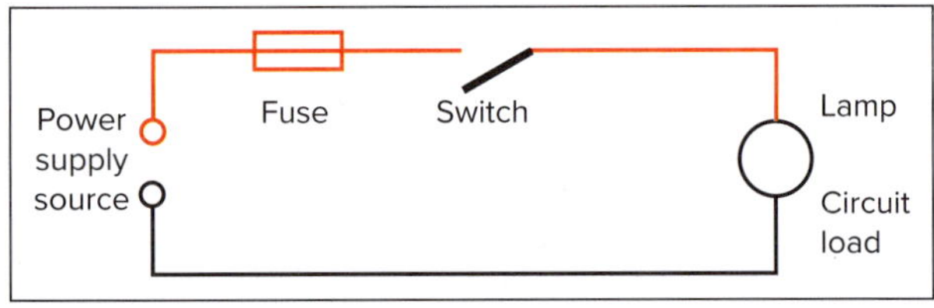

FIGURE 1.4 **Circuit diagram of a final subcircuit**

In terms of complexity, the circuit in Figure 1.4 is a simple series circuit. The switch symbol shown in Figure 1.4 only shows the electrical contact for ON-OFF control, and not how it is operated. There are five levels of circuit complexity: simple, series, parallel, compound and complex. Regardless of its level of complexity, an electrical circuit contains the components listed in the first column of Table 1.1.

TABLE 1.1 **Circuit components, symbols and functions**

Component	Symbol	Function
Supply	a.c. d.c.	Power source for a circuit. It can be either: direct current (e.g. a battery); or alternating current (e.g. a socket-outlet)
Circuit protection	Fuse Circuit-breaker	Devices for quick disconnection of supply to protect the wiring against excessive current

TABLE 1.1 Circuit components, symbols and functions (continued)

Component	Symbol	Function
Switch		Control device electrical contact to turn the load ON and OFF
Wires/cables	Active/Positive Neutral/Negative	Provide a path for electric current between other circuit components
Lamp/light/load	⊗ or ○	Load device for a circuit

CHECK YOUR UNDERSTANDING

1.1 Give three examples of appliances that usually require the installation of dedicated circuit load devices.

1.2 What is another term for the circuits in a switchboard that supply end-user equipment?

1.3 List the five components of an electrical circuit (it is not necessary to state their levels of complexity).

1.2.3 Protection to meet safety requirements

Circuit protection

All circuit components, devices and circuit wiring need to be protected. In electrical installations, circuit design, load devices and wiring protection are critical factors when it comes to the safety of people, livestock and premises: they minimise the risk of electric shock, fire and physical injury.

Examples of circuit-protective devices are miniature circuit-breakers (MCBs), residual current devices (RCDs) and residual current-operated circuit-breakers with overcurrent protection (RCBOs). An RCBO acts as a residual current device and a circuit-breaker to provide both overcurrent and earth leakage protection.

While a type of fuse called an HRC (high rupturing capacity) fuse remains common in commercial and industrial equipment applications, MCBs, RCDs and now RCBOs provide protection for new installations in domestic and commercial environments.

Protective devices in electrical equipment and electronic devices are located at the beginning of the circuit path, between the power source and the power ON switch, as shown in Figure 1.5. Using a light switch in a lighting circuit enables ON-OFF switching control.

In Figure 1.5, circuit protection is provided by a circuit-breaker. (The **ammeter** and the **voltmeter** shown in the diagram are test measurement devices.)

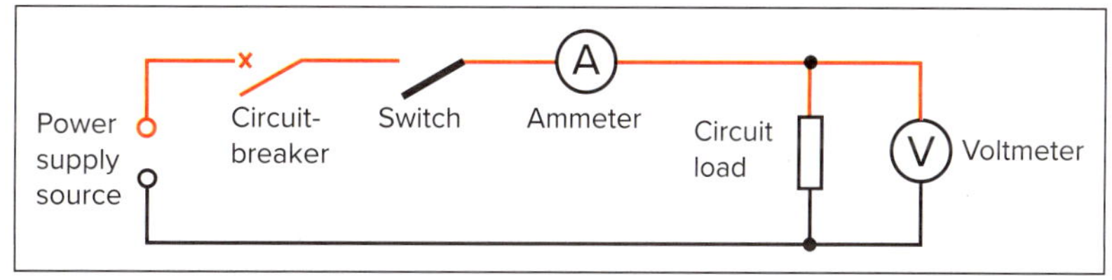

FIGURE 1.5 Protective circuit-breaker located at the start of the circuit

Circuit components, devices and accessories will usually have a specified voltage, current or wattage (or combination of those). If these specifications are exceeded, the component, device or accessory will fail. Fuses, switches and accessories are rated in amps; component ratings for lamps and resistors are usually specified as voltage and wattage. Extension cords, power boards, toasters and motors have a specified current rating in amps and a power rating in watts or kilowatts. These ratings are the basis for the selection of components in extra-low voltage electrical and electronic circuits.

Circuit-breakers can be sourced with different current ratings. While these ratings are important for overcurrent protection and coordination between conductors, another important consideration is the *kA protection rating* against short-circuits, transient voltage spikes, external faults and the effects of lightning strikes. The kA rating for protective devices is determined by how far the supply transformer is from the installation, with the resistance based on cable length. A device with a 6 kA rating can deal with fault currents up to 6000 amps and 10 kA devices can handle fault currents up to 10 000 amps.

CHECK YOUR UNDERSTANDING

1.4 Name four types of devices that are used for circuit protection.

1.5 What does 'RCBO' stand for?

1.6 Explain the purpose of using a light switch in a lighting circuit.

1.2.4 The difference between alternating current and direct current

Electricity comes in two forms, alternating current (a.c.) and direct current (d.c.). The latter is a flow of charge in only one direction while alternating current is a flow of charge whose direction changes periodically.

Alternating current is generated when a *loop conductor* (or *coil*) is rotated within a uniform magnetic field. It is used in preference to d.c. for power transmission, transformers and effects that require **self-** or **mutual inductance**. The electrical supply and distribution network is primarily a.c. and supplies the majority of equipment in domestic, commercial and industrial environments. Alternating current is commonly used in many types of electric motors.

Direct current is produced by several sources, including batteries, fuel cells and solar cells. (Solar panels produce d.c., which is then converted to a.c. via an *inverter*.) More commonly, d.c. is used to supply portable equipment such as torches, laptops, phones and electric power tools.

Alternating current has a repeating **sine wave** (or **sinusoidal waveform**) which cycles through peaks and troughs. The peaks are also referred to as the 'maximum' or 'V_{max}' values for voltage. Direct current has a steady (or almost flat) line representing a constant d.c. value. Terms such as V_{PEAK}, V_{RMS} and V_{AV} are associated with sine waves. The V_{RMS} voltage of a waveform has to be measured and displayed by a meter called a 'true-RMS' multimeter, whereas most low-cost meters measure a.c. voltage as V_{AV} (refer to Figure 1.6).

An alternating current waveform is shown in Figure 1.6(a) as one cycle of a sine wave. The peak, RMS and average values are also shown. The value of either a voltage or current waveform at one instant of time is called the 'instantaneous value'. The RMS (root-mean-square) value in Figure 1.6(a) and (b) is the d.c. equivalent value and is shown as a constant or steady value. Under ideal conditions, a d.c. voltage attains a steady value and remains constant at that value.

The sinewave in Figure 1.6 is one cycle (in Australia, the electrical supply and distribution is maintained at 50 cycles per second, which is called a 'frequency' of 50 hertz). The frequency of a wave is the number of times it repeats itself in one second.

In Figure 1.7(a), d.c., as shown by the waveforms, is the usual energy source for torch and car batteries. Electron movement or 'flow',

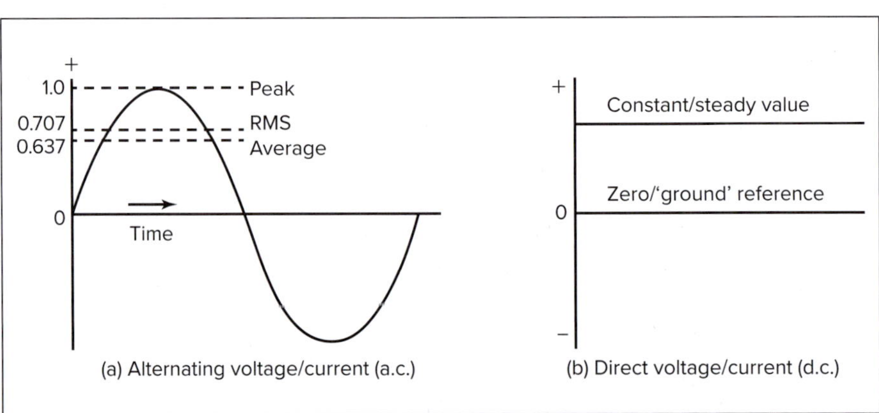

FIGURE 1.6 RMS d.c. value superimposed on a continuous, repeating a.c. sinewave waveform

a phenomenon that is often used in electronics, is from negative (−ve) to positive (+ve). Before the discovery of electrons, the original convention (which is still used today) was a uni-directional movement or flow of current from positive (+ve) to negative (−ve). Figure 1.6(a) shows one cycle of an a.c. waveform, while Figure 1.7(b) shows this as a continuous, repeating a.c. sinusoidal waveform.

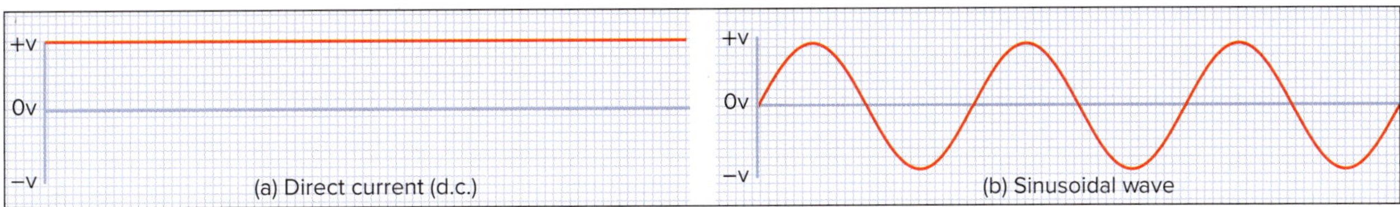

FIGURE 1.7 Waveforms

Figure 1.8 shows a series circuit with a steady d.c. supply from a 12 V car battery to a lamp as the load.

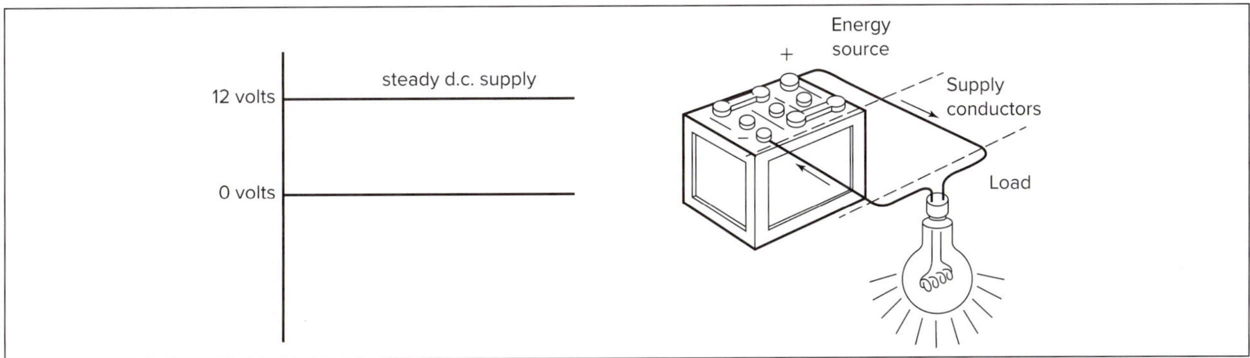

FIGURE 1.8 Series circuit with steady d.c. supply

CHECK YOUR UNDERSTANDING

1.7 State what the electrical supply and distribution frequency is in Australia.

1.8 Explain what is meant by 'wave frequency'.

1.9 List five electrical items which operate on a.c.

1.2.5 Calculation and measurement of voltage, current, resistance and power in practical circuits

Calculation of voltage, current, resistance and power in practical circuits

Generally, the four parameters of a practical circuit that need to be calculated are voltage (V), current (I), resistance (R) and power (P). The acronym VIRP (voltage, current, resistance and power) is useful in representing symbols or values in calculations. If you know two of the values, a third unknown value can be calculated.

The circuit in Figure 1.9(a) shows a supply voltage (V) from an energy source, a load device resistance (R) and the measured voltage across the load, which in this case is the same as the supply voltage. The load current (I) is measured using an ammeter. Figure 1.9(b) shows the battery as the energy source that provides a supply voltage (V), and the lamp as the load device (R), which draws a current (I).

When a **potential difference** (or voltage) is applied to a *load* (resistance), a flow of electricity results in the form of current ($I = \frac{V}{R}$). This flow will result in an alternating current (a.c.) for an a.c. power source and a direct current (d.c.) for a d.c. power source.

The measured current being drawn by the load device can be calculated using Ohm's Law, whose formula forms the basis of many circuit calculations. Ohm's Law states that the current flowing between any two points in an electric circuit is directly proportional to the potential difference between these points and inversely proportional to the resistance of the circuit between these points. Ohm's Law is represented as $I = \frac{V}{R}$.

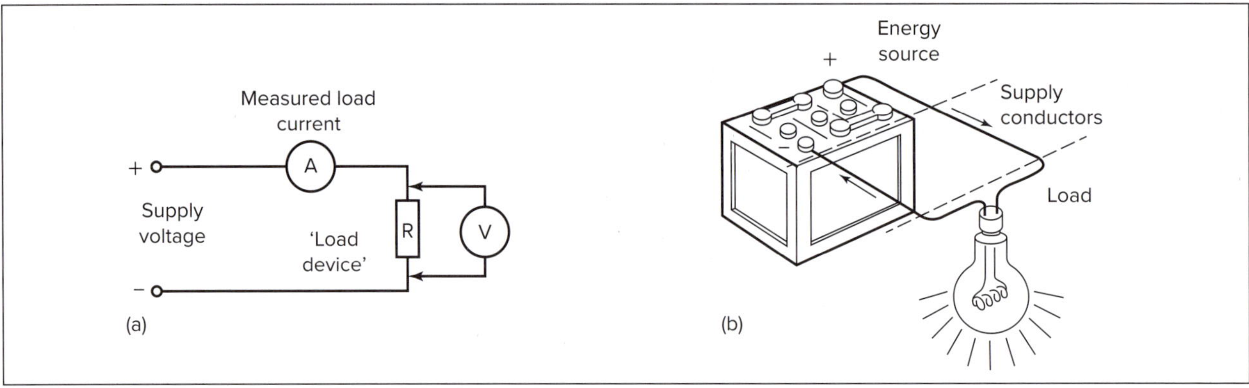

FIGURE 1.9 **(a) Simple device circuit (b) Lamp connected to a car battery**

The power (P) consumed by the circuit can also be calculated by multiplying the voltage by the current, which is represented as Power (P) = Voltage (V) × Current (I).

Loading of a circuit is measured in two values, as maximum current in amps or as power in watts. Power usage over time is referred to as 'energy consumption' and is measured in kilowatt hours (kWh).

EXAMPLE 1.3

Power boards are devices that are commonly used in a domestic environment as a means of providing additional socket-outlets for electrical appliances. A four-socket-outlet power board is shown in Figure 1.10.

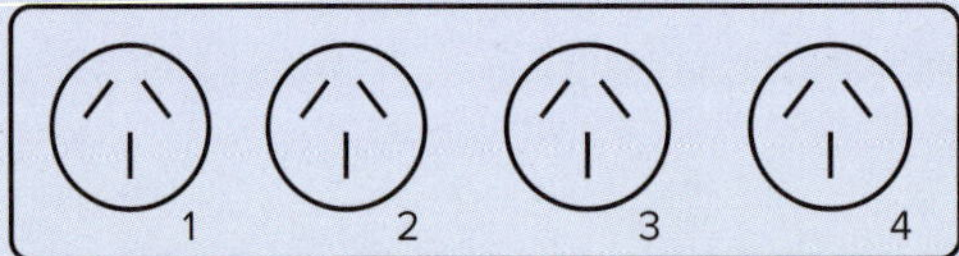

FIGURE 1.10 **Four-socket-outlet power board**

Domestic electrical power boards are rated at a maximum current of 10 amps. Markings on these are sometimes labelled as 240 V, but normally the value of voltage calculations is 230 V (as per AS/NZS 3000). In the following example, if the supply voltage is nominally 240 volts, calculate the power rating in watts:

Power (P) = Voltage (V) × Current (I)
Power (P) = 240 V × 10 A
Answer: P = 2400 W

Plug-in electrical appliances include televisions, electric toasters, kettles, refrigerators and vacuum cleaners. Each of these appliances has a nameplate that states a rating, and this rating is based on how much power is consumed by the appliance. This is referred to as the 'wattage'. Other details are provided on these nameplates, usually including voltage and current.

If two, three or four appliances are plugged into a power board (such as that shown in Figure 1.10), the total wattage and total current drawn under load conditions can be calculated by simple addition, as follows:

$$P_{Total} = P_1 + P_2 + P_3 + P_4 \text{ (kW)}$$
$$I_{Total} = I_1 + I_2 + I_3 + I_4 \text{ (A)}$$

Measurement of voltage, current, resistance and power in practical circuits

Circuit voltage and load current can be calculated or given as values. Single-phase electrical installations have a nominal voltage of 230 volts (previously this was 240 volts, and many pieces of electrical equipment are still labelled thus). But for calculation purposes (unless stated otherwise), a value of 230 volts is used. Circuit values can also be measured using ammeters and voltmeters for extra-low voltage and d.c. circuit applications.

Using multimeters and test instruments

Test instruments are used to measure circuit values such as voltage, current, resistance and power. Taking measurements, using the appropriate test instruments and correctly connecting them into electrical circuits without causing component damage, short-circuits and the risk of burns and electric shock requires specific knowledge and skill.

Figure 1.11 shows two test meters. An ammeter is connected in series with the load (to measure the load current drawn from the supply), and a voltmeter is connected in parallel with the load resistance to measure voltage drop across the resistor. The current in each component of a series circuit is the same. Simple and series circuits have only one current path but parallel circuits can have any number of current paths, and these are referred to as 'branches'.

In Figure 1.11, the lamp symbol (circuit load) shown in Figure 1.4 has been replaced with the resistor symbol.

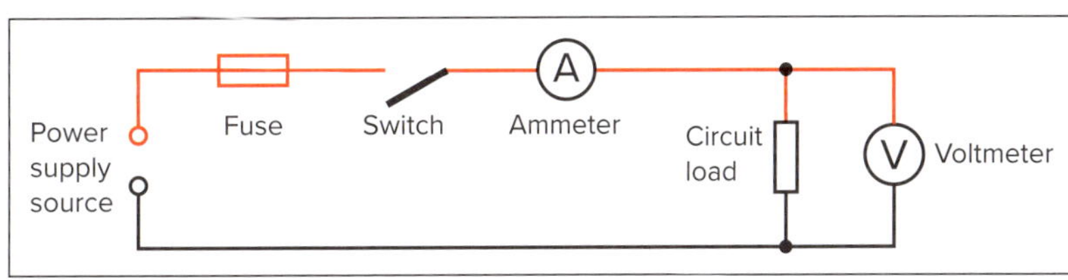

FIGURE 1.11 Measuring current and voltage with an ammeter and a voltmeter

Safety considerations are crucial in the use of test instruments or meters. For example, an ammeter connected incorrectly across a voltage source can cause a *dead-short* hazard or a short-circuit in another part of the circuit. This would at best trip the breaker or blow the fuse. Other considerations in such a situation would be personal safety and whether the equipment was damaged or circuit components stressed.

Whenever using a test instrument such as a multimeter, it is important to ensure that the test leads are inserted into the correct meter terminals and that the correct range and meter setting has been selected. Figure 1.12 shows two examples of how the mode selected on a multimeter results in an illumination of the correct meter terminals to use. An incorrect connection of the leads would result in the sounding of an alarm.

FIGURE 1.12 Meter mode selection for (a) a.c. current and test probe connections and (b) d.c. current and test probe connections
Tony Jones

Figure 1.13 in Workplace Scenario 1.1 shows the same measuring probe connection for five different measurement readings. In Figure 1.13(a), the a.c./d.c. voltage readings can be taken without changing the test probe position. Figure 1.13(b) shows that four possible tests (diode check, continuity, resistance and capacitance) can be selected without changing the test leads.

The meter test mode is selected via a function button which allows the user to toggle between the four modes that are available. Knowledge and experience are required to take live measurement readings safely. Human error such as a wrong contact with the circuit under test could result in damage to the meter, to the circuit and possibly to the user. Apprentices under strict supervision during Stage 3 (the third year of training in a four-year apprenticeship) will be taught how to perform these tests; however, being taught how to perform them and actually being permitted to perform them are different things.

WORKPLACE SCENARIO 1.1

A licensed electrician with metering endorsement has been authorised to install a replacement meter. Figure 1.13(a) shows the correct meter setting when testing for supply (voltage). The electrician needs to focus her mind on the job. If she were to become distracted and incorrectly set the meter to the setting shown in Figure 1.13(b), an electrical incident could result.

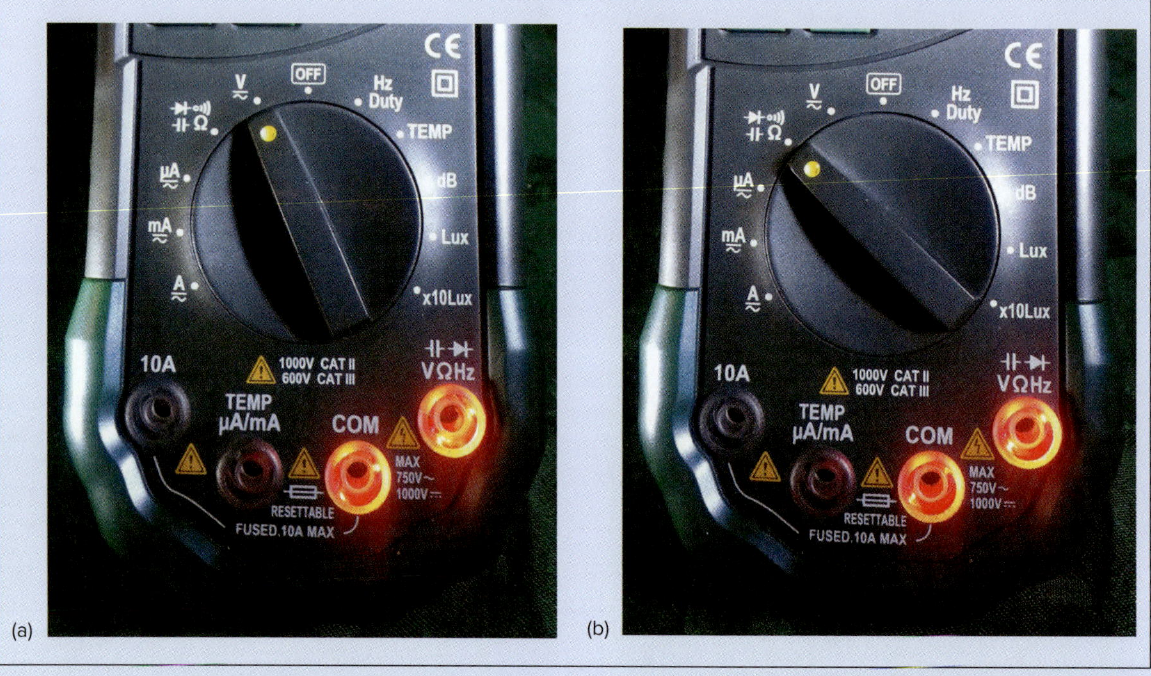

FIGURE 1.13 **(a) Meter mode selections for a.c./d.c. voltage and (b) meter mode selections for resistance**
Tony Jones

CHECK YOUR UNDERSTANDING

1.10 A four-year apprenticeship usually covers four stages of learning. In which stage is electrical testing taught?

1.11 What are the dangers of incorrectly connecting a test instrument to an electrical circuit?

1.12 Identify the three main safety considerations for 'live' electrical circuit testing.

1.2.6 Applications of magnetism and electromagnetic induction

Magnetism, **electromagnetism** and **magnetic induction** provide useful insights into electrical installations, power distribution, electrical transformers, generators and motors.

One way of conceptualising magnetism and electromagnetism is to consider all materials as consisting of domains. When a magnet or electromagnet influences an unmagnetised (or demagnetised) **ferromagnetic** material, that material changes from being in a randomly oriented domain state to being magnetised. That is to say, its domains line up magnetically. The term for this is 'magnetic induction'.

Figure 1.14 illustrates material domains in magnetised and unmagnetised/demagnetised states.

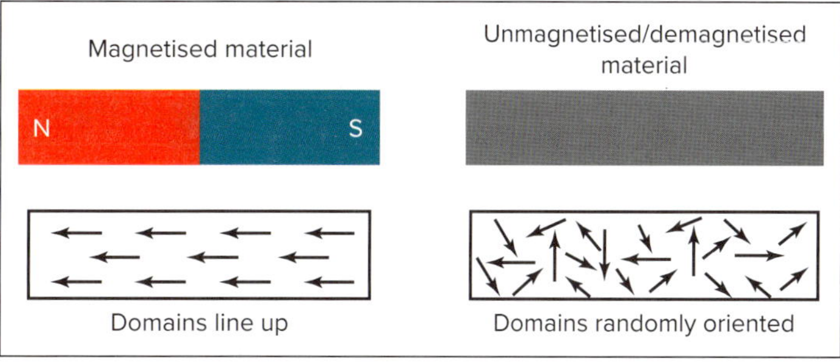

FIGURE 1.14 Material domains in unmagnetised/demagnetised and magnetised states

Magnetism is an invisible force that influences ferromagnetic materials made of iron, cobalt or nickel. When materials become influenced, they become magnetised and form a North (N) and a South (S) pole. This is known as a 'dipole' (meaning two poles). A magnetic field of force (also known as 'magnetic flux') radiates external to the magnet itself. By convention, this field is often referred to as 'unbroken magnetic lines of force'. The lines radiate from the North pole of a magnet and return via its South pole, as illustrated in Figure 1.15.

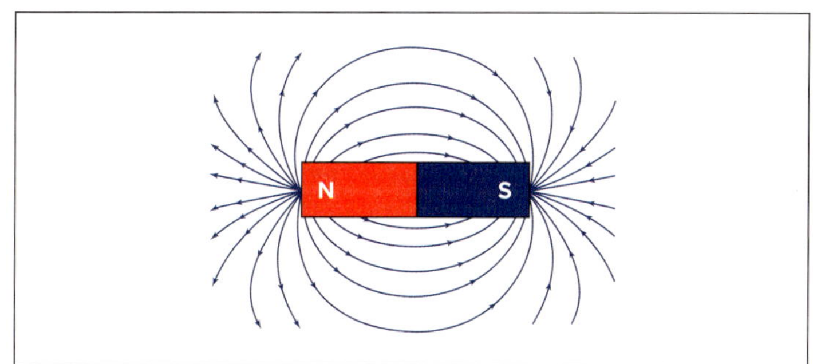

FIGURE 1.15 Bar magnet with lines of force

When a current flows through a conductor or coil, a magnetic field results. There is a relationship between the magnitude of current and the magnitude of the magnetic field produced such that an increase in the former causes an increase in the latter. Similarly, when current decreases, the magnetic field also either decreases in strength or collapses. When magnetic fields come into contact with one another, this magnetic force causes physical movement via either an attraction or a repulsion. This can be applied as a principle of operation to a range of electrical devices and test instruments, as illustrated in Figure 1.16.

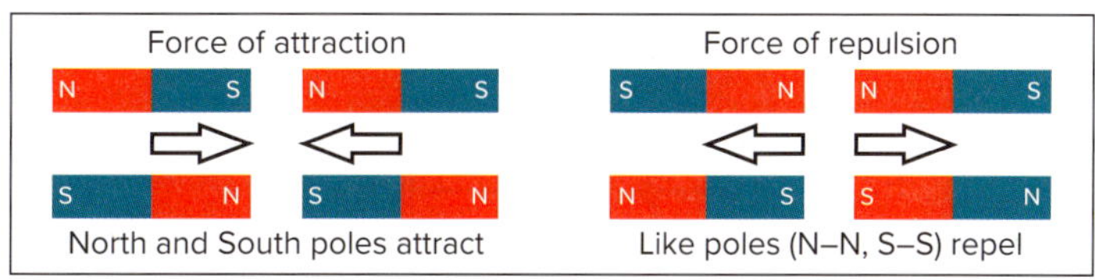

FIGURE 1.16 Magnetic field attraction and repulsion

While a magnet cannot be turned on or off, a temporary magnet (or electromagnet) can be created by using electricity. When current flows through a conductor or coil, a magnetic field forms around a wire or component. This is called 'electromagnetism'.

The advantage of an electromagnet is that the magnetic field can be controlled; it can be switched on and switched off when required. Due to the relationship between current and magnetic flux, an increase in current flow through a conductor or coil results in an increase in the magnitude of the magnetic field created. The same principle applies to coils and windings. Just as the unlike magnetic poles of a bar magnet attract and the like poles repel, the interaction of these fields can be utilised in other devices such as motors and generators.

Examples of the application of the force of attraction are the operation of reed switches in door and window security, electric motor contactors and relays required for smaller load current and circuit control applications.

Circuit-breakers used for circuit protection operate on the principle of repulsion under fault conditions. A short-circuit will often result in large fault current values. Two separate coils wound around the contacts of a miniature circuit-breaker (MCB) form part of its internal construction. As an increase in current produces a magnetic field, the short-circuit current through the MCB produces two powerful and opposing magnetic fields of force, rapidly pushing (blowing) the contacts apart, disconnecting the supply and protecting the final subcircuit from damage. The two coils produce magnetic fields which exhibit a force of repulsion like the bar magnets shown in Figure 1.16.

Electrical devices that operate on magnetic and electromagnetic principles include residual current devices (RCDs), circuit-breakers, relays and contactors. Variable speed drives need to use cables specifically designed to overcome the effects of electromagnetic interference.

CHECK YOUR UNDERSTANDING

1.13 What is the reaction between like poles?

1.14 What is the reaction between unlike poles?

1.15 What is magnetism?

1.2.7 Transformer operating principles and their application

A **transformer** is an alternating current (a.c.) electromagnetic device which has no moving parts in its construction and operates on the principle of mutual induction. Transformers are designed to be capable of *stepping up* or *stepping down* the voltage to a load device. The voltage at the secondary terminals of a transformer can be higher, lower or the same voltage as the primary supply voltage.

Transformers have an *input winding* and an iron core magnetically coupling to an *output winding*. Multi-tap transformers have multiple output windings. The input winding is usually referred to as the 'primary side' and is connected to an electrical supply. Output windings are known as the 'secondary side' and are connected to the load. The primary coil is energised by an a.c. supply, which magnetises the iron core.

As shown in Figure 1.17, the constantly changing a.c. input supply magnetises and demagnetises the iron core at a rate of 50 cycles per second, a frequency of 50 hertz. This ongoing process converts electric energy from the primary to magnetic energy in the iron core, inducing a voltage in the secondary coil.

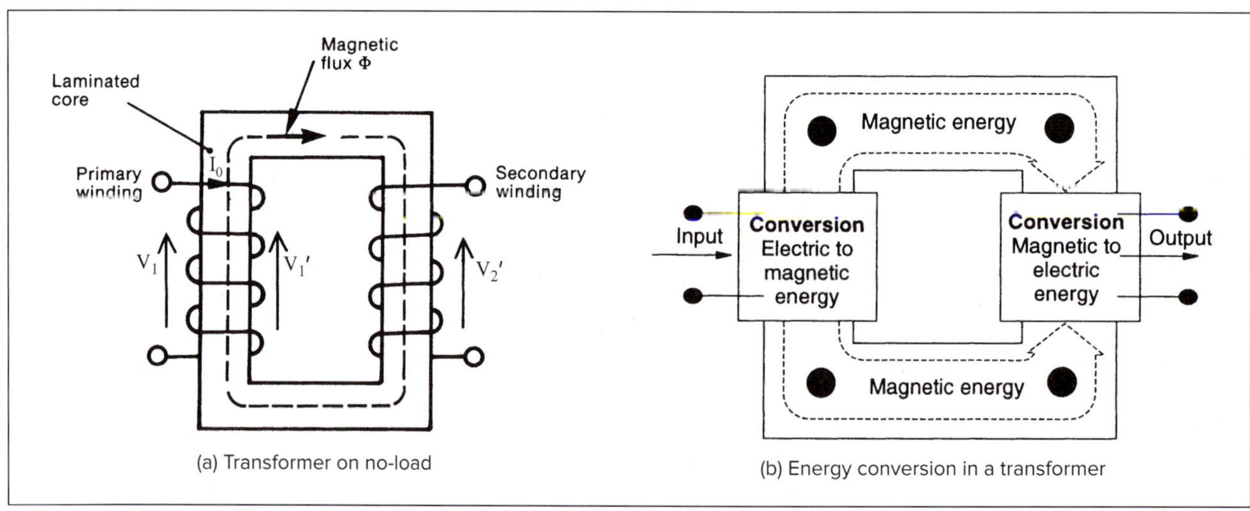

FIGURE 1.17 **The conversion of electric energy into magnetic energy**

The induced voltage on the transformer's secondary side can be higher, lower or the same value as the primary supply voltage, depending on the number of coil turns.

Figure 1.18 shows how transformers are used in the supply system to change the voltage to the required level. See text discussion for explanation of numbers.

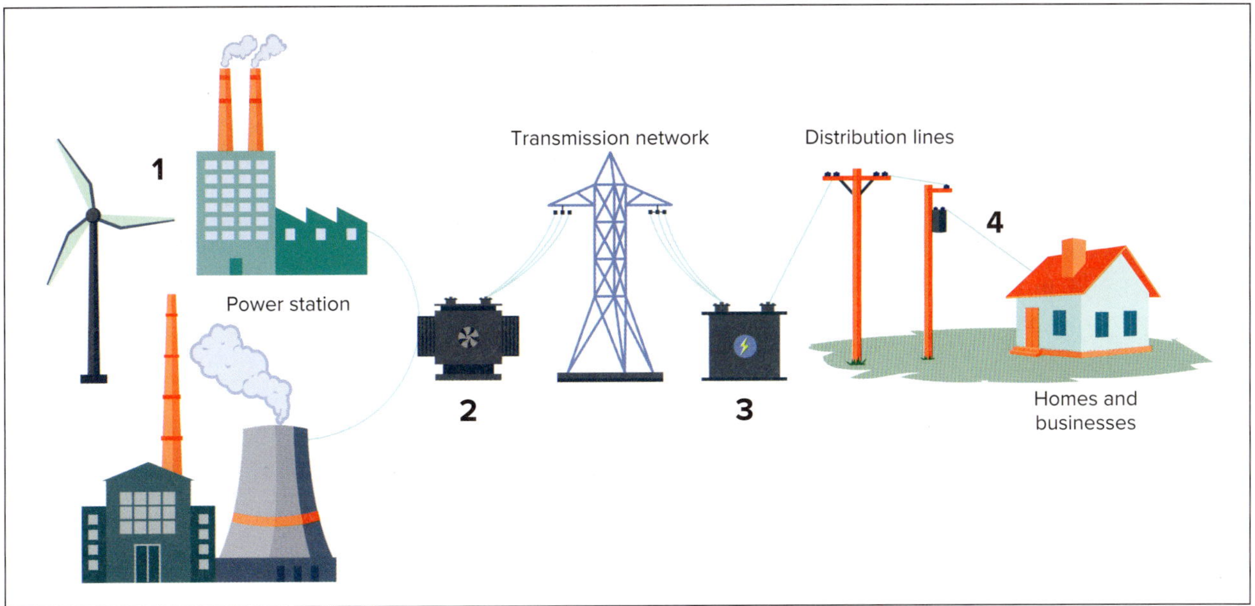

FIGURE 1.18 **How transformers are used in the supply system to change the voltage level**

The process represented in Figure 1.18 can be described in four steps:

1. The power station or plant, whether coal-fired, wind, solar, hydroelectric or another energy source, generates electricity.
2. The transformer steps up the voltage for transmission to reduce the current, and therefore the power losses due to heat (the greater the value of current, the greater the power losses in transmission). Transmission lines transport the electricity over long distances.
3. Substation transformers step down the transmission voltage to a lower value. However, the value at the substation is still much higher than those needed for homes and businesses.
4. Distribution lines carry electricity via overhead lines, in this example between neighbourhood distribution transformers and pole transformers, to further step down voltages to end users for electricity supply to homes and business.

Electrical installations also use special-purpose step-down transformers in (or associated with) the main switchboard arrangement or load centre to monitor incoming supply voltage and current to an electrical installation. Working on a principle of induction, potential transformers (PTs) monitor voltage while current transformers (CTs) monitor the current on each supply phase of an installation.

Common applications of step-down transformers include supplying power for laptop computers, architectural lighting or downlights in homes, retail and commercial installations and also in electronics, industrial machinery and equipment for printed circuit boards (PCBs), transforming voltage levels from 240 V a.c. down to 24 V, 12 V and 5 V a.c. values.

It is important to be aware that transformers in old installations may contain PCBs (polychlorinated biphenyls). These chemicals are a safety hazard and can cause skin rashes, chemical-induced acne and even liver cancer.

CHECK YOUR UNDERSTANDING

1.16 Briefly describe what a transformer is.

1.17 Give the full names of the two devices that are known respectively by the abbreviations PT and CT.

1.18 Give two examples of the application of step-down transformers.

1.19 What is the operating principle of a transformer?

1.2.8 Hazards associated with electrical systems and apparatus

Electric shock

Electric shock is the result of electrical energy causing a disruption to living tissue function. The effects of various amounts of current on the human body are as follows (the values are given in milliamps, or mA, 1 mA being equal to one-thousandth of an amp).

- 10 mA current causes muscle pain and shaking.
- 10 mA–30 mA causes severe muscle contractions, pain and internal organ stress.
- Over 30 mA there is a risk of ventricular fibrillation of the heart.
- Over 100 mA causes the body tissue, muscle and organs to be burnt, torn or strained to an extent that necessitates hospitalisation, could render a person unconscious or even result in death.

Any work on circuits and equipment rated as low voltage (greater than 50 V a.c. and 120 V d.c.) or high voltage (greater than 1000 V a.c. and 1500 V d.c.) must only be carried out by a licensed electrician or suitably qualified person as those voltages could be lethal. (Clause 1.4.128 of AS/NZS 3000:2018 provides definitions for what constitutes extra-low, low and high voltage.)

Equipment operators are trained in the starting-up and shutting-down procedures of equipment, plant and machinery. These procedures differ from correct electrical isolation practice and are usually documented as standard operating procedures or safe operating procedures (SOPs). Performing an isolate, lockout, tagout procedure (sometimes abbreviated to 'ILOTO') on electrical equipment, machinery or apparatus de-energises the equipment in terms of electrical energy, but that may not be the only type of energy present. Other energy sources could include:

- battery and/or capacitor banks
- fuels
- heat
- fluids or gases under pressure (e.g. water, air, steam, hydraulic oil)
- stored mechanical energy (e.g. compressed springs)
- gravity (potential energy)
- solar panels.

The companion textbook *Electrical Wiring Practice* provides more guidance on the identification of hazards associated with working with electricity, how to isolate electrical system apparatus and supply, and the effects of electric current on the human body.

The early signs of an electrical fault

The ability to recognise the early signs of an electrical fault (and thus prevent a serious situation from developing) does not require qualifications, extensive work experience or a high level of electrical knowledge. It simply requires taking some care in using electrical cords, power boards and appliances, being aware and using common sense.

The early signs of a fault include:

- a burning smell, which could indicate overloading or that an appliance or plug is heating up
- discolouration, brown patches or blackened marks around plugs, sockets or cables
- evidence of melting or the occurrence of sparks when inserting or removing an appliance plug
- plugs or appliance leads with fraying, cuts, damage or exposed wires.

The following measures can prevent a fault from developing:

- Never connect multiple extension leads together (making what is referred to as a 'daisy-chain').
- Avoid multi-way block/double adaptors. Do not plug them into extension boards.
- Avoid using extension leads or power boards as a long-term electrical solution.

Working safely with electricity

Working safely in any business environment, workplace, installation area or building and construction site requires everyone to have knowledge, skill, training and a site induction. Working safely with electricity requires taking a significant amount of care, for the stark reason that it can kill.

Table 1.2 lists ways of working safely with electricity and reducing the likelihood of electric shock.

TABLE 1.2 Working safely with electricity

Work scenario	How to reduce the likelihood of electric shock
Portable equipment	Where possible, replace corded power tools with battery-powered tools. Check tools before and after use, according to standard operating procedures. Plug portable equipment/appliances into an RCD (safety switches can fail, so check their operation before use). Always switch electrical equipment off before inserting or removing a plug.
Extension cords and power boards	Do not overload cords or power boards. To prevent power reels or cords from overheating, use single lengths and unwind them fully. For power boards, use individually switched socket types and have overload protection. Plug appliances into a dedicated wall socket wherever possible. Always switch electrical equipment off before inserting or removing a plug. Electrical cords and leads are trip hazards. Avoid: ▶ running electrical cords across walkways ▶ pulling out equipment by the cord ▶ creating extension cord chains.
Water and electricity	The key principle is: keep it dry. Keep electrical equipment and leads well clear of water. Keep yourself dry, have dry hands and use dry gloves.
Using corded tools and welding equipment	Check equipment before and after use: get it repaired if it is damaged. Tag equipment: identify any faulty items and remove them from service to be repaired. Plug corded tools into an RCD-protected supply. Look after your mate: keep others clear of the work area and run leads safely. Keep yourself, your equipment and the job dry. Replace gloves if they get wet.
Keep your distance	Do not access unauthorised areas or cabinets in commercial buildings, factories or on building sites. If they are locked, they are locked for a reason. Plan and check: do not puncture walls or the ground before checking what is behind or beneath. Look up and live: avoid working near powerlines and electrical network assets.

CHECK YOUR UNDERSTANDING

1.20 What is a non-electrical safety hazard presented by cords, leads and power boards?

1.21 Name the device required for electrical protection when using corded tools.

1.22 What are the types of procedures that relate to checks before and after tool, appliance and equipment use?

Non-electrical hazards associated with electrical systems, equipment and apparatus

There are many non-electrical hazards associated with working on equipment and machinery. These include becoming entangled in machinery or getting cuts from tools and rotating equipment such as circular saws and grinders. Other possible hazards are friction and getting abrasions from belt-sanding machines, as well as shearing and amputation and crushing or trapping hazards.

Process and industrial electricians working in factories and around machinery will encounter rotating machine and equipment hazards involving belts, pulleys, chains, sprockets and rollers. Injuries can occur if a part of the body is drawn into a 'nip-point' formed by counter-rotating parts, as well as by rotating and fixed parts. Technicians working on photocopiers and other office equipment will encounter smaller versions of the same type of hazard.

Figure 1.19 illustrates the drawing-in hazards posed by the counter-rotating parts of various types of machinery. The solid red arrows indicate where a part of the body could be drawn into a 'nip-point' and the white arrows show the direction of movement of machine parts.

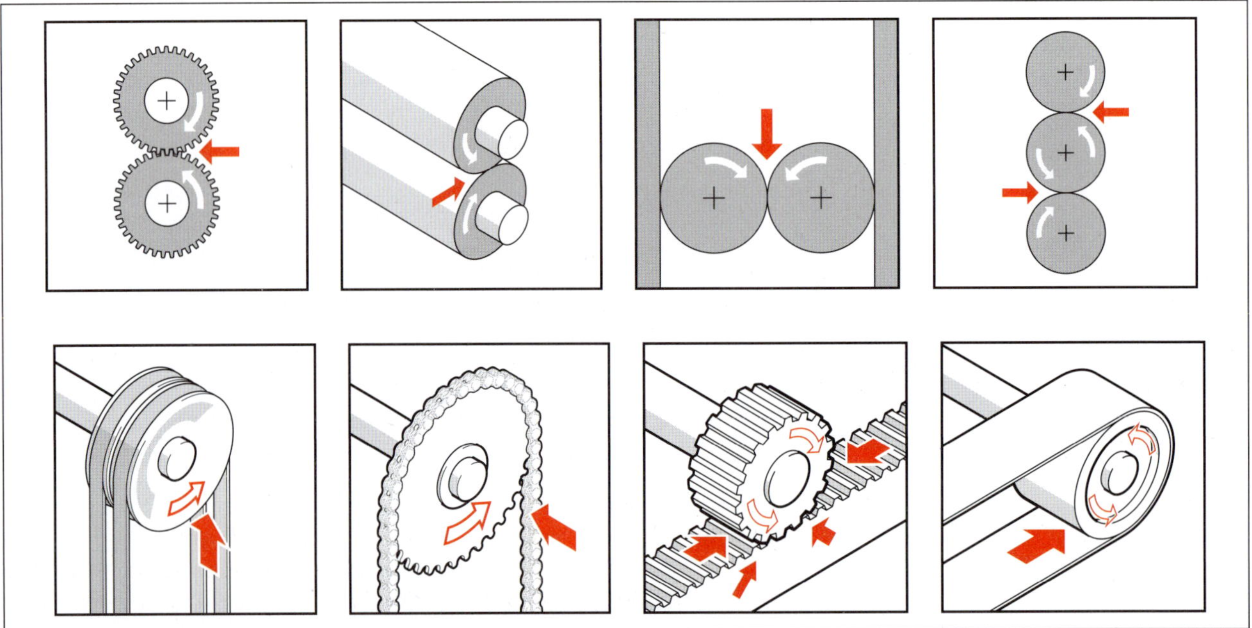

FIGURE 1.19 Counter-rotating machinery and roller 'nip point' hazards

Flying objects from grinding wheels or moving and rotating parts such as drills and drill press equipment can also cause puncture or stabbing injuries. Figure 1.20 illustrates equipment impact, crushing, flying object and moving parts hazards.

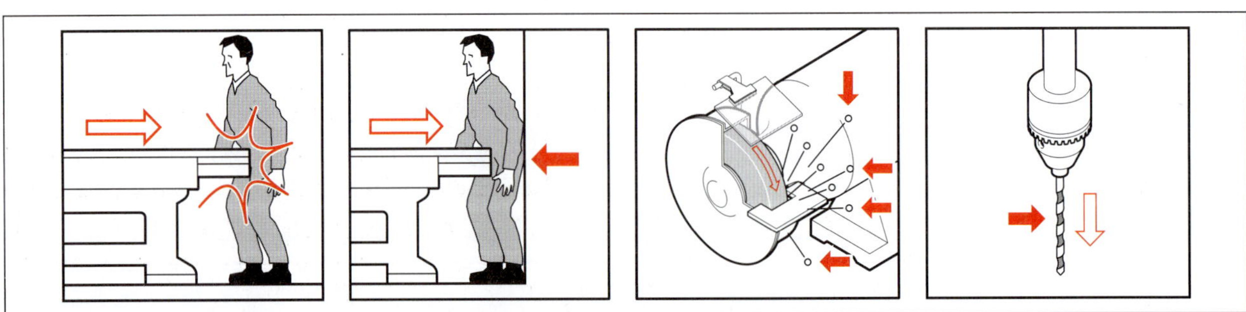

FIGURE 1.20 Equipment impact, crushing, flying object and moving parts hazards

Injury from machine and equipment hazards can be avoided or minimised by wearing appropriate personal protective equipment (PPE) and having administrative controls such as signage, safe operating procedures and training in place. An even more effective safety measure is having engineering controls in the form of protective mechanical guards and covers. These would protect workers from severe cuts and lacerations while using the table and band saws and portable grinder in the examples shown in Figure 1.21.

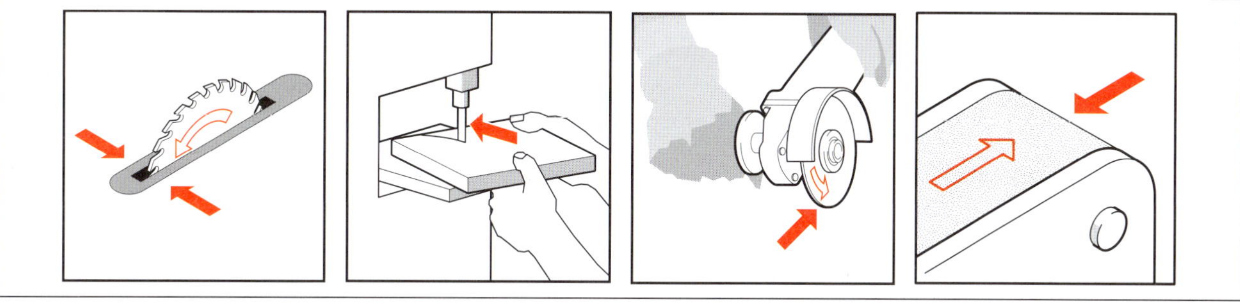

FIGURE 1.21 Cutting tool and abrasion hazards in the absence of safety guards and covers

Safety guards and covers are provided and installed for normal operating conditions. However, fault and abnormal operating situations (including maintenance mode activities) may result in interlocks being bypassed and a machine or piece of equipment being able to operate without safety guards. This can result in severe injury.

Figure 1.22 illustrates how a combination of physical guards and a photoelectric light curtain provide safe operation of a paper-cutting guillotine. Not having them would result in the possibility of accidents such as crush injuries by the clamp or amputation by the blade. From an operational perspective, the two push buttons and foot pedal contacts would be physically operated to electrically change state from *normally open (NO)* to closed, thus completing the circuit to operate. From a safety perspective, the guillotine operator would need to have their hands and arms clear of the clamp and blade area. To depress the foot switch, they would need to be standing and not leaning into the machine. The *normally closed (NC)* photoelectric sensor contacts open when the light beam presence detection is 'broken' by the operator's hand or arm. The light curtain safety detection, if operated, breaks the circuit electrically and prevents guillotine blade or clamp operation.

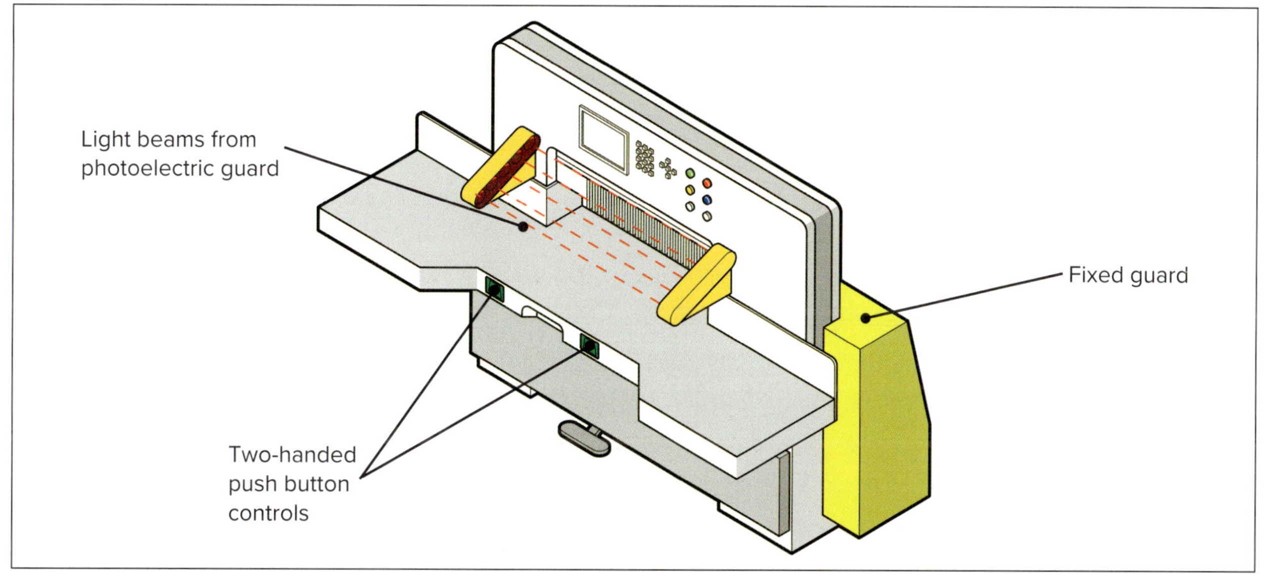

FIGURE 1.22 Paper-cutting guillotine

1.3 Useful terms, definitions and industry knowledge

1.3.1 Workplace documentation and activities

All electrotechnology workers have to meet the *work health and safety (WHS)* and *occupational health and safety (OHS)* requirements that are relevant to the environment in which they are operating. Satisfying these requirements involves:

- applying risk control measures
- observing workplace procedures and instructions
- following sustainable energy principles and practices
- maintaining a clean and tidy worksite
- checking and maintaining tools and equipment
- keeping correct records
- following workplace documentation.

Workplace documentation includes the following:

- workplace procedures
- regulations
- equipment specifications
- industry standards
- operation and manufacturer manuals
- work records
- reports on work activities.

It also includes codes of practice, which provide the relevant WHS/OHS requirements, job safety assessments and risk-mitigation processes.

Apprentices, trainees and industry workers are given workplace documentation and information to ensure that they perform work activities and tasks in a safe and timely fashion. Workplace documentation helps people to work to a standard of quality and with minimum waste. This is achieved by the application of the skills and knowledge required to safely and correctly use routine tools, equipment, plant, technologies and appropriate PPE. The skills and knowledge are subject to regulations and are directly related to contracts of training.

Work instructions are direct, general or broad directions, under supervision, given to the apprentice as guidance to perform work activities.

Work activities are any work activity or task to be carried out. Tasks include sweeping and cleaning the work area, getting tools and materials as work preparation, stripping cable, and drilling, cutting and fabricating activities.

Safe work practices include:

- following work instructions to carry out work activities safely by observing the SOPs for maintaining and using tools and equipment correctly
- using correct lifting, carrying and manual handling techniques to prevent injury
- wearing PPE such as safety glasses, earmuffs/plugs, clothing, shoes/boots and gloves.

Before being allowed to perform any work tasks, in addition to attending mandatory site inductions, electrotechnology professionals may be required to undertake training in first aid (including cardiopulmonary resuscitation, or CPR), height safety, manual handling, working in confined spaces and asbestos awareness. Undergoing this training may be a prerequisite to gaining work licences and permits in some jurisdictions.

1.3.2 Energy sector tools, equipment and technology

A tool is an instrument or simple piece of equipment that can be used to perform a particular task. A hand tool is an instrument that is manipulated without an a.c.- or d.c.-powered motor or other energy source. Hand tools include screwdrivers, pliers, knives, snips, saws and cutters, files, bench vices, clamps and striking tools.

Power tools can be corded (electric) or battery powered. Examples are drill/impact drivers, hammer/rotary drills, hand and bench grinders and drill presses.

Plant and equipment includes machinery, appliances, containers, implements, tools, PPE, ladders, lifts, cranes, computers, machinery, conveyors, forklifts, vehicles, power tools and mobile plant. Equipment and technology also includes computers, test instruments and multimeters.

1.3.3 Energy sector industry standards and codes of practice

While standards and codes of practice can both be categorised as safe work practices, they are not the same thing. A standard is a written document of how a process should be carried out. AS/NZS 3000:2018 is an example of a standard. A code of practice is a practical guide, providing specific information on work situations and activities that helps meet legal requirements, duties of work health and safety laws. Codes of practice state relevant industry standards, requirements for training, instruction, supervision and risk-management processes. Examples of codes of practice that are relevant to the energy sector include:

- Manual Handling Operations Regulations
- Work Health and Safety (Confined Spaces) Code of Practice 2015
- Compliance Code: Demolition
- Model Code of Practice: Managing the Risk of Falls at Workplaces
- Construction and Operation of Solar Farms Code of Practice
- Electrical Safety–Works (2020)
- Code of Practice–Safeguarding of Machinery and Plant
- Model Code of Practice: Managing Risks of Plant in the Workplace.

1.3.4 Operating instructions for tools, equipment and technologies

Operating instructions include SOPs which are the recommended start-up and shut-down procedures for plant, equipment and machinery. They also include manufacturer guidelines, instructions and specifications and industry codes of practice. SOPs may also be in the form of an observation checklist for workers to follow. These checklists cover such subjects as inspection before use for damage, safe operation tips and what to do after use. (Once a work task has been completed, all tools, equipment and other items involved in the execution of that task need to be inspected for damage, cleaned and packed up.)

1.3.5 Sustainable work practices

Sustainable work practices aim to recycle, reuse and minimise the waste of materials. They promote the efficient use and preservation of resources. Simple actions such as switching off lights in the daytime or using technology such as light sensors are examples of sustainability aimed at conserving energy. The correct disposal of waste products and material is an example of sustainable work practice to protect the environment. Sustainable practices include programs, initiatives and actions relating to the so-called 'three pillars of sustainability': environmental sustainability, social sustainability and economic sustainability.

When *roughing in* wiring or *fitting off* switches and electrical accessories, economic sustainability can be achieved by reducing the amount of cabling being cut off, discarded and wasted. If you were to pull through 500 mm cable excess to need at each socket-outlet in an apartment complex, the total length of wasted cable would add up; and that wastage could equal the cost of a drum of cable.

1.3.6 Workplace policies, procedures and instructions

Safe work environments require policies and procedures to deal with unwanted workplace behaviour. These policies and procedures are supported by legislation designed to protect everyone from bullying, harassment and discrimination and encourage equal employment opportunities (EEO).

CHECK YOUR UNDERSTANDING

1.23 Identify four examples of workplace documentation that are required for the safe performance of work.

1.24 Briefly state the knowledge, skill and work requirements for using equipment, plant and technologies.

1.25 Give a brief description for each of the following terms:

(a) work instructions

(b) work activities

(c) safe work practices.

INDUSTRY IN FOCUS

Work competency was once defined as the intersection of knowledge, skills and attitude. Instruction, training (on and off the job) and job experience provide workers with the skills and knowledge required to perform a range of tasks. This includes the knowledge of what the tasks are and why they are carried out.

The term 'work attitude' is defined in terms of how we apply ourselves to perform work. This encompasses work ethic, employability skills and adaptation to culture (or 'the way we do things around here'), WHS legislative requirements and organisational policies and procedures.

'Performing work' is a term that includes preparing to work, doing the work, cleaning up and carrying out job documentation. For example, performing work using battery and corded tools requires inspection before and after use, following a safe operating procedure and completing job documentation. Technical staff and operators working on plant, equipment and machinery need to know how those assets operate, the relevant start-up and shut-down procedures and maintenance mode functions. Equipment training is often provided to operators and technical staff for these activities.

WORKPLACE SCENARIO 1.2

Andrew and Elizabeth are in the electrotechnology industry. Andrew is a first-year apprentice and Elizabeth has just completed her fourth year and is now a tradesperson. Both work for the same organisation and, along with their colleague Stu, who is also a tradesperson, they carry out installation, maintenance and repair work in the building and construction industry. They also do factory and commercial jobs, regularly moving from site to site. Each site they visit has a mandatory induction for contractors and all work staff which includes details of operational requirements.

To be able to do the kind of work they do, Andrew, Elizabeth and Stu had to obtain their 'White Card' (general construction induction card). They also had to undergo first-aid, cardiopulmonary resuscitation (CPR) and asbestos awareness training. The company that Andrew, Elizabeth and Stu work for has also provided training in manual handling, working at heights, elevated work platform (EWP) work and isolate, lockout and tagout procedure. This means that all three are allowed to work at heights and operate the EWP. Being licensed electricians, Elizabeth and Stu are authorised to isolate and energise electrical equipment and supplies. Together with Andrew, they are trained in, and permitted to carry out, operator start-up and shut-down procedures.

As they move from site to site, their job tasks include testing and isolation before performing maintenance and repair work. Precautions need to be taken to identify and isolate all energy sources, such as:

- sources of electricity, including mains, solar panels and generators
- fuels
- heat
- steam
- fluids under pressure, such as water, air or hydraulic oil
- stored energy
- gravity
- radiation.

In industrial and production environments, qualified technical staff must be trained in using equipment in maintenance or test mode. They also have to be authorised to operate it. This may involve removing protective guards and covers and opening access panels, thus exposing moving parts and rotating machinery.

CHECK YOUR UNDERSTANDING

1.26 Describe what is meant by 'work attitude'.

1.27 Briefly explain why Andrew, Elizabeth and Stu require a site induction.

1.28 List possible energy-source hazards that could be involved with the operation of equipment and machinery in factory and production work environments.

SUMMARY

- The electrical industry consists of four subsectors: generation, transmission and distribution; industrial and mining; commercial construction, installation and maintenance; and domestic construction, installation and maintenance.
- The electronics industry's focus is on electronic components; the electrical industry deals with wiring and larger electrical equipment and apparatus.
- 'Electrical supply' refers to the energy source supplying a network.
- Circuits connecting the main switchboard to another distribution board are called 'submains'.
- Circuits supplying end-user equipment are called 'final subcircuits'.
- Circuit diagrams are a pictorial method of recording circuit connections in a simple, easy-to-understand way.
- Final subcircuits in an installation normally have five main components: an energy source or supply; circuit protection; control devices such as switches and isolators; a conducting path; and the load.
- Protective devices are necessary for electrical installations, equipment, machinery and electronic devices to protect all circuit components, devices and circuit wiring.
- Power usage over time is referred to as 'energy consumption' and is measured in kilowatt hours (kWh).

- Alternating current (a.c.) shows a repeating sine wave, and a steady or almost flat line represents a direct current (d.c.) value.
- When a current flows through a conductor or coil, a magnetic field results.
- A temporary magnet or electromagnet can be created by using electricity.
- Transformers are devices that have an input winding and an iron core magnetically coupling to an output winding.
- Belts, pulleys, chains, sprockets and rollers are all rotating machine and equipment hazards.
- Injury from machine and equipment hazards can be avoided or minimised by wearing appropriate personal protective equipment (PPE) and having engineering and administrative controls.
- Workplace documentation helps support workers in performing work in an agreed timeframe, to a quality standard and with minimum waste.
- Codes of practice are industry 'standards' which set out requirements for training, instruction, supervision and risk management.
- Sustainable work practices are aimed at achieving the effective and efficient use and preservation of resources.
- Proper disposal of waste products and material is an example of an environmentally sustainable work practice.
- Safe work environments require policies and procedures to deal with unwanted workplace behaviours.

END-OF-CHAPTER QUESTIONS

1.1 Name the four subsectors of the electrotechnology industry.

1.2 The industrial and mining subsector includes instrumentation. Describe what instrumentation is.

1.3 Where are current transformers (CTs) and potential transformers (PTs) usually found in an installation?

1.4 New domestic installations will have at least two lighting circuits. Give two practical reasons for this.

1.5 State three ways in which a component, device or power board is rated.

1.6 Where in a final subcircuit would you place the protective device?

1.7 What value of an a.c. waveform would be equivalent to a steady, constant d.c. value?

1.8 Your supervisor has asked you to identify some observation steps (things you would look for) as part of a corded drill standard operating procedure (SOP) checklist. List at least three.

1.9 List five hand tools.

1.10 List five other items of plant and equipment.

1.11 Provide three unwanted results of the incorrect use of a multimeter.

1.12 Name the AS/NZS 3000:2018 clause that provides a definition of voltage.

1.13 What injuries are prevented by safety devices and systems fitted to an industrial guillotine?

1.14 A simple transformer is made up of two coils and a core. Which part magnetises?

1.15 What is the difference between shutting down a piece of equipment and performing an isolate, lockout, tagout (ILOTO) procedure?

1.16 List the possible energy sources to check for when performing ILOTO.

1.17 Identify six hazards which can cause injury to equipment technicians and maintenance electricians performing work tasks.

1.18 What are codes of practice?

1.19 What are sustainable work practices aimed at achieving?

1.20 Give one example of sustainability practice that is aimed at conserving energy.

CHAPTER 2

Electrical fundamentals

LEARNING OBJECTIVES

- Understand the relationship between the structure of atoms and the creation of electricity
- Explain static electricity
- Explain dynamic or current electricity
- Understand how electricity is produced by both renewable and non-renewable sources
- Explain how electrical power is transported from a generation source and distributed to end users
- Understand how electricity can be used by basic electrical load devices such as resistors, inductors and capacitors
- Demonstrate how to perform basic calculations involving properties of electricity, speed and velocity

PREREQUISITE KNOWLEDGE

Year 10 mathematics

2.1 Static and current electricity

2.1.1 Atoms, molecules and matter

To understand the nature of electricity, it is important to understand the nature of atoms. Atoms are the building blocks of all matter. If an atom attaches to one or more atoms through **chemical bonding**, they form a **molecule**. A molecule is the smallest fundamental unit of matter into which a pure substance can be divided and still retain its composition and chemical properties.

2.1.2 Atomic structure

Atoms are made up of particles called 'protons', 'neutrons' and 'electrons'. Their almost unimaginably small size means that, historically, much of the extensive scientific study of the atom that has been carried out has been focused on the conceptual level. This endeavour, along with related experimentation, has yielded much in the way of understanding, a prime example of which is Bohr's Model of the atom. Developed by the Danish physicist Niels Bohr, the model is crucial to our understanding of how the properties of atoms give rise to electricity.

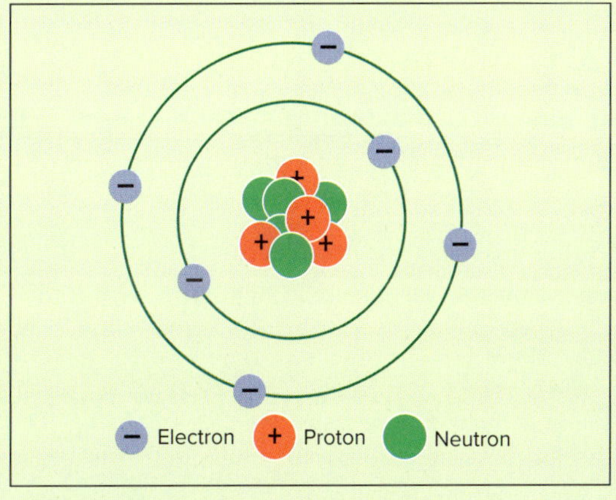

FIGURE 2.1 **Bohr's Model of the atom**

Bohr's Model, which is illustrated in Figure 2.1, is planetary in form in that it presents the structure of an atom as being like a small solar system, with a nucleus in the centre occupying the position of the sun. The nucleus represents most of the atom's mass and is made up of protons and neutrons. Electrons orbit the nucleus. They are about three times smaller than protons, and much lighter.

Two phenomena keep the electrons in orbit:

1. the force created by their momentum, which stops them from crashing into the nucleus
2. the attraction between the electrons and the nucleus, which stops them from flying off into space.

Atoms of different elements differ by the number of protons, neutrons and electrons contained within them. The **atomic number** is the number of protons contained within an atom, and this number defines each *chemical element* in the **Periodic Table**. For example, copper, which is an excellent conductor of **electric current**, has an atomic number of 29. That means it contains 29 protons and 29 electrons. It is the distribution of the electrons around the atom that makes copper a very effective conductor.

Inside the nucleus, the protons contain a **positive (+ve) charge** while the neutrons contain no charge. The electrons orbiting the nucleus contain a **negative (−ve) charge**. Atoms in their pure state have no overall charge as they contain the same number of protons and electrons. All the positive charge from the protons is balanced by the negative charge from the electrons, which causes the atom to be electrically balanced.

Atoms that have more than two electrons have them arranged into separate orbits (or **shells**) at varying distances from the nucleus. Each shell has its own energy level of electrons. The major shells are identified by number or letter, starting with K, which is the shell closest to the nucleus, and proceeding outwards alphabetically, as in Figure 2.2. There is a maximum number of electrons that each shell can contain. The closer the electrons are to the nucleus, the tighter they are bound to the atom and the greater the amount of external energy required to move them. Electrons in shells further away from the nucleus require less energy to be moved.

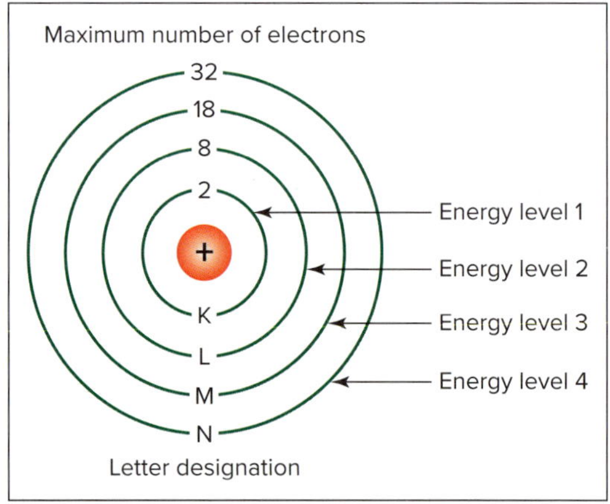

FIGURE 2.2 Electron shells

If the total number of electrons for a given atom or element is known, determining how they are arranged within the shells is easy. Each shell layer, beginning with the first, is filled to its maximum number and this is repeated for each corresponding layer. For example, copper has 29 protons and 29 electrons. Its electrons are arranged as follows:

Shell K (or shell 1) = 2 electrons (full shell)

Shell L (or shell 2) = 8 electrons (full shell)

Shell M (or shell 3) = 18 electrons (full shell)

Shell N (or shell 4) = 1 electron (incomplete shell).

2.1.3 Charged particles

If a force external to an atom causes the atom to lose or gain an electron, the atom is no longer electrically balanced and gains an electrical charge. This process is known as 'ionisation' and the atom involved becomes an **ion.** An ion with a negative charge is an atom that has gained an electron, resulting in a larger negative charge within the atom as there are now more electrons than protons. An ion with a positive charge is an atom that has lost an electron, resulting in a larger positive charge within the atom as there are now more protons than electrons.

2.1.4 Movement of electrical charge

An atom's outermost shell (the one furthest away from the nucleus) is known as the 'valence shell', and the electrons contained within it are called 'valence electrons'. The number of valence electrons determines an atom's ability to gain or lose an electron. This in turn determines the electrical properties of an atom (and thus an element). If an atom has a valance shell with only one valance electron, it will easily give up that electron to a neighbouring atom. Conversely, it will readily accept an electron from a neighbouring atom.

When a valance electron becomes free of its orbit due to an external force, it is known as a 'free electron'. This external force can be applied by an **electromotive force (EMF)** or voltage across a material. Friction is another type of force that can move electrons. When a free electron leaves an atom, it creates a space for a new electron. The movement of free electrons within a material is known as 'electron flow'. **Electrical current flow** is the movement of electrical charge, also known as 'drift'.

When a material allows electrons to flow within it, the material is said to 'conduct' electricity. Materials that contain only one valance electron readily lose or gain electrons and thus conduct electricity best. Such materials are known as 'electrical conductors'. Materials with atoms containing a full valance shell are less likely to lose or gain electrons and thus make the material resist current flow. These materials are known as 'electrical insulators'. It is important to remember that some insulators can still conduct electricity if a large enough force or voltage is applied.

CHECK YOUR UNDERSTANDING

2.1 What are the three particles that make up an atom?

2.2 Which of those particles contains a positive charge, which contains a negative charge and which contains no charge?

2.3 Describe the process by which an atom can gain or lose an electron.

2.4 What is the movement of free electrons within a material known as?

2.1.5 Static electricity

Static electricity is electrical energy that gathers in one location and thus creates an imbalance of electric charge within or on the surface of a material. The electric charge will remain until it can be discharged. Most people have experienced the effects of static electricity in one form or another, common examples being lightning or the 'zapping' experienced when shuffling along carpet or removing nylon clothing. The zapping sensation is what you feel when the electric charge is discharged.

Production of an electrical charge

The ancient Greeks observed that the material we call amber, when rubbed with cloth, attracted light particles such as feathers. (The Greek word for amber, *elektros,* was used to name this phenomenon, and it is from this word that the term 'electricity' is derived.) The same effect can be created when glass is rubbed with silk or when vulcanite (ebonite) is rubbed with flannel or fur. The cause of this is static electricity (whose effect is illustrated in Figure 2.3), and the material it acts on is said to be 'charged' with electricity.

The following experiments show the effects of charges of electricity on rods of glass and ebonite.

1. If an ebonite rod that has been rubbed with fur or flannel is suspended by a dry silk thread and a second ebonite rod that has also been rubbed with fur or flannel is brought near the first, the suspended rod will move away from the other rod. It is said to be 'repelled' (Figure 2.4(a)). A glass rod rubbed with silk will 'attract' the suspended ebonite rod (Figure 2.4(b)).

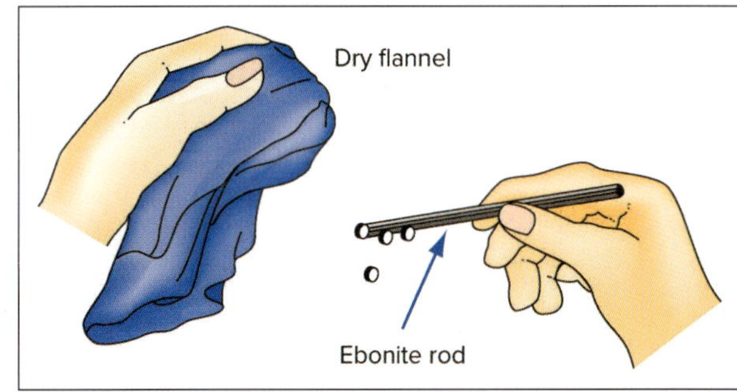

FIGURE 2.3 Electrostatic charge caused by friction

2. If an electrically charged glass rod is suspended from a silk thread, a second glass rod that has been rubbed with silk will repel it when brought nearby. An ebonite rod that has been rubbed with fur will attract it.
3. If the suspended electrostatically charged rod is replaced with an uncharged rod, the uncharged rod will be attracted to either an electrified glass rod or an electrified ebonite rod when either is brought near to it.

The charge exhibited by glass or glass-like materials when rubbed with silk is positive, and the state of **electrostatic charge** of ebonite or resinous materials when rubbed with fur is negative. Electrical charges of the kind indicated by the movements of the rods are described as 'static' because there is no electron drift – the electrons are not continuously moving but are stored on the surface of the materials.

The friction of fur on an ebonite rod causes a transfer of electrons from the fur to the surface atoms of the rod. The rod is now negatively charged because it carries a larger number of electrons than usual. In the process, the fur becomes positively charged because it has lost electrons.

The electrons involved in the transfer come from the outermost electron orbits of the atoms, which leaves the atoms temporarily ionised. In an attempt to return to an electrically neutral stable state, the ions exert an attracting force on the electrons in orbit around adjacent atoms. The charges on the rod remain there (as static charges) because the rods used are very poor conductors of electrical charges.

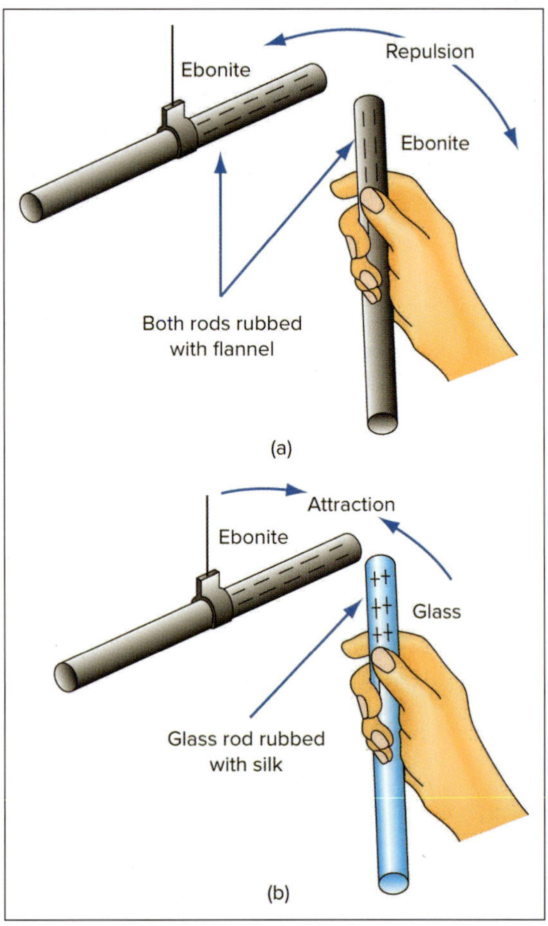

FIGURE 2.4 (a) Repulsion and (b) attraction between electrical charges

The electrostatic charges created on the rods and the resulting attraction and repulsion demonstrate an important concept about electricity and electrical charges: like charges repel and opposite charges attract. For example, two materials with negative charges will repel one another, while two materials will attract if one is positively charged and the other is negatively charged. This fundamental electrical concept is explored further in Chapter 6.

Electrical potential

When two different materials are rubbed together, the electrons from the surface of one are transferred to the surface of the other. Experiments have shown that the intensity of the electric charge is different for different combinations of materials.

Displaced electrons and their 'parent' atom (now an ion) have an electrostatic charge that produces an **electrostatic field**. The strength of this electrostatic field increases as the number of displaced electrons increases. A material with an electrostatic charge is said to possess an 'electrical potential', meaning that the build-up of electrical charge gives the material the potential for electron flow. The difference in electrical potential between two bodies is referred to as the 'potential difference' (or 'pd'), and this is measured in volts. An example is the electrostatic pressure required to remove electrons from one insulated surface and store them on another insulated surface.

The return of electrons from the negatively charged body to the positively charged body results in the potential difference reducing to zero. Equilibrium is restored and the two materials are once again neutral.

Partial charge

In a perfect (and thus theoretical) world, a charged body would always discharge to an exactly neutral state where the number of protons is equal to the number of electrons. What happens in the real world is that bodies of materials that come close to one another share what they have until they have an equal charge, at which point they are said to have reached 'local equilibrium'.

In dry summer weather, a person may experience an electric spark by touching a car door. The car has picked up an electrostatic charge from being driven against the hot, dry wind. Touch the car a second time and a second spark will often result. This is because the first spark made the charge on the driver equal to the charge on the car. But the driver, standing on the ground, has already lost much of that charge to the ground. So, the second spark comes from a mass that is still charged (the vehicle) to the mass that is now less charged (the driver).

Figure 2.5 illustrates three materials with a small number of atoms each. In the first row, the materials are charged +6 on the left, neutral in the middle and −5 on the right. This means that the material on the left is missing six electrons, the centre material is neutral and the material on the right has five extra electrons.

In the second row, the centre material touches the material to the left and, to equalise the charge between the two, it gives up three electrons. The result is that each of the two materials is now positively charged +3.

In the third row, the centre material, now with a +3 charge, moves to the right and touches the material there. To equalise the charge, the centre material takes three electrons from the material on the right. This leaves the material on the right with a −2 charge. In consequence, it gives up another electron and the two materials each become negatively charged −1.

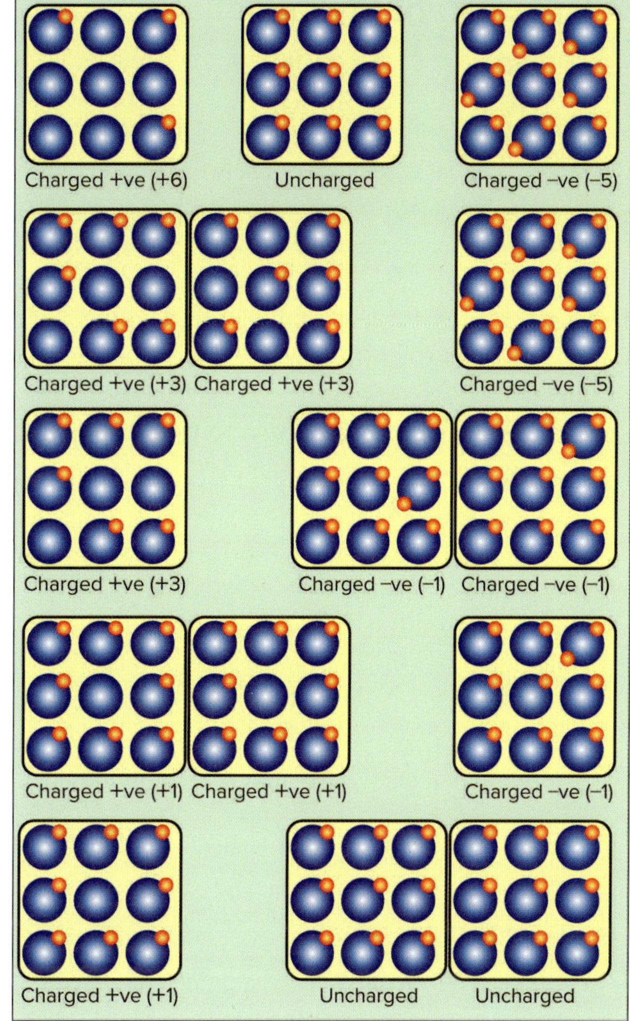

FIGURE 2.5 **The transfer of electrons between materials**

In the fourth row, the centre material again touches the material on the left, which first gives up one electron to the material on the right to neutralise its charge. This results in the material on the left having a positive +2 charge. To equalise the charge between the two materials, the material on the left gives up one electron and the two materials become positively charged +1.

In the last row, the centre material touches the material on the right. The material on the right gives up its one remaining free electron to the centre material and both neutralise their charge.

As to the final state of each material, nothing will happen if there is a difference of just one electron; all three will remain in their present charged and uncharged states.

CHECK YOUR UNDERSTANDING

2.5 Describe what is meant by the term 'static electricity'.

2.6 How does a material become either positively or negatively charged?

2.7 What does the term 'potential difference' mean?

2.1.6 Current (dynamic) electricity

The process of electrostatic discharge illustrates that electricity is the flow of electrons from a place that has a surplus of them to a place with fewer of them – that is, from a more negatively charged point to a more positively charged point.

Large electrostatic charges can exist without necessarily causing electricity to flow. An electron flow can only occur if a conducting path exists between two bodies and if there is a sufficient difference in potential between them. The force that causes the electrons to flow is electromotive force (EMF). The strength of EMF is measured by the potential difference in volts between two points.

Electrons can flow directly from one body to another when the bodies touch. If points of different potential are joined by an electrical conductor, the electrons will be forced away from the point where there is a surplus of them towards the point where there is a deficiency (a positively charged point).

2.1.7 Electron flow vs conventional current flow

As shown above, electrical current flows from negative to positive. In the electrical trade, this is termed 'electron flow'. However, in 1752 (before the nature of atoms and electrons was correctly understood), Benjamin Franklin theorised that electrical current flowed from a positively charged terminal to a negatively charged terminal, that is from positive to negative. We now know that Franklin's theory (known as 'conventional current flow') was incorrect. Despite that, the electrical trades still use it for the study of circuits. It is important to remember that the difference between electron flow and conventional current flow does not affect practical circuits. When studying those circuits, your calculations will be fine as long as you choose one convention to follow and stick with it.

Electrostatic discharge or static electricity is not much use as a power source because the electron flow ceases when it runs out of charge. To make it more useful, that and the following three issues must be addressed.

1. Potential difference (source): Once the potential difference between two points has been reduced to zero, the flow of electrons stops. An irregular flow of electricity is unsatisfactory for power-supply purposes; to ensure a continuous flow, a continuous source of electromotive force is required.
2. Electrical materials (path): Electrical energy needs to be guided from where it is generated to where it can be used effectively. One type of material (a conductor) guides the electrical energy to where it can be used, while another (an insulator) prevents any unwanted flow from the electric circuit to other materials.
3. Use (load): Electricity is energy, and that energy can serve many purposes. But a way must be found of changing the electrical energy into another form so it can do the work required of it. A device that converts electrical energy is called either an 'appliance' or a 'load' (the latter being the more common term).

CHECK YOUR UNDERSTANDING

2.8 What is the difference between electron flow and conventional current flow?

2.9 How is current electricity more useful than static electricity?

2.10 What three things are required to make current electricity work?

2.2 Production of electricity by renewable and non-renewable energy sources

Electricity has historically been mainly produced from non-renewable sources. A non-renewable energy source is one that will eventually be depleted, coal, petrol and gas being examples. As climate change and environmental concerns put pressure on these technologies, renewable sources of electrical energy are increasingly being researched and are becoming more widespread in their use. Renewable sources are those that can be replenished or that come from a source that will not run out in our lifetime.

Types of renewable electricity include solar, wind, hydroelectric, bio-fuel and geothermal. Developing technology and its associated efficiencies have led to increased viability, and consequently renewable energy is beginning to replace conventional fuel in certain areas. Renewable energy markets are projected to grow significantly in the near future.

Another type of electricity that is non-renewable but is debatably clean (meaning that it creates little or no carbon dioxide emissions) is nuclear. But the (few) devastating accidents, both created by humans and naturally occurring, that have occurred at nuclear reactors have led to a decline in its popularity. Nuclear fission reactors are being decommissioned across the globe. Technological breakthroughs have raised the possibility of nuclear power being utilised again, but the wealth of other clean renewable resources mean that it is highly unlikely to be a favoured option in Australia in the foreseeable future.

2.2.1 Power generation

Physicist Michael Faraday's experiments with induction led to the development of rotating machinery to produce dynamic electricity. Rotating machinery was initially of open construction and was primitive by today's standards.

Figure 2.6 shows an early dynamo (as generators were once called). It can be seen that there has been little change in the basic parts of direct current generators during the last hundred years—they still consist of an armature rotating within a magnetic field. There has, however, been extensive refinement of the construction methods employed and the materials used.

FIGURE 2.6 An early high-current dynamo
FLHC 5/Alamy Stock Photo

Inventor and electrical engineer Nikola Tesla realised that alternating current (a.c.) offered many advantages over direct current (d.c.) in both the generation and transmission of electrical power. However, the changeover from direct current to alternating current took many years to implement, and both systems were in use for a considerable time.

Early methods of producing mechanical energy to generate electrical power remained the same as the old steam-driven reciprocal engines that had originally been used to drive low-speed d.c. generators came to be used to drive a.c. generators instead. (Generators that produce an alternating current are usually called 'alternators'.)

2.2.2 Non-renewable energy sources

The bulk of electrical energy is still generated in the same way as it was a hundred years ago, through electromagnetic generation. Fossil fuel is burnt to generate the heat required to turn water into steam to drive mechanical turbines which then drive electrical generators. These generators employ the principle of electromagnetism by forcing conductors through a magnetic field to generate a current and a voltage in them. The only real variations in the process concern the type of prime mover used. The driving unit may be a steam, water or gas turbine or a diesel engine. Each has its advantages and disadvantages.

The electromechanical method is used wherever large quantities of electrical power are supplied from a common source. The operating efficiency of these various processes has gradually been improved but is still a subject of great concern to engineers and environmentalists.

Thermal steam turbines

Steam engines were slow and had low efficiency. They gave way to steam turbines, which offered higher speed and efficiency. A steam turbine is similar to a jet engine in that escaping hot gases cause a number of fans to rotate, producing power for the alternator. The most common method of generating power in the mainland states of

Australia is with steam turbines. In the majority of cases, steam is produced by burning coal, oil or gas, as governed by local conditions. Coal-fired steam-powered generating stations are often sited adjacent to large quantities of coal and water for cooling and condensation purposes.

Nuclear power

Nuclear energy has many advantages, including higher efficiency, lower pollution levels and greater reliability. It does not generate the levels of greenhouse gases that coal-fired power stations release. Nuclear reactors produce steam, which is then fed to turbines to drive alternators, as in thermoelectric power stations.

Nuclear reactors already drive power stations, ships and submarines in many countries, but the safety of nuclear power generation and the storage of nuclear waste fuel are a concern to the general public. As stated earlier, many nuclear power plants have been (or are due to be) decommissioned.

Engine-driven alternators

In remote areas, gas or diesel engine-driven alternators are common. The power is generated in a somewhat similar fashion to that of a portable generator, but on a larger scale. A comparatively expensive method of generating electrical power, it is used when other fuel supplies such as water and coal or furnace oil are not available to generate steam at an economical rate. Many small communities use diesel alternators for peak periods and battery back-up at other times.

Gas turbine engines

Gas turbines, which are a type of jet engine that drives a shaft, are noted for their efficiency and their ability to produce small quantities of electricity quickly. They can run up alternators and get them online rapidly at peak times and they make excellent standby plants for emergencies. Gas turbines are high speed, which means that they can drive high-speed alternators efficiently and with good speed regulation.

Electrochemical sources

Most electrochemical sources come in the form of batteries of electric cells. These are discussed in detail in Chapter 3. In general terms, secondary or rechargeable cells are used to store energy from other generating sources when it is plentiful, and they return it when those sources fail. For example, a wind generator charges the batteries when it is windy so the user can have electricity when the wind drops.

A less common method is chemical bonding, which is the combination of chemical elements such as hydrogen, carbon and oxygen to form compounds such as methane, water or carbon dioxide. Chemical bonding involves ions and the transfer of electrons from one atom to another, energy being either generated or absorbed in the process. Adaptations of this method lead to several types of fuel cells. One of these types consists of two chambers with two porous electrodes separated by an electrolyte. Hydrogen and oxygen are supplied to the two chambers and, in the presence of a catalyst, they react to provide ions and free electrons. (A by-product of this process is water.)

Fuel cells were used for the rockets that went to the moon, as well as for many subsequent space craft. They provide high efficiencies and, under ideal conditions, can last for many years. A fuel cell has no moving parts, gives off no noxious fumes and, in some cases, can produce drinkable water. This is an important reason for their use in space travel. However, some fuel cell systems require extensive (and expensive) auxiliary equipment, and their use is accordingly restricted.

2.2.3 Renewable energy sources

Hydroelectric power

In many parts of the world, hydroelectric power is generated by releasing large volumes of water through low-speed turbines. The water is collected in high-level reservoirs (as high as possible) and a power station is then built at as low a level as possible to maximise the energy provided by gravity. Figure 2.7 shows the Murray 1 Hydroelectric Power Station in the Snowy Mountains, NSW.

FIGURE 2.7 **The Snowy Mountains Scheme hydroelectric power station**
Cephas Picture Library/Alamy Stock Photo

The energy in the water is used to drive turbines (which are like medieval waterwheels, only far more efficient). The release can be controlled, as can the amount of electricity generated.

Geothermal steam

Italy and New Zealand have been generating electrical energy for many years by harnessing steam that emerges naturally from the earth. The steam is cleaned and fed to low-pressure turbines that drive alternators. Precautions have to be taken to ensure that fine solids and wet steam are prevented from reaching the turbine blades and eroding them. Large volumes of steam are required due to the relatively lower temperatures and pressure that are involved in the process.

Tidal movements

The principle is to harness the energy of tidal flows by using directly driven propellers to generate electricity as water moves in and out of a channel. This provides maximum power at peak tidal flow but no power at the top and bottom of the tides.

France and the United Kingdom have done innovative work in collecting seawater by tidal movement for generating power with hydroelectric turbines. Some coastal areas of those two countries have tidal movements of between 6 m and 9 m. The north-west coastal regions of Australia have similar variations in sea level but, because of low population densities, this kind of power generation is currently not a viable proposition here.

FIGURE 2.8 **A bank of solar cells**
Xavier Lorenzo/Shutterstock.com

Photoelectric sources

A solar cell is basically a large semiconductor diode. When light falls on the junction, light energy is converted directly into electrical energy. The cells are usually joined together into banks or panels (as in Figure 2.8). Although they theoretically have an unlimited lifespan, their efficiency is affected if they are not cleaned regularly and coated with protective chemicals.

At 12 noon, when the sun is directly overhead (or at its zenith), the value of the energy that it shines on the earth is around 1000 W/m^2 (watts per square meter). As most commercially available solar panels range from 15% to 25% efficiency, they will generate between 150 W/m^2 and 200 W/m^2 at this time. Efficiencies greater than 40% or 400 W/m^2 are available but are generally cost prohibitive and complex to achieve.

As efficiency increases and costs decrease, houses can be built with solar arrays mounted on their roofs that not only supply the household energy requirement but also feed into the grid to help meet industrial needs.

Photoelectric sources are also known as 'solar-electric' sources.

Electrostatic sources

Electrostatic generators produce very high voltages at very low current capabilities. Usually, they store an electrostatic charge either on their surface or in Leyton jars. The charge is released quickly and dramatically.

Two common ways of producing substantial electrical charges are via the Wimshurst machine and the Van de Graaff generator. The Wimshurst machine (see Figure 2.9) consists of two parallel insulating plates rotating in opposite directions, the charge being conducted away by two contacts. The Van de Graaff generator has a motor-driven rubber belt which rubs lightly against a flexible conducting comb. At the other end of the belt, another comb conducts the charge to a hollow ball. This type of generator is mainly used in the study of high-voltage effects and in testing insulators.

The use of static electricity in industry is growing. Typical applications are dust precipitation and spray painting. In dust precipitation, dust particles are charged negatively by passing them first through a charged grid and then through a positively charged collector. Rather than being discharged into the atmosphere, the charged particles are attracted to the collector, where they congregate into a mass which is periodically shaken into a hopper for disposal.

For spray painting, the object to be painted is made negative with respect to earth and the paint particles are attracted to it. The result is an evenly distributed coat of paint. The paint is even attracted around the sides of the object and into areas that are difficult to reach.

It is important to remember that very high voltages (typically 50 kV) are used and precautions have to be taken when working on electrostatic units.

FIGURE 2.9 A Wimshurst machine
Roland Bouvier/Alamy Stock Photo

CHECK YOUR UNDERSTANDING

2.11 What do the terms 'renewable' and 'non-renewable energy' mean?

2.12 List three forms of both renewable and non-renewable energy.

2.13 What is an alternator?

2.14 How does a solar cell produce electricity?

2.3 Transportation of electricity from the source to the load via transmission and distribution systems

2.3.1 Early power-distribution systems

The first power-distribution systems supplied consumers with direct current (d.c.) power. The simplest system consisted of two conductors, with the current flowing outwards from the generator in one conductor and returning via the other. As with any long conductors, the voltage gradually became lower and lower the further the conductors were located from the supply.

To try to minimise the size and length of conductors in street lighting, the circuits in some towns initially had the lamps connected **in series.** The overall voltage of a circuit had to be quite high so that each lamp would receive enough power to light it. Only one conductor was required but it would have to go up one street and return via another, going around the block. If any lamp failed, the whole block went out. If lamps had to be added, the voltage had to be increased to compensate for the resulting voltage drops.

As distribution systems grew in size and the number of customers increased, cable sizes became larger. In some systems, the ground was used as the return path. This practice sometimes had serious consequences such as the early deterioration of metal pipes buried in the ground. Another consequence was unwanted voltages, for example between iron fences and metal pipes entering the ground. The two-wire d.c. series system needed to be replaced with something better.

2.3.2 Modern power-distribution systems

Almost without exception, electrical energy is today distributed to consumers with a three-phase, four-wire system that uses distribution transformers to change the voltages.

Power-distribution systems are divided into two network groups:

1. Transmission networks, which transport electricity from generators or power stations to distribution networks.
2. Distribution networks, either regional or metropolitan, which comprise interconnecting substations and allow the end customer to connect to the network or grid.

Power stations generate at voltages ranging from 6.6 kV (kilovolts) to 33 kV, which are then transformed in a switching yard into higher voltages suitable for the transmission stage.

Long-distance transmission of power to distribution networks usually means that the voltage is converted to very high levels so that the loss of electrical energy in the transmission lines that naturally occurs is minimised. In Australia, transmission networks operate at voltages above 220 kV. In Europe and some parts of the USA, voltages are even higher.

At the consumer end, the transmission lines are connected into distribution networks. This is achieved by the transmission lines connecting into a substation where the voltages are reduced to lower levels for distribution. A result of this transmission/distribution arrangement is that few power stations are needed, but there is an increased need for substations to convert voltages to suitable levels. These are often cross-linked to other substations so that alternative routes for supplying them are available.

Figure 2.10 shows a stylised layout for a transmission system. In this example, the power station generates 23 kV and the voltage is increased to 220 kV, which is then transmitted to distribution substations. The voltage is then reduced to 11 kV for distribution to suburban substations where it is again transformed to 400/230 V for supply to households and small businesses.

In Australia, the transmission and distribution networks are owned and maintained mainly by state government-owned businesses and are regulated by the Australian Energy Regulator. Table 2.1 shows the companies responsible for transmission lines across Australia, as well as their owners.

Regional and local distribution networks are also mainly owned and operated by state government-owned companies. For example, the three companies that own and manage the distribution network assets in New South Wales are Ausgrid, Endeavour Energy and Essential Energy. The companies whose distribution networks facilitate

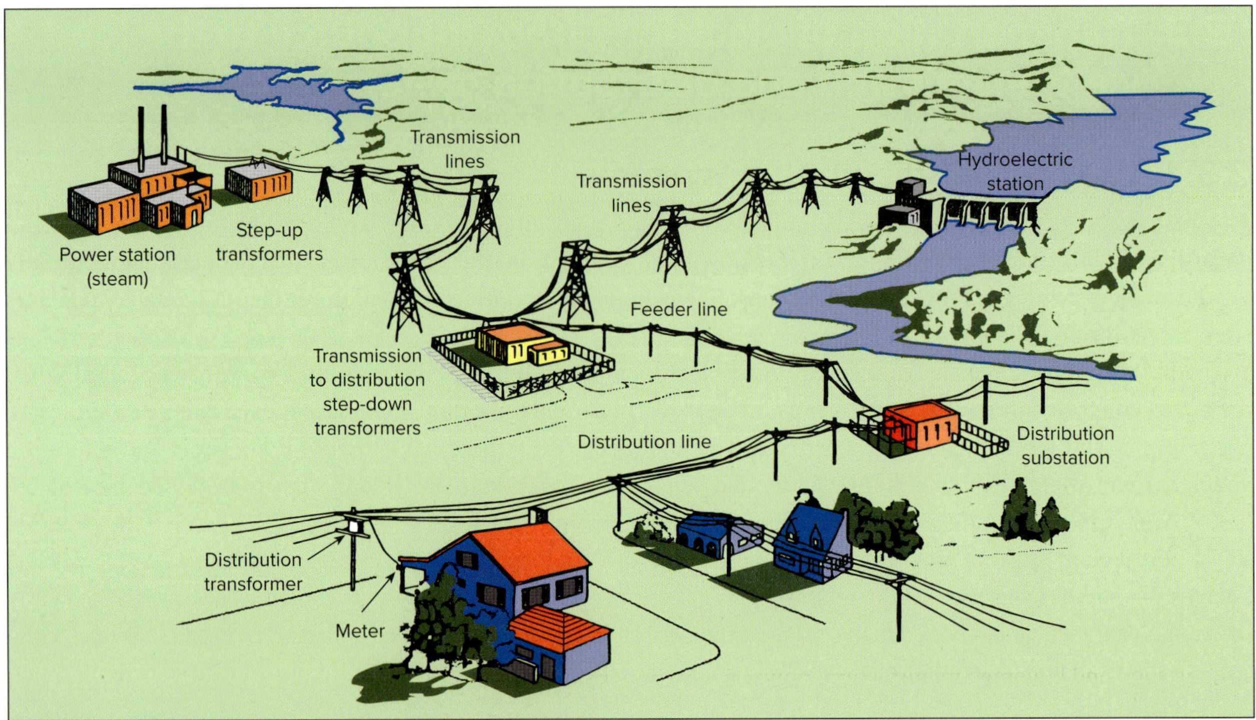

FIGURE 2.10 Stylised transmission and distribution system

TABLE 2.1 Companies responsible for transmission lines across Australia and their owners

Region	Company responsible	Owner
Queensland	Powerlink	Queensland Government
New South Wales and ACT	TransGrid	New South Wales Government
Victoria	AusNet Services	Singapore Power International
South Australia	ElectraNet	Powerlink, YTL Power Investment, Hastings Utility Trust
Tasmania	TasNetworks	Tasmanian Government
Western Australia	Western Power	Western Australian Government
Northern Territory	Power and Water Corporation	Northern Territory Government

end customers' receiving electricity are also known as 'wholesalers' as they sell their electricity at wholesale price to electricity retailers. End customers buy their electricity from the electricity retailers. There are many electrical retailers as the Australian Government has opened this segment of the market to competition. Added competition is an advantage to customers as it gives them more choice over who to buy their electricity from. This encourages retailers to add benefits such as tariff discounts.

 CHECK YOUR UNDERSTANDING

2.15 Why is electrical energy transmitted at high voltage?

2.16 What are the main components of transmission and distribution networks?

2.17 What is the difference between an electrical wholesaler and a retailer?

2.4 Utilisation of electricity

2.4.1 Electrical appliances

Electrical appliances convert electrical energy into other forms of useful energy. Those found in homes (often known as 'home appliances', 'domestic appliances' or 'household appliances') are mostly used either as labour-saving devices or for entertainment and security. Some home appliances are portable; they are connected to the electrical supply via an electrical lead and a plug that goes into an electrical socket (also called a 'socket-outlet' or 'power point'). This type of electrical appliance gives flexibility around where it is located and how it is used. Examples include vacuum cleaners, toasters, kettles, TVs and computers.

Other home electrical appliances are fixed; they are wired with a permanent connection to the electrical supply and cannot be relocated or moved freely. Examples of fixed appliances are ovens, range hoods, cooktops, hot water systems, heated towel rails and pool pumps.

Many household appliances can be categorised as 'white goods'. These are large, heavy appliances that usually connect to a power point and do not lend themselves to being moved around. Examples of white goods are refrigerators, washing machines and clothes dryers.

Industrial and commercial settings have specialised electrical machines for specific processes such as those used in food and beverage manufacture, power generation, raw material refining, transport and logistics, large-scale computer servers and electronic data storage. The variety of specialised machines and the industries they serve is immense.

2.4.2 Appliances that convert electricity to heat energy

When electrical current flows through a circuit, heat is always produced—even if in such a small quantity as to be undetectable. This is usually detrimental to circuits as excess heat can damage electrical accessories and cable insulation. However, it is possible to design circuits in such a way as to stop that happening.

Heat energy has many beneficial uses, such as in cooking and space heating. To produce heat, electricity is connected to some form of electrical resistance which, in this context, is called an 'element'. Electrical household appliances that produce heat energy, such as hair dryers, toasters, ovens and cooktops, all contain some type of heating element.

2.4.3 Appliances that convert electricity to light energy

One of the very first uses of electricity was for powering street lighting. Traditional light globes had a *resistive element* in a glass container. When electric current flowed through the light element, it would glow white-hot, producing visible light. This method of producing light is very inefficient and better ways have been developed. Advances in lamp technology have led to the development of halogen, fluorescent, CFL (compact fluorescent) and LED (light-emitting diode) lamps. LED lamps produce more light and use much less electrical power than other lamps and are thus the default choice for most electrical installations.

2.4.4 Appliances that convert electricity to kinetic energy

Kinetic energy is motion or 'movement energy'. Electrical appliances that convert electrical energy into motion fall mostly into the category of motors (however, there are other devices that perform the same function, such as solenoids and relays). Electrical generators link rotary movement with electricity, although they convert kinetic energy to electrical energy. Electrical motors are used extensively in industrial installations. In residential or domestic installations, motors can be found in many appliances such as washing machines, vacuum cleaners, hair dryers and refrigerators.

2.4.5 Resistors

Resistors are components that resist the flow of electric current and cause a drop in voltage and radiate heat. If enough heat is generated, a resistor glows with incandescent light.

Resistors usually consist of a length of conductor which is sometimes wound into a coil or laid into a grid so that the heat can escape. In electronics, resistors can be as small as $\frac{1}{8}$ W and measure 2 mm by 1.5 mm (see Figure 2.11). Even smaller resistors are found in microelectronics. Equally, they can be as large as necessary for their particular function.

Resistors are used to:

- restrict current flow
- develop a voltage drop
- generate heat (see Figure 2.12)
- generate light (see Figure 2.13).

A resistor's ability to control current flow is used in electronics and some electrical applications. For example, an operator can change settings such as volume and tone in an amplifier by adjusting resistance: a variable resistance device such as a **rheostat** or **potentiometer** is mounted on a front panel, and turning the device changes the resistance.

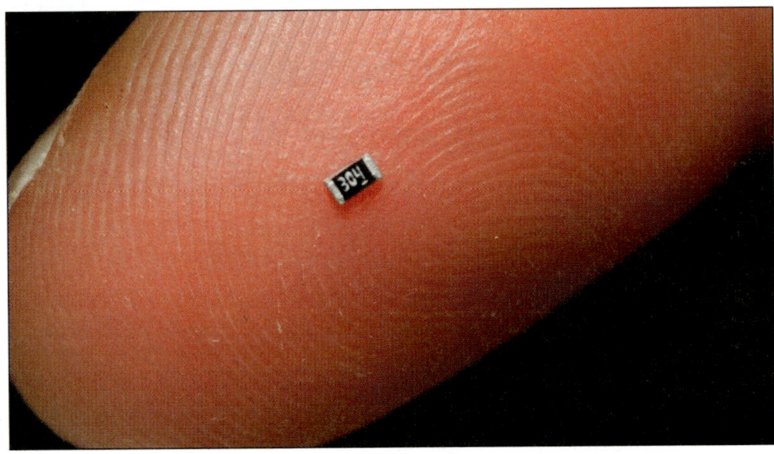

FIGURE 2.11 Small chip resistor electronic component on finger
Adam Hayes/Shutterstock.com

FIGURE 2.12 The heating elements in an electric toaster
Solaris/Alamy Stock Photo

FIGURE 2.13 A conventional lamp, a compact fluorescent lamp (CFL) and an LED lamp
Botastock images/Alamy Stock Photo

2.4.6 Inductors

Put simply, an **inductor** is a coil of wire that uses magnetism for its operation. Inductors are used to create other electrical devices such as solenoids, relays, transformers and fluorescent light ballasts. They are also used extensively in electronic circuits. The operation of inductors is explained in more detail in Chapter 6.

FIGURE 2.14 **Capacitors**
KPixMining/Alamy Stock Photo

2.4.7 Capacitors

Capacitors are devices for storing an electrical charge.

They come in a large range of sizes and shapes, depending on the manufacturer and their intended use. A capacitor is constructed of two conductive surfaces separated by an insulator so that an electric charge can be stored between those surfaces. Capacitors are used in many appliances and electronic circuits (see Figure 2.14). Their operation is explained in more detail in Chapter 5.

CHECK YOUR UNDERSTANDING

2.18 What is the purpose of a resistor?
2.19 What is the purpose of an inductor?
2.20 What is the purpose of a capacitor?

2.5 Basic calculations

2.5.1 Terms and units

SI units

The electrotechnology industry uses a standard set of units of measurement, quantity and associated calculations. Those units of measurement are contained within a comprehensive system known as the 'International System of Units' or 'SI units'. SI units are part of the metric system and are the agreed international standard for measurement.

The building blocks of this system are the seven SI base units, which form the basis of all physics calculations. These base units are listed in Table 2.2.

From these base units, a set of derived units has been formulated to measure physical quantities and attributes. Those that are relevant to the electrotechnology industry are described in Table 2.2.

TABLE 2.2 SI base units

Name	Unit symbol	Quantity name	Definition	Dimension symbol
Metre	m	length	Originally corresponding to 1/10 000 000 of the distance between the Equator and the North Pole, but now the distance travelled by light in a vacuum in 1/299 792 458 second.	L
Kilogram	kg	mass	Originally the weight (mass) of a cube of pure water measuring 0.1 m × 0.1 m × 0.1 m at its freezing point, but now the international prototype kilogram.	M
Second	s	time	Originally based on 1/86 400 of a day (24 hours × 60 minutes × 60 seconds), but now based on the change in energy levels of the caesium-133 atom.	t
Ampere	A	electric current	The constant current that produces a force equal to 2×10^{-7} Newtons per metre of length between two conductors placed 1 m apart.	I
Kelvin	K	temperature	Originally, temperature SI units were based on the centigrade scale of 0°C being the freezing point of water and 100°C being the boiling point of water; but now the Kelvin is used (it is the theoretical point where molecular movement stops).	T
			0° Kelvin is approximately −273°C	
Mole	mol	amount of substance	The amount of substance based on atoms in 0.012 kg of carbon-$12_{[n\ 4]}$.	N
Candela	cd	luminous intensity	The intensity of light in any direction from a source of light equal to $5.4 \times 10_{14}$ hertz at 1 m from the source.	J

Time

Time is a universal constant and a base SI unit that is used in many electrical calculations to define the rate at which events are occurring. Unless stated otherwise, this textbook presents time in increments of seconds. Time is represented by a lower case 't' and measured in seconds (s).

Electric current

Electric current is described as a flow of electric charge or electrons past a point. This relates in the context of the electrotechnology industry to electric circuits and electrons moving through a wire or conductor. As has been discussed, electric current flows from a place of excess electrons to a place of insufficient electrons. For convenience, these small electric charges are grouped into larger units called 'coulombs'.

The SI unit for electric current is the **ampere** (or 'amp' for short). It is one of the seven base SI units and is defined as the rate of flow of electric charge in coulombs across a surface or through a conductor per second. Electric current is represented by an upper case 'I' and is measured in amperes (A).

A current flow of 1 amp transfers a charge of 1 coulomb or 6.24×10^{18} electrons per second.

Electrical charge

The unit of electrical charge is the coulomb. A coulomb is defined as the amount of electrical charge transferred through a point in an electrical circuit when a current flow of 1 amp is maintained for 1 second. This amount of flowing electrical charge equals 6.24×10^{18} electrons per second. In electrical calculations, electrical charge is represented by the symbol (Q) and is measured in coulombs (C). The amount of charge can be calculated by multiplying the current in amps by the time the current is flowing in seconds, i.e. $Q = I \times t$.

Electromotive force (EMF)

Electromotive force (EMF) is the electrical pressure between two points that causes a current to flow. It is the force that moves electrons. Electric circuit sources are measured by their EMF in units known as 'volts' (V), after the physicist Alessandro Volta. Electromotive (ε) force is usually abbreviated to EMF, or simply E. As this E is easily confused with energy (which is also represented by E), the term 'EMF' is used in this textbook, except where there is no doubt as to what E represents.

Voltage and potential difference

When electrons are stored, their total charge or quantity of electricity is measured in coulombs (C). The energy (E) required to store them is measured in joules (J). The ratio of work energy to the quantity stored is measured in volts and is calculated from $V = \frac{E}{Q}$ (this is not a required formula). So, the more energy it takes to store a given amount of electricity, the greater the measured voltage.

Stored electricity has the potential to do work and is therefore a form of potential energy. If two bodies are charged with different potential energies, there is a potential difference (or a difference in potential). In physics, energy is expected to flow from a body with high potential energy to a body with low potential energy. In electrical contexts, the term 'potential difference' is generally called 'voltage' as it is measured in volts.

One volt is the EMF that causes 1 watt to be dissipated in a circuit when 1 amp is flowing.

Electrical work and energy

Electrical energy is quantified by the unit of work or energy and is known as the **joule** (J). In electrical terms, a joule is defined as either the:

- work required to produce 1 watt of power for 1 second; or
- energy used when 1 watt of power is used for 1 second.

One joule of electrical energy is an extremely small value, so a more manageable and practical unit is used, the kilowatt-hour (kWh). One kilowatt-hour equals 3.6 megajoules. The amount of electrical work or energy (W) can be calculated by multiplying the electrical power (P) required or used multiplied by the time taken to use the power, i.e. $W = P \times t$ (note: for a kilowatt-hour, the power is in kilowatts and the time is in hours).

Power

'Power' is defined as the rate of doing work or expending energy. In physics, power is calculated from the amount of work done divided by the time it takes to do the work.

The symbol used to represent power is (P) and the unit is watts (W).

One watt is defined as 1 J of energy expended in one second, i.e. $P = \frac{W}{t}$.

Electrical power

Electrical power (P) is calculated from electromotive force and the current that flows due to that force. As current is already a rate (coulombs per second), power can be calculated from the force involved (the EMF) and the displacement (current flow). Therefore, power equals the voltage multiplied by the current.

$$P = V \times I$$

Electrical resistance

Electrons passing between the atoms of a conductor is known as 'current flow'. During this process, some electrons collide with ionised atoms and give up their energy as **photons** of heat or light. The more collisions, or the greater the energy lost in each collision, the greater the heat generated by the conductor. The energy source supplies more energy to lift electrons out of their orbit and back into the current flow. The amount of energy required to release the electrons differs between materials, and so different materials are said to have different **resistance** or **resistivity**.

Resistance is the property that opposes current flow in a conductor. The unit of resistance is the Ohm. It is represented by the Greek letter omega (Ω).

2.5.2 Distance and displacement

Distance and displacement are similar, but they are not the same thing. Distance is a **scalar** measurement, meaning it refers to magnitude only. Displacement is a **vector** measurement, meaning that it refers to magnitude and direction. For example, a runner sprinting around a 400 m track will have travelled a distance of 400 m on the completion of one lap. But their displacement will be 0 m since they started and finished in the same position.

The symbol used to represent distance and displacement in a formula is generally (d), or sometimes (s), from the Latin word *spatium,* meaning 'space'. Rotating objects use either (d) for degrees or (r) for radians. Length can also be represented by the symbol (*l*). The base unit for distance or length used in most formulas is the metre (m).

2.5.3 Speed and velocity

As with distance and displacement, the difference between speed and velocity is one of direction. Speed is a scalar quantity, while velocity is a vector quantity. Velocity could also be called 'directional speed'. If the runner in the in the example in the previous section ran the 400 m track in 100 seconds, the speed would be 14.4 km/h. But the average velocity would be zero since the displacement is zero metres. However, if the runner were instead to run the 400 m in a straight line, the velocity would be 4 m/s.

The symbol used to represent average velocity is (v).

Velocity equals distance divided by time, i.e. $v = \frac{d}{t}$ and is measured in meters per second (m/s).

CHECK YOUR UNDERSTANDING

2.21 What are SI units?

2.22 What is the difference between distance and displacement?

2.23 What is the difference between speed and velocity?

THE CURRENT WAR: EDISON VS WESTINGHOUSE

The late nineteenth century saw a race to develop a form of electrical lighting that could compete with the gas- and oil-based systems that were already in use.

Winning that race would necessitate the development of an electrical distribution system. Thomas Edison, an American inventor and businessman, realised that improving the reliability of electrical lighting was also going to be crucial. In 1878, he succeeded in creating an incandescent light globe for indoor use. He also devised a distribution system that could compete against the gas and oil utilities. However, his was a direct current system and had the drawback that, due to electrical losses, it required many d.c. power stations to be installed near customers. It also required the use of large, expensive electrical distribution wiring. This made Edison's system inefficient and economically costly to maintain.

Enter Nikola Tesla, a visionary Serbian-American engineer and inventor who had formerly worked for Edison. Tesla's pioneering efforts with alternating current power systems and, later, a.c. induction motors, helped Edison's major competitor George Westinghouse develop a distribution system that made it possible to transmit electrical power long distances over thinner, cheaper wires. This reduced electrical losses, making the system more efficient and economical than Edison's. Westinghouse's a.c. system used high voltage and a network of transformers to step it down to end users, enabling the supply of power over vast distances. It thus became the obvious choice to handle the increased demand for electrical power over the continental US. By contrast, Edison's system was only viable for small, large-density customer bases such as big cities. When it came to supplying lighting across the whole country, it simply could not compete.

In 1892, the current war was won by Westinghouse when Edison was forced out of control of his own company, Edison General Electric, by the stockholders. Later in that same year, the company became part of a new organisation called General Electric. General Electric subsequently controlled three-quarters of the US electrical business and continued to vie with Westinghouse for dominance of the US alternating current market.

SUMMARY

- Matter is composed of molecules, and molecules contain atoms.
- Atoms consist of three main particles: protons (positive), neutrons (neutral) and electrons (negative).
- All atoms can be ionised.
- If an electron is removed from an atom, the atom becomes a positive ion.
- If an atom gains an electron, the atom becomes a negative ion.
- A material that has a large number of excess electrons is considered to have a negative electrostatic charge.
- A material that has a large number of electrons missing is considered to have a positive electrostatic charge.
- The difference in the charge levels of two materials is called the 'potential difference'.
- Materials with no potential difference are considered to have reached equilibrium.
- Like charges repel; unlike charges attract.
- A neutral mass (no charge) is attracted by both positively and negatively charged rods.
- Static electricity is a charge that has gathered in one location but has no current flow.
- Current or dynamic electricity is when an EMF forces electrons to flow through a conductor.
- Electron flow is when electrons move from a negative potential to a positive.
- Conventional current flow is regarded as being from positive to negative. This is the direction in which early electrical pioneers believed current flowed and is still the convention most used.
- Renewable and non-renewable resources produce electricity.
- A non-renewable energy source is one that will eventually be depleted.
- Renewable energy sources are those that can be replenished or those that will not run out in our lifetime.
- Early power-distribution systems used d.c. but, as use of this system increased, it became inefficient and, ultimately, prohibitively expensive.
- Modern power-distribution systems use high-voltage a.c. as this is a more efficient means of transporting power to many users and reduces costs.
- There are many types of household appliances that use electrical energy by converting it into other forms of energy, such as light, heat and kinetic (or movement) energy.

- Household appliances can be small and portable or large and permanently positioned.
- Units of measurement are contained within a comprehensive, widely adopted system known as the 'International System of Units' (or SI units).
- The SI units contain seven base units from which most electrical units are derived.
- The SI units are used in most electrical calculations.

END-OF-CHAPTER QUESTIONS

2.1 What is the difference between static and current electricity?

2.2 List three different ways of producing electricity from renewable sources and three different ways of producing electricity from non-renewable sources. Briefly explain each one.

2.3 What is one effect of electricity that is always present when an electric current is flowing?

2.4 Describe the phenomenon that is termed 'EMF' and state how it is used in modern electrical systems.

2.5 What are the differences between the terms 'electrical work' and 'electrical energy'?

CHAPTER 3
Electrical circuits

LEARNING OBJECTIVES

- Describe a simple d.c. circuit
- State the components of a simple d.c. circuit
- Describe a short-circuit, an open circuit and a closed circuit
- Understand Ohm's Law
- Apply Ohm's Law calculations to simple d.c. circuits
- Understand power in a simple d.c. circuit
- Understand the relationship between voltage, current, resistance and power in a series circuit
- Understand the relationship between voltage, current, resistance and power in a parallel circuit
- State the effects of electrical current
- Describe the practical uses of electrical current
- Describe the various methods of producing an electromotive force (EMF)

PREREQUISITE KNOWLEDGE

- Understand the concepts of voltage, current and resistance
- The ability to transpose mathematical equations

3.1 Symbols used to represent an electrical energy source, a load, a switch and a circuit-protection device in a circuit diagram

3.1.1 Electrical circuits

Understanding *electrical circuits* is the essence of an electrical worker's job. People in other occupations may also need to understand them: for example, automotive mechanics find that there is an ever-increasing electrical aspect to automobile technology, and electronics technicians cannot understand or repair electronic circuits without understanding electrical circuits.

An electrician's most valuable tool is the ability to calculate the electrical values in a circuit before it is even turned on.

One way to categorise electrical circuits is by their complexity, which means in practical terms whether they have just one load or many, one current path or many and one or more sources of energy. The five types of circuits can be categorised as:

1. simple
2. series
3. parallel

4. compound
5. complex.

There is another type of circuit, the three-phase circuit, which has three alternating current (a.c.) power sources. The three-phase circuit is discussed in Chapter 10.

3.1.2 Drawing a circuit

Circuit diagrams are a pictorial method of recording the connections of a circuit quickly, simply and clearly. Standard symbols are used to represent the parts of a circuit—the components and connections—and a standard is also applied to labelling a circuit. A circuit drawn by one person must be able to be easily read by others who know the standards.

These standard symbols generally represent a concept rather than a picture of reality. The figures in this chapter contain several common symbols, most of which might be somewhat familiar.

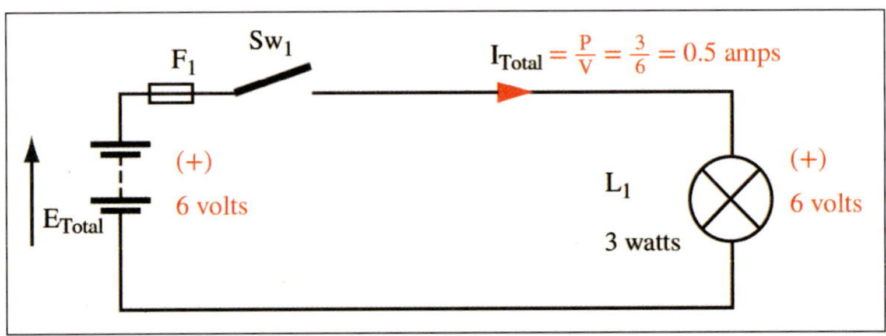

FIGURE 3.1 Labelling a simple circuit diagram

3.1.3 Labelling a circuit

Components are labelled from left to right and from top to bottom. They have a capital letter and a subscript number or letter that is used in equations and calculations. For example, resistor 1 could be labelled 'R_1' while the power supply might be labelled simply 'E'. A switch might be 'S_1' and a fuse 'F_2'. The total current can be labelled 'I_{Total}'. These symbols and their names (or 'nomenclature') quickly become familiar.

Circuits for analysis should also be marked to show the **polarity** of the voltage source and the direction of current flow. Voltage sources should have an arrow beside the symbol, as shown in Figure 3.1.

COMMON ELECTRICAL ABBREVIATIONS

Circuits are often labelled with abbreviated terms to simplify them and make reading them quicker. Although you will notice many abbreviations (some of which will be unique to their own place of work), the following are commonly used:

- +ve = positive—labels the positive terminal of a component or supply
- −ve = negative—labels the negative terminal of a component or supply
- V = voltage in volts (V)
- pd = potential difference—the difference in potential between two points
- VD = voltage drop—the voltage difference between two points, e.g. across a component
- cct = circuit (used mainly in example questions)
- I = current in amps (A)
- R = resistance in ohms (Ω)
- P = power in watts (W)
- E, EMF or emf = electromotive force. E is also used to represent energy, but what it represents in a particular situation is usually obvious from the context.

Current direction should also be noted on a circuit by placing an arrowhead (a filled triangle) as a pointer on the conductor in the direction of current flow. A label is usually drawn beside the symbol to name the current, e.g. 'I_{Total}' (see Figure 3.1).

CHECK YOUR UNDERSTANDING

3.1 What is the main use of a circuit diagram?

3.2 What are some of the conventions used to label circuit diagrams?

3.3 What do the abbreviations 'E', 'VD' and 'P' represent when used on a circuit diagram?

3.2 Purpose of each component in the circuit

3.2.1 Parts of electric circuits

A basic circuit has to contain three parts: a source, a load and a path, as shown in Figure 3.2. A circuit may also contain control devices, protection devices and measuring instruments.

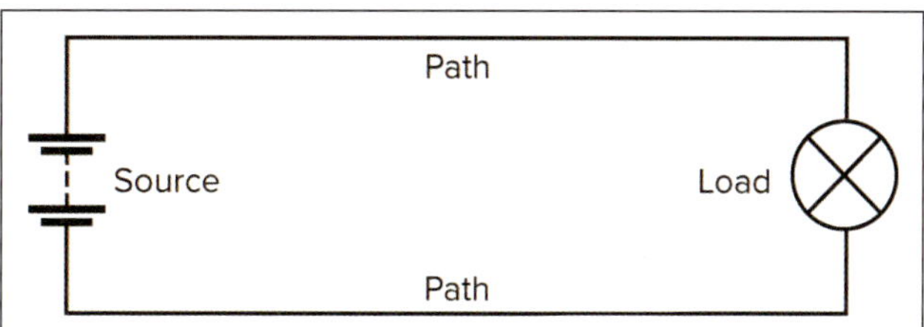

FIGURE 3.2 Basic circuit parts

The source

The source is simply where a circuit's energy comes from. In many cases, the source is a battery, but most often it is the power-supply distribution network. The source may also be a generator or solar cell, or one of many other means of generating electricity.

The load

The load is the 'working' part of the circuit and converts electric current into heat and/or light, or uses the magnetic effect of electric current to produce motion.

The path

The word 'circuit' is derived from the Latin word 'circuitus', whose meaning relates to 'going around'. Electricity will only flow if a circuit is complete, which can also be referred to as the 'current path' or simply 'the path'. A circuit needs a conductor to carry the current and the conductor is the path of the electricity.

Control devices

Paths often include control devices such as switches and electronic devices that are neither a source nor a load; they control current flow within a circuit. A switch is simply a mechanical device that opens or closes a circuit. When a circuit is open, no current flows and the switch is said to be 'off'. When the switch is closed, current flows and both the switch and the circuit are said to be 'on'.

Protective devices

Protective devices are comprised of circuit-breakers and fuses, and there is a wide variety of both. They are not essential for the operation of the circuit but are there to prevent shocks to people and animals, and to ensure that all components (including the current path conductors) are not damaged or become a fire threat. When a circuit-breaker is closed or a fuse is intact, the circuit can operate as usual, but when the circuit-breaker 'trips' and opens or a fuse 'blows', the circuit is opened and no current flows.

Measuring instruments

Voltmeters and ammeters are measuring instruments that are commonly shown in circuit diagrams to represent the value of voltage or current at a given point.

Ammeters must be placed in line (in series) with a conductor, as a replacement for a part of the conductor, for the meter to read the current. This is because ammeters must pass the current through themselves to read the value.

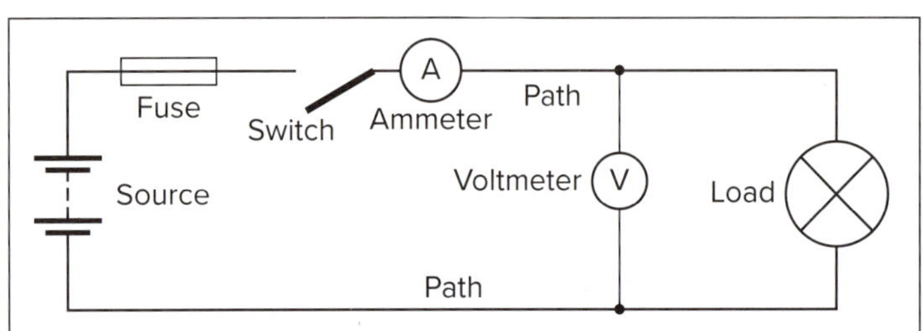

FIGURE 3.3 Basic circuit, including circuit protection and measuring instruments

Voltmeters, on the other hand, must be placed across a component, or across two points in a circuit, to measure the voltage difference or potential difference between those two points. Ammeters do not measure anything when the circuit is open; voltmeters may, however, measure voltages when the circuit is open, depending on where they are placed.

Figure 3.3 shows a basic circuit that includes protection, control and measuring instruments.

CHECK YOUR UNDERSTANDING

3.4 What are the three components of a basic circuit?

3.5 What is the purpose of a circuit's protective device?

3.6 Where should the circuit-protective device be positioned in the circuit?

3.7 Which measuring instruments are placed in series and which measuring instruments are placed in parallel?

3.3 Effects of an open circuit, a closed circuit and a short-circuit

3.3.1 Electric circuits

Electric circuits can usually be turned on or off, although they can at times have a fault which changes them entirely. In electrical circles, the term 'circuit' may often be abbreviated to 'cct'.

3.3.2 Open circuit

If the current path is not continuous between two points of potential difference, the circuit is referred to as an 'open circuit' (see Figure 3.4(a)). The circuit has a break in it and the switch is said to be 'open'. No current can flow and the lamp will not light. This condition occurs in a normal circuit when it is switched off, but may also occur when a wire is broken.

3.3.3 Closed circuit

If the circuit is complete, as in Figure 3.4(b), current can flow and the lamp lights up (assuming it is a lamp meant for 12 V and the battery is not flat). This condition occurs in a normal circuit when it is switched on and the switch is 'closed'.

A closed circuit is an essential condition for current flow. For a continuous current flow, a continuous source of electrical energy must be provided.

In Figure 3.4, a car battery is shown as the source of energy, but other devices such as generators can be used when greater quantities of energy are needed. Static electric charges generated by friction are not usually able to carry out this function. The circuit in Figure 3.4(b) is known as a 'simple circuit'.

FIGURE 3.4 Lighting circuits based on a car battery

3.3.4 Short-circuit

This type of circuit is one that has a 'fault', a condition that is to be avoided. In Figure 3.4(c), the lamp is bypassed by a conductor connected directly from one terminal of the battery to the other. The circuit is now shorter than it was before and the lamp is said to be 'shorted' out of the circuit; or the circuit is said to have a 'dead short' in it.

'Shorted' means that the current now has a shorter path. Because of this, the current flow through the circuit's component or components is either limited or entirely prevented and so the circuit does not function as expected. 'Dead short' refers to an almost-zero-resistance short as compared with a partial short-circuit, where the resistance is low but not zero. A dead short is a dangerous situation where large amounts of current can flow. Usually, a protection device will operate to open the circuit and prevent further current flow.

In a short-circuit, no current flows through the lamp but excessive current flows directly from one terminal of the energy source to the other. The current flow is not limited by the load, as it is in Figure 3.4(b). In this particular case, unless the energy source (such as a battery) is protected against excessive current flow internally, the battery and the conductors can easily be damaged and may melt before the current stops. The battery may even explode.

Short-circuits are not always bad. Sometimes they are introduced deliberately to cause a particular effect. When a load is short-circuited, the high current generated by the low-resistance current path is what activates the circuit-breaker or fuse. For example, if a wire was to break off from the circuitry in an appliance and touch the exterior metal frame, it would be dangerous for someone to touch it. However, a low-resistance circuit path would be created by connecting an earth wire to the frame; as soon as the wire touched the frame, a high current would be produced, activating the circuit-breaker and protecting the circuit.

CHECK YOUR UNDERSTANDING

3.8 Which of the following is generally a fault condition: an open circuit, a closed circuit or a short-circuit?

3.9 When can the use of a short-circuit be beneficial?

3.10 What are the terms that are used for a circuit that contains approximately zero resistance due to a fault condition?

3.4 Multiple and sub-multiple units

3.4.1 Prefixes and multipliers

While most of the calculations in this textbook use simple values such as 5 Ω, 10 V and 0.5 A, the values of units in the real world can also be very large or very small. As a result, other systems of communicating such values simply have been devised. These include *multiples* and *sub-multiples* and *scientific* and *engineering notation*.

In the multiple and sub-multiple system, large or small units have abbreviated prefixes. Examples of prefixes for large values are 'kilo', 'mega' and 'giga', while 'milli', 'micro' and 'nano' are examples of those used for small values. These are then abbreviated further through the designation of a single letter for the particular unit. For example, 1000 watts can be abbreviated to 1 kilowatt, which can be further abbreviated to 1 kW.

Table 3.1 lists the commonly used prefixes, symbols and multipliers.

TABLE 3.1 Summary of common prefixes, symbols and multipliers

Prefix	Symbol	Multiplier
pico	p	10^{-12}
nano	n	10^{-9}
micro	μ	10^{-6}
milli	m	10^{-3}
kilo	k	10^{3}
mega	M	10^{6}
giga	G	10^{9}
tera	T	10^{12}

Extremely large and extremely small values can also be represented using either scientific or engineering notation. While both use the same principles, engineering notation uses values that are easily translated into multiples and sub-multiples.

Scientific notation writes the number in two parts:

1. The first part is the digits, with a decimal point placed after the first digit.
2. The second part is × 10 to the power that puts the decimal point to where it needs to be.

For example, 200 metres can be expressed as 2×10^2 metres (or, better still, 2×10^2 m) as 10^2 equals 100 and 2×100 equals 200. Alternatively, a small number such as 0.02 metres can be expressed as 2×10^{-2} metres (2×10^{-2} m) as 10^{-2} equals 0.01 and 2×0.01 equals 0.02.

Engineering notation only uses powers of 10 that are multiples of 3 (see Table 3.1). For example, 2000 metres can be expressed as 2×10^3 metres (or, better still, 2×10^3 m), which can then be easily converted to 2 kilometres (2 km) since 2×10^3 m equals 2000 metres. Alternatively, a small number such as 0.002 metres can be expressed as 2×10^{-3} metres (2×10^{-3} m), which can then be easily converted to 2 millimetres (2 mm) since 2×10^{-3} equals 0.002 metres. With engineering notation, the digits before the exponent will always be a number between 1 and 999.

The difference between scientific and engineering notation can be explained using 25 000 metres as an example. In scientific notation this value would be simplified to 2.5×10^4 metres, whereas it would be expressed as 25×10^3 metres in engineering notation.

3.4.2 Scientific calculators

There are many different brands of scientific calculators available today, including those that are applications on mobile phones and tablet devices. However, each brand generally only uses one of several types of function buttons to enter scientific and engineering notation.

The function button for entering the exponents is generally labelled as 'EXP', '× 10^{X}' or sometimes 'EE'. For the sake of simplicity, let's call the button 'EXP'.

To enter a number (e.g. 4.15×10^6) expressed in scientific or engineering notation, do the following:

Step 1. Enter the digits required for the number (4.15).

Step 2. Use the 'EXP' button to enter the exponent (6). *Do not use the 'X' function button* (this is already assumed by the use of the 'EXP' button).

Step 3. Use the '+/−' function button to change the exponential from positive to negative, if required.

Most calculators also have a function button to allow the resultant numbers to be displayed in normal, scientific or engineering notation. This button is typically labelled 'SCI' or 'ENG' and, generally, multiple presses or the use of a 'SHIFT' button will change the values back and forth.

For a more detailed explanation of how to use a scientific calculator, refer to the manual for the particular device that you use.

CHECK YOUR UNDERSTANDING

3.11 What does the term 'scientific notation' mean?

3.12 What does the term 'engineering notation' mean?

3.13 What is the prefix for 10^{-12}, 10^6 and 10^{-6}?

3.5 Direct current (d.c.) single path circuit

3.5.1 Simple circuits

Simple circuits have only one source and one load. They also have only one unbroken path of conducting material for the electric current to flow along. An example of a typical simple circuit is a standard torch. It has a battery (source), a lamp (load) and a path which includes a switch. A desk lamp is the same circuit, except that the source is now the power-distribution network, via a three-pin plug. The three-pin plug is not the actual source but, as it connects to a known source, most people accept that a socket-outlet that a three-pin plug connects to is the source when connected and switched on.

Circuits are not usually shown as images or drawings but as circuit diagrams that replace the components with symbols. The simple circuit pictured in Figure 3.5(a) is drawn as the circuit diagram shown in Figure 3.5(b). Each component has its own symbol.

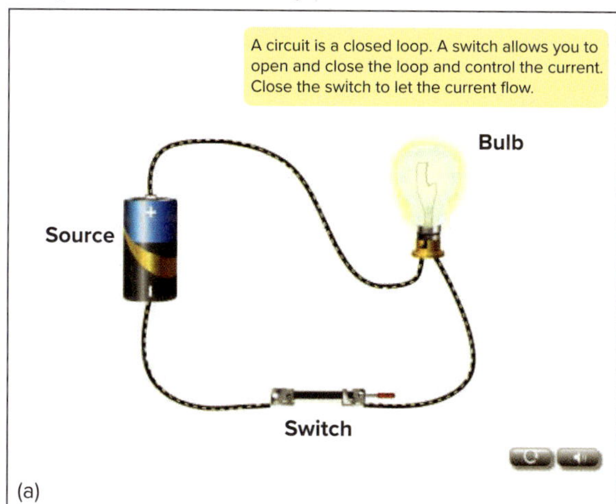

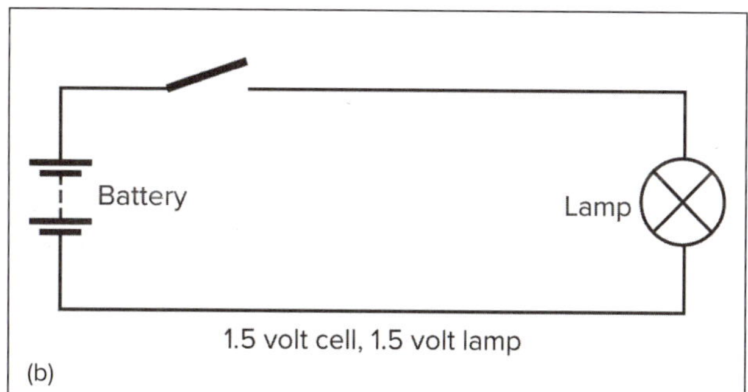

FIGURE 3.5 Simple circuit in pictorial and diagrammatic form
(a) McGraw-Hill Education

To be able to work safely and effectively with electricity, professionals in the electrotechnology industry must be aware of all the electrical properties of electrical circuits. Electrical workers must know what devices are used to take measurements in electric circuits. They also need to understand the relationship between current flow, voltage and resistance so that they can determine the correct operation of electrical circuits.

3.6 Voltage and current levels in a basic d.c. single path circuit

3.6.1 Electrical units

Electrical current

As stated in Chapter 2, electrical current is the flow or movement of electric charge through conductive materials from a point of higher electrical potential to a point of lower electrical potential. The unit of electric charge is the coulomb (C), which represents a total charge of 6.24×10^{18} (6 240 000 000 000 000 000) electrons. This number may appear to be very large, but remember that an individual electron is incredibly small, as is its charge. Current flow is electric charge that is moving. It is represented by the symbol (I), which stands for 'intensity' and is measured in amperes (A). One ampere is defined as the flow of one coulomb of charge in one second, i.e. $I = \frac{Q}{t}$ (coulombs per second).

It is important to understand that when electric current begins to flow, its effect is felt instantly. This is due to the fact that the electrons only have a very small distance to travel before they change orbit to a neighbouring atom. This transfer of energy repels another electron out of orbit, thus transferring its energy to the ejected electron. The second electron repeats the process, which continues through the conductive material. The overall effect of this process is that, while the individual electrons move through the conductor at a very slow rate that is called the 'drift velocity' (typically in the order of millimetres per hour), the electric current travels through the conductor at almost the speed of light, which is approximately 300 000 km/second.

Voltage

Electric current will not flow around a simple circuit unless the electrons have some form of pressure pushing them. Voltage can be thought of as the electrical pressure required to do that. It is a difference in electrical charge between two points and can also thus be called 'potential difference' (because the difference in charge has the potential to do work). Therefore, a voltage source is a form of potential energy which is not realised until it is connected to an electrical load via conductors and a switch. The unit of voltage is the volt and the symbol that represents it is usually (V), although other symbols such as (E), (U) and (EMF) are used in different electrical contexts. Voltage is defined as 1 volt required to move 1 ampere of current through 1 ohm of resistance.

A d.c. voltage source (e.g. a battery) is drawn with a '+' on one side and a '–' on the other side. These marks indicate the direction in which the voltage will attempt to push the electric current, and this potential direction is referred to as 'polarity'. It is important to remember that voltage is always relative between two points. This is very useful when trying to find a fault, for example a drop in voltage (or 'voltage drop') around a circuit. This is discussed in more detail in Chapter 10.

Resistance

The opposition to current flow is called 'resistance', which can be thought of as a measure of friction in a material when electrons attempt to flow through it. It is measured in ohms and is represented by the Greek letter omega (Ω).

When electric current flows, the movement of the electrons from atom to atom is resisted by the attraction that the negative electrons have to the positive nucleus of the atom. This resistance increases as countless electrons collide within the conductive material as part of electrical current flow. The increase in resistance creates heat within the conductor. It is therefore important to note that one of the effects of current flow is to generate heat within the conductor; heat is always generated from current flow, even if it is too small to be sensed. The higher the opposition to current flow a material has, the greater its value of resistance will be. For example, the heating element in an oven will become very hot when electric current passes through it. This is due to the fact that it has much higher resistance than the conductors that transfer the electrical current to it. When considering electrical circuits and electrical components, it is important to remember that *everything* contains some value of resistance. So, for example, when considering the electrical continuity of an electrical conductor, we must not think of the conductor as having zero resistance; it will still have a very small value of resistance. Resistance and all the factors that affect it are discussed in detail in Chapter 4.

3.6.2 Electrical measuring devices

Galvanometers

Dating back to 1820, the galvanometer (named after the scientist Luigi Galvani) was one of the earliest electrical instruments. The original galvanometer was simply a compass within a coil of copper wire. Any deflection of the compass needle indicated that an electric current flowed in the coil, and the position of the galvanometer in the circuit determined whether the experiment was measuring voltage or current.

Before electronics, highly accurate instruments were generally based on the galvanometer, with a very light mirror used in place of the needle and a reflected light beam indicating deflection, perhaps on a far wall in a darkened room.

Galvanometers are still used in some laboratories (Figure 3.6 shows an example). However, they have generally been replaced by ammeters and voltmeters with electronic amplification to allow very small voltages and current to be measured. Although the principles of both are based on the moving coil galvanometer, their movements are optimised for the task they are intended to perform.

FIGURE 3.6 A galvanometer
Pavel L Photo and Video/Shutterstock.com

Ammeters

Ammeters are devices that measure current flow. Since current flows through a circuit, an ammeter must always be connected **in series** with a circuit. Analogue or moving needle ammeters must have their positive (+) or red terminal always connected to the positive side of the circuit, otherwise the meter needle will attempt to move backwards against the meter stops and the instrument may be damaged. Ammeters have a very low internal resistance so that their presence in the circuit will not restrict the current flow and produce a voltage drop within the circuit; that would alter the correct operation of the circuit. This means an ammeter should never be placed in parallel with any points in a circuit that have a voltage or potential difference across them. A meter connected in that way could be seriously damaged.

The current flowing through the windings of the ammeter in Figure 3.7 (which is an analogue meter), when connected to a small source voltage (for example, 5 V) and a high resistance load will be very small. However, if the same analogue meter were connected across a 230 V supply, the current would be over four-and-a-half times as great. Although that is a small current, it is beyond the value for which the meter winding was designed, so the meter could be damaged.

FIGURE 3.7 An analogue ammeter
Michael Burrell/istockphoto/Getty Images

FIGURE 3.8 **An analogue voltmeter for 0 to 600 V of alternating current**
Shiyan Sergiy/Shutterstock.com

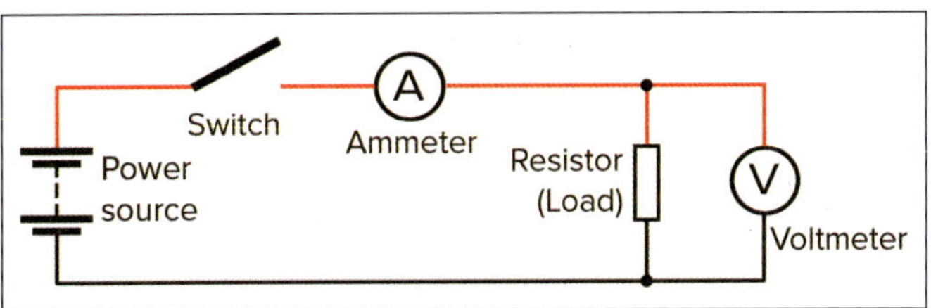

FIGURE 3.9 **Measuring instrument connections**

Voltmeters

Voltmeters measure voltage, which can be measured across any two points in a circuit. They must therefore be connected *in parallel* with those two points. To avoid excessive current flow, the internal resistance of a voltmeter must be very high. Common digital voltmeters are at least 10 megohms (MΩ), and many have an internal resistance of 200 MΩ. Due to a voltmeter's very high resistance, if one were accidentally connected in series with a circuit, it would look like a high-resistance joint or an open circuit and would prevent the circuit from functioning correctly. Figure 3.8 shows an analogue voltmeter.

Meters can only be used to measure values within their capacity. The maximum capacity of a meter is called its 'full-scale deflection' (FSD) value. If any meter pointer moves rapidly across the scale and beyond its FSD (the highest division marked on the scale), the power should be switched off immediately, before the meter is damaged.

Figure 3.9 shows the correct symbols and connections for an ammeter and a voltmeter when connected to read current and voltage in a simple circuit. Note that the ammeter is connected in series and the voltmeter is connected in parallel.

CHECK YOUR UNDERSTANDING

3.14 What are the three electrical units that are used with basic electrical circuits?

3.15 How are current and voltage measured in a circuit?

3.16 Why must an ammeter be connected in series and a voltmeter in parallel?

3.7 Relationship between voltage and current from measured values in a simple circuit

3.7.1 Ohm's Law

In 1826, the German physicist Georg Simon Ohm carried out a series of experiments in which he connected various pieces of material (some conductive, some resistive and some insulating) as the resistance in a simple circuit. He then measured the amount of current that passed through those materials and the voltage that was required to cause the current to flow.

Plotting his results, he was able to deduce from the resulting straight-line graphs that the current was directly proportional to the voltage.

He also saw that, in all of his experiments, as the voltage approached zero, the current also approached zero, and the plotted line of the voltage and current values was straight (or linear). The angle of this line on a graph only changed when the material he was testing changed. He was then able to state with confidence that in any electric circuit the current is directly proportional to the voltage and inversely proportional to the circuit resistance.

From that came Ohm's Law, which can be stated as: the current flowing between any two points in an electric circuit is directly proportional to the potential difference between these points and inversely proportional to the resistance of the circuit between these points.

The term 'directly proportional' means here that, as one quantity increases (in this case, the voltage), the quantity that depends on it (the current) will also increase. When two quantities are said to be 'inversely proportional', it means that, as one quantity increases (in this case, the resistance), the dependent quantity (the current) decreases.

Ohm's Law can be expressed by an equation:

$$I = \frac{V}{R}$$

Table 3.2 shows the elements of Ohm's Law. Electrical students should learn Ohm's Law and be able to explain it in clear, concise terms. It is the most important rule for any electrical worker, forming as it does the basis of many circuit calculations. Its importance in later studies of electrical theory, right through to electronics and engineering, cannot be overemphasised.

TABLE 3.2 The elements of Ohm's Law

Quantity	Symbol	Unit	Unit symbol	Effect in circuit
Voltage	V	volt	V	Electrical pressure
Current	I	amp	A	Electron flow
Resistance	R	ohm	Ω	Opposition

EXAMPLE 3.1

Given that a lamp draws 3 A when powered by a 6 V battery, determine the resistance of the lamp.

In this example, the voltage and the current are known so the equation to be used is $R = \frac{V}{I}$.

Step 1. List the known and unknown values:

$V = 6\text{ V}$
$I = 3\text{ A}$
$R = ?$

Step 2. Select a suitable equation to solve the problem:

$$I = \frac{V}{R}$$

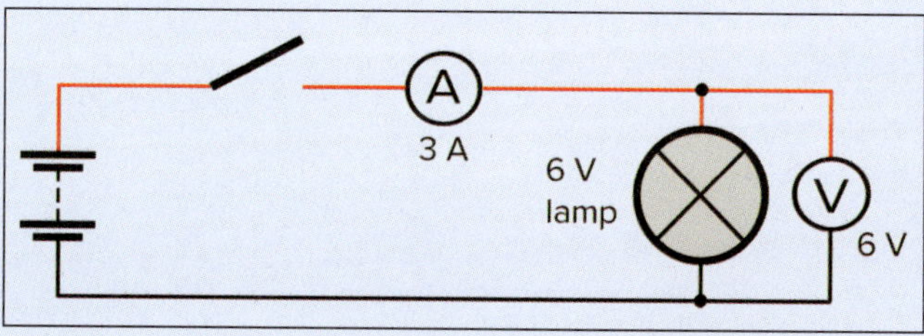

FIGURE 3.10 6 V battery connected to a lamp

Step 3. Transpose the equation, if necessary:

$$R = \frac{V}{I}$$

Step 4. Enter the values into the equation:

$$R = \frac{V}{I} = \frac{6}{3} = 2\ \Omega \text{ (ans)}$$

CHECK YOUR UNDERSTANDING

3.17 Explain Ohm's Law.

3.18 What three electrical units are used in Ohm's Law?

3.19 Why is Ohm's Law highly significant for electrical workers?

3.8 Determining voltage, current and resistance in a circuit, given any two of these quantities

The values of voltage, current and resistance can be determined in any circuit if two of the three values are known. This can be done by using Ohm's Law in either its given form or through transposition.

For example, to find the resistance required in a circuit that has 12 V applied and an allowable current of 2 A:

$$R = \frac{V}{I}$$
$$I = \frac{12}{2}$$
$$2 = 6\ \Omega$$

If the allowable current is known and the value of resistance is already determined, a transposed version of Ohm's Law would apply.

For example, if the allowable current is 3 A and the resistance is 20 Ω, the required applied voltage can be determined with:

$$\begin{aligned} V &= I \times R \\ &= 3 \times 20 \\ &= 60\ V \end{aligned}$$

Similarly, the current that will flow through a known resistance with a known value of voltage can also found by using the third version of Ohm's Law (by transposition).

For example, if a resistor of 10 Ω has a voltage of 24 V applied to it, the current that flows through the resistor can be determined with:

$$\begin{aligned} I &= \frac{V}{R} \\ &= \frac{24}{10} \\ &= 2.4\ A \end{aligned}$$

By using the principles of *direct inverse proportionality* (as shown in Figure 3.11), these values can also be quickly determined by understanding their relationships.

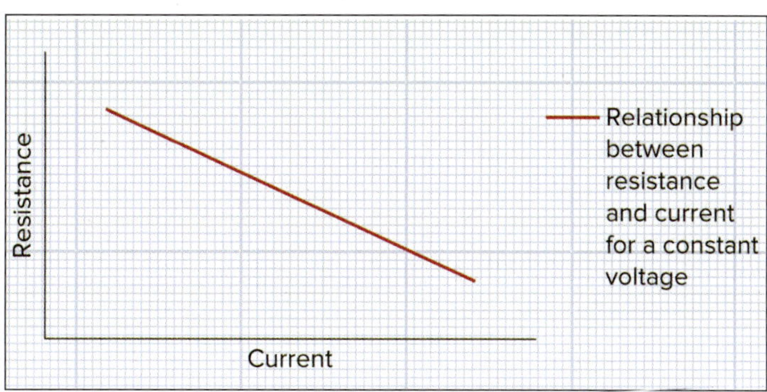

FIGURE 3.11 **Relationship between current and resistance for a constant voltage**

As resistance is equal to the voltage divided by the current, it can also be expressed as: resistance is directly proportional to voltage and inversely proportional to current. These relationships are shown in Figures 3.12 and 3.13 respectively.

Increasing current in a fixed resistance, such as drawing more current through a conductor, has the direct consequence of increasing the voltage drop across the conductor since V = IR.

FIGURE 3.12 **Relationship between voltage and resistance for a constant current**

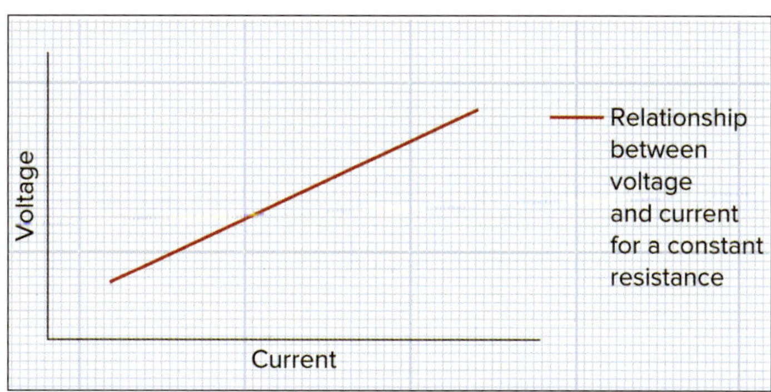

FIGURE 3.13 **Relationship between voltage and current for a constant resistance**

CHECK YOUR UNDERSTANDING

3.20 In a circuit with a fixed voltage, what will happen to the value of current if the circuit resistance is increased?

3.21 What will happen to the voltage drop across a fixed resistor if the current through the resistor is increased?

3.9 The relationship between force, power, work and energy

3.9.1 Force

To move an object from one position to another requires an amount of 'force'. Force is the mass of the object multiplied by the change in velocity, or acceleration, of the object. This can be either from standstill to moving or while the object is moving. For example, to get a vehicle to move from standstill to a velocity of 10 m/s takes a certain amount of force. Similarly, to reduce the vehicle's velocity from 10 m/s to 5 m/s takes a certain amount of force in the other direction.

The symbol used to represent force is (F) and the unit for force is the Newton. The unit symbol for the Newton is (N).

$$F = m \times a$$

When an object is being raised or lowered, it is subjected to gravity, which has a constant acceleration of 9.81 m/s^2. As this value is constant for earth, it has its own symbol in calculations (g).

$$F = m \times g$$

3.9.2 Energy and work

Energy and work are the same quantity, regardless of context (which could, for example, be mechanical, chemical or electrical). Energy can either be used to do work or can be stored. For instance, pumping water up to higher dams in a hydroelectric power system requires work to be done and energy to be expended. Once the dam is full, it provides a store of energy which can be directed through generators later to produce energy.

$$W = F \times d$$

3.9.3 Power

As seen in Chapter 2, 'power' is defined as the rate of doing work or expending energy.

$$P = \frac{W}{t}$$

CHECK YOUR UNDERSTANDING

3.22 What are the symbol and unit that are used to represent energy?

3.23 Explain the relationship between energy and work.

3.24 What are the symbol and unit that are used to represent power?

3.25 How are power and work different?

3.10 Power dissipated in a circuit from voltage, current and resistance values

3.10.1 Electrical power

Electrical power is calculated from electromotive force and the current that flows due to that force. As current is already a rate (coulombs per second), power can be calculated from the force involved (i.e. the EMF) and the displacement (i.e. the current flow). Therefore, power equals the voltage multiplied by the current.

$$\text{Power} = V \times I \qquad (1)$$
$$\text{i.e. } P = VI \qquad (2)$$

Sometimes, the available information includes the voltage and resistance, or current and resistance. The power can be found from those values using Ohm's Law and a process of substitution.

We know that:

$$P = VI \qquad (3)$$
$$\text{and } I = \frac{V}{R} \qquad (4)$$

Therefore, we can substitute the I component of the P = VI equation, giving us a new power equation:

$$P = V \times \frac{V}{R} \qquad (5)$$
$$= \frac{V^2}{R} \qquad (6)$$

Also:

$$P = VI \qquad (7)$$
$$\text{and } V = IR \qquad (8)$$

Therefore, we can substitute the V component of the P = VI equation, giving us a third power equation:

$$P = IR \times I \qquad (9)$$
$$= I^2R \qquad (10)$$

Electrical students should practise substitution by finding these two equations from Ohm's Law and the power equation. The equations above show why power is said to be 'proportional' either to voltage squared or current squared.

In a circuit with a fixed resistance, any increase in voltage will cause a proportional increase in current, and therefore the change in power will equal the change in voltage squared.

Likewise, in a circuit with a fixed resistance, any increase in current will cause a proportional increase in voltage drop, and therefore the change in power will equal the change in current squared. This is in line with the equations above.

Finally, in a circuit where the resistance can be changed, any increase in resistance will cause a decrease in current flow and, therefore, a decrease in power; and a decrease in resistance will allow more current to flow and, therefore, cause an increase in power.

So power is said to be 'inversely proportional' to resistance. For example, given that the load resistance in a 12 V circuit is 4 Ω, the power consumed is:

$$P = \frac{V^2}{R} = \frac{12^2}{4} = \frac{144}{4} = 36 \text{ W}$$

If the load resistance is doubled to 8 Ω, the current flow drops to half and the power consumption becomes 18 W. (Try proving this yourself using Ohm's Law.)

Real-world electrical calculations rarely correspond exactly to the basic equations. As learning an equation for every situation would require learning thousands of them, students must be prepared to master a basic set of equations and then convert them via algebra and transposition into what they require for the particular problem at hand.

EXAMPLE 3.2

A car with a 12 V battery has two headlamps rated at 50 W each. When both lamps are alight, what current will be drawn from the supply?

$$P = VI$$

Therefore, by transposition:

$$\begin{aligned} I &= \frac{P}{V} \\ &= \frac{(2 \times 50)}{12} \\ &= \frac{100}{12} \\ &= 8.33 \text{ A (ans)} \end{aligned}$$

EXAMPLE 3.3

A 12 V portable car cabin heater has 4 × 45 W elements.

Calculate the:

(a) resistance of the circuit
(b) value of current drawn from the supply.

(a) $P_{Total} = 4 \times 45 \text{ W} = 180 \text{ W}$

$$P = \frac{V^2}{R} \therefore R = \frac{V^2}{P} = \frac{12^2}{180} = 0.8\ \Omega \text{ (ans)}$$

(b) $P = VI \therefore I = \frac{P}{V} = \frac{180}{12} = 15 \text{ A (ans)}$

Check: $P = I^2R = 15^2 \times 0.8 = 180 \text{ W}$

3.10.2 Electrical energy

Electrical energy (like mechanical energy) is a product of power and time. Energy is usually measured in joules, but as power is the common unit for the rating of appliances, electrical energy is instead measured by the power of the appliances multiplied by the time that they are in operation.

Electrical energy is typically sold to consumers in kilowatt-hours (kWh). The kilowatt hour refers to the power in thousands of watts multiplied by the length of time the power was used in hours. One kWh is equal to 3.6×10^6 J or 3.6 MJ.

EXAMPLE 3.4

How much power does a 100 W light bulb use in a month, assuming it is used for 6 hours a day and there are 30 days in a month?

First as measured in joules:

Step 1. Time in seconds:

The lamp would operate for 30 days × 6 hours × 60 minutes × 60 seconds

$\therefore t = 648\ 000$ seconds

Step 2. Energy is power multiplied by time: $W = P \times t$

$\therefore W = 100 \times 648\ 000 = 64\ 800\ 000$ J or 64.8 MJ

Now as measured in kilowatt hours (kWh):

1 kWh = 3.6 MJ

$$\therefore \text{kWh} = \frac{64\,800\,000}{3\,600\,000} = 18 \text{ kWh (ans)}$$

EXAMPLE 3.5

An electric heater rated at 2400 W is switched on for an average of 6 hours per day for the whole of winter, from 1 June to 31 August.

How much electricity is used in that period by the heater?

Step 1. Calculate the time in hours:

$$\text{Days} = 30 + 31 + 31 = 92$$
$$\therefore \text{t} = 92 \text{ days} \times 6 \text{ hrs}$$
$$\therefore \text{t} = 552 \text{ hrs}$$

Step 2. Calculate the power in kilowatts:

$$\text{Power} = \frac{2400 \text{ W}}{1000} = 2.4 \text{ kW}$$

Step 3. Calculate kilowatt hours:

$$\text{W} = \text{P} \times \text{hr}$$
$$= 2.4 \text{ kW} \times 552 \text{ hrs}$$
$$\text{W} = 1324 \text{ kWh (ans)}$$

CHECK YOUR UNDERSTANDING

3.26 What are the methods and units that are used to calculate electrical power?

3.27 If electrical current passing through a lamp is doubled, what will happen to the electrical power used by the lamp?

3.28 What is the commercial reason for using the kilowatt hour?

3.11 Series circuit analysis

3.11.1 Series-connected circuits

In a *series-connected circuit* there is only one path. The components are connected one after the other like carriages in a train, and the current from the source flows through each. Therefore, the current in each component is the same as in every other component, and is also the same as in the path and in the source. Figure 3.14 shows a diagram of a series circuit.

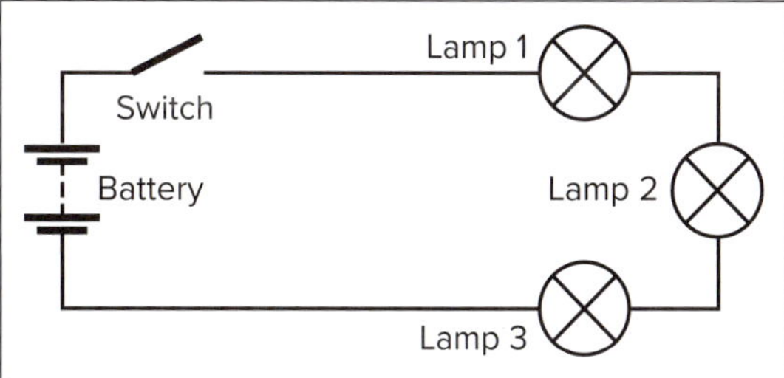

FIGURE 3.14 **Diagram of a series circuit**

3.11.2 Voltage in series circuits

As current flows through every component, every component must, according to Ohm's Law, have a voltage drop. Since the circuit cannot drop more voltage than it has to start with, it makes sense that the voltage drops must add up to the total applied voltage. This simple but important fact was documented by Gustav Kirchhoff in 1845. It became known as **Kirchhoff's Voltage Law (KVL)**.

Kirchhoff's Voltage Law states:

In any given circuit, the algebraic sum of the applied EMFs is equal to the algebraic sum of the voltage drops. This is written mathematically as:

$$\Sigma E = \Sigma V$$

Or, put simply:

$$E_{Total} = V_1 + V_2 + V_3 + \ldots V_n$$

3.11.3 Current in series circuits

A **series circuit** is defined as having only one current path or loop. That means that all the current coming from the source must pass through every component before getting back to the source. Therefore, every component will have the same current. This is written mathematically as:

$$I_{Total} = I_1 = I_2 = I_3 = \ldots I_n$$

So, a current of 5 A flowing into a series circuit will flow through every component, and the current measured passing through every component will be 5 A.

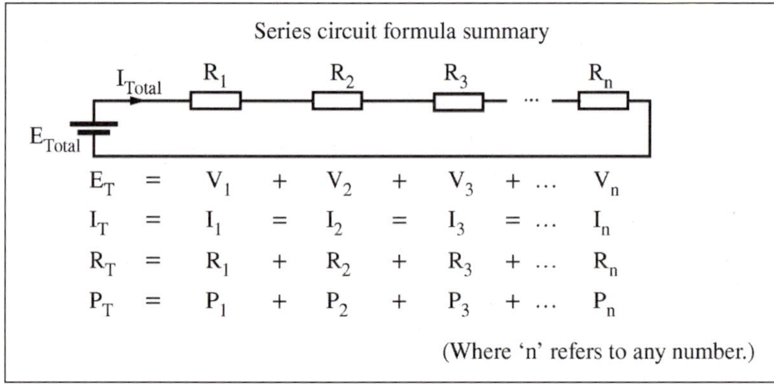

FIGURE 3.15 Resistors in series

Note: The equivalent resistance of a number of series resistors is always greater than any individual resistance.

3.11.4 Resistance in series circuits

Kirchhoff's Voltage Law, Ohm's Law and simple algebra dictate that the resistors in a series circuit such as that shown in Figure 3.15 must also be added to find the total resistance (R_{Total}) or equivalent series resistance (R_E).

3.11.5 Power in series circuits

Electrical power is calculated from the voltage and current, voltage and resistance or current and resistance. Power is the energy being used at a given rate to perform work. Therefore, it makes sense to assume that any work done requires power, which adds up as it is used. This is written mathematically as:

$$P_{Total} = P_1 + P_2 + P_3 + \ldots P_n$$

3.11.6 Faults in series circuits

When several components are in series, the current must pass through each to return to the source. Therefore, should any one component, conductor or joint become open circuit, the current cannot flow and the whole circuit will fail.

If a component becomes shorted, the voltage on all the remaining components must increase, with a resulting increase in current flow. Therefore, the components in series circuits must be carefully chosen to be suitable for both normal and fault conditions. Series circuits are rarely used for distribution systems as adding or removing loads would change the voltage requirements of every component in the circuit.

3.12 Parallel circuit analysis

3.12.1 Parallel-connected circuits

In a *parallel-connected circuit*, there is one path for each load. The separate current paths each have a load. The current flowing from the source in Figure 3.16 is divided into three, with a part of the total current flowing through each lamp. After the current passes through the lamps, the current recombines to become the total current that returns to the source.

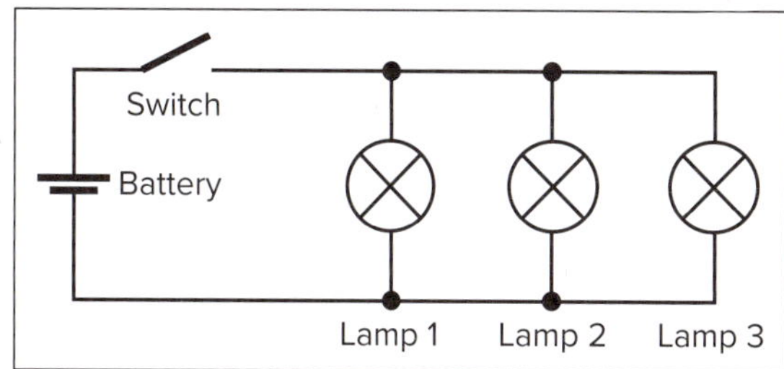

FIGURE 3.16 Parallel circuit

3.12.2 Voltage in parallel circuits

Voltage is measured across the terminals of a component by placing a voltmeter in parallel with the component.

Therefore, the voltage across any number of components connected in parallel will always be the same, regardless of where it is measured.

Another way to look at it is to recognise that a parallel circuit requires only two nodes and, as the voltage at a node is the same as everywhere on that node, the voltage between the two nodes must be the same anywhere between those two nodes, across any component.

This is written mathematically as:

$$V_{Total} = V_1 = V_2 = V_3 = \ldots V_n$$

3.12.3 Current in parallel circuits

Simple and series circuits have only one current path, but parallel circuits can have any number of current paths, sometimes referred to as 'branches'. The voltage across each branch causes a current to flow in accordance with Ohm's Law and the value of resistance of each branch. Kirchhoff noticed this and stated it in what we now call **Kirchhoff's Current Law (KCL).**

Kirchhoff's Current Law states:

The sum of the currents entering a junction equals the sum of the currents leaving that junction.

This is written mathematically as:

$$\sum I_{in} = \sum I_{out}$$

Or, put simply:

$$I_{Total} = I_1 + I_2 + I_3 + \ldots I_n$$

In other words, current cannot accumulate in a junction and, as there are no losses, what goes in must come out!

3.12.4 Resistance in parallel circuits

Since the resistors in a parallel circuit such as that shown in Figure 3.17 allow current to flow, the more resistors that are connected in parallel, the more current is allowed to flow. As the current is inversely proportional to resistance, more current means less total resistance. The formula for total resistance can be derived from the current and voltage formulas as follows:

$$\frac{1}{R_{Total}} = \frac{1}{R_1} + \frac{1}{R_2} + \frac{1}{R_3} + \ldots \frac{1}{R_n}$$

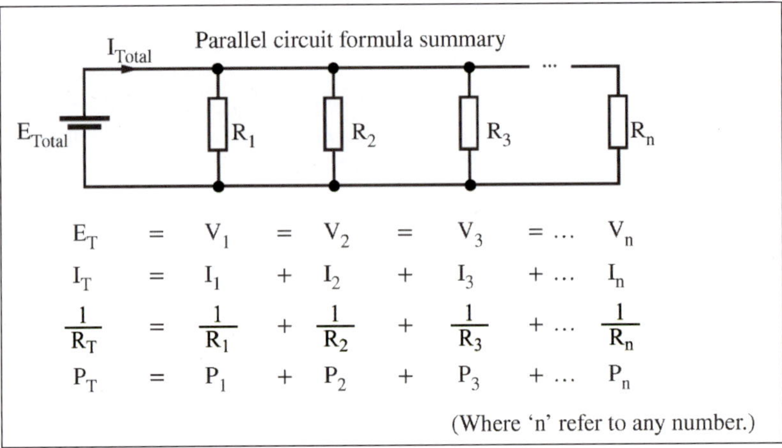

FIGURE 3.17 Resistors in parallel
Note: The equivalent resistance of a number of parallel resistors is always smaller than any individual resistance.

3.12.5 Power in parallel circuits

As in series circuits, power is used in individual resistors according to the voltage and current in the individual resistor. The total of that power is the sum of the power used in each resistor, i.e. the power formula is the same as for series circuits.

$$P_{Total} = P_1 + P_2 + P_3 + \ldots P_n$$

3.12.6 Faults in parallel circuits

Parallel circuits are preferred to series circuits in distribution systems because one branch can become open circuit without affecting the other branches. More importantly, new loads can be added to a circuit without any apparent effect on the voltage to other loads, provided that the maximum capacity of the source is not exceeded. All devices and appliances can be manufactured to the same rated voltage, as the voltage is the same in parallel circuits. For Australian use, 230 volts a.c. is the current standard voltage.

If one branch becomes open circuit, the other branches are not affected as they each have the same voltage applied. The total current is reduced by the amount that no longer flows in the open circuit branch.

If one branch becomes short-circuited, the voltage on the other branches will also initially drop. However, assuming branches have their own protection devices fitted, the remaining branches will be back to normal as soon as the offending one is disconnected by its circuit protection (i.e. when its fuse blows or its circuit-breaker opens).

3.13 Compound circuit analysis

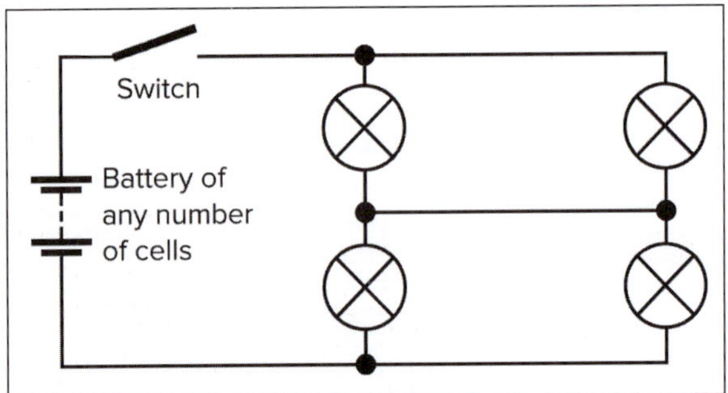

FIGURE 3.18 A compound circuit

3.13.1 Compound (series/parallel) circuits

Compound circuits (see Figure 3.18) are made up of both series and parallel parts, but the components are always either clearly in series with another component or else clearly in parallel with another component. The series or parallel components can always be replaced by a single component to simplify the circuit. Compound circuits are the most common type of circuit.

3.13.2 Equivalent resistance

Before delving into circuits of multiple components, it is important to be clear on what is meant by the term 'equivalent resistance'. When resistors are connected together, they can be replaced by a single resistor with the same overall resistance as the set of resistors. That is the way that compound circuits are analysed (on paper at least)—by replacing groups of resistors with a single resistor. A resistor that has the same value as a group of combined resistors is called an 'equivalent resistance' (R_E).

3.14 Power ratings of devices

All electrical devices are rated in terms of their output power, or the power that they can dissipate without being destroyed. The output power rating is the power that a device performs its rate of work at or the amount of energy the device can convert.

For example, a motor with a rating of 1 kW lifting a weight of 1000 kg up to 1 m:

$$\begin{aligned} F &= m \times g \\ &= 1000 \times 9.81 \\ &= 9810 \text{ N} \end{aligned}$$

$$\begin{aligned} W &= F \times d \\ &= 9810 \times 1 \\ &= 9810 \text{ J} \end{aligned}$$

$$\begin{aligned} P &= \frac{W}{t} \\ \therefore t &= \frac{W}{P} \\ &= \frac{9810}{1000} \\ &= 9.81 \text{ s} \end{aligned}$$

So the 1 kW motor can lift the 1000 kg up 1 m in 9.81 s.

The power rating of a device is the amount of power needed for the device to work. However, the amount of energy or electricity the device uses depends on how long the device is powered on for. It is also an indication of the amount of heat that the device can dissipate per second without suffering damage. If this value is exceeded for a certain period of time, the device will burn out. This time frame gets exponentially shorter as the amount of exceeded power increases.

CHECK YOUR UNDERSTANDING

3.29 How do input power and output power differ?

3.30 What does the term 'efficiency' mean?

3.31 Why can a machine never be greater than 100% efficient?

3.15 Methods for measuring electrical power in a d.c. circuit

The simplest method of measuring power in a d.c. circuit is by using a voltmeter and ammeter and multiplying the resulting two values together.

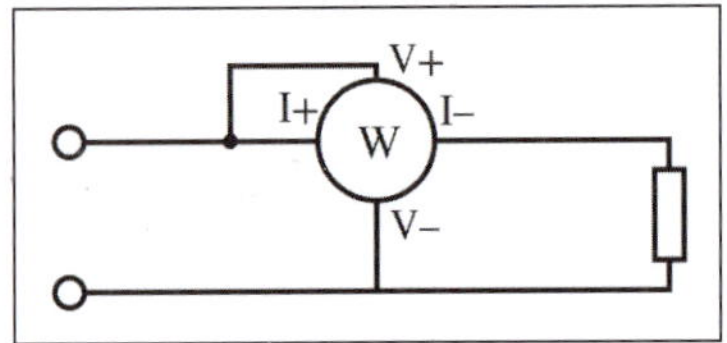

FIGURE 3.19 Connections for a wattmeter

Power can also be measured directly with an instrument known as a 'wattmeter' or 'power meter'. The method for connecting a wattmeter is shown in Figure 3.19. The meter has two measuring circuits: one to measure the current flow (I+ to I−), the other to measure the supply voltage (V+ to V−).

The two measuring circuits control the instrument needle in such a way that its position indicates power instead of current or voltage.

3.16 Physiological effects of current

3.16.1 Electric shock

As explained in Chapter 1, electric shock is the result of electrical energy causing a disruption to living tissue function. A human being who sustains an electric shock may stop breathing, and their heart may either stop beating or enter a condition known as 'ventricular fibrillation'. This is a state of rapid, uncoordinated muscle spasms. Blood stops flowing around the body, and in around four minutes the brain starts dying. If the person dies, they are said to have been electrocuted. Happily, advances in electrical protection and safe working practices have dramatically reduced the incidence of that.

In cases where electric shock has caused a person's heart to enter a stage of ventricular fibrillation, trained medical attendants may use a defibrillator to attempt to restart its normal pumping action. A defibrillator is an electrical device in which a capacitor is charged to a pre-determined value. Electrodes connected to the device are placed at appropriate places on the body and a capacitor is discharged in an attempt to shock the heart into resuming its normal rhythm. All training in the electrotechnology industry incorporates basic first-aid training in CPR (cardio-pulmonary resuscitation), and full first-aid training is highly desirable.

Electricians and the general public should be aware that delayed-onset symptoms can occur long after an electric shock, and internal damage may not become obvious for many hours.

3.16.2 The fundamental principles of AS/NZS 3000:2018 Wiring Rules

AS/NZS 3000:2018 is the standard that is required by law under electrical safety acts in Australia and New Zealand. Clause 1.5 of AS/NZS 3000:2018 outlines the fundamental principles of the particular standard that ensures the safety of persons, livestock and property from the dangers and damage that can result from reasonable use of electricity. It identifies three major types of danger and potential damage.

1. *Electric shock*: this is defined in AS/NZS 3000:2018 as the shock current that can arise from contact with parts that are live during normal service and those that are live when a fault exists. Contact with parts that are live under normal service, such as live wires inside an appliance, is avoided by the use of basic protection (protection from direct contact). Contact with parts that are live under fault conditions, such as a live frame on a faulty appliance, is avoided by the use of fault protection (protection from indirect contact).
2. *Excessive temperatures*: electricity flowing through conductors produces heat which, when it becomes excessive, can result in a hazard by either burning the insulation and other parts of the wiring or appliance or burning other materials in close proximity to the wiring or appliance. This could result in a fire and smoke hazard, physical burns for people or livestock and/or impairment of the safe operation of equipment.
3. *Explosive atmospheres*: electricity and its resulting heat can cause the explosion of gases and dust. Special care must be taken when installing electrical equipment in areas where such hazardous atmospheres are present. Examples of hazardous areas are mines, petrol stations and grain silos.

CHECK YOUR UNDERSTANDING

3.32 What is AS/NZS 3000:2018?

3.33 What are the three major types of danger and potential damage that are mentioned in AS/NZS 3000:2018?

3.34 What are some of the possible consequences of a person getting an electric shock?

3.17 Principles by which an electric current can produce heat, light, motion and a chemical reaction

3.17.1 Electrostatic effects

Although the electrostatic effects of electricity are less common than other types, their uses are increasing. Chimney stacks in power stations, for example, have an electrostatic generator that energises the smoke particles so they become attached to an electrostatic grid. The collected particles cluster and then fall into a chute to be taken away as pozzolanic ash for use in making cement. This reduces pollution and acid rain. Electrostatic processes are also used in painting and even in making sandpaper.

3.17.2 Electrochemical effects

Electrochemical processes occur in electric storage cells (accumulators) and batteries, and in **electroplating** and **anodising**. In fact, many chemical processes use electricity to separate, combine or refine chemicals, and to separate gases.

Manufacturing methods also commonly use electrical processes to etch and *electroform*, or for *electroerosion* (processes where electrical current removes material from—or shapes—the product being manufactured).

3.17.3 Heating effects

When the opposition of a conductor (its resistance) is overcome by an applied voltage and a current flows, work is done and energy is expended. When heat is produced, this raises the temperature of the conductor and increases its resistance.

If a conductor of high resistance is concentrated in a small area with suitable safeguards and insulation, a source of heat is available, as in the domestic radiator. This is a desirable outcome, but consider the end result if great quantities of heat are produced in the conductors supplying the electrical energy to the radiator. The conductors could be inside the hollow walls of a building and, if the temperature around the enclosed conductors were raised sufficiently, there would be a possibility of fire.

Steps must be taken to prevent such an undesirable event. The current flow must be reduced or the conductor's cross-sectional area increased. Both steps lead to reduced power loss and heating effect in the supply conductors.

The heating effect is always present when electric current flows.

3.17.4 Magnetic effects

Magnetism is created around any conductor with a current flowing through it. If this magnetism is unwanted, consideration must be given to rerouting the conductors to diminish it. Magnetic compasses are subject to stray magnetic fields and care must be taken to ensure that they are not affected in any way. With large currents, considerable magnetic forces can be set up between conductors, in which case precautions have to be taken.

When the magnetic effect is a desirable one, steps are taken to concentrate the magnetism in specific locations. The magnetic effect is increased by creating coils of conductors called 'solenoids'. Many turns may be added to the coils to enhance the effects. For more information about magnetism, see Chapter 6.

The magnetic effect is always present when electric current flows.

CHECK YOUR UNDERSTANDING

3.35 How does electric current produce light?

3.36 How does electric current produce heat?

3.37 What are the two effects of electricity that are always present when an electrical current is flowing?

3.18 Typical uses of the effects of current

3.18.1 Heating

Heat is a fundamental effect of electric current, and one of the most used. Electrical heating is highly efficient—any loss in efficiency is given off as heat. Heating devices range from radiant panel and oil heaters to fan heaters and large electric furnaces used in industrial processes.

A comparatively more energy- and cost-efficient method of heating and cooling is the use of refrigeration and air-conditioning. This method's comparative efficiency is due to the operating principle of transferring heat from one space to another through the use of a refrigerant gas, an electric compressor pump and electric fans.

3.18.2 Lighting

As the most noticeable effect of electric current after heat, light has been used for over a hundred years (although it was actually the introduction of electric refrigerators that accelerated the use of domestic electricity in the middle of the last century).

More recently, lighting has progressed from the old-fashioned incandescent type to gas-discharge lighting such as fluorescent, sodium and neon and then to highly efficient LED technology.

3.18.3 Producing motion

Just as electricity can be produced by moving or rotating a conductor through a magnetic field, motion can be produced by reversing the process. That is, when a conductor with a current flowing in it is placed in a magnetic field, it will move as a result of both the field around it and the external magnetic field.

This process can be used to move everything from small servos in robotics and compact disc drives to large machines such as those used in mining and rail. It can also be deployed to open and close switches in relays and contactors. Similarly, solenoids can be used in valves and locks to open and close mechanical devices.

3.18.4 Sound and radio frequencies

Similar to the way that electromagnetism can create motion, it can also create sound through the use of a small solenoid attached to a diaphragm or cone. When an alternating or oscillating signal is applied to the solenoid, the cone vibrates and the result is sound. This principle can also be reversed to create a microphone. A variation of this circuit is used for transmitters and receivers for radios, televisions and mobile phones.

The **piezoelectric effect** is also used in microphones to convert sound into electric signals. It is also used in ultrasonic cleaners and similar devices.

CHECK YOUR UNDERSTANDING

3.38 What are electrical appliances?

3.39 What are some common household appliances that use electrically produced heat?

3.19 Mechanisms by which metals corrode

3.19.1 Electrolytic corrosion

Dissimilar metals

The best-known effect of destructive corrosion occurs when two different metals are in contact in the presence of an electrolyte. An increasing problem in Australia is the connection of copper conductors to aluminium conductors. The copper and aluminium act as electrodes when an electrolyte is present due to condensation or rain that contains dissolved sulphur products. A simple cell is created and is effectively short-circuited on itself, as illustrated in Figure 3.20.

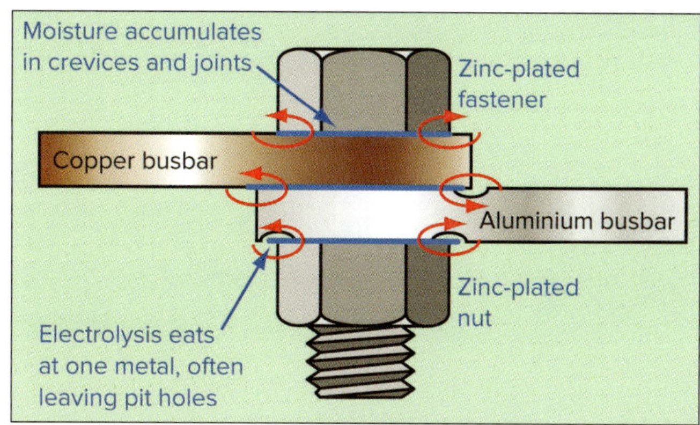

FIGURE 3.20 Corrosion of dissimilar metals

The metal with the higher potential becomes the anode that will tend to go into solution in the electrolyte (i.e. will corrode away). The result is a progressively deteriorating electrical connection, the generation of heat and the eventual failure of the connection. In some circumstances, the heat produced can set fire to an installation—and the problem then is more significant than just the failure of an electric circuit.

One way to combat this type of corrosion is to minimise the number of connections where dissimilar metals are joined. Another method is to prevent the entry of moisture to the joint by applying paint or another covering such as a paste that goes onto both metals prior to termination. Alternatively, all the materials that are being joined can be electroplated with a common metal, e.g. cadmium.

Stray electric currents

Underground pipes and cables can become corroded by the electrolytic action of stray underground currents. Dissimilar metals that are adjacent to each other in damp ground may produce these stray currents, or they may result from a nearby faulty electrical installation.

The electrical bonding of tracks in electric traction systems helps to reduce underground stray currents. Covering underground pipes and conductors with plastic sheaths minimises the corrosive effects. Corrosion tends to be concentrated at sharp bends in underground pipework, and appropriate measures to avoid this should be taken where practicable.

Electrolytic protection: sacrificial anodes

Electrolytic corrosion can be minimised by implementing sacrificial anodes. Anodes made of zinc blocks are bolted inside boilers or to the metal hulls of ships, adjacent to bronze propellers. The intention is to sacrifice a metal block, which can be replaced comparatively easily, rather than risking the corrosion of the boiler or ship's hull.

Electrolytic protection: cathodic protection

With this method, an external d.c. voltage is applied between the equipment to be protected and earth such that the equipment is at a lower potential than the surrounding soil. In practice, it has been found that most forms of corrosion can be prevented when the pipes or other metallic structures are approximately 0.5 V negative in relation to the surrounding soil.

AS/NZS 3000:2018, Clause 1.5.14 states that all parts of an electrical installation shall be protected from external influences that can occur under normal operation. These include corrosion and galvanic action.

CHECK YOUR UNDERSTANDING

3.40 What does the term 'electrolytic corrosion' mean?

3.41 How can electrolytic corrosion be minimised?

3.42 What detrimental effects can stray underground currents cause?

3.20 Fundamental principles listed in AS/NZS 3000:2018 for protection against the damaging effects of current

AS/NZS 3000:2018 is founded on the principle that electrical installations must be safe. The two general methods of achieving this are called 'basic protection' and 'fault protection'.

3.20.1 Basic protection

Basic protection is the means of protecting persons and livestock from shock hazards that would arise from direct contact with parts of the installation that are live under normal operation.

The four basic protective methods described in Clause 1.5.4.2 of AS/NZS 3000:2018 are:

1. *Insulation*: the most common method, this involves covering the live conductors with a non-conductive material such as the thermoplastic insulation used for electrical cables.
2. *Barriers or enclosures*, such as equipment housing and switchboards that can only be opened by a key or tool.
3. *Obstacles*, such as the safety fences surrounding high-power transformers.
4. *Placing out of reach*, as with, for example, overhead power lines.

The clause also notes that residual current devices (RCDs or, as they are more commonly known, safety switches) are only to be used as an additional means of protection and are now mandatory on all circuits in residential installations—the other means of protection listed above must be used as primary protection.

3.20.2 Fault protection

Fault protection is the means of protecting persons and livestock from shock hazards that would arise from indirect contact with parts of the installation that are not live under normal operation. These parts are only live when there is a fault, but they present just as big a risk of shock or injury as parts that are usually live.

The four protective methods described in Clause 1.5.5.2 of AS/NZS 3000:2018 are:

1. *Automatic disconnection of supply*, which is provided by circuit-breakers, fuses or a combination of both. Rewireable fuses were used in the past and, although they are still in use in old installations, they are now prohibited.
2. Preventing the fault current passing through a body by the use of *Class II (double-insulated) equipment*, as defined in Clause 1.4.32. This type of equipment is identified by the double square symbol.

3. Preventing the fault current passing through a body by the use of *electrical separation* where a transformer or other source of electrical supply (such as a battery) is used and connection to earth does not complete the circuit.
4. *Limiting the value of shock current* so that it does not reach a dangerous level. This is generally done through electrical circuitry that senses the output current and lowers the output energy if it becomes excessive.

Clause 1.5.5.2 also notes that, of these four methods, automatic disconnection of supply is the most commonly used.

CHECK YOUR UNDERSTANDING

3.43 What are the two main methods of protecting against the effects of current?

3.44 List the four basic methods of protection described in Clause 1.5.4.2 of AS/NZS 3000:2018.

3.45 List the four fault-protection methods described in Clause 1.5.5.2 of AS/NZS 3000:2018.

3.21 Electromotive force (EMF) sources and conversion of electrical energy

All systems of energy conversion—mechanical, chemical, magnetic or light/heat—that produce work exhibit some kind of energy loss. This loss is generally in the form of friction, resulting in the production of heat. Examples include heating due to resistance (from the friction of electrons in a conductor), mechanical friction in motor bearings, and air drag and noise loss, which in turn creates friction heat from the movement of air.

3.21.1 Input, output and losses of electrical systems

In any work-related system there are three main forms of energy or power. These are:

1. *input energy* or *input power*
2. *output energy* or *output power*
3. *losses* (which equal the difference between input and output).

For example, in an old-fashioned electric lamp there is the electrical input power, the losses as a result of the heat produced by the resistance of the lamp filament and the output power in the form of light energy.

Mathematically, this can be expressed by the equation:

$$\text{Output} = \text{input} - \text{losses or } P_{out} = P_{in} - P_{losses}$$

Efficiency is the ratio of the output to input and is generally expressed as a percentage. To determine the percentage efficiency of a system, the output is divided by the input and multiplied by 100.

Mathematically, this can be expressed by the equation:

$$\text{Efficiency } \% = \frac{\text{output}}{\text{input}} \times 100 \qquad \text{or} \qquad \eta\,\% = \frac{P_{out}}{P_{in}} \times 100$$

It is important to remember that, in real-world systems, the output will never be more than the input and the efficiency will never be greater than or equal to 100%; there are *always* losses.

Most significant losses in machines result in dissipation of heat, i.e. heat losses. A reduction in the amount of these losses results in an increase in the efficiency of the system. In the electrical industry, a low level of loss or high efficiency are generally considered desirable because of the economic and environmental advantages. Losses in an electrical system result in higher costs of energy production/supply on the input side and increased emissions from electrical energy produced from non-renewable sources. There are many other negative effects of low efficiency in an electrical system.

Primarily, low efficiency means increased power, and increased power at the same voltage equates to higher current drawn. When higher current is flowing in circuit wiring, the heat resulting from the resistance of the wiring leads to breakdown of the wiring insulation and eventually short-circuits between conductors. Alternatively, overheated conductors can also become open circuits when they melt.

This issue with overheating and its effect on conductors can also apply to machines. Transformers or motors that have increased losses (perhaps from damaged bearings or cooling systems) will eventually fail if the internal insulation breaks down or conductors open circuit.

EXAMPLE 3.6

An electric motor rated at 5 kW output is driving a machine. If the efficiency is 89%, calculate the power input required from the electrical supply.

Calculate the input power:

$$\text{Efficiency \%} = \left(\frac{\text{power output}}{\text{power input}}\right) \times 100$$

$$\text{Transposed, power input} = \text{power output} \times \left(\frac{100\%}{\text{efficiency \%}}\right)$$

$$\text{Power input} = 5 \times \left(\frac{100\%}{89\%}\right)$$

$$\text{Power input} = 5.62\ \text{kW (ans)}$$

If this machine runs at full load for four hours, how much energy is used?

Calculate the energy used:

$$\begin{aligned}\text{Energy used} &= \text{power input} \times \text{time (in hours)}\\ &= 5.62\ \text{kW} \times 4\ \text{hrs}\\ &= 22.5\ \text{kWh (ans)}\end{aligned}$$

3.21.2 Sources of electromotive force

There are six common sources of electromotive force:

1. **Electromagnetic**—electricity generated from electric conductors moving through a magnetic field.
2. **Electrochemical**—electricity generated from the reactions between chemicals, as used in the cells of batteries.
3. **Electrostatic**—electricity resulting from friction between appropriate materials, which is used to generate very high voltages at low power levels.
4. **Photoelectric**—electricity resulting from the conversion of light into electricity, e.g. solar panels.
5. **Thermoelectric**—electricity generated by dissimilar metals being exposed to heat, e.g. thermocouples used to measure high-temperature ovens.
6. **Piezoelectric**—the electric potential generated by placing a crystal under stress or releasing it from stress, e.g. gas lighters.

CHECK YOUR UNDERSTANDING

3.46 What does the term 'EMF' stand for?

3.47 What are the six methods of generating an EMF?

3.22 Principles of producing an electromotive force

3.22.1 Electromagnetic sources

The bulk of electrical energy is still generated in the same way as it was a hundred years ago, through an electromagnetic method. Fossil fuel is burned to produce the heat required to turn water into steam to drive mechanical turbines, which then drive electrical generators. However, renewable energy is beginning to replace conventional fuel in certain areas, and the market for it is projected to grow significantly in the near future.

The burning of fossil fuels results in pollution and causes global warming, so a clean, green replacement for them is both needed and inevitable. For more information about sustainable sources of energy, see Chapter 8.

See Chapter 2, sections 2.2.2 and 2.2.3, for more information on the production of electricity by renewable and non-renewable energy sources.

3.22.2 Thermoelectric sources

The thermoelectric method of voltage generation is mostly used as a means of temperature measurement. A **thermocouple** consists of two dissimilar metals joined at a point where the heat is applied (a 'hot junction'). A potential difference is created and appears at the other two ends of the conductors, where it is measured with a meter that is usually calibrated as a thermometer (see Figure 3.21).

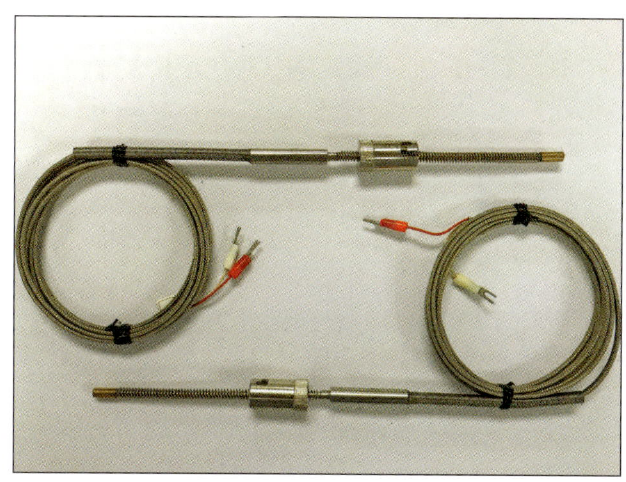

FIGURE 3.21 Thermocouples
Meaw_stocker/Shutterstock.com

The electromotive force produced is very small, and reading it requires a sensitive meter. A typical thermocouple made of copper and constantan (an alloy) would produce an EMF of around 9 millivolts for a temperature rise of 200°C. For greater voltages, thermocouples are sometimes connected as a group so their voltages add up to a higher value. The unit is then called a 'thermopile'.

Other combinations of metals allow temperatures to be measured from levels ranging from very low to very high (−273°C to >2000°C).

3.22.3 Piezoelectric sources

Some materials, when subjected to mechanical stress, generate a voltage between their faces. This is called the 'piezoelectric effect'. One common use of this phenomenon is the striking of an arc (or spark) to ignite the gas in a gas lighter, gas barbecue or gas stove (see Figure 3.22).

FIGURE 3.22 Piezoelectric striker ignition system
donikz/Shutterstock.com

The voltage generated is also directly proportional to the pressure, so piezo microphones, which are very small, are often used in devices such as mobile phones. Audiophiles might notice that very-high-frequency speakers are also piezo devices, and the speaker in any mobile phone is also likely to be a piezo. In these cases, however, the piezo crystal is working the other way, converting a voltage into a mechanical deflection of the very thin piezo wafer. Piezo speakers can also be found as the initiating element in echo sounders and ultrasound diagnostic machines in hospitals. They are used for the generation of very-high-frequency sound waves in ultrasound cleaning machines and distance-measuring devices such as electronic (ultrasonic) rulers.

The materials most commonly used are naturally occurring crystals such as quartz and Rochelle salts. These are cut into chips (also called 'crystals') with their two faces parallel to one of the three axes in the main crystal, and then mounted in holders before being connected into circuits.

When a voltage is applied to the opposite faces of the chip, the chip distorts and the movement can be measured. When the chip is distorted, a voltage appears at the opposite faces. At certain frequencies, the applied voltage causes a mechanical distortion, which causes a voltage to be generated in the crystal, which causes a distortion and so on.

This results in a natural rate of resonance, depending mainly on the angle of the cut and the physical dimensions of the crystals. This has led to one of the most common uses of crystals being to determine the operating frequencies in radio transmitter and receiver circuits.

CHECK YOUR UNDERSTANDING

3.48 Describe the six main methods of generating an EMF.

3.49 What is a renewable energy source?

3.50 What is a non-renewable energy source?

3.23 Principles of producing an electrical current from primary cells, secondary cells and fuel cells

3.23.1 Introduction

The term 'electrochemistry' refers to the effects that electricity has on chemicals or the electrical effect of chemicals. The common process of corrosion is an electrochemical effect, and electrochemical processes are widely used to prevent rusting.

The battery in a car or handheld torch or mobile phone is an electrochemical component. The human brain is an electrochemical organ.

Many manufacturing processes use electrochemical effects as a means of protection, or even of electromachining metal parts. Electroplating is used to protect items from corrosion or make them more appealing, such as its use in silver-plating brass tableware to create the appearance of a more expensive item. More recently, electroplating technology has been improved to an atomic level through nanolaminating, where coats of dense and ultra-fine materials can be used to add strength and other physical properties to weaker metals.

3.23.2 Electrochemistry

Electrochemistry in general refers to two opposite processes, one using electrical energy to create a chemical effect, the other using a chemical effect to create electrical energy.

The first process is often used in electroplating, where an external voltage is applied across a pair of electrodes, causing a current to flow through the electrolyte. The electrode connected to the positive polarity of the supply is called the 'anode' and the electrode connected to the negative is called the 'cathode'. The process of metal ions being removed from the anode and deposited onto the cathode has been adapted for electroplating and the refining of metals, as well as machining by electroerosion.

The second process converts chemical to electrical energy, which is produced within a cell by the disassociation of chemical molecules. This energy will flow through an external circuit.

3.23.3 Electrochemical energy sources

Primary cells

Primary cells are electrochemical devices that convert chemical energy into electric energy but stop when the chemicals are depleted. Until generators were invented, primary cells were the only source of electricity other than static electricity. Primary cells are fully charged when they are assembled and cannot be recharged.

Many types of primary cells have been invented, but the one that has held the field for the longest time is the Leclanché cell. Its modern form still has many uses, although other types of cells have been developed with greater efficiency, longer lifespan and greater energy density.

Secondary cells

Another type of cell is the secondary or rechargeable cell. Secondary cell chemistries allow the cell to be recharged by reversing the current flow using another energy source. Secondary cells have a zero or low potential difference (voltage) when assembled and need to be charged. They can subsequently also be recharged.

Many types of secondary cell have been invented and most remain in use. As with primary cells, modern examples have many improvements and new types are still being created.

Fuel cells

Fuel cells are electrochemical devices that are similar to primary and secondary cells in that they have an anode, cathode and an electrolyte. However, there is a significant difference in that they use a 'fuel'. This is typically hydrogen or hydrocarbons such as methane and LPG (liquefied petroleum gas) that combine with oxygen in the air to produce electricity with a waste product of water (for hydrogen) or water and carbon dioxide (for hydrocarbons). As long as this fuel is supplied, the fuel cell will deliver electricity. An essential part of this process is that the electrolyte is made of a specific material that only allows the passage of positive ions, forcing the electrons to seek an alternative route, thus creating electricity.

There are several types of fuel cell, the most common being the proton exchange membrane fuel cell (PEMFC). The applications of fuel cells include energy storage and management for peak electrical supply periods (peak levelling), cogeneration (generation of both electricity and heat in the form of hot water from cooling the cell) and fuel cell electric vehicles (FCEV).

Flow batteries are similar to fuel cells in their operation but have two sets of electrolytes stored outside them.

Voltaic cells

Figure 3.23 shows a simple voltaic cell consisting of two electrodes, one of copper and the other of zinc, immersed in a solution of dilute hydrochloric acid. The electrolyte need not be hydrochloric acid; other acids such as chromic, acetic or sulphuric acid can be substituted. Salt solutions such as common salt or copper sulphate can also be used.

To obtain a difference of potential between the two electrodes, only two conditions are necessary: the electrodes must be of different metals and they must be immersed in an electrolytic solution comprising an acid, alkali or salt. These cells have come to be known as 'galvanic' or, more popularly, 'voltaic' cells after Luigi Galvani and Alessandro Volta, who were pioneers in this area.

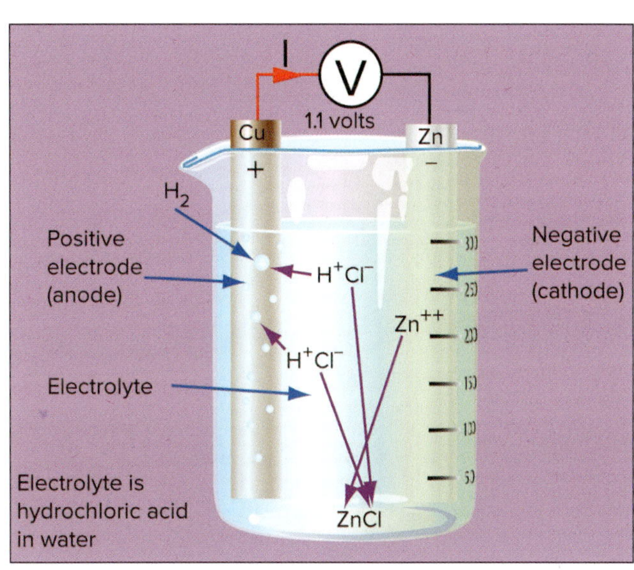

FIGURE 3.23 Voltaic cell

The container is made from a non-metallic material (e.g. glass) that will not be affected by the acid. It must not be made of a conducting material, and care should be taken that the electrodes do not touch each other.

Dissolving an electrolyte of hydrochloric acid (HCl) in pure water allows the acid to separate out into positive hydrogen ions (H+) and negative chlorine ions (Cl−). The chlorine ions combine with the zinc atoms, creating positive zinc ions (Zn++). This leaves free electrons on the zinc electrode, giving it a negative charge. The positive zinc ions combine with free chlorine ions, forming zinc chloride (ZnCl), a solid which sinks to the bottom of the cell.

Positive hydrogen ions move across to the copper electrode, where they combine with surface electrons from the copper electrode and become neutralised hydrogen atoms.

The hydrogen collects as bubbles that rise to the surface of the electrolyte and escape as free hydrogen gas. The removal of electrons from the copper electrode causes it to have a positive charge, and the cell has a potential equal to the difference between the negative charge on the zinc electrode and the positive charge on the copper electrode. This potential difference is usually around 1.1 V.

This simple cell is not very practical. The copper electrode becomes covered with hydrogen gas, preventing hydrogen ions from taking further electrons from the surface. This effect is known as 'polarisation', and it increases the internal resistance of the cell, reducing the output voltage.

CHECK YOUR UNDERSTANDING

3.51 What does the term 'electrochemistry' refer to?

3.52 How do primary and secondary cells differ?

3.53 What is a fuel cell?

SUMMARY

- The essence of an electrical worker's job is to understand electric circuits.
- There are five basic types of electrical circuits: simple circuit, series circuit, parallel circuit, compound circuit and complex circuit.
- Electrical circuit diagrams use standard symbols and labelling techniques so that electrical workers find them easy to understand.
- There are many common electrical abbreviations used with electrical circuits. These include 'V', 'I', 'R' and 'P'.
- A basic circuit has three main parts: source, path and load.
- Simple circuits are basic circuits that may also contain control devices such as a switch and protective devices such as a fuse.
- Electrical workers use measuring devices to measure the values of electrical properties in circuits.
- An ammeter is an electrical measuring device that is used to measure current.
- A voltmeter is an electrical measuring device that is used to measure voltage.
- Ammeters are connected in series.
- Voltmeters are connected in parallel.
- Generally speaking, there are three conditions of electric circuits: open circuit, closed circuit and short-circuit.
- Open and closed circuits are normal conditions of circuits, whereas a short-circuit is normally a fault condition.
- When dealing with very large and very small values in electrical circuits, engineering notation and electrical prefixes are used to make those values easier to represent.

- Ohm's Law describes the relationship between electric current, voltage and resistance in electric circuits.
- Ohm's Law states: the current flowing between any two points in an electric circuit is directly proportional to the potential difference between these points and inversely proportional to the resistance of the circuit between these points.
- A major advantage of using Ohm's Law is that it can determine how an electric circuit should function before power is applied.
- Electrical power used in a circuit is measured in watts and can be determined by using any two values from Ohm's Law.
- In the electrical context, the terms 'work' and 'energy' refer to the amount of electrical power used or supplied over time, its unit being the joule (J).
- The commercial unit of electrical energy is the kilowatt hour (kWh).
- The electrical measuring device used in circuits to measure power is the wattmeter.
- AS/NZS 3000:2018 states the electrical wiring rules that all electrical workers must follow to ensure all electrical work is performed to the required standard to keep persons and livestock safe from the dangers of electrical current.
- Electrical current can produce light, heat and motion, and these effects are used in electrical appliances.
- Electromotive force (or voltage) can be produced by electrical circuits by the following six methods: electromagnetic, electrochemical, electrostatic, photoelectric, thermoelectric and piezoelectric.
- Electrochemical circuits are the basis of batteries and fuel cells.
- Electrical batteries are grouped into two types, primary and secondary cells.

END-OF-CHAPTER QUESTIONS

3.1 State the five main components that are required to construct a basic electrical circuit.

3.2 Explain what is meant by 'open circuit', 'closed circuit', 'short-circuit' and 'fault'.

3.3 State Ohm's Law in your own words.

3.4 If the voltage in a basic d.c single path circuit is increased, will the current increase, decrease or stay the same?

3.5 Explain what is meant by the 'power rating' of a device and what the effects of exceeding this value would be.

3.6 If the resistance of a basic d.c. single path circuit is reduced, will the output power of the circuit increase, decrease or remain the same?

3.7 Explain the terms 'basic protection' and 'fault protection', as described in AS/NZS 3000.

3.8 What are the two effects of electricity that are always present when an electrical current is flowing?

3.9 What is the difference between a primary and a secondary cell?

3.10 Is the output power of an electrical device equal to the input power plus the lost power or minus the lost power?

3.11 In a series circuit, what is the relationship between the voltage drop across all the components and the supply voltage?

3.12 Give an example of where a series circuit is used in the electrotechnology industry.

3.13 In a parallel circuit, what is the relationship between the current in each parallel branch and the current drawn from the supply?

3.14 Give an example of where a parallel circuit is used in the electrotechnology industry.

3.15 Give an example of where a series/parallel (compound) circuit is used in the electrotechnology industry.

3.16 Using Ohm's Law, calculate the missing numbers from the following list:

12 V, 1 A, ? Ω
6 V, 0.5 A, ? Ω
24 V, ? A, 100 Ω
12 V, ? A, 10 Ω
? V, 100 mA, 240 Ω
? V, 3 A, 8 Ω

3.17 Calculate the power used by each of the following:

12 V, 1 A
6 V, 0.5 A
24 V, 100 Ω
12 V, 10 Ω
100 mA, 240 Ω
3 A, 8 Ω

3.18 How much power is required to run an electric motor if the motor requires 8 A at 230 V? How much energy will the motor use in a week if it runs for 8 hours per day?

3.19 A motor draws 15 A from a 2.5 mm^2 circuit over a distance of 80 m (each way). If the resistance of the 2.5 mm^2 cable is 0.7 ohms per 100 m, how much energy is lost in the cable over a 24-hour period?

3.20 An electric heater is rated at 3.6 kW when used on 230 V. How much current will it take?

3.21 Define a series circuit, a parallel circuit and a compound circuit.

3.22 A simple circuit has a 12 V battery supplying current to a 20 Ω resistor. Calculate the current drawn and the power used by the resistor.

3.23 What value resistor must be added in series with a 12 Ω resistor to give the circuit a total resistance of 28 Ω?

3.24 Find the total resistance of the following groups of resistors, each of which is connected in series:

12 Ω, 15 Ω and 21 Ω
6 Ω, 7.5 Ω, 10 Ω and 13.5 Ω
25 Ω, 22 Ω, 42 Ω and 55 Ω
1.8 kΩ, 2.7 kΩ, 3.3 kΩ and 4.7 kΩ
900 Ω, 1.2 kΩ, 82 kΩ and 1.5 MΩ

3.25 Find the total resistance of each of the following groups of resistors, each of which is connected in parallel:

180 Ω, 150 Ω, 120 Ω and 100 Ω
6 Ω, 8 Ω, 12 Ω and 15 Ω
9 Ω, 9 Ω, 12 Ω and 12 Ω
36 Ω, 24 Ω and 18 Ω
1.2 MΩ, 1.8 MΩ and 820 kΩ

3.26 In a series circuit as shown in Figure 3.24, the four resistors are 2 Ω, 4 Ω, 6Ω and 8 Ω. If the potential difference across the 4 Ω resistor is 10 V, find the:

(a) current in the circuit
(b) p.d. across the other resistors and the supply voltage
(c) power consumed by each resistor and the total power used.

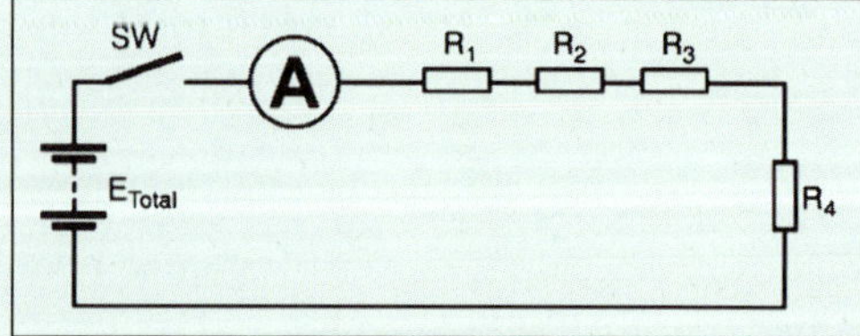

FIGURE 3.24 Series circuit

3.27 The parallel circuit in Figure 3.25 has the following values known. The resistance of the first branch is 12 Ω. The current in the second branch is 300 mA. The power taken in the third branch is 1.5 W and the fourth branch has a voltage of 10 V. Calculate all the VIRP values in the circuit if the total current is 1.5 A.

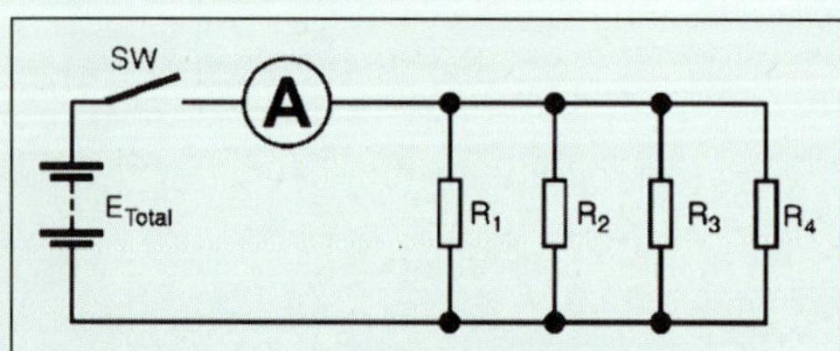

FIGURE 3.25 Parallel circuit

CHAPTER 4

Resistors and resistance measurement

LEARNING OBJECTIVES

- Identify the types and applications of fixed and variable resistors
- Understand the characteristics of temperature-, voltage- and light-dependent resistors
- Understand why resistors have a power rating
- Understand how power loss occurs in resistors
- Apply the resistor colour-code tables to determine resistor values
- Understand the four factors that affect resistance
- Apply techniques for calculating the resistance of conductors
- Understand how to use digital and analogue meters to measure resistance
- Understand techniques for measuring continuity and insulation resistance for AS/NZS 3000:2018 verification purposes

PREREQUISITE KNOWLEDGE

- Understand the concepts of voltage, current and resistance
- Be able to transpose mathematical equations
- Understand Ohm's Law

4.1 Types and applications of fixed and variable resistors used in the electrotechnology industry

4.1.1 Resistance

Resistance is the opposition to current flow, or the restriction caused by the atomic attraction between protons and electrons. The unit of electrical resistance is the ohm (Ω).

Resistance was first quantified by Georg Ohm, who established the relationship between resistance, voltage and current:

$$R = \frac{V}{I}$$

One ohm is defined as the electrical resistance between two points in a conductor when a constant electric potential of one volt is applied across the two points, producing a current flow of one ampere through the conductor.

Two main practical effects of resistance must be considered in electrical work. The first is when current flowing through a resistor results in a voltage drop; while that can be detrimental to electrical appliances, it can be used advantageously in electrical circuits. The second effect of resistance is that of heat generation, which is mostly used for heating and cooking. The most common usage of resistors is to bring about these two effects.

4.1.2 Resistor types

Resistors are chosen according to several parameters. These include resistance, current-carrying capacity, voltage rating and power dissipation. Generally, the size of a resistor is determined by all of those factors. Resistors can be broadly classified into three types: fixed, variable and adjustable. Adjustable resistors can be adjusted by a moving tap, while variable resistors can be operated from a control panel. In between those two types are resistors that are adjusted by using a tool such as a screwdriver. They are known as 'trimmer', 'trimpot' or 'pre-set' resistors.

Fixed resistor

A fixed resistor is a resistor whose resistance value cannot be varied and does not change with fluctuations in voltage, current or temperature.

Figure 4.1 shows the IEC (International Electrotechnical Commission) standard symbol for a fixed resistor.

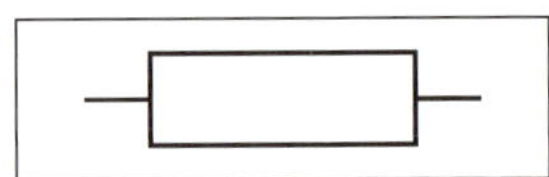

FIGURE 4.1 IEC symbol for a fixed resistor

Fixed resistors are the most-used resistors in electronic circuits. They come in various shapes and sizes and can be manufactured using a range of materials. Types of fixed resistor include:

- wire-wound resistor
- carbon-composition (also called 'carbon-composite') resistor
- carbon-film resistor
- metal-film resistor
- metal oxide film resistor
- metal glaze resistor
- foil resistor.

Variable resistor

A variable resistor can have its resistance value changed. This is an easy process that involves a rotary or linear sliding contact moving across a resistive element or track. Variable resistors are mostly used where the user does not know exact values, and they allow the user to change the resistance value while the circuit is either de-energised or energised. The volume control on a conventional radio is an example of a variable resistor. Two-terminal variable resistors that are usually used to control a current are known as 'rheostats'. Variable resistors with three terminals are usually called 'potentiometers' as their main function is to control a voltage. Often, a three-terminal potentiometer will have the middle terminal connected to one end to make it into a rheostat, even though it may still be called a potentiometer (or 'pot' for short).

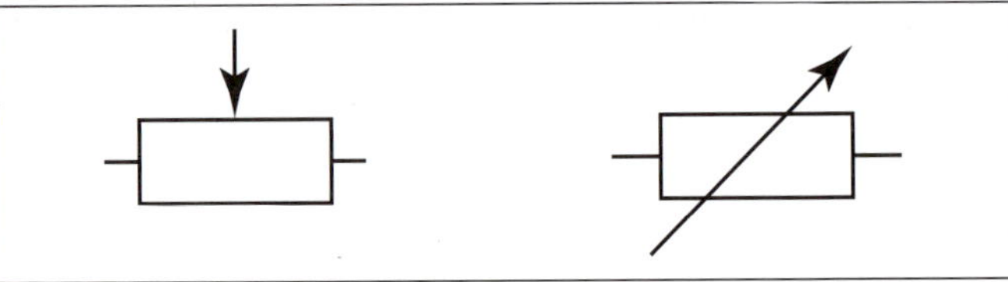

FIGURE 4.2 IEC symbols for variable resistors

Figure 4.2 shows the IEC standard symbols for variable resistors while Figure 4.3 shows what a resistor looks like, along with its diagrammatic representation.

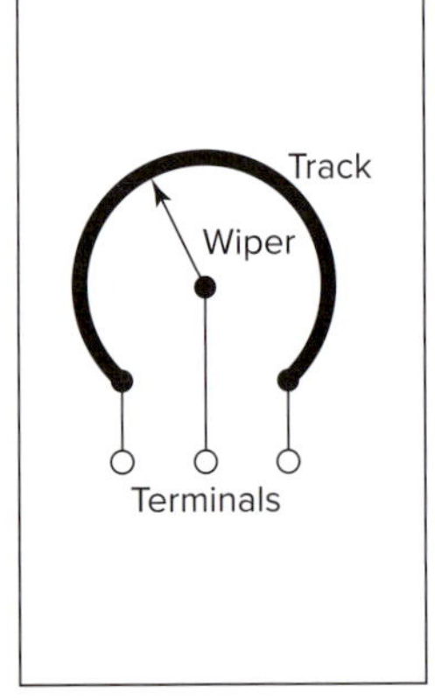

FIGURE 4.3 Variable resistor in diagrammatic and actual forms
(photo) Oasishifi/Shutterstock.com

Variable resistor types include:

- potentiometer
- rheostat
- thermistor
- magneto resistor
- photoresistor
- humistor
- force-sensitive resistor.

Adjustable resistor

An adjustable resistor is a type of variable resistor that is used by a technician to adjust, calibrate or fine-tune circuit parameters. (They are not intended to be adjusted by the device user.) Trimpots or trimmer potentiometers are adjustable resistors. If used as a rheostat, they are called 'pre-set' resistors.

Figure 4.4 shows the IEC standard symbol for trimpot or 'trimmer' adjustable resistors while Figure 4.5 shows what they actually look like.

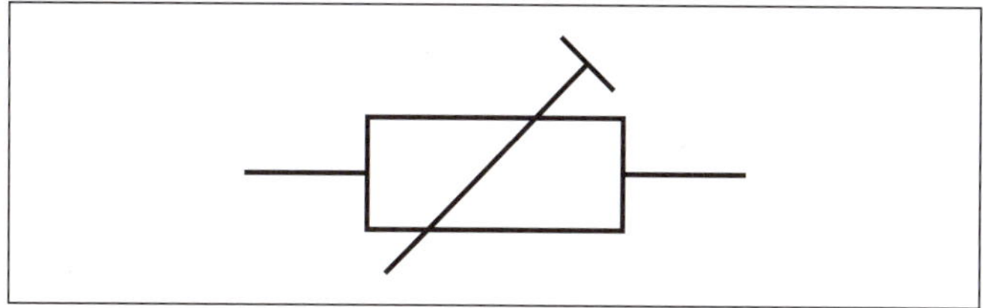

FIGURE 4.4 IEC symbol for a trimmer adjustable resistor

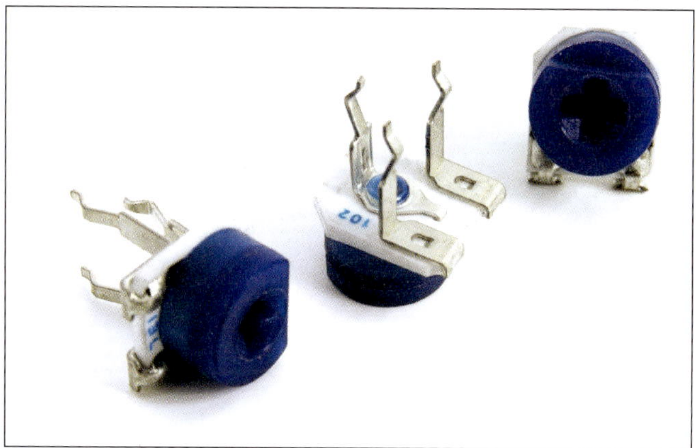

FIGURE 4.5 Various trimmer resistors
MilanStock.com/Alamy Stock Photo

Linear vs non-linear resistors

Variable and adjustable resistors can be classified as either 'linear' or 'non-linear'. The classification depends on how the resistance value changes in response to the movement of the adjustable control. Figure 4.6 is a graphical representation of both linear and non-linear variable resistance responses.

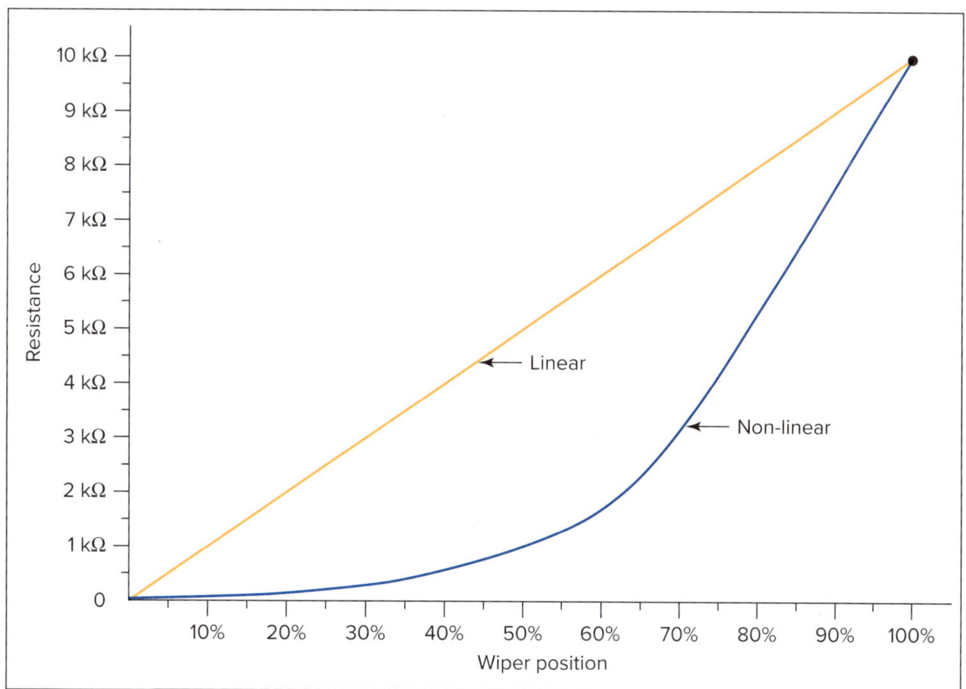

FIGURE 4.6 Linear and non-linear resistance responses

Linear resistance

For linear resistors the change in resistance value is directly proportional to the movement or position of the adjustable control. For example, if a variable resistor with a resistance range of 0 to 10 kΩ has its adjustable control moved halfway of its travel (i.e. 50%), the resistance value would be set at 5 kΩ, which is 50% of 10 kΩ. If the same variable resistor had its adjustable control moved to 85% of its travel, the resistance value would be set to 8.5 kΩ, which is 85% of 10 kΩ.

Non-linear resistance

For non-linear resistors the change in resistance value is *not* directly proportional to movement or position of the adjustable control. For example, if a variable resistor with a resistance range of 0 to 10 kΩ has its adjustable control moved to 40% of its travel, its resistance value changes only slightly to 500 Ω. However, if the adjustable control is moved another 40% so that its total movement is now at 80%, the resistance value changes considerably to be approximately 5.5 kΩ.

Cast grid resistors

Large high-power resistors were once commonly cast from iron or an iron alloy. They may be cooled by having forced air blown over them or by a liquid such as water flowing through a hollow core. Cast grid resistors are generally used only where an extremely high current needs to be controlled, such as for speed control of winding motors for large lifts and cranes and traction motors. Figure 4.7 shows a cast grid resistor.

FIGURE 4.7 A cast grid resistor
Simon Dand

Co-axial sheathed elements

The heating elements used in electric stoves, electric fry pans and strip heaters are commonly made from a resistive conductor inside an insulating ceramic powder, which is in turn within an earthed metal sheath. As used in stoves, the element heats the sheath to a temperature of around 660°C (hot enough to melt aluminium if an empty saucepan were to be left on the element unattended). Figure 4.8 shows a diagram of the three parts of a co-axial sheathed element beside a picture of its application in a domestic oven.

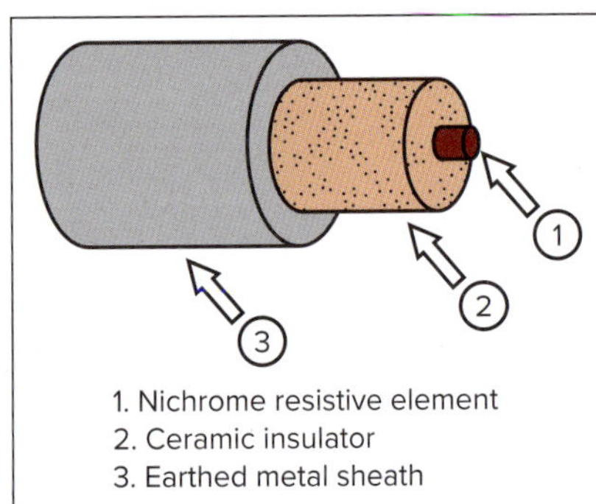

FIGURE 4.8 Co-axial sheathed element
(photo) Simon Dand

Wire-wound resistors

Smaller resistors from five to several hundred watts are commonly made by winding a resistance wire around a ceramic former. An insulating layer is generally used to cover the wire, not only as insulation but also to protect it from damage and corrosion. Large resistors, which are designed to dissipate a lot of heat, generally have a large surface area or some other means of conveying the heat away from them. Some resistors come in their own finned heatsink (which looks and acts like a radiator).

Wire-wound resistors are generally large enough to have their resistance and power rating printed on their body. Some are made to be tapped or adjusted via a panel-mounted shaft and knob. Figure 4.9 shows a wire-wound resistor.

FIGURE 4.9 Wire-wound resistor
Vladislav12/Shutterstock.com

Carbon-composition resistors

Carbon-composition resistors are physically smaller and have lower power ratings than other types of resistor. Common ratings are 1/4 W, 1/2 W, 1 W and 2 W. Figure 4.10 illustrates some different kinds of resistors, with the construction of a carbon-composition resistor shown at the top.

With carbon-film resistors, a resistive carbon material is placed on a ceramic tube and laser cut into a spiral to attain the exact resistance required (within the tolerance required). During manufacture, the thickness and composition of the compound can be varied to make resistors of different values between 0.01 Ω and 10 MΩ.

Carbon resistors can age and their resistance can change as they do so, necessitating the development and use of a more practical type of small resistor. The metal-film resistor is like the carbon type in size and appearance, and only its base colour identifies it as not being a carbon resistor. (Carbon resistors have a brown or cream base colour, while metal-film resistors are usually blue or green.)

Carbon-film variable resistors are manufactured by applying a conductive paint to a non-conductive base material. A wiper contact slides over the conductive paint to allow the terminal resistance to be controlled.

The resistors described above are too small for their resistance and power rating to be printed on them. This has led to the development of a standard colour-coding identification scheme. (The coloured bands on fixed-value resistors are placed closer to one end of the body during manufacture.)

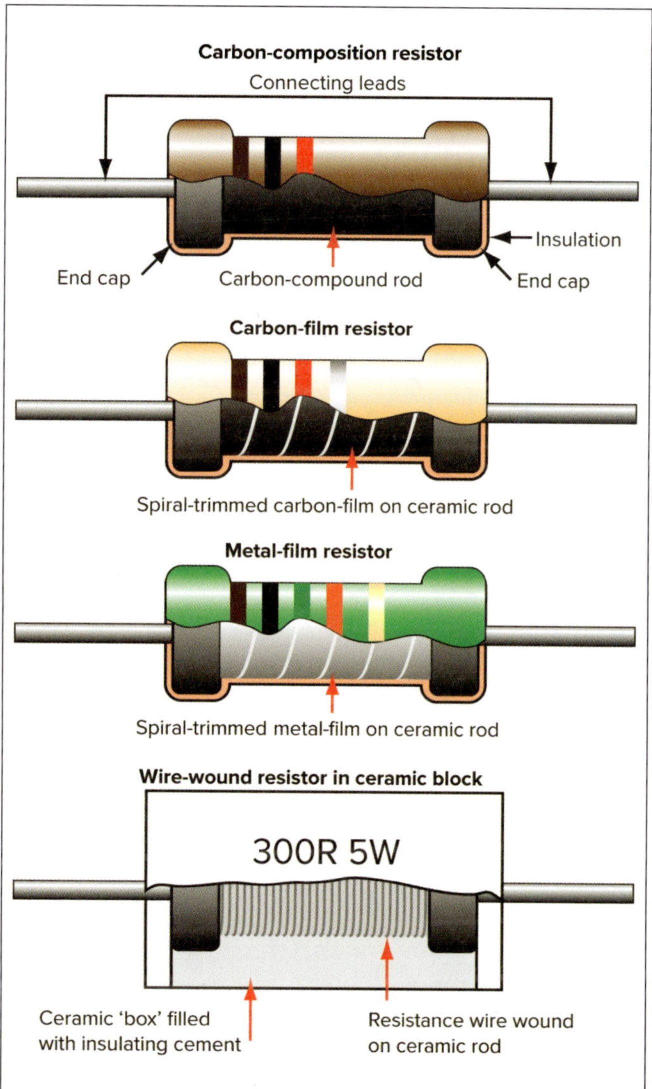

FIGURE 4.10 Resistor types

CHECK YOUR UNDERSTANDING

4.1 What are four resistor types used in the electrotechnology industry?

4.2 What class of resistors are adjusted by using a tool such as a screwdriver?

4.3 What are the common power ratings of carbon-composition resistors?

4.4 What are the standard symbols used in Australia to represent fixed, variable and adjustable resistors?

4.5 What is the difference between a rheostat and a potentiometer?

4.2 Special-purpose resistors

4.2.1 Resistors as transducers

Non-ohmic (non-linear) resistors exist that change resistance according to other physical parameters such as voltage, temperature and ambient light. These resistors are useful as sensors in instrumentation, and when used as a sensor are known as 'transducers'.

4.2.2 Thermistors

A **thermistor** is a type of resistor whose resistance value varies with small changes in temperature. (The word 'thermistor' is derived from the combination of 'thermal' and 'resistor'.)

Thermistors can be grouped into two types, based on how their resistance value changes in response to temperature.

Negative temperature coefficient (NTC) thermistors

The resistance value of NTC thermistors decreases with an increase in temperature. Therefore, current flowing through an NTC thermistor will increase with an increase in temperature. The IEC standard symbol for NTC thermistors is shown in Figure 4.11.

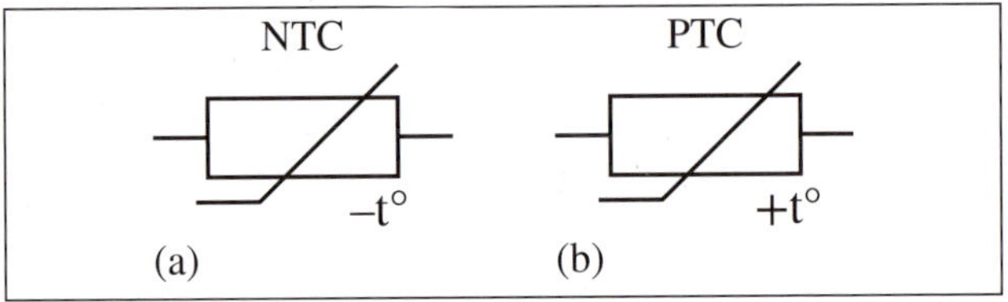

FIGURE 4.11 IEC symbol for (a) an NTC thermistor and (b) a PTC thermistor

Positive temperature coefficient (PTC) thermistors

The resistance value of PTC thermistors increases with an increase in temperature. Therefore, current flowing through a PTC thermistor will *decrease* with an increase in temperature.

Thermistors are used for:

- motor winding protection
- hot ends of 3D printers
- home appliances such as ovens, hair dryers, toasters and refrigerators
- computers
- temperature sensors.

Figure 4.12 shows the typical characteristics of NTC (blue) and PTC (red) thermistors. Usually, the resistance values are shown on a logarithmic scale, but in order to illustrate the sharp 'knee' of the curve as the PTC heats up, a representative curve has been drawn from test

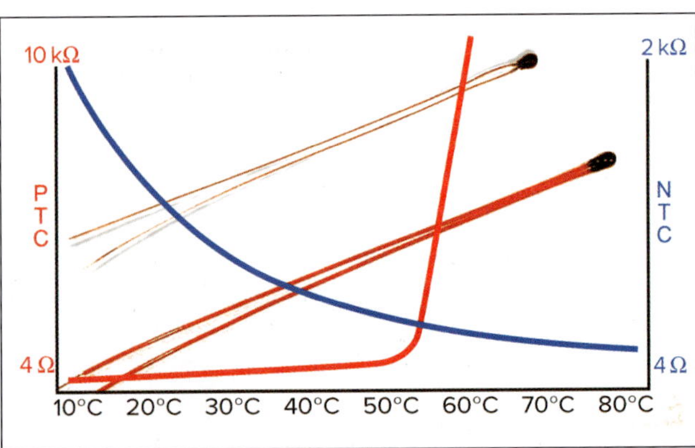

FIGURE 4.12 Typical thermistor characteristics

results on actual thermistors. The resistance of the PTC thermistor is only 4 Ω at room temperature, but rises rapidly once the temperature goes above around 50°C. The NTC thermistor on the other hand has a high resistance at low temperatures and drops in a non-linear fashion as temperature rises.

FIGURE 4.13 Voltage-dependent resistors (VDRs)
badamar/Shutterstock.com

4.2.3 Low temperature coefficient resistors

In some circumstances, a resistor is required to retain its value over a range of temperatures. An ideal material in such circumstances is manganin, which has a temperature coefficient of resistance given as 0.00001 (10^{-5}). Its resistance change is generally so small compared with other metallic elements that it is often listed as zero. Manganin is used where resistors of high temperature stability are required, for instance in measuring instruments.

4.2.4 Voltage-dependent resistors (VDRs)

A **voltage-dependent resistor** (**VDR**) or **varistor** is a component that has its resistance varied with the applied voltage. VDRs are manufactured from a mixture of materials to have a very high resistance at lower voltages but a very low resistance above a certain critical value. A VDR is usually presented in disc form, with two leads for connection into a circuit, as shown in Figure 4.13.

The VDR is connected in parallel with the voltage supply, close to the circuit it must protect. Usually it does not conduct but when the supply voltage exceeds a designed limit, it breaks down and conducts the excess energy away from the circuit it is protecting. Modern VDRs are constructed using sintered ceramic metal-oxide materials and are commonly known as *metal-oxide varistors (MOVs)*. Figure 4.14 shows how a VDR's resistance varies with applied voltage.

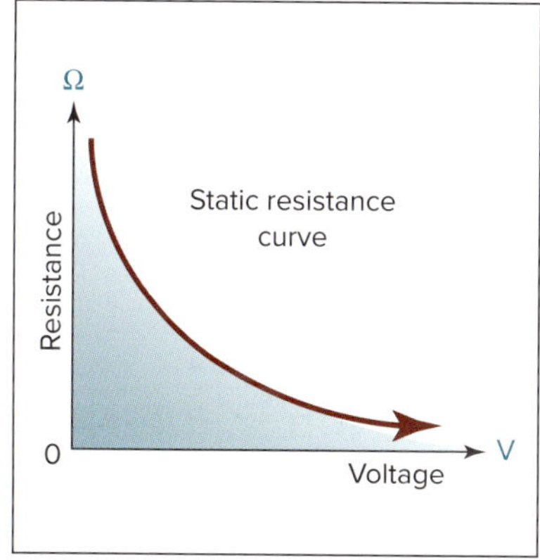

FIGURE 4.14 VDR's response to applied voltage

The primary purpose of a VDR is to protect equipment against voltage surges such as lightning strikes. The response time is extremely quick—an obviously necessary characteristic.

In some circuits, VDRs are intended to blow a fuse or trip a circuit-breaker to isolate the electric circuit being protected. Sometimes a VDR is built into a package called a 'surge protector', which can be installed in the main switchboard or on the consumer's terminals. Depending on the individual device and the severity of the surge, a VDR might have to be replaced after it has operated.

The IEC standard symbol for a voltage-dependent resistor is shown in Figure 4.15.

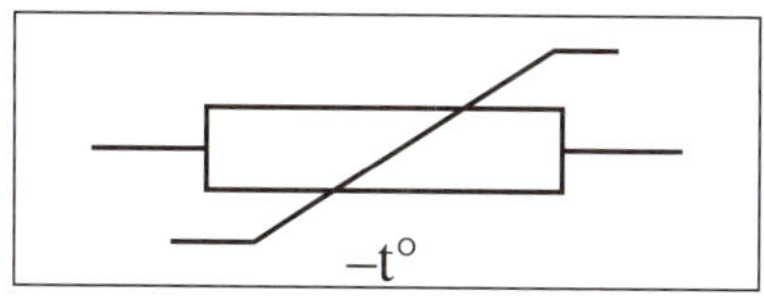

FIGURE 4.15 IEC symbol for a voltage-dependent resistor

4.2.5 Light-dependent resistors (LDRs)

Light-dependent resistors (LDRs) are used to detect light levels such as in PE (photo-electric) cells on power poles to turn streetlights on and off, or to control other night lighting. When light falls on the resistor, the resistance changes and an electronic circuit senses that change to turn a relay or solid-state switch on or off. Figure 4.16 shows the relationship between resistance and illumination in an LDR.

Light-dependent resistors are usually made from cadmium-sulphide film mounted on a ceramic or phenolic plate and covered with a conductive grid. To ensure there is no build-up of contamination and to protect the active

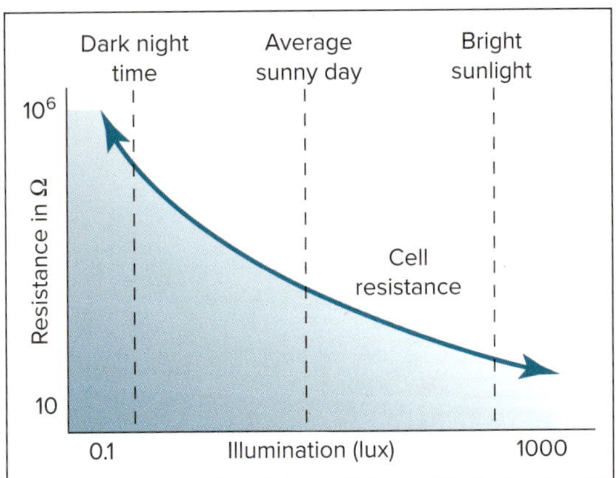

FIGURE 4.16 Resistance and illumination in an LDR
Jumaat, S & Othman, M. (2018). Solar Energy Measurement Using Arduino. MATEC Web of Conferences. 150. 01007.10.1051/matecconf/201815001007. Fig 3 CC BY 4.0; https://creativecommons.org/licenses/by/4.0/

FIGURE 4.17 LDRs
Alexander Konradi/Shutterstock.com

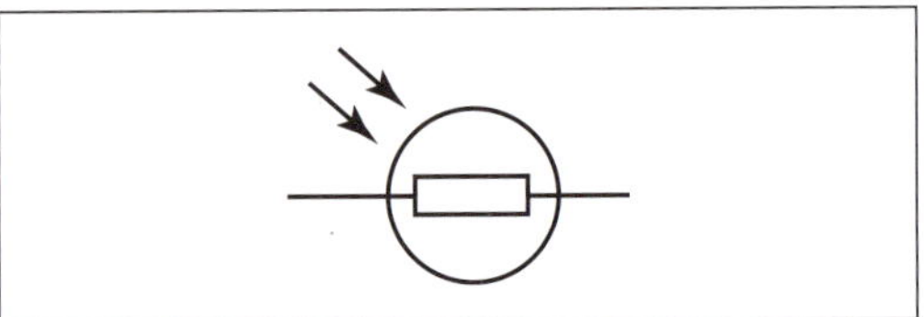

FIGURE 4.18 IEC symbol for a light-dependent resistor

material, the resistor is mounted in an evacuated glass envelope or covered with a clear plastic encapsulation. Figure 4.17 shows a variety of LDRs.

Figure 4.18 shows the IEC standard symbol for a light-dependent resistor.

The resistance of the device varies considerably with the amount of light received on the surface. Typically, the resistance in complete darkness can be as high as 10 MΩ, reducing to possibly 100 Ω in sunlight.

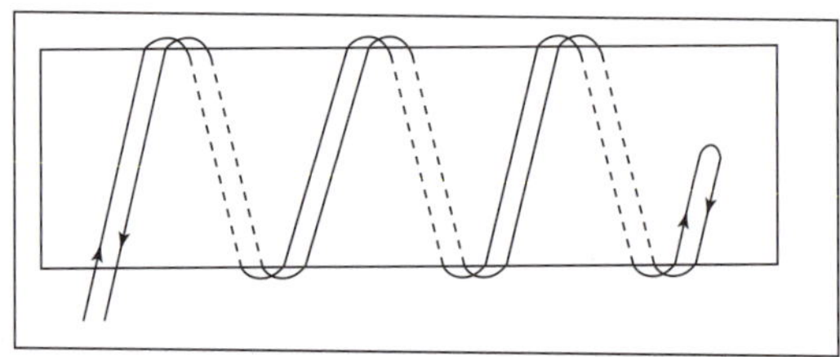

FIGURE 4.19 Bifilar resistor construction

4.2.6 Non-inductive resistors

A critical point to keep in mind when designing switching or high-frequency circuits is that the construction of most resistors is similar to that of inductors, i.e. they contain a length of wire wound like a coil around a ceramic former. This means that resistors may add unwanted inductance to a circuit. To avoid inductance and make the resistor as purely resistive as possible, the wire-wound resistor is wound back on itself, which has the effect of cancelling out any inductance produced. This type of wound resistor is also called a 'bifilar resistor', the construction of which is shown in Figure 4.19.

4.2.7 Liquid resistors

Resistances for motor starters use a resistive liquid which is stored in a tank, with two electrodes conducting the power to it. The resistance is free to cool through convection and conduction through the tank surface, so the tank size can be designed to suit the required dissipation.

One advantage of liquid resistance is that the resistance value decreases as the temperature rises (exhibiting a negative temperature coefficient or NTC), which is useful for motor starters. Figure 4.20 shows a liquid resistance wound rotor induction motor starter.

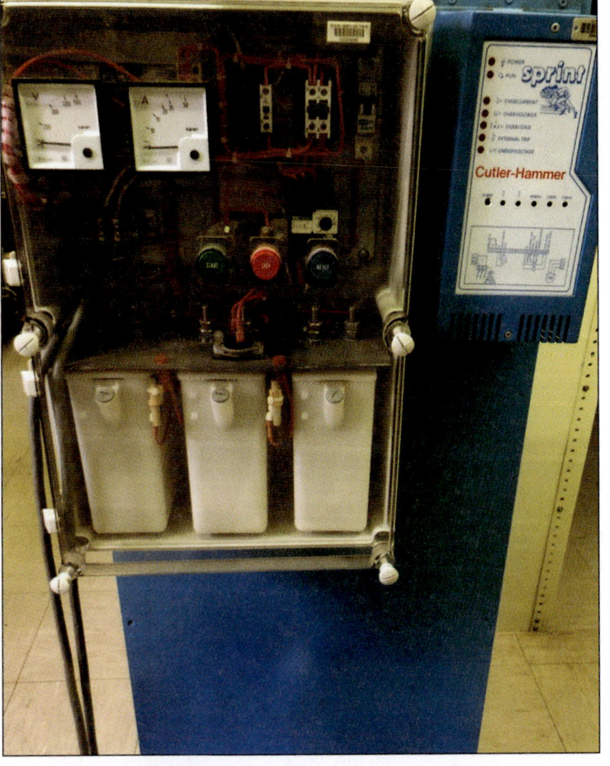

FIGURE 4.20 Liquid resistance rotor starter
Simon Dand

CHECK YOUR UNDERSTANDING

4.6 What is a thermistor?

4.7 What is the difference between an NTC and a PTC thermistor?

4.8 What is an MOV, and what is it commonly used for?

4.9 How does resistance in an LDR change in response to light?

4.10 How can resistor design remove unwanted inductance?

4.3 Power ratings of a resistor

All components in an electric circuit have a limit to the amount of heat they can dissipate. This level of heat is expressed as power in watts. The power limit is known as the 'power rating' of the component, and if it is exceeded the component will fail.

In the case of resistors that are specifically used to either limit current flow or produce a voltage drop, this rating is the maximum power that the resistor can dissipate continuously. Care must be taken in circuit design not to exceed this level. In the case of resistors that are used to produce light and heat, this rating is the amount of energy that the device will output when it has the rated voltage and (as a result) current applied to it. As an example, Figure 4.21 shows three 100 Ω resistors. All have the same resistance value, but each has a different power rating. It can be clearly demonstrated that the physical size increases as the power rating increases, to give the resistor more surface area to dissipate the heat generated.

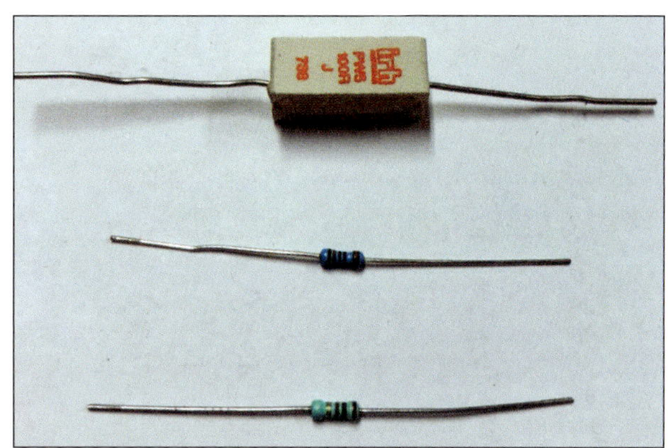

FIGURE 4.21 Various 100 Ω resistors
Simon Dand

4.4 Resistor labelling and colour codes

4.4.1 Resistor labelling

Resistors with a power rating of 5 W or greater usually have their resistance value printed on their body. The 'Ω' symbol is often dropped in favour of a letter based on the multiplier value, with 'R' used for ohms, 'k' used for kilohms and 'M' used for megohms (e.g. 100 R, 33 k, 1 M).

Where there is a decimal point in the value, to avoid it being mistaken for a scratch, the decimal point is omitted and the letter placed within the value (e.g. 4.7 Ω would be labelled '4R7', 6.8 kΩ would be labelled '6k8' and 1.2 MΩ would be labelled '1M2').

In the past, some components were too small to have numerical values printed on them, so a colour code was devised. Ironically, as component sizes decreased when 'surface-mount devices' became common in modern electronic circuits, writing their value on the component was reintroduced since very thin coloured lines proved to be more difficult to read than written values, especially in the case of devices as small as 0.8 mm × 1.2 mm.

4.4.2 Resistor colour codes

In the very early days of electronics, resistor labelling was problematic (and very difficult to read). As carbon-filled resistors became more popular, a new standard for labelling resistors was created. It employed a system where the value and tolerance of the resistor is made up of a series of colour bands printed on the body of the resistor. Each

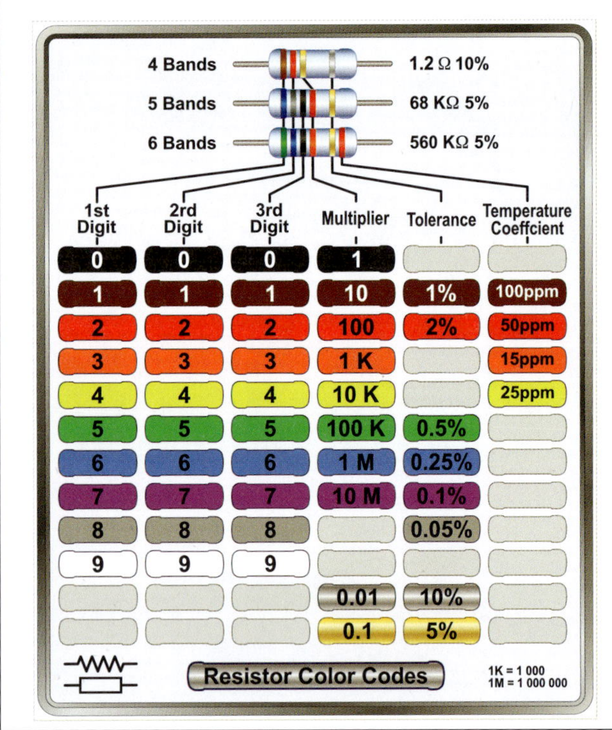

FIGURE 4.22 **Resistor colour-code chart**
Fouad A. Saad/Shutterstock.com

band has an associated number and the combination of numbers is used to represent the resistance value. Figure 4.22 shows a resistor colour-code chart.

To read a resistor, look at it and identify which end the main group of colours is nearest to. On four-band resistors, the colours are usually all together at one end. Make that end the left end by turning the resistor into that position if necessary. Assume a resistor has four colour bands—red, violet, yellow and gold. Reading from the band nearest to the end, the first band indicates the first digit of the value and the second band the second digit. In our example, these are red and violet, which are 2 and 7 respectively. That makes the value 27. The third band is the 'multiplier', which effectively specifies the number of zeroes to write after the value. The fourth band is yellow, which means four zeroes. So, the resistance is 270 000 Ω (or 270 kΩ).

The final band on our resistor is the tolerance. Gold represents a tolerance of 5%, which indicates that the resistor has been manufactured to have an actual value of 270 kΩ +/– 5%, or +/– 13.5 kΩ. The value will be between 256.5 kΩ and 283.5 kΩ.

The gold and silver colours are mainly used in the fourth band (tolerance), but on small values of resistance (less than 10 Ω) gold and silver are also used as multipliers of 0.1 and 0.01 respectively.

EXAMPLE 4.1

Determine the resistance values for these resistors:

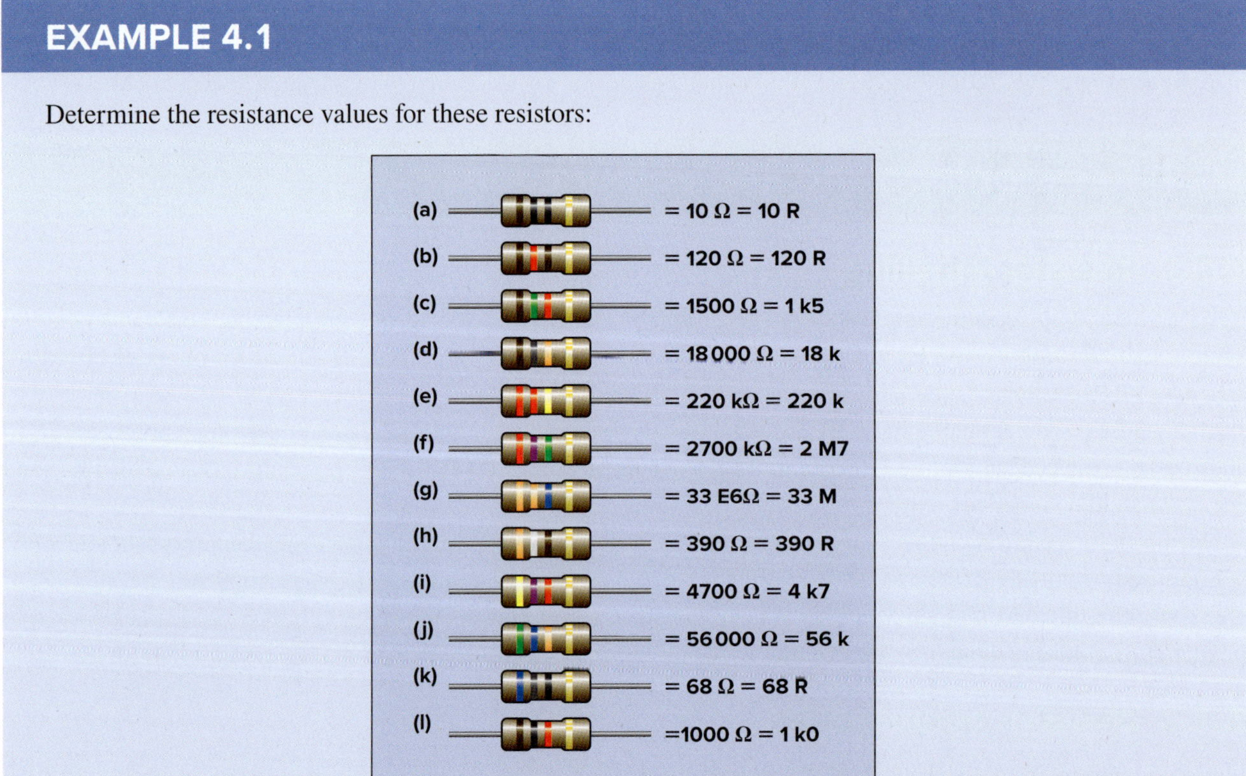

4.4.3 Preferred resistor values

Manufacturers cannot be expected to produce resistors in every possible value, so initially only those of the most common values were made. Later, a plan was adopted to make specific values with overlapping (or very nearly overlapping) tolerances.

For example, on the 10% tolerance scale, 10% of 10 Ω is 1 Ω. Approximately twice 10% (1 Ω) distant there should be another standard value resistor, and that means 12 Ω. The tolerance of the 12 Ω resistor is 1.2 Ω, and 12 Ω plus twice that, 2.4 Ω, is almost 15 Ω. The tolerance of the 15 Ω resistor is 1.5 Ω, so 15 Ω plus twice 1.5 Ω is 18 Ω. The tolerances almost overlap.

Therefore, the E12 scale very nearly covers the whole range from 10 to 100 Ω, and then starts over again in another decade. The next decade has values from 100 to 1000 Ω and so on. Table 4.1 shows the preferred values for three E series of tolerances. There is also an E48 range which, naturally enough, has 48 values.

Figure 4.23 shows the E12 scale series.

TABLE 4.1 Preferred range of resistor values

E6	E12	E24
20% tolerance	10% tolerance	5% tolerance
10	10	10
		11
	12	12
		13
15	15	15
		16
	18	18
		20
22	22	22
		24
	27	27
		30
33	33	33
		36
	39	39
		43
47	47	47
		51
	56	56
		62
68	68	68
		75
	82	82
		91

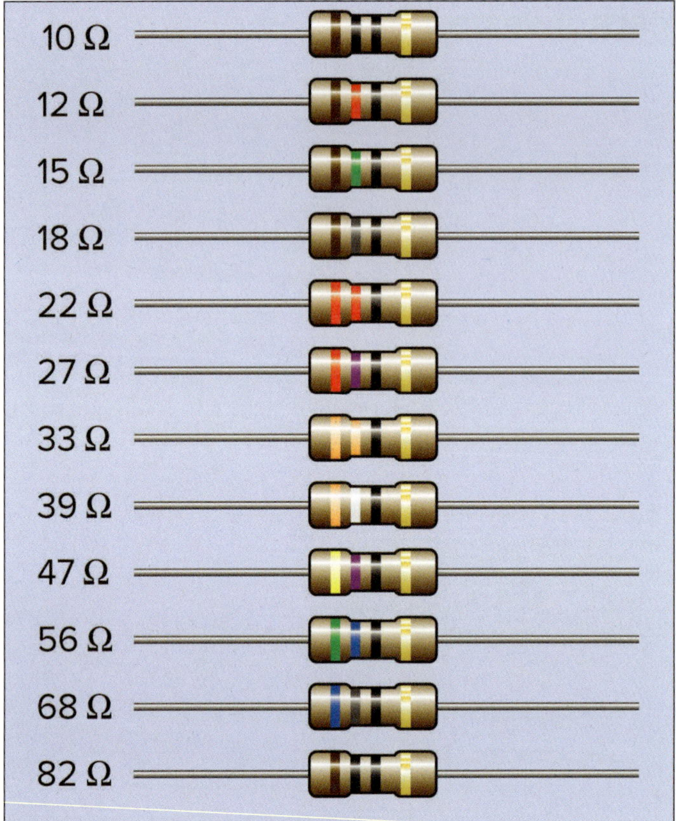

FIGURE 4.23 E12 series from 10 Ω to 82 Ω 10%

CHECK YOUR UNDERSTANDING

4.11 What is the resistor colour chart used for?

4.12 What do the gold and silver colours on resistor bands stand for?

4.13 What labelling convention is used for large wire-wound resistors?

4.5 Specifying a resistor for a particular application

4.5.1 Selecting a resistor

The primary factor in the selection of a resistor is its purpose in the circuit. As stated earlier, specific types of resistors can be used for specific tasks such as measuring light, heat, voltage across their terminals or, in the case of variable resistors, varying voltage and/or current flow in a circuit for a desired output (such as volume control on a speaker). The most common application of a component resistor is to control voltage and/or current to a required level to suit circuit operation.

Resistors of this type are generally limited by calculated value. For example, if a current of 100 mA was to be maintained to a device with negligible resistance that requires a 12 V supply, a 120 Ω resistor would have to be placed in the circuit to limit the current. The precise value required for this resistor would be determined by the resistance and tolerance. Any variation from the required value could be made up for with an adjustable resistor (e.g. a trimpot).

Once the value is determined, the next factor to consider is the power rating of the resistor, which in turn determines its physical size. Smaller resistors have a lower power rating (up to 1 W) and larger resistors have higher power ratings. If the power applied to a resistor exceeds this value, the resistor will burn out.

EXAMPLE 4.2

A resistance of 100 Ω is required to carry 100 mA of current. What value of power dissipation is required?

$$P = I^2 R \quad (1)$$
$$= 0.1^2 \times 100 \quad (2)$$
$$= \underline{1\ W} \quad (3)$$

4.6 Factors affecting resistance

4.6.1 Four factors that affect the resistance of a conductor

Ohmic materials are those that have a fixed resistance regardless of the applied voltage. The voltage-versus-current graph will be a straight line showing a constant resistance value. (As most common resistors are ohmic resistors, the word 'ohmic' is generally not used.)

In electronics, non-ohmic resistors change their resistance according to some other parameter such as applied voltage, light or temperature. (Again, the term 'non-ohmic' is not generally used.) Examples of non-ohmic resistors are tungsten lamp elements, photo-sensitive resistors, voltage-dependent resistors and temperature-sensitive resistors.

All parts of an electric circuit (the supply source, the conductors and the load) resist current flow. Even the chemicals inside a battery have resistance that changes with temperature, state of charge and the gases dissolved in the electrolyte. To understand resistance, and therefore electric circuits, it is necessary to understand that resistance is determined by four factors:

- length
- cross-sectional area
- type of material
- temperature.

4.6.2 Length

Electrical resistance is associated with the collisions between moving electrons (the electric current) and the atoms of the conducting material. As with driving down a road, the risk of having a collision increases the further one travels. In fact, travelling twice the distance doubles the risk. So, the resistance of a conductor is proportional to its length:

$$R \propto l$$

This can be proved quite simply by measuring the resistance of the active conductor in a 100 m roll of 1 mm^2 cable (it should be about 1.7 Ω). Then measure the resistance of the neutral conductor (again, about 1.7 Ω). Finally, measure the resistance of both the active and neutral conductors joined together at one end of the roll (now 200 m of wire). The resistance should be twice the resistance of either wire, as the total length is twice as long (approximately 3.4 Ω, or twice what was measured for one wire).

4.6.3 Cross-sectional area (CSA)

The roll of cable used in the previous section should be labelled 'CSA = 1 mm^2'. That is the area of the end of the wire if it is cut across at 90°, which is known as the 'cross-section' of the wire. The area of the cross-section is called the 'cross-sectional area' or 'CSA' (and is not to be confused with the diameter).

If the CSA is doubled, it is easy to imagine that more electrons can pass easily down the wire, and therefore it will have less resistance. The resistance is said to be proportional to the inverse of the CSA. In other words, as the CSA increases, the resistance decreases, so resistance is inversely proportional to area:

$$R \propto \frac{1}{A}$$

Using the roll of cable again and knowing the resistance of the conductors, connect both ends (which is the same as increasing the CSA to 2 mm^2, or double what it was). The resistance of 100 m of 2 mm^2 wire should be half the resistance of one wire (approximately 0.85 Ω).

4.6.4 Type of material (resistivity)

The copper used to make electrical wire (annealed copper) has a known value of resistance. A cube of copper 1 m on each face has a resistance of approximately 1.72×10^{-8} Ω.

The standard of one square metre of material one metre long is based on the SI system, but no lab would attempt to measure the resistance of such a large lump of material. Instead, a sample such as 100 m of 1 mm^2 is used and the resistivity is calculated mathematically. (Resistivity is given the Greek letter 'rho', which is written as ρ and pronounced 'roe'.)

Resistance is directly proportional to resistivity:

$$R \propto \rho$$

Resistivity of a material is defined as the resistance between the opposite faces of a 1 m cube at a specified temperature (e.g. 20°C) and is measured in ohm-metres (Ωm).

$$\text{So } \rho = \frac{RA}{l}$$

or transposing to define R:

$$R = \frac{\rho l}{A}$$

where ρ = resistivity.

Knowing the resistivity of any material, the resistance of any conductor can be calculated (with allowances being made for temperature differences where necessary). In Table 4.2, some electrical materials are listed, together with their resistivity values.

The values in the table are given in ohm-metres because the formula $R(\Omega) \times \frac{A(m^2)}{l(m)}$, when simplified, becomes:

$$\frac{\Omega m^2}{m} = \frac{\Omega m^{\not 2}}{\not m} = \Omega m$$

Resistivity also changes depending on whether the material is mechanically hard or soft. Annealed copper (copper which has been heated to make it more flexible) has a higher resistivity than hard copper.

The resistivity of a conductor also depends on the purity of the material and the nature of any gaseous inclusions within it. Hi-fi speaker installers pay higher prices for 'oxygen-free' speaker leads.

Table 4.2 shows that silver has the least resistance, closely followed by copper; but copper is less expensive than silver, so it is used extensively as an electrical conductor.

The four materials listed at the end of the table are alloys that are generally used for making resistors—that is, they restrict the flow of electricity far more than those above them.

When calculating the resistance of a solid material, there are 1000 × 1000 square millimetres in a square metre ($1 \text{ m}^2 = 1 \times 10^{-3} \times 1 \times 10^{-3} = 1 \times 10^{-6} \text{ m}^2$).

TABLE 4.2 Resistivity of selected materials

Conductor	Resistivity (ρ) @ 20°C	Use
Aluminium	2.83×10^{-8} Ωm	Pure metals used for conductors
Copper	1.72×10^{-8} Ωm	
Gold	2.44×10^{-8} Ωm	
Lead	2.04×10^{-8} Ωm	
Platinum	10.09×10^{-8} Ωm	
Silver	1.63×10^{-8} Ωm	
German silver	33×10^{-8} Ωm	Alloys used as resistance wire
Advance	49×10^{-8} Ωm	
Manganin	48×10^{-8} Ωm	
Nichrome	112×10^{-8} Ωm	

EXAMPLE 4.3

Find the resistance of a copper cable 500 m in length if it has a cross-sectional area of 2.5 mm^2. Take the resistivity of copper to be 1.72×10^{-8} Ω/m.

Note that 2.5 mm^2 is 2.5×10^{-6} m^2.

$$R = \frac{\rho l}{A} \quad (1)$$

$$= \frac{1.72 \times 10^{-8} \times 500}{2.5 \times 10^{-6}} \quad (2)$$

$$= \underline{3.44\ \Omega} \quad (3)$$

EXAMPLE 4.4

To manufacture a 15 Ω resistor from 0.2 mm^2 diameter manganin wire, what length of wire is required?

$$A = \frac{\pi d^2}{4}$$

$$= \frac{\pi \times (0.2 \times 10^{-3})^2}{4}$$

$$= 0.0314\ \text{mm}^2$$

$$R = \frac{\rho l}{A}$$

$$\therefore l = \frac{RA}{\rho}$$

$$= \frac{15 \times 0.0314 \times 10^{-6}}{48 \times 10^{-8}}$$

$$= 0.981\ \text{m}$$

$$= 981\ \text{mm}$$

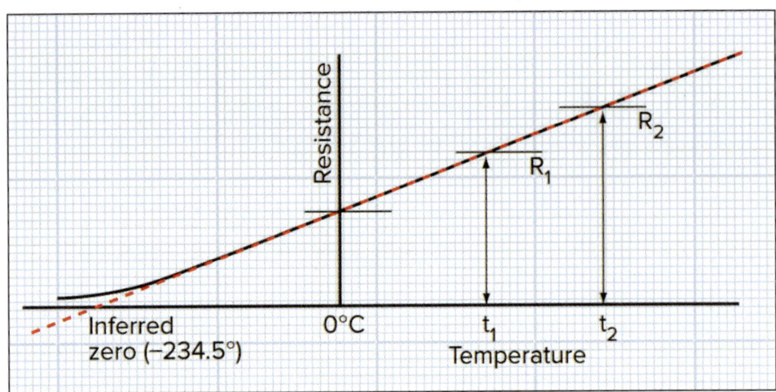

FIGURE 4.24 Effect of temperature on resistance

4.6.5 Temperature

In all these calculations, the resistance value is accurate only at 20°C. As the temperature increases or decreases, allowances may have to be made for a change in resistance.

For some materials, an increase in temperature causes an increase in resistance. These materials are said to have a 'positive temperature coefficient' (PTC). When a material has a lower resistance at higher temperatures, it is said to have a 'negative temperature coefficient' (NTC). Some resistors are made from specific materials to take advantage of these characteristics. The temperature coefficient of resistance is defined as the change in resistance per ohm per degree Celsius (or Kelvin).

Resistivity values are specified at a particular temperature because resistance can change with temperature. The resistance of most metallic conductors increases with temperature (PTC) and over a limited range. The increase is a linear function of temperature, or very close to it within that range. This leads to what is called the 'inferred zero' method of calculating the resistance of conductors at another temperature. The inferred zero value varies for different materials, but the method is illustrated in Figure 4.24.

Copper has an inferred zero resistance at −234.5°C, and the increase of resistance plotted against temperature is basically linear. The resistance of a length of copper wire can be given as a resistance of R0 at 0°C, and the increase in resistance per degree C will continue to be linear through R1 and R2. Therefore, at any temperature significantly above −234.5°C, the resistance can be calculated from any other known resistance at a known temperature. The calculation is a simple ratio:

$$\frac{R_2}{R_1} \alpha \frac{t_2}{t_1}$$

Temperatures are usually taken as relative to 0°C, so the formula needs to recognise the inferred zero resistance at −234.5°C plus the resistor temperature.

$$R_2 = R_1 \frac{234.5 + t_2}{234.5 + t_1}$$

where:

R_1 = resistance at temperature t_1
R_2 = resistance at temperature t_2.

An electric motor may be tested to see how hot the windings become in full load use. To measure the temperature directly would require the motor to be disassembled for a temperature probe to be inserted into the windings, but another method is often used.

EXAMPLE 4.5

The resistance of a coil of copper wire is 34 Ω at 15°C. What would be its resistance at 70°C?

$$R_2 = R_1 \frac{234.5 + t_2}{234.5 + t_1} \quad (1)$$

$$= 34 \times \frac{234.5 + 70}{234.5 + 15} \quad (2)$$

$$= 34 \times \frac{304.5}{249.5} \quad (3)$$

$$= \underline{41.49\ \Omega} \quad (4)$$

The motor winding resistance is measured when the motor is cold and then re-measured immediately after it is shut down. The temperature can be calculated from the cold temperature and the resistance change of the windings using the formula shown before.

EXAMPLE 4.6

A motor at 20°C has a winding resistance of 16 Ω. After running up to temperature at full load, the resistance is measured as 24.8 Ω. What is the temperature of the windings?

$$R_2 = R_1\left[\frac{234.5 + t_2}{234.5 + t_1}\right] \quad (1)$$

by transposition: (2)

$$t_2 = \frac{R_2}{R_1} \times (234.5 + t_1) - 234.5 \quad (3)$$

$$= \frac{24.8}{16} \times (234.5 + 20) - 234.5 \quad (4)$$

$$= 1.55 \times 254.5 - 234.5 \quad (5)$$

$$= \underline{160°C} \quad (6)$$

Resistance values can also be calculated from the temperature coefficient of resistance, which is defined as the change in resistance per ohm per degree change in temperature (symbol α—alpha).

$$\alpha = \frac{\text{change in resistance per °C}}{\text{resistance at } t_1}$$

Table 4.3 lists a selection of conductors and the temperature coefficients of resistance for those conductors at 0°C and 20°C.

For most metals, the change in resistance per ohm per °C is relatively constant, but the coefficient changes as temperature changes. The temperature at which the value of the coefficient is effective is usually given by a subscript to the symbol (e.g. α_0 and α_{20}, indicating the temperature coefficients at 0°C and 20°C respectively).

TABLE 4.3 Temperature coefficients of resistance

Temperature coefficient of resistance (/°C)		
Conductor	α_0	α_{20}
Aluminium	0.00423	0.0039
Copper	0.00427	0.00393
Gold	0.00368	0.00343
Lead	0.00411	0.0039
Platinum	0.00367	0.0039
Silver	0.004	0.004
Zinc	0.00402	0.004
German silver	0.0004	0.0004
Advance	0.00002	0.00002
Manganin	0.00001	0.00001
Nichrome	0.0002	0.0002

Mathematically, the new resistance can be calculated by adding the change in resistance to the original resistance, which in turn is the change in temperature multiplied by the temperature coefficient.

$$R_2 = R_1 [1 + \alpha(t_2 - t_1)]$$

where:

R_1 = resistance at temperature t_1
R_2 = resistance at temperature t_2
α = temperature coefficient of resistance.

EXAMPLE 4.7

A 2.5 mm^2 copper conductor has a resistance of 0.241 Ω at an ambient temperature of 20°C. Find its resistance at 75°C. Do the sums in brackets first.

$$\begin{aligned}
R_2 &= R_1[1 + \alpha(t_2 - t_1)] && (1)\\
&= 0.241 \times [1 + 0.00393 \times (75 - 20)] && (2)\\
&= 0.241 \times [1 + 0.00393 \times 55] && (3)\\
&= 0.241 \times [1 + 0.21615] && (4)\\
&= 0.241 \times 1.21615 && (5)\\
&= \underline{0.293\ \Omega} && (6)
\end{aligned}$$

EXAMPLE 4.8

Copper conductors of 2.5 mm^2 cross-sectional area are supplying a current of 15 A to an air-conditioner. If the conductors have a total resistance of 0.43 Ω, calculate the power lost in the conductors.

$$\begin{aligned}
P &= I^2 R && (1)\\
&= 15^2 \times 0.43 && (2)\\
&= 225 \times 0.43 && (3)\\
&= \underline{96.75}\ \text{W} && (4)
\end{aligned}$$

4.6.6 Superconductors

Many materials produce an effect known as 'superconductivity' when they are cooled below a certain temperature. Even lead is a superconductor at around −256.8°C. At the critical temperature, electrons can pass through the material with seemingly zero resistance. Other materials, mostly pure metals and some special alloys, have different critical temperatures but exhibit the same total lack of resistance below that temperature.

Since a superconductor has no resistance, once a current flow is initiated, the current will continue to flow at the same value without an applied potential. If a conductor has no resistance, then current flow through it generates no heat. If no heat is generated in a conductor, the amount of current passed through it can be increased far beyond normal values.

The current flowing through a superconductor creates a magnetic field that becomes equal and opposite to any applied magnetic field, with the result that extremely powerful electromagnets can be constructed.

Research has been carried out for many years to try to produce a superconductor effect at higher temperatures. Although breakthroughs could occur at any time, at the moment superconductors still need to be chilled to temperatures well below freezing point, below the temperatures where most gases become liquids.

Although materials may be made superconductive, it has been discovered that their current capacity is not unlimited and current over a certain level destroys the superconductivity.

The next three paragraphs discuss applications for superconductors.

Magnetic levitation

Japan and other countries are investigating the use of superconductors to levitate electric trains. The use of ceramic magnets operating in liquid nitrogen allows very strong elcctromagnetic fields to support the trains so that they float above their tracks. With no friction, they can travel much faster and more quietly than conventional trains.

Magnetic resonance imaging (MRI)

Powerful electromagnets are used to excite atoms, which then give off tell-tale radio frequencies that are used to generate images of the human body in far greater detail than any X-ray technology.

Particle accelerators

Extremely strong magnets are used to accelerate atomic particles to very high speeds and energy levels in order to smash atoms into their parts for study by physicists.

CHECK YOUR UNDERSTANDING

4.14 What are the four factors that affect resistance?

4.15 What is a superconductor and what are they used for?

4.7 Effects of resistance on current-carrying capacity, voltage drop and power loss in cables

4.7.1 Current-carrying capacity

For each size of conductor, there is a value of resistance ($R = \frac{\rho l}{A}$), so the power loss per unit length will depend on the current passing through the cable. The heat produced causes the conductor temperature to rise, and if it exceeds the temperature rating of the cable, the insulation can be damaged. Depending on the installation, poor heat dissipation can result in a fire.

As the current flowing through the conductor generates heat, standards have been determined which govern the maximum amount of current that can be allowed to flow in a conductor in an installation. A conductor in the open air can dissipate heat more readily than if it were one of several conductors in a conduit, all generating heat. Standards Australia stipulates maximum current-carrying capacities for conductors based on how cables are installed. These are listed in the Australian/New Zealand Wiring Rules and cable selection standards that are generally adopted throughout Australia.

4.7.2 Voltage drop in conductors

AS/NZS 3000 stipulates that the maximum conductor voltage drop is not to exceed 5% of the supply voltage. This means that, on a 230 V supply, 11.5 V is the maximum allowable voltage drop.

EXAMPLE 4.9

In Example 4.8, the current is 15 A and the conductor resistance is 0.43 Ω. The voltage drop in the cable is therefore equal to V = IR = 15 × 0.43 = 6.45 V. This is less than 11.5 V so the circuit is acceptable.

If the circuit run is 25 m long, what is the voltage drop per metre? How long could the cable run be before the allowable voltage drop was exceeded on a 230 V supply?

$$V_{dm} = \frac{V_d}{\text{length}} \quad (1)$$

$$= \frac{6.45}{25} \quad (2)$$

$$= \underline{0.258 \text{ V/m}} \quad (3)$$

$$V_d = V_{dm} \times \text{length} \quad (4)$$

$$\text{by transposition:} \quad (5)$$

$$\text{length} = \frac{V_d}{V_{dm}} \quad (6)$$

$$= \frac{11.5}{0.258} \quad (7)$$

$$= \underline{46.5 \text{ m}} \quad (8)$$

4.7.3 Power loss in a conductor

In a conductor supplying a load there is an inherent resistance and a current flowing through that resistance. Accordingly, the power consumed in the conductor is 'lost' as far as the load is concerned. The power lost in conductors is commonly referred to as 'copper loss'. This shows up as heat by raising the temperature of the conductor and its surrounding insulation. Heat lost in the conductors is waste, which reduces the efficiency of the circuit. The calculation for copper loss is:

$$P_{CU} = I^2 \times R$$

To reduce copper losses, either the current or the resistance must be reduced. To reduce current but still obtain a usable amount of power out of the cable, the voltage needs to be increased. This method is used in high-voltage transmission lines to reduce loss.

Two methods are generally used to reduce the resistance in the cable:

1. shortening the length of the cable
2. increasing the cross-sectional area in the cable.

By decreasing the resistance in the cable, both copper losses and voltage drop over the length of the cable are reduced.

4.7.4 Resistance tables

Conductor resistance can be established comparatively easily by reference to appropriate tables such as those in AS/NZS 3008.1. Nominal resistance is given in ohms per km of cable and, by proportionate scaling either up or down, the resistance of any length of conductor can be obtained. For example, for 10 mm^2 cable, the resistance of 1000 m is 1.79 Ω. The resistance of 100 m would therefore be 0.179 Ω.

4.7.5 Resistance faults

When testing and fault-finding electrical circuits, resistance measurement results can tell an electrician a lot about the health of a circuit. One of the most useful tools an electrician has is the ability to accurately predict how a circuit will perform before it is powered up. This skill can also be applied to a faulty circuit to determine where the fault lies and how to fix it. It is important to remember that when performing resistance measurements on electrical circuits, the circuits need to be isolated and tested dead *before* attempting to measure for resistance. *Never* perform resistance measurements on a live circuit. There are generally two resistance fault conditions that can be tested for—high resistance and low resistance.

High resistance

High resistance readings in an electrical circuit indicate that sections of the circuit are preventing current from flowing. The worst case of high resistance would be a reading of infinity ohms, indicating an open circuit, a location in a circuit where no current will flow and, most likely, the circuit will fail to operate completely. An open circuit can be thought of as an open switch. A device for circuit protection such as a fuse or circuit-breaker is designed to open a circuit and switch it off.

Other common high resistance or open circuit fault conditions can be the result of:

- heater elements failing
- poor electrical connections
- burnt electrical cables and terminals
- incorrectly wired relay or contactor contacts
- motor windings failing.

If testing a faulty circuit for open circuits, an electrician needs to be mindful of the many open circuit conditions (and their root causes) that can result in an electrical circuit not working.

Low resistance

Low resistance readings in an electrical circuit indicate sections of the circuit where extremely high current may flow due to a low resistance value. To take a worst-case example of the possible consequences of this, in a domestic circuit where a dead short across the load existed, the only resistance in the circuit would be provided by the electrical cable. This resistance value could be as low as 0.01 Ω. If we apply Ohm's Law to a circuit with a dead short where the total resistance is 0.01 Ω, the potential current flow will equate to 230 V divided by 0.01 Ω. That equals 23 000 A. If this amount of current were allowed to flow, it would greatly exceed the safe current-carrying capacity of the cable and its associated electrical accessories and fittings. This would irreversibly damage the electrical components and probably cause a fire or explosion. This example shows why circuit protection is critically important. Circuit protection such as a fuse or circuit breaker would operate very quickly in a fault state like this and thus protect the circuit.

Electricians must be able to detect low resistance faults. Understanding Ohm's Law and resistance enables the accurate prediction of how a low resistance reading will affect a circuit. This means that action can be taken to fix it. If a circuit-breaker or fuse has activated and isolated the circuit, it is important to ascertain why the circuit-protection device operated and what conditions in the circuit could have caused this to happen. Then it should be possible to repair the electrical fault.

Common low-level resistance faults include:

- moisture ingress into electrical equipment
- earth faults (i.e. active electrical wires connected somehow to earth)
- failure of any component or machine that includes a coil or winding, for example:
  - motors
  - transformers
  - relays
  - contactors
  - fluorescent light ballasts.

Testing on electrical circuits is mandatory to confirm there are no high or low resistance faults in them before they are placed into service.

4.8 Resistance measurement

4.8.1 Digital ohmmeters

An ohmmeter is an electrical test instrument that is used to measure resistance. Ohmmeters can be either analogue (with a moving needle display) or digital. The ohmmeter of choice for today's electrical professional would come as part of a high-quality digital multimeter, as shown in Figure 4.25.

Generally combining a voltmeter, ammeter and ohmmeter, a multimeter has the following features:

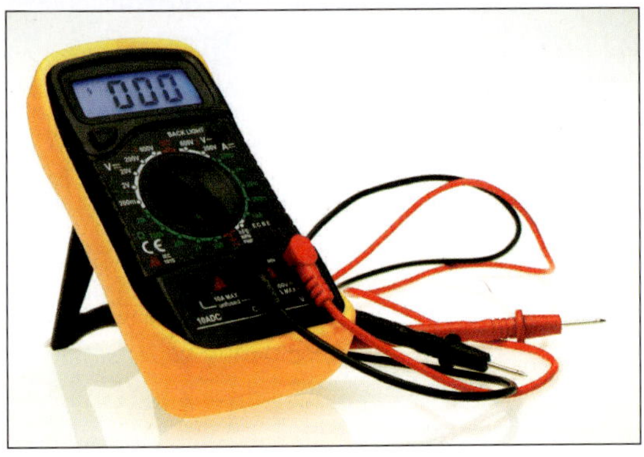

FIGURE 4.25 **Digital multimeter**
David John Abrams/Shutterstock.com

- Function selection—voltmeter, ammeter and ohmmeter (with a.c. or d.c. values).
- Range—selects the range or full-scale range for the meter. The range needs to be correctly selected to match the values of the electrical quantity being measured. Some meters have an auto-ranging feature so that the correct value is always displayed without any intervention from the user.

When using an ohmmeter or the ohmmeter function of a multimeter, the circuit should always be isolated and tested dead before connecting the meter into it. *Never* use an ohmmeter on a live circuit as the meter could be damaged and could even explode.

4.8.2 Earth continuity testing

AS/NZS 3000:2018 requires that all parts of the earthing system must be installed and tested to ensure that protective devices (such as circuit-breakers and RCDs) operate in the event of a fault and that exposed conductive parts (such as the enclosures on electrical equipment and plumbing pipework) do not reach unsafe voltage levels when these faults exist. The requirements for earth continuity are documented in Section 8 of AS/NZS 3000. Earth continuity testing basically requires a test between all parts required to be earthed (e.g. the earth pin of each socket-outlet) and the main earthing conductor in an installation's switchboard, with an expected value no greater than 0.1 Ω.

4.8.3 Insulation resistance testing

AS/NZS 3000:2018 requires the insulation resistance on all parts of an electrical installation to be high enough to prevent electric shock from contact in normal use, fire hazards from short-circuits and equipment damage. To check that the insulation of an installation (including all wiring and equipment insulation) has not been damaged during installation or deteriorated over time, an insulation resistance tester should be used. The requirements of this check are also documented in Section 8 of AS/NZS 3000.

When applied between live conductors (all active and neutral conductors) and earth, the insulation resistance tester must be able to output 500 V d.c. to stress the insulation. The resistance reading when performing this stress test between live conductors and earth must be not less than 1 MΩ.

4.8.4 Care in the use of insulation resistance testers

A 500 V insulation resistance tester can cause an electric shock unless care is taken. Inadvertent contact with the test leads can give rise to a shock, which could affect other technicians handling the same conductors.

Underground cables are often tested with higher voltages, often by IR testers generating 3000 V. Apart from direct electric shocks from the instrument, there is an additional danger created by the capacitance effects of the cable. This applies particularly to armoured or metal-sheathed underground cables as well as to mineral-insulated metal-sheathed (MIMS) cables.

The capacitance created by the method of construction of the cable enables an electric charge to be stored on the cable. The charge is created from the d.c. of the IR tester and generally exists between the conductor and the armoured metal sheath protecting the cable. Because the actual capacitance can vary widely from cable to cable, it is usually expressed in microfarads per unit length. A typical cable has a capacitance of approximately 0.2 μF/300 m. Others may have higher or lower capacitances.

At 3000 V, this capacitance relates to an energy storage of around 9 J. This quantity of charge at this voltage can cause enough of a shock to immobilise a technician for a time. Sometimes medical attention is needed, and this is a reportable incident by law.

On an aerodrome, for example, there can be many kilometres of underground cable, so the scope for an electric shock is considerable. Even with a comparatively short length of underground cable, there is an energy content at a voltage that can kill. Therefore, it is a good idea to discharge the cable after testing. Typically, electricians will do this by shorting a screwdriver or something similar across the sheath and conductor. However, care should be taken in doing this as it may present other risk hazards, and so a slower discharge method such as using a high-ohm-value bleed resistor with a high wattage rating is safer.

Before relying on readings taken by an IR tester on an installation that contains capacitance, the operator should ensure that the installation is charged up to the voltage of the IR tester. This is generally done by extended testing on any one conductor for a period. The meter reading usually indicates that this has been achieved when the reading stabilises at one value.

For example, when reading the resistance of one conductor in an underground cable to its sheath, the IR tester may show a reading which indicates a low resistance path to earth. On persisting with the test, the IR tester reading will generally climb to a much higher, more satisfactory, reading.

SUMMARY

- Resistance is the opposition to current flow.
- Resistance is measured in ohms.
- Resistance is defined using Ohm's Law.
- There are generally three types of resistors: fixed, variable and adjustable.
- Fixed resistors' value does not change.
- Variable resistors can have their value changed by an appliance operator.
- Variable resistors are potentiometers and rheostats.
- Adjustable resistors can have their value changed by technicians for calibration and fine circuit adjustment.
- Linear resistors change value in proportion to the movement of the adjustment control.
- Non-linear resistors change value in a non-proportional way.
- Other resistor types include thermistors, varistors, light-dependent resistors and liquid resistors. Each has different applications.
- Resistors have different power ratings and varying sizes for a given power rating.
- There is a resistor colour chart used for labelling resistors to identify their resistance value.
- Four factors affect resistance value: length, cross-sectional size, resistance material and temperature.
- Resistance value increase is directly proportional to length.
- Resistance value increase is inversely proportional to cross-sectional area.
- Materials have a property called 'resistivity' which varies for different materials.

- Resistance value increase is directly proportional to a material's resistivity.
- For some materials, resistance value can increase with temperature increase.
- For some materials, resistance value can decrease with temperature increase.
- Resistance in circuits is responsible for power losses.
- Ohmmeters are used to measure resistance values.
- High resistance readings may be open circuit faults.
- Low resistance readings may be short-circuit faults.
- Earth continuity testers are used to measure resistance values in earthing systems.
- Insulation resistance testers are used to measure the resistance value of wiring insulation.

END-OF-CHAPTER QUESTIONS

4.1 List two fixed and three variable types of resistors used in the electrotechnology industry.

4.2 What does the term 'preferred value of resistance' mean and what device is required to match the circuit to the required resistance?

4.3 Explain why the length of a conductor affects its resistance.

4.4 What other three factors affect the resistance of a conductor?

4.5 What is the purpose of an insulation resistance tester?

4.6 When using an analogue ohmmeter to measure the resistance of an earth connection, what must be done to ensure accuracy before taking the measurement?

4.7 A 1 mm^2 TPS (thermoplastic-sheathed) cable has a resistance of 1.7 Ω per 100 m per conductor. If a load takes 10 A, what voltage would you estimate is lost per metre of cable, remembering that 1 m of cable is 2 m of conductor?

4.8 What three values are specified when resistors are ordered?

4.9 What is the resistance of a full roll (100 m) of 2.5 mm^2 copper cable, based on a resistivity of 1.72 Ωm?

4.10 How much cable is left of that 2.5 mm^2 roll when the resistance of the conductor is only 0.1032 Ω?

CHAPTER 5
Capacitors

LEARNING OBJECTIVES

- Identify various types of capacitors that are used in the electrotechnology industry
- Know which standard symbol to use when including capacitors in electrical drawings
- Understand capacitance, charge and time constants
- Describe the factors that affect capacitance
- Know how to connect capacitors in series and how to calculate total series capacitance
- Know how to connect capacitors in parallel and how to calculate total parallel capacitance
- State the safety hazards involved in working with capacitors
- Understand the correct handling techniques when working with capacitors
- Identify and test for faults in capacitors

PREREQUISITE KNOWLEDGE

- Understand the concepts of voltage, current and resistance
- Be able to transpose mathematical equations
- Understand Ohm's Law
- Understand series and parallel circuits

5.1 Techniques for identification of various types of capacitors commonly used in the electrotechnology industry

Components used in electrical circuits are categorised into two main groups, **passive components** and **active components.** Passive components are comprised of resistors, inductors and capacitors that do not require any source of external signal or power other than the circuit that they are contained in. Active components are comprised of semiconductor devices such as diodes, transistors and integrated circuits that supply power and/or signals from other external sources to control voltage and/or current in the main circuit.

A capacitor is a sandwich consisting of an insulator between two conductors. If a voltage is applied, electrons from one conductive side are transported to the other conductive side via a circuit. (This is called 'charging' the capacitor.) When the capacitor has been charged, the supply voltage may be removed and the charge will remain in place until leakage causes it to drain away.

In real circuits, the capacitor acts like a very fast battery, storing electrical energy when there is an excess and giving it back when the voltage falls. Capacitors can appear to have no resistance when a sudden high voltage appears on a circuit and, if correctly placed, can safely dump the current from surge voltages to ground.

5.1.1 Capacitors

A capacitor consists of two conducting surfaces called 'plates' that are separated by an insulating material called the 'dielectric' so that an electrostatic field can be stored between them. Each of the plates has a separate conductor that joins it to a set of connecting leads. Capacitors come in a large range of sizes and shapes, depending on their manufacturer and intended use.

A capacitor stores energy in the form of an electric charge contained by an electrostatic field. This is like a magnetic field, except that a magnetic field results from current and an electrostatic field results from potential difference.

Capacitors can store electrical energy when the voltage is high and return it when the voltage is low. They are commonly used in power supplies to 'smooth out' the voltage. Capacitors are also used in single-phase electric motors to help them start and develop full torque (see Chapter 12). Inductors are very common in *electric* circuits while capacitors are more common in *electronic* circuits. Figure 5.1 shows capacitors on an electronic board.

FIGURE 5.1 Capacitors
Tomáš Hašlar/Alamy Stock Photo

CHECK YOUR UNDERSTANDING

5.1 What is the purpose of a capacitor?

5.2 How is a capacitor constructed?

5.3 What are some of the uses of capacitors?

5.4 What is the difference between a passive and an active electrical component?

5.2 Capacitor types and circuit symbols

5.2.1 Capacitor types

The different types of specialised capacitors that exist are too numerous to detail here. The following are the most common.

Ceramic

Ceramic capacitors (see Figure 5.2) are made by silverplating both sides of a ceramic insulator, which is usually a circular disc or square 'chip'. The capacitor is then 'encapsulated' or enclosed in an insulating material such as ceramic, enamel or epoxy resin. Encapsulation protects the conducting surface and reduces leakage between the conducting surfaces. Ceramic capacitors have values from 0.5 pF to 0.1 μF and voltage ratings exceeding 6 kV.

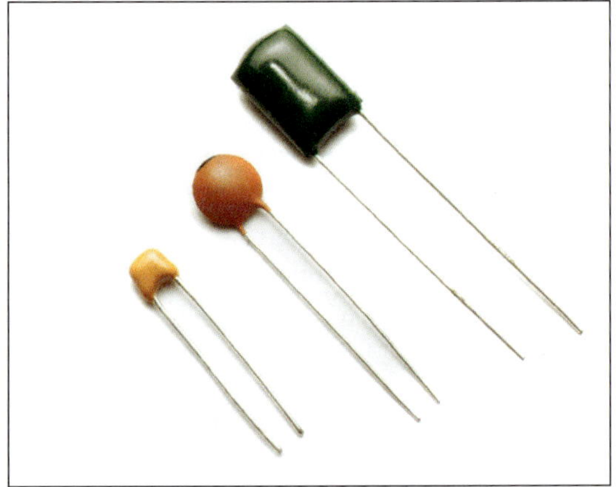

FIGURE 5.2 Three different types of ceramic capacitors
Cristian Storto/Shutterstock.com

Ceramic capacitors are often constructed for low-capacity high-voltage use in electronic circuits. They are common at very high frequencies that other types of capacitor cannot handle.

For electrical work, the main uses of ceramic capacitors are voltage-surge suppression and arc suppression. Some ceramic capacitors are quite large, and some are meant for very high voltages and very high surge currents. Capacitors for distribution systems are often tubular ceramic.

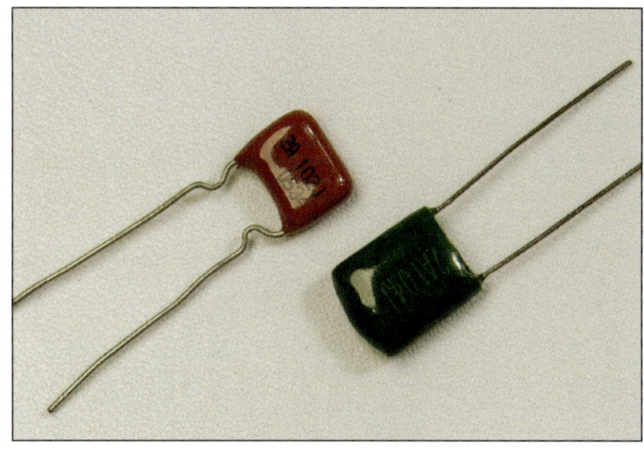

FIGURE 5.3 Stacked-plate capacitors
Hendrik Sejati/Shutterstock.com

Stacked-plate

In stacked-plate capacitors (see Figure 5.3), the conductive plates and dielectric insulators are stacked in a pile, and alternate plates are joined to form two large surface areas. Mica capacitors are a common example of stacked-plate capacitors (and are so named because of the mica dielectric material).

Stacked silver-mica capacitors are good for high-stability, high-voltage, low-capacity applications such as radio transmitters and medical equipment. In the electrical field, their main use is for the suppression of voltage surges and contact arcing.

Rolled

An insulating material such as polyester or polycarbonate is made into a thin sheet which is then 'aluminised' on one side by exposure to an aluminium plasma cloud. Two sheets are rolled together to make the double layer conductor/insulator/conductor/insulator set. This roll is then cut into sections, wires are attached and the whole assembly encapsulated to make a tightly bound small capacitor with medium-size capacitance that ranges from 1 pF to several μF (see Figure 5.4).

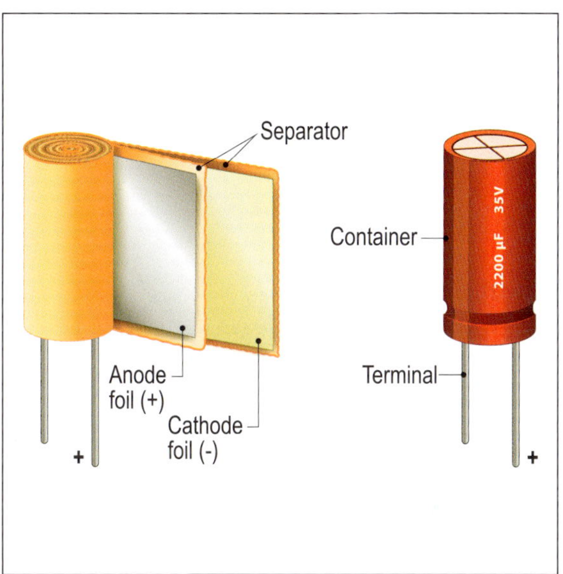

FIGURE 5.4 Rolled capacitors
Designua/Shutterstock.com

The main uses of large rolled capacitors are in motor starting and running capacitors, as well as for general power-factor correction purposes. Small polymer (plastic) capacitors have replaced paper in most applications, especially in electronics.

Polyester, polycarbonate, polypropylene or mylar capacitors are used extensively in electronics and consumer products, as well as in industrial instrumentation and control.

Electrolytic

Electrolytic capacitors (see Figure 5.5) are similar to rolled capacitors, except that the dielectric is an absorbent material impregnated with a borax electrolyte solution. The plates are etched to increase their surface area. Electrolytic capacitors provide a very large capacitance in proportion to physical size but are **polarised** because of the electrolyte. This is a disadvantage in that a voltage of the wrong polarity can destroy the capacitor. Electrolytic capacitors range from 1 μF to 1 F. Voltage ratings are generally from around 10 V to 500 V.

Electrolytic capacitors have the greatest capacity of all capacitor types and are therefore popular as d.c. filters in power supplies and large amplifiers. Electrolytic capacitors are used in switch-mode power supplies that power most consumer electronics and in an increasing number of industrial applications. Electrolytic capacitors also find use in audio and lower-frequency circuitry.

Electrolytic 'super-capacitors' are available with 25 F capacity, which is sufficient storage to replace batteries in electronic circuits.

(a) Aluminium electrolytic (b) Tantalum electrolytic

FIGURE 5.5 Two types of electrolytic capacitors
(a) Lipowski Milan/Shutterstock.com; (b) David J. Green - electrical/Alamy Stock Photo

Oil-filled

Oil-filled capacitors (see Figure 5.6) are generally for high-power/high-voltage applications. They are typically constructed with a wound metalised polypropylene film so that they can adequately store large amounts of energy. They are frequently found in a.c. circuits such as air-conditioners, single-phase motors, florescent lighting and compressors.

Variable

Variable capacitors (see Figure 5.7) are rarely used today, but at one time they had many applications in the electronic communications industry such as tuning radios, antennas, amplifiers and test equipment. In the normal variable capacitor, one set of capacitor plates can be moved within the other set so the surface area between the plates changes. Other variable capacitors are adjusted by changing the space between the plates or by sliding a dielectric in or out of that space. Most variable capacitors range from several pF to perhaps 1000 pF.

FIGURE 5.6 Oil-filled capacitors
High Energy Corp.

FIGURE 5.7 Variable capacitors for radio tuning
David J. Green - electrical/Alamy Stock Photo

Trimmer capacitors (see Figure 5.8) are small adjustable capacitors that are usually adjusted by means of a screwdriver. They are used to initially set and/or calibrate variables in a circuit. Should the values drift over time, these trimmer capacitors allow technicians to re-calibrate equipment.

Many older types have been replaced in the past few decades by modern plastic-insulated types that are both more reliable and smaller for a given value.

FIGURE 5.8 **Trimmer capacitor**
khak/Shutterstock.com

5.2.2 Symbols used in electrical drawing for capacitors

The IEC (International Electrotechnical Commission) standard symbols used in electrical drawings for capacitors are as follows.

For a fixed capacitor:

For an electrolytic or polarised capacitor:

For variable and trimmer capacitors:

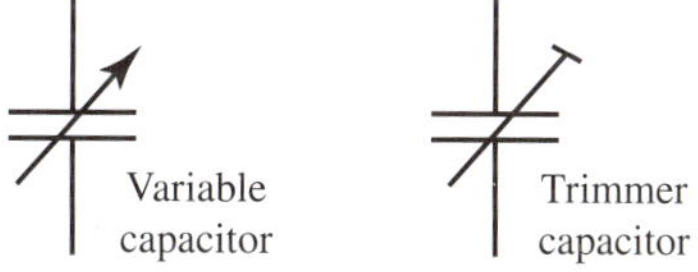

CHECK YOUR UNDERSTANDING

5.5 List the common types of capacitors.

5.6 What is an electrolytic capacitor?

5.7 What is a trimmer capacitor?

5.3 Terms and units for capacitance and electric charge

5.3.1 Capacitance

Capacitance is the measure of the ability of a capacitor to hold an electric charge. The size of a capacitor is known as the 'capacity'. In the automotive world, capacitors are often called 'condensors', which goes back to a time when they were thought to 'condense' electricity.

Unit of capacitance

The unit of capacitance is the farad (F), which is defined as the capacity of a capacitor that stores a charge of one coulomb at a potential difference of one volt.

A coulomb, represented by the letter Q, is defined as the charge that passes a point in one second when a current of one ampere flows.

$$\text{So, } C = \frac{Q}{V}$$

$$\therefore Q = CV \qquad (1)$$

$$Q = It \qquad (2)$$

$$\therefore Q = CV = It \qquad (3)$$

where:

Q = charge in coulombs
V = voltage
C = capacitance in farads
I = current in amperes
t = time in seconds.

This is an important set of relationships.

Up until recently, the farad was considered to be a very large unit (sub-multiples were, and still are, commonly used). Advances in technology have made capacitors of up to several thousand farads possible.

The two most common sub-multiples of the farad are:

1. microfarads (μF), $1\ \mu F = 1 \times 10^{-6}\ F$
2. picofarads (pF), $1\ pF = 1 \times 10^{-12}\ F$

The quantity of charge held in a capacitor is dependent on both capacitance and the voltage across the capacitor. The same quantity of charge can be held in a large capacitor at a low voltage and in a small capacitor at a high voltage.

EXAMPLE 5.1

A 10 μF capacitor is charged to a potential difference of 100 V. Calculate the charge.

$$\begin{aligned} Q &= CV \\ &= 10 \times 10^{-6} \times 100 \\ &= 0.001\ C \\ &= 1\ mC \end{aligned}$$

A 1000 μF capacitor is charged to a potential difference of 1 V. Calculate the charge.

$$\begin{aligned} Q &= CV \\ &= 1000 \times 10^{-6} \times 1 \\ &= 0.001\ C \\ &= 1\ mC \end{aligned}$$

5.3.2 Capacitance parameters

Capacitance varies according to these physical parameters.

The effective area of the plates

Capacitance is directly proportional to the effective area of the plates and is increased by increasing the number of plates (as in stacked-plate capacitors) or the total area of the plates (as in rolled capacitors).

'Effective area' means the surface area adjacent to a plate of the opposite polarity. The outsides of the plates at either end of the stack do not count. (Actually, what is known as 'stray capacitance' exists between any two conductors, but this is not generally considered in calculations.)

Electrolytic capacitors are etched to increase their surface area via the bumps and hollows that are formed.

The distance between the plates

As the separation decreases, the electrostatic field strength increases, so more charge can be maintained, thus increasing the capacitance.

The permittivity of the dielectric

Capacitance is affected by the type of material used as the dielectric. For example, if glass is used instead of air, the capacitance increases approximately six times. Glass also increases the breakdown voltage of the capacitor considerably, so much higher voltages can be used.

The ratio by which the dielectric can increase the charge, relative to air, is called the 'dielectric constant'. The measure of the dielectric effect is known as 'permittivity'. Even a vacuum has permittivity (known as 'the permittivity of free space' or 'absolute permittivity').

For a capacitor consisting of two parallel plates, the capacitance can be found from the following equation:

$$C = \frac{\varepsilon_0 \varepsilon_r A}{d}$$

where:

C = capacitance in farads
ε_0 = absolute permittivity ($= 8.85 \times 10^{-2}$)
ε_r = relative permittivity
A = area of plates in square metres
d = distance between the plates in metres.

Today, 2.5 V, 25 F 'supercapacitors', although still rare, can be bought from electronics suppliers. The value of most electrolytic capacitors is usually expressed in microfarads, even when the figure is 10000 microfarads. The terms 'farad' and 'millifarad' are still not commonly used, presumably due to tradition.

5.3.3 Dielectric constants

The **dielectric constant** signifies the degree to which capacitance can be increased by replacing the air between the plates with a dielectric. For example, if two parallel plates in air had a capacitance of 120 pF and then the air was replaced with glass (the area and distance between the plates being unchanged), the capacitance would increase to 720 pF.

An insulating material with a high dielectric constant is used to increase capacitance without an increase in physical size.

To increase capacitance further (or to reduce physical size), the dielectric is made very thin. However, if the voltage across the dielectric material is increased beyond the material's breakdown voltage, the insulation fails and becomes useless as an insulator between the plates. The same result occurs in variable capacitors if the voltage is kept constant and the thickness of the material is reduced.

The voltage-per-unit thickness necessary to cause breakdown is called the 'dielectric strength' of the insulating material (selected values of this are shown in Table 5.1). The maximum voltage applied across a dielectric depends on both the type of material and its thickness.

TABLE 5.1 Dielectric constants

Dielectric	Dielectric constant (e_r)	Dielectric strength (kv/mm)
Air/vacuum	1	3
Paper	2	40
Transformer oil	4	15
Mica	5	100
Glass	6	30
Porcelain	6	7

CHECK YOUR UNDERSTANDING

5.8 What does the term 'capacitance' mean?

5.9 What factors affect capacitance?

5.10 What is the dielectric?

5.11 What units are used to measure capacitance?

5.4 Behaviour of a series d.c. circuit containing resistance and capacitance components—charge and discharge curves

5.4.1 Capacitors in a d.c. circuit

Figure 5.9 illustrates a capacitor connected to a battery. When first connected, the capacitor would have no charge, meaning that the number of free electrons on either side of it would be approximately equal. When the switch is closed, the capacitor would charge instantly to a voltage equal to the battery voltage, with the positive plate of the battery attracting some of the free electrons from the capacitor, thus causing the connected capacitor plate to become positively charged.

The other plate of the capacitor, connected to the negative of the battery, would receive the free electrons displaced from the other side of the capacitor and therefore become negatively charged.

Capacitors are insulators, so the current measured in any circuit containing them is the movement of the free electrons from the positive side of a capacitor to the negative side of that capacitor or another capacitor.

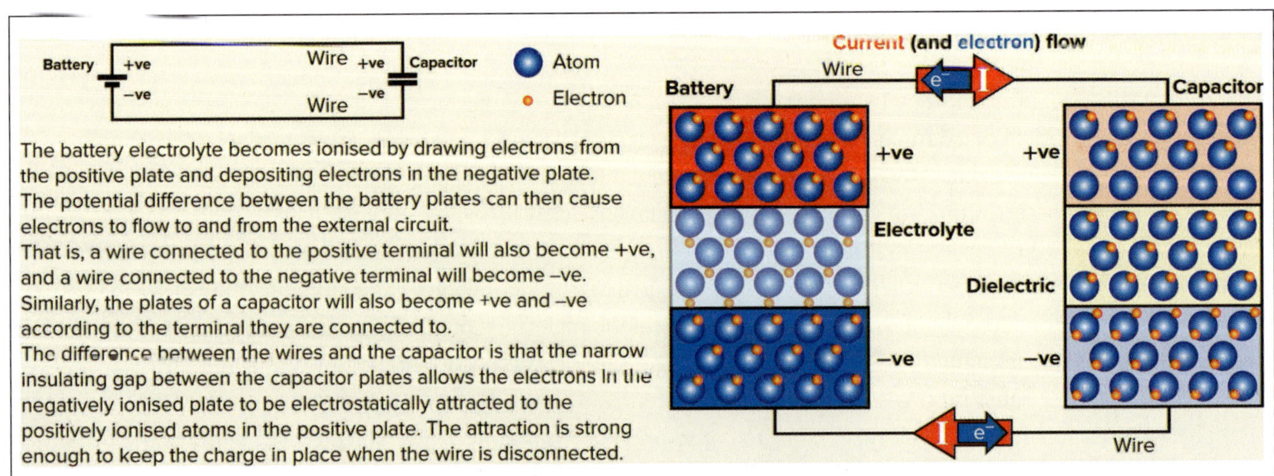

FIGURE 5.9 Capacitor connected to battery

The current does not flow through the capacitor, as current does not flow through insulators. When the capacitor voltage equals the battery voltage, there is no potential difference, the current ceases to flow and the capacitor is considered to be fully charged. If the voltage is increased, a further migration of electrons from the positive to negative plate results in a greater charge and a higher voltage across the capacitor. With no resistance in the circuit, the transfer of charge (electrons) happens almost instantaneously.

5.4.2 The time constant

The rate at which a capacitor is charged depends on its capacitance and the circuit resistance. The equation has already been used several times:

$$Q = CV = It$$

Therefore:

$$t = \frac{CV}{I}$$

and you know that:

$$R = \frac{V}{I}$$

Therefore:

$$t = RC$$

where:

t = charge time for the capacitor in seconds (one 'time constant')
R = resistance in ohms
C = capacitance in farads.

The time constant is usually denoted by the Greek letter 'tau' or τ (pronounced 'tow', rhyming with 'cow'). This is the time the capacitor takes to charge up to 63% of the applied voltage.

Therefore, to increase the charging time, either the capacitance or the resistance needs to be increased. Likewise, decreasing either value decreases the time constant.

Notice that the equation does not include voltage or current. The supply voltage does not affect the charging time for any given capacitor. Doubling the supply voltage doubles the charging current, but the quantity of electric charge being pushed into the capacitor is also doubled, so the charging time remains the same. Plotting the values of voltage against time for any capacitor charging from a constant voltage results in an exponential curve increasing towards the applied voltage (see Figure 5.10).

Discharging a charged capacitor into a fixed resistance creates another exponential curve, this time reducing towards zero.

The discharge current is shown as a negative value because of the reversal of current flow. The charge is now flowing out of the capacitor.

The curves in Figure 5.11 show that the current is at a maximum when the voltage is changing most rapidly (i.e. at the start of charging and discharging). The current is zero when the voltage is steady, but technically the voltage never

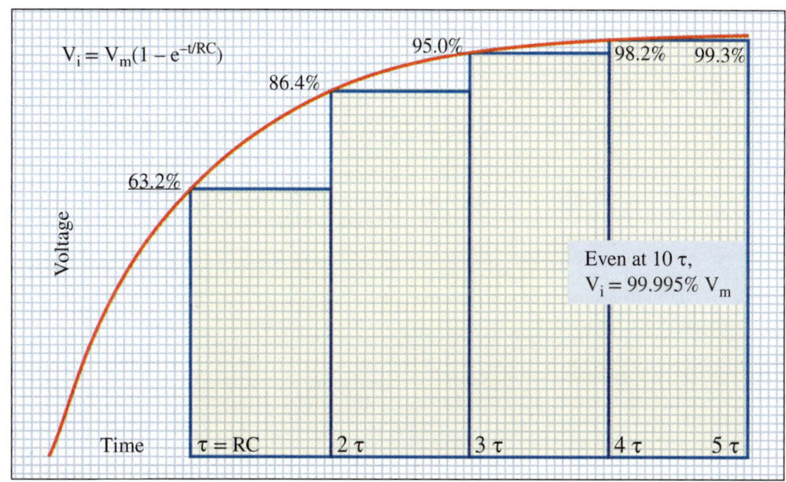

FIGURE 5.10 Capacitor charging voltage

quite reaches maximum as exponential curves continue to rise for infinity (i.e. the voltage never reaches a steady value).

When the circuit resistance value is very small, extremely high values of current can result and the charging time may be reduced to millionths of a second. By Q = VC = It, when the discharge time is small, high currents must flow. This high discharge current can be dangerous.

As no dielectric is a perfect insulator, a charged capacitor will slowly lose its charge as current leaks from one plate to another. However, a really good capacitor may hold its charge for a very long time.

Therefore, to reduce the risk of an electric shock, many high-voltage, high-power circuits have a high-value 'bleed resistor'. This is a resistor connected across the capacitor to reduce the charge to a safe limit (see Figure 5.12). AS/NZS 3000:2018 requires that the terminal voltage of capacitors greater than 0.5 μF discharges to a voltage equal to or less than 50 V after:

- 1 minute, where capacitors are rated up to and including 650 V
- 5 minutes, where capacitors are rated above 650 V.

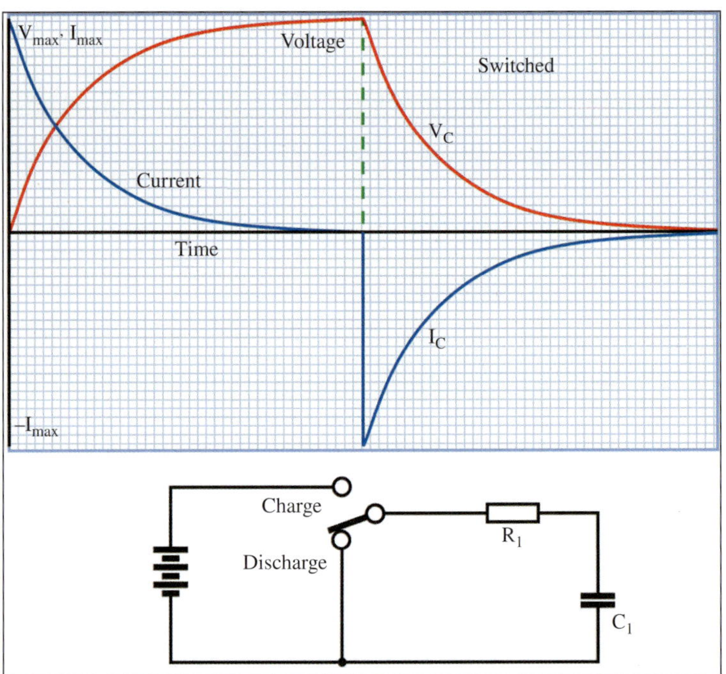

FIGURE 5.11 Capacitor charge/discharge

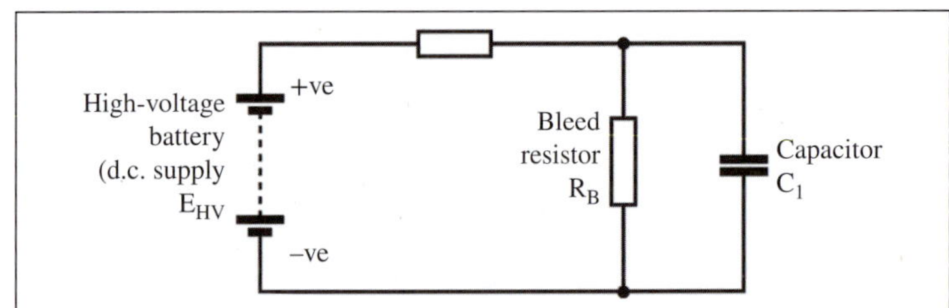

FIGURE 5.12 Capacitor charging circuit

To meet AS/NZS 3000:2018 requirements, the bleed resistor's value must be carefully selected for high-voltage, high-power circuits containing capacitors greater than 0.5 μF.

Note: a bleed resistor is not required where a capacitor naturally discharges through other components, for example a motor winding.

5.4.3 Harnessing the time constant

As the time a capacitor takes to charge (or discharge) to a set voltage can be calculated from resistance and capacitance, a circuit can be designed that operates at that value, perhaps to turn a light on or off or to control how long a motor runs or takes to start. This circuit is commonly referred to as an RC circuit.

In electronics, the circuit known as a 'voltage comparator' is used. Although an electrician may never need to design or build one, they may fail or behave erratically, so it is necessary to remember and understand time constants in order to be able to repair them.

5.4.4 Energy stored in a capacitor

When a capacitor is charged, a static electric field exists between the plates. This is a result of the electrons being drawn from the positive plate to the negative plate, and of the attraction between these electrons and their respective positive ions. The actual value of stored energy depends on the capacity of the capacitor and the voltage on the capacitor.

Unlike an inductor, which must have a dynamic flow of electrons (a current) to maintain its charge, a capacitor needs only a stored (static) charge of electrons. The attraction between the electrons and the positive ions keeps the electrons in place and the capacitor remains charged until leakage allows the charge to escape.

The actual value of energy stored in the field depends on the applied voltage and the capacitance of the capacitor. The energy stored in a capacitor can be found from the equation:

$$W = \frac{1}{2}CV^2$$

where:

W = work or energy in joules
C = capacitance in farads
V = voltage in volts.

CHECK YOUR UNDERSTANDING

5.12 Describe what happens when a power source is connected to a capacitor.

5.13 In the context of resistor/capacitor circuits, what does the term 'time constant' mean?

5.14 The energy stored in a capacitor depends on which two values?

5.5 Dangers of a charged capacitor and the consequences of discharging a capacitor through a person

5.5.1 Dangers of a charged capacitor

Electrical workers should always take care when dealing with capacitors. The value of energy stored in them can be low but, because the potential difference across the terminals can be over 300 V, an unpleasant (if not dangerous) electric shock can occur.

Also, capacitors can store the charge for a long time after the supply has been disconnected. A capacitor that has been used on three-phase line voltages can have a charge exceeding 500 V. Electric circuits such as modern switch-mode welders can have large capacitors that are charged to well above the supply voltage and still very much alive, even after the plug has been removed from the socket.

Depending on actual circumstances, a high discharge current can flow from a charged capacitor. This current lasts for only a very short period of time before deteriorating to a much lower value, but it can cause damage to a circuit if the circuit or the capacitor are not built to withstand such current surges.

A circuit of this type is the basis of photographic flash guns, which generate a high current for a few milliseconds to produce the bright electric arc that is seen as a camera flash. A capacitor is charged up to between 200 and 500 V and discharged into a xenon-gas-filled tube.

5.5.2 Safe handling procedures when working with capacitors

Before handling capacitors or working on circuits where they are used, it is a sensible precaution to ensure that they have been discharged. Capacitors with a small value of capacitance can be discharged directly with a short-circuit but, where there is a safety issue, larger values might need a resistor to control the value of the current during discharge.

As mentioned earlier, some circuits have high-value 'bleed' resistors permanently connected across a capacitor to ensure discharge occurs when the power goes off. This avoids the dangerous possibility of a handler touching a charged capacitor, particularly when working with higher-voltage circuits.

High-voltage cables should also be treated as capacitors as cables have capacitance and can store a great deal of energy. Therefore, when working on high-voltage cables, a grounding hook and cable must be used to ensure that they are grounded and any stored energy in the cables can be dissipated into the mass of earth.

CHECK YOUR UNDERSTANDING

5.15 What should you do to a capacitor before handling it?

5.16 What are some of the dangers involved in working with capacitors?

5.17 What device is used to ensure there is no stored energy remaining in disconnected high-voltage cables?

5.6 Capacitors connected in parallel

Placing two or more capacitors in parallel is the same as increasing the area of the plates. As each capacitor is added, the effective area of the plates is increased and thus capacitance of the group is increased. The actual dimensions do not matter, and the method of calculating parallel capacitors is easy—simply add them together; the total capacitance in a parallel circuit is the sum of the individual capacitances, as shown in Figure 5.13.

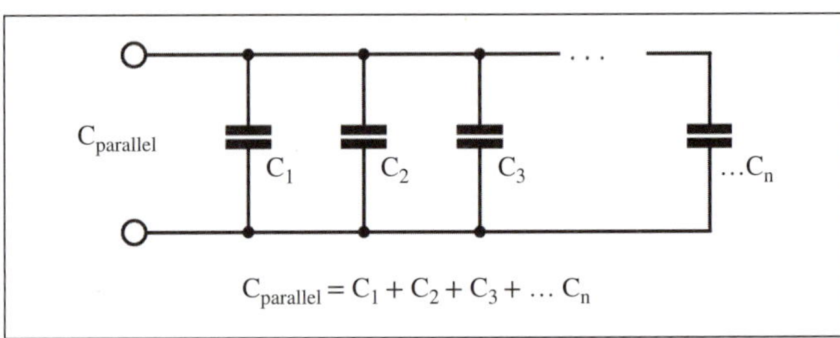

FIGURE 5.13 Capacitors in parallel

Capacitors in parallel are subject to the same rules as other components in parallel circuits. They have the same voltage across them. Since the voltage is the same across each capacitance, the total charge is equal to the sum of the charge on each capacitor:

$$V_T = V_1 = V_2 = V_3$$

$$Q_T = Q_1 = Q_2 = Q_3$$

The total charge can also be calculated from the sum of the capacitances and the applied voltage:

$$Q_T = C_T V_T$$

EXAMPLE 5.2

Three capacitors with values of 100 μF, 150 μF and 220 μF are connected in parallel to a 120 V d.c. supply. Calculate the:

- charge stored by each capacitor:

$$Q_1 = C_1 \times V_1 = 100 \times 10^{-6} \times 120 = 0.012 \text{ C or } 12 \text{ mC}$$
$$Q_2 = C_2 \text{ x } V_2 = 150 \text{ x } 10^{-6} \times 120 = 0.015 \text{ C or } 18 \text{ mC}$$
$$Q_3 = C_3 \times V_3 = 220 \times 10^{-6} \times 120 = 0.0264 \text{ C or } 26.4 \text{ mC}$$

- total charge on the circuit:

$$C_T = C_1 + C_2 + C_3 = 100 + 150 + 220 = 470 \text{ μF}$$
$$V_T = V_1 = V_2 = V_3 = 120 \text{ V}$$
$$Q_T = C_T \text{ x } V_T = 470 \text{ x } 10^{-6} \times 120 = 0.0564 \text{ C or } 56.4 \text{ mC}$$

or:

$$Q_T = Q_1 + Q_2 + Q_3 = 0.012 + 0.018 + 0.0264 = 0.0564 \text{ C or } 56.4 \text{ mC}$$

CHECK YOUR UNDERSTANDING

5.18 What overall effect does adding capacitors in parallel have on the capacitor plates?

5.19 What calculation is used to determine the total capacitance of capacitors connected in parallel?

5.7 Capacitors in series

When capacitors are connected in series, as shown in Figure 5.14, the effect is the same as adding the distances between the plates of each capacitor. The total distance between the plates is greater and therefore the total capacitance is less.

This can be proven with maths (the equation has been used for over a century). The total series capacitance is found by using the equation shown in Figure 5.14.

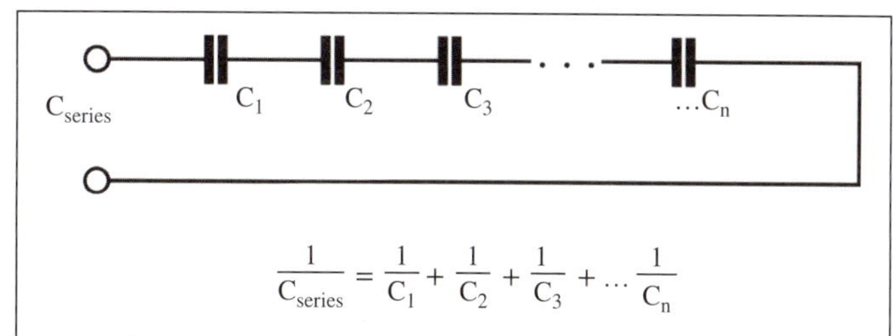

FIGURE 5.14 Capacitors in series

The total capacitance will always be less than the smallest sized capacitor in a parallel group. Also, the potential differences across the capacitors must add up to the applied voltage (according to Kirchhoff's Voltage Law).

Remember that, by Kirchhoff's *Current* Law, the current in each of the capacitors is the same as the total circuit current. Charge is dependent on current and time by the equation Q = It. Therefore, the charge in each capacitor (and in the whole circuit) is the same; only the voltage across each capacitor changes. So:

$$Q_T = Q_1 = Q_2 = Q_3$$

$$V_T = V_1 + V_2 + V_3$$

The total charge can also be calculated from the sum of the capacitances and the applied voltage:

$$Q_T = C_T V_T$$

EXAMPLE 5.3

Three capacitors with values of 100 µF, 120 µF and 220 µF are connected in series with a 120 V d.c. supply. Calculate the following:

- total capacitance

$$C_T = \frac{1}{\left(\left(\frac{1}{C_1}\right) + \left(\frac{1}{C_2}\right) + \left(\frac{1}{C_3}\right)\right)}$$

$$= \frac{1}{\left(\left(\frac{1}{100} \times 10^{-6}\right) + \left(\frac{1}{120} \times 10^{-6}\right) + \left(\frac{1}{220} \times 10^{-6}\right)\right)}$$

$$C_T = 43.71\ \mu F$$

- total charge on the circuit

$$C_T = 43.71\ \mu F$$

$$V_T = 120\ V$$

$$Q_T = C_T \times V_T = 43.71 \times 10^{-6} \times 120 = 0.00525\ C \text{ or } 5.25\ mC$$

or

$$Q_T = Q_1 = Q_2 = Q_3 = 5.25\ mC$$

- voltage across each capacitor

$$V_1 = \frac{Q_1}{C_1} = \frac{5.25 \times 10^{-3}}{100 \times 10^{-6}} = 52.5 \text{ V}$$

$$V_2 = \frac{Q_2}{C_2} = \frac{5.25 \times 10^{-3}}{120 \times 10^{-6}} = 43.7 \text{ V}$$

$$V_3 = \frac{Q_3}{C_3} = \frac{5.25 \times 10^{-3}}{220 \times 10^{-6}} = 23.8 \text{ V}$$

and

$$V_T = V_1 + V_2 + V_3$$
$$= 52.5 + 43.7 + 23.8$$
$$V_T = 120 \text{ V}$$

CHECK YOUR UNDERSTANDING

5.20 What overall effect does adding capacitors in series have on the capacitor plates?

5.21 What calculation is used to determine the total capacitance when connected in series?

5.8 Common faults in capacitors

5.8.1 Capacitor faults

As the size of a capacitor decreases and greater capacity is expected of it, greater stress is placed on both the insulation and the conductive foils. If any point in the insulation should fail, there is no insulation and the whole capacitor fails.

The applied voltage, which causes great stress on the insulation, might be better appreciated when it is considered that a voltage of 100 V applied across an insulator just 0.1 mm thick represents a voltage gradient of $100/0.1 \times 10^{-3} =$ 1 million volts per metre.

Current flow can also be a problem as capacitors discharge at a rate determined by the resistance of the circuit, and sudden discharges such as the shorting of the terminals of the capacitor result in very high internal currents. A short is theoretically zero ohms, so the current would theoretically be infinite.

Fortunately, capacitors have internal resistance, although its value may only be in milli-ohms. The resulting current can easily fuse the thin foil conductors. For example, a 1000 microfarad capacitor charged to 1000 volts represents a charge of 1 coulomb. If that charge is discharged in 1 millisecond, up to 1000 amps would flow internally.

5.8.2 Dielectrics drying out

Electrolytic and oil-filled capacitors can leak or dry out, with the result that their capacity falls to a lower level than they were designed to contain. Eventually the capacitor will fail to meet its design criteria.

Electrolytic capacitors in switch-mode power supplies reach higher temperatures than the same capacitors on d.c. or 50 hertz. The higher temperature is the result of dielectric heating, which occurs in most capacitors, especially at higher frequencies.

CHECK YOUR UNDERSTANDING

5.22 What parts of a capacitor are subject to physical faults?

5.23 What are some of the problems that may occur with a capacitor's dielectric?

5.9 Techniques for testing capacitors to determine serviceability

5.9.1 Testing capacitors

To test a large capacitor, an ohmmeter is placed across the terminals of a known discharged capacitor. If the resistance is low at first but increasing, there is capacity in the capacitor. However, in order to know whether there is enough capacity, a good capacitor of the same value should be checked on the same resistance range and thc result compared with the suspect capacitor.

Some multimeters have a capacitance range that will allow direct measurement of capacitance up to around 200 microfarads. The testing procedure will be given in the instruction book for the instrument.

An instrument that is currently gaining favour is the effective series resistance (or ESR) meter, which measures the internal series resistance of electrolytic capacitors and some other components. The reading gives an indication of the condition of the electrolyte within the capacitor.

Another simple testing method requires the capacitor to be removed from the circuit and placed in series with a resistor of known value across a d.c. power supply. The capacitor is charged to 63% of the applied voltage and the charge time noted. Using the time constant and the resistance value, the capacitance can be calculated.

Small capacitors may be tested at their expected operating frequency against a capacitor that is known to be functioning well in a circuit called a 'capacitance bridge' or an 'LCR bridge'.

Capacitors with a bulging bottom or leaking electrolyte should be replaced, even if they are still working.

CHECK YOUR UNDERSTANDING

5.24 What test devices can be used to test a capacitor?

5.25 What does an ESR meter do?

5.26 How can a resistor and d.c. power supply be used to test a capacitor?

5.10 Application of capacitors in the electrotechnology industry

Capacitors serve many functions in the electronics industry, including **coupling** (blocking d.c. from a.c. signals) and **decoupling** (separating signals from other circuits), high- and low-pass filtering (allowing only certain audio such as bass or treble or radio frequencies through), noise filters, oscillators (that produce sound and radio waves) and radio tuning circuits.

In the electrical sector of the electrotechnology industry, capacitors can be used for energy storage (especially super capacitors whose capacitance is greater than 1F), power factor correction, motor starting and 'snubbing' relay circuits (especially d.c.), where opening the relay results in arcing across the contacts.

As capacitors have the ability to discharge a large amount of energy in an almost instantaneous time, they are also used in many types of weapons, ranging from tasers to railguns.

CHECK YOUR UNDERSTANDING

5.27 Name some of the uses of capacitors in the electrical sector of the electrotechnology industry.

SUMMARY

- Capacitors are passive electrical devices that are used to store an electric charge.
- Capacitors include two conductive plates to store the charge that are separated by an insulating medium called a dielectric.
- There are various capacitor types, such as ceramic, rolled, stacked-plate and electrolytic.
- Capacitors can have a fixed value or can be variable.
- A capacitor's capacity is measured in capacitance and its value is the farad (F).
- The factors that affect the capacitance of a capacitor are plate area, distance between the plates and permeability of the dielectric.
- When connected in series with a resistor, a capacitor takes time to reach a level of charge. This RC circuit allows timing circuits to be created.
- Capacitors store energy based on their capacitance value and the voltage applied to it.
- In some applications, the amount of energy stored in a capacitor is very large and can be dangerous—or even fatal—if that energy is discharged into a body.
- Safe handling practices must be used when working with capacitors and high-voltage cables that store large amounts of electrical energy.
- Capacitors are used extensively in the electrical industry for power-factor correction, motor starting and filter circuits.
- Various devices and techniques are used to test the electrical integrity of a capacitor.
- For capacitors connected in series, the effective distance between the plates is increased, and thus the overall capacitance value is decreased.
- For capacitors connected in parallel, the effective area of the plates is increased, and thus the overall capacitance value is increased.

END-OF-CHAPTER QUESTIONS

5.1 Describe the construction of a capacitor.

5.2 What three factors affect the capacitance of a capacitor?

5.3 State two applications for capacitors in the electrotechnology industry.

5.4 What are the hazards associated with capacitors in electrical circuits and how can they be minimised?

5.5 Find the equivalent capacitance of a 7 μF capacitor and a 33 μF capacitor connected in series.

5.6 What is the equivalent capacitance of a 7 μF capacitor and a 16 μF capacitor connected in parallel?

5.7 Determine the amount of charge held in a 100 μF capacitor at a voltage of 100 V d.c.

5.8 If the quantity of charge held on a 1000 μF capacitor is 1 coulomb, what would be the voltage across its terminals?

5.9 Calculate the energy stored in a 33 μF capacitor at a voltage of 100 V.

CHAPTER 6

Electromagnetism: solve problems in magnetic and electromagnetic devices

LEARNING OBJECTIVES

- Understand natural magnetism
- Use fields to predict reactions between magnets
- Understand the characteristics of magnetic materials
- Compare magnetisation and hysteresis curves for various materials
- Relate magnetic effects to electric current
- Describe a magnetic circuit, magnetic leakage and magnetic fringing
- Predict magnetic fields around conductors
- Apply the right-hand (grip) rule to determine polarity
- Calculate electromagnetic values
- Explain Faraday's Law of Induction
- Understand Lenz's Law
- Describe inductor types
- Explain inductance as an electrical effect
- Understand what 'time constant' means
- Describe the construction and operation of moving coil and moving iron meters
- Apply magnetic principles to simple machines

PREREQUISITE KNOWLEDGE

- Understand how to apply work health and safety regulations, codes and practices in the workplace
- The ability to solve problems in direct current circuits

INTRODUCTION

In 1820, the Danish physicist Hans Christian Ørsted (also spelled 'Oersted') discovered that electric currents exerted forces on magnets. In 1831, the English scientist Michael Faraday observed that magnetic fields could induce electric currents. Up until then, electricity and magnetism had been thought of as unrelated. However, the scientific study of electricity, and subsequently of atoms, led to the belief that magnetism was an electric effect of an atomic particle such as an electron moving at a very high velocity.

Following discoveries by Einstein that built on the work of the Scottish mathematician James Clerk Maxwell, many scientists continue to pursue what is called the 'Unification Theory'. This theory suggests that electromagnetism, nuclear weak force, nuclear strong force and gravity may all be related.

6.1 Magnetism

Magnetism that can be artificially produced using electricity is termed 'electromagnetism'. It can be turned on and off and controlled using electrical principles. Every electric motor, generator, alternator, transformer, microphone, loud speaker, solenoid, relay and contactor relies on magnetism to work.

At one time, magnets were considered to be magical, the words 'magnet' and 'magic' both being derived from the Latin word 'magus', which relates to a magician or priest. The phenomenon of magnetism was believed to be a mechanical property of a material that caused it either to attract or be attracted to other materials with similar properties.

A material called 'lodestone' or 'magnetite' is a natural form of a magnet whose use goes back over two thousand years. One of its applications was as a compass for sailing ships. A ship's compass needs a point of reference for navigation. By convention, the compass needle is said to be 'north seeking' (meaning that it seeks or points to the north). Since unlike poles attract and like poles repel, the north of the compass needle is attracted to the unlike south magnetic pole. The Earth itself acts as a magnet, as if there was a giant upside-down magnet buried underground (see Figure 6.1).

The orientation of compass needles with the *magnetic lines of force* of the Earth gave rise to navigation as well as a number of conventions for magnetism. The compass needle is a dipole magnet (meaning that it has both a north and a south pole), which orientates to the geographic North and aligns with the Earth's magnetic field. This scientific convention led to an assumption that magnetic lines of force represent vectors.

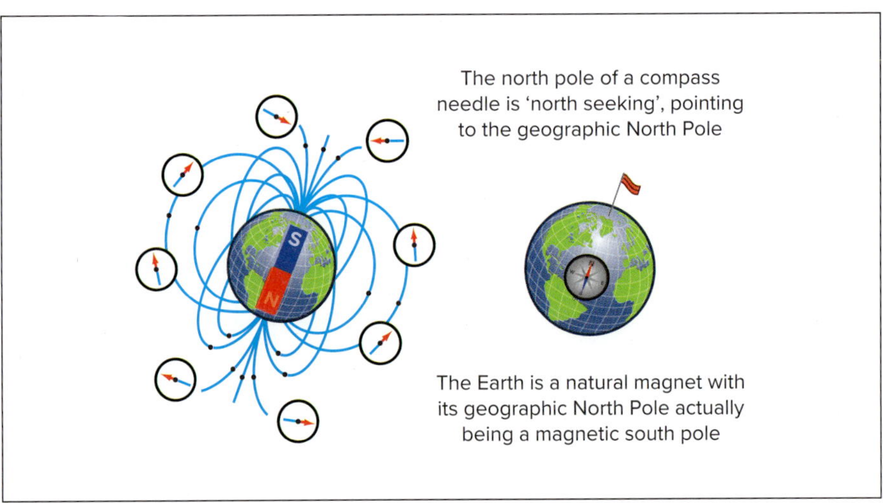

FIGURE 6.1 **Earth conceptualised as a magnet with magnetic lines of force**

6.1.1 Common magnetic and non-magnetic materials and their groupings

The three most important magnetic materials are iron, nickel and cobalt, together with their alloys. Of these, iron produces the greatest magnetic effect and was, until the end of the last century, the most widely used magnetic material. Alloys of magnetic materials have produced **permanent magnets** of superior properties, an example of which is an 'Alnico' magnet, which is an alloy of aluminium, nickel and cobalt, with a few minor impurities.

Magnetic and non-magnetic materials are grouped or classified as 'ferromagnetic', 'paramagnetic' or 'diamagnetic'. (Iron and iron alloys are ferromagnetic.)

Ferromagnetic materials that are easily 'influenced' (temporarily magnetised) are termed 'magnetically soft'. As easily magnetised as they may be, they tend to demagnetise themselves when the magnetising force is removed. Any magnetism that remains is called 'residual magnetism'. Materials whose magnetic properties are difficult to alter can be permanently magnetised, and are said to be 'magnetically hard'.

Diamagnetic materials are substances that we think of as being non-magnetic; however, they are actually repelled when brought near a magnetic field. Examples include wood, paper, plastic, glass, marble, antimony, bismuth, brass, zinc, silver, gold and copper.

Paramagnetic materials have little or no magnetic properties and are thus weakly attracted by the magnetic field. Examples include aluminium, platinum, chromium, sodium, lithium, tungsten, magnesium and caesium.

Soft magnetic materials such as iron or steel can be machined with little difficulty and may be rolled into a sheet from which shapes can be punched and stacked to form what are known as 'laminations'. Laminations are usually made from an alloy of silicon and iron called 'silicon steel'.

A magnetically hard material is mechanically hard. Some cannot be machined with any degree of success, although others can be surfaced by grinding. Shaping these materials usually involves casting them into the required shape and magnetising them while they are still above their Curie temperature (the point at which a material loses its magnetic property). Higher temperatures make magnets weaker.

Some modern magnetic materials are formed by placing a mixture of powdered magnetic materials and a ceramic binder under high temperature and pressure to form ferrite, a very high-permeability magnetic material. This process is known as 'sintering' and produces sintered ferrite cores. Many of the materials used today are called 'rare earth materials' because of the difficulty in isolating the element from the surrounding material. These elements are from the lanthanide (or lanthanoid) group of the periodic table.

Neodymium (element 60) appears to make the strongest magnets currently available. It has a very high melting point, which adds to the difficulty of working it. Neodymium needs an extremely high force to magnetise it but has the advantage of requiring a much higher magnetic force to de-magnetise it. Neodymium magnets tend to be rather brittle and can shatter if roughly handled.

All materials have some form of magnetism or magnetic effect. However, paramagnetic and diamagnetic materials (such as glass, paper or plastic) under everyday use and with normal temperatures and pressures have a net zero result magnetically. They do not exhibit the same internal alignment or a north-south pole orientation like ferromagnetic materials.

Materials such as copper and aluminium make good non-magnetic conductors. Brass, copper and plastic have been used in the electrotechnology industry for cable glands, couplings, fittings and accessories (as non-moving fixtures and fittings) due to their diamagnetic properties.

6.1.2 Magnetic field pattern of bar and horse-shoe magnets

Magnets are ferromagnetic substances with dipoles, meaning they have two poles (monopoles do not exist in nature). By convention, these terms are called a 'north pole' and a 'south pole'. Figure 6.2 shows a bar magnet and a horse-shoe magnet, each with a north pole and a south pole. Invisible lines of force are drawn to show the pole-to-pole magnetic field path, north to south.

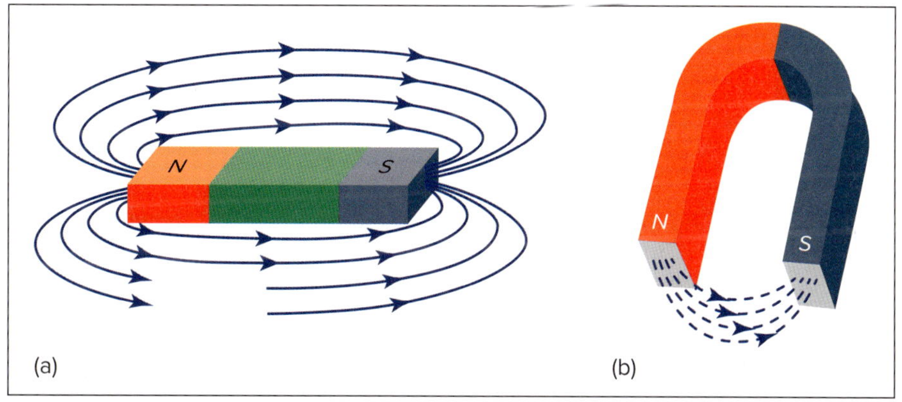

FIGURE 6.2 (a) Dipole bar and (b) horse-shoe magnets' north-to-south lines of force

Experiments in magnetism: the case for iron filings

A series of iron filings experiments that were carried out over two hundred years ago influenced conventions that are still in use today.

Figure 6.3 shows the magnetic field produced by a magnet. The filings arrange themselves in a series of lines that commence and end near opposite poles of the magnet.

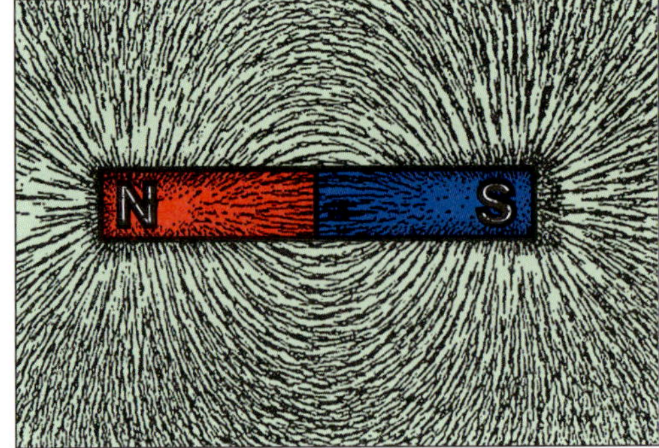

(a) Iron filing pattern on flat surface

(b) 3-D iron filing pattern

FIGURE 6.3 Iron filings in magnetic fields
(photo) Tony Jones

Magnetic field patterns can be made by placing a sheet of paper, thin card or Perspex over the magnet to prevent the filings from sticking to it. When iron filings are sprinkled onto the paper or Perspex over and around a magnet, a strange pattern emerges which looks like lines coming from one pole and going towards the other pole. These lines appear to spread out from the first pole and converge at the other pole. Early experimenters called this pattern a 'magnetic field' that was made up of what Michael Faraday referred to as 'lines of force'. This term is still used in explaining magnetic fields.

Practical and safety tips for experiments involving iron filings

Iron filings are elongated particles of iron and produce better field patterns than iron powder particles, which are spheroidal. If iron filings or powder get onto a magnet, they are extremely difficult to remove and present a metal splinter hazard for experimenters' fingers. Field demonstrators are easy to use and are quick to set up. Magnetic fields (or lines of force) convention states that a magnetic force always acts outwards from the north-seeking pole of a magnet and inwards at the south-seeking pole (see Figure 6.4).

(a) Bar magnet

(b) Horse-shoe magnet

FIGURE 6.4 Lines of force traced using magnets
Tony Jones

Figure 6.4 shows the magnetic field produced by bar and horse-shoe magnets. A magnetic compass is a magnet, and if one is slowly moved along a line of force as an experiment, the north pole of the compass will indicate the path in which the magnetic force is acting.

Permeability and reluctance

A relationship exists between the **permeability** and the **reluctance** of a particular material, as shown in Table 6.1.

A magnetic field tends to take the path of least magnetic resistance, which means taking the shortest possible path between the north and south poles of a magnet, as shown in Figure 6.5(a). The field will pass directly through materials that cannot be magnetised (non-magnetic materials).

TABLE 6.1 Relationship between permeability and reluctance

Permeability	Reluctance
$\mu = \mu_o \mu_r$ where: μ = actual permeability μ_o = permeability of free space (vacuum) μ_r = relative permeability.	$R_m = \dfrac{l}{\mu_o \mu_r A}$ where: R_m = magnetic reluctance l = length μ_o = permeability of free space (vacuum) μ_r = relative permeability A = the area or cross-sectional area (m^2).

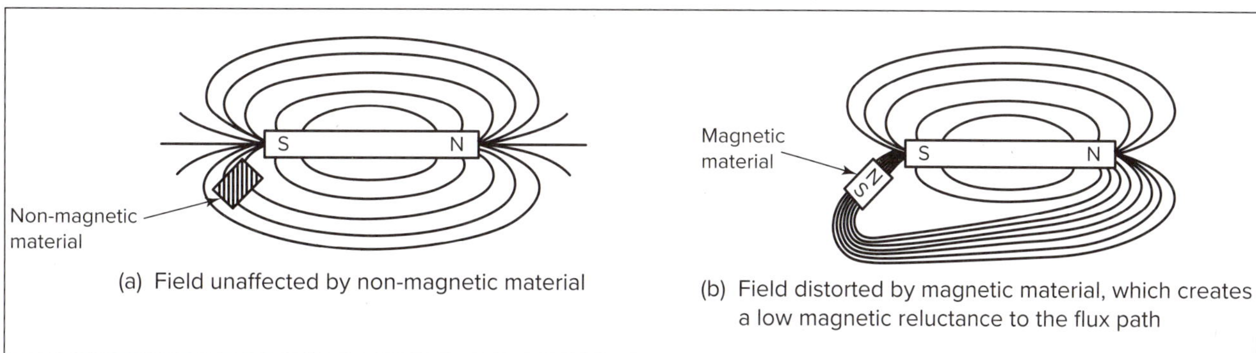

FIGURE 6.5 Effect of magnetic field on magnetic and non-magnetic materials

Figure 6.5(b) shows that when a material that *can* be magnetised (a magnetic material) is placed within the magnetic field, the field is distorted—and perhaps even lengthened—in order to pass through this magnetic material rather than through air. This is because a magnetic material offers much less opposition to the field than a non-magnetic material. The magnetic flux passes through, altering the material and resulting in an effect called 'magnetic induction'.

Both materials in Figure 6.5(a) and (b) have the magnetic property that is referred to as 'relative permeability'. The term 'permeability', which was coined in 1885 by the mathematician and physicist Oliver Heaviside, is defined in this context as the ease with which a magnetic flux can be created in a material. The non-magnetic material in Figure 6.5(a), which could be paper, glass or plastic, has a very low relative permeability (μ_r) and its physical property is unaffected. The material in Figure 6.5(b) has ferromagnetic properties and a very high permeability value. (Note that the material is shown with north and south poles due to a magnetic field influence called 'magnetic induction'.)

An effective way of comparing magnetic and non-magnetic materials is to consider the following scenario. You are having cereal for breakfast. Milk is the magnetic flux, and the breakfast table has a very low permeability. You spill some milk on the table. The milk does not get absorbed because of the table's low permeability and spreads over the table. If you apply an absorbent cloth, the milk is soaked up by the cloth. The breakfast table, like the non-magnetic material shown in Figure 6.5(a), is unaffected by the magnetic flux. But the absorbent cloth, like the magnetic material in Figure 6.5(b), absorbs the milk (the magnetic flux) because it is highly permeable.

A material that has a high reluctance also has a low permeability. Similarly, a material with high permeability will have a low reluctance. Although air has a high reluctance, we know there are no insulators or barriers to the phenomenon of magnetism. (Magnetic fields are peculiar in that they have the ability to pass through almost any material.) Figure 6.6 shows that, like electric currents which travel by the path of least resistance, magnetic flux will travel by the path of least reluctance (magnetic resistance).

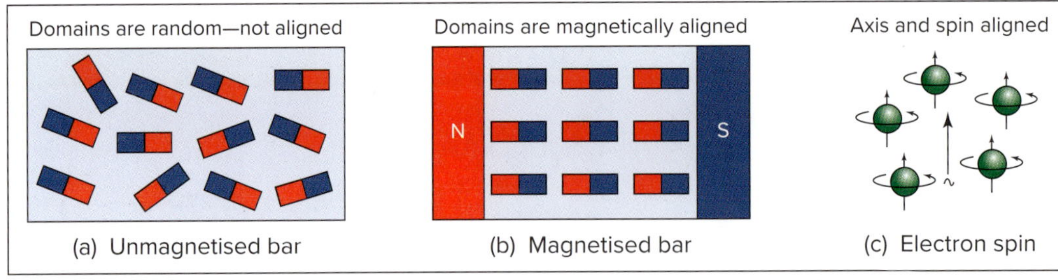

FIGURE 6.6 Domain and electron alignment

CHECK YOUR UNDERSTANDING

Think back to the earlier iron filings experiment, contrasting it with the breakfast cereal example.

6.1 What would be the magnitude (high or low) of relative permeability (μ_r) for the iron filings and for air?

6.2 What are the effects of lines of force through different materials in terms of relative permeability and reluctance?

Magnetic induction

When a magnet is attracted to a piece of magnetic material, the magnetic field passes through the material, turning it into a temporary magnet and creating a north and south pole. The magnetic field is said to be induced into the (ferromagnetic) material by a process called 'magnetic induction'. A number of theories attempt to explain this physical change in ferromagnetic materials.

The molecular theory of magnetism was developed by the German physicist Wilhelm Weber in 1852 and advanced by Scottish engineer James Ewing in 1890. It is known as the 'Weber-Ewing molecular theory'. 'Magnetic domain theory' was developed by the French physicist Pierre-Ernest Weiss in 1906. The 'electron theory' of magnetism suggests that magnetic materials tend to have their electrons spinning in the same direction (see Figure 6.6). The spinning effect creates a magnetic field while the field polarity is determined by the direction of electron spin. Electrons of non-magnetic material spin in different directions, cancelling the magnetic effect.

Magnet 'keepers' are a ferromagnetic material. They join the ends of a magnet (a practical application of magnetic induction) and are a means of safely storing magnets. The keepers, as shown in Figure 6.7(a), serve to prolong the life of a magnet and reduce magnetic leakage and fringing.

Magnetic induction always takes place when attraction occurs between a magnet and a material that is not magnetised. The green items in Figure 6.7(b) are both ferromagnetic while the yellow item is diamagnetic, which means that its permeability is less than that of free space.

(a)

(b)

FIGURE 6.7 (a) Magnetic keepers and (b) magnetic induction

(photo) Tony Jones

6.1.3 Magnets' attraction and repulsion when brought into contact with each other

If two straight (bar) magnets are placed with their unlike poles together (N–S and N–S), the magnets will attempt to click together and remain attached due to the attraction between their poles. The field pattern between the magnets, if kept apart, will show a magnetic 'fringe' between their co-joined poles, which means that the magnetic field spreads out between the poles (see Figure 6.8).

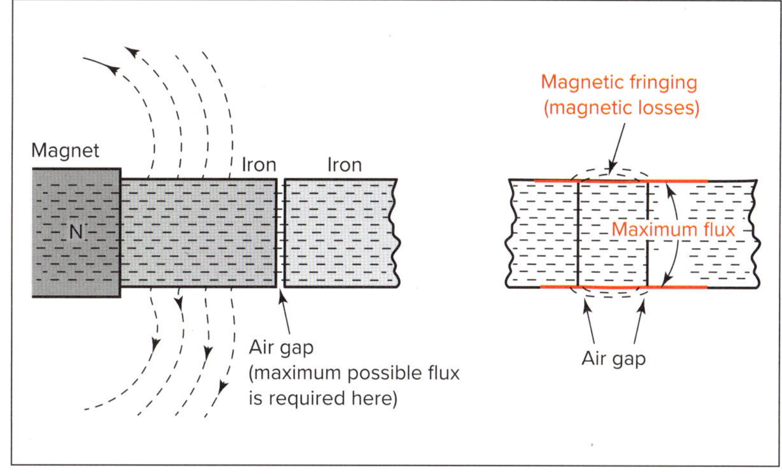

FIGURE 6.8 Air gap magnetic flux 'fringing'

If two bar magnets are placed alongside one another with their like poles together, the magnets will tend to move apart, owing to the force of repulsion between their like poles. The two fields will remain separate, unable to form a single magnetic field, as shown in Figure 6.9(b).

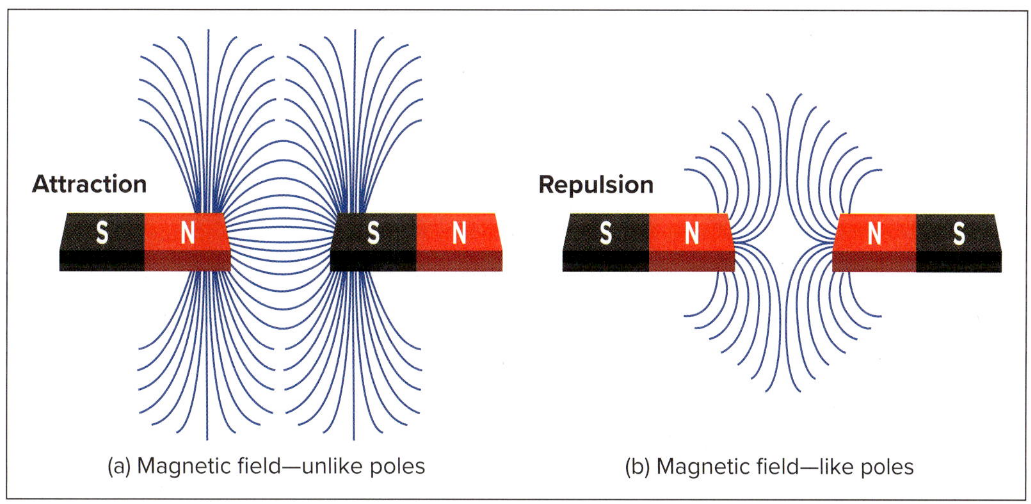

FIGURE 6.9 Magnetic fields—attraction and repulsion
OSweetNature/Shutterstock.com

6.1.4 Practical applications of magnets

In the past, the permanent magnet concept was used to create data on computer discs, as well as on audio and video tapes. Permanent magnets are still used in computer hard disc drives (although they are giving way to solid-state drives) and are also used to provide a constant magnetic flux for some of the test and measurement instruments used in industry.

Probably the most common applications of permanent magnets are the magnetic compass (for those who do not use GPS) and the rare earth magnets in audio headphones. Rare earth magnets can be manufactured in any shape by casting or sintering. Other applications for these magnets include electric motor fields for permanent magnet motors such as those used in cordless tools.

Due to their smaller size and stronger field effect (purportedly ten times the performance of ferrite magnets), rare earth magnets are replacing electromagnets in industry. In equipment, machinery and production environments, they are used in rotary actuator and linear actuator high-performance a.c. servo motors.

Today, the permanent magnet has a wide variety of applications. These include audio loud speakers and microphones, the magnetic tape used in credit cards, fridge door seals, proximity relays, alarms, small generators, debris-collecting sump plugs and printed-circuit motors.

Magnetic chuck

A common and important application in modern machining practices is the magnetic chuck, as shown in Figure 6.10. This uses the holding power of magnets to retain magnetic materials firmly in position on the work table of a machine during machining processes. No electrical connections are needed and, therefore, in the event of an electrical failure, the material being machined is not accidentally released or allowed to move or cause damage.

FIGURE 6.10 Magnetic chuck
Rito Succeed/Shutterstock.com

As it is not possible to 'switch off' the magnetism of a permanent magnet, some other means must be used to release articles held by the chuck. This is achieved by shunting (bypassing) the magnetic flux through low-reluctance bridges.

Construction, operation and applications of reed switches

Reed switches consist of two ferromagnetic wires and specially shaped contact blades enclosed in a hermetically sealed glass capsule with a gap between them. The glass capsule is filled with inert gas to prevent activation of the contacts (see Figure 6.11).

They are operated using a magnetic field created by either a permanent magnet or current-carrying coil located close to the switch. The magnetic field will force the reed contacts to close. When the magnetic field is removed, the switch will open. Figure 6.11(b) shows a cutaway of an industrial-style reed switch while Figure 6.11(c) shows two types of switches used for alarm and security applications on doors and windows.

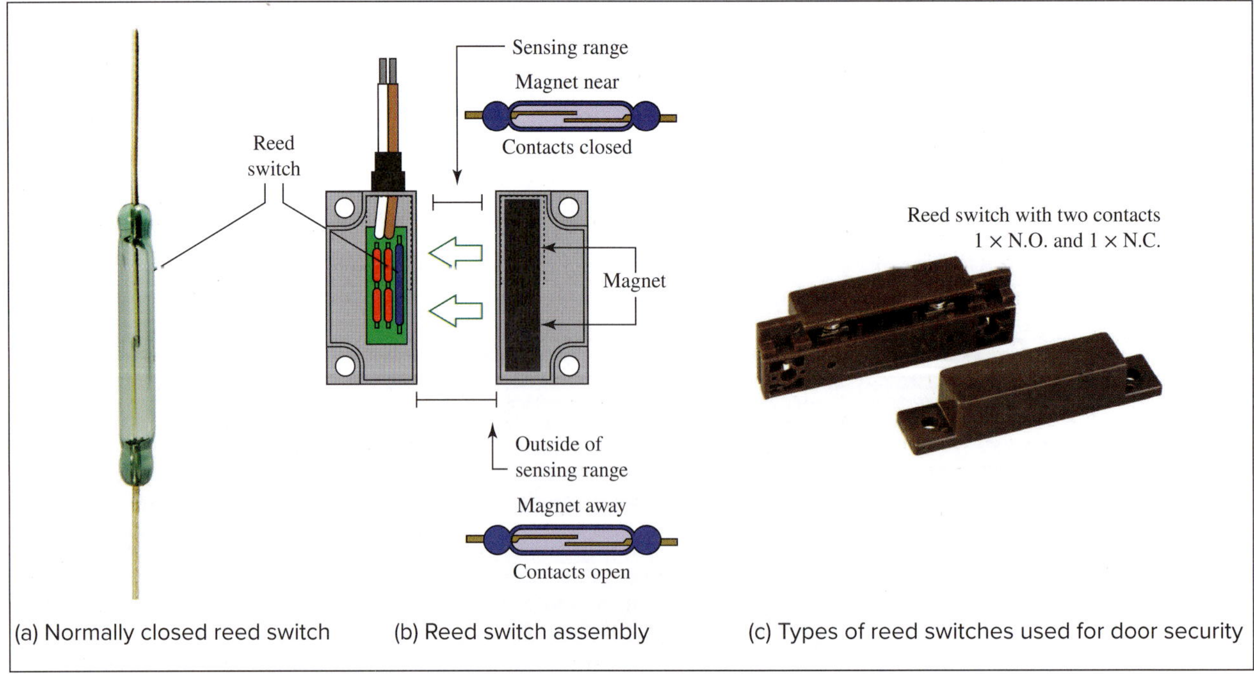

FIGURE 6.11 Reed switches
(a) Darkdiamond67/Shutterstock; (c) OKcamera/Shutterstock.com

6.1.5 The principle of magnetic screening (shielding) and its applications

Electricians often have to deal with the separation of data cables from power cables on cable trays in commercial and industrial installations and in data centres. They also have to manage the separation of extra-low-voltage (ELV, such as TV, telephone and data cabling) and low-voltage (LV, such as light and power) cables within domestic installations. Data, telephone, TV installation technicians and electrical contractors should undertake installation measures to ensure electrical separation for safety and to avoid electrical interference.

One way to reduce the effect of unwanted magnetic interference (including induced voltages) is the application of electrical installation methods used for data and power cables. Magnetic screening or cable shielding works on the physical magnetic properties of ferromagnetic and other materials. Shielding and screening does not block magnetism. Using material permeability, magnetic flux can be diverted or 'shunted' like current in electrical circuits, via the path of least magnetic resistance and around whatever needs screening or shielding.

The purpose of shielding is to prevent unwanted magnetic fields and signals produced by electrical devices or cabling from interfering with other electrical devices, signals and cables. This interference can be either electromagnetic interference (EMI) or radio frequency interference (RFI). Aluminium and metallised plastics are normally used when radio frequency shielding is required to stop high-frequency fields (>100 kHz).

Magnetic shielding is effective for a.c. and d.c. applications and typically used in the 30–300 Hz a.c. range. Shielding is also used for fields produced by magnets, variable frequency drive motors and other 50–60 Hz electric power equipment.

All ferromagnetic metals (anything containing iron, nickel or cobalt) can be used for magnetic flux screening and shielding. In some situations, ferromagnetic steel is used because of its relatively low cost and availability. Steel is commonly referred to as either *mild steel* or *stainless steel*. However, not all steel is 'stainless', nor is it necessarily magnetic. Magnetic properties of stainless steel are highly dependent on the elements added into the alloy. Type 316 stainless steel, a molybdenum-alloyed steel, is also negligibly responsive to magnetic fields and can be used in applications where a non-magnetic metal is required.

Steel alloys containing more than 10.5% chromium can be classified as stainless steel. The addition of nickel renders steel non-magnetic, as in the case of 304 stainless steel, for example. Stainless steel can be either Martensitic or Austenitic steel. Austenitic steel contains nickel and is non-magnetic while Martensitic steel has a ferritic microstructure, which makes it magnetic.

Magnetic shielding does not block a magnetic field. Magnetism (and electromagnetic radiation) cannot be stopped and there are no insulators. However, magnetic shields can divert the magnetic field by providing a path of much greater permeability than free space. Just as in Figure 6.12(a), the magnetic material creates a low resistance to the magnetic flux. This concept is applied in Figure 6.12(b) to the wire braid shielding. Shielding is used in cabling with foil or a metal braid wrapped around the conductors. One end is terminated to 'ground' any interference to an earth potential.

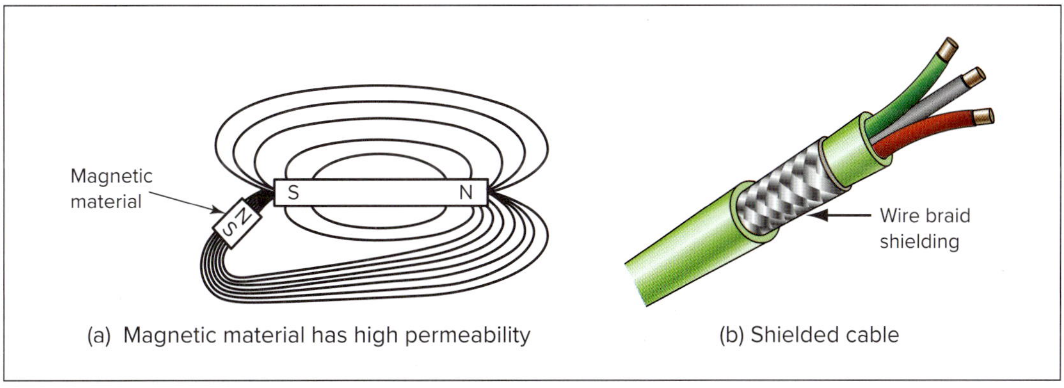

(a) Magnetic material has high permeability

(b) Shielded cable

FIGURE 6.12 Material permeability and shielding

Computers and electrical and electronic devices and equipment can be sensitive to EMI or RFI affecting their operation. When such devices need to be protected from interference, a magnetic shield like the one in Figure 6.13 can be used.

The magnetic field flows *around* the device (via shielding) instead of through it. The shielding also needs to have greater permeability than the material encasing or surrounding the device to be protected.

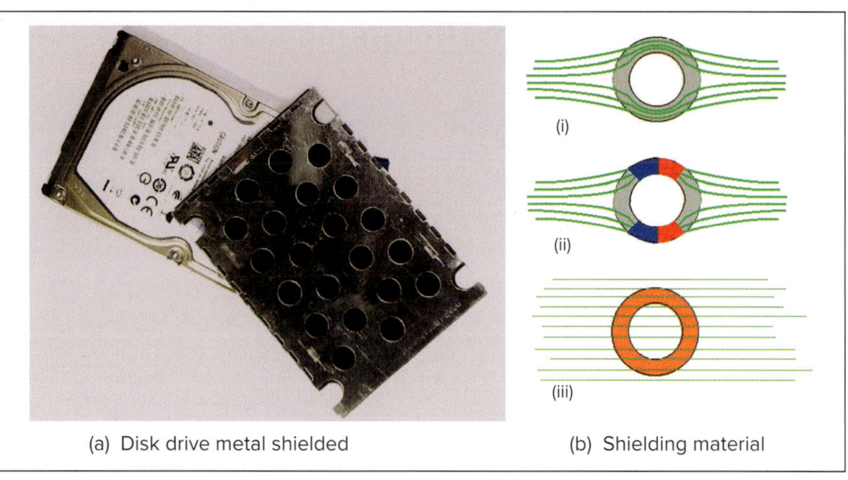

FIGURE 6.13 Magnetic shielding
Tony Jones

Magnetic shields in sensitive instruments are often made of a material known as MuMETAL®, a nickel-iron alloy that is widely used for magnetic shielding purposes. It has a composition of 80% nickel (Ni), 4.5% molybdenum (MO) and 15.5% iron (Fe). This gives MuMETAL® highly permeable properties and it saturates at 0.76 tesla (T). There are several compositions of MuMETAL® available, another being Ni 77%, Fe 14%, Cu 5% and MO 4%.

Figure 6.13(b) depicts the cross-section of a cylindrical magnetic shielding material. Magnetic field lines are directed around the area within the shield material as shown in Figure 6.13(b)(i) and (ii). The central regions for Figure 6.13(b)(i) and (ii) indicate the field-free region. Figure 6.13(b)(iii) shows the magnetic field effect through unshielded, non-ferromagnetic material.

6.2 Electromagnetism

When current travels along a conductor, a magnetic field surrounds the conductor (Figure 6.14(a)). The magnetic field around the conductor, represented as B, is 'north seeking' and the direction depends on which way current flows through the conductor. This magnetic field increases if the current increases and decreases if the current decreases. If the conductor is made into a loop, the density of the magnetic field increases inside the loop due to compression of the field inside the curve. Multiple turns further increase the field strength as if greater current flows within the single loop. Stretching the field outside the loop reduces the magnetic field density. Figure 6.14(b) shows field lines spreading out.

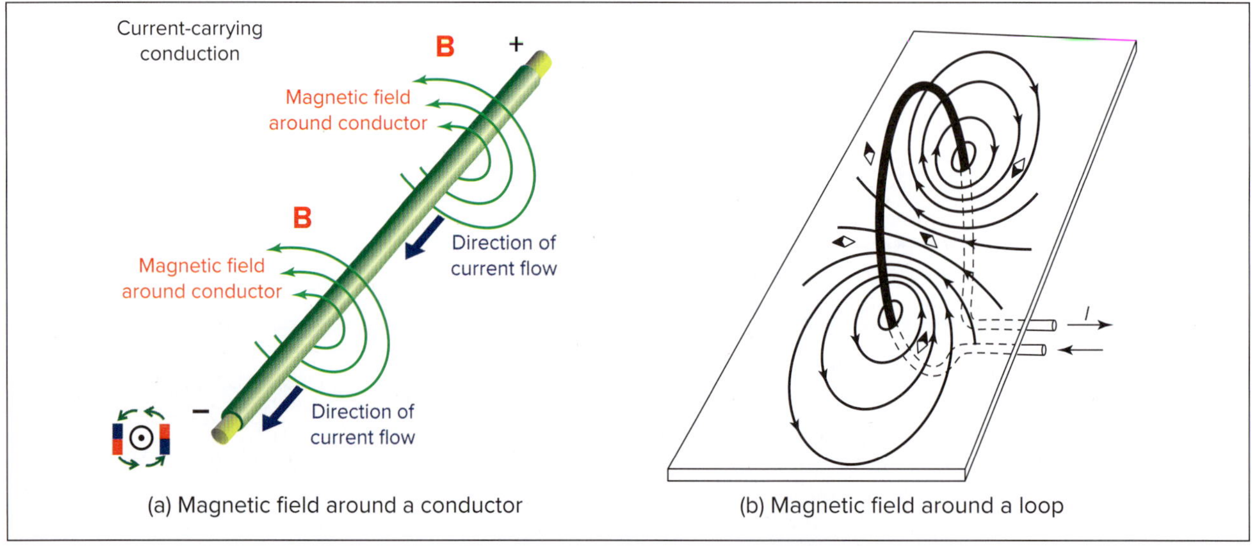

FIGURE 6.14 Magnetic fields

6.2.1 Conventions representing direction of current flow in a conductor

Figure 6.15 illustrates the direction of current flow in a conductor or coil. Two symbols are commonly used to indicate the direction of current flow in any conductor seen end-on. A circle with a dot in the centre represents a current flowing *towards* the viewer, as if it were the point on the end of an arrow. A circle with a cross inside it represents a current moving *away from* the viewer, with the cross as the feather flights of the arrow.

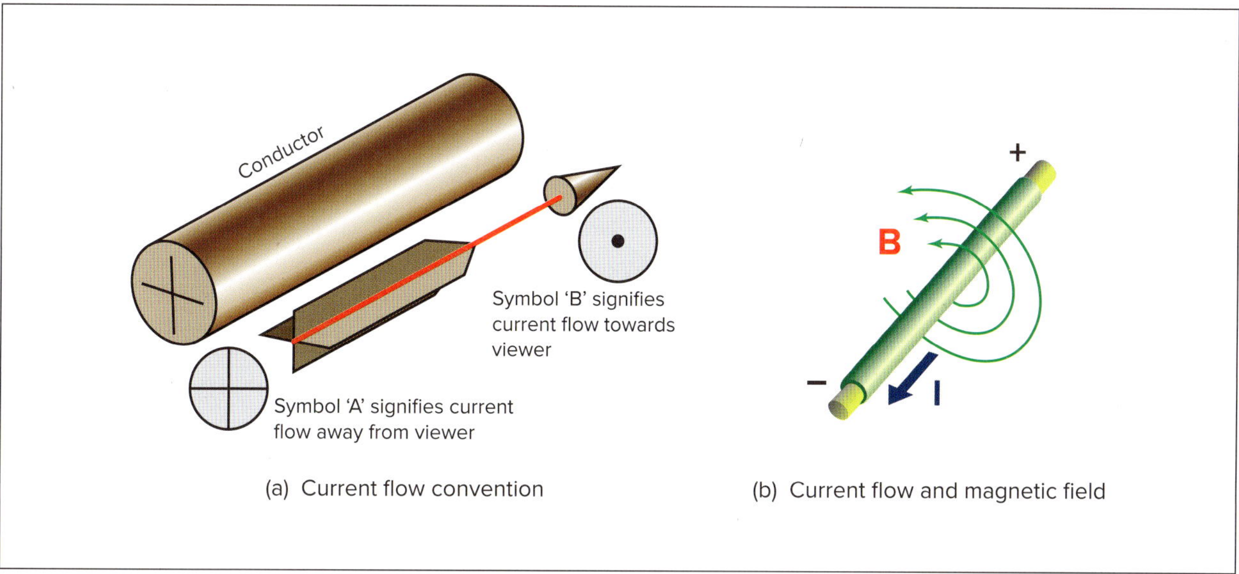

FIGURE 6.15 Current flow

In relation to conventional current flow and polarity, Figure 6.15(b) shows that the arrow tail is positive (+ve) and its tip negative (−ve); convention states that current flow is from positive to negative. The tip and tail symbols can be thought of as being like battery terminal symbols—the tail is a positive potential and the dot is a negative potential.

There is a simple rule to help determine the direction of the field for a given direction of current flow—the *right-hand (grip) rule*. Figure 6.16(a) shows the magnetic field flux around a straight current-carrying conductor. Figure 6.16(b) shows the right-hand (grip) rule. With a straight conductor, if the conductor is grasped (only if it has been established that it is not energised) in the right hand with the thumb pointing in the direction of the current flow, the fingers point in the direction of the magnetic field.

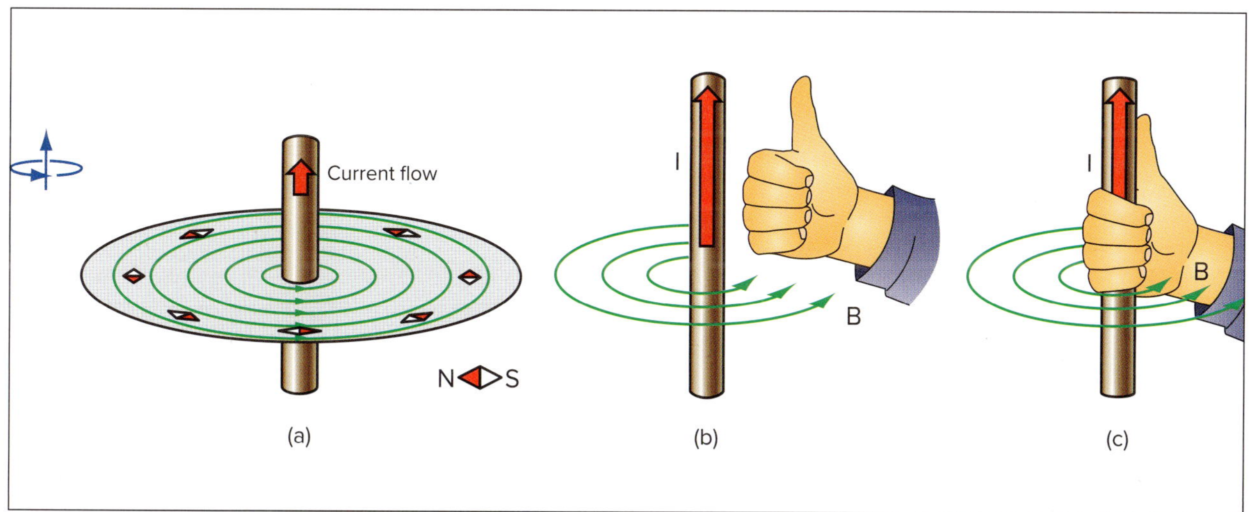

FIGURE 6.16 Magnetic field around a straight conductor as determined by the right-hand (grip) rule

The strength of the magnetic field around a straight conductor depends on the value of the current in the conductor. Doubling the current results in double the field strength—the field strength is proportional to the current. As illustrated in Figure 6.17, compass needles act as small magnets to show the effect of the magnetic field around a current-carrying conductor.

A single conductor passing vertically through a flat plane (provided by a sheet of cardboard or clear plastic) is given a large current of perhaps 100 amps. If iron filings are sprinkled on the plane, a pattern appears, circling the conductor. That pattern is the magnetic field that surrounds every current-carrying conductor.

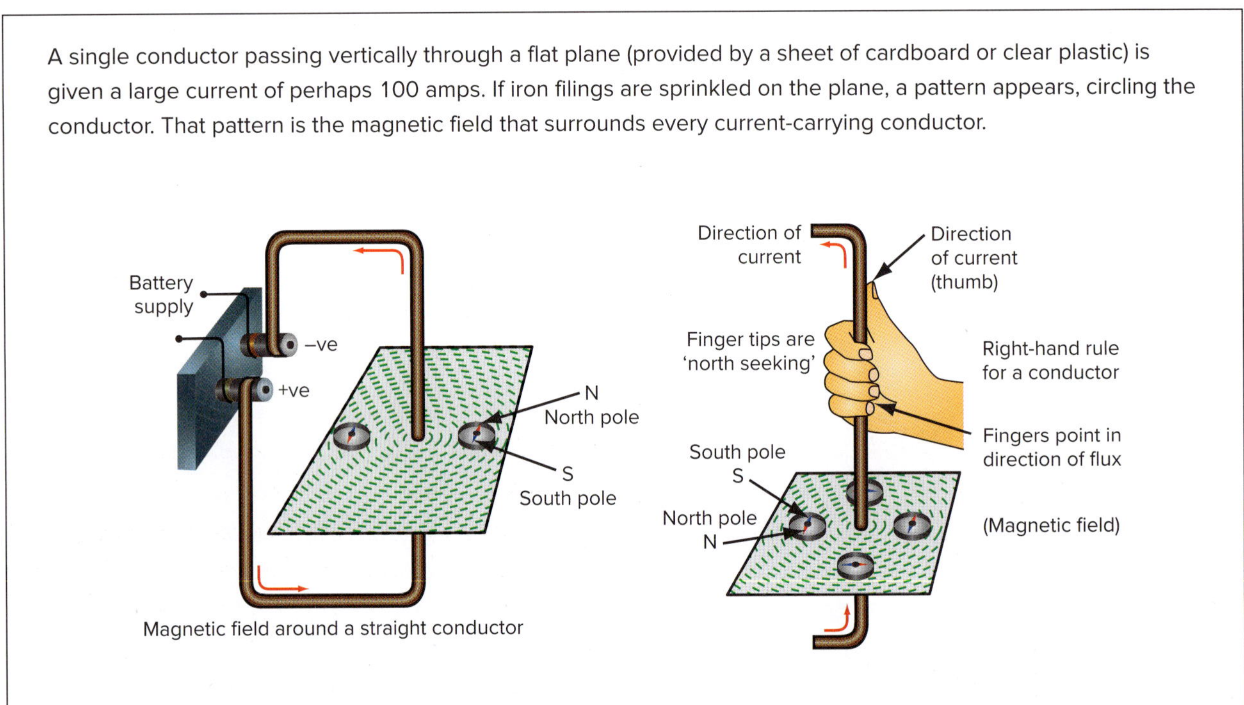

FIGURE 6.17 **Magnetic field around a conductor and the right-hand (grip) rule**

6.2.2 Direction of force between adjacent current-carrying conductors

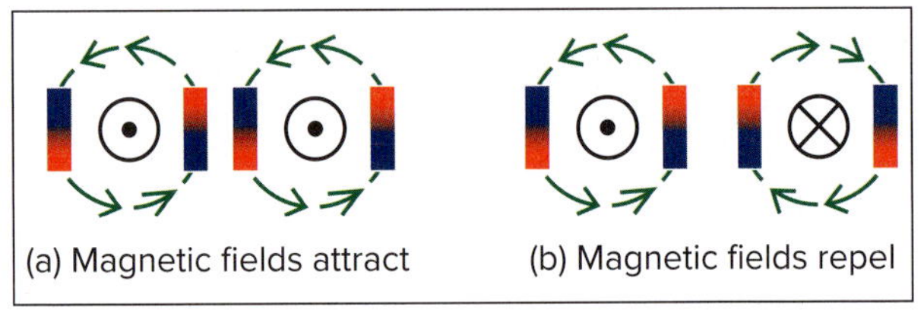

FIGURE 6.18 **Magnetic fields attracting and repelling**

Two adjacent current-carrying conductors (or busbars) will each have a magnetic field around them. These magnetic fields will be a result of the direction of current flow. The interaction of these fields will be either one of attraction or repulsion. Figure 6.18 shows the interaction between two adjacent magnetic fields as (a) attraction and (b) repulsion like that between two bar magnets. Applying the right-hand (grip) rule for conductors, the finger tips are 'north seeking'. If we show the north-seeking magnetic fields as bar magnets, Figure 6.18(a) illustrates the interaction of unlike magnetic fields. This results in a force of attraction. Figure 6.18(b) shows the interaction of like magnetic fields. This results in a force of repulsion.

Figure 6.19(a) shows the flux around two straight conductors carrying current in the same direction. *The flux of each conductor unites to form a single flux around both conductors.* A flux tends to take the shortest possible path, which in this case will tend to pull the two conductors together.

If the current in each conductor flows in opposite directions, as shown in Figure 6.19(b), the flux between the two conductors is acting in the same direction. *Flux in the same direction tends to spread out, forcing the two conductors apart.*

Figure 6.19(b) shows conductor movement will occur when the magnetic force of repulsion exceeds the physical forces holding the conductors in position. This situation can occur in practice with heavy-current switch gear and

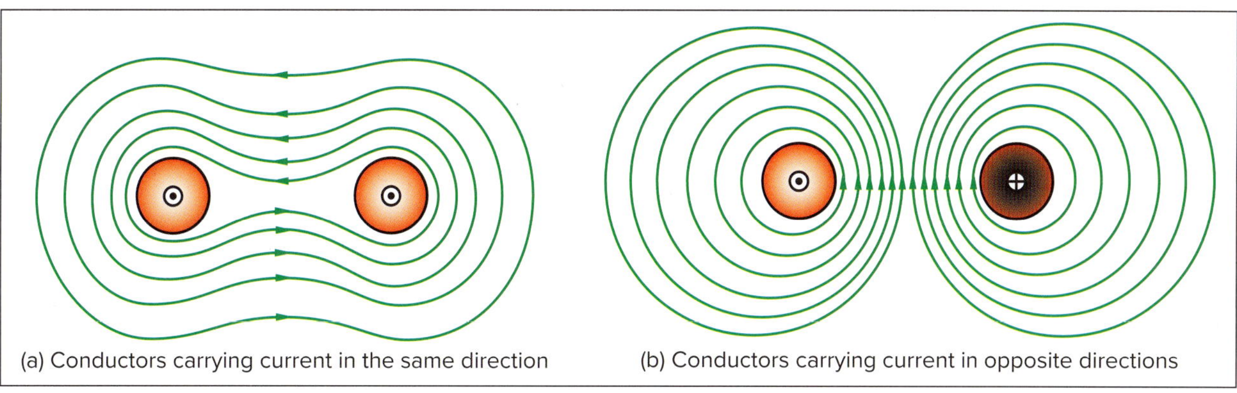

FIGURE 6.19 Flux around straight conductors

machinery, however the effects are not always undesirable. Conductors within electric motors and measuring instruments, for example, depend on these forces for their operation.

6.2.3 The effect of current, length and distance apart on the force between conductors (including forces on busbars during fault conditions)

An ampere is the current flowing in each of two parallel conductors of infinite length and negligible cross section (area). The ampere is defined as the current that would cause a force of 2×10^{-7} newtons per metre between two conductors placed one metre apart. Therefore, the force between two conductors with a known current flow in each conductor and with a known distance separating the conductors can be calculated. This force of 2×10^{-7} newtons per metre, its relationship to coils, solenoids and their relationship to the magnetic flux field and flux density are examined in section 6.3.5. For now, only the force and attraction or repulsion effects between two magnetic fields are being considered.

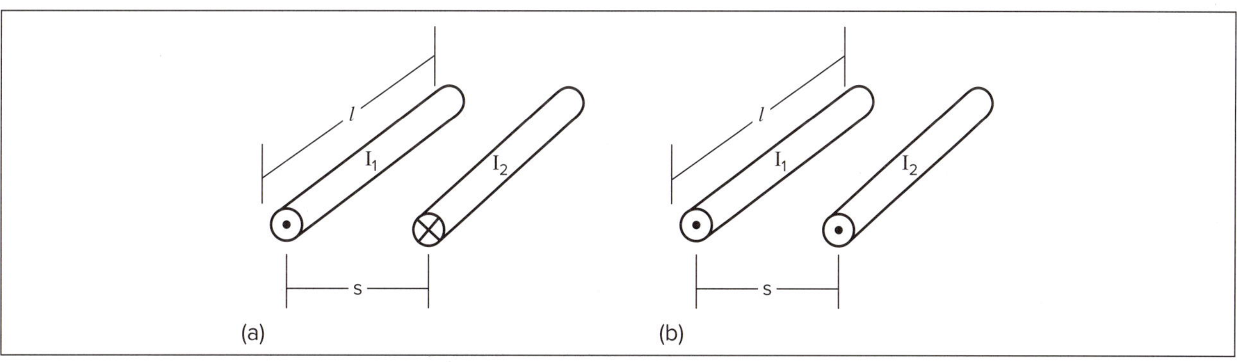

FIGURE 6.20 Parallel current-carrying conductors

EXAMPLE 6.1

Force between conductors

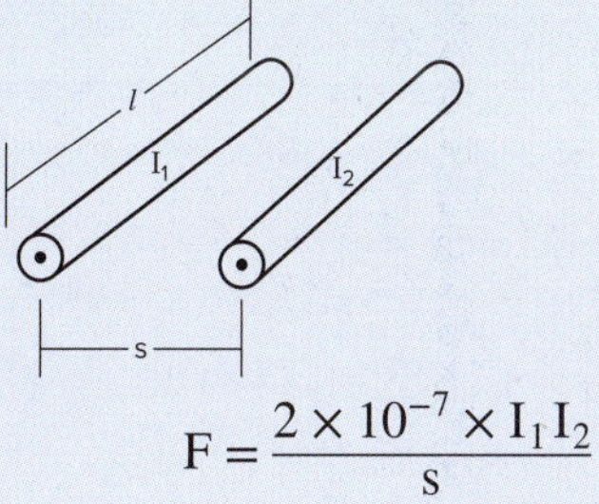

$$F = \frac{2 \times 10^{-7} \times I_1 I_2}{s}$$

Two long parallel conductors 0.1 m apart each carry a current of 100 A in opposite directions. Calculate the force between them.

$$F = \frac{2 \times 10^{-7} I_1 I_2}{s} \quad (1)$$

$$= \frac{2 \times 10^{-7} \times 100 \times 100}{0.1} \quad (2)$$

$$= 0.02 \text{ N} \quad (3)$$

where:

F = force between conductors in newtons
I_1 = current in the first conductor in amps
I_2 = current in the second conductor in amps
s = distance separating the conductors in metres.

Press the following buttons on your calculator:

2 × 10x (–) 7 × 1 0 0 × 1 0 0 ÷ . 1 =

CHECK YOUR UNDERSTANDING

Review Figures 6.18, 6.19, 6.20 and the sets of parallel conductors.

6.3 What is the direction of magnetic flux around each conductor (using the right-hand (grip) rule for conductors)?

6.4 What is the force of either attraction or repulsion for each pair of conductors (using the right-hand (grip) rule for conductors)?

6.2.4 Magnetic field around coils and electromagnets

A length of wire wound into a number of loops forms a coil. These loops of wire are also known as 'coil windings' or 'coil turns', each complete winding being one turn (N). Just as a magnetic field exists around a current-carrying conductor, when a current flows through the coil windings, a magnetic field is created.

Figure 6.21(a) shows how the current in each adjacent turn of a coil flows in the same direction. The magnetic fields around each loop will combine to form a single magnetic flux embracing all the turns of the coil. The strength of the resultant flux will be equal to the total of all the separate fields set up by each coil turn. As a result, the flux (Φ) inside the coil is proportional to the number of turns (N) in the coil.

Figure 6.21 shows that the flux flowing through the centre of the coil establishes a north pole at the end where it leaves the inside of the coil and a south pole at the end where it enters the coil. The field around the outside of the coil is similar to the field around a bar magnet as in Figure 6.21(b). Such a coil is commonly referred to as a 'temporary magnet' or 'electromagnet' because a magnetic field is produced by the current flowing through it.

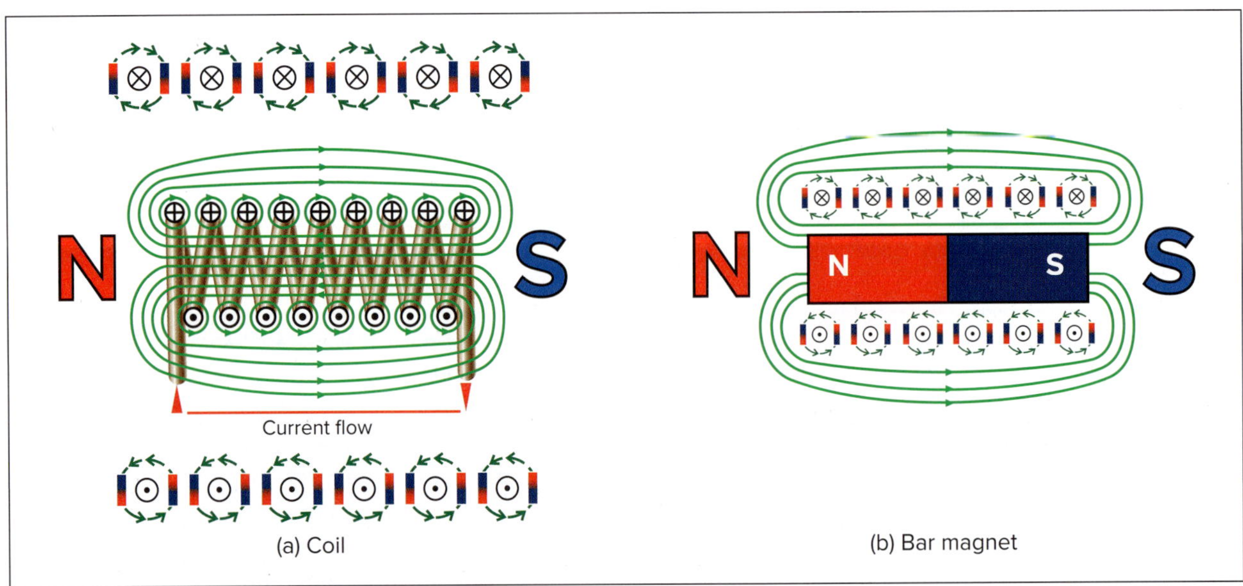

FIGURE 6.21 Flux loop around adjacent conductors in a coil

An electromagnet can be defined as a coil producing a magnetic field that is concentrated on the core or centre of the coil. The core of an electromagnet is usually a high-permeable material such as iron or mild steel but in some applications can be non-magnetic or an air core. The magnetic field generated by a solenoid can also be determined by the right-hand (grip) rule, as shown in Figure 6.22(a) and (b).

The right hand is placed over a solenoid or coil, with the fingers pointing in the direction of the current flow through the windings (shown by the red arrows). The thumb points at a 90-degree angle from the fingers, in the direction of the north-seeking magnetic field. The thumb points to the north end of the coil.

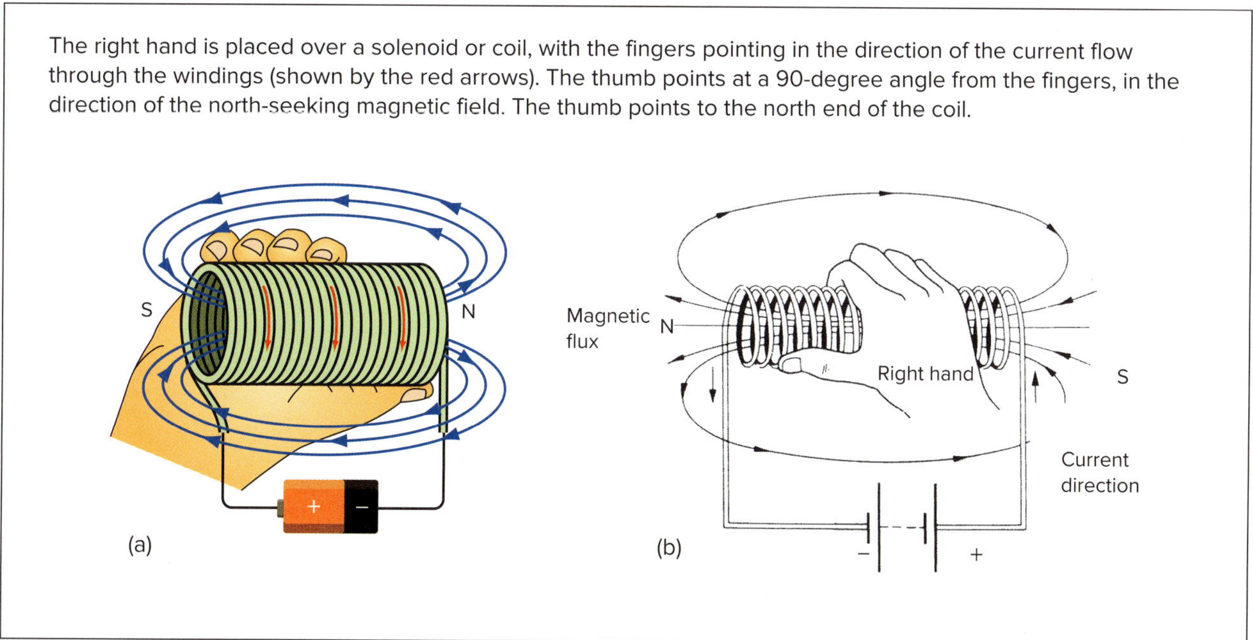

FIGURE 6.22 **Right-hand (grip) rule for a coil**

6.2.5 Magnetomotive force (MMF) and its relationship to the number of turns in a coil and the current flowing in the coil

If a straight conductor is bent to form a loop, the strength of the flux inside the loop is due to the magnetic field generated so the field concentration is doubled (as shown earlier in Figure 6.14(b)). By winding the conductor into a coil of many turns, the magnetomotive force (MMF) is increased in proportion to the number of turns in the coil, as illustrated in Example 6.2. More detailed information about flux density and calculations can be found in section 6.3.5.

EXAMPLE 6.2

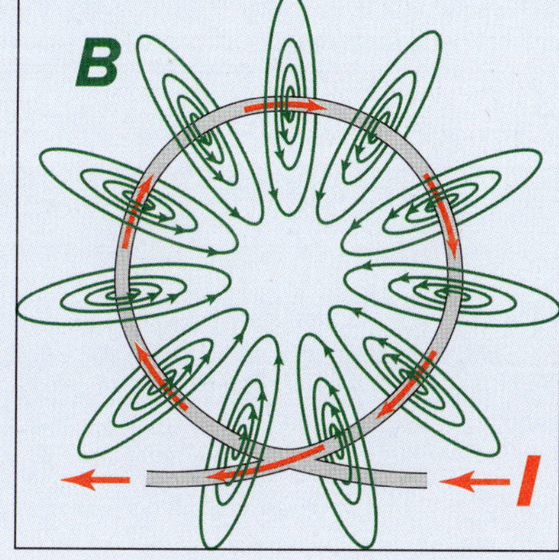

Magnetic flux density inside a loop (B)

$$B = \frac{\mu_0 I}{2r}$$

where

B = flux density
μ_0 = permeability of free space (vacuum)
I = current in the coil loop
r = radius of the loop.

The MMF is dependent on the current flowing in a conductor while the magnetic field of a coil is dependent on the number of turns of the coil. The field strength of a coil is therefore proportional to the product of the current (I) and the number of turns (N) in the coil (see Figure 6.23).

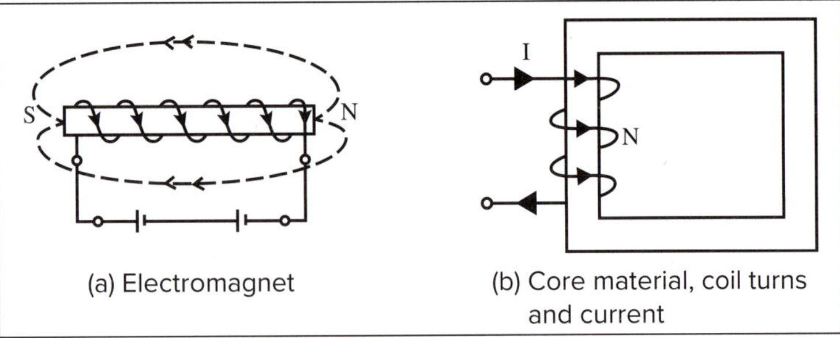

FIGURE 6.23 **Magnetomotive force (MMF)**

In pure SI units, the MMF unit is the ampere because the number of turns in a coil is considered to be dimensionless. In calculations, however, the number of turns has to be included. A conductor wound around a former to form a coil is called a solenoid. A solenoid with a magnetic core becomes an electromagnet. The magnetic field strength depends upon current strength (I), number of turns (N) and the core material.

EXAMPLE 6.3

Magnetomotive force (MMF)

$F_m = IN$

where:

F_m = magnetomotive force
I = current
N = number of coil turns.

If a current of 5 A is flowing in a coil of 120 turns, find the value of MMF creating a magnetic flux.

$$F_m = IN \qquad (1)$$
$$= 5 \times 120 \qquad (2)$$
$$= 600 \text{ At} \qquad (3)$$

6.2.6 Practical applications of electromagnets

Electromagnets can replace permanent magnets or temporary magnets in most applications, with the advantage that they can be turned on and off and varied in magnetic strength. Electromagnets can also have alternating fields when an alternating current is applied. Permanent magnets cannot be made to alternate, especially at the line frequency of 50 Hz or 100 times per second.

The electromagnetic effect is used in inductors (chokes and ballasts). Operating coils can be used to create an electromagnetic effect in devices such as contactors relays and solenoids. Section 6.8 covers the operation and application of magnetic devices in the electrotechnology industry.

There are three types of devices:

1. Tractive—the electromagnet attracts an armature to operate attached electrical contacts (contactors and relays). Section 6.8.2 gives further detail on the operation and construction of relays and contactors.
2. Solenoid—the coil surrounds a sliding plunger which is usually drawn into the coil (or extend) when energised (e.g. hot and cold water valves on a washing machine).
3. Lifting—the poles attract magnetic material for transportation or magnetic material separation from waste.

Electromagnets are used in motors and generators, transformers, microphones, loud speakers, indicators such as doorbells and car-ignition coils. Figure 6.24(a) shows an audio loud speaker and (b) an example of a D'Arsonval galvanometer. Moving coil meters and speakers have some common electrical and magnetic operating principles.

In terms of how audio speakers work, when the magnetic fields of coils interact with the magnetic fields of permanent magnets, the principle of attraction or repulsion applies. Speakers are constructed with two magnetic field components which interact with one another. An effect is created within a speaker which is similar to that with parallel conductors where, depending on the direction of current flow in each, the resulting magnetic fields interact, causing a force of either attraction or repulsion.

On the speaker base is a permanent magnet with a constant magnetic field. Attached to the speaker cone is a coil. When current flows through the speaker cone coil, a magnetic field is created which interacts with the magnetic field of the permanent magnet. The audio signal is a varying analogue signal, like a rapidly changing sine wave with positive peaks and negative troughs. These peaks and troughs have the effect of reversing the current flow through the coil, changing the coil field direction.

The audio signal from an amplifier varies the amount of current flowing through the speaker cone coil, which, in turn, varies the magnetic field. This interaction between the field of the permanent magnet and the constantly changing magnetic field of the speaker coil causes the speaker cone to physically move back and forth. It is this rapid back-and-forth cone movement or vibration that creates the sound we hear from a speaker.

Knowing the magnetic principles of how a loud speaker works aids the understanding of how other devices like transformers and measurement instruments operate.

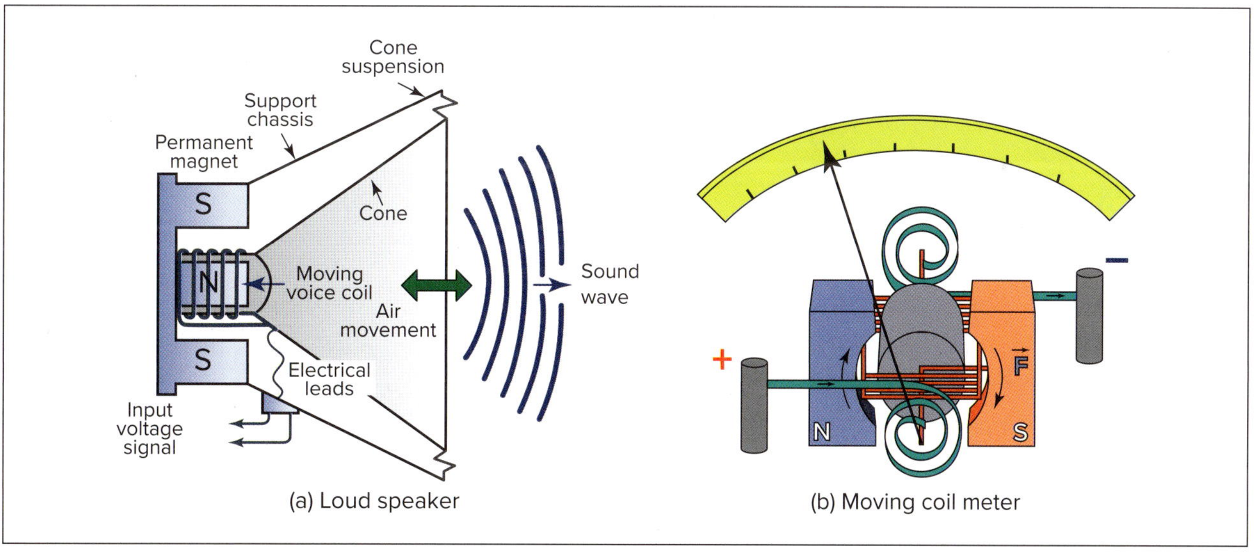

FIGURE 6.24 **Loud speaker and moving coil meter**

CHECK YOUR UNDERSTANDING

6.5 List two devices which operate on magnetic attraction.

6.6 List two devices which operate on magnetic repulsion.

6.7 Briefly explain how you think the D'Arsonval galvanometer shown in Figure 6.24(b) might operate.

6.3 Magnetic circuit types and associated terminology

Electrical circuits and magnetic circuits have similarities and differences in how they operate. These can be shown via a comparison between Ohm's Law for electrical circuits (voltage, current and resistance) and the equivalent rule for magnetic circuits (magnetomotive force, magnetic flux and reluctance).

Electrical circuits and magnetic circuits also use different terminology, as shown in Table 6.2.

TABLE 6.2 Comparison of electrical and magnetic circuit terminology

Term	Electrical	Magnetic
Pressure	Electromotive force (EMF)	Magnetomotive force (MMF)
Flow	Current (I)	Flux (Φ)
Opposition	Resistance (R)	Reluctance (R_M) or (S)

6.3.1 Comparing electrical and magnetic circuits

An electromotive force (EMF) can exist without producing a current flow, but a magnetomotive force (MMF) cannot exist without producing a magnetic flux. Electric circuits can be turned on and off, conduct electricity and be in either an 'open circuit' or a 'short-circuit' condition.

The magnetomotive force (either MMF or Fm) creates the magnetic field (flux) in a coil, causing the magnetic flux orientation from north to south around an iron core circuit (magnetic circuit). It acts like the electromotive force in an electric circuit and is the driving force necessary to set up a magnetic flux in a magnetic circuit. Magnetic flux is measured in webers (Wb) and has the symbol Φ (phi, pronounced 'fye'). The opposition to the magnetic flux is called 'reluctance'.

TABLE 6.3 Comparing electrical and magnetic circuits using Ohm's Law

Electrical circuits	Magnetic circuits
Ohm's Law, when applied to electrical circuits, gives the following equation: $I = \frac{V}{R}$ where: I = current flow in amperes V = electromotive force R = circuit opposition or resistance.	A similar law can be applied to magnetic circuits: $\Phi = \frac{IN}{R_m}$ where: Φ = magnetic flux in webers (Wb) IN = magnetomotive force in ampere-turns (At) R_m = magnetic reluctance in ampere-turns per weber (At/Wb).

6.3.2 Reluctance as the opposition to the establishment of magnetic flux

Reluctance is comparable with resistance in an electric circuit. Some 'hard' materials such as steel require high magnetising forces to align their atomic dipoles in the same direction while others are readily magnetised by small forces. All materials offer some opposition or resistance (referred to as 'magnetic reluctance') to being magnetised.

As stated earlier, materials with high permeability offer less opposition to magnetic flux. A higher permeability value results in a lower material reluctance. Permeability $\mu = \mu_o \mu_r$ is used in equations for determining the reluctance (Rm).

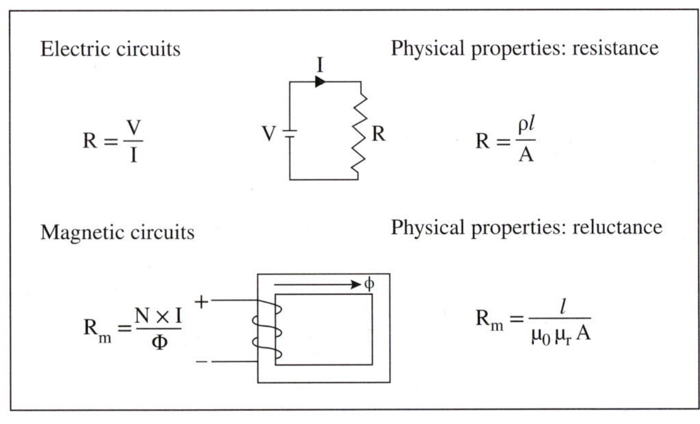

FIGURE 6.25 Comparing electrical and magnetic circuits

Figure 6.25 compares electrical and magnetic circuits and shows the effect of changes in permeability (μ), length (l) and area (A) for reluctance.

The symbol used for reluctance can be either (S) or (R_m). For consistency, (R_m) is used throughout this chapter. Also, when comparing electrical and magnetic circuits, electrical resistance (R) and magnetic reluctance (R_m) are more easily identified.

EXAMPLE 6.4

Reluctance

$$R_m = \frac{l}{\mu_o \mu_r A}$$

where:

R_m = magnetic reluctance

μ_o = permeability of free space (vacuum)

μ_r = relative permeability

A = cross-sectional area (m^2).

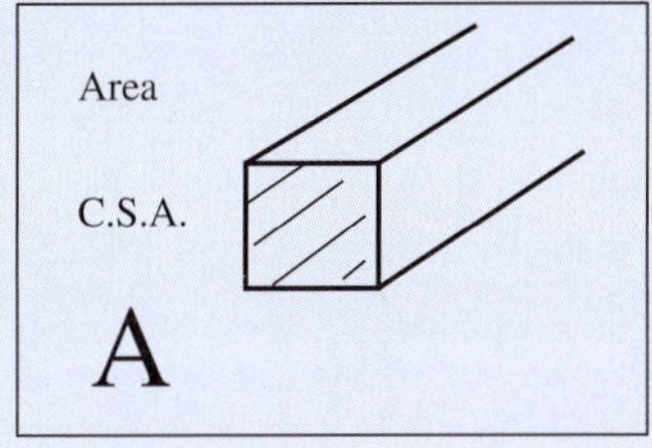

The total mean length of the path of an iron core is 200 mm. The core is rectangular in cross section, with dimensions of 15 mm × 10 mm. If the core has a relative permeability of 830 at the designed flux density, calculate the reluctance of the core.

$$R_m = \frac{l}{\mu_o \mu_r A} \quad (1)$$

$$= \frac{200 \times 10^{-3}}{4 \times 10^{-7} \times \pi \times 830 \times 15 \times 10^{-3} \times 10 \times 10^{-3}} \quad (2)$$

$$= \frac{1.278 \times 10^{6}\,\text{At}}{\text{Wb}} \quad (3)$$

In Example 6.4, the permeability of free space (a vacuum) is included in the calculation and is the reference against which the permeability of other materials is compared. If there is no material, such as in a vacuum, the value can be shown to be $4\,\pi \times 10^{-7}$. Some calculators may give a 'syntax error' when these values are entered. A way around this problem would be to enter the values as $4 \times 10^{-7}\,\pi$, or as an approximation, 12.57×10^{-7}.

For air and other non-magnetic materials, μ_r has the value of unity ($\mu_r = 1$). However, in Example 6.4 the relative permeability valve for the iron core is 830. This has the overall effect of reducing the magnetic reluctance.

Remember that, when performing magnetic circuit device calculations, the coil or solenoid ferromagnetic core will have a very small reluctance value compared with the very high reluctance value for the circuit air gap. The total circuit reluctance is obtained by adding the individual reluctances for the iron core and air gaps together. Examples of this are shown in Figure 6.26.

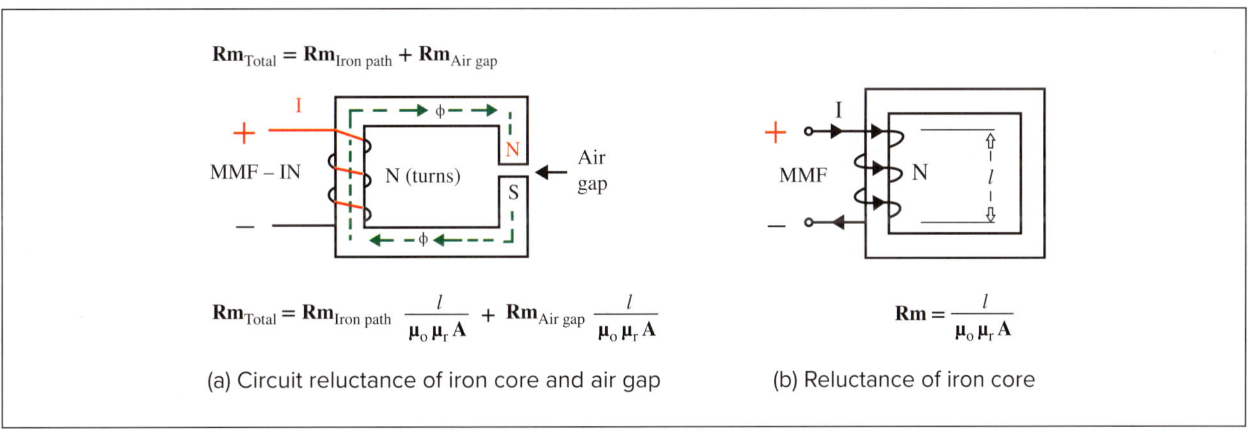

FIGURE 6.26 Circuit reluctance (Rm or S)

6.3.3 Effect of an air gap in a magnetic circuit

With the exception of transformers, magnetic circuits must have a gap in the soft iron of the circuit to do any useful work. This air gap has a very high reluctance. Examples include contactors and relays (see Figure 6.27). Magnetic circuit devices such as motors, generators and moving coil meters such as the D'Arsonval galvanometer shown in Figure 6.24 have two air gaps. Unlike electrical circuits, the magnetic flux path of a magnetic circuit can 'jump' across these high-reluctance air gaps.

Improvements in machine efficiency can be achieved through precision engineering and design to reduce the size of the air gap in a magnetic circuit. This promotes flux paths with lower reluctance values and reduces the effects of flux leakage and fringing.

6.3.4 Magnetic leakage and magnetic fringing

In practical magnetic circuits there is a tendency for flux to leak through the surrounding air, bypassing the intended path. Unfortunately, the flux tends to spread out so the flux density in the intended path of the magnetic circuit is reduced to a lower value.

Figure 6.27 shows examples of the magnetic field path, flux leakage and fringing in a magnetic circuit. The lines of force near the centre line of the flux path are straight, providing maximum flux. Lines of force at the edges of the field tend to curve outwards in the air gap and spread according to the rule that flux travelling in like directions repels. As a result, the area of the flux path in the air gap is greater than in the material of the magnetic circuit. The flux density of the air gap will be less than that within the magnetic material on either side. This effect is known as 'magnetic fringing' and must be allowed for when designing magnetic circuits with air gaps. Some magnetic leakage also occurs at the sides of the air gap, but this is normally counted as a part of the fringing. Circuit calculations use the material area as the area for the air gap.

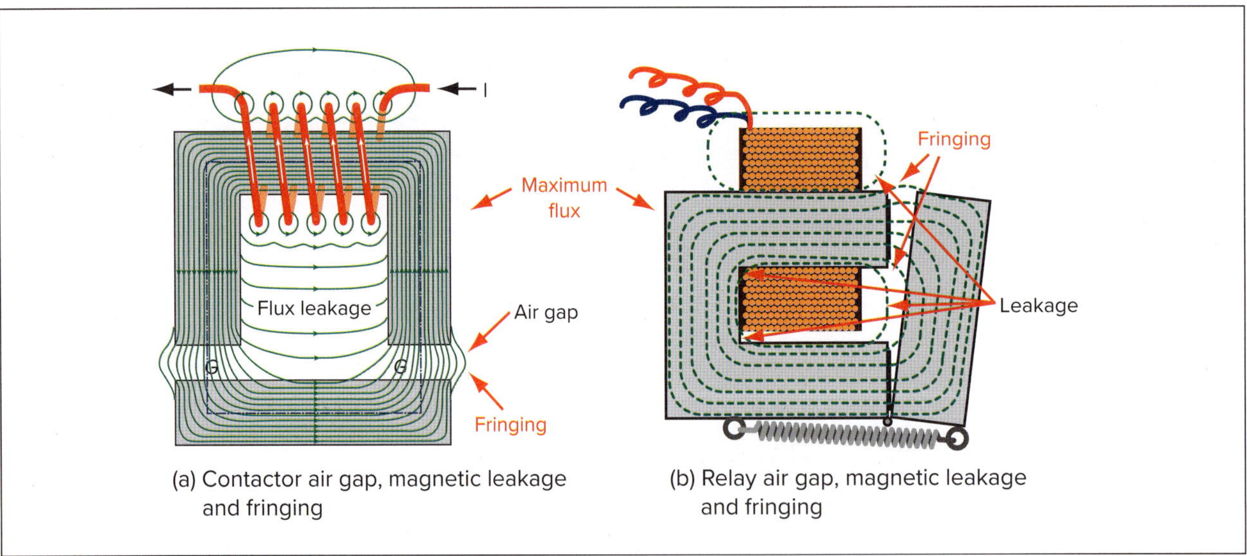

FIGURE 6.27 **Magnetic flux leakage and fringing**

CHECK YOUR UNDERSTANDING

6.8 Coils can have an air core, a non-magnetic material core or a ferromagnetic core. What would be the effect of replacing the non-magnetic core of a solenoid with a magnetic material?

6.3.5 Magnetic circuit properties

Physical, mathematical and graphical relationships exist between permeability (μ), flux density (B) and magnetising force (H) in magnetic circuits.

Flux density

Here, 'density' refers to the quantity of flux per unit area and is not a measure of field strength.

EXAMPLE 6.5

Flux density (B)

$$B = \frac{\Phi}{A}$$

where:

B = flux density in tesla (Wb/m^2)
Φ = total flux in weber
A = cross-sectional area in m^2.

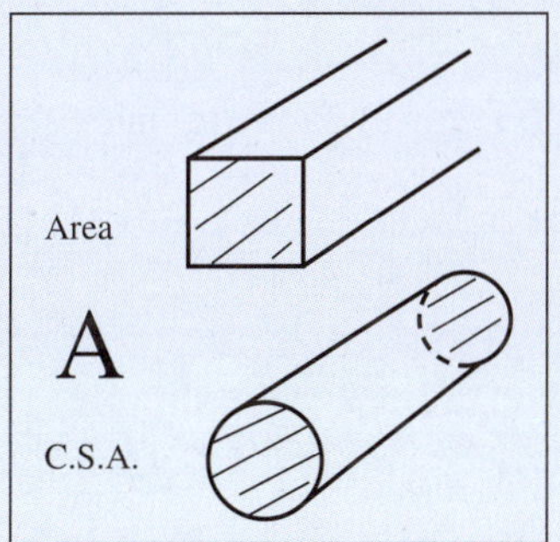

A magnetic circuit has a cross-sectional area of 100 mm^2 and a flux density of 0.01 T. Calculate the total flux in the circuit.

$$B = \frac{\Phi}{A}$$

Transpose the above equation to find Φ (1)

$$\Phi = BA \quad (2)$$
$$= 0.01 \times 100 \times 10^{-6} \quad (3)$$
$$= 1 \times 10^{-6}\,Wb \quad (4)$$

CHECK YOUR UNDERSTANDING

6.9 A contactor coil has an iron core, rectangular in section and 20 mm × 30 mm. The flux density in the magnetic circuit is 1.2 T. What is the total flux?

Magnetising force

The MMF required to magnetise a length of magnetic path is termed the 'magnetising force' for that portion of the magnetic circuit. Magnetising force (H), which is ampere-turns-per-metre (At/m), should not be confused with magnetomotive force (MMF or IN), which is ampere-turns.

Magnetising force is applicable only to that section of the magnetic circuit path that is defined by the letter l (for length in metres) in the equation $H = \frac{IN}{l}$ as shown in Example 6.6. While magnetising force and flux density are closely related, they are different. Neither quantity represents the magnetic circuit field strength. That is determined by the relationship between permeability, flux density and magnetising force.

EXAMPLE 6.6

Magnetising force (H)

$$H = \frac{IN}{l}$$

where:

H = magnetising force
l = length
N = number of conductors
I = current flowing in amperes.

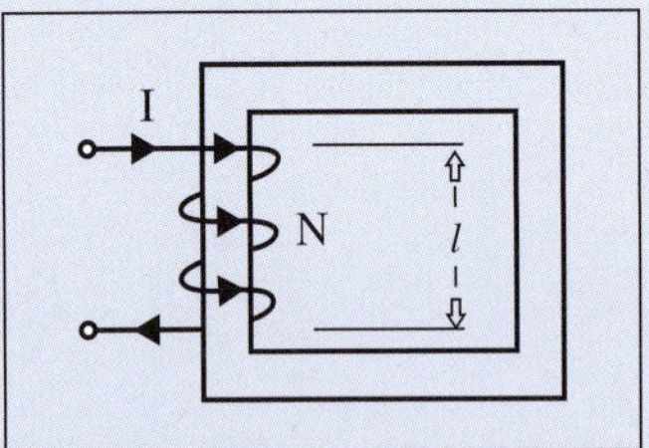

Calculate the magnetising force for a coil 0.2 m in length, with 500 turns when a current of 1.5 A flows through the coil.

$$H = \frac{IN}{l}$$

$$H = \frac{1.5 \times 500}{0.2}$$

$$H = \frac{750}{0.2}$$

$$H = 3750 \text{ At/m}$$

6.3.6 Magnetisation and magnetic characteristic curves—the B/H curve

When values of flux density (B) are plotted against values of magnetising force (H) for a magnetic material, the resulting graph is in the form of a curve. Table 6.4 shows figures for an iron sample.

TABLE 6.4 Magnetisation curve for a magnetic material (iron)

Magnetising force H (At/m)	Flux density B (Wb/m^2)	Permeability $\mu = B/H$	Relative permeability $\mu_r = \mu/\mu_o$
100	0.04	0.00040	318
200	0.12	0.00060	477
300	0.40	0.00130	1058
400	0.90	0.00225	1790
500	1.00	0.00200	1591
600	1.06	0.00177	1408
700	1.11	0.00159	1265
800	1.15	0.00144	1146
900	1.18	0.00131	1042
1000	1.21	0.00121	963
1200	1.25	0.00104	828
1400	1.29	0.00092	732
1600	1.32	0.00083	660
2000	1.36	0.00068	541

Table 6.4 gives typical values of B and H for iron. Therefore, it is possible to calculate permeability for each particular flux density (B) and magnetising force (H). In columns 3 and 4, values for μ and μr have been calculated from the given values of B and H.

A graph plotted from these figures is shown in Figure 6.28. Since values of B are plotted against values of H, the graph is known as a 'B/H curve'. These curves are commonly used as a means of comparing the magnetic characteristics of different types of magnetic materials.

The permeability of ferromagnetic materials changes with differing values of flux density. For a given flux density, permeability (μ) is equal to the ratio B/H.

The graph shows that the permeability curve rises steeply to a peak. Beyond this point of maximum permeability, the curve slopes away quite rapidly.

This indicates that permeability becomes progressively less as H is increased beyond the value that causes magnetic saturation.

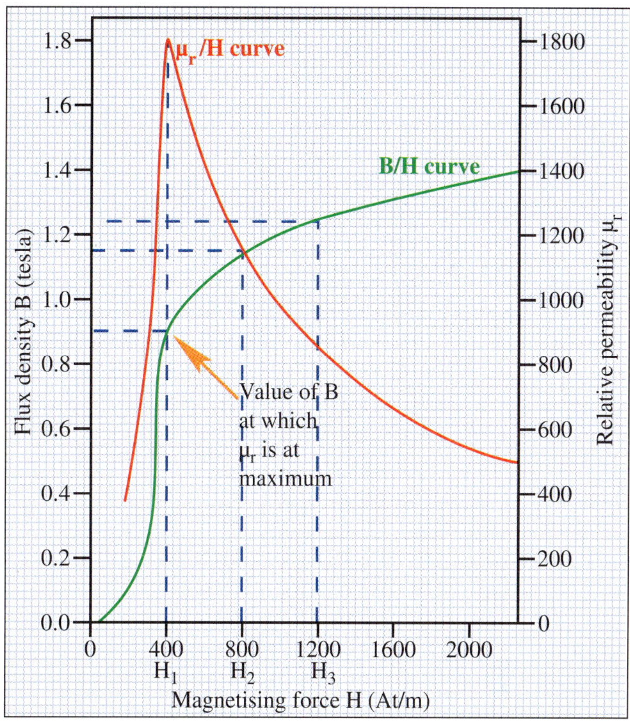

FIGURE 6.28 Flux—magnetisation curve

Magnetic saturation

Saturation is said to occur at a flux density near the centre of the 'knee' of the B/H curve. The values of μr have been plotted against the values of H in Figure 6.28 to give the μr/H curve. We can see from the graph that the saturation point is at the tip of the μr/H curve with a B value of 1.8 tesla and an H value of 400 At/m.

In practice, it is not economical to magnetise steel to a flux density much beyond the point of magnetic saturation. A large increase in magnetising current produces only a small increase in flux density, resulting in a waste of electrical power without achieving any useful increase in flux.

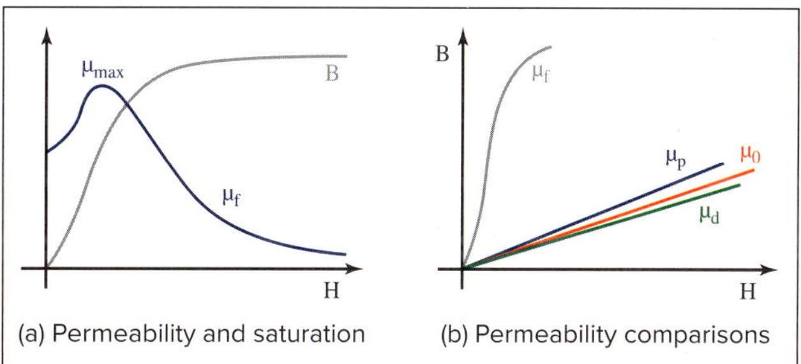

FIGURE 6.29 Permeability and saturation comparisons

Permeability of a ferromagnetic redundant material and its saturation point (μ_{max}) are shown in Figure 6.29(a). It should be noted that this is the 'knee point' on a B/H curve for a particular ferromagnetic material.

Figure 6.29(b) shows a simplified comparison of permeability for ferromagnets (μ_f), paramagnets (μ_p), free space (μ_o) and diamagnets (μ_d). Figure 6.30 shows B/H curve regions.

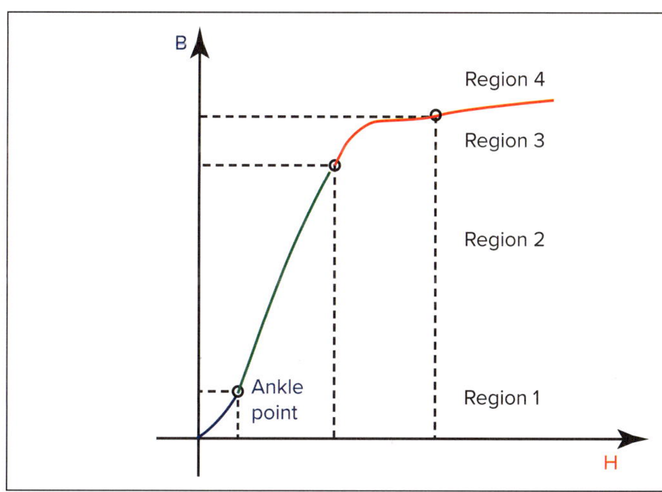

Region 4: Saturation region—no further increase in flux with increase in current.

Region 3: Knee point of curve—referred to as the 'saturation point'. Magnetic material has less flux change compared with changes in current.

Region 2: Ideal range for most magnetic materials to operate. Small increases in current produce significant increases in flux.

Region 1: Small increases in flux with change in current.

FIGURE 6.30 B/H curve regions

6.3.5 Comparison of B/H magnetisation curves

B/H magnetisation curves plot the values of flux density (B) against magnetising force (H). The effects of magnetising materials is shown as four distinct regions in Figure 6.30. Figure 6.31(a) shows B/H magnetisation curves for nine ferromagnetic materials:

1. sheet steel
2. silicon steel
3. cast steel
4. tungsten steel
5. magnet steel
6. cast iron
7. nickel
8. cobalt
9. magnetite.

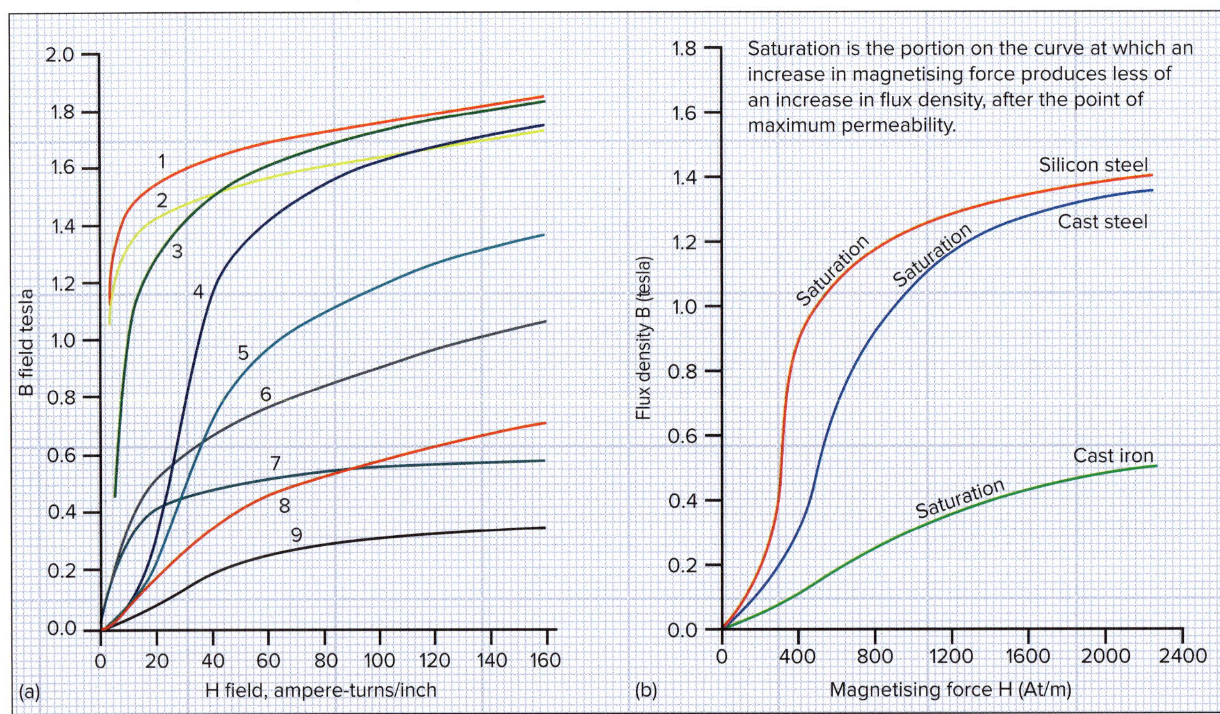

FIGURE 6.31 Magnetisation curves

Note that in Figure 6.31(a) the H scale has been expanded to show 0–160 in inches. Normally SI units are used and the correct scale is usually shown as ampere-turns per metre (At/m).

Figure 6.31(b) illustrates the magnetisation curves for silicon steel, cast steel and cast iron. The following points should be noted:

- The materials become magnetically saturated in the region that corresponds with the centres of the 'knees' of the respective curves.
- When the value of H is in the lower ranges, much greater flux density will be produced in silicon steel compared with that produced in cast steel or cast iron.
- Silicon steel saturates at a slightly lower value of flux density (0.8–0.9 tesla) than cast steel (approximately 1 tesla), based on the graph in Figure 6.31(b).
- Cast iron saturates at much lower values of flux density than either silicon steel or cast steel.

Hysteresis loops

The term 'hysteresis' means a lag. Materials that are magnetised, demagnetised and remagnetised show hysteresis in the form of residual magnetism. Residual magnetism is that portion of the flux that remains in the ferromagnetic material when the magnetising force is removed. This can be seen in Figure 6.32. The force used to remove residual magnetism is known as a 'coercive force'.

Magnetisation curves are plotted to show variation in flux density of ferromagnetic materials for increasing and decreasing values of magnetising force.

In Figure 6.32 the initial magnetisation curve OA begins at the origin (O) for both the B and H axes—no flux or magnetic force exists. The magnetising force (+H) increases to point A, well past the knee of the curve, and results in a flux (+B) that is in the saturation region.

When the magnetising force is reduced back to zero (from point A to point B), the flux density reduces, but not to zero. Point B on the curve shows the flux density as the residual flux (Br).

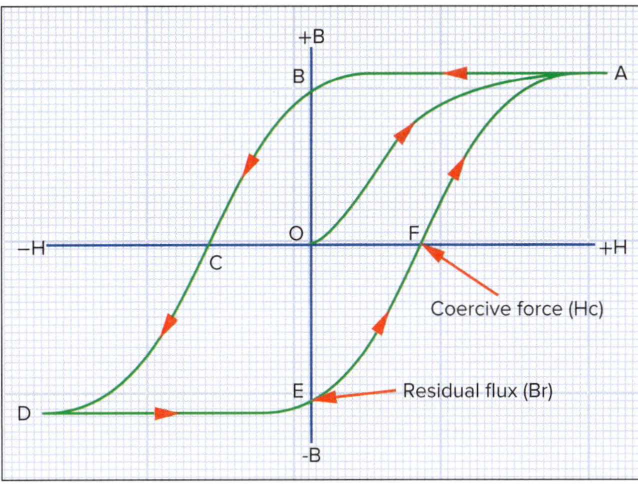

FIGURE 6.32 Magnetic hysteresis

In order to remove the residual magnetism, a reverse magnetising force or coercive force (Hc) is required (shown between OC). This causes the flux density to be reduced to zero (see point C). Part BC of the curve shows the coercive force in effect.

Curve DEFA is an exact reflection of ABCD. Thus, the combined magnetisation curves form a closed loop, ABCDEFA. This loop is commonly known as a 'hysteresis loop' for a ferromagnetic material.

6.4 Electromagnetic induction

An EMF can be induced in a conductor if the conductor cuts across (or is cut across by) a magnetic field. That is to say, *when there is relative motion between a conductor and a magnetic field, a voltage will be induced in the conductor and an induced current will flow if there is a complete circuit.*

Three factors are needed for electromagnetic induction to produce an EMF:

1. a length of conductor or coil (an inductor) (L)
2. a magnetic field (B)
3. relative motion between the inductor and the magnetic field.

Figure 6.33 demonstrates these three factors with an experiment involving a cable moving through a magnetic field. Fleming's right-hand rule for generators can be used to determine the relationship between the direction of motion, field flux and induced voltage.

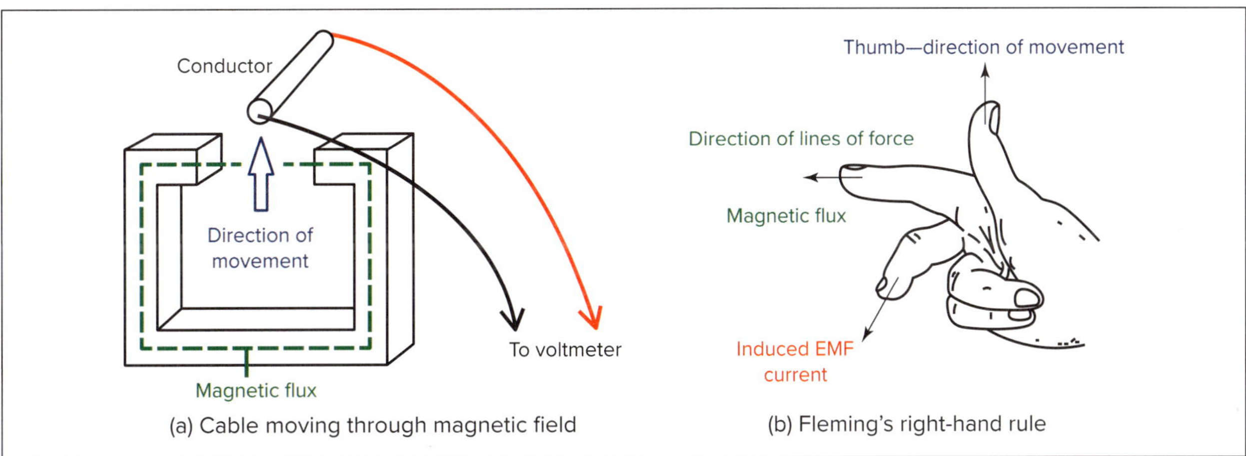

FIGURE 6.33 Fleming's right-hand rule

Skill: Calculation of induced EMF in a conductor, given the conductor length, flux density and velocity of the conductor.

EXAMPLE 6.7

Generated voltage (induced EMF)

$$e = Blv \sin\theta$$

where:

e = induced EMF in volts (V)
B = magnetic flux field in tesla (T)
l = length of conductor moving through field
v = velocity in metres per second (m/s).

Note: the conductor cuts through the magnetic field at right angles, that is at 90 degrees (sine 90 = 1).

A 200 mm conductor is rotated within a magnetic field of a constant density of 1 T. If the velocity relative to the field is 10 m/s, what voltage will be generated at an angle of 60 degrees?

$$e = Blv \sin\theta$$
$$e = 1 \times 0.2 \times 10 \sin(60^{\circ})$$
$$e = 1 \times 0.2 \times 10 \times 0.866$$
$$e = 1.73 \text{ volts}$$

CHECK YOUR UNDERSTANDING

6.10 A 200 mm conductor is rotated within a magnetic field of a constant density of 1 T. If the velocity relative to the field is 10 m/s, what voltage will be generated at an angle of 85 degrees?

6.4.1 Faraday's Law

The value of the EMF induced in a circuit depends on the number of conductors in the circuit and the rate of change of the magnetic flux linking the conductors.

Michael Faraday found that the voltage generated in a coil due to a changing magnetic field is directly proportional to the change in flux and inversely proportional to the change in time. He also noted that increasing the number of conductors (or turns) of a coil is equivalent to increasing the current.

EXAMPLE 6.8

Faraday's Law—(induced) voltage

$$V - N\frac{\Delta\Phi}{\Delta t}$$

where:

V = induced voltage
N = number of turns
$\Delta\Phi$ = flux change in weber
Δt = time change in seconds.

A coil of 500 turns has a permanent magnet moved into it such that 0.2 Wb cuts across the coil in four seconds. Find the average induced voltage.

$$E = N\frac{\Delta\Phi}{\Delta t} \quad (1)$$
$$= 500 \times \frac{0.2}{4} \quad (2)$$
$$= 25 \text{ V} \quad (3)$$

Experiments by Ørsted and Faraday

Figure 6.34 combines experiments carried out by Faraday and Ørsted. In 1820, Ørsted conducted an experiment involving a magnetic field caused by an electric current in a wire. This experiment requires a current-carrying conductor and a compass. The compass is positioned so that the wire is placed across it in the same orientation as north-south. The coil, bar magnet and multimeter are the apparatus for Michael Faraday's magnetic field induction experiment (1831).

Whenever a steady direct current passes through a conductor, a magnetic flux is set up around the conductor. Ørsted's Law for a straight wire carrying a steady direct current was based on the following five observations:

1. Magnetic field lines are concentric, encircling the current-carrying conductor.
2. Magnetic field lines of force will be perpendicular (at a right angle) to the conductor.
3. If the direction of the current is reversed, the direction of the magnetic field reverses.
4. The field area is directly proportional to the magnitude of the current.
5. The field strength will decrease with distance from the conductor.

Figure 6.34 shows the compass needle deflecting when current flows through the conductor. As the bar magnet is removed from the coil, the current direction changes, as does the direction of the compass needle deflection. These experiments and observations were taken up by André-Marie Ampère and James Clerk Maxwell as the basis of the right-hand rule for straight conductors/redundant, as a busbar is a conductor.

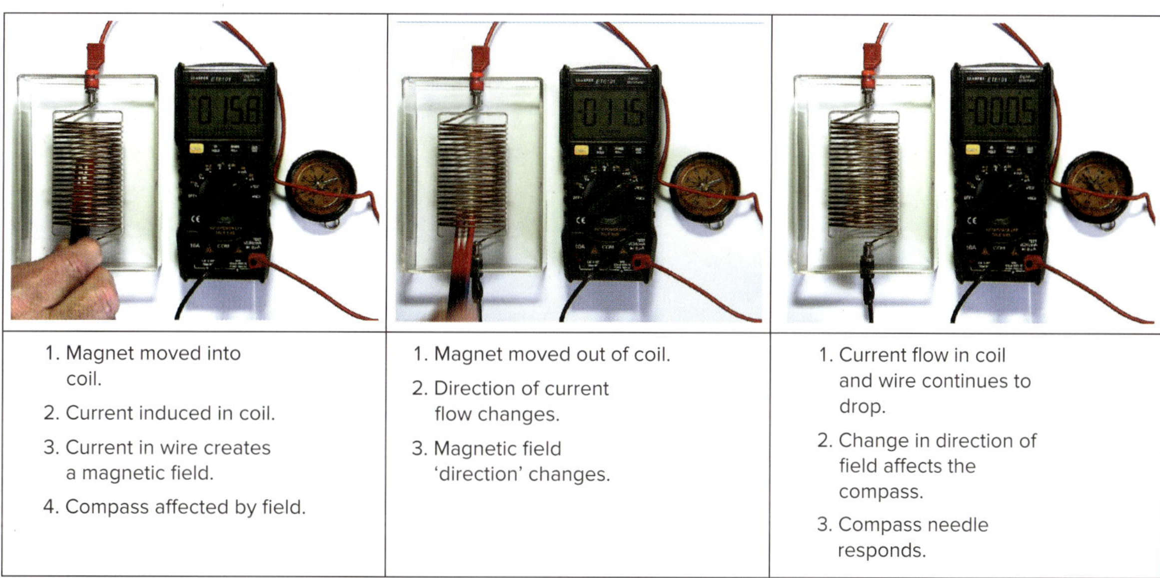

FIGURE 6.34 **Ørsted's and Faraday's experiments**
Tony Jones

For the experiments shown in Figure 6.34, the meter sampling rate is three times per second. This means there is a slight lag between the measurement value displayed and the physical response by the magnetic needle of the compass. This type of experiment was originally conducted before the invention of multimeters and digital electronics.

Ørsted's experiment shows a practical consideration which applies to data cable installation and the application of $e = Blv$. The direction of the magnetic field around the conductor is shown in Figure 6.34 using a compass. When two conductors are run in parallel to one another, a changing magnetic field produced by one conductor would induce a voltage in the adjacent conductor. This is problematic with power cables laid in parallel with telecommunications cables.

When there is relative movement between a conductor and a magnetic field (expanding and collapsing), a voltage will be induced. We also know that if this movement between conductor and magnetic field is at 90 degrees, a maximum voltage is induced.

If the power and data cables are spaced apart, the induced voltage is minimised (which results in lower interference).

The experiment shown in Figure 6.34 shows that by crossing another conductor over the first, in the same direction as the compass needle, no voltage would be induced in the second conductor. This is because there would be no relative movement between the second conductor and the magnetic field (of the first conductor). Therefore, an induced voltage will not be created in a conductor placed in the same direction as the magnetic field.

If the cables need to cross one another, they should cross at 90 degrees. This will minimise any 'relative movement' between field and conductor. This too will minimise induced voltages in the telecommunications cable, ensuring they operate with minimal interference.

A practical application of this concept is to use a compass to detect whether a car alternator is charging the battery. Place a compass across the battery cable. Deflection of the compass needle after starting the vehicle will show the resulting magnetic field from alternator current flow to recharge the battery. In the past, a galvanometer (see Chapter 3) was used. This gives a quicker needle movement response. Figure 6.35 shows Faraday's experiment using a galvanometer.

Figures 6.35(a) and (b) provide a similar result as shown in Figure 6.34 using the north pole of the bar magnet. Figure 6.35(c) and (d) show the experiment using the south pole.

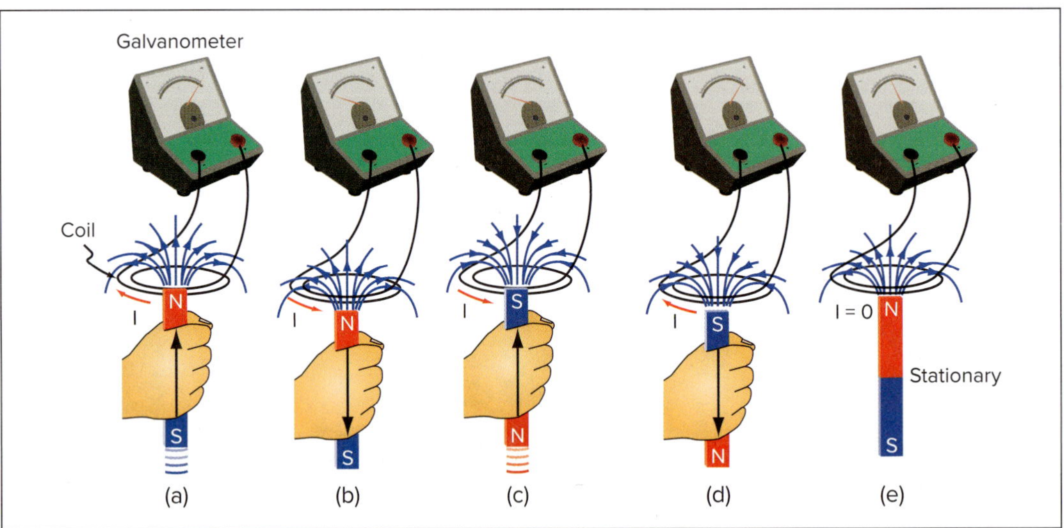

FIGURE 6.35 **Faraday's experiment using a galvanometer**

CHECK YOUR UNDERSTANDING

6.11 When a current flows through a conductor or coil, the moving charges create a magnetic field. How could a compass needle help in determining whether the charging system in a transportation vehicle is working?

The experiments shown in Figures 6.34 and 6.35 involve a number of variables. We can see that for a current to flow in the circuit, a voltage is created. The permanent bar magnet possesses magnetic field properties and the coil has a particular number of 'turns' (turns are the number of cable loops or windings). A relative motion between the field flux and the coil creates this voltage. The variable shown here is how fast or slow the magnet is inserted and removed from the coil.

EXAMPLE 6.9

Calculate for induced voltage (EMF)

$$V = N\frac{\Delta\Phi}{\Delta t}$$

List the known values:

N = number of turns
ΔΦ = flux change in webers
Δt = time change in seconds

Steps to find Δt:

Φ1 = 80 μWb
Φ2 = 30 μWb
ΔΦ = 50 μ Wb

Find for V (average induced voltage)

A coil of 600 turns has a flux of 80 μWb passing through it. If the flux is reduced to 30 μWb in 15 ms, find the average induced voltage.

$$V = N\frac{\Delta\Phi}{\Delta t} \quad (1)$$

$$= 600 \times \frac{80 \times 10^{-6} - 30 \times 10^{-6}}{15 \times 10^{-3}} \quad (2)$$

$$= 600 \times \frac{50 \times 10^{-6}}{15 \times 10^{-3}} \quad (3)$$

$$= 2\text{ V} \quad (4)$$

6.4.2 Applications of electromagnetic induction

Electromagnetic brake

Electromagnets are often used in conjunction with an electric motor, especially a lifting motor as used on a crane. When the lifting motor is engaged, a solenoid releases a brake, allowing the lifting drum to rotate and wind up the load. Disconnecting the motor releases the solenoid and automatically applies the brake. The system is considered failsafe as the loss of power does not permit the load to fall.

Electromagnetic brakes can also be made to operate when power is applied, either when a control circuit applies energy or the operator presses a button or foot switch.

A typical brake mechanism is shown in Figure 6.36.

FIGURE 6.36 **Brake mechanism—car brake part at a garage**
kpakook/Shutterstock.com

Electromagnetic clutch

At times, an electric motor cannot speed up or slow down fast enough for a given load. Take a big metal-working guillotine, for example: the electric motor runs a large flywheel up to speed, and when the press is operated an electromagnetic clutch engages the flywheel to the cutting blade for one revolution; the flywheel provides most of the energy for the cut and, when it is completed, the motor drives the flywheel back up to its idling speed.

Many other machines use an electromagnetic clutch to engage or disengage various functions—often multiple functions from a single driving motor.

FIGURE 6.37 **Electromagnetic chuck**
Nordroden/Shutterstock.com

Electromagnetic chuck

An electromagnetic chuck (see Figure 6.37) is an electrically operated device for holding work on a machine. They provide a consistent clamping pressure which improves working safety for the user. Applications for electromagnetic chucks include drilling, cutting, milling, turning and grinding machines. On a grinding machine, for example, an iron part might be held steady by placing it on the surface of the chuck and applying power to the magnet. This aids in reducing vibration to provide accuracy and precision during the machining process. A disadvantage of this is that the power supply must be very reliable as loss of magnetism can result in major damage to the machine. The system is therefore mostly used to add security to a holding mechanism where the job is unlikely to move anyway. Often the machine is a computer numerical controlled (CNC) machine.

Lifting electromagnet

Scrap metal dealers often need to move large amounts of jagged and dangerous iron and steel. To handle the material manually would be risky, and ropes and lifting straps might be damaged by it. The answer is to use electromagnets to pick it up (see Figure 6.38).

FIGURE 6.38 **Lifting electromagnet**
Â©Photolibrary/Age Fotostock

The magnet is lowered onto the scrap, energised and lifted away to wherever the scrap is intended to be placed. It is then de-energised so that it drops the scrap. Some electromagnets can lift many tonnes of iron in one lift, including large vehicles.

Electromagnetic separator

As electromagnets only work on magnetic materials, they can be used to separate them from non-magnetic materials. As shown in Figure 6.39, a conveyor belt typically carries materials to a dumping bin; but the end drum is magnetised so magnetic materials fold around to the underside of the drum, where they drop off into a separate bin. Other methods of separation exist, depending on the type of materials involved.

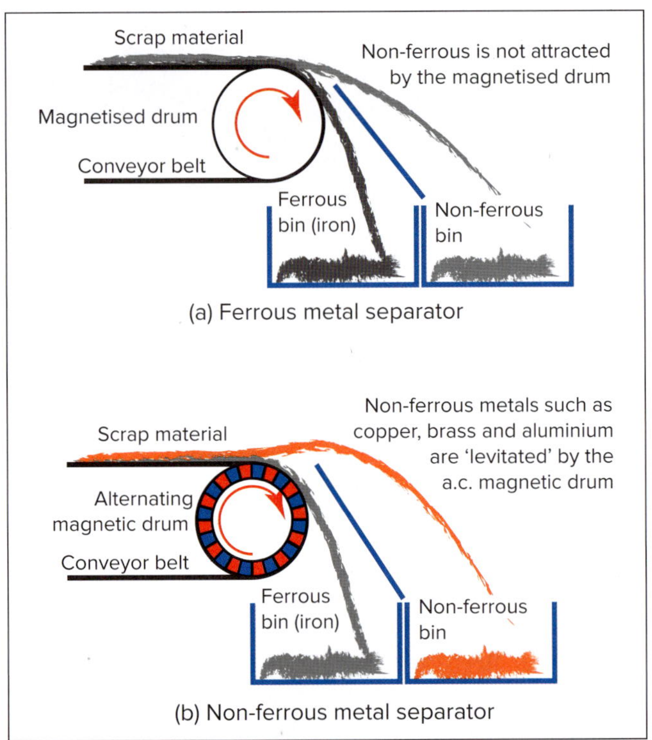

FIGURE 6.39 Electromagnetic separator

6.4.3 Lenz's Law

Newton's Third Law of Motion states that for every action there is an equal and opposite reaction. In 1834, the Russian physicist Heinrich Friedrich Emil Lenz formulated its electrical counterpart, that action and reaction are equal and opposite.

Current flowing in a conductor creates a magnetic field around it. The magnetic field increases (grows) as the current increases, and decreases (collapses) as the current decreases. This change in flux ($\Delta I \propto \Delta\Phi$) as a result of a change in current is not immediate—there is a delay (or lag). Lenz's Law relates to why this lag occurs.

When a changing magnetic field cuts across a conductor, a voltage or EMF is induced in that conductor, and if a circuit exists, a current will flow as shown in Figure 6.40. This was also shown earlier in Figure 6.34 with Faraday's induced-voltage experiment.

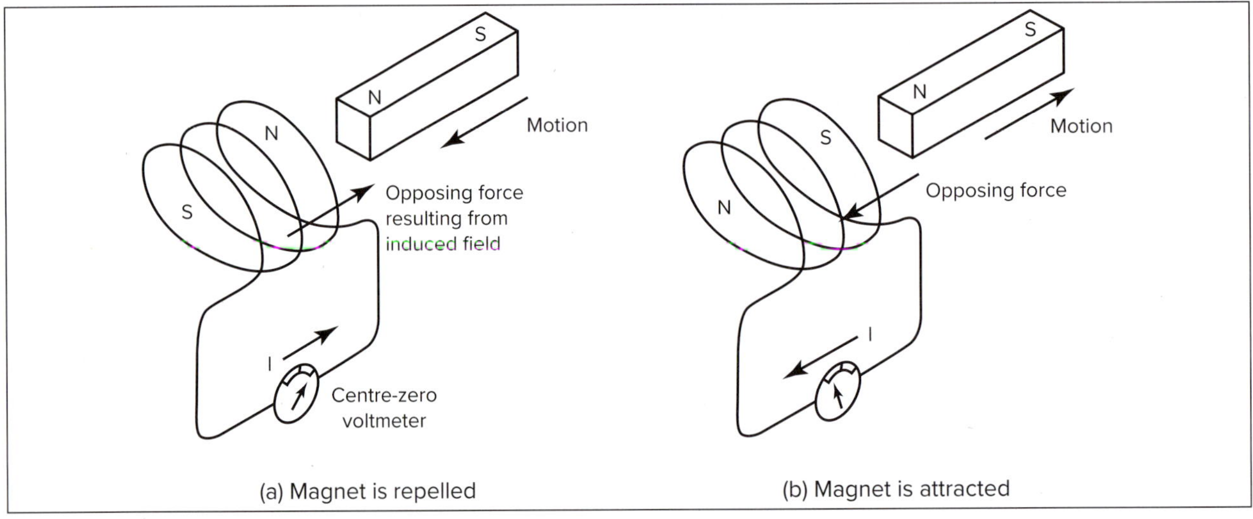

FIGURE 6.40 Lenz's Law—field moving

While Faraday showed that a moving magnetic field (in the form of a bar magnet) could induce a coil voltage (measured using a galvanometer), the curious effect of removing the magnet resulted in both the magnetic coil field poles and the direction of current flow through the coil reversing. The induced voltage in the coil can be calculated and measured, and this opposing force can be demonstrated as a physical phenomenon, as illustrated in Figure 6.41(a).

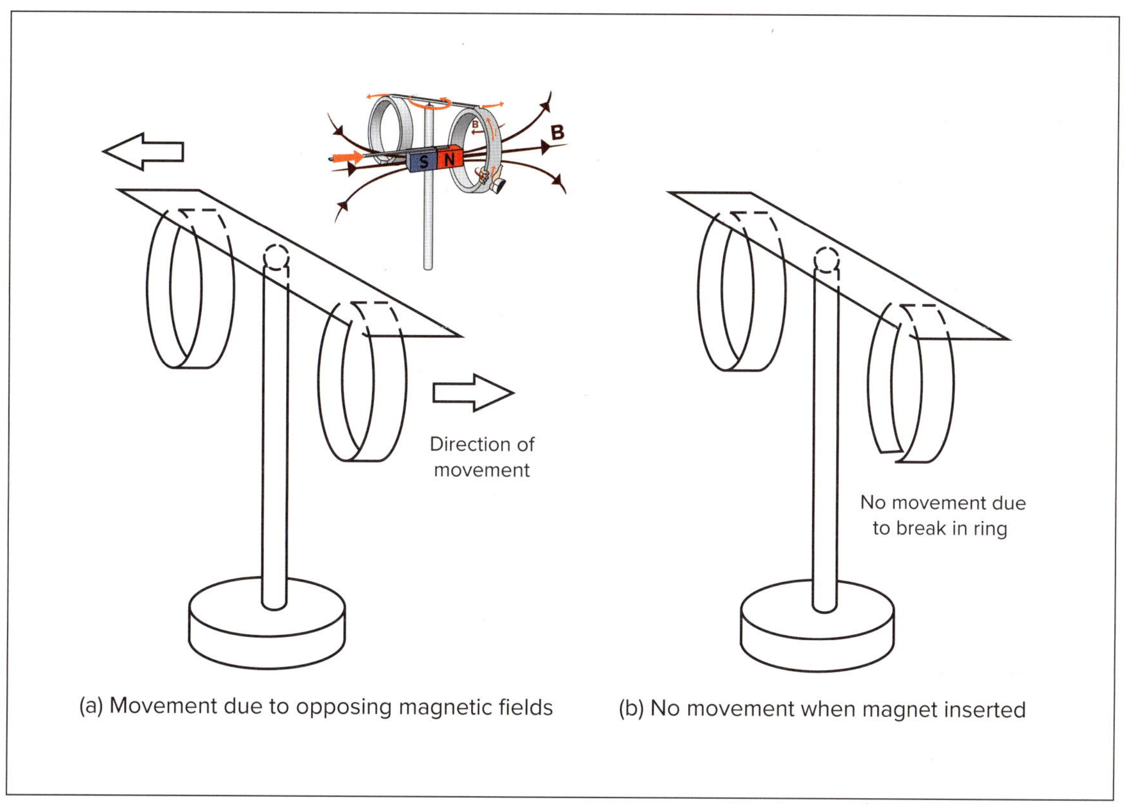

FIGURE 6.41 Lenz's Law demonstrator

The experiment in Figure 6.41(a) shows that the rings (which are non-magnetic) move when a bar magnet is inserted into their open spaces. This is because, like the Faraday coil experiment, an EMF is induced in the conducting ring. This EMF causes an eddy current to circulate within the ring, creating a magnetic field which opposes the field of the bar magnet. The two magnetic fields repel, causing the rings to move on the apparatus pivot. Figure 6.41(b) shows that the ring has a split. When the bar magnet is inserted into the open space of the ring, the break or split creates an air gap, preventing the circulation of the eddy current. No magnetic field is created to oppose the bar magnet field so no movement results. This principle has been used on electrical switchboards to prevent circulating eddy currents between conduit entry points into metal boards.

Lenz's Law states that the direction of an induced EMF is such that the resulting current flow will produce a magnetic field which tends to oppose the original motion that caused the induced EMF. This means that the polarity of the generated voltage is opposite to that of the applied voltage during a current increase, and the generated current direction opposes the original current direction.

Calculation of an induced voltage (through the use of Faraday's Law) determines the magnitude of that value, the minus sign in the equation signifying what Lenz described as the 'opposing force'.

TABLE 6.5 Induced voltage equation—Faraday and Lenz

Faraday's Law—(induced) voltage	Lenz's Law
$V = N\frac{\Delta\Phi}{\Delta t}$ where: V = induced voltage N = number of turns $\Delta\Phi$ = flux change in webers Δt = time change in seconds.	$V = -N\frac{\Delta\Phi}{\Delta t}$ where: the minus sign in the equation relates to the opposing polarity of the induced EMF (or current flow in the circuit).

The magnitude and direction of the induced EMF can be established using Lenz's Law (the induced current will appear in such a direction that it opposes the change that produced it). Lenz's Law tells us three things about current in a conductor:

1. A current that is increasing produces an opposing EMF that causes the increase to be slower than expected.
2. A stable current in a conductor maintains a magnetic field of stored energy.
3. A current that is decreasing converts the stored energy back into an EMF that assists the applied voltage such that any decrease in current is slower than expected.

It will be shown that in a.c. circuits the current is constantly changing in both strength and direction.

Figure 6.42 represents the downwards movement of a conductor through a magnetic field as a result of mechanical force acting downwards. This movement through the field will induce an EMF in the conductor. The resulting current, in a closed circuit, will create a circular magnetic field around the conductor.

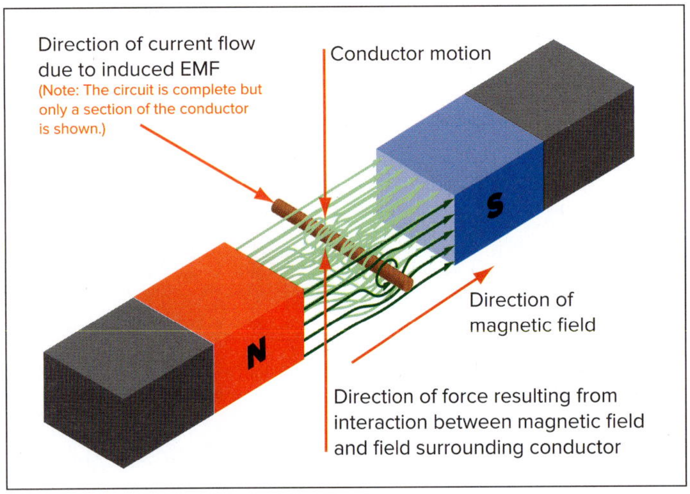

For Lenz's Law to apply, the induced field must oppose the motion. The direction of the induced field must be such that the following reactions will occur between the main and induced fields:

1. The main field will be compressed in front of the moving conductor and expanded behind it.
2. The compressed magnetic field is a stronger force than the expanded magnetic field; therefore, the difference in the forces acts in opposition to the motion of the conductor.

FIGURE 6.42 **Forces on a conductor in a magnetic field**

CHECK YOUR UNDERSTANDING

6.12 Which 'hand' rule can be applied for the experiments shown in Figures 6.40 and 6.41?

6.13 When the bar magnet north pole is inserted into the inductor ring of Figure 6.41(a), what does the ring do?

6.14 What happens when the bar magnet south pole is inserted into the inductor ring of Figure 6.41(a)?

6.5 Inductors, inductance and induction

Faraday experimented with magnets and coils to induce a voltage. The coil was an **inductor** and the EMF created was a result of electromagnetic induction. He found that increasing the number of coil turns also increased the magnitude of the voltage, as did how fast the magnet moved in relation to the coil.

So, the rate of change in time (Δt) was also found to be an important quantity in determining EMF. When the magnet was removed from the coil, there was an opposition and a change in the direction of current flow in an inductor, as supported by Lenz's Law.

An inductor is a passive electrical component, usually taking the form of a coil of wire wound around an iron or other magnetic core. Inductors have physical properties which we can use to calculate a value for inductance (L). Inductance is a magnetic circuit property of all inductors that enables an EMF to be induced. When current flows through a coil, an electromagnetic field is created around the inductor. This interaction between magnetism and electricity with the inductor is called 'inductance'. Inductors have the effect of storing energy as an electromagnetic field whenever the current increases and giving it back when the current decreases.

'Inductance' is also the name given to the component property that opposes any change of current flowing through it, when either generating a magnetic field or being influenced by a magnetic field.

Therefore, an inductor is often used in electrical circuits to smooth the current value. This is called 'filtering', and in this case the inductor is called a 'choke'. Inductance coils have a wide variety of names, depending on their expected uses. For example, the inductance coil used in automotive ignition systems is known as an 'ignition coil'.

Inductance is the property itself, while induction is the process by which inductance operates.

'Induction' is the process by which this voltage is produced by the electric current. That is, the voltage is 'induced' in the conductor as the result of an electric current in another (or the same) conductor.

The definition of unit of inductance is as follows:

The unit of inductance is the henry, which is defined as the inductance of a closed circuit in which an EMF of 1 V is produced when the electric current flowing in the circuit varies uniformly at the rate of one ampere per second.

$$V = L\frac{\Delta I}{\Delta t}$$

where:

L = inductance in henrys (H)
V = voltage or EMF (electromotive force) in volts (V)
I = current in amperes (A)
t = time in seconds (s).

While inductance can be calculated on the basis of fixed physical properties, the value of inductance can also be influenced by a magnetising force (H) with changes in current and the resulting changes in magnetic flux.

EXAMPLE 6.10

Inductance

$$L = N\frac{\Delta \Phi}{\Delta I}$$

where

L = inductance in henrys
N = number of conductors
$\Delta\Phi$ = change in flux in webers (Wb)
ΔI = change in current in amperes (A).

The current in a 100-turn coil changes from 5 A to 0 A, causing a flux change of 300 mWb. What inductance does the coil have?

$$L = N\frac{\Delta \Phi}{\Delta I} \quad (1)$$

$$= 100 \times \frac{300 \times 10^{-3}}{5} \quad (2)$$

$$= 6 \text{ H} \quad (3)$$

CHECK YOUR UNDERSTANDING

6.15 The current in a 200-turn coil changes from 5 A to 0 A, causing a flux change of 200 mWb. What inductance does the coil have?

6.5.1 Applications of the different types of inductors

Straight or line type

At radio frequencies, an antenna is often a straight conductive tube of a specific length that makes the inductance just right for the antenna to work. Inductors made from straight conductors are also used inside UHF radios. The effect of frequency on inductors is covered in Chapter 12.

Loop

When a conductor is bent around to form a loop, the magnetic field induced inside it is compressed into the area of the loop, while the magnetic field outside the loop is expanded. Therefore, the magnetic effect is greater inside the loop.

Solenoid coil—air-core

Adding more loops to an inductor by adding turns has the same effect as increasing the current flow—that is, adding two turns makes the current go around twice and therefore has the effect of twice the current. Equations will often use 'IN' or current multiplied by the number of turns to take account of the multiple turns in a coil.

Coils wound with multiple turns in one or more layers are often called 'solenoids'.

Solenoid coil—magnetic core

To further increase inductance, a coil may have a magnetic core added (perhaps laminated iron or even ferrite). Other man-made materials are also formed into magnetic cores to increase the inductance of the coil.

Relays and plunger solenoids use a magnetic core to increase the force that can be applied, and also to reduce current flow when the relay or plunger is closed.

Toroidal core

Although toroidal cores have existed since Faraday made his experimental transformer from a toroidal core of iron wire, toroids were 'rediscovered' in the 1970s. This was the result of needing higher efficiencies from smaller and lighter inductors.

Toroidal cores of ceramic materials have made inductors of very high value easy to produce and are therefore popular in many areas of electronics and electrical engineering.

Multi-coil—transformer

Whenever more than two coils are placed together, usually on a common core, their inductance interacts, causing what is known as 'mutual inductance', so that the rising or collapsing field in one coil causes a voltage to be generated in every other coil. This is the transformer effect—a very important effect in electrical engineering.

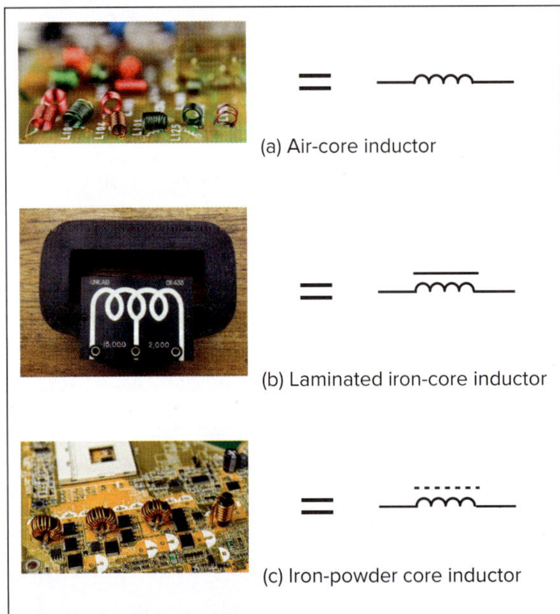

FIGURE 6.43 Inductor types and symbols
(a) KPixMining/Shutterstock.com; (b) sciencephotos/Alamy Stock Photo; (c) Ladislav Kubeš/istockphoto/Getty Images Plus

6.5.2 Australian Standard circuit diagram symbols for the four types of inductors

Normally, inductors are made as a coil of wire, sometimes wound around a core of a magnetic material such as iron. The symbol for a single coil is shown in Figure 6.43(a). A line drawn next to a symbol for a coil indicates that a magnetic core has been used. If there is no bar, this often means that it is an air-core coil which has no magnetic core. However, the general symbol for a coil as described in Standards Australia publications does not require the bar to show the type of core. The bar is included only when necessary, such as when some inductors in a circuit have a core and others do not.

Inductors can also have multiple coils with the same symbol used for each coil. Generally, a core is shown as a solid line (see Figure 6.43(b)) or as a dashed line to represent a ferrite (ceramic iron) core (see Figure 6.43(c)).

The symbol for a *bifilar winding* can be seen in Figure 6.44. A bifilar coil contains two closely spaced insulated parallel windings. The current flow through these coils can be in the same direction to magnify the effect of a resultant magnetic field. They can also be connected so that the current flow produces opposing magnetic fields, the purpose of which is to cancel the resulting magnetic fields produced in each coil. Bifilar windings are also used to suppress *back EMF*.

FIGURE 6.44 Bifilar symbol

6.5.3 Common types of inductor cores

Air cores

An air-core inductor is simply a coil of wire that is sometimes wound around a plastic former but often simply wound into a coil, as shown in Figure 6.43(a). Air-core inductors have no core and therefore have no appreciable core losses. This means they can be used at any frequency. However, they can only convey a small amount of energy at lower frequencies; to convey more power at low frequencies, a magnetic core must be used.

Ferrite cores

Sometimes called 'iron-powder' cores, ferrite core inductors (see Figure 6.43(c)) are manufactured from very fine iron powder mixed with an insulating bonding medium moulded into the required shape. Owing to the small size of the particles, iron losses are greatly reduced.

Ferrite cores can be used above 30 MHz, depending on the application. The central core iron-powder position is often adjustable so the inductor can be tuned by altering the inductance (by changing the position of the core).

Iron-powder cores are also used in small transformers in switch-mode power supplies and in special-purpose inductors operating at higher frequencies. As frequency increases, the size of a transformer decreases. Switch-mode power supplies use higher frequencies to allow the size and weight of power supplies to be drastically reduced.

Iron cores

Iron cores are available in two types—laminated and solid (see Figures 6.43(b) and 6.45). Solid cores are used for d.c. magnetic pole pieces. The inductance of the coils associated with solid iron poles is usually so high that solid iron poles are restricted to frequencies below about 5 Hz. The iron losses are prohibitively high if the magnetic polarity of the pole changes at any higher frequency.

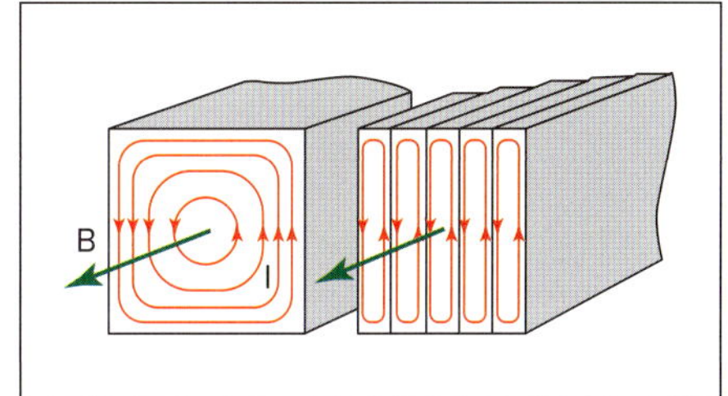

FIGURE 6.45 Solid and laminated cores

Laminated cores are made from stacked thin sheets of iron to reduce what are known as 'eddy currents' and to reduce iron losses at power-line frequencies. Special types of ferrous materials have been developed to reduce these losses further.

Plain iron sheeting can be used to make laminations for small, low-energy transformers; but for larger distribution transformers, a more expensive, electrical silicon-steel alloy is used. The special steel allows higher flux densities and a smaller iron core for the same power output. Lower iron losses result in less heat being generated in the cores.

6.5.4 Construction of an inductor

The unit of inductance is the henry, which is defined as the inductance of a closed circuit in which an EMF of 1 V is produced when the electric current flowing in the circuit varies uniformly at the rate of one ampere per second. See Table 6.6. for the physical construction of an inductor (coil).

TABLE 6.6 Physical construction of an inductor (coil)

Inductance	
$$L = \frac{N^2 \mu_o \mu_r A}{l}$$ where: L = inductance in henrys; N = number of turns; μ_o = permeability of free space; μ_r = relative permeability; A = area of the core in metres-squared (m^2); l = length of the coil in metres (m).	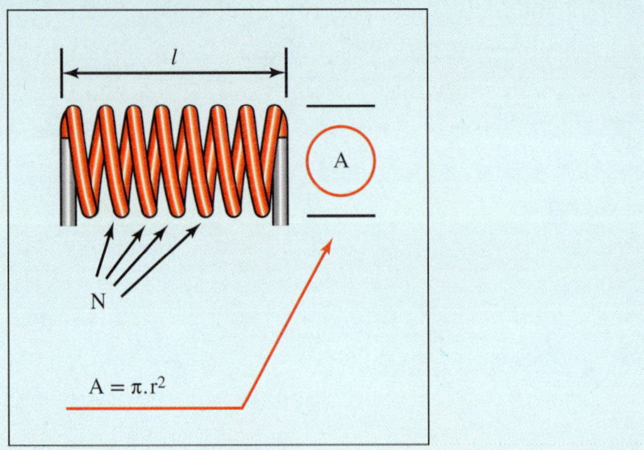

The relationship between reluctance, permeability and inductance due to physical properties including μ_o μ_r area, length and coil turns is:

$L = N^2 \frac{\mu_o \mu_r A}{l}$	$R_M = \frac{l}{\mu_o \mu_r A}$	$R_M = \frac{N^2}{L}$	$L = \frac{N^2}{R_M}$

CHECK YOUR UNDERSTANDING

6.16 What are the four factors that determine the inductance of a coil?

6.17 Two air-core coils have the same number of turns and cross-sectional areas. Coil A is shorter in length than coil B. What is the effect on inductance for coil A?

6.18 Two coils of similar-length core material and turns have different core areas. Coil A has a smaller CSA than coil B. Which coil has a higher inductance?

6.5.5 Self-inductance and mutual inductance

In an electric circuit, inductance occurs when a current in one conductor causes a voltage to be induced in that same conductor or in a nearby conductor via a magnetic field.

Self-inductance

The term 'self-inductance' is used when a conductor or a coil has a voltage induced in it by its own magnetic field.

Figure 6.46(a) shows the field that surrounds a conductor in which current flows at a steady value. Because the current flow is steady, the field is also steady and there is no relative movement between the field and the conductor.

An increasing current will cause an expanding magnetic field which is moving outwards in relation to the conductor. This relative movement induces a voltage within the conductor opposing the existing current (Lenz's Law). When current flow in a conductor decreases, flux density also decreases and the magnetic field collapses. The relative movement is opposite in direction and is shown in Figure 6.47.

When the current from the power source reaches a steady value as in Figure 6.47(a), the relative movement of the field ceases and no induced voltage is generated.

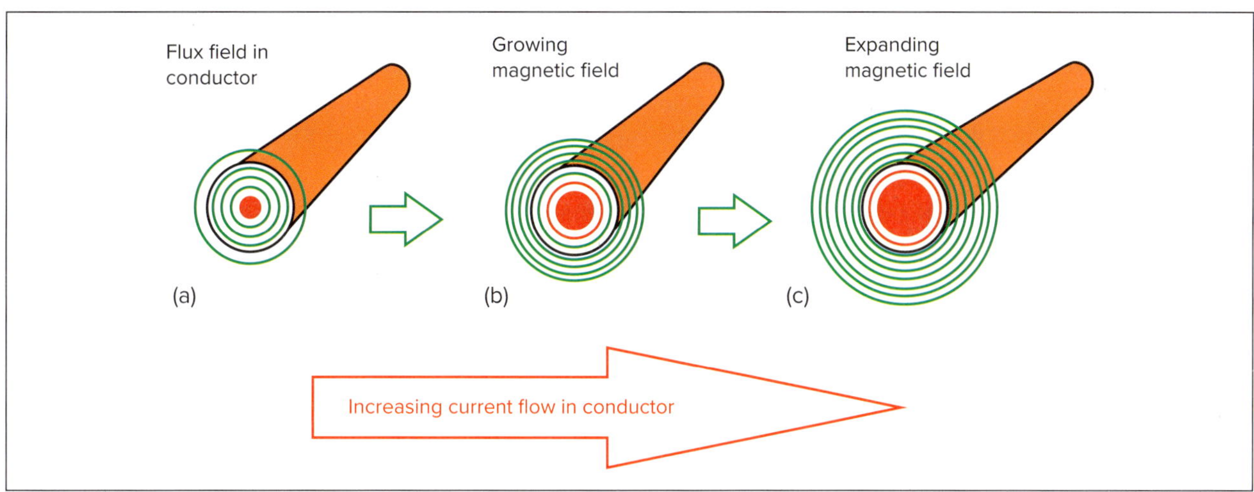

FIGURE 6.46 Relationship between conductor current and magnetic flux

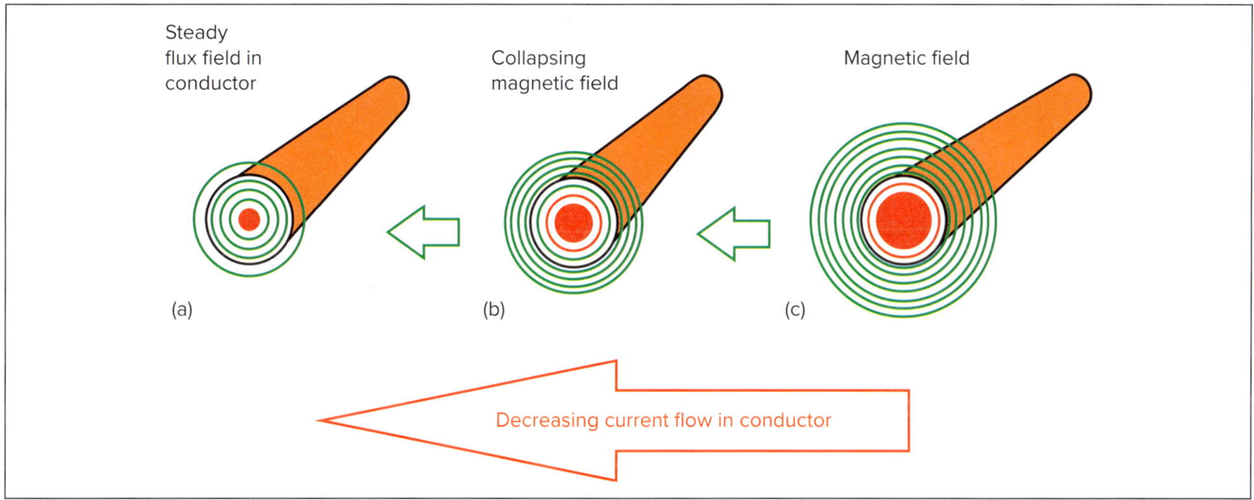

FIGURE 6.47 Collapsing magnetic field

EXAMPLE 6.11

Self-induced EMF

$$V = L\frac{\Delta I}{\Delta t}$$

where:

V = voltage generated
L = inductance in henrys
ΔI = change in current
Δt = change in time.

If the current through an inductor of 1.5 H is reduced uniformly from 5 A to 1 A in 0.5 s, find the value of the induced EMF.

$$V = L\frac{\Delta I}{\Delta t} \quad (1)$$

$$= 1.5\frac{5-1}{0.5} \quad (2)$$

$$= 1.5\frac{4}{0.5} \quad (3)$$

$$= 12\ \text{V} \quad (4)$$

Mutual inductance

The term 'mutual inductance' is used to describe the effect when variations of current flow in a conductor cause an induced EMF in a neighbouring conductor. It is not necessary for there to be an electrical connection between the conductors.

Section 6.5 introduced the concept of induction, Fleming's right-hand rule and the three factors required for inducing an EMF. Figure 6.48 shows the relevant motion between the magnetic field and the second conductor. When the field grows and collapses, the relevant motion is actually the second conductor and not the field. The key phrase in understanding this is 'relative motion'. This can be confirmed by applying Fleming's right-hand rule to the second conductor.

Two inductors placed together sharing a magnetic field produced by one of the inductors will experience mutual inductance to some extent. Figure 6.48 shows two parallel conductors. AB is connected in series with a switch and battery. The conductor CD and a centre-zero millivoltmeter form a separate circuit.

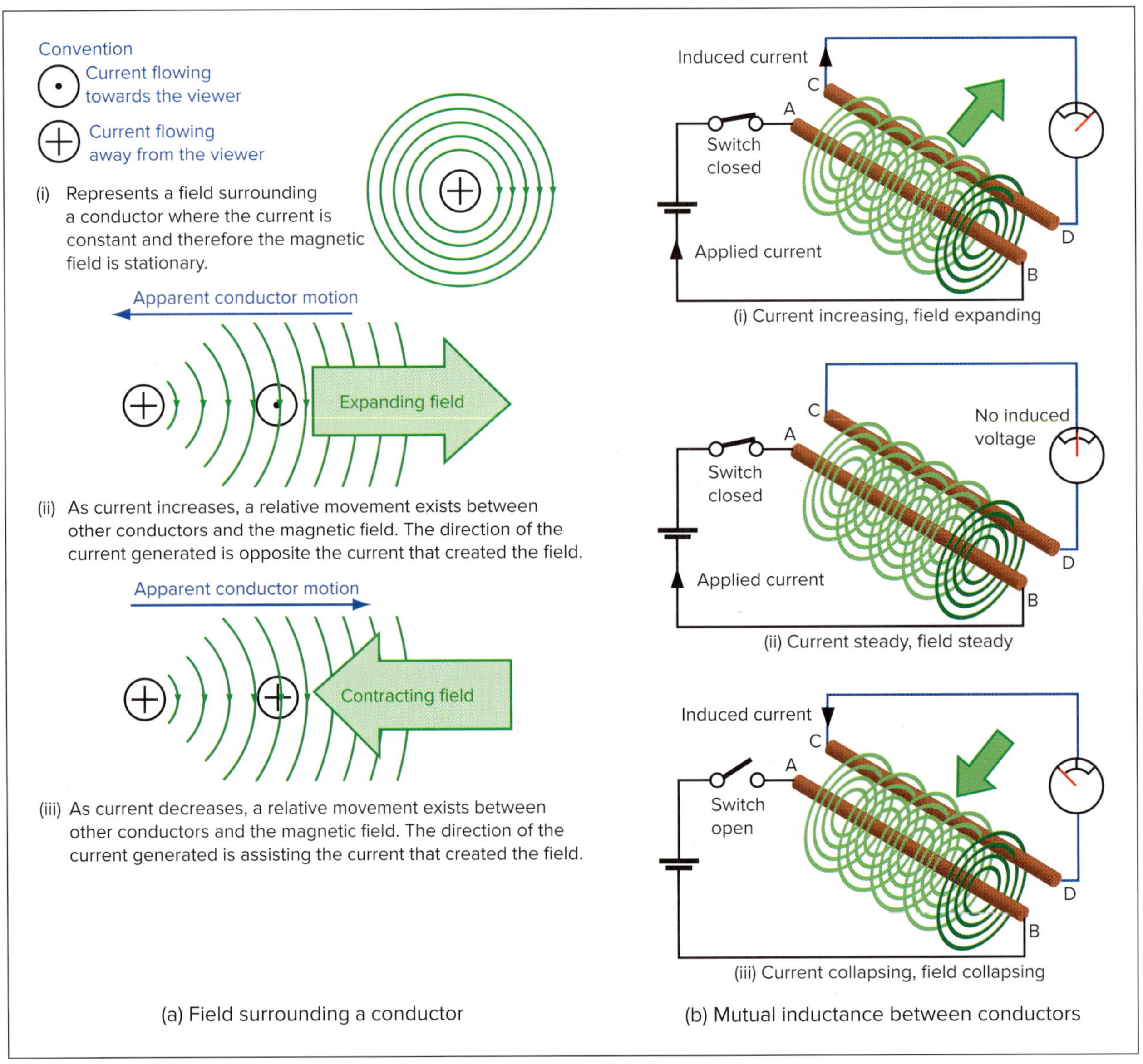

FIGURE 6.48 **Mutual induction of parallel conductors**

Figure 6.48(a)(i) shows a constant magnetic field when there is a steady current flow. Relative movement between a coil or conductor with an expanding or collapsing magnetic field creates an induced voltage. Figure 6.48(a)(ii) illustrates the effect of an expanding magnetic field.

When the switch is closed, as shown in Figure 6.48(b)(i), a circuit path is through conductor AB, creating an expanding magnetic field (by the green arrow). While current is building up in AB, the field around AB increases in strength.

The expanding magnetic field causes relative movement between it and conductor CD, and this will produce an induced voltage in conductor CD. This induced voltage will act in the direction shown by the centre-zero millivoltmeter gauge in part (b)(i). The needle pointer moves from the at the 12 o'clock position (zero volts) to the 2 o'clock position (indicating a +ve voltage).

Figures 6.48(a)(i) and 6.48(b)(ii) both show the effect when the current flow in AB remains at a steady value. In Figure 6.48(b)(ii), the field around AB will be at a fixed strength because there is no relative movement between the field and conductor CD. There is no induced voltage or current flow in CD. This is represented by a zero voltage on the millivoltmeter gauge.

Figures 6.48(a)(iii) and 6.48(b)(iii) show what happens immediately when the switch is opened. The current flow in AB will stop and, in doing so, will cause the field around AB to collapse. This contracting or collapsing field (shown by the green arrow) causes relative movement between the magnetic field and conductor CD. This will also induce a voltage in CD but in the opposite direction as shown by the needle pointer of the millivoltmeter.

The change in direction of movement causes a change in the polarity of the induced emf. The needle pointer of the millivoltmeter moves to the 10 o'clock position (indicating a −ve voltage).

Conductor AB is termed the 'primary winding' because it is connected to a source of power. Conductor CD is termed the 'secondary winding' because it has an EMF induced in it by the magnetic flux from the primary.

These terms are still applicable if multi-turn coils are used instead of single conductors. If conductor CD had the power connected to it, it would be the primary winding and conductor AB would be the secondary winding because it would then generate the induced EMF.

Mutual induction occurring between two coils

When a powered coil (called the 'primary' coil) is energised, a magnetic field is generated that expands through the second coil (named the 'secondary' coil). The flux in the field generates a voltage in the secondary coil, which causes a current to flow if the circuit is complete, as seen in Figure 6.49(a). It is important to note the polarities in the secondary coil.

When the inductor has fully charged, the field has expanded fully and there is no longer any self-induced voltage. There will also be no secondary-induced voltage and the current in the secondary coil will fall to zero. For as long as the primary coil is energised, the field will be maintained at a steady strength and there will be no self-induced voltage or secondary-induced voltage. Therefore, there will be no current in the secondary coil, as shown in Figure 6.49(b).

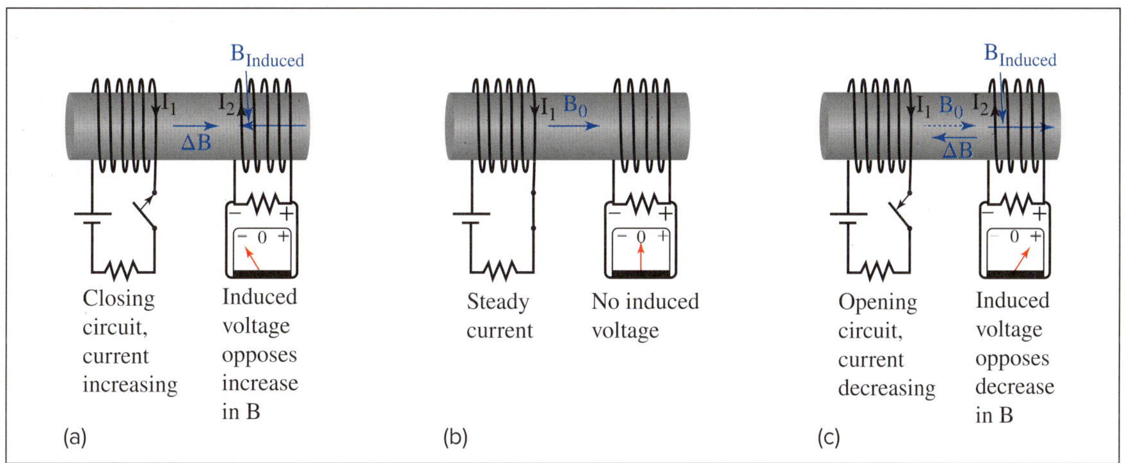

FIGURE 6.49 Mutual inductance between coils

When the current is switched off, the primary field collapses, cutting across the conductors of the secondary coil. The relative direction of movement will be opposite to what occurred during charge-up and the polarity of the induced voltage in the secondary coil will also be reversed (see Figure 6.49(c)).

If the current could be continuously turned on and off, a current would continue to flow in the secondary coil. This is approximately what happens in a transformer connected to an a.c. source.

Energy stored in a magnetic field

The energy that can be stored in an inductor can be calculated from the equation given in Example 6.12. This energy is stored in the electromagnetic field while the current is flowing but is released very quickly if the circuit is turned off or power is lost.

Large inductors driven by a source such as an automotive battery can deliver a lethal voltage across their terminals when discharged. The sudden discharge can cause a very high EMF to be generated.

Figure 6.50 illustrates the relative directions and values of induced voltage during *circuit make* and *circuit break*. The following points should be noted:

- Self-induced voltage opposes current build-up.
- The maximum value of self-induced voltage during switching on is less than the applied voltage.
- Self-induced voltage also opposes current collapse at switching off.
- The value of self-induced voltage is greater at circuit break than circuit make, and can be many times the value of applied voltage.
- The greatest values of self-induced voltage occur at points where the current curves have the steepest slopes. At these points, the rate of change of current is greatest.
- Where there is no current change, there is no self-induced voltage.

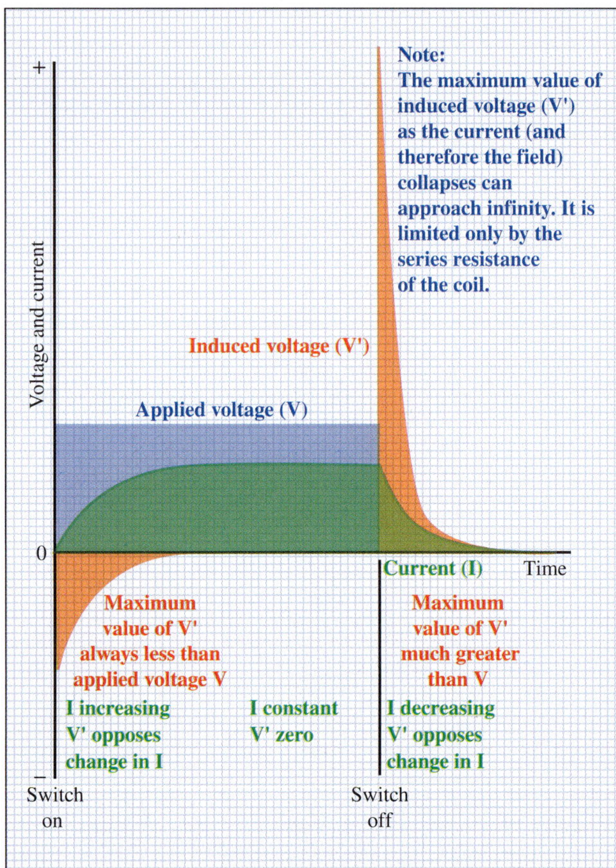

FIGURE 6.50 Inductor charge and discharge

Figure 6.50 shows that the current flow takes some time to reach its maximum value. The voltage across the inductor falls as the current rises until a value of approximately zero is reached. The current flow through the inductor will increase until the only limiting factor is the circuit resistance, and the maximum current can be calculated by Ohm's Law, $I = \frac{V}{R}$.

EXAMPLE 6.12

Energy stored in a magnetic field

$$W = \frac{1}{2}LI^2$$

where:

W = energy in joules (J)

L = inductance in henrys (H)

I = current flow in amperes (A).

A 10 H electromagnet has a current flowing through it of 5 A. Find the energy stored in the fully charged magnetic field.

$$W = \frac{1}{2}LI^2$$

$$W = \frac{1}{2} \times 10 \times 5^2$$

$$W = 125 \text{ J}$$

CHECK YOUR UNDERSTANDING

6.19 A 12 H electromagnet has a current of 3 A flowing through it. What is the energy stored in the fully charged magnetic field?

Time constants

When a circuit containing an inductor and a resistor is first energised, the flow of current is opposed by the back EMF created in the inductor. The expected maximum current can be calculated using Ohm's Law. However, it will take five time periods for the current to reach a maximum. The percentage of current existing in the circuit at each of these time periods can be determined using a universal time constant chart (see Figure 6.51).

In one time constant, the current will increase to 63% of the value of the maximum current. As this increase is exponential, in theory the current will never reach its maximum value but becomes 63% closer with each time period. When the supply is removed, the current will take five time periods to reach 0 A.

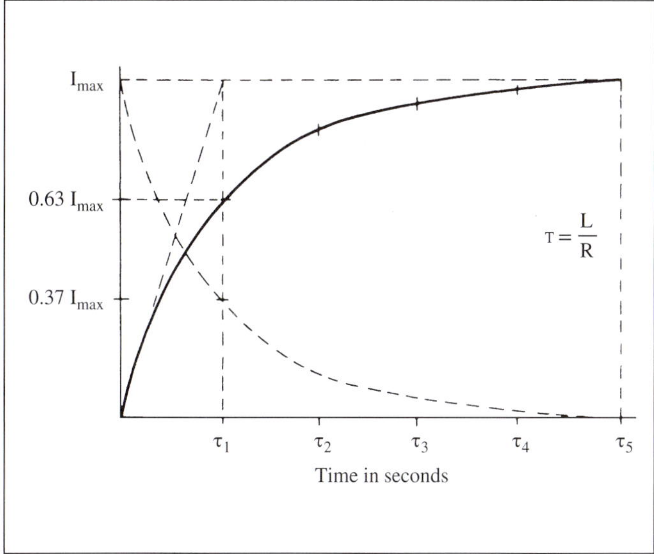

FIGURE 6.51 The time constant as a chart

The voltage across the inductor will reduce as the current in the inductor increases, taking five time periods to decrease from a maximum voltage to 0 V. When the current in the circuit is at a maximum, the supply voltage will appear across the resistor.

Calculating the instantaneous values of current at any of these time periods is achieved by multiplying the maximum current value by the percentage identified by the time constant chart. For example, a the time constant chart, if the maximum current in the circuit (determined by Ohm's Law) is 10 A, there would be 6.32 A at the first time period, 8.64 A at the second time period, 9.5 A at the third, 9.82 A at the fourth and 9.93 (or 100%) at the fifth.

EXAMPLE 6.13

Time constant (tau)

$$\tau = \frac{L}{R}$$

where:

τ = time constant in seconds
L = inductance in henrys
R = resistance in ohms.

A 1 H choke has an internal resistance of 25 Ω. Find the time constant of the choke.

$$\tau = \frac{L}{R} \quad (1)$$

$$= \frac{1}{25} \quad (2)$$

$$= 40 \text{ ms} \quad (3)$$

This expression shows that a greater inductance and/or a lower resistance will cause a longer time constant.

CHECK YOUR UNDERSTANDING

6.20 A 10 H electromagnet with an internal resistance of 50 Ω has a current flowing through it of 5 A. Find the discharge time to turn-off (5 τ).

6.6 Magnetic principles in measurement instruments

The skills and knowledge associated with testing and using test equipment are beneficial to apprentices and qualified electricians alike. The following information provides a foundational awareness of test instruments and equipment in the context of the application of magnetic principles.

6.6.1 Moving coil meters

Figure 6.52 is an exploded view of a moving coil movement. It can be seen that a coil that is free to rotate is suspended in the field of a permanent magnet. The coil ends are connected to a suspension system so that current can be passed through the coil.

The current is governed by the value of the applied voltage. The coil sets up its own magnetic field. This field reacts with that of the permanent magnet and causes the coil to rotate. A pointer attached to the coil gives a voltage reading against a scale.

The meter movement can only work satisfactorily on direct current. If a.c. is applied to the movement, it tries to turn the coil rapidly in opposite directions, with the result that the coil effectively remains stationary.

The meter can only operate on a.c. if the a.c. is rectified to d.c. before it flows through the meter. Because of these factors, a moving coil meter always reads average values of current and voltage.

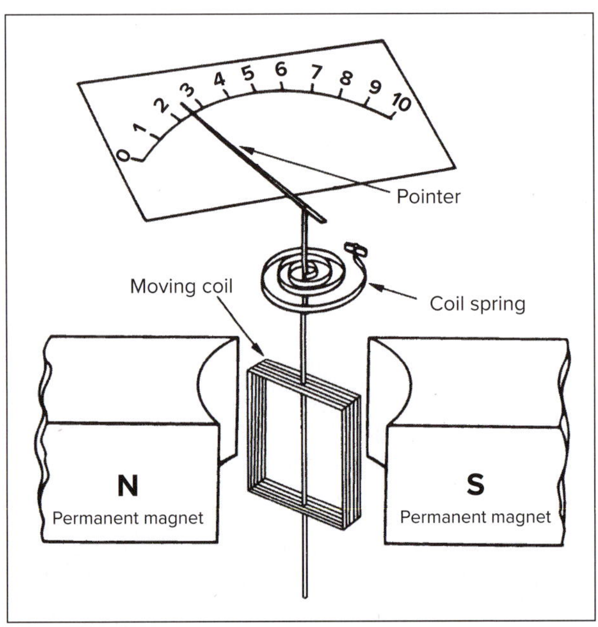

FIGURE 6.52 Exploded view of a moving coil movement

6.6.2 Moving iron meters

Figure 6.53(a) is an exploded view of a moving iron meter to illustrate its operating principle. In practice, the construction is slightly different and is shown in Figure 6.53(b).

There are two magnetically soft iron vanes in the movement. One vane is fixed and the other is pivoted and free to rotate. A pointer attached to the moving vane moves across a scale as an indicator.

When an electric current is passed through the coil, both the fixed and the moving vanes are magnetised and have like poles at adjacent ends. As like poles repel each other, the movable vane moves away from the fixed vane. The attached pointer then indicates a value against a calibrated scale. A restraining spring provides opposing torque so that the vane movement can be stabilised. It should be noted that the scale of a moving iron meter is non-linear.

6.6.3 Dynamometer meter movement

An exploded view of a dynamometer movement is shown in Figure 6.54(a). The meter has two circuits—one for voltage, the other for current. The model illustrated has a soft iron core, around which is wound a low-resistance coil to carry the circuit current. This coil produces a magnetic flux proportional to the current flowing in a circuit. Not all dynamometer-movement meters have an iron core; some are air cored.

The meter's second circuit consists of a coil with a series resistance of high value. This is the voltage circuit and it produces a magnetic field proportional to the applied voltage.

The direct multiplication of voltage and current values in an a.c. circuit to obtain a power value can at best be only an approximation. With some electrical components, the voltage and current can be out of step with each other. This type of meter construction, with its two magnetic fields, takes into account any displacement between voltage and current and gives a 'true power' reading.

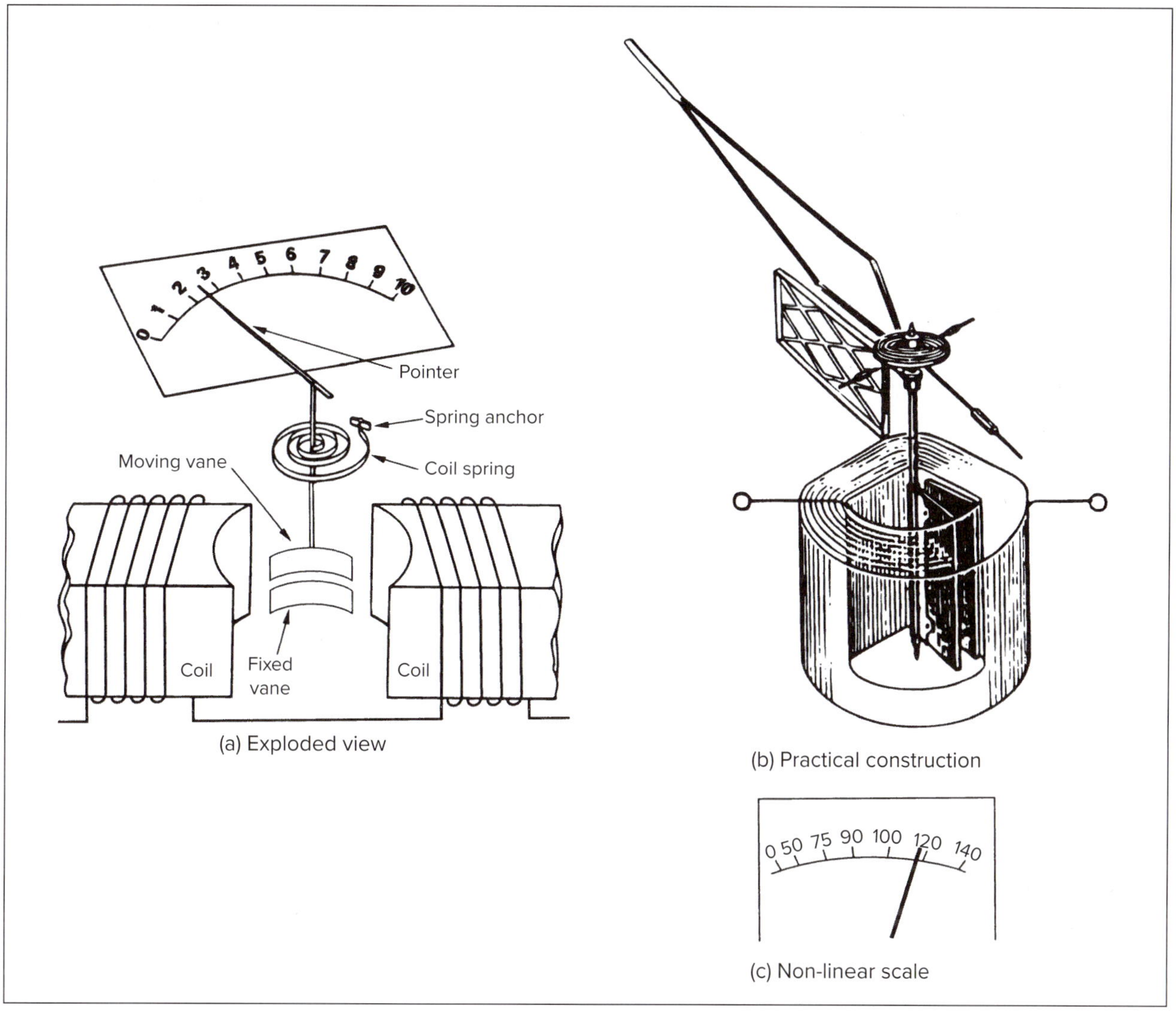

0 1 2 3 4 5 6 7 8 9 10
Pointer
Spring anchor
Coil spring
Moving vane
Coil
Fixed vane
Coil
(a) Exploded view
(b) Practical construction
0 50 75 90 100 120 140
(c) Non-linear scale

FIGURE 6.53 Moving iron meter

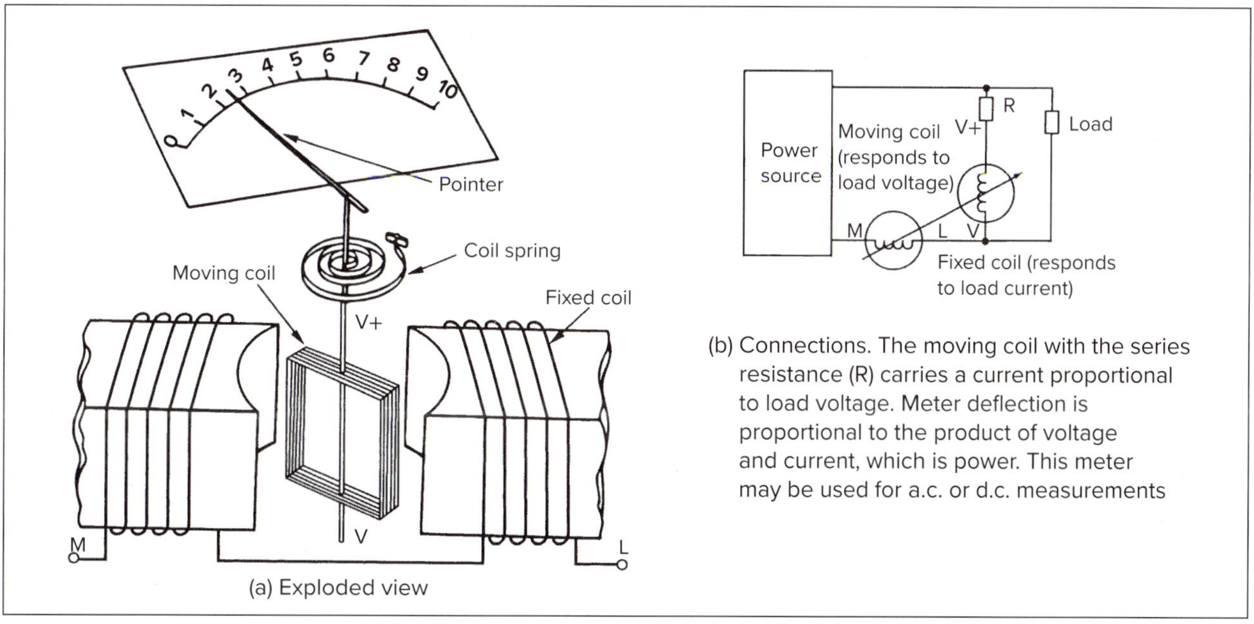

0 1 2 3 4 5 6 7 8 9 10
Pointer
Coil spring
Moving coil
Fixed coil
V+
V
M
L
(a) Exploded view
Power source
Moving coil (responds to load voltage)
V+
R
Load
M
L
V
Fixed coil (responds to load current)
(b) Connections. The moving coil with the series resistance (R) carries a current proportional to load voltage. Meter deflection is proportional to the product of voltage and current, which is power. This meter may be used for a.c. or d.c. measurements

FIGURE 6.54 A dynamometer movement

6.6.4 Voltage testers

These testers can detect either 'moving charges' (magnetic field indicator) or electric fields associated with a.c. voltages. Figure 6.55 shows the electric field and magnetic field. There are two types of voltage tester, *capacitive coupled* and *inductively coupled*. The inductive coupled tester has a sensor winding in its tip. A voltage is induced within the winding in the presence of an electromagnetic field. This type of tester works for wires in operational conditions such as circuits under load. They will not detect voltage or live conductors when there is no current flow ('moving charge' is needed to produce a magnetic field).

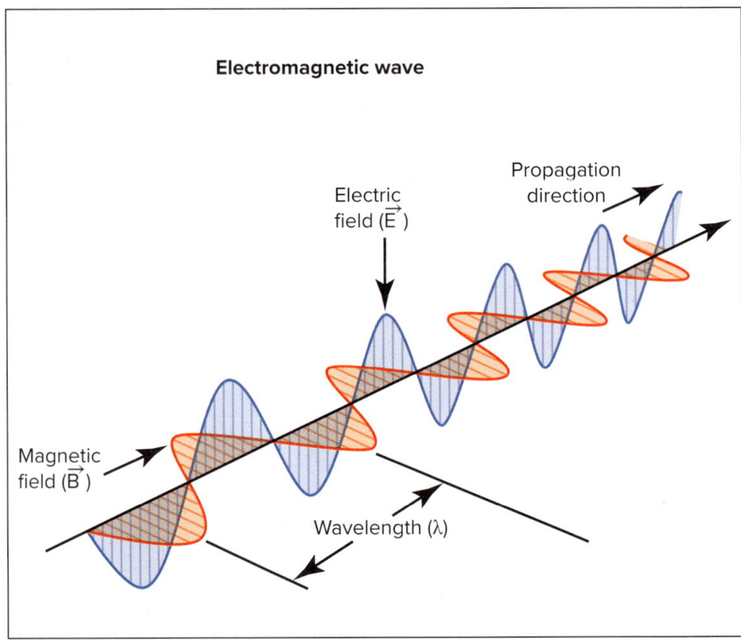

FIGURE 6.55 **Electric and magnetic fields testing**

Voltage is detected by a 'capacitive coupled' device that is sensitive to the electric fields produced by the circuit voltage. It is the basis of the 'finder' used to locate live conductors buried in walls up to a depth of 2.5 cm. Most units give both audible and visual signals.

6.6.5 Supply current and voltage—monitoring instruments

Some meters for testing voltage and current are designed to make no electrical contact with the circuit under test. The majority of these (for a.c.) work on the basis of mutual induction. Several types are available, either as fixed monitoring instruments or as portable test equipment used by electricians. They essentially operate on a similar principle to transformers.

Instrument transformers are used in electrical installations for the safe monitoring of supply voltage and current. Voltage or potential transformers (VTs or PTs) step down the supply voltage to a lower, safer level for monitoring (typically 110 V). Current transformers (CTs) step down the current to either 1 amp or 5 amps. This transformation ratio is determined by the number of turns of the primary and secondary windings.

6.6.6 Current testers

Current testers are marketed under various names, most of which are trade related—'tong testers', 'clamp meters', 'clip-on testers', 'link test meters' and so on. Clamp action meters are generally used to measure current without having to interrupt the circuit being tested.

The jaws of the instrument are opened with a lever and then placed around the chosen conductor. They are then allowed to close. The magnetic field around the conductor enters the low-permeability path of the iron, and the meter movement responds according to the strength of that magnetic field.

Originally there were only two types of current tester. One worked on the repulsion principle of the moving iron meter while the other used a transformer combined with a switch to select the desired current ranges.

Repulsion-type movement (a.c. and d.c.)

The operating principle was that of the moving iron meter. Plug-in modules catered for different current ranges. The magnetic field created by the current produced a repulsion between the meter elements and caused the moving section with the pointer attached to rotate. They were capable of use on both a.c. and d.c.

Transformer operated (a.c. only)

Different current ranges were catered for by using a transformer with tappings connected to a range switch. The transformer prevented it from being used on d.c. The basic principle is shown in Figure 6.56.

The indicating meter could be a d.c.-operated one if a rectifying unit was connected between the transformer output and the meter movement. With a d.c. meter, the scale then became linear (with a moving iron meter, the scale was non-linear).

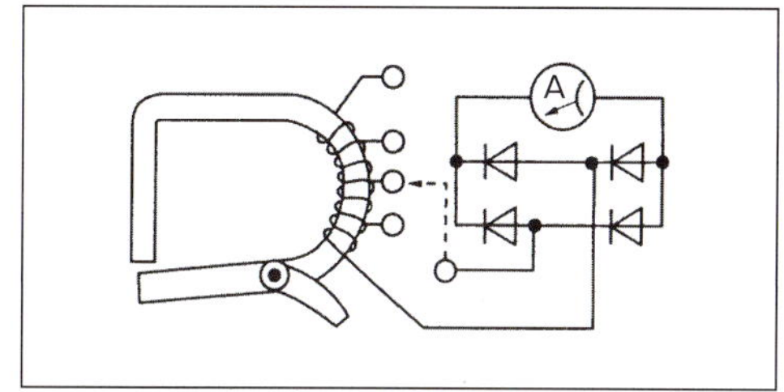

FIGURE 6.56 Internal circuit arrangement of a current transformer

6.6.7 Multifunction clamp meters

The repulsion- and transformer-type current testers described above have been superseded by technological innovations and improvements incorporated into multifunction clamp meters. Three types that are now available are:

1. Current transformer clamp meters—only measure alternating current.
2. Flexible clamp meters (Rogowski coil)—only measure alternating current.
3. Hall-effect clamp meters—measure both alternating and direct current.

CTs, PTs and a.c. clamp meters work on a similar principle to that of transformers. A cable under test acts as the primary winding while the clamp meter jaws act as a secondary. The jaws are made of ferrite iron and are individually wrapped by coils of wire. The jaw tips are flush when closed (as shown in Figure 6.57(b)), but when open reveal bare metal core faces. The jaws form a magnetic core during measurements.

Flexible clamp meters which also only measure a.c. operating on the Rogowski coil principle do not have an iron core. They work on the principle of an air gap between coil windings for measurements.

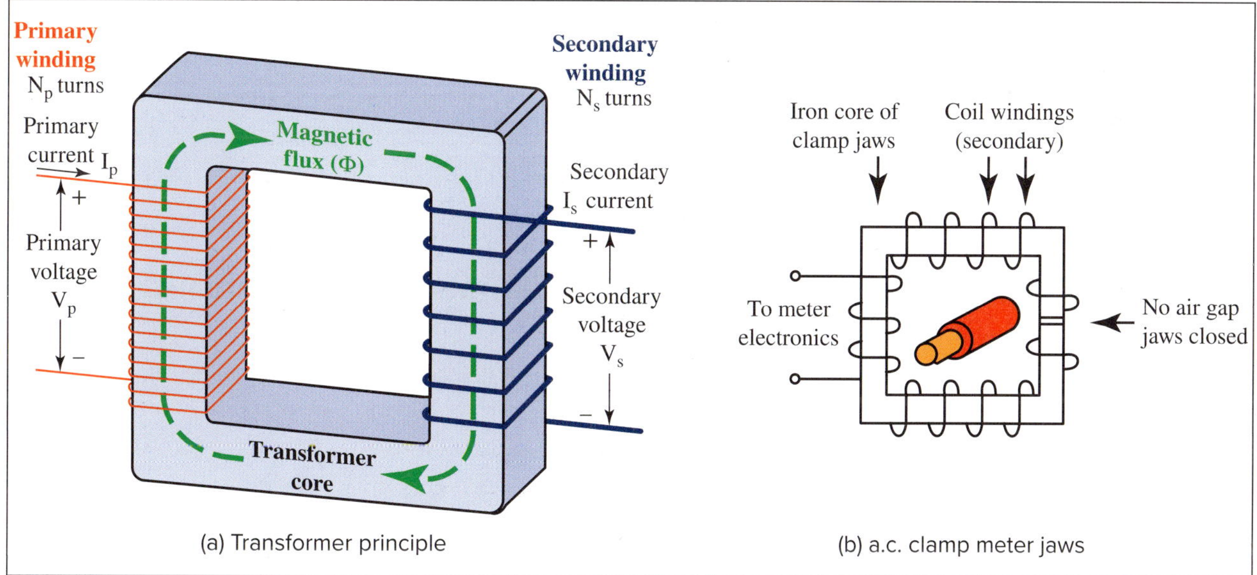

FIGURE 6.57 Transformer principle for a.c. clamp meters

Rogowski coils are sometimes called 'air-cored coils' or 'flexible current probes' because they use a wound, helix-shaped coil for measuring the rate of change of a conductor's magnetic field.

The Hall-effect sensor is a transducer which converts the changing magnetic field of the conductor to a voltage proportional to the magnetic flux in the jaw tip's air gap. These meters have a Hall-effect semiconductor encased by a thin plastic moulding where the jaws meet. This establishes an air gap which the magnetic flux field passes through, completing the magnetic circuit. This concept of an air gap and iron core construction is shown in Figure 6.58(a). The iron core jaws themselves permit the magnetic flux to pass through more easily than air. The air gap ensures that the core is not saturated by limiting the magnetic flux.

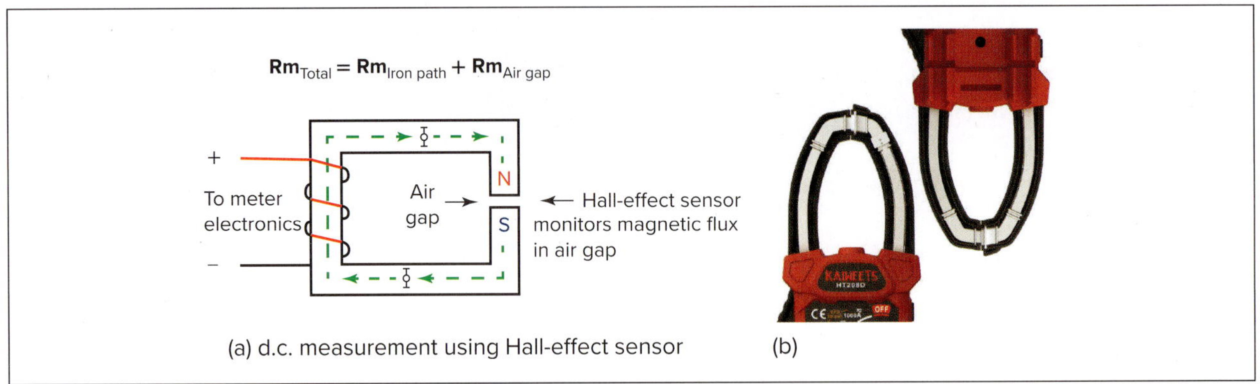

FIGURE 6.58 Clamp meter jaw construction
(photo) Courtesy of Kaiweets

Figure 6.58(b) shows the physical jaw construction for an a.c. clamp meter (on the left) and a d.c. clamp meter (on the right).

Hall-effect clamp meters are used for measuring d.c. current in electrical cables and conductors. Current flow in a conductor produces a magnetic field, and *Hall elements* respond to the field flux created by producing a voltage proportional to the current in the conductor under test. If the cable current is d.c., the magnetic field will be steady and the *Hall voltage* constant. When the cable current is a.c., the flux field will vary and the Hall voltage will also vary to match the field changes.

6.7 Magnetic devices

Solenoids, relays and contactors operate on the principle of magnetism. Magnetic devices used for sensing, detection, control and braking functions include eddy current, Hall-effect and magnetostriction devices.

Magnetic sensor technology is a rapidly growing industry. These sensors provide a quick response, sensitive detection and are low cost. They have become increasingly popular for their versatility, non-contact operation, low maintenance and hardy build.

6.7.1 Construction, operation and application of magnetic sensing devices

Magnetic sensors are solid-state devices that detect moving ferrous metal and can be used in applications for sensing position, velocity or directional movement. Magnetic and inductive sensors detect changes and alterations in magnetic fields, working on the properties of magnetic flux, magnetic induction, inductance or reluctance.

Magnetic sensors in the automotive industry are particularly useful in determining position, distance and speed. They are helpful in determining the position of car seats and seat belts for air-bag control, ignition timing, tachometers and wheel rotation speed detection for anti-lock braking systems (ABS).

Sensors are also used in conjunction with conveyor belts, chain drives and equipment cycles for position measurement in manufacturing, industrial and processing environments. They monitor presence, rotation and direction for machinery as well as equipment functions and product pieces.

Hall-effect devices

The Hall effect was discovered in 1879 by the American physicist Edwin Hall. Hall-effect sensors are transducers that convert magnetic flux to electrical voltage. A Hall-effect integrated circuit includes a Hall sensing element, linear amplifier and emitter-follower output stage. The sensor output voltage levels are dependent on the magnetic flux levels.

Hall-effect sensors are activated in response to an external magnetic field or when a predetermined level of field flux is exceeded. These sensors are used to detect the position of permanent magnets, the presence of a magnetic field or to measure the magnitude of a changing magnetic field.

Hall sensing applications for motion, proximity and gear teeth are shown in Figure 6.60. Note that the magnet polarities associated with the notch sensor and gear tooth sensor are different due to their respective applications. The notch sensor detects ferrous material and the gear tooth sensor detects the absence of ferrous material (gaps between teeth).

When a magnetic field is detected as in Figure 6.59, these sensors generate an output voltage (a Hall voltage) which is instantaneous and directly proportional to the magnitude of flux density (B) and magnetic field polarity (north and south poles).

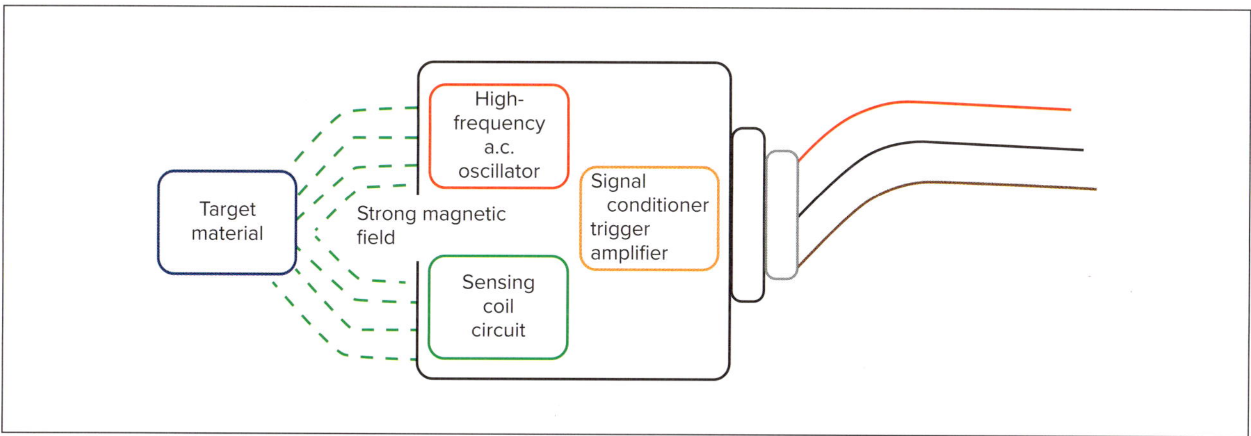

FIGURE 6.59 Solid-state device with integrated circuitry

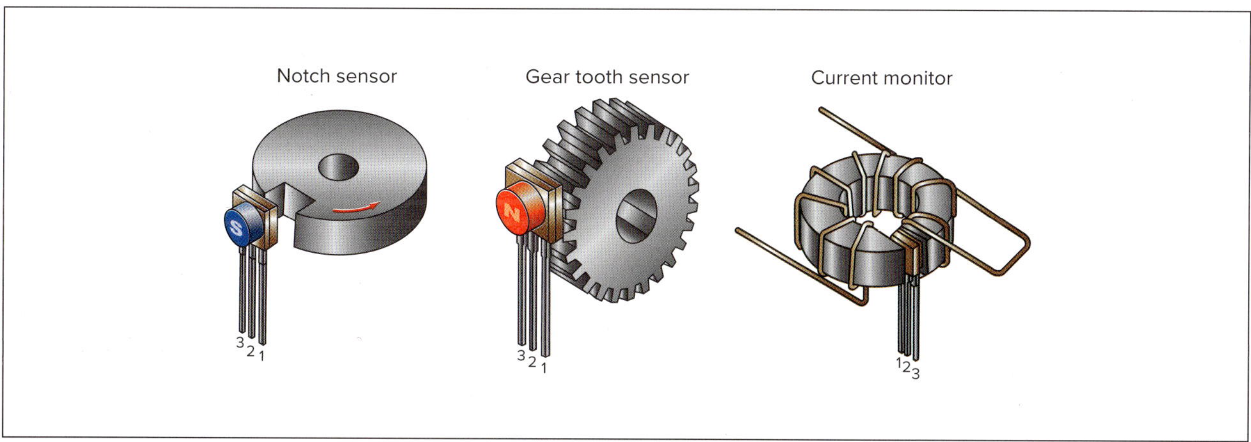

FIGURE 6.60 Allegro ratiometric, linear Hall-effect sensor applications

Hall-effect devices are used in speed control, motion, proximity, positioning systems and current-sensing applications. They are also used in clamp meter testers for contactless measurement of current.

Magnetostriction equipment

Magnetostriction is a process in which a ferromagnetic material changes shape when introduced to a magnetic field. Ferromagnetic materials include iron, nickel, cobalt, gadolinium, pure nickel, nickel-iron alloys and alloys of iron with chromium, cobalt and aluminium. These ferromagnetic materials are used to build actuators and sensors like the one in Figure 6.61. Magnetostrictive devices include audio-frequency oscillators and transducers such as that shown in Figure 6.62.

The best known magnetostrictive material is Terfenol-D, a compound of Terbium (Te) and iron (Fe).

When a ferromagnetic material is magnetised, changes in physical dimensions occur. These are caused by the shifting of boundaries between domains within the material, which affects the orientation of the domains.

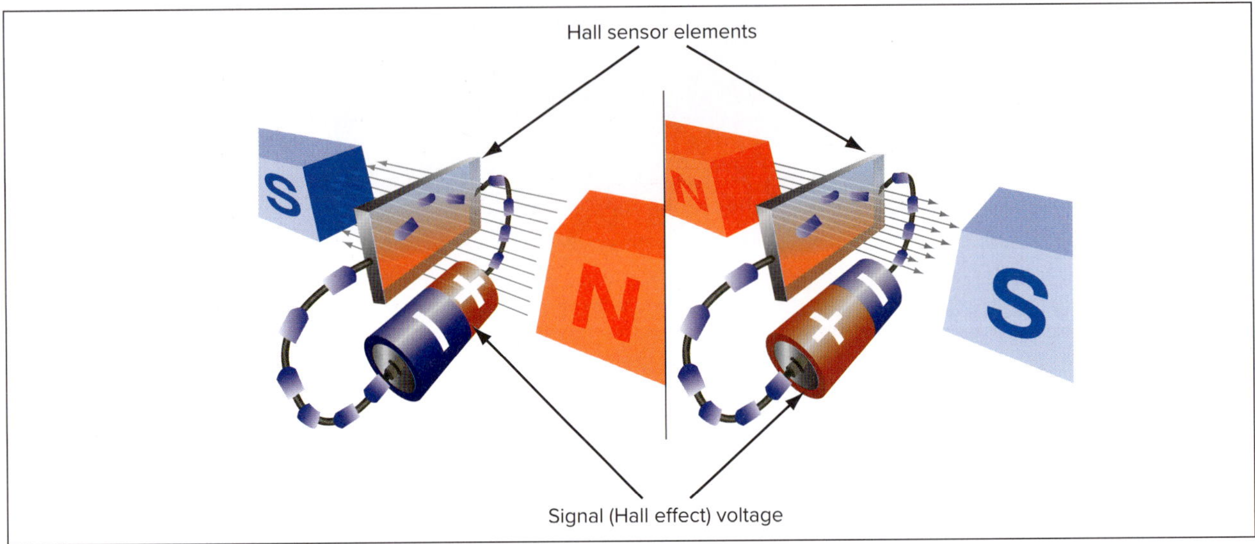

FIGURE 6.61 Hall sensor elements

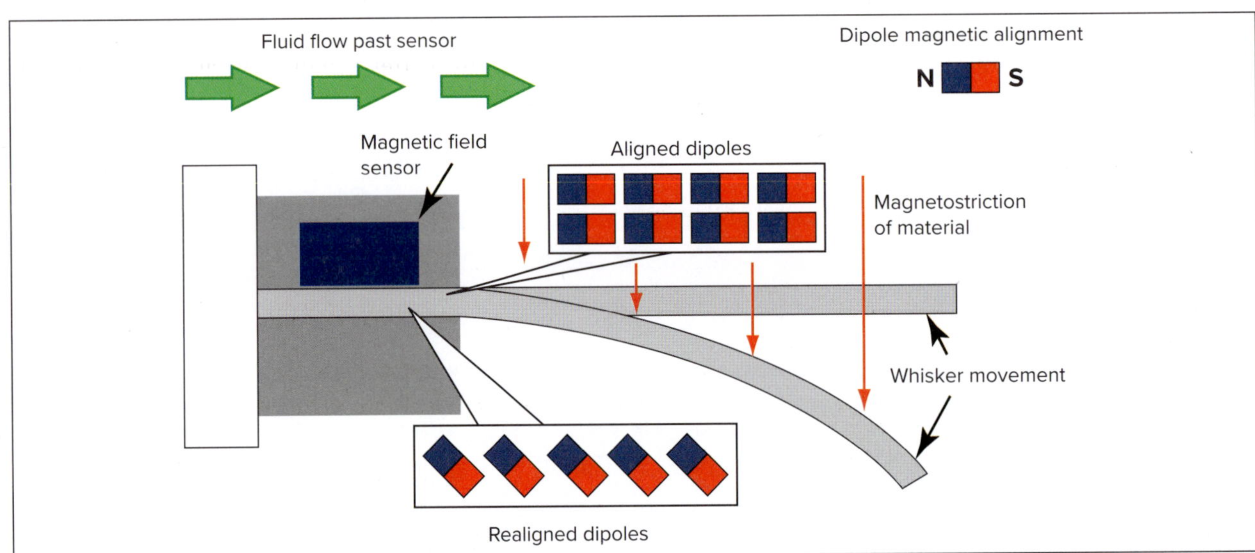

FIGURE 6.62 Magnetostrictive transducer construction

Magnetostriction changes in material dimension can be of three types:

1. longitudinal changes in the direction of applied field
2. transverse changes perpendicular to applied field
3. volume changes perpendicular and parallel to applied field.

Magnetostrictive devices are commonly used in security systems where labels made of ferromagnetic material are secured to, for example, goods in shops and books in libraries. When the label signal and the transmitter pulse interact, an alarm will sound (see Figure 6.63).

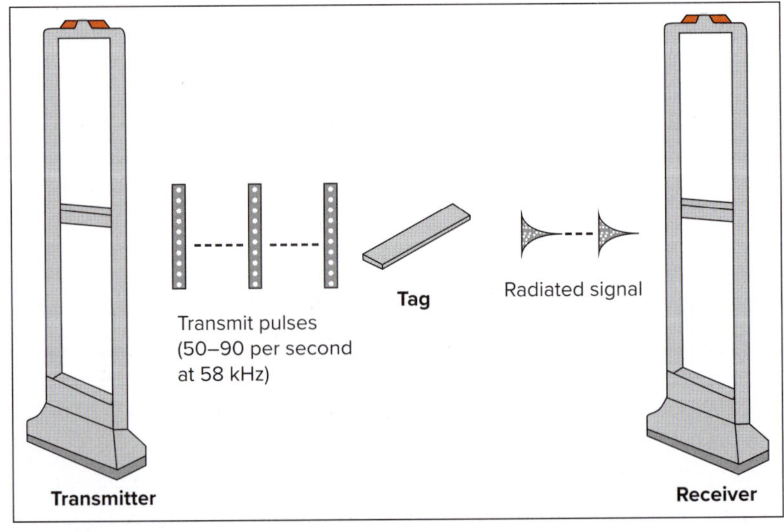

FIGURE 6.63 Magnetostrictive equipment

Eddy current sensors

These sensors are based on the principle that eddy currents circulate inside the conductive material under inspection when induced by an alternating magnetic field (which is in accordance with both Faraday's and Lenz's laws). Eddy current sensors are used for measuring conductivity in electrically conductive materials, surface and near-surface defects and flaws, thickness measurement of materials, displacement and the position measurement of ferromagnetic and non-ferromagnetic (electrically conductive) materials.

Eddy current sensors do not register non-conductive material like oil, dirt, dust or moisture, which makes them suitable for use in harsh and industrial environments. They cannot sense through metals so are manufactured in either non-metallic enclosures or have non-metallic caps. Inductive sensors have a coil wound around a ferromagnetic core while eddy current sensors incorporate an air-core coil.

An alternating current is applied to the coil and this produces a magnetic field, which acts as the primary field as seen in Figure 6.64(a). When the coil is brought near a conductive material, an electrical (eddy) current is induced inside the material. Induced eddy currents circulate within the conductive material, producing a secondary magnetic field which opposes the primary field (Lenz's Law).

The interaction of these magnetic fields produces a signal as the sensing coil is moved towards or along the material being measured or inspected, as shown in Figure 6.64(b). The signal varies due to the distance between the probe and the target material or imperfections in the material.

Current flows by the path of least resistance and, similarly, eddy currents will flow around cracks and imperfections. The changes in eddy current flow cause signal changes which are detected by the receiving (sensing) coil in Figure 6.64(a).

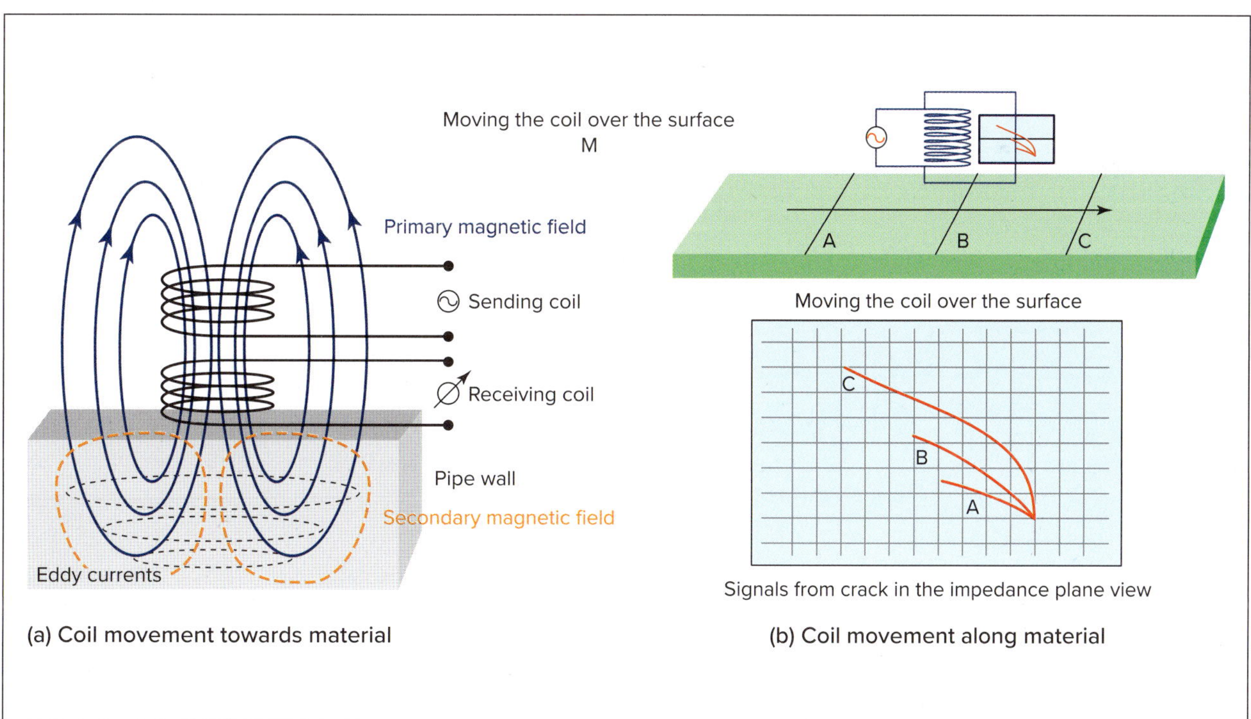

FIGURE 6.64 Sensing coil movement producing eddy currents

Eddy current depth and penetration of a material depends on the frequency of the alternating field, material permeability and conductivity. Figure 6.65 illustrates how higher frequencies of a.c. supply oscillation will result in eddy currents concentrating near the surface of the material. Lower frequencies result in eddy currents being produced deeper in the material.

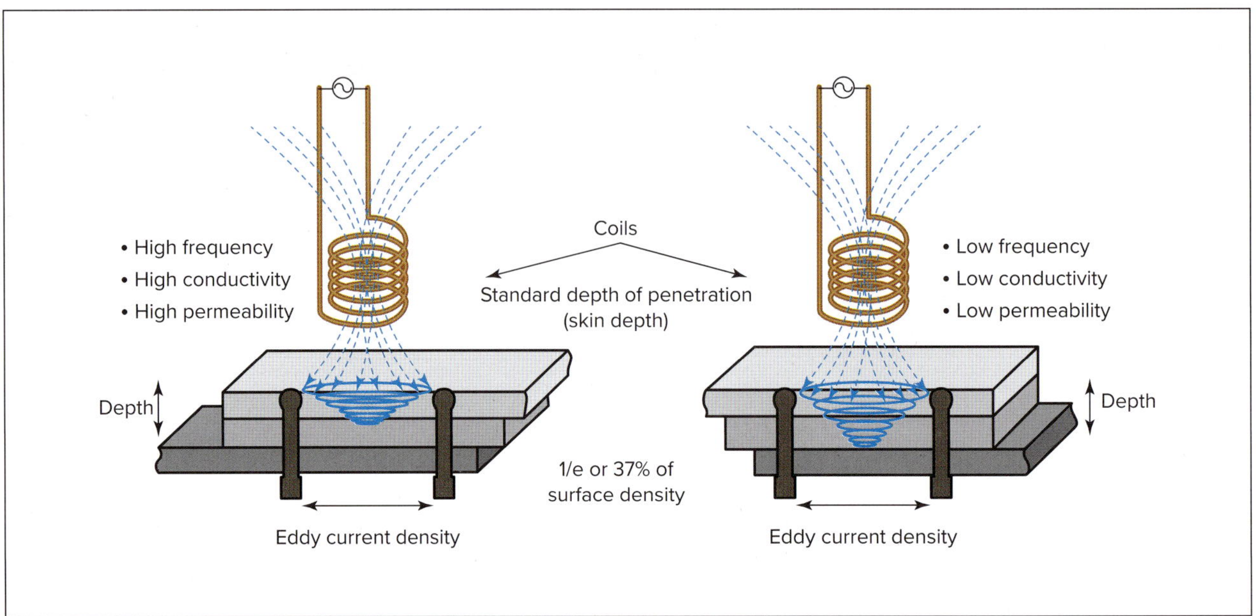

FIGURE 6.65 Effect of higher frequencies of a.c. supply oscillation on eddy currents

6.7.2 Relays and contactors

Relays and contactors are similar in their operating principles. They are devices that are electromagnetic and their operation is controlled by energising and de-energising a coil. Relays are used for switching and control applications while contactors generally switch larger-current-load devices such as heaters and motors.

The magnetic field created by energising the coil causes a fixed iron stator to become magnetised and thus attract the moving iron armature.

Typical relay construction

Magnetic relays switch electrical contacts electromagnetically by a solenoid coil. The electrical contacts, one set fixed and attached to the stator, the other moveable and attached to the armature, open or close when the coil is energised, allowing for opening and closing of these contacts and control of electrical circuits.

It consists of a fixed iron core (a **stator**) and a part that is moveable (an **armature**). A return spring is often used to hold the armature away from the stator or core when the coil is de-energised. When the coil is energised, the armature is pulled against the stator, closing the magnetic circuit and mechanically closing (or opening) one or more sets of contacts.

When current flows in the operating coil, a magnetic flux (shown as dotted green lines in Figure 6.66) is created in the soft iron core and around the magnetic circuit, including the armature and the air gap. The iron core domains align. (You may recall that magnets and electromagnets have a north pole, a south pole and similar magnetic field patterns.) There is a force of attraction between the fixed stator and armature, overcoming the spring tension so that the armature will close. The force exerted on the armature in the closed position (minimum or no air gap) will be many times greater than when the armature is in the fully open position (maximum air gap).

The iron core and air gap have different reluctance (R_m) properties. Reluctance of air (or magnetic resistance) is much greater than that of iron. Air and iron also have different permeability (μ). The ease with which the iron core and stator can be magnetically influenced is much greater than it is for air.

Typical contactor construction

A contactor operates sets of contacts electromagnetically by a solenoid coil. These contacts are usually meant to operate an a.c. circuit, particularly a three-phase or high-voltage/current circuit. It should be noted that d.c. contactors are also used in industry.

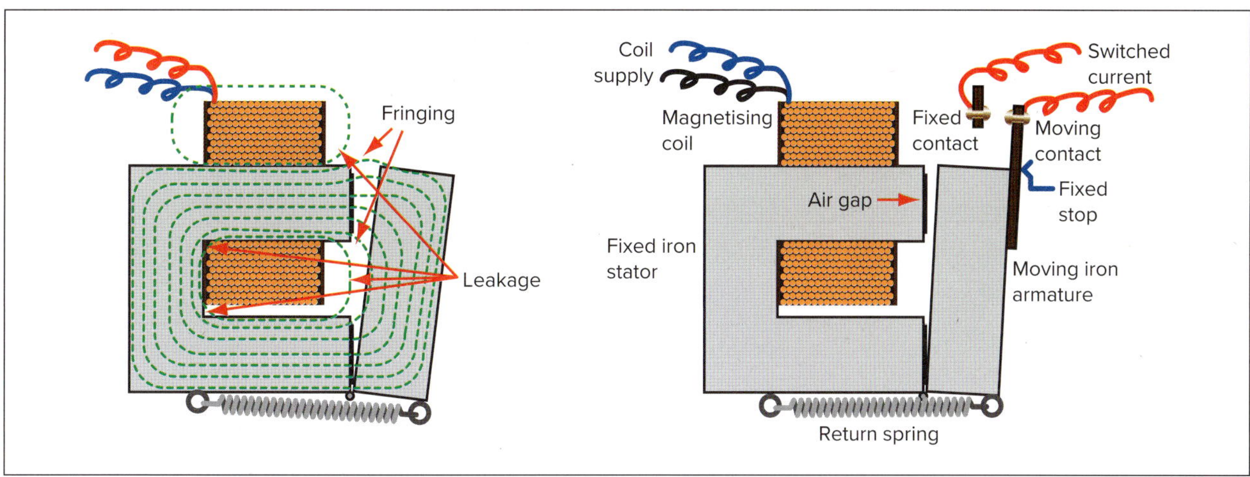

FIGURE 6.66 Armature relay/contactor core

Relays have normally open (N.O.) and normally closed (N.C.) electrical contacts for switching and controlling circuits. Contactors generally operate higher current loads with normally open contacts. Additional auxiliary contacts (normally open and closed) are used for other circuit operations such as hold-in contacts and control circuit switching, once the contactor coil has been energised.

In Figure 6.67(a) the contactor is in a de-energised state, the yellow rectangle highlighting the open contacts. Once energised, the moveable part (armature) is attracted magnetically to the fixed-base stator (housing the coil), pulling against the spring tension. The coil, when energised, magnetises the stator (magnetic induction), causing the electrical contacts to change state from 'open' to 'closed'. This is shown in Figure 6.67(b) inside the red rectangle. When the contactor energises, the electrical supply connected to the 'line' side of the contactor is connected to the 'load' side contacts.

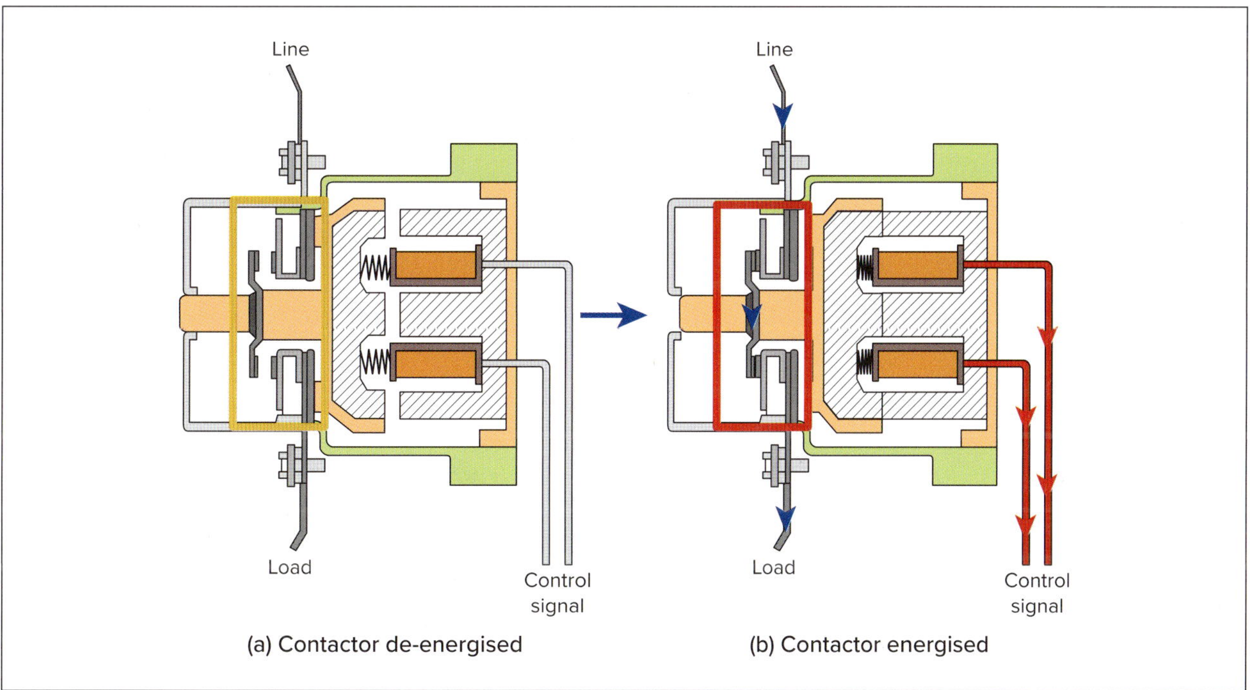

FIGURE 6.67 Contactor operation

There are many types of relays and contactors (see Figures 6.68 and 6.69).

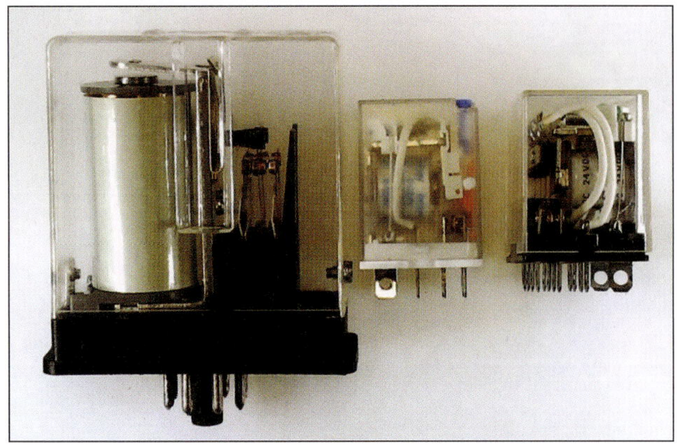

FIGURE 6.68 Typical small relays
Tony Jones

FIGURE 6.69 Relays and contactors
Tony Jones

6.7.3 Solenoids

A solenoid is a type of electromagnet comprising a coil, housing and a moveable plunger (armature). Solenoids convert electrical energy into mechanical work and are categorised as solenoid actuators and solenoid valves. The linear electromechanical actuator (LEMA), produces a straight-line linear movement and the *rotary solenoid* produces a rotational movement.

Linear actuators consist of an electrical coil wound around a cylindrical tube with an actuator or 'plunger'. The plunger is free to move or slide in and out of the coil, and this enables it to open or close doors and valves, as well as to move and operate mechanisms.

Figure 6.70 shows the various parts of a solenoid valve. Around the coil is a housing made from iron or steel. The ferromagnetic properties concentrate the magnetic flux produced by the energised coil for better magnetic attraction of the valve plunger. The valve plunger serves to block the inlet air or fluid from flowing through the outlet of the valve. The solenoid valve may be for water supply to a domestic washing machine or a commercial dishwasher, or may act as a control device for an industrial application in machinery and equipment.

The valve is controlled by the solenoid coil for opening and closing. When the coil is energised, a magnetic field draws in the valve plunger, opening the valve. While the coil is energised, the plunger is pulled, which permits the valve to open. When the coil is de-energised, spring tension forces the plunger to seal the valve.

In a dishwasher or washing machine application, mains water pressure is sufficient for the movement of water present at the valve inlet to flow through the valve outlet. Industrial applications may also incorporate pumps to provide pressure.

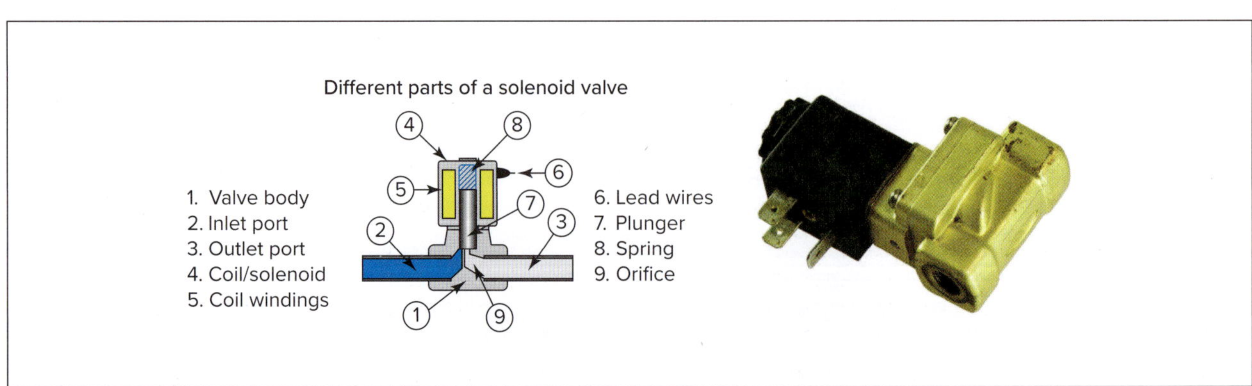

FIGURE 6.70 Solenoid valve
(photo) iKhai_TH/Shutterstock.com

6.7.4 Magnetic methods used to extinguish the arc between opening contacts

An arc-suppression blowout coil produces a magnetic field in an electrical switching device (such as a contactor or circuit-breaker) for the purpose of lengthening and extinguishing the electric arc that is formed as the contacts of the switching device open to interrupt the current. The magnetic field produced by the coil is approximately perpendicular to the arc. The interaction between the arc current and the magnetic field produces a force driving the arc into an 'arc chute', which assists in extinguishing the arc (see Figure 6.71).

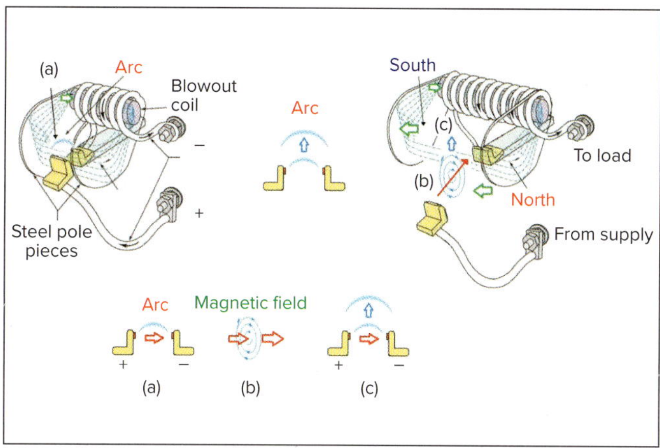

FIGURE 6.71 Magnetic blowout coil

INDUSTRY IN FOCUS

Induction loops may not be a new idea, but their application within the electrotechnology industry and the wider world remains important. Electromagnetic principles of operation apply to metal detectors, traffic-light control and hearing-loop systems.

A notable example of the use of induction loops as metal detectors was for submarine detection in ports and harbours (this was introduced between 1915 and 1918, during the First World War). The passage of a submarine past a cable that was connected as an induction loop could induce a voltage (in millivolts) in the cable that could be detected by a sensitive galvanometer. In the Second World War, a submerged inductive loop cable gave first warning of the 1942 attack on Sydney Harbour by detecting a midget submarine as it passed over it.

Other examples of the application of inductive loops are vehicle detection, classification and traffic light control systems. These systems (and other forms of metal detectors) use the wire loops as inductive elements. When ferrous metals such as a car chassis form part of the magnetic loop circuit, the circuit inductance value changes.

Hearing augmentation—hearing loop system

According to Australian research, more than one in six people suffer from hearing loss. Many people who work in industry are exposed throughout their career to vibration and high sound levels from the operation of tools, equipment and machinery.

Hearing augmentation systems can help people who experience hearing loss. There are three types:

1. hearing loop
2. radio frequency (commonly referred to as 'FM')—analogue or digital
3. infrared (IR) light transmission system.

Hearing loops, also referred to as 'audio-frequency induction loops' (AFILs), magnetically induce the sound from a microphone, TV or audio signal directly to hearing aids and cochlear implants without interference, via a hearing-loop amplifier and loop cable. An electromagnetic field is created that users with hearing aids can connect to by switching on a telecoil in their hearing device. A *telecoil* used in conjunction with a hearing loop assists with hearing in challenging listening situations.

A telecoil, sometimes called a 'T-switch', 'T-coil' or 'telephone switch', is a tiny coil of wire (wound around an iron core) inside a hearing device (see Figure 6.72). The presence of a changing magnetic field will induce an electric current in the coil.

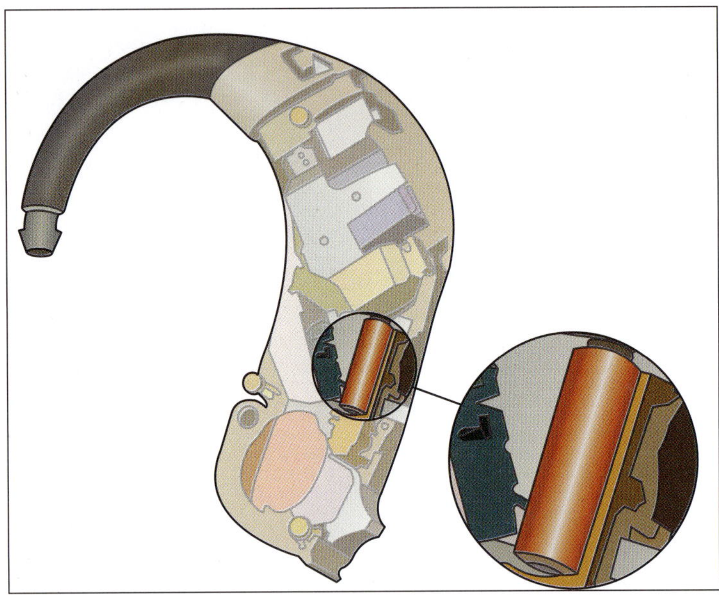

FIGURE 6.72 T-coil inside a hearing device

Hearing loop systems have four components: the listener (with a hearing aid or cochlear implant); the loop cable; a loop amplifier; and an audio input (signal source), as shown in Figure 6.73.

A hearing loop involves the installation of a physical loop to provide a magnetic, wireless signal. The loop can be flat copper tape or copper wire run in the floor, under the carpet, around walls or in the ceiling (often along the perimeter of a room). Installation is ideally where no metal is present in the building structure. A typical hearing loop system room layout is shown in Figure 6.74.

The strength of the electromagnetic field within a hearing loop can vary and is determined by the signal current flow in the coil. You may recall the relationship between current and magnetic flux, i.e. that an increase in current results in an increase of flux. The hearing loop signal can be measured using a field strength meter. Areas more than 6 metres wide on steel-reinforced concrete will require multiple loops and a phase shifter.

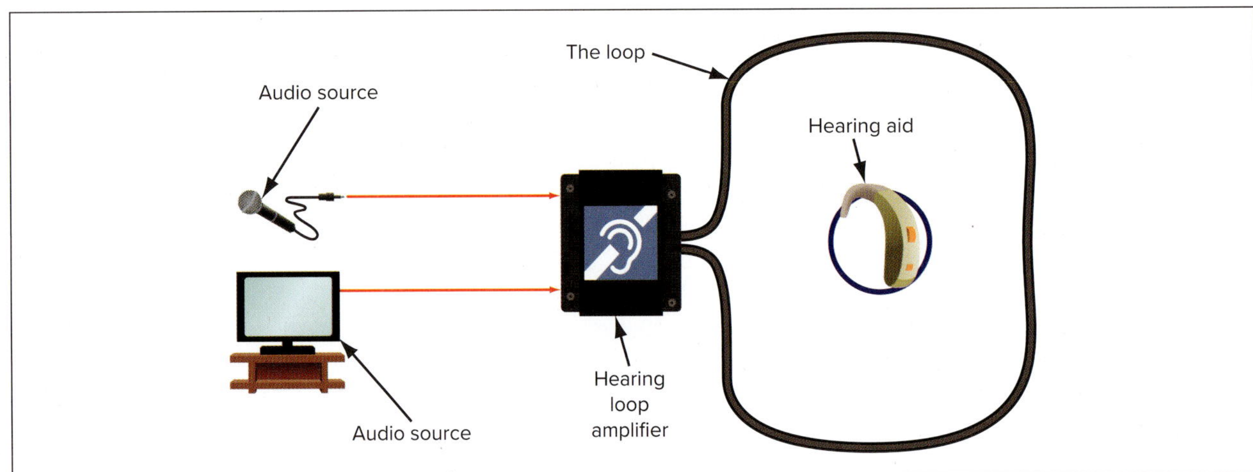

FIGURE 6.73 Basic hearing loop system

Hearing loops are beneficial in a wide range of environments, from large venues such as theatres and conference facilities to teaching spaces, meeting rooms and one-to-one communication points such as customer and reception counters. Hearing loops are an effective solution for users in transport environments, vehicles, terminals and stations (see Table 6.7).

FIGURE 6.74 Induction loop system room layout where no steel reinforcement is present

TABLE 6.7 Applications of hearing augmentation systems

Environment	Applications
Safety	Public address systems, voice alarm systems and help points.
Transport	Airports, stations and transport networks, elevators, help points and carpark access points.
Point of service	Counters, intercoms and entry phones, drive-throughs and help points.
Work	Meeting rooms, video conference facilities, desks and offices.
Education	Lecture halls and classrooms.
Vehicles	Taxis and private cars, minibuses, coaches, trains, trams and boats.
Venues	Theatres, cinemas, concert halls, stadia, sports venues, places of worship and conference and lecture halls.

Counter loops consist of three main components that are required for installation—microphone, amplifier and loop, as shown in Figure 6.75.

Microphone type, selection and positioning are all important in capturing a good, clean audio signal that is free of background noise. An amplifier and loop are normally attached underneath the counter. There are different methods of providing the loop.

Electrotechnology workers, including technicians, electrical contractors and professional specialists who are involved with the planning, installation and testing of hearing augmentation systems, are required to follow Australian Standard AS 1428.5, Design for access and mobility, Part 5: Communication for people who are deaf or hearing impaired.

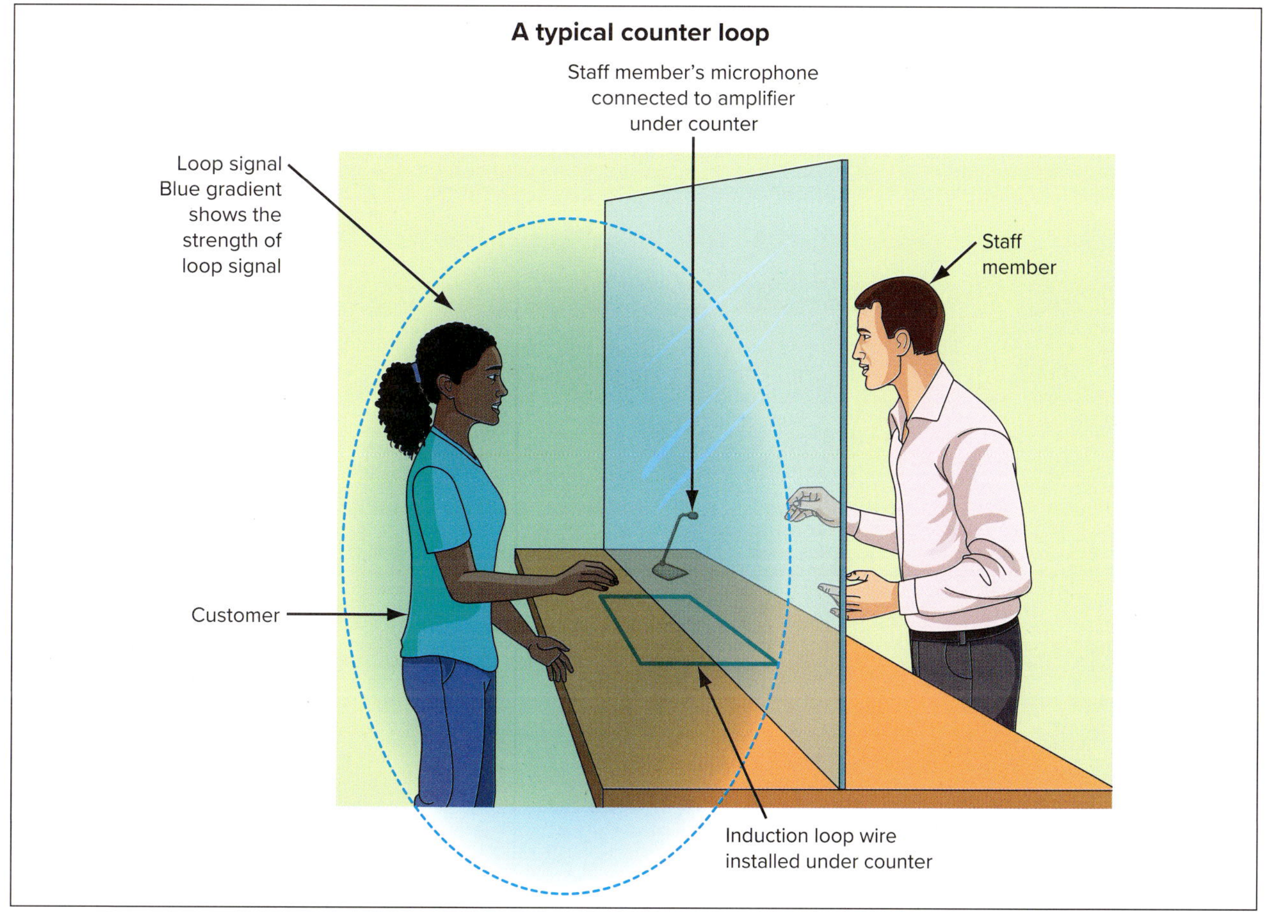

FIGURE 6.75 Counter hearing loop

SUMMARY

- A magnetic field acts outwards at the north pole and inwards at the south pole.
- A magnetic field tends to expand to fill the available space, producing a field of flux that will extend to infinity in a vacuum.
- Like poles repel and unlike poles attract.
- Lines of force existing outside the desired magnetic path are called 'leakage flux'.
- Induction is a process whereby magnets induce a magnetic field in other magnetic materials.
- The permeability of a magnetic material indicates the ease with which magnetic induction can occur.
- Paramagnetic and diamagnetic materials have low permeability and are generally termed 'non-magnetic' materials.
- Ferromagnetic materials have high permeability and are generally termed 'magnetic' materials.
- Materials that are used for permanent magnets are called 'hard' materials.
- Materials that can easily be induced to exhibit magnetic properties but lose the induced magnetic field when the magnet is removed are called 'soft' materials and are useful in electromagnetism.
- Magnetism apparent in a material after the magnetising force has been removed is called 'residual magnetism'.
- Magnetic fields can be shielded by highly permeable materials such as MuMETAL®.
- The right-hand (grip) rule for magnetism can be used for a straight conductor or a solenoid.
- The force created by conductors in a magnetic field is proportional to the flux density, the length of the conductor and the current in the conductor. It can be found from the equation $F = BIl$
- Magnetomotive force (MMF) creates a magnetic field and can be found from the equation $Fm = IN$ (ampere-turns)
- Magnetising force can be found from the equation $H = \frac{IN}{l}$ (ampere-turns/metre)
- Flux density is the flux per unit of area and can be found from the equation $B = \frac{\Phi}{A}$ (webers/m^2)
- Permeability of free space can be found from the equation $\mu_0 = 4 \times 10^{-7}\pi$ (or $4\pi \times 10^{-7}$)
- Permeability (actual) can be found from the equation $\mu = \mu_r \times \mu_0$ (for air, $\mu = 1$) $= \frac{B}{H}$
- Relative permeability is the permeability of a material relative to free space and can be found from the equation $\mu_r = \mu/\mu_0$
- The reluctance of a magnetic circuit can be found from the equation

$$R_m = \frac{l}{\mu_o \mu_r A}$$

- The reluctance in a magnetic circuit is related to the MMF and the flux in much the same way as resistance is related to voltage and current. Therefore, reluctance can be found by the equation $R_m = \frac{IN}{\Phi}$
- The magnetisation curve for a non-magnetic material is a straight line.
- The magnetisation curve for magnetic materials is a curve with a pronounced 'knee' at saturation.
- Magnetic hysteresis is the difference between the increasing and decreasing magnetisation curves resulting from residual magnetism and the coercive force required to remove it.
- The area within a hysteresis curve is proportional to the losses in a magnetic material.
- Permeability can be found using the straight portion of the B/H curve and the equation $\mu = \frac{B}{H}$
- Magnetic leakage is a magnetic flux that passes outside the intended magnetic circuit.
- Magnetic fringing is a magnetic flux that expands over a wider area than intended when passing through an air gap in a magnetic circuit.

- Electromagnetic relays have a fixed magnetic part called a 'stator', a moving magnetic part called an 'armature', a coil and a set of contacts that switch on or off.
- An inductor is a component that generates a magnetic field when current passes through it, and that magnetic field in turn causes a current to be generated in the inductor while the magnetic field is changing.
- The inductance of an inductor is determined by the:
  - material in the core
  - number of turns of wire in the coil
  - cross-sectional area of the coil
  - length of the coil.
- The core of an inductor is usually air, solid iron, laminated iron or iron powder.
- The unit of inductance is the henry (H). A henry is the inductance of a closed circuit in which an EMF of 1 volt is produced when the electric current flowing in the circuit varies uniformly at the rate of 1 ampere per second.
- When a conductor cuts through a magnetic flux, a voltage is induced that is proportional to the number of turns, the strength of the flux and the inverse of the time taken to cut across the flux (i.e. the induced voltage is proportional to the velocity of the conductor). Induced voltage is therefore found by the equation $E = N\frac{\Delta\Phi}{\Delta t}$
- The voltage induced into an inductor can be calculated using $e = Blv \sin\theta$
- Lenz's Law states that the induced current will appear in such a direction that it opposes the change that produced it.

END-OF-CHAPTER QUESTIONS

6.1 How would you determine whether a piece of iron was magnetised?

6.2 Describe what is meant by the term 'magnetic field'.

6.3 Show how a sensitive instrument can be protected against an external magnetic field.

6.4 Draw the general arrangement of an electromagnetic relay switch, labelling all parts.

6.5 What effect do electromagnetic forces have on magnetic and non-magnetic materials?

6.6 Draw a single wire formed into a coil with 20 turns. One end is attached to a battery positive terminal and the other to a battery negative terminal. (*Do not actually try this, as the current would be very high.*) Draw the current flow in the wire and show the expected magnetic field around the wire.

6.7 Draw a solenoid. Show a direction of current flow and the magnetic polarity of the solenoid (N and S).

6.8 In street lighting, two wires carry the current up the street and back again. Will the current cause the wires to be attracted or repelled? (Use a diagram to help visualise this.)

6.9 How can the force of a solenoid be calculated? What values must be known?

6.10 What is the important difference between the permeability of a ferromagnetic material and that of air?

6.11 Sketch a typical B/H curve for a magnetic core made from silicon steel. Label the axes with the correct units. Mark the estimated saturation region on the curve. What is the significance of this saturation region?

6.12 What is the value of the permeability of free space?

6.13 Briefly explain what is meant by the terms 'magnetic leakage' and 'magnetic fringing'.

6.14 State Lenz's Law in your own words and explain what it describes.

6.15 What is electromagnetic force?

6.16 When and where does magnetic leakage occur?

6.17 Where does magnetic fringing occur?

6.18 State typical applications of electromagnets.

6.19 What is an electromagnetic relay? Why is it used?

6.20 What is a contactor? How is it different from a relay?

6.21 Name the unit of inductance and define the unit.

6.22 What is meant by the terms 'primary coil' and 'secondary coil'?

6.23 What is a time constant? What symbol is used to represent it?

6.24 What is the value of current and voltage after one time constant?

6.25 What happens when a highly inductive circuit is opened quickly? What adverse effect could result from this?

6.26 List types of inductors and their common uses.

6.27 Two conductors placed 10 mm apart carry a current of 35 A to and from a d.c. motor 50 m away. What force acts between the conductors per-metre of length? Are they attracting or repelling?

6.28 What magnetomotive force (MMF) is generated by an electromagnet with 150 turns when a current of 12 A flows in the coil?

6.29 Determine the MMF necessary to create a flux of 0.2 Wb in a magnetic core which has a reluctance of 2000 At/Wb.

6.30 A magnetic circuit has a reluctance of 750 At/Wb. The coil, having 800 turns, carries 0.5 A. Find the total flux produced.

6.31 The mean length of a magnetic path is 600 mm. The cross-sectional area is 800 mm^2. The relative permeability is 600. Determine the reluctance of the magnetic path.

6.32 The magnetising force in an iron ring is 1500 At/m. It creates a flux density of 0.95 T. Find the relative permeability for these conditions.

6.33 A long solenoid of 0.8 m has a current of 2 A flowing through a coil of 2000 turns. What is the magnetising force?

6.34 Calculate the voltage that would appear across the terminals of a 10 H inductor if the current of 5 A is reduced to zero in 0.2 seconds.

6.35 What is the inductance of an inductor that has a d.c. resistance of 35 Ω and takes 0.8 s to charge to 63.2% of full current?

6.36 A 0.8 H inductor has an internal resistance of 10 Ω. What is the maximum current it will take when supplied from a 12 V battery?

CHAPTER 7

Direct current machines

LEARNING OBJECTIVES

- Understand how a direct current (d.c.) machine is constructed
- Identify all the parts of a d.c. machine
- Apply magnetic principles to simple machines

PREREQUISITE KNOWLEDGE

- Understand electromagnetism
- Understand d.c. theory

INTRODUCTION

A direct current (d.c.) machine is one that either produces or consumes electricity as direct current, acting therefore as an energy converter. The same d.c. machine can be either a motor or a generator, although the design is generally optimised as one or the other. When electricity is applied to the machine, it can function as a motor and convert electrical energy to mechanical energy, causing it to rotate. When the same machine is driven by mechanical energy so that it rotates, it generates electrical power.

7.1 Rotating machine construction, testing and maintenance

7.1.1 Components of a d.c. machine

The main parts of a typical d.c. machine (a motor or generator) are:

- field frame or yoke
- end-shields and bearings
- field poles
- field coils
- armature and commutator
- brush gear and brushes.

Some of these components are shown in the exploded view in Figure 7.1.

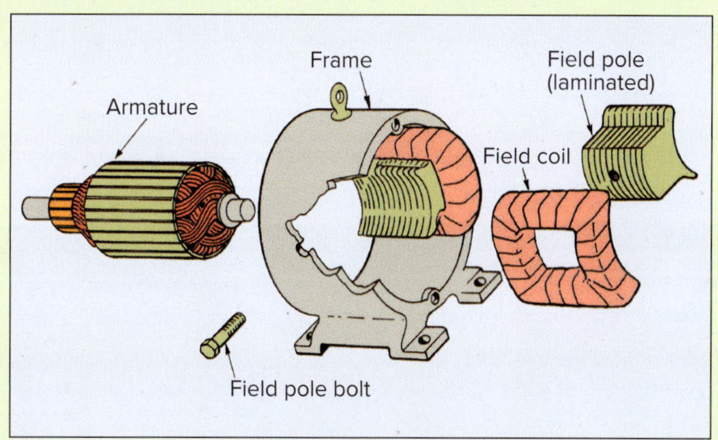

FIGURE 7.1 **Major components of a d.c. machine**

Field frame or yoke

In a d.c. machine, the field frame (or yoke) holds the field poles and provides a magnetic path between them. As a material of high magnetic permeability is required to provide the magnetic path, cast iron or steel are often used. The yoke has high mechanical strength and brings the advantage of retaining residual magnetism, a required factor for generating a voltage.

FIGURE 7.2 **Dismantled d.c. machine**
theLIMEs/Shutterstock.com

End-shields and bearings

End-shields support the bearings in which the **armature** rotates. Also, one of the end-shields usually has provision for mounting the brush gear for electrical connection to the rotating armature.

The design of the end-shields and bearings varies according to the desired use of the machine. They might be waterproof for one situation or completely open for another.

The end-shields and bearings can be seen in the dismantled view of a d.c. machine shown in Figure 7.2. The bearings are attached to the armature shaft.

Field poles

The field poles provide the magnetic field inside the d.c. machine. This magnetic field is required to either induce an EMF or to be used to create a force to move and spin the conductors placed in the armature. The field poles can be thought of as magnets—either permanent magnets or temporary electromagnets. Their high permeability provides a high concentration of magnetic flux at a particular position in the d.c. machine.

These field poles (essentially magnets) can usually be separated from their respective field coils so that they can be placed into the machine, as in Figure 7.1. The outer surface of the poles is shaped to fit tightly against the inner surface of the field frame, while their inner surface is curved to closely follow the shape of the armature. The pole tips have been extended to contain the magnetic fringe. This shaping permits a greater area in the flux path close to the armature and reduces the magnetic reluctance of the air gap. An incidental advantage is that the tips can be used to hold the field coils in position when the pole piece is bolted into place in the yoke.

Field coils

Field coils are wound with insulated copper wire to create an electromagnet. They can be connected to the conductors placed in the armature in several combinations. The way they are connected—i.e. series, shunt (parallel) or a combination of both—determines the type of machine.

If the field coils are connected in series with the armature conductors, they are called 'series coils'. These are made of thick wire or even copper bar. If the field coils are connected in parallel with the armature conductors, they are called 'shunt coils'. Shunt coils are usually wound with much thinner wire.

The actual construction of the coil will vary accordingly. Series coils must carry the high armature current, so they are wound with heavier conductors than shunt coils.

Typical shunt and series field coils are shown in Figure 7.3. The conductor sizes and comparative number of turns on the field coils are typically shown in circuit diagrams as three turns for series coils and four turns for shunt coils.

Some magnetic fields are electrically excited. This means that the field poles become magnetised from the field coils or windings. When electrical current flows through the field coils, a magnetic field is created that magnetises the field poles. The field coils are defined by the way in which they are connected to the armature of the motor. Series field coils are connected in series with the armature, and shunt field coils are connected in parallel with it. If a d.c. machine contains both series and shunt field coils, it is considered to contain compound windings. The other types of magnetic fields are created by non-electrical means such as permanent magnets.

The magnetomotive force (MMF) created must be great enough to produce the required magnetic flux through the yoke, field poles, air gaps and armature in a series-parallel magnetic circuit (see Figure 7.4).

FIGURE 7.3 Field coil construction
Simon Dand

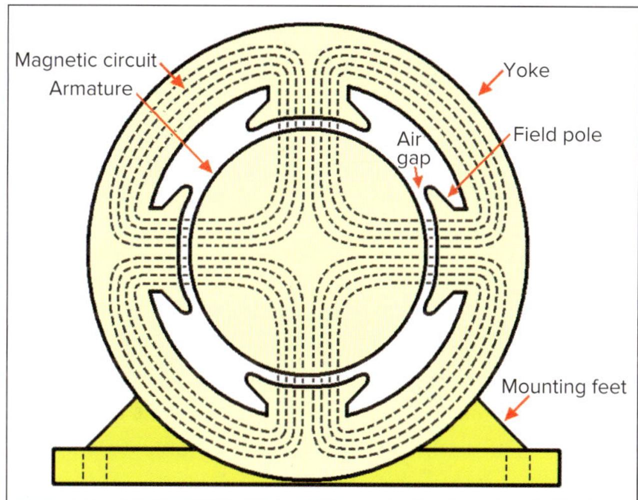

FIGURE 7.4 Magnetic circuit in a four-pole d.c. machine

Armature

The armature is the part of the machine that holds the conductors which are used to induce a voltage. It spins around inside the machine. The armature core is assembled from many steel laminations that are stacked together, which reduces eddy currents. The steel used is called 'electrical steel', and it contains silicon, which helps to reduce hysteresis loss. Figure 7.5(a) illustrates a typically shaped lamination. Equally spaced slots are stamped around the circumference to hold the armature conductors. Their shape influences the winding method used for any armature. Laminations for larger armatures also have punched ventilating holes. An unwound armature is shown in Figure 7.5(b).

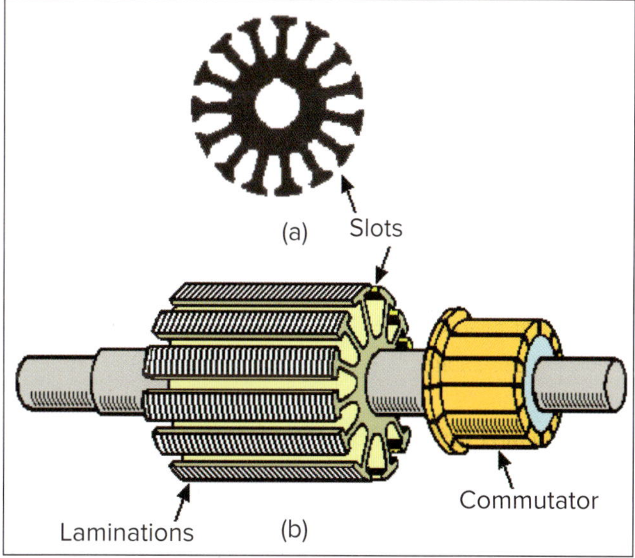

FIGURE 7.5 Typical lamination and armature before being wound

Commutator

As the armature spins past the magnetised field poles, an induced voltage will be produced in the armature conductors. This voltage will be an alternating value (a.c.), meaning that the EMF will alternate between a maximum positive and a maximum negative value. The commutator connects the armature with the circuit outside the machine. It is constructed in such a way that it converts the alternating current flowing within the armature into direct current flowing in the external circuit; without it, the machine will only work with an alternating current external circuit.

The commutator is placed on the shaft of the armature. The individual segments are insulated from each other, and from the armature shaft and the retaining rings, by mica insulation (see Figure 7.6(a)). Each pair of commutator segments is then connected to the ends of an armature coil. The many pairs of commutator segments (and thus armature coils) improve the d.c. machine's efficiency.

When assembled, the segments lie parallel to one another to form a cylinder. A sectional view of a commutator is illustrated in Figure 7.6(b). In exceedingly small commutators, the fixing of segments is achieved by clamping under pressure and rolling over or crimping the edges of the clamping rings.

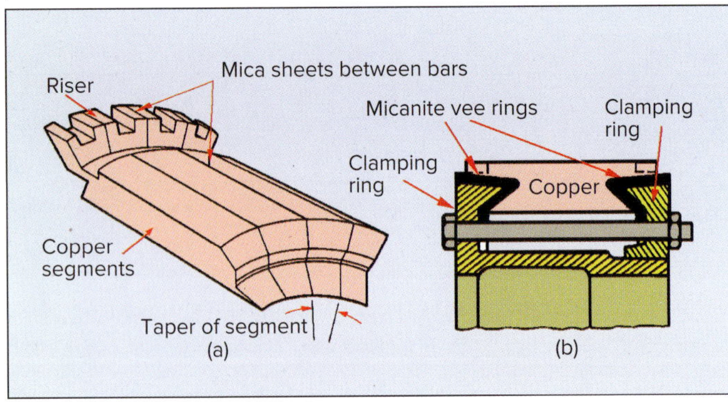

FIGURE 7.6 Commutator construction

The commutator may be pressed directly onto the shaft or bolted to the armature spider. Its segments connect to the external circuit via stationary brushes that push against the segments but have a sliding contact as the commutator rotates. The commutator acts as an automatic switch, swapping the current flow through the armature loops to the external circuit by exchanging the connections between the loop and the brushes.

The commutator is designed to produce a d.c. voltage output or accept a d.c. voltage input.

Brush gear and brushes

The brushes are used to transport the current into or out of the armature conductors, depending on whether the machine is a motor or a generator.

Graphite or carbon brushes, sometimes mixed with copper dust, are used to provide a low-resistance electrical connection between the armature windings and the external circuit. The brush makes a sliding contact with the commutator.

Figure 7.7 illustrates the brush gear for a four-pole machine. The carbon brushes must be able to move freely within the brush holder to ride up and down over any irregularities on the commutator surface, and to adjust for wear.

The brush is usually connected to the brush holder with a flexible copper lead called a 'pigtail' and is held down by a spring. The same connection terminal is used to connect to the external circuit.

The brushes must maintain contact with the spinning commutator. The brush gear must be suitably insulated from the frame, and in many cases the position of the brush around the commutator must also be adjustable. Figure 7.7 shows the brush mounting on an adjustable ring, which permits positioning of the brush gear.

FIGURE 7.7 Brush gear assembly of a d.c. machine
Sharomka/Shutterstock.com

7.1.2 Difference between a generator and a motor in terms of energy conversion

As a d.c. machine can be used as either a motor or a generator, the construction of each is identical. A d.c. machine operating as a generator will convert mechanical energy to electrical energy. A d.c. machine operating as a motor will convert electrical energy to mechanical energy.

7.1.3 Machine nameplates

Figure 7.8 shows a typical nameplate for a d.c. machine. It provides the manufacturer, year of manufacture, serial number, output power rating, rated speed, field excitation voltage, rated armature current and mass.

FIGURE 7.8 d.c. machine nameplate
Simon Dand

7.1.4 Using electrical equipment to take electrical measurements and comparison of readings with nameplate ratings

It is important to be able to quickly and safely measure the voltage, current and resistance of a d.c. machine when trying to identify operational issues. The nameplate in Figure 7.8 shows that the armature has a voltage of 495 V and a current of 29.9 A. The excitation field has 300 V and 2.18 A, which, when measured with an ohmmeter, would give a reading of 137 Ω. It should be possible to identify the armature and field windings and to measure the voltages across each. Current can be measured using a d.c. ammeter to verify correct operating values.

7.1.5 Identification of faults in a machine from electrical measurements

There are several tests for identifying a fault in a d.c. machine. First, ensure that the machine is safe to touch. The casing should be tested for voltage using a voltmeter from the frame to a known earth. The meter should be tested on a known live source after testing for voltage. Then a resistance test should be carried out from the frame to the known earth to ensure that the frame is earthed correctly.

The case should be opened and checked for voltage (+ve to −ve, +ve to earth). If there is voltage at the terminals and the machine is not working, the machine will have to be investigated. If there is no voltage, the issue is between the machine and the switchboard.

The supply should be correctly isolated and tagged, and the resistance of the field windings and the armature checked. As mentioned earlier, the nameplate in Figure 7.8 shows that the armature has a voltage of 495 V and a current of 29.9 A. The excitation field has 300 V and 2.18 A, which would give a reading of 137 Ω. An infinite reading would mean an open circuit; a smaller reading on the excitation field (perhaps 30 Ω) would indicate that there is a short-circuit in the windings. Other possible issues could include worn brushes not making contact with the commutator.

7.1.6 Care and maintenance processes for rotating machines

It is important to create a preventative maintenance program for rotating machines. The type of program and regularity of the maintenance will be determined by the time of use and general environment that the machines are operating in.

Some of the tasks carried out during these maintenance checks would be:

- visual inspection for condition, safety and fitness for purpose
- general clean of machine body, vents and screens
- check commutator and brushes
- test field coils and armature windings
- lubricate bearings.

7.1.7 Safety risks associated with using plant

Plant is a major cause of workplace death and injury. Risks associated with its unsafe use include:

- limbs being amputated by unguarded moving parts of machines
- being crushed by mobile plant
- fractures from falls while accessing, operating or maintaining plant
- electric shock from plant that is not adequately protected or isolated
- burns or scalds due to contact with hot surfaces or exposure to flames or hot fluids.

Other risks include hearing loss due to noise, and musculoskeletal disorders caused by manually handling or operating plant that is poorly designed.

Safe Work Australia has released a code of practice for managing the risks of plant in the workplace. It is available at www.safeworkaustralia.gov.au. The information in the code of practice covers plant installation, commissioning and use through to decommissioning and dismantling.

CHECK YOUR UNDERSTANDING

7.1 List all the major parts of a d.c. machine.

7.2 State the difference between a d.c. generator and a d.c. motor in terms of energy conversion.

7.3 What is the purpose of a commutator?

7.4 List the three types of field coils found in d.c. machines.

7.5 State some of the safety risks associated with rotating machinery.

7.2 Generators

7.2.1 Basic operating principle of a generator

A basic generator converts mechanical energy to electrical energy, and to generate a voltage there needs to be a magnetic flux, conductors and movement. The d.c. machine achieves this effectively and efficiently. When a conductor passes through a magnetic field, a voltage is induced into the moving conductor. This can be determined by the equation:

$$e = Blv$$

This is done in a generator by spinning the conductors through a magnetic field. This spinning conductor is placed into the armature.

7.2.2 Applying Fleming's right-hand rule for generators

The direction in which the induced voltage in a conductor causes current to flow can be determined by using Fleming's right-hand rule (see Figure 6.33 in Chapter 6).

When conductor end 'a' in Figure 7.9 is forced past the north pole, the right-hand rule for generators determines that the current in conductor end 'a' will be flowing towards the viewer. When conductor end 'b' is forced past the south pole, the current will be flowing away from the viewer.

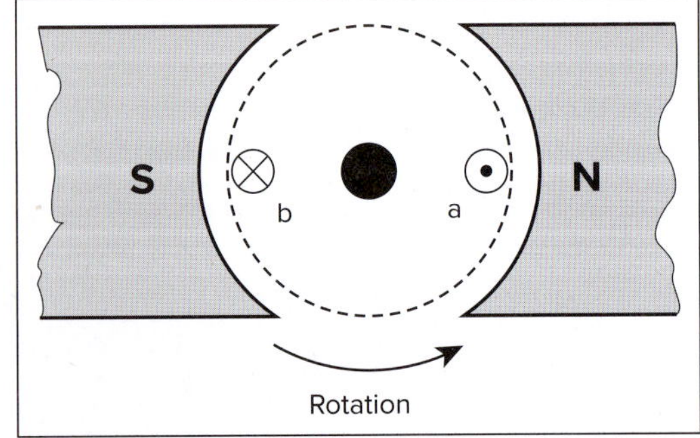

FIGURE 7.9 Generator rotation

7.2.3 Calculation of generated and terminal voltage of a d.c. shunt generator

The value of the voltage generated in an armature winding depends on the following three factors:

1. strength of field flux
2. number of effective armature conductors connected in series (influenced by the number of coils and the turns in each)
3. relative speed between conductors and magnetic field.

The generated voltage for a d.c. machine is found from:

$$E_g = \frac{p\Phi nZ}{60a}$$

where:

E_g = generated voltage
p = number of poles in the machine
Φ = magnetic flux per pole in webers
n = revolutions per minute (rpm)
Z = number of effective armature conductors
a = number of parallel paths in the armature.

A lap-wound armature has as many parallel paths as there are poles. A wave-wound armature always has two parallel paths.

7.2.4 Voltage regulation

Voltage regulation is the difference between the no-load and full-load values compared with the full-load value as a percentage. That is:

$$V_{reg}\% = \frac{V_{NL} - V_{FL}}{V_{FL}} \times \frac{100}{1}$$

where:

V_{NL} = generator terminal voltage at no load
V_{FL} = generator terminal voltage at full load.

7.2.4 Prime movers, energy sources and energy flow used to generate electricity

In order to generate a d.c. supply, the generator needs to be mechanically driven. There are many ways to provide the mechanical input. A prime mover (motor) connected to the shaft of a generator can be fuelled by diesel, gas or electricity. More sustainable methods are being developed to provide the drive for generators, with many smaller installations successfully using hydro, geothermal, tidal and wind technologies as a fuel source.

7.2.5 Types of d.c. generators and their applications

There are two main methods of voltage control for d.c. generators:

1. *Field flux control*
   The general concept is of resistance in series or parallel with field:
   (a) permanent magnet—shunt of soft iron to bypass flux
   (b) separately excited—resistance in series with field to limit current
   (c) shunt excited—resistance in series with field to limit current
   (d) series excited—resistance in parallel with field to bypass field
   (e) compound excited—resistance in shunt field to limit current.

2. *Speed control*
   Use is limited mostly to special cases, the concept being to adjust the speed of the prime mover.

Uses

1. Separately excited:
   (a) a permanent magnet—instruments (e.g. tachogenerator)
   (b) wound field—process control, rotary amplifiers
2. Shunt excited—small, cheap generators
3. Series excited—almost nil
4. Compound excited—all general-purpose work.

Excitation

1. Separately excited—permanent magnet or power from outside source
2. Shunt, series and compound—builds up own field; self-excited.

The factors affecting self-excitation of shunt, series and compound generators are:

1. residual magnetism needed
2. resultant field flux to assist residual magnetism
3. machine connected correctly
4. machine electrically in good order
5. direction of rotation.

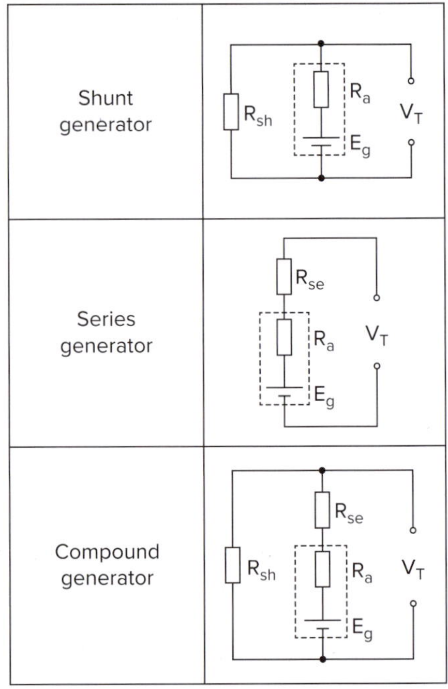

FIGURE 7.10 Equivalent circuits for shunt, series and compound generators

Figure 7.10 shows the equivalent circuits for shunt, series and compound generators. Figure 7.11 is a summary of the key points about all types of d.c. generators.

Type	Field winding	Circuit diagram	Voltage/speed	Voltage/load	Characteristics	Voltage control
Permanent magnet	–	G	Volts, Speed	Volts, Load	1. Volts directly proportional to speed 2. Output drops slightly on load	Use magnetic shunts on field system
Separately excited	Many turns fine wire	d.c. supply, G	Volts, Speed	Volts, Load	1. Volts directly proportional to speed 2. Output drops slightly on load	d.c. supply, G, Control field current
Shunt generator	Many turns fine wire	G	Volts, Speed	Volts, Load	1. Voltage is not linear with speed 2. Voltage drops more than permanent magnet on load	G, Control field current
Series generator	Few turns heavy wire	G	Volts, Speed, With a set load	Volts, Load	Rising voltage characteristics	G, Diverter
Cumulative compounding	Series assists shunt	G	Volts, Speed	Volts, over, flat, Load	Voltage characteristic can be made flat or increase with load, depending on compounding	G, Control shunt field
Differential compounding	Series demagnetises shunt	G	Volts, Speed	Volts, Load	Voltage drops quickly on load	G, Control shunt field

FIGURE 7.11 Key points for all types of d.c. generators

7.2.6 Methods of excitation used for d.c. generators

Separately excited generators

The term 'separately excited generator' indicates that the field windings have current passing through them from an external source; the current is not supplied by the generator itself. The field windings can have a few turns with a comparatively high current flowing or many turns with a much lower value of current, depending on the external source used.

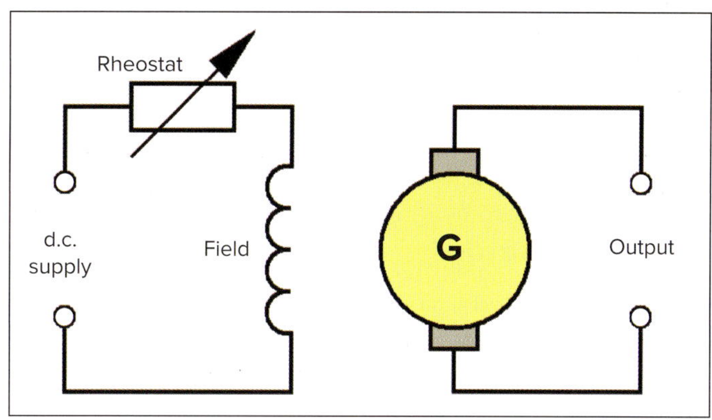

FIGURE 7.12 Separately excited generator circuit diagram

The connection diagram is shown in Figure 7.12, and the characteristics in Figure 7.13. As with the permanent-magnet generator, the terminal voltage versus speed characteristic of the separately excited generator is linear.

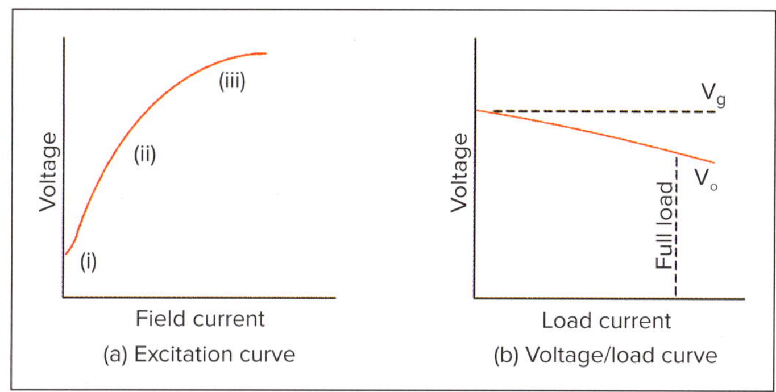

FIGURE 7.13 Separately excited wound field generator characteristics

The usual method of voltage control is by series resistance to regulate the current flowing through the field. A stronger field gives a higher voltage output at the same generator speed. The excitation curve showing the variation in generated voltage for varying values of field current is illustrated in Figure 7.13(a). At zero field current, there may be a small generated voltage due to residual magnetism in the field pole, in region (i). In region (ii), the curve is effectively linear, showing a steady increase in output voltage as the field current is increased. Once the saturation region (iii) is reached, large changes in field current are needed to produce small changes in voltage. This indicates that there is not enough iron in the generator for this level of excitation. The generator may be designed to work in either region (ii) or region (iii), according to the application.

Figure 7.13(b) illustrates the relationship between the generator terminal voltage and load current drawn from the generator. When the generator is on no load, there is no current flowing through the generator armature. Consequently, the full voltage generated by the armature can be measured at the generator terminals. However, when a load is connected and a closed circuit is created between the load and the armature, current will then flow through the armature and out to the load. The armature has resistance, and Ohm's Law tells us that when current flows through a resistance, voltage drop is created. Therefore, when current flows through the armature, a voltage drop is created across it. This is an internal voltage drop within the generator and is known as 'armature voltage drop'. The armature voltage drop is subtracted from the armature's generated voltage and the remaining generated voltage is available at the generator terminals. Figure 7.13(b) shows that this voltage drop is proportional to the load current and, therefore, as the load current increases to the generator's rated full-load current, the terminal voltage also falls to the generator's full-load terminal voltage. Most publications express armature voltage drop as the product of the armature current and the armature resistance. The following equation is used to calculate armature-generated voltage:

$$E_g = I_a R_a + V_T$$

where:

E_g = armature-generated voltage in volts
I_a = armature current in amps
R_a = armature resistance in ohms
V_T = terminal voltage in volts.

Open circuit characteristic (**OCC**) shows the relationship between generated EMF at no load and the field current at a given fixed speed. The OCC curve mimics the magnetisation curve and is similar for all types of generators. OCC curve data is obtained by operating the generator at no load and keeping a constant speed. It is also known as 'magnetic characteristic' or 'no-load saturation characteristic'.

The OCCs for a separately excited d.c. generator can be seen in Figure 7.13(a).

Shunt generators

In a *shunt generator*, the armature voltage is applied to the generator field that is connected in parallel with the armature. The connection is shown in Figure 7.14. The machine is said to be 'self-excited' because, when running, it does not rely on an external power source for field excitation. The shunt generator does, however, require a field from some source in order to start generating.

The open circuit characteristics for a shunt d.c. generator can be seen in Figure 7.15(a) which shows the relationship between field current and generated voltage. In the load characteristic curve shown in Figure 7.15(b), the decreasing voltage characteristic that is typical of the separately excited machine can still be seen, although in the shunt machine the voltage drop is more pronounced at full load. This is because the load is also in parallel with the field and a decreasing output voltage means a decreasing field current, which further decreases the output voltage.

If the generator is loaded to the point where the armature is short-circuited, then the field is also short-circuited because it is in parallel with the armature. The only magnetic flux left in the machine is that due to residual magnetism, and the result is a small generated voltage, which causes a circulating current to flow between the armature and the short-circuit. This is shown in Figure 7.15(b) as a value of current when there is no apparent voltage being generated.

The shunt generator can be safely run under short-circuit conditions, but it is the only type of generator that can.

Under the usual operating conditions, terminal voltage control is by means of a rheostat connected in series in the field circuit (see Figure 7.14). The practical range of voltage control is limited; the upper voltage limit is governed by field pole saturation while the lower voltage is determined by residual magnetism. Voltage control can also be achieved by speed control, but the method has limitations. These include maximum design speed, physical size and the need for a variable-speed drive source.

The uses for a shunt generator are generally limited to smaller, less costly machines or cases where the load and speed are practically constant. The shunt generator is very seldom used for larger machines.

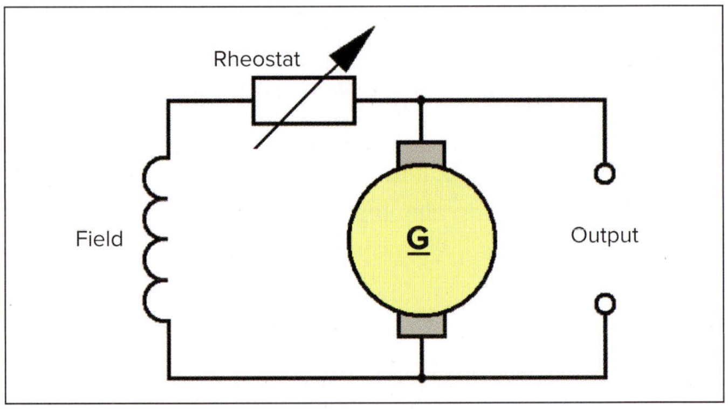

FIGURE 7.14 Shunt generator circuit diagram

Series generators

The *series generator*, being a self-excited machine, also requires certain conditions to be fulfilled before it can generate a satisfactory output voltage. These are identical to the requirements of the shunt generator in that there must be residual magnetism present, and the magnetism produced by the fields must assist the residual magnetism. As with the shunt machine, any loss of residual magnetism can be rectified by the application of a direct current to the fields for a short time.

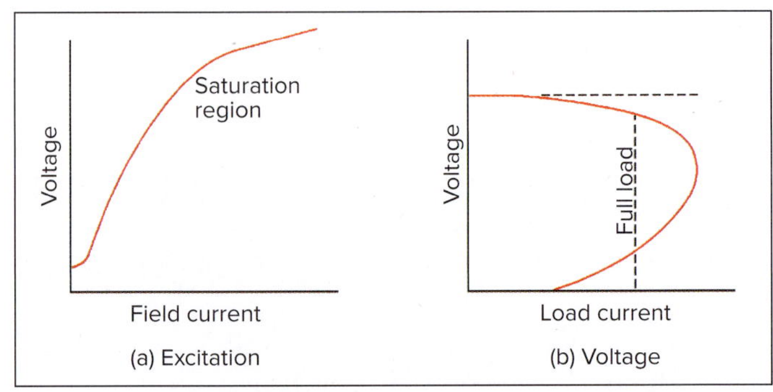

FIGURE 7.15 Characteristic curves for a shunt generator

The OCCs for a series d.c. generator can be seen in Figure 7.16(b).

In the series generator, the full load current also flows through the field winding. Consequently, a series field winding must have a low resistance so as not to restrict the load current or cause excessive terminal voltage drop. In the no-load condition, no current flows through the field; the only generated voltage is that from the existence of residual magnetism. As load is applied to the series generator, the field current and the magnetism in the field poles gradually increase. The increasing field strength gives rise to a generator voltage/load current characteristic that is also an excitation characteristic for the field magnetic circuit as shown in Figure 7.16(b). It has the same typical shape as the excitation curves for separately excited and shunt machines (see Figures 7.13(a) and 7.15(a)). Once the fields have saturated, any increase in load results in decreasing output voltage because the field flux remains almost constant, while losses in the machine continue to increase.

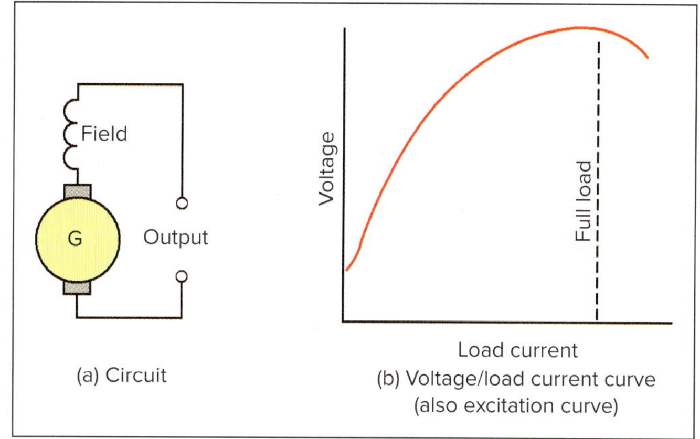

FIGURE 7.16 Series generator connections and load characteristics

Output voltage control can be achieved by speed variations within certain limits, but the most usual method is by connecting a resistor in parallel with the field to divert current around the field. When used in this manner, the resistor is called a 'diverter resistor'.

Compound generator

A *compound generator* has both shunt and series field windings. As with series and shunt generators, a compound generator begins to generate by residual magnetism. The field flux is generated in the voltage-driven shunt windings and the current-driven series windings. In the no-load condition, the primary operational field is the shunt field, and because of that the machine excitation curve is the same as that for any shunt machine (see Figure 7.15(a)). The series field, however, only becomes operational as the load increases. Therefore, the series field is useful for adding to the excitation as load is applied. The circuit for the compound generator is shown in Figure 7.17(a).

If the shunt field is connected across the armature as shown in Figure 7.17(a), the generator is called a 'short-shunt compound machine'; if connected across the output terminals, the generator is called a 'long-shunt compound machine'. From a theoretical viewpoint, there are minor differences in losses and voltages—but in practice there is almost no difference in performance, whether as a motor or a generator, and either connection can be used.

In the compound d.c. generator, the rising voltage characteristic of the series field is used to compensate for the falling voltage characteristic of the shunt field. When the series field overcompensates for the falling voltage characteristic of the shunt field, the terminal voltage rises as load is applied and the generator is said to be 'over-compounded' or 'over-compensated', as shown by the appropriate curve in Figure 7.17(b). An overcompensated generator is often used on long feeder lines, where line losses due to voltage drop result in a low terminal voltage at the load end at full load. The generator would be adjusted so the terminal voltage at the load end of the feeder lines is approximately constant from no load to full load. The final adjustment is by means of a diverter resistance adjusted to the level required on the actual installation. As with the shunt machine, initial voltage adjustment is with a series field resistor. It is sometimes combined with a measure of speed control.

The under-compounded generator is seldom used because the prime function of a generator is to supply varying loads at a constant voltage. The level-compounded generator, sometimes called the 'critically compounded' generator, is often used as a power source for installations where the machines it supplies are close to the generator, thereby eliminating voltage-drop problems. Examples are shipboard installations, machinery where a high degree of control is necessary (as with rolling mills), large, automated planers and multiple-lift installations where a converter is needed.

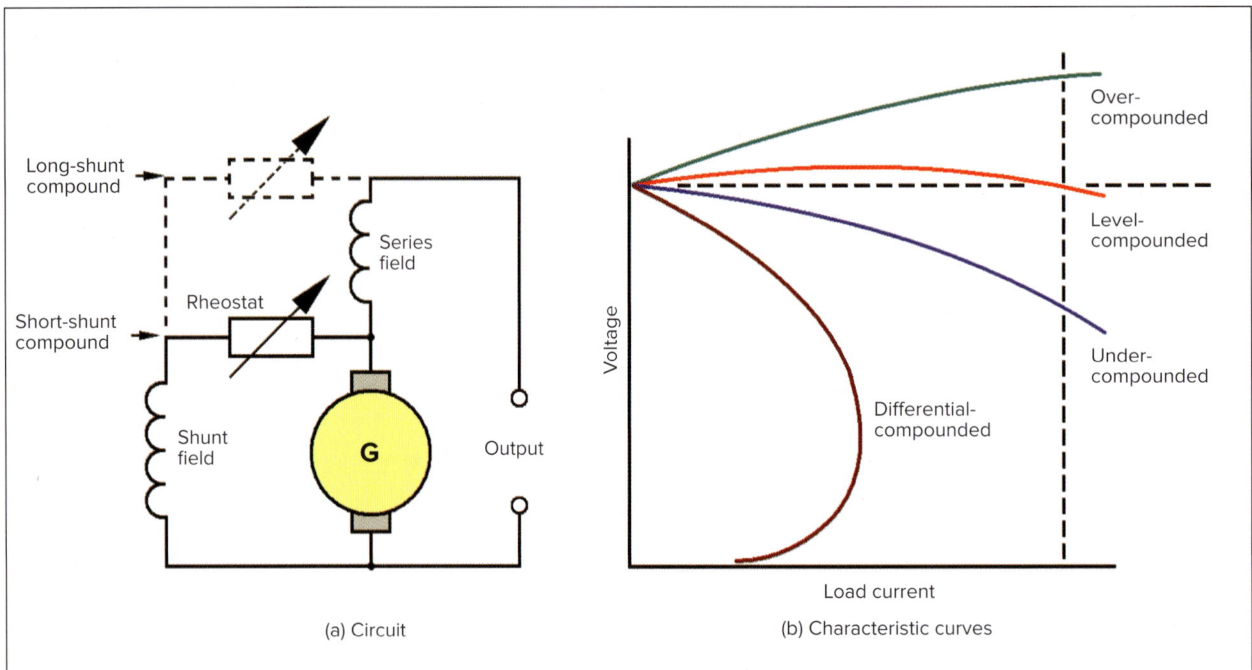

FIGURE 7.17 **Compound generator connections and load characteristic curves**

A series field that opposes a shunt field can be used to create a differentially compounded machine. The characteristic of the differentially compounded generator is to drop terminal voltage dramatically under load. Figure 7.17(b) also shows the characteristics for a differentially compounded machine. The differentially compounded connection is seldom used but often occurs when one of the field windings is mistakenly connected in reverse after repairs. Differentially compounded generators have been used in the past as electric welding machines (with varying degrees of success).

7.2.7 Importance of residual magnetism for a self-excited generator

Self-excited generators rely on a magnetic field supplied by the magnetism that remains in the iron field poles from the last time the generator was operated. It is a result of the hysteresis of iron and is known as 'residual magnetism'. When the generator armature conductors start to cut this residual flux, a small voltage is generated, causing a current to flow in the field windings.

If the residual magnetic field has the same polarity as the field generated by the field poles, the amount of magnetism increases, thus increasing the field—and so on as the generator comes up to full speed and full excitation.

If the residual magnetism opposes the field flux, the two will cancel and no energy will be generated. If the two magnetic components oppose each other, the machine cannot build up a voltage.

Self-excitation therefore only occurs when:

1. a residual field exists
2. the residual field acts in the same magnetic polarity as the wound field
3. the generator is rotated in the correct direction for the excitation to generate a supporting field.

If any of these three factors are missing or are of incorrect polarity, the machine will not generate a voltage. If these three factors are correct, the small current circulating in the field winding provides a strengthening of the magnetic flux. This leads to an increased generated voltage, more field current, more magnetic flux and so on until the field current is high enough to provide a magnetic flux great enough to send the field poles into saturation. When this is achieved, the generated voltage stabilises at some design value. Control of output voltage is achieved by a field resistance, as with the separately excited machine.

7.2.8 Reversing the polarity of a d.c. generator

To reverse the polarity of a generator, reverse either the field or armature leads, but not both.

7.2.9 Connecting and testing a d.c. generator on no load and load

To ensure proper operation, d.c. generators should undergo both the open circuit test and the load test. The open circuit test determines the OCC or magnetising characteristics by giving the MMF and thus the excitation or field current necessary to produce the required voltage on no load at a fixed speed. The OCC curve shows the variation of induced EMF as a function of field current at constant speed and zero load current.

The load test should be carried out to establish the rating of a d.c. generator. Some energy is lost while running a machine, converting into heat and other losses. If too much heat is produced it can affect the operation of the machine and eventually lead to it breaking down. The load must be tested to ensure it is able to operate safely within the temperature limit.

7.2.10 Identifying safety risks associated with using generators

As mentioned in section 7.1.7, Safe Work Australia has released a code of practice for managing risks of plant in the workplace. It is important to be familiar with this document prior to working on rotating d.c. machines.

CHECK YOUR UNDERSTANDING

7.6 List the different types of d.c. generators.

7.7 Why does a d.c. generator's voltage decrease as the generator approaches full load current?

7.8 How is the polarity of a d.c. generator reversed?

7.9 What is the difference between a long-shunt and a short-shunt compound generator?

7.3 Motors

7.3.1 Basic operating principle of a motor

Motors and generators have the same basic construction and a standard machine can be used as either.

For a motor to be useful, it must develop **torque** or a turning effect. When electrical energy is applied to a d.c. machine, current flows in the armature conductors and produces a magnetic field that interacts with the main magnetic field, causing a rotational torque to be developed (see Figure 7.18).

When the two magnetic fields interact, an attractive (axial) force is created on one side of the conductors and a repulsive (radial) force is created on the other side. The resulting magnetic field produces a force that acts on the conductors, as indicated in Figure 7.18.

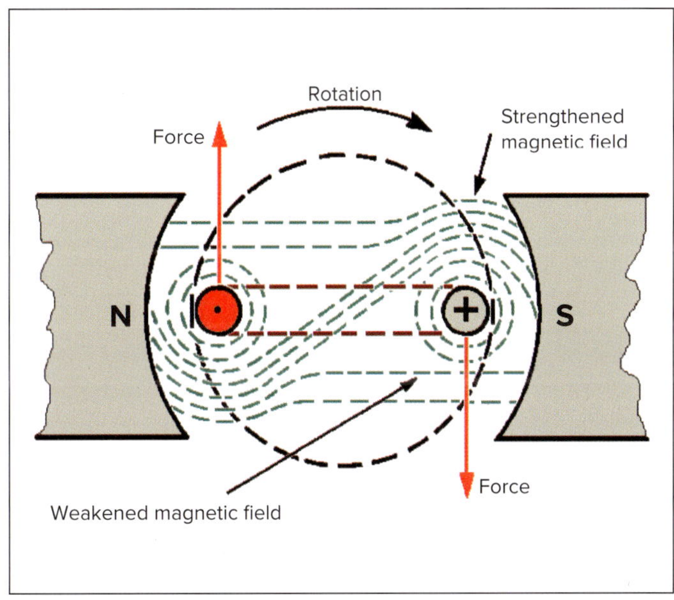

FIGURE 7.18 Motor effect produced by an electric current

7.3.2 Applying Fleming's left-hand rule for motors

The direction in which the force acts can be determined by using Fleming's left-hand rule (see Figure 7.19). This works in a similar way to the right-hand rule for obtaining the direction of an induced voltage in a conductor.

The rule works as follows:

1. Arrange the thumb, first and centre fingers of the left hand at right angles to each other.
2. Point the first finger in the direction in which the lines of force (flux) are acting.
3. Point the centre finger in the direction in which the current is flowing.
4. The thumb will then point in the direction in which the force is acting on the conductor.

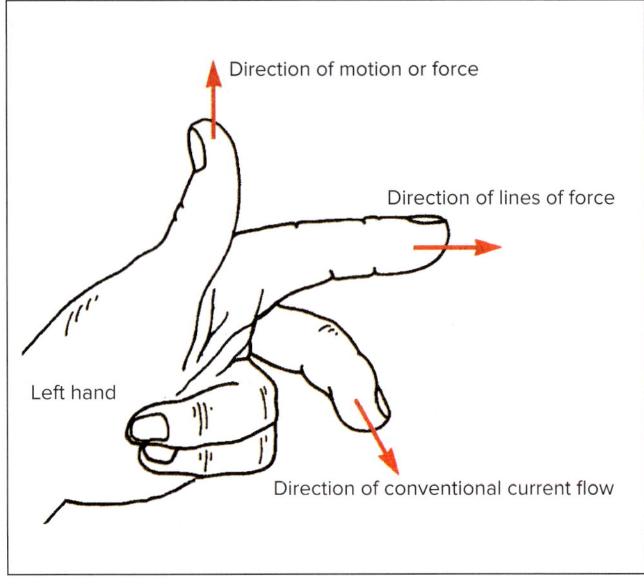

FIGURE 7.19 **Fleming's left-hand rule**

The left-hand 'conductor' in Figure 7.18 has a tendency to be forced upwards, and the right-hand conductor downwards. As these are normally embedded in an armature core at a fixed distance from the centre of rotation, the effect is to create a turning movement, or torque. In a practical d.c. motor, there is a large number of conductors and the torque produced is sufficient to drive the load connected to the motor, unless it is severely overloaded.

7.3.3 Calculation of force and torque developed by a motor

The magnitude of the thrust exerted on a conductor carrying a current when located in a magnetic field is given by the equation:

$$F = BlI$$

where:

F = force on conductors in newtons (N)
B = flux density of main field in tesla (T)
l = length of the conductor in the field in metres (m)
I = current in the conductor in amperes (A).

In a practical case, where an armature has many conductors and also may have several paths, the equation becomes:

$$F = \frac{BlIZ}{a}$$

where:

I = total armature current
Z = number of armature conductors
a = number of parallel paths in the armature.

Since T = Fr where r is the radius of rotation of the armature conductors, the equation becomes:

$$T = \frac{BlIZr}{a}$$

where:

T = torque in newton metres.

The equation can be further developed to:

$$T = \frac{p\Phi IZ}{2\pi a}$$

where:

T = torque in newton meters (Nm)

p = number of poles

Φ = flux per pole in webers (Wb)

I = total armature current in amperes (A)

Z = number of armature conductors

a = number of parallel paths in the armature.

7.3.4 Operation of a motor and its energy flow

A direct current machine is one that either produces or consumes electricity as direct current. The machine therefore acts as an energy converter. The same d.c. machine can be either a motor or a generator, although the design is generally optimised as one or the other—when electricity is applied to the machine, it can function as a motor and convert electrical energy to mechanical energy, causing the machine to rotate.

A d.c. motor converts electrical energy to mechanical energy.

7.3.5 Effect of back EMF in d.c. motors

When the armature of a d.c. motor rotates under the influence of the driving torque, the armature conductors move through the magnetic field and EMF is induced in them, just as in a generator. According to Lenz's Law, the induced EMF acts in the opposite direction to the applied voltage (V) and is known as 'back EMF' or 'counter EMF'. It is always smaller than the applied voltage.

7.3.6 Types of d.c. motors and their applications

The most common d.c. motor of any type is the permanent-magnet electric motor. In this motor, the field has no winding but consists of a permanent magnet. Everyday use of the permanent-magnet motor is in small applications such as the vibrating motor in mobile phones, cordless tools, toys, radio-controlled aeroplanes and boats. This category of both motors and generators was once reserved for smaller machines, but the present stage of development in permanent magnets allows the construction of much larger motors with permanent magnet fields.

FIGURE 7.20 View of the Alnico magnets of a permanent-magnet motor
Bosca78/E+/Getty Images Plus

As they have no wound field, there is a growing tendency towards a reduced unit cost and an increase in efficiency. Motors up to approximately 7.5 kW use ceramic magnets. While highly resistant to demagnetising, they have a relatively low flux level and are therefore limited in application and size.

For larger motors, Alnico magnets are used and the motor is easily adaptable to extreme applications such as in steel mills (furnace electrode drives and live table drives, for example).

Normally, the magnet is moulded and set in the motor frame and used for low-speed applications (such as machine tools). The magnets can be seen in Figure 7.20. These magnets have been fixed in position by a high-temperature bonding agent.

Speed control is traditionally achieved by electronically varying the voltage applied to the armature, which is a very effective method. Torque is comparatively linear throughout the normal load range. Figure 7.21 shows characteristic curves for a permanent-magnet motor.

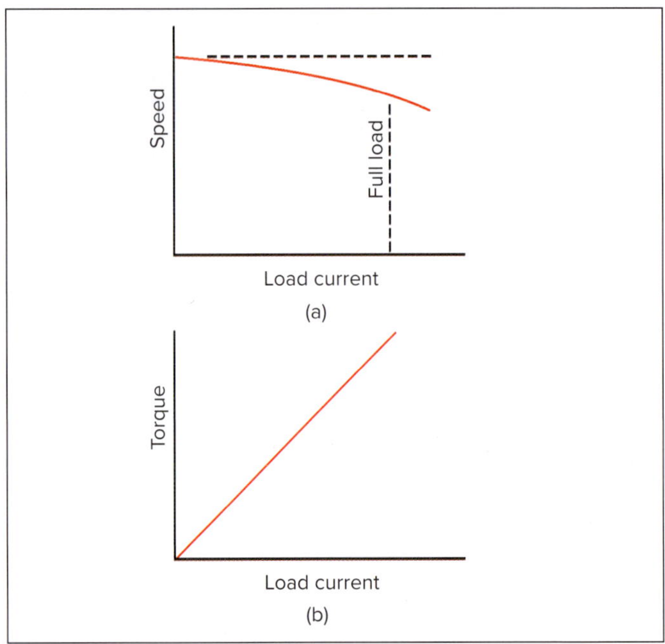

FIGURE 7.21 Characteristics for a permanent-magnet motor

Printed circuit motors

Printed circuit motors are a variation in construction of the permanent-magnet motor. The armature itself has no iron in its construction and is consequently an air-cored armature. Figure 7.22 shows the construction of the motor in its simplest form. Several circular magnets are fastened to a casting that forms the basis of field support and motor end-shield in one piece.

The motor end has provisions to mount a bearing so that the armature can be maintained in its correct position within the air gap. The second end-shield consists of magnetic material to concentrate the magnetic paths within the motor and provides brush and bearing mounting facilities. For increased torque, motors may have a set of magnets each side of the armature, and the material used in end-shield construction may be non-magnetic.

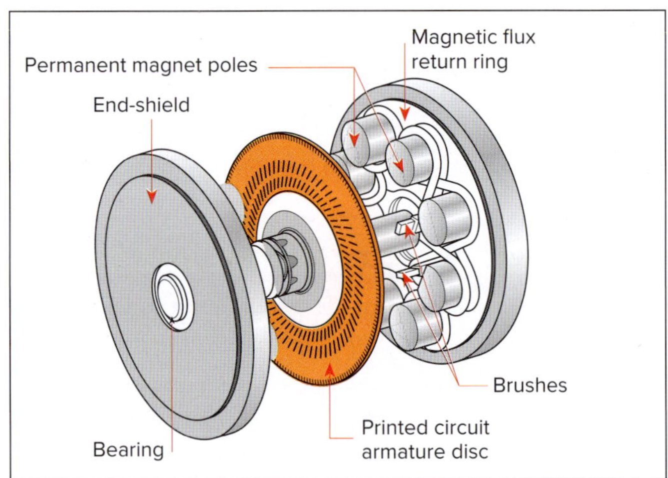

FIGURE 7.22 General construction features of a printed circuit motor

The armature coils are created by an etching process on a non-conducting substrate. Often of Bakelite or fibreglass, the base material is initially coated with a film of pure copper deposited on it by an electroplating process and may be single or double sided. Where an armature coil might need to consist of more than one turn, double-sided material may be used. The conductor shapes are outlined on the copper and the material in between is removed by photo-etching.

Printed circuit motors are light-weight, have a short shaft length and run at low speed and low voltage. Because of the armature construction where no iron is used, output torque is limited and efficiency is low. Commutation problems are never encountered because of low coil inductance.

Separately excited motors

The wound-field separately excited motor has no limitations in size. Like the separately excited generator, the motor is used mainly in process-control work. This type of field connection is adaptable to a wide range of speed control, and the torque is also linear with respect to the applied load.

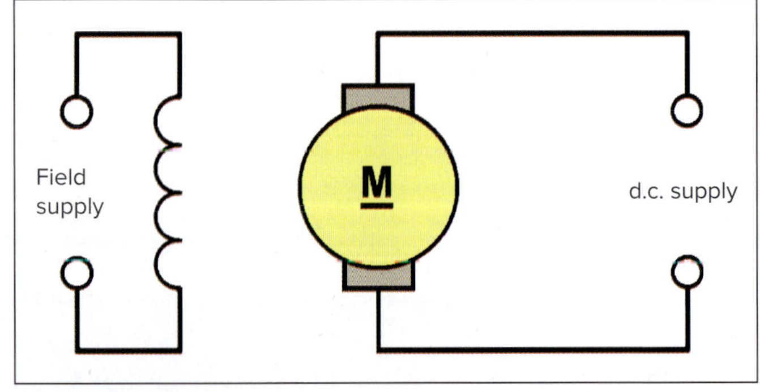

FIGURE 7.23 Circuit for a separately excited wound-field motor

Figure 7.23 shows the circuit of the wound-field type and can be compared with the generator circuit shown in Figure 7.12.

As the machine has similar characteristics to the permanent-magnet motor, its uses are also similar. It has the advantage of being used in other circuits, such as *Ward-Leonard control systems* and rotary amplifiers, where a small change in field current can be made to cause a large change in speed. A special construction of the separately excited wound-field motor makes it suitable for positioning work, as in long-distance readings of anemometers and engine speed governors.

The normal or rated speed of a separately excited d.c. motor is obtained with rated voltage applied to both the field and the armature. This speed is termed the 'base speed' of the motor. If the voltage applied to the armature is reduced, the motor will slow down. So, for speeds below base speed, armature voltage control is used.

If the voltage applied to the field is reduced, with rated voltage on the armature, the motor speed will increase. This method is called 'field weakening' and is used to give speeds above base speed.

The field-current control method is normally preferred, due to lower currents in the control device resulting in less electrical power wastage. However, modern electronic methods of armature voltage control are very efficient.

When a motor armature rotates in its magnetic field, a voltage is generated that opposes the applied voltage. The current flowing in the armature is, based on Ohm's Law, proportional to the difference in these two voltages divided by the armature resistance.

If a rheostat is connected in series with the field, any reduction in resistance results in an increase in field current and, in turn, an increased field strength. At a constant armature speed, an increase in field strength leads to an increased generated voltage (back EMF), which tends to reduce the armature current, and the motor therefore produces less torque. Consequently, the motor slows down as the armature current increases and it stabilises at a lower speed.

Conversely, a decrease in field current leads to an increase in speed. Therefore, precautions must be taken to ensure that the field current does not decrease below a certain level that permits excess armature currents and dangerous speeds to result.

Shunt motors

The shunt-field connection is commonly used in smaller-sized motors. In larger-sized motors, the shunt connection is found less frequently, but is still used because of its constant speed characteristic. Of the several types of motor connections, the shunt-excited motor has the best speed regulation throughout the normal speed range. Like the shunt generator, the shunt-field of the motor is connected in parallel with the armature and the motor speed can be controlled by a series resistor regulating the current flow through the field. The basic motor circuit is shown in Figure 7.24, together with the characteristic curves for speed and torque.

Under normal operating conditions, the speed of the motor is set with the field rheostat. Decreasing the resistance value increases the field current and causes the same sequence of events with speed adjustment for the separately excited machine. Below normal speed operating ranges, the voltage to the armature is varied with a series armature

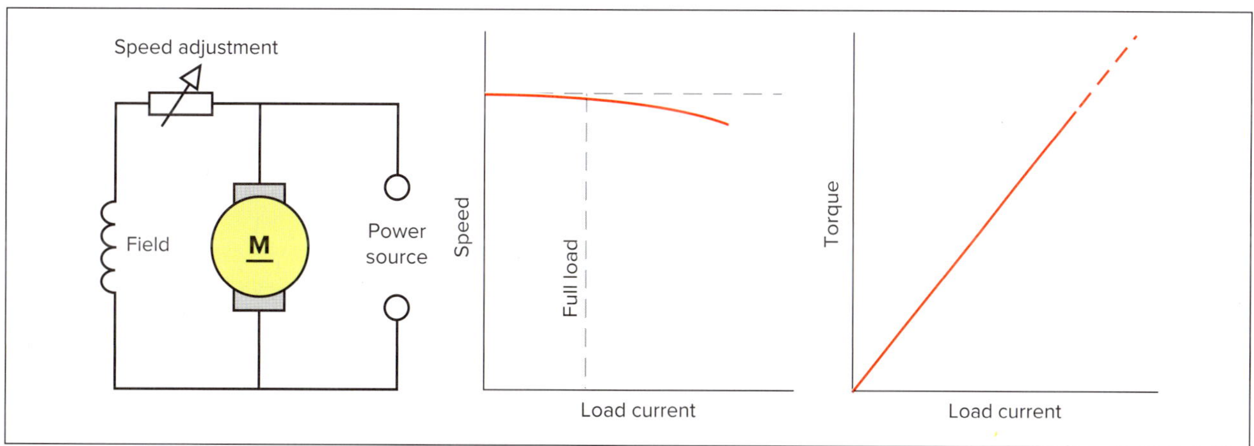

FIGURE 7.24 Shunt motor characteristics

resistor to give speed variations. In a similar way to the separately-excited motor, precautions must be taken to ensure that the field current is never reduced below a certain value in order to prevent excessive armature speeds and current.

Reversal of rotation is by the same method as the separately-excited machine in that either the armature or the field current flow is reversed—but not both.

Series motors

In series motors, the field and armature are in series with the supply in a circuit similar to the series generator. The circuit and characteristics are shown in Figure 7.25. The series motor is subject to wide changes in speed as its load is varied because of the changing field current. With a series motor on full load, both the armature and field current are at comparatively high values.

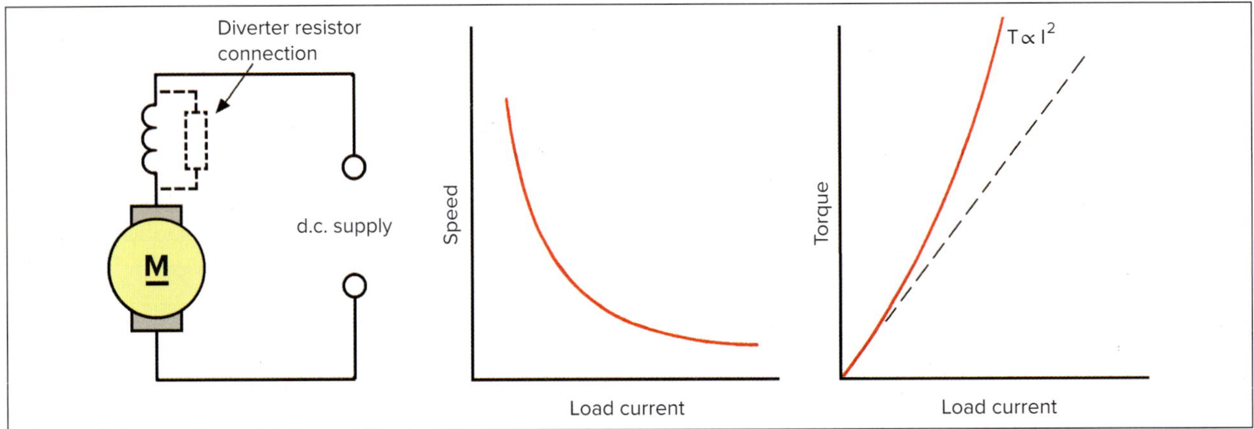

FIGURE 7.25 Series motor characteristics

As explained previously, increasing the field current of a motor reduces its speed and decreasing the field current increases its speed. Thus, with full load (and field) current, the speed of a series motor is low; and, as the mechanical load is removed from the motor, the armature current (also the field current) is reduced. When the magnetic field becomes weaker, the motor speeds up.

With larger-sized series motors, speeds can be attained under no-load conditions that are sufficiently high to cause damage to them. A normal precaution is to have a minimum load permanently connected by direct coupling or similar means to prevent the possibility of removing all the mechanical load.

The torque characteristic of a series motor is non-linear because, as the load applied to the motor increases and the motor slows down, both armature and field current increase together.

Torque is:

$$T = \frac{p\Phi IZ}{2\pi\alpha}$$

For any one machine, the number of poles 'p', the number of conductors 'Z', the number of parallel paths 'a' and the value of 2π will remain constant—so the equation can be written in the form $T \propto \Phi I$. Since Φ, the field flux, is proportional to the armature (field) current, $T \propto I^2$ for a series motor.

Inspection of the characteristic curves shows that an armature current increase is associated with a decrease in speed and an increase in torque. This is the big advantage of the series motor (and its most common application): starting against heavy loads.

Typical uses are traction motors in electric trains, cranes, anchor winches, lifts and—the most common application—as the starter motors for motor vehicles.

To reverse the direction of rotation, reverse either the field or armature leads—but not both. Speed-control methods are not applicable to series motors because of their inherent wide range of speeds, but at any one load and voltage the speed can be increased with the use of a diverter resistor to bypass some of the armature current around the field.

Compound motors

The compound motor is the general workhorse among d.c. motors. It is especially suited to loads that require a reasonably high degree of starting torque but do not need a series motor. The shunt winding allows the motor to run at exceptionally light loads, a factor that gives it an advantage over the series motor.

In the cumulatively compounded connection, the series and shunt windings assist one another, thereby increasing magnetic field strength. This is the connection normally used. As with compound generators, compounding can be under-compounded, level-compounded, over-compounded or differentially compounded. Differential compounding is possible but has little practical use. If a differentially compounded motor is loaded beyond a certain point, its current increases rapidly and the motor abruptly changes its direction of rotation. This can create a dangerous situation as, depending on the size and torque of the motor, a sudden reversal can cause damage to couplings and gearboxes or even twist off a motor shaft.

The cumulatively compounded motor combines the characteristics of both the series and shunt motors. Its speed regulation is not as good as in the shunt motor but is superior to that of the series motor. While the torque of the shunt motor is approximately linear, the torque of the compound motor increases more rapidly, but at the cost of some loss in speed. Its torque, however, is less than that of the series motor.

Applications include punches, shears, rolling mills and drive motors for machines subject to sudden or shock loads and reversals such as large metal-planing machines. Characteristics of the compound motor are shown in Figure 7.26.

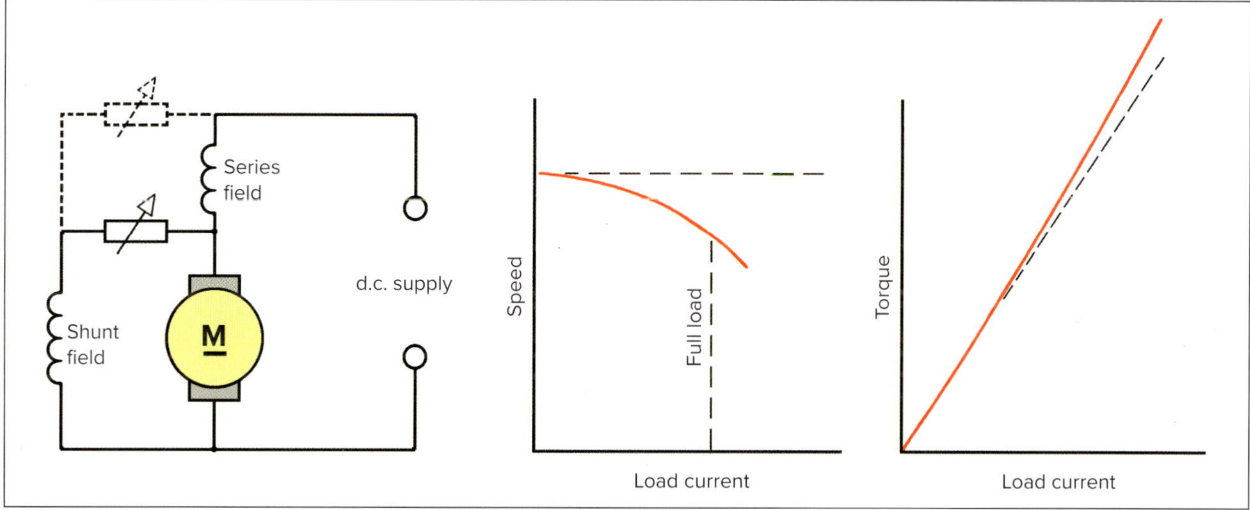

FIGURE 7.26 Compound motor characteristics

With large traction loads such as diesel electric drives in ships, and particularly diesel electric locomotives with multiple bogie drives, the basic drive motor is referred to as a 'series motor'. (In fact, it is probably a compound motor with the shunt field open-circuited during starting.) Several series motors may also be connected in series with one another. As the load becomes mobile, it is normal practice to reconnect the series motors in parallel and then to convert the series motor connection to a shunt or a cumulative compound type.

While a series motor (or a shunt motor) is applicable to certain jobs, the motor can be electrically reconnected while still rotating to suit conditions that have changed since the job started. Similarly, a d.c. motor in motion can be reconnected to behave as a generator supplying a resistive load to form part of a braking sequence to reduce wear on brake shoes (e.g. in an electric train travelling downhill).

One form of emergency braking for electric trains is to reverse the electric motors and apply full voltage. When reversing the cumulatively compounded motor, it must be remembered that there are two field windings. As is often the case, the machine might have interpoles as well and the reversing must be handled more carefully. Both the shunt and series windings must be reversed together—but not the armature or the interpoles. Reversing either the shunt or series windings on their own merely changes the machine's connection from cumulative to differential compounding.

Figure 7.27 provides a summary of the key characteristics of all types of d.c. motors, including the circuit diagrams for each type. The circuits are identified by the excitation used by the machine.

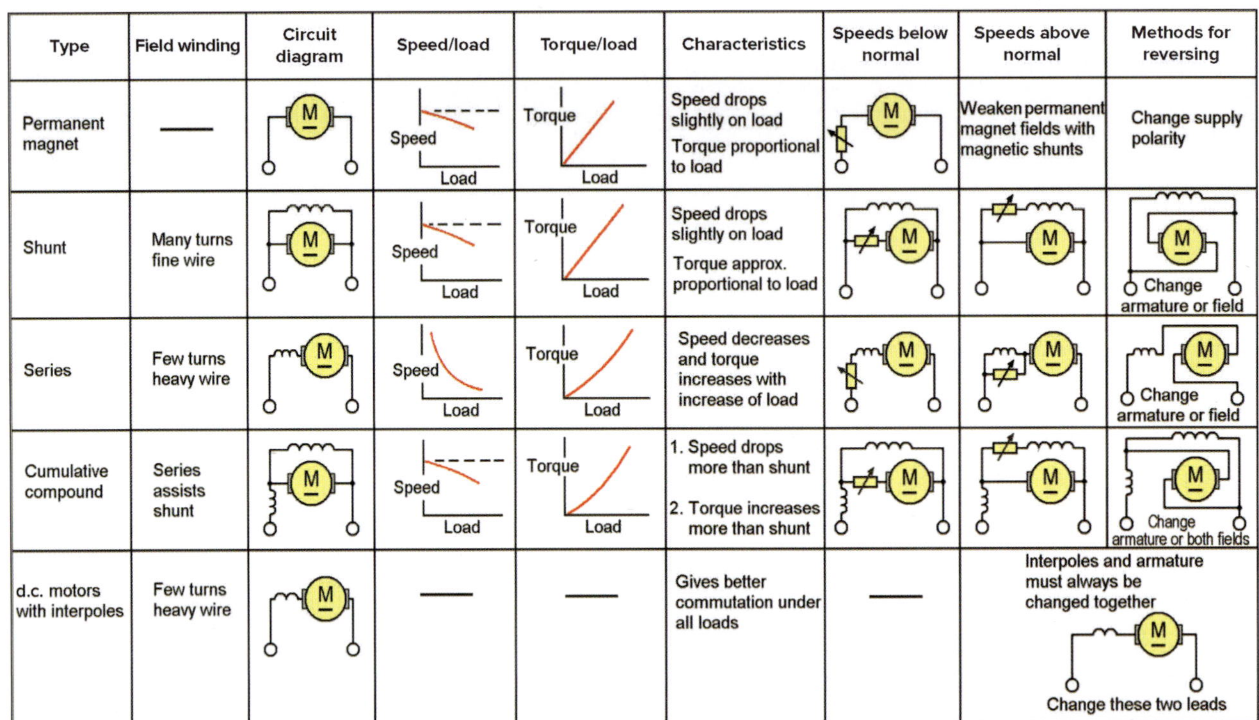

Type	Field winding	Circuit diagram	Speed/load	Torque/load	Characteristics	Speeds below normal	Speeds above normal	Methods for reversing
Permanent magnet	—	M	Speed, Load	Torque, Load	Speed drops slightly on load Torque proportional to load	M	Weaken permanent magnet fields with magnetic shunts	Change supply polarity
Shunt	Many turns fine wire	M	Speed, Load	Torque, Load	Speed drops slightly on load Torque approx. proportional to load	M	M	M Change armature or field
Series	Few turns heavy wire	M	Speed, Load	Torque, Load	Speed decreases and torque increases with increase of load	M	M	M Change armature or field
Cumulative compound	Series assists shunt	M	Speed, Load	Torque, Load	1. Speed drops more than shunt 2. Torque increases more than shunt	M	M	M Change armature or both fields
d.c. motors with interpoles	Few turns heavy wire	M	—	—	Gives better commutation under all loads	—	Interpoles and armature must always be changed together M Change these two leads	

FIGURE 7.27 Summary of the characteristics of d.c. motors

7.3.7 Calculation of power output of a motor

The output power of a motor is based on the machine's mechanical performance. It can be calculated using:

$$P = \frac{2\pi NT}{60}$$

where:

P = output power in watts (W)
N = speed in revolutions per minute (rpm)
T = torque in newton-metres (Nm).

7.3.8 Connecting and testing a d.c. shunt motor on no load and load

The connection and testing of a d.c. motor on load or no load is conducted in the same way as it is with d.c. generators (as described in section 7.2.9).

7.3.9 Reversing the direction of rotation of a d.c. motor

Motor reversal is achieved by altering the polarity of the field windings or the armature. The basic principle of reversal of rotation for a d.c. motor requires that the direction of current flow through either the field or the armature windings be reversed—but not both. Reversing the polarity of both results in no change in rotation, as the direction of current flow in both the armature and field windings is reversed.

Figure 7.28 illustrates one method of reversal of rotation. In both diagrams, the field polarity is unchanged while the direction of current flow through the armature is changed (see Figure 7.28(b)). The torque created is equal in both cases, but the resultant direction of rotation has changed.

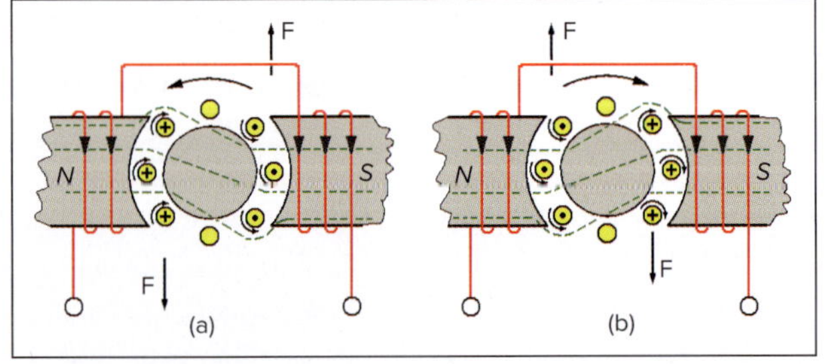

FIGURE 7.28 Reversal of rotation of a d.c. motor

7.3.10 Safety risks associated with using motors

It is important to be familiar with Safe Work Australia's code of practice relating to managing risks of plant in the workplace prior to working on rotating d.c. machines.

CHECK YOUR UNDERSTANDING

7.10 List the five basic types of d.c. motors.

7.11 How is the direction of a d.c. motor reversed?

7.12 How is torque developed in a d.c. motor?

7.13 What is the back EMF developed in a motor and how is it useful?

7.14 How is the output power for a d.c. motor calculated?

7.4 Machine efficiency

7.4.1 Losses that occur in a d.c. machine

Losses in machines include friction, windage and electrical losses. Friction is present in all rotating machinery. Windage is present because of air resistance to rotating components and in fans added to ensure forced circulation of air for cooling purposes. In electrical machines, the term 'losses' comprises copper losses, iron losses, magnetic leakage and other, lesser losses. Together, the losses represent wasted energy, which should be reduced as much as possible, often by simple good maintenance.

Copper power losses are due to the resistance of electrical windings, while iron power losses are caused by hysteresis and eddy currents in the iron core of the armature. While the iron loss is almost constant from no load to full load, the copper loss varies considerably due to load current. These two are the main electrical losses in a motor and are added to obtain the total electric power loss. The power loss in copper conductors varies as the square of the current flowing ($P = I^2R$). At light loads, the small current flow means the copper loss is at a minimum. If the armature current is doubled, the copper loss becomes four times as great, and four times as much heat is generated; this heat has to be removed, usually by air circulation, which adds a further loss to the system.

7.4.2 Methods used to determine the losses in a d.c. machine

For the purpose of analysis, it is usual to assume that all the armature resistance is concentrated into one component and not distributed throughout the windings. Figure 7.29 shows a shunt-connected generator separated into its various conceptualised parts, while the broken lines indicate actual components.

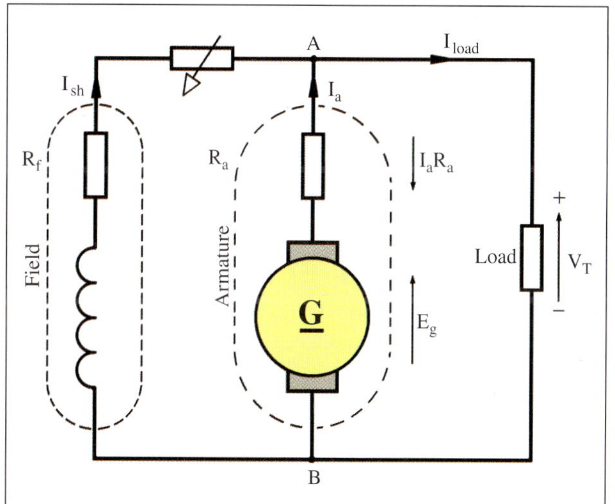

FIGURE 7.29 Equivalent circuit of a shunt generator

7.4.3 Calculation of losses and efficiency of a d.c. machine

If the designed generated voltage of the generator in Figure 7.29 is 200 V, and the armature has a resistance of 0.5 Ω, for every ampere of current being supplied by the armature there is an internal voltage drop of 0.5 V due to the armature resistance.

For every 2 A of load current, 1 V will be dropped internally; and if 200 V is required at the generator terminals, the generating section will have to generate a higher voltage in the windings. That is, for a 10 A load, the generated voltage will have to be 205 V to give a terminal voltage of 200 V between points A and B.

The armature current I_a will also include the field current I_f as well as the load current I_{load}, that is:

$$I_a = I_f + I_{load}$$

The voltage drop due to internal resistance is equal to I_aR_a (V = IR) and the generated voltage Eg is equal to the terminal voltage V_T plus the I_aR_a voltage drop. So for a generator:

$$E_g = I_aR_a + V_T$$

With a series field winding, the resistance of the field must be added to the armature resistance.

The theoretical approach to the efficiency of a motor is similar to the generator method. Armature resistance is considered as one component and the winding as another. This is shown in Figure 7.30, and the similarity to Figure 7.29 is evident.

The back EMF generated is subject to the armature voltage drop (I_aR_a) and is equal to the difference between the supply voltage and the I_aR_a voltage drop. That is, for a motor:

$$V_T = I_aR_a + E_g$$

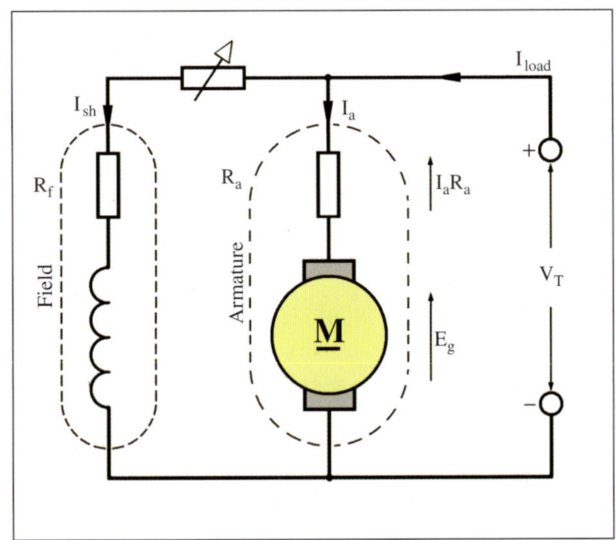

FIGURE 7.30 Equivalent circuit of a shunt motor

The only variation from the equation for generators is the polarity of the armature voltage drop (I_aR_a). This is illustrated in Figure 7.30, with arrows indicating the directions of the applied and generated voltages and showing them opposing each other. The result is that the effective voltage causing a current to flow through the armature circuit is smaller than the applied voltage. That is:

$$I_a = \frac{V_T - E_g}{R_a}$$

The overall efficiency of a motor can be found in a similar manner to that of a generator, that is:

$$P_{input} = P_{output} + P_{losses}$$

Whereas the input to a generator is mechanical power and the output electrical power, the input to a motor is electrical power and the output mechanical power. The losses are shown in Figure 7.31.

7.4.4 Efficiency characteristics of a d.c. machine and the conditions for maximum efficiency

The efficiency of a d.c. machine will change with a change of load and will always be at a maximum when the variable losses equal the constant losses.

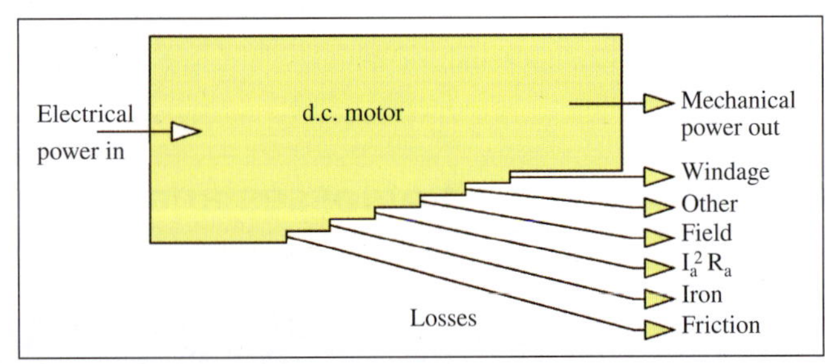

FIGURE 7.31 Losses in a d.c. motor

7.4.5 Application of Minimum Energy Performance Standards (MEPS)

Minimum Energy Performance Standards (MEPS) are used for all three-phase **squirrel cage motors** over 0.75 kW. MEPS specify the minimum level of energy performance that appliances, lighting and electrical equipment (products) must meet or exceed before they can be offered for sale or used for commercial purposes. It should be noted, however, that d.c. motors are not currently subject to MEPS requirements.

7.4.6 Methods used to maintain high efficiency

Machine efficiency is affected by several losses. In order to improve or maintain the efficiency of a d.c. machine, it is important to understand what the losses are and their impact on the machine.

Heat plays a role in poor efficiency. The frame is designed to assist in the transference of heat via forced movement of air. The fan design will affect the cooling of the machine. Ensuring that the vents are not blocked or up against a barrier will allow the air to move freely.

Copper losses in the form of electrical resistance will contribute up to 60% of the losses in a machine. The design of the field system and sizing of field windings needs to take into consideration the impact that I^2R losses have on the overall efficiency of the machine.

Approximately 20% of the losses are iron (magnetic) losses. Hysteresis and eddy current losses can be reduced by using high-grade iron and steel containing small amounts of silicon to be used in laminations.

Mechanical losses can be minimised by preventative maintenance. Check bearings and brushes regularly and ensure that the machines are free of grit and dirt.

CHECK YOUR UNDERSTANDING

7.15 How is the efficiency of a d.c. motor calculated?

7.16 What are the losses in a d.c. motor?

7.17 How do Minimum Energy Performance Standards (MEPS) apply to d.c. motors?

SUMMARY

- The basic construction of a d.c. machine comprises the following components:
  - field frame or yoke
  - end-shields and bearings
  - field poles
  - field coils
  - armature and commutator
  - brush gear and brushes.
- The basic operation of a d.c. machine requires two opposing magnetic fields, one generated in the field poles, the other generated in the rotating armature.
- The magnetic field in the field poles can arise from permanent magnets or from field coils.

- The field coils used in creating a magnetic field are defined by their connection to the armature. They can be series, shunt (parallel) or compound (series and shunt) field coils.
- The magnetic field is a catalyst required to obtain either generator action or motor action. All the electrical power is either created or used within the armature.
- Generator action is where electromagnetic induction within the rotating armature generates an EMF that provides a source for current to flow to a connected load.
- Motor action is electrical power supplied to a rotating armature, creating torque, which allows a motor to drive a load.
- A commutator is mounted on the rotating shaft next to the armature winding and, via a sliding brush gear, is used to connect the armature winding to the external circuit. The commutator also allows external d.c. circuits to be connected to the alternating current circuit of the armature.
- The brushes are stationary contacts that slide against the commutator as it rotates and allow current from the external circuit to conduct through them into the commutator and then into the armature conductors.
- Traditional d.c. machines are categorised by the field coils used within them, i.e. series, shunt (parallel) and compound (series/shunt).
- A d.c. machine converts energy from one form to another, i.e. mechanical energy to electrical energy in a d.c. generator and electrical energy to mechanical energy in a d.c. motor.
- A d.c. machine has numerous losses within it. Some are electrical in nature and some are mechanical. The relationships between a machine's input power, output power and its losses are used to determine how efficient the machine is.

END-OF-CHAPTER QUESTIONS

7.1 What is the effect of the magnetic field that is produced when electrical energy is applied to a d.c. machine and current flows in the armature conductors?

7.2 What condition can prevent the torque produced in a d.c. motor from driving the motor?

7.3 Which of the following is NOT a typical part of a d.c. machine: a yoke; slip-rings; an armature; a field coil?

7.4 What is the main reason for shaping field poles?

7.5 Which type of d.c. generator is safe to operate with a short-circuited load?

7.6 What is the effect on the field current and the magnetism in the field poles when load is applied to a series generator?

7.7 What is the most common type of d.c. motor?

7.8 What will happen to the copper losses if the armature current is doubled?

7.9 What percentage of the total losses in a machine is contributed by copper losses in the form of electrical resistance?

7.10 In large traction loads such as diesel electric drives in ships, and particularly diesel electric locomotives with multiple bogie drives, what is the basic drive motor referred to as?

7.11 How is the commutator connected to the external circuit?

7.12 Which of Fleming's visual rules can be applied for generators?

7.13 Which three factors determine the value of voltage generated in a d.c. generator's armature?

7.14 What does the term 'voltage regulation' mean in regard to a generator?

7.15 Why is residual magnetism important in a self-excited generator?

CHAPTER 8

Apply environmentally sustainable procedures in the energy sector

LEARNING OBJECTIVES

- Understand matters that are relevant to environmentally sustainable work practices, including:
  - the concept of sustainable work practice
  - the consequences of neglecting sustainable work practices
  - the greenhouse effect—causes and consequences
  - national and international greenhouse imperatives
  - the role of regulators and similar bodies
  - legislative requirements
  - the economic benefits of applying sustainable initiatives.
- Understand techniques for reducing carbon-produced energy and hence greenhouse gases, including:
  - domestic, commercial and industrial strategies
  - trade-related technologies and methods
  - energy-efficient retrofits
  - renewable energy technologies.

INTRODUCTION

In 2015, the United Nations member states adopted 17 Sustainable Development Goals (SDGs) as part of the organisation's 2030 Agenda for Sustainable Development. This is a 15-year plan to stimulate action to end poverty, protect the planet and improve the lives and prospects of everyone who lives on it.

This chapter will cover aspects of the SDGs, including:

- affordable and 'clean' energy
- industry, innovation and infrastructure
- responsible consumption and production
- climate action.

Figure 8.1 illustrates the 'Three Pillars of Sustainability'—social equity, environmental protection and economic viability. These are commonly referred to as 'the three Ps'—people, planet and profit. The diagram shows that sustainability occurs at the point at which the three Ps meet.

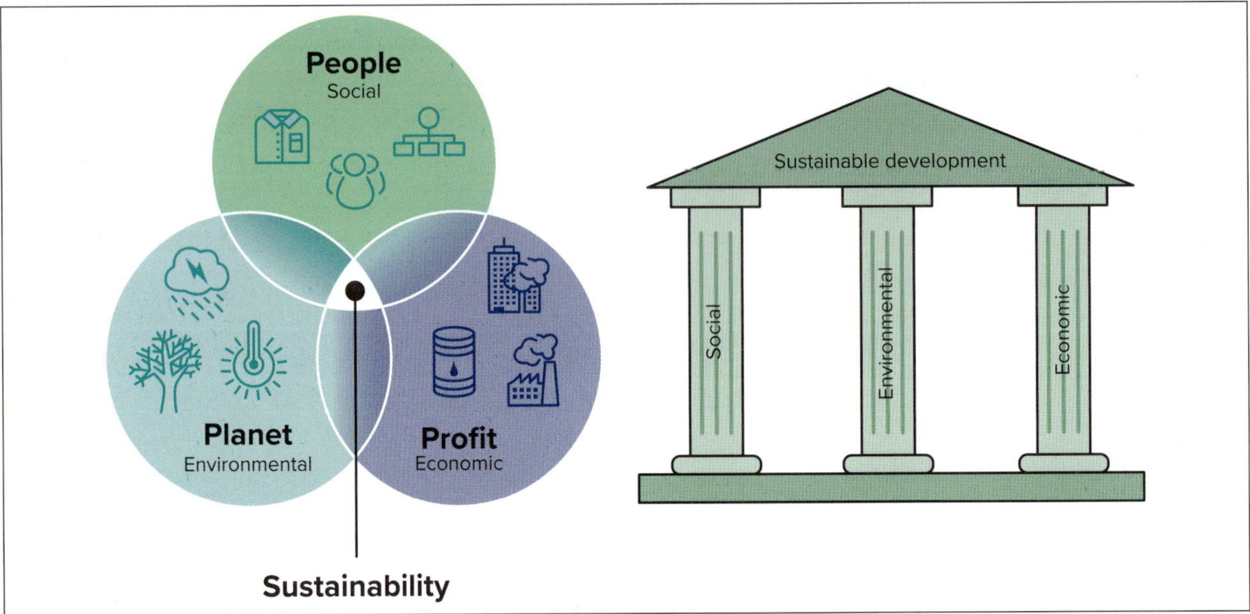

FIGURE 8.1 The Three Pillars of Sustainability

8.1 Sustainability

The term 'sustainability' means the ability of something to be maintained at a certain level. In the global environmental context, it specifically means the ability of the environment in which human beings live (Earth) to allow our quality of life—and way of life—to be maintained. If, for example, the generation of power from a solar panel installation is sufficient to keep a battery bank charged at a useable level (taking into account the current drawn from the batteries), it can be said to have sustainability. Or, if the amount of water consumption of a particular household is low enough for rainfall to be able to keep the supply tanks from emptying, the water-supply system has sustainability.

An example of a way of gauging global sustainability would be to ask whether the recycling of finite metals such as copper and gold is sufficient to allow us to continue to produce the cables and electronic components that use them. We might also ask:

- What strategies will we adopt when copper and gold resources are depleted?
- How will we continue to provide the function of a copper cable when there is no copper left to mine or recycle?
- How will we continue to produce electronic components such as microprocessors when there is no gold left to plate their electrical contacts?

Answers to these questions include finding ways of minimising or avoiding the use of resources that are under threat. For example, a wireless transmission medium can be used in place of copper cabling, and alternative materials can replace the gold coatings on the contacts of many electronic components. However, alternative materials are not always ideal. Aluminium can be used in place of copper, for example, but it has a lower conductivity and is too finite a resource itself. Materials such as silver and tin can be used instead of gold but they tend to corrode more easily.

The concept of sustainability includes acknowledgement of the detrimental effects that resource usage has on our planet. When population growth is factored in, at some stage there will simply not be enough to go around. It is important not only to conserve the Earth's resources but also to use them in such a way that the planet does not become uninhabitable for human beings.

Environmental sustainability is a required state for human existence. The decade from 2010 to 2020 was the hottest in recorded history, with the seven years 2014 to 2020 inclusive also being the hottest years on record. Figure 8.2 shows the seven hottest years prior to 2021.

These stark facts raise questions about how we can reduce greenhouse gas emissions to a level that will avoid the Earth's temperature rising so high as to threaten its ability to support human life.

Refrigerants and other chemicals are also relevant to sustainability and the electrotechnology industry. They are pollutants that damage the Earth and its atmosphere and affect global warming. Chemicals leak into waterways, causing damage to plants and animals, while man-made refrigerants (used for refrigeration and air-conditioning) affect the ozone layer, allowing harmful ultraviolet radiation from the sun through to the Earth.

Environmental sustainability strategies usually involve the minimisation of harm and/or resource usage. These, it is hoped, will buy us enough time to make alternatives such as renewable energy wholly viable as a replacement for energy derived from fossil fuels.

The production of energy through the use of gas, coal and petroleum and the extraction of metals such as copper and gold are examples that show that the electrotechnology industry must implement sustainable work practices. Copper and gold are resources that affect sustainability as they are limited and their use and/or manufacture creates **greenhouse gases** (**GHGs**) that contribute to **global warming,** a phenomenon that could make life on Earth unsustainable.

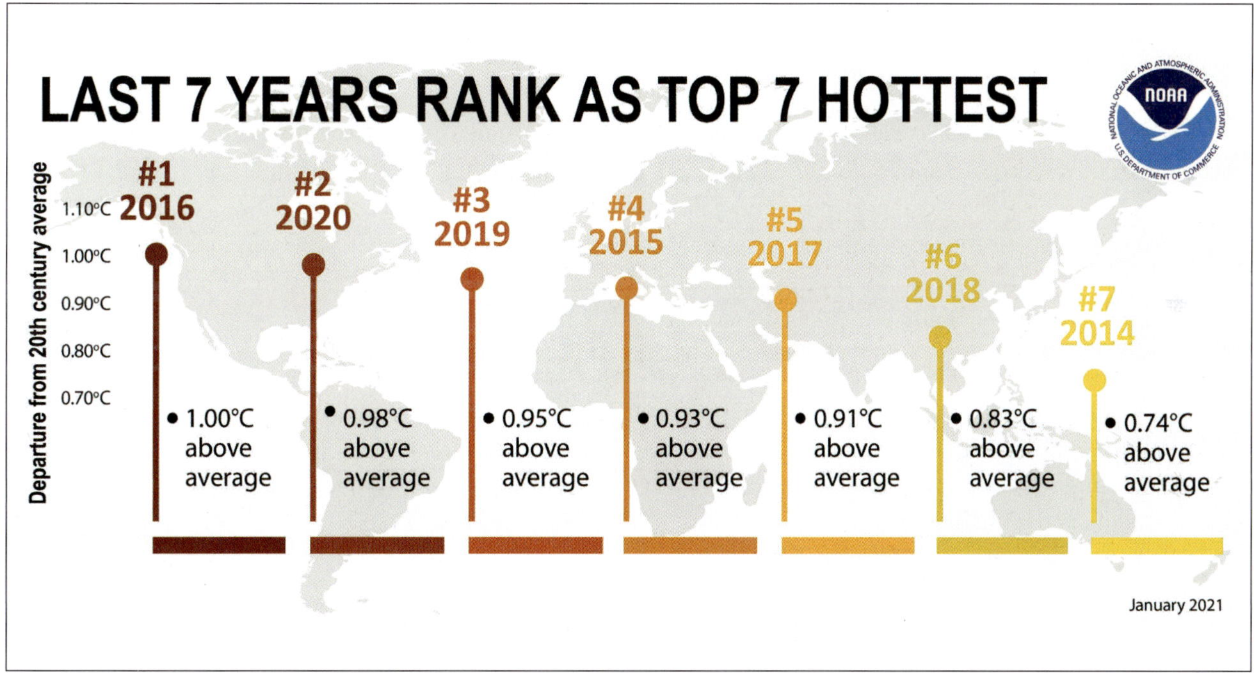

FIGURE 8.2 The seven hottest years prior to 2021
NOAA

CHECK YOUR UNDERSTANDING

8.1 Name the Three Pillars of Sustainability.
8.2 Identify the three Ps of sustainability.
8.3 Define sustainability.
8.4 Explain why environmental sustainability is required.
8.5 List the six hottest years in recorded history.

8.2 Sustainable work practices

The activities of businesses, companies and corporations affect the environment in many ways through manufacturing processes, waste material, pollution, land use, location, use of raw materials and work practices.

Sustainable work practices are minimisation and/or replacement strategies to reduce the impact of resource usage on the Earth and its inhabitants and maintain the current environment. They aim to bring about behaviour change in the workplace and prompt, for example, the selection of environmentally friendly materials, minimisation of offcuts, waste recycling and the use of energy-efficient tools and equipment. Sustainable work practices should be applied across the whole electrotechnology industry, and there are many well formulated guiding principles that organisations can adopt. These include the:

- 3 Rs of waste management—Reduce, Reuse and Recycle.
- 5 Rs of waste management—Refuse, Reduce, Reuse, Recycle and Repair.
- 5 Rs to save the environment—Refuse, Reduce, Reuse, Repurpose and Recycle.

At the individual level, everybody would do well to follow one of these approaches. The 5 Rs of waste management provide a hierarchy of sustainable work practice to save the environment. The fundamental aspects of sustainability are summed up in the phrase 'reduce, reuse, recycle, repair'. These concepts are examined in the next sections.

8.2.1 Reduce

Construction and maintenance activities are labour- and material-intensive, so job sites are places where the consumption of resources is above average. An effective way of reducing this consumption is to consider options for more efficient use of machinery and equipment. These include using low energy settings or simply turning machinery and apparatus such as vehicle engines, taps and electric motors off when they are not in use.

Refrigeration, cooling and space-heating applications operate more efficiently through thermostatic control to maintain a constant temperature rather than being switched on and off as needed.

Sustainable work practices aimed at reduction apply to all workplaces. They include:

- reducing waste resulting from disposable products
- reducing the use of paper products
- only using recycled print products
- reducing paper use by printing on both sides
- introducing paper-recycling schemes
- reducing the need for printing by using electronic files.

In domestic, office, commercial, industrial and hospitality environments, sustainable reduction practices include only using washing machines and dishwashers when there is a full load and, where health and hygiene does not require hot water settings, using cold water settings to conserve energy.

Some organisations and businesses that are starting to focus on sustainability may find that there is a trade-off between practicality, operational efficiency, cost-effectiveness and protection of the environment; it is advisable to balance any short-term inconvenience involved in adopting sustainability measures with their long-term impact.

8.2.2 Reuse

When upgrading, replacing or changing infrastructure or equipment, organisations often find that some items have become surplus to requirements. Donating such items to other organisations allows them to be reused rather than being consigned to the tip.

Materials that can be reused in the electrotechnology industry include (but are not limited to) cabling, cable tray, ducting and even saddles. PVC conduit can also be reused, and it is important to do so where possible as it cannot easily be recycled. Although it is predominantly plastic, it contains additives such as lead, cadmium, tin, barium and zinc that cannot readily be separated from it. As with all plastics, its manufacture requires oil.

8.2.3 Recycle

Most materials used in the electrotechnology industry are recyclable. Metals like copper and aluminium can be sold to scrap metal yards, while paper, glass and plastic can be delivered to recycling plants. The resource cost offset by recycling rather than producing new materials is significant.

There are three main types of recycling—primary, secondary and tertiary. *Primary (or closed loop) recycling* is the process of turning one thing into more of the same thing, for example turning paper into more paper products. *Secondary recycling* refers to turning something into other things made of the same material, for example car tyres and fluorescent tubes. *Tertiary recycling* is the breaking down of a product using a chemical process and then turning it into a completely new product. Recycling in these ways uses energy.

The recycling of fluorescent tubes involves separating the glass, aluminium, phosphor and mercury and salvaging them as reusable materials. This process represents a closed loop, with no waste to landfill. That is significant as mercury in landfill converts to toxic methylmercury and spreads through the wider environment via the air, water and soil, creating risks to the environment and human health. It is estimated that in Australia 95% of mercury-containing lamps are currently sent to landfill.

Recycling as a process (see Figure 8.3) can range from sorting usable items to be reused or repurposed to identifying damaged items that are repairable or can be used in the refurbishment of other items.

Batteries can also be recycled. When they are dumped into landfill, the chemicals they contain can leak and contaminate groundwater, posing a health hazard to humans and animals. Batteries contain hazardous waste but also recyclable material such as lead; delivering them to a recycling plant is worth the trip. The Australian Battery Recycling Initiative (ABRI) was created to promote the responsible disposal of batteries. ABRI's website shows where batteries can be recycled and is a useful, informative resource. It can be found at http://www.batteryrecycling.org.au/

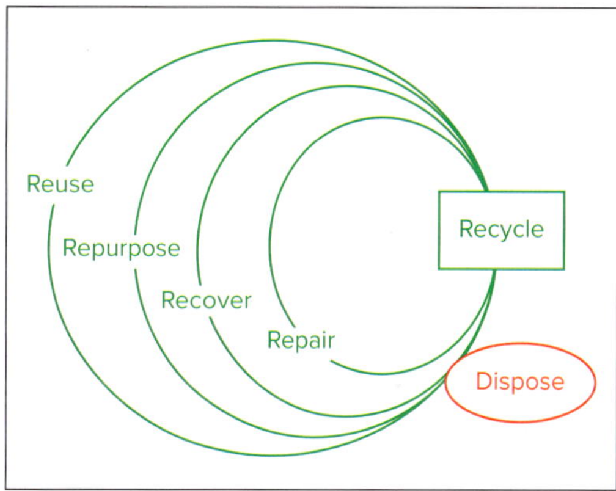

FIGURE 8.3 Recycling—reuse, repurpose, recover and repair

8.2.4 Repair (or replace)

Many products are cheaper to replace than repair. As stated earlier, one of the three sustainability pillars is 'profit'—businesses need to establish financial practices to remain competitive and viable. Business and financial structures will vary across industries and determine different approaches to asset management, depreciation, procurement and disposal processes. For larger businesses, accrual accounting and depreciation can lead to the disposal of an item appearing to be a better option than repair. Sustainable approaches to the disposal of equipment which cannot be reused, repurposed or repaired might include donation or gifting as an alternative to landfill. Smaller businesses with a different approach to asset management may prefer the repair option.

Repairing items may be the best option in terms of environmental responsibility. Reselling or recycling are preferable to simply throwing an item out. Sometimes, however, factors such as design, age or efficiency mean that certain items cannot be modified, updated or repaired. They may not be economically fit for purpose, and they may need to be replaced.

For example, an inefficient fleet of vehicles with poor fuel consumption will, over time, cost increasingly more to run, which will affect their economic viability. Inefficient cooling and heating systems and high-energy-consumption devices are examples of items whose replacement by more efficient, environmentally friendly models can cut a business's costs.

Families and individuals may also consider replacing standby electronic items like televisions and white goods with models that are more water- and energy-efficient. Energy Rating Labels (see section 8.5) can guide consumers to energy-saving benefits and decreased expenditure due to lower operating costs.

8.2.5 Responsible consumption, production and waste management

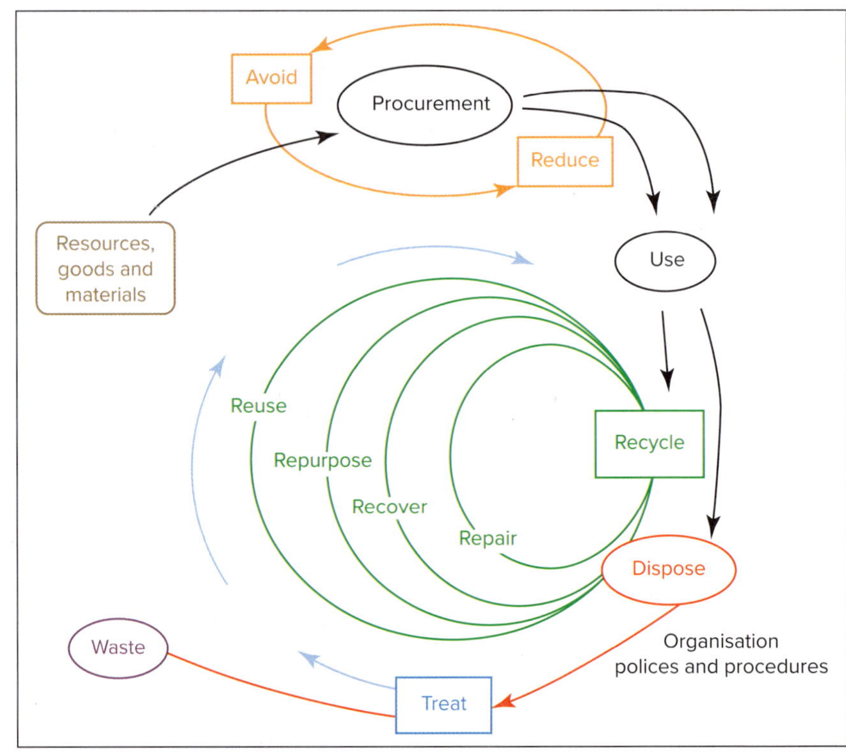

FIGURE 8.4 Responsible consumption, production and waste management

The notion of sustainable work practice includes finding innovative methods of responsible consumption, production and managing waste. A relevant framework to consider (building on 'reduce', 'reuse', 'recycle' and 'repair') is 'avoid', 'reduce', 'reuse', 'repurpose', 'recover', 'recycle' and 'repair'. From the viewpoint of the 'profit' element of the three Ps, sustainability can be built into the procurement phase for businesses through cost savings, responsible consumption and minimising waste (see Figure 8.4).

In this context, 'avoid' means reducing waste through planning, procurement and supply-chain processes. It is in many ways a similar concept to 'refuse' as it promotes choosing, for example, products with comparatively less (or even no) packaging. It also means avoiding disposable products such as certain cups and bags and choosing products that are more efficient in their water and/or energy requirements. Another aspect is avoiding purchasing items that are inferior, have a comparatively short lifespan or are not fit for purpose.

Consumables advertised as being 'recycled' and 'environmentally friendly' should be approached with caution unless their quality can be ascertained. For example, low-quality paper and print cartridges should not be purchased if they are not recommended for use on the type of equipment you have; they could eventually damage it. This could lead to repairs being necessary, lost production time, premature equipment disposal and unwanted replacement costs—all of which affect profit. It is advisable to check out the viability of recycled products by referring to equipment user and manufacturer specifications. As for paper, the paper-production industry is the fifth-largest energy consumer in the world, but recycled paper produces half the waste of traditional paper-making processes, using 45% less energy and reducing overall consumption.

'Recover' is a waste-management concept that links recycling goods and materials to avoiding and reducing waste. When replacing items or equipment, the concept of 'recover' requires considering the use of existing 'fit-for-purpose' items as an option. In the case of office furniture and equipment, as well as cable tray, ducting and other electrical items and accessories, this could mean using repaired or refurbished products rather than purchasing new ones.

Sustainable work practice on job sites includes waste minimisation. Recovering materials that are appropriate for use or repurposing on another job can be part of the pack-up and clean-up phase of the work day.

Manufacturers and suppliers of equipment and machinery may have a repair-and-replace exchange program as support for customers. For example, removable printed circuit cards could be exchanged for a set fee. Damaged or faulty circuit cards could be returned to the supplier for testing, repair and refurbishment. This type of recovery arrangement minimises waste as the circuit card is repaired rather than thrown out. Repurposing circuit cards through a flat-rate exchange for, say, \$500 per card as opposed to spending between \$2000 and \$5000 on a new one is an example of an environmentally sustainable work practice in the electrotechnology industry. Often the life cycle of circuit cards can be extended through a repair-and-exchange program with warranty and customer support. Under this type of arrangement, whether a card is new or refurbished is irrelevant: functionality and operation are the critical factors, minimising production downtime and sustaining businesses' profitability.

CHECK YOUR UNDERSTANDING

8.6 List the five steps of the extended waste management hierarchy.

8.7 Give five examples of recycling/recovery.

8.8 Explain how fluorescent lamp tubes in landfill could be environmentally hazardous.

8.9 What is the difference between being usable and being 'fit for purpose'?

8.3 Effects of neglecting sustainable work practices

One catastrophic long-term consequence of neglecting sustainable work practices is that the Earth's resources are being consumed at a faster rate than they can be replaced; and some are finite and cannot be replaced once they've been exhausted. A lack of resources, combined with damage to the environment, could cause widespread famine and disease. Something that is happening now is the loss of habitats for animals, resulting in certain species becoming extinct and changes to complex ecologies that will have far-reaching, but as yet unknown, impacts.

The immediate consequences of failing to practice sustainability are considerably less speculative and include an increase in:

- air pollution, causing a decrease in air quality and an increase in respiratory problems and illness
- water pollution, causing a decrease in water quality, potentially making it unsuitable for drinking
- contaminated land, causing potential sickness and a reduction in usable land
- global warming, making life on Earth increasingly difficult to sustain.

The causes of pollution as a result of neglecting sustainable work practices include (but are not limited to):

- chemical and oil spills
- generation and inappropriate disposal of waste
- asbestos (handling and removal)
- air emissions (toxic and greenhouse gases)
- noise and vibration
- dust generation
- PCB (polychlorinated biphenyls) management
- electromagnetic radiation (EMR) from electric and magnetic fields
- greenhouse gas emissions.

Neglecting sustainable work practices in the long term will exact a heavy toll on future generations. One person's excess waste is not enough to doom humanity, but as each resource is squandered the cost to the environment is increased. The resources that we currently take for granted may not be available in five or 50 years' time. The environment we live in today is unlikely to remain as healthy as it is now if the demand for raw materials increases. If sustainable practices are not introduced and followed, future generations may well have a reduced quality of life. They may need to adapt to an inhospitable environment, with fewer resources available and more health issues emerging.

8.4 The greenhouse effect

8.4.1 The greenhouse effect—definition

A greenhouse is a plastic or glass structure in which plants are grown. Its purpose is to counteract the effects of cold weather. Greenhouses stay warm because heat from the sunlight which shines through the transparent glass or plastic is partly trapped inside.

The term 'greenhouse effect' is often used in connection with the Earth and sustainability. A natural greenhouse effect is created by water vapour and greenhouse gases such as carbon dioxide in the atmosphere reflecting back to the Earth some of the sunlight that rebounds from it (see Figure 8.5); the water vapour and greenhouse gases have a similar effect to the plastic or glass of a greenhouse.

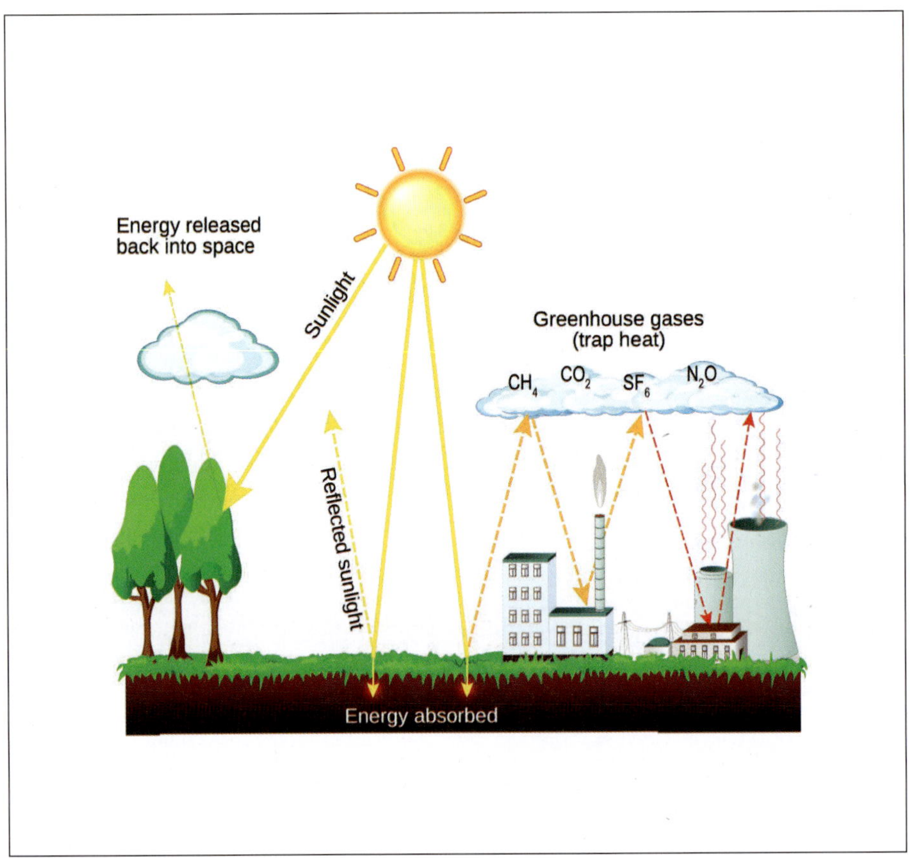

FIGURE 8.5 The greenhouse effect
Designua/Shutterstock

8.4.2 Causes of the greenhouse effect

Although the greenhouse effect occurs naturally, some industrial activities increase it by burning fossil fuels to generate electricity and power vehicles. Making concrete for construction produces large amounts of carbon dioxide (CO_2), increasing the greenhouse effect. Other man-made chemicals and the effects of pollution and land clearing increase the warming effect.

Reports of the major greenhouse gasses produced in Australia are updated quarterly. These are carbon dioxide (CO_2), methane (CH_4), nitrous oxide (N_2O), perfluorocarbons (PFCs), hydrofluorocarbons (HFCs) and sulphur hexafluoride (SF_6). Emissions of the greenhouse gas nitrogen trifluoride (NF_3) are considered negligible and are not estimated.

Figure 8.6 shows the different types of industrial emissions in Australia for the year to March 2021. The sector shown in the figure as 'LULUCF' represents land use, land-use change and forestry.

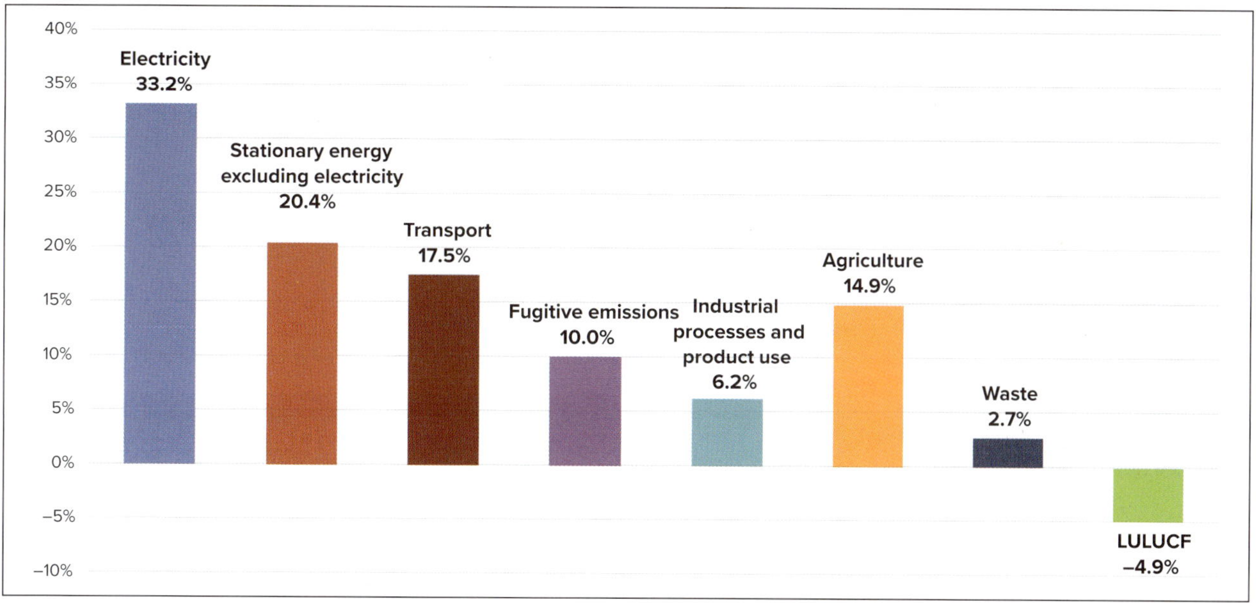

FIGURE 8.6 Share of total Australian industrial emissions for the year to March 2021
Quarterly Update of Australia's National Greenhouse Gas Inventory: March 2020, Australian Government Department of Industry, Science, Energy and Resources, p. 9 © Commonwealth of Australia 2020 CC BY 4.0, https://creativecommons.org/licenses/by/4.0/

There are various *feedback loop mechanisms* which also increase the greenhouse effect. For example:

- Burning fossil fuels creates more CO_2 in the atmosphere.
- More CO_2 in the atmosphere prevents heat from escaping, further warming the Earth.
- The Earth becomes hotter, melting the ice at the polar caps.
- White ice reflects sunlight/heat away from the Earth and, as the ice melts, less sunlight/heat is reflected and is instead absorbed by the Earth. Consequently, the Earth heats up even more, melting more ice and continuing the feedback loop.

The main feedback loop that contributes to the greenhouse effect involves water vapour. As the Earth warms, more water vapour is produced, which traps more heat and heats the planet further, thereby creating even more water vapour. This traps more heat and the Earth becomes still warmer—and the cycle continues.

Feedback loops such as these may lead to very sudden 'avalanche' effects rather than the gradual changes that have been noticed since the beginning of the Industrial Revolution in the late 1700s.

8.4.3 Consequences of the greenhouse effect

The fundamental consequence of the greenhouse effect on the Earth is an increase in the global temperature. The ten warmest years on record have been 1998, 2005 and 2013 to 2020 inclusive; the eight warmest have occurred since 2005. Based on global average temperatures, 2016 was the hottest year recorded to date, with 2020 being the second-hottest ever recorded. 2020 was the second-hottest year after 2016. Although a certain amount of global warmth is required for the Earth and its inhabitants to thrive, too much causes the following:

- an increase in the Earth's atmospheric temperature
- changed patterns and amounts of precipitation
- reduced ice and snow
- raised water levels
- increased ocean acidity
- increased frequency, intensity and/or duration of extreme weather events
- shifting characteristics of ecosystems.

These issues increase health risks for humans and animals, as well as the need to migrate to maintain acceptable nutritional, climatic and other living conditions.

CHECK YOUR UNDERSTANDING

8.10 State three consequences of neglecting sustainable work practices.

8.11 Explain what is meant by 'the greenhouse effect'.

8.12 Identify three causes of the greenhouse effect.

8.13 Name the major greenhouse gasses that need to be monitored.

8.14 State three consequences of the greenhouse effect.

8.5 International and national greenhouse imperatives

Australia has been active in some ways in limiting greenhouse gas emissions since 1988, when the Australian Government adopted a reduced CO_2 emissions target in response to the World Conference on the Changing Atmosphere in Toronto. Since then, the country has continued to be a part of a worldwide commitment to climate change, participating in various national and international initiatives such as the Kyoto Protocol (1992) and the Paris Agreement (2015).

However, according to the 2018 Climate Council of Australia Limited's working paper *Australia's rising greenhouse emissions,* "Emissions reductions above 26% in the electricity sector are realistic and achievable. The electricity sector can and should do more". Two key findings of the working paper were:

1. Australia's emissions are rising and projected to continue to do so in the absence of credible and comprehensive climate and energy policies.
2. Australia's emissions reduction target is woefully inadequate to protect Australians from intensifying climate change.

The report stated further that if other countries were to adopt climate policies similar to Australia's, the global average temperature rise could reach over 3°C and up to 4°C.

Global CO_2 emissions and the climate are concerns for all governments. Tellingly, in December 2020 leaders from Australia, Brazil, Saudi Arabia, Mexico, Turkey and Poland were not invited to speak at a global climate summit involving 70 other countries. This was because invitees had to be able to produce 'concrete and ambitious' plans to reduce carbon emissions and the use of fossil fuels, and to promote climate change commitments. However, at the 2021 United Nations Climate Change Conference in Glasgow (also known as COP26), the Australian Government announced a plan to achieve zero net carbon emissions by 2050.

Australia has instituted various policies, programs and partnerships that provide tools and programs to help reduce climate change. Examples include the:

- National Landcare Program—aimed at creating a more sustainable and productive agriculture industry.
- Asia-Pacific Rainforest Partnership—formed to reduce greenhouse gas emissions from deforestation.
- International Partnership for Blue Carbon—targets the restoration of coastal ecosystems.

Long-term strategy and domestic initiatives incorporate emissions monitoring and accountability systems, including the Renewable Energy Target (RET) and the National Greenhouse and Energy Reporting (NGER) scheme.

8.5.1 The Renewable Energy Target (RET)

The Renewable Energy Target is an Australian Government scheme that was introduced to bring about the production of a higher level of renewable energy. The RET offers incentives for small- and large-scale renewable energy installations. It has two components:

- The Small-scale Renewable Energy Scheme (SRES)—covers small-scale generation and residentially based power generation (solar hot water heaters and photovoltaic panels).
- The Large-scale Renewable Energy Target (LRET)—covers commercial-scale renewable power generation.

Examples of large-scale systems are power stations such as wind and solar farms, while smaller systems include solar water heaters and small-scale wind farms. In 2015, the LRET was reduced from 45 000 gigawatt-hours (GWh) to 33 000 gigawatt hours of Australia's electricity coming from renewable energy by 2020. In 2019 this target was achieved.

8.5.2 The Emissions Reduction Fund (ERF)

The Emissions Reduction Fund provides businesses, land owners, government departments, community groups and individuals with incentives to reduce emissions via new technologies and practices. It aims to reduce 2020 emissions to 5% below 2000 levels and 2030 emissions by 26%–28%.

To benefit from the ERF, a business needs to create a project that reduces emissions beyond its current activities. Once that is done, an administrative body called the Clean Energy Regulator will issue one Australian Carbon Credit Unit (ACCU) for each tonne of reduced emissions. One ACCU is earned for each tonne of carbon dioxide equivalent (CO_2-e) stored or avoided by a project. ACCUs can then be sold to the Government or to private-sector purchasers who can offset their emissions with them. The Australian National Registry of Emissions Units (ANREU) tracks the location and ownership of ACCUs.

8.5.3 The Clean Energy Innovation Fund

The Clean Energy Innovation Fund supports the growth of emerging technologies and businesses that target clean energy, renewable energy, energy efficiency and low emissions. Suitable initiatives might include large-scale solar with storage, off-shore energy, biofuels and smart grids.

8.5.4 National Greenhouse and Energy Reporting (NGER) scheme

The National Greenhouse and Energy Reporting scheme is a framework for the reporting and dissemination of company information about greenhouse gas emissions, energy production and energy consumption. The information is used to inform policy and meet international reporting requirements.

8.5.5 Other national initiatives

Other Australian schemes and initiatives include:

- The Carbon Neutral Program—certifies an organisation as having met certain carbon offset standards.
- Hydrofluorocarbon (HFC) management—controls manufacturing, importing and exporting of ozone-depleting substances and synthetic greenhouse gases.
- Taxation measures—for example, luxury cars that are not fuel efficient incur higher tax penalties than those that are.
- 20 Million Trees—a successful scheme to plant 20 million trees by 2020 to help reduce greenhouse gas emissions by creating a carbon sink.
- The Solar Communities Program—provides funding for community groups in some regions to install rooftop solar PV (photovoltaic systems), solar hot water and solar-connected battery systems.
- Energy Efficiency Schemes—aim to reduce greenhouse gas emissions by lowering energy consumption for households and businesses. These include a number of state and territory schemes:
  - Energy Efficiency Improvement Scheme (EEIS)—ACT
  - Energy Savings Scheme (ESS)—NSW
  - Retailer Energy Efficiency Scheme (EERS)—South Australia
  - Victorian Energy Upgrades program.
- The Carbon Farming Initiative—allows farmers who store carbon or reduce greenhouse gas emissions to earn Australian Carbon Credit Units that can be sold to businesses who want to offset their emissions.
- Product Stewardship (Oil) Scheme—pays incentives to industry to encourage the environmentally sustainable management and refining of used and recycled oil. Used oil (sump oil) can be cleaned of

contaminants and recycled again and again. Industry and the community generate 250 million litres of used oil in Australia each year. Recycled oil as re-refined base oil can be used for lubricant, hydraulic or transformer oil.
- Minimum Energy Performance Standards (MEPS)—ensure that electrical products meet a minimum performance level before they can be supplied or sold in Australia. MEPS is a mandatory regulation.
- Energy Rating Labels—a scheme (part of the Equipment Energy Efficiency program) to ensure that electrical appliances have an energy rating label attached, enabling consumers to make informed decisions about purchases. Appliances covered by the scheme include washing machines, fridges, freezers and air-conditioners (see Figure 8.7).

FIGURE 8.7 Star energy rating icon on appliances
Tony Jones

Australia's various initiatives fulfil binding requirements set by The United Nations Framework Convention on Climate Change (UNFCCC). Compared with many other countries, Australia is a relatively small contributor to overall global emissions. However, its per-capita emissions are among the highest in the world.

8.5.6 The United Nations Framework Convention on Climate Change (UNFCCC)

In 1992, Australia, along with 153 other nations, signed The United Nations Framework Convention on Climate Change. The aim of the framework was to ensure the protection of the atmosphere. All of the signatory nations recognised that the continual production of harmful greenhouse gases was contributing to climate change, and therefore that the activities of human beings are contributing to global warming.

The UNFCCC created a formal agreement to ensure that participating countries took steps to limit the amount of greenhouse gasses (GHGs) and to allow ecosystems to adapt to the ever-changing climate. Two major initiatives have come from the UNFCCC—the Kyoto Protocol and the Paris Agreement.

These UNFCCC initiatives set Australia two targets for greenhouse gas emission reduction. These were that emissions should be:

1. 5% below 2000 levels by 2020 (under the Kyoto Protocol)
2. 26–28% below 2005 levels by 2030 (under the Paris Agreement).

8.5.7 The Kyoto Protocol

The Kyoto Protocol was an international agreement created in Kyoto, Japan, in 1997. It aimed to reduce the world's collective greenhouse gas emissions by binding developed countries to emission-reduction targets, but it left developing countries to join voluntarily. The latter category included China and India, both of which had significant emissions due to their large populations, despite being relatively under-developed. The developed/under-developed issue led to some large-emissions countries dropping out of the agreement, including the USA and Canada. Figure 8.8 illustrates Australia's progress towards its Kyoto Protocol targets.

In 2015, the Kyoto Protocol was effectively replaced by the Paris Agreement.

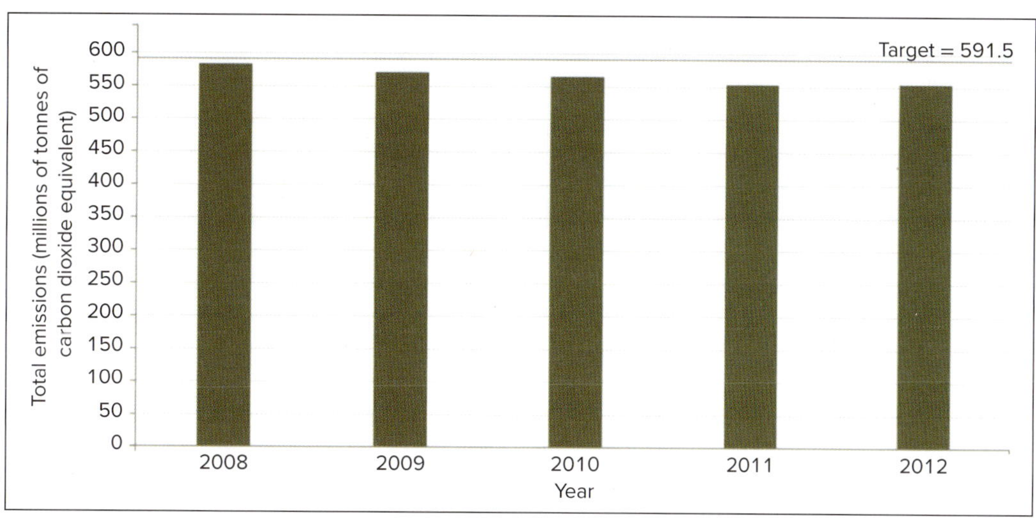

FIGURE 8.8 **Australia's progress towards the Kyoto Protocol target from 2008 to 2012**
Data: Australian Government, Department of Department of Industry, Science, Energy and Resources, National Inventory Report 2012

8.5.8 The Paris Agreement

The Paris Agreement aims to reduce climate change but, unlike the Kyoto Protocol, it binds both developed and developing countries to it, which has led to more countries signing up to it.

The Paris Agreement aims to ensure that the Earth's 'pre-industrial' temperature is not exceeded by more than 2°C this century, and to endeavour to limit the actual temperature increase to no more than 1.5°C.

To put this into perspective: as of 2018, the world was at about 1°C above pre-industrial temperatures and, predictions suggest, if emissions are allowed to continue unabated, the temperature could rise to about 4°C above pre-industrial temperatures by the year 2100. It is important to note that this refers to the *mean* global temperature; the temperature in some areas could rise well above that, and in other areas it could fall below that. The consequences of a 4°C mean global temperature increase would be catastrophic. According to the Climate Council, the impacts of a 1.5°C and a 2°C rise would be as follows:

1. Impact of 1.5°C rise:
   - 8% of plants will lose half of their habitable area
   - 6% of insects will lose half of their habitable area
   - 70–90% further decline in coral reefs
   - 14% of global population exposed to severe heat every one-in-five years.
2. Impact of 2°C rise:
   - 16% of plants will lose half of their habitable area
   - 18% of insects will lose half of their habitable area
   - 99% further decline in coral reefs
   - 37% of global population exposed to severe heat every one-in-five years.

8.6 The role of regulators and similar bodies

There are various regulators nationally and internationally that oversee policies, legislation and changes relevant to emissions and the climate. Federal, state and local governments jointly administer environmental protection laws in Australia.

8.6.1 Climate Change Authority (CCA)

The Climate Change Authority provides expert advice about Australia's climate-change policies. It has a particular role in reviewing and making recommendations on the Carbon Farming Initiative and the National Greenhouse and Energy Reporting scheme.

8.6.2 Clean Energy Regulator

The Clean Energy Regulator's role is to administer Australian Government schemes for measuring, managing, reducing or offsetting the country's carbon emissions. The Regulator is responsible for the National Greenhouse and Energy Reporting scheme, the Emissions Reduction Fund, the Renewable Energy Target and the Australian National Registry of Emissions Units.

8.6.3 The Environment Protection Authority (EPA)

Each state and territory in Australia has an EPA. Their purpose is to protect the environment and human beings by taking action on pollution and waste.

8.6.4 Department of the Environment and Energy

The Department is responsible for the protection of the environment, water and heritage. It implements policies, procedures and, through a regulatory framework, supports the environment, community and economy. This is achieved by protecting, conserving and improving environment and heritage, ensuring that energy is affordable, reliable and sustainable.

8.6.5 International bodies

The peak world body for climate change information is the Intergovernmental Panel on Climate Change (IPCC). Formed in 1988, it regularly releases reports that assess the impacts of climate change. The IPCC coordinates thousands of experts from around the world to gather information to help understand climate science, including meeting the challenges posed by climate change.

8.6.6 Australian Energy Regulator

The Australian Energy Regulator (AER) is responsible for monitoring and regulating the wholesale electricity and gas markets in Queensland, New South Wales, South Australia, Tasmania (electricity only) and the ACT.

8.6.7 Australian Renewable Energy Agency

The Australian Renewable Energy Agency (ARENA) funds innovation and shares knowledge about renewables. Its purpose is to improve the competitiveness of renewable energy technologies and increase the supply of renewable energy through innovation that benefits Australian consumers and businesses.

8.6.8 Climate Council

The Climate Council supplies independent, authoritative climate-change information to the Australian public.

8.6.9 Australian Solar Council

The Australian Solar Council is a not-for-profit, membership-based organisation concerned with the advancement of solar power and complementary technologies in Australia. Council activities include the promotion of research, development, public advocacy and adoption of solar energy through education.

8.6.10 Solar Energy Industries Association (SEIA)

Members of the SEIA include installers, retailers, researchers and lobbyists, as well as manufacturers and importers of equipment relating to solar energy.

8.6.11 Smart Energy Council

In 2017 the Australian Solar Council and the Energy Storage Council combined to form the Smart Energy Council, which is now the peak body for the solar, storage and smart energy industries. It is committed to promoting clean, efficient, cheap and smart energy solutions.

8.6.12 Australian Photovoltaic Institute (APVI)

The institute focuses on data analysis, information and collaborative research, nationally and internationally. It participates in two International Energy Agency programs, PVPS (Photovoltaic Power Systems) and SHC (Solar Heating and Cooling).

8.7 Legislative requirements

Legislation relevant to sustainability comes from acts, regulations and policies generated by government. There are examples relating to climate change, air pollution, water, chemicals and ozone protection, as well as renewable energy. The key pieces of legislation are dealt with in the next sections.

8.7.1 *Climate Change Act 2017*

All Australian states and territories have created Acts of Parliament, regulations and policies relevant to climate change. For example, Victoria's *Climate Change Act 2017* is consistent with the Paris Agreement and is designed to manage climate-change risks.

8.7.2 *Ozone Protection and Synthetic Greenhouse Gas Management Act 1989*

The Act was introduced to control the manufacture, import and export of all ozone-depleting substances (ODSs) and synthetic greenhouse gases (SGGs). Controls have been introduced in Australia to guide the use of synthetic refrigerants and to ensure that only licensed persons can work with them.

8.7.3 *Environment Protection and Biodiversity Conservation Act 1999*

The Environment Protection and Biodiversity Conservation (EPBC) Act 1999 is the major single piece of legislation relating to the environment, and it applies to all states and territories of Australia. It provides a legal framework to protect and manage important flora, fauna, ecological communities and heritage areas, wetlands, threatened species and marine areas.

8.7.4 *Greenhouse and Energy Minimum Standards (GEMS) Act 2012*

On 1 October 2012, the *Greenhouse and Energy Minimum Standards (GEMS) Act 2012* came into effect in Australia. The GEMS Regulator replaced the previous state regulators and is responsible for administering the legislation in Australia. The GEMS Regulator oversees a number of programs, including the Equipment Energy Efficiency program in conjunction with the Energy Efficiency Advisory Team. The Advisory Team members include NSW, Victoria, Queensland, Western Australia, South Australia, Tasmania and the ACT. Known also as the 'E3' program, it is a cross-jurisdictional program through which the Australian and New Zealand Governments collaborate on energy efficiency standards and product labelling. The program was devised to improve energy efficiency and reduce energy consumption.

8.7.5 Administration of laws, regulations and policies

Each state and territory's acts, regulations and policies are administered by its Environment Protection Authority (EPA). When businesses, government departments and land owners lodge an application for works, the EPA must consider the potential impacts of those works in light of the requirements set out in all legislation. If it is established that the works will either not meet or will be in contravention of legislative requirements, they are not allowed to proceed.

CHECK YOUR UNDERSTANDING

8.15 Name the two main international agreements that relate to climate change.

8.16 Name three key regulators of matters relating to the environment and sustainability in Australia.

8.17 State the greenhouse gas emission-reduction targets set for Australia by international agreements.

8.18 Identify three key pieces of legislation that are relevant to sustainability.

8.19 Explain briefly the purpose of the *EPBC Act 1999*.

8.8 Economic benefits of sustainable initiatives

The electrotechnology industry in Australia contributes significantly to both the national economy and to the production of greenhouse gases. Table 8.1 shows the actual annual emissions by sector for the year to March 2019 and the year to March 2020.

Incorporating sustainable work practices in the electrotechnology industry can bring significant benefits, not simply in terms of the longevity and wellbeing of the environment but also in terms of the economic welfare of electrical businesses and their customers. Various economic initiatives provide direct economic benefits and incentivise businesses, land owners, state and local governments, community organisations and individuals to adopt new practices and technologies which reduce emissions. These include the Emissions Reduction Fund (ERF) and the two core components of the Renewable Energy Target (RET).

TABLE 8.1 Emissions by sector 2019–20

	Annual emissions (Mt CO_2-e)		
Sector	**Year to March 2019**	**Year to March 2020**	**Change (%)**
Energy–electricity	180.5	172.9	−4.2
Energy–stationary energy excluding electricity	100.0	102.7	2.7
Energy–transport	100.1	99.7	−0.4
Energy–fugitive emissions	55.7	55.8	0.2
Industrial processes and product use	34.8	34.6	−0.6
Agriculture	72.0	68.0	−5.5
Waste	13.0	13.1	1.0
Land use, land use change and forestry	−19.7	−18.1	8.0
National inventory total	**536.4**	**528.7**	**−1.4**

Australian Government, Department of Industry, Science, Energy and Resources, Quarterly Update of Australia's National Greenhouse Gas Inventory: March 2020, Table 3

Adopting practices to reduce waste can enable businesses and organisations to reduce costs, spend more efficiently and achieve greater profits. If practices to reduce carbon emissions are adopted, businesses or individuals may be eligible to seek monetary compensation. For example, a person or business who invests in small-scale renewable power systems can access credits as part of the Small-scale Renewable Energy Scheme.

8.8.1 HFC Refrigerant Levy

As part of the Clean Energy Future Plan (2012), synthetic greenhouse gas refrigerants attract a levy. The size of the levy is proportional to the gas's Global Warming Potential (GWP). Although it is the refrigerant importers who pay the levy, the costs are passed down the supply chain, ending up with the customer. The HFC refrigerant levy aims to bring about these benefits:

- reduced refrigerant leakage through improved system design
- reduced emissions through improved maintenance
- systems that use a smaller charge of refrigerant being preferred for future use
- lower GWP refrigerants and systems being more likely to be implemented.

The electrotechnology industry is under constant pressure to move towards the use of work practices and procedures that minimise the use of electrical energy and synthetic greenhouse gases.

8.9 Techniques for reducing carbon-produced energy and greenhouse gases

Burning fossil fuels such as natural gas, coal and oil causes the level of carbon dioxide in the atmosphere to rise, which is a major contributor to the greenhouse effect and global warming. Individuals can take action by using energy wisely and thus reducing the demand for fossil fuels. Table 8.2 shows methods of reducing greenhouse gas emissions.

TABLE 8.2 Reducing greenhouse gas emissions

Method	Application and results	
Reduce, reuse, recycle, repair	Buying products with minimal packaging helps to reduce waste. Over 1000 kilograms of CO_2 can be saved annually if each person recycles half of their household waste.	nd3000/Shutterstock
Use less heat and air-conditioning	Adding insulation to walls and installing double-glazed windows can lower heating costs by more than 25% by reducing the amount of energy needed to heat and cool a home. Heating can be turned down while people are sleeping at night or are out during the day. Temperatures should be kept moderate at all times, and set no higher than 19°C in winter and no lower than 25°C in summer. By doing this, Australian businesses could save $100 million and 300 000 tonnes of carbon every year.	Kate Kunz/Glow Images

(continues)

TABLE 8.2 Reducing greenhouse gas emissions (continued)

Method	Application and results	
Replace light bulbs	Traditional inefficient incandescent light bulbs have been phased out in Australia. There are other more efficient types of lighting available, including light-emitting diodes (LEDs) and compact fluorescent lamps (CFLs). It is estimated that the phasing out of incandescent light bulbs is saving the average household 300 kilowatt hours and $75 per annum.	Nathan Mead/Age Fotostock
Drive less and drive wisely	Less driving means fewer emissions. People could use public transport or consider carpooling to work. If people decide to drive, they could make sure that their cars are running efficiently—having tyres over- or under-inflated by 1 bar or 15 psi from the recommended inflation pressure could lead to a 5% difference in rolling resistance, which may result in a significant fuel cost increase. A reduction of 10% of rolling resistance on a complete vehicle results in approximately 3% reduced fuel consumption. Every litre of petrol saved keeps 10 kilograms of CO_2 out of the atmosphere.	Don Mason/Blend Images LLC
Buy energy-efficient products	Home appliances come in a range of energy-efficient models. Compact fluorescent bulbs, along with LED lamps, use far less energy than standard light bulbs.	Tavis Wright/Image Source, all rights reserved (left); Tony Jones (right)
Use less hot water	An average household can use around 25% of its total energy on heating water. Using a more efficient hot water system reduces energy costs.	Elnur/Shutterstock

TABLE 8.2 Reducing greenhouse gas emissions (continued)

Method	Application and results	
Remember to switch lights and appliances off	People can switch off lights when they leave a room, and use only as much light as they need. It is also important to turn off televisions and computers that are not in use.	Alexey Rotanov/Shutterstock
Plant a tree	One tree will absorb approximately one tonne of carbon dioxide during its lifetime.	Shutterstock/nBhutinat
Request an energy audit from utility companies	Many utility companies provide free home energy assessments to help consumers identify areas in their homes that may not be energy efficient. Utility companies may also offer rebates to help pay for the cost of energy-efficient upgrades.	Purestock/SuperStock
Encourage others to conserve energy	People can share information about recycling and energy conservation with friends, neighbours and colleagues, and take opportunities to encourage public officials to establish programs and policies that are good for the environment.	Caia Image/Image Source

8.10 Domestic, commercial and industrial strategies

The electrotechnology industry as a whole could adopt sustainable work practices to reduce greenhouse gas emissions and fossil fuel use. Three areas where changes could be made are in its use of copper, electricity, and oil and gas.

8.10.1 Copper

The electrotechnology industry uses large amounts of copper. Recycling is beneficial as copper can be recycled over and over again with no loss in quality. In fact, a process of electrolysis can be refined to upgrade it.

Recycling copper means there is less of it in landfill and less of it being mined (it is cheaper to recycle copper than to mine it). This in turn means there is less environmental damage due to a reduction in the waste gases produced when mining and refining copper. Further benefit from recycling copper comes from a reduction in the use of the resources used to manufacture it: oil, gas and coal are used as fuel sources in its production; not having to use them is good for the environment.

Reducing the use of copper is not as easily achieved. However, reductions can be made in hard-wired copper-LAN cabling by using wireless communication technology when possible. This could include an entire network, particularly for small installations, and could also apply to a single run only. For example, rather than copper cabling to an outside IP camera, perhaps a wireless IP camera would suffice. Installation is also more economical for the customer and, with future innovation, could eventually lead to increasing data-throughput speeds.

The diameter of copper cable used is important and is indeed mandatory in relation to safety, functionality and reliability. Australian Standards mandate that cabling of a minimum specific diameter is used in various situations. However, it is not uncommon for larger diameters to be used, for example, in some commercial lighting installations. However, as copper is a finite resource it is important that this practice is curbed within the limits of safety, reliability and functionality.

The overall benefits of recycling copper and reducing its use include:

- minimised environmental damage
- reduction in landfill
- copper resource conservation
- financial savings.

INDUSTRY IN FOCUS: COPPER

The peak body for the country's copper industry is the International Copper Association Australia (ICAA). The organisation has committed itself to supporting the industry's efforts towards the reduction of the carbon and environmental footprint of the global copper mining sector.

The ICAA's initiatives align with all 17 of the United Nations Sustainable Development Goals, in particular:

- affordable and clean energy (UN Goal 7)
- sustainable cities and communities (UN Goal 11). McKinsey & Company, a management consulting firm that advises on strategic management to corporations, governments and other organisations, has estimated a 43% potential increase in copper demand by 2035 in comparison with today's 22-million-tonne demand
- Climate Action (UN Goal 13). By 2030, recycling copper could reduce the world's carbon footprint by 16%. This is likely to be viable as:
  - Two-thirds of the 550 million tonnes of copper mined since 1900 is still in productive use.
  - Recycling copper requires up to 85% less energy than primary production. Around the world, this saves 40 million tonnes of CO_2, the equivalent emissions of 16 million cars. Of all the copper needed across the world, 34% comes from recycling (41% in Europe). On average, copper products contain 35% recycled content.
  - 25% of all copper produced is used in buildings for plumbing, roofing and cladding.
  - Copper has the highest electrical conductivity of any metal apart from silver.
  - Approximately 70% of copper produced worldwide is used for electrical/conductivity applications and communications.
  - 60% of total copper use is for electricity and heat applications. Products containing copper tend to operate more efficiently because it is the best non-precious conductor of heat and electricity.
  - Up to 12 times more copper is being used in renewable energy systems than in traditional systems to ensure efficiency.
  - Electric vehicle manufacturing uses four times as much copper as combustion engine vehicles. It is used for batteries, motors, charging stations and supporting infrastructures.
  - Copper is essential in cutting-edge fields such as nanoparticles, wearable technologies, health, space exploration and renewable-energy technologies including solar panels, wind turbines and geothermal sources.

8.10.2 Electricity

Australia is one of the three highest exporters and fifth largest miner of coal on the planet after the US, China, Russia and Saudi Arabia. In 2019, 76% of Australia's electricity came from burning fossil fuels and 24% from renewables.

Significant change has been made in moving towards renewable energy through hydropower, wind, rooftop solar and bioenergy. Between 2019 and 2020, the contribution to electricity generation made by renewables rose from 24 to 27.7%. According to statistics from the Australian Government Department of Industry, Science, Energy and Resources, about 14% of Australia's electricity was generated outside the electricity sector by business and households in 2018–19. Table 8.3 shows states' and territories' contributions to electrical generation from renewable energy sources in 2020 with 2019 figures shown in brackets.

TABLE 8.3 Generation of electricity from renewable energy by state and territory in 2020 (2019)

ACT	NSW	Victoria	Tasmania	South Australia	Western Australia	Northern Territory	Queensland
100%	21%	27.7%	99.2%	59.7%	24.2%	8%	16.6%
(100)	(17.1)	(23.9)	(95.6)	(52.1)	(20.9)	n/a	(14.1)

The following are strategies for reducing electricity consumption:

- Turn off appliances at the wall when not in use rather than putting them on standby.
- Replace standard devices and lights with more efficient or lower-consumption versions. Examples include replacing fluorescent lights with LEDs and replacing desktops with laptops or tablets.
- Turn off lights when they are not required or installing motion detectors to control lighting in infrequently used rooms.
- Install blinds and shutters on windows to block out direct sun, reducing the need for air-conditioning in the summer and allowing sun and light in during winter.
- Keep doors closed to minimise the need for air-conditioning and setting the temperature to avoid overheating or overcooling (19°C during winter and 25°C during summer).
- Turn air-conditioning off for the last hour or so of the day.
- Clean air-conditioning filters and condenser coils regularly.
- Use signs, posters and stickers to keep the idea of saving energy in people's minds.
- Incorporate building improvements that maximise natural heating and cooling systems.
- Hire an energy-efficiency expert to analyse work areas and develop an energy savings plan.

Environmentally friendly alternatives to meeting current electricity consumption include:

- minimising the use of fossil fuel by moving towards renewable energy sources
- choosing green energy plans from energy providers
- installing solar power systems.

8.10.3 Oil and gas

The use of fossil fuels such as oil and gas is one of the major contributors to the production of greenhouse gases. Oil is distilled to create petrol and diesel and, along with gas, is used to fuel vehicles. Gas in a different form is also used for heating.

Sustainable work practices for minimising oil and gas use include:

- being aware that gas heating is similar to heating with electricity. Certain points that apply to electricity, such as making building improvements, keeping doors closed and not setting the temperature too high, also apply to gas

- using energy-efficient vehicles that consume less fuel. These include hybrid electrical/fuel vehicles as well as vehicles that use less petroleum-based fuel
- minimising the purchase of plastic products (crude oil is used in the production of plastic)
- using non-synthetic work clothes and boots where safe to do so rather than those made from oil-based nylon and polyester plastics
- switching to soy-based printing inks rather than the standard petroleum types
- using natural cleaners rather than oil-based cleaners
- choosing plastic products made from recycled material that can be re-recycled
- using tap water rather than bottled water to reduce the use of petroleum-based plastic bottles
- purchasing products made locally rather than overseas to reduce long-distance transportation

Of these sustainable work practices, reducing vehicle fuel use and the use of plastics would yield the most beneficial results.

8.11 Trade-related technologies and methods

There are various technologies and methods for reducing energy consumption in the electrotechnology industry.

8.11.1 Technologies

Greater energy efficiency can be achieved by using these technologies:

- programmable timer clocks—switch appliances on and off at preferred times and during holiday periods
- motion sensors—enable ventilation systems to operate during holiday periods if regular motion is detected
- LED lighting—more energy efficient and cost effective than standard lighting
- building design and position—can maximise weather conditions such as sunlight
- solar energy systems
- energy monitoring systems—can be used to assess the energy efficiency of an installation
- building management systems (BMS)—can control plant and equipment. For example, a BMS can be set to use air from outside rather than conditioned air
- C-Bus microprocessor systems—capable of controlling lighting, security, air-conditioning and other residential and commercial functions.

8.11.2 Methods of reducing energy consumption

Methods of reducing energy consumption include:

- switching devices and machines off when not in use, including via the use of automation such as sensors and timers
- matching generators to the load
- replacing ageing, low-efficiency equipment with newer, more efficient versions
- using blinds and window shutters
- wearing clothing to keep warm rather than using air-conditioning
- adjusting the thermostat for air-conditioning 1 degree warmer in summer and 1 degree cooler in winter
- using induction cooktops rather than gas
- using renewable power sources and batteries rather than diesel generators
- installing variable speed drives (VSD).

In addition to using technology and other ways of reducing electricity consumption, the regular servicing of plant and equipment is necessary to keep all systems working as efficiently as possible.

Sustainable work practices should aim to minimise waste, reduce energy consumption and avoid damage to the environment. Work practices and methods that contribute to sustainability include:

- aiming to reduce electricity bills by conducting a site survey of the plant equipment and air-conditioning equipment and measuring power consumption
- ensuring that any visible faults such as leaks and damaged switches are repaired or replaced
- replacing worn or damaged cables, seals etc.
- selecting the most efficient electrical appliances for each installation
- fitting electronic timers to systems to ensure they are not accidentally left running in an empty building or office
- evaluating proposed and existing installations and testing the installation before leaving the site.

Electrical installations are often evaluated for proposed alteration or additions to improve energy efficiency and investigate functional problems.

8.12 Trade-related retrofits

In the context of sustainability, retrofitting is the process of replacing old technologies with newer ones to improve energy efficiency. The many examples of these technologies include:

- solar photovoltaic systems and battery storage systems—these respectively generate and store power so that fossil-fuel-based power does not need to be used
- electric vehicles—reduce the use of fossil-fuel-based vehicles, reducing fossil fuel use and greenhouse gas emissions
- variable speed drives that match working conditions—can cut energy use by up to 50% in comparison with constant-speed drives (this includes applications in pumping, fans and HVAC systems)
- low-power LEDs—can replace compact fluorescent lighting
- programmable controllers—enable the efficient use of resources by activating equipment only when needed, based on factors such as time and temperature
- electric water heaters and induction cook tops—these are energy-efficient replacements for standard electric appliances
- improving power quality—reduces electromagnetic disturbance from inferior equipment such as lighting systems and electric motors and decreases power and energy losses
- replacing synthetic refrigerants with natural refrigerants such as ammonia (R717), carbon dioxide (R744) and hydrocarbons (R290 and R600). There are dangers associated with these natural refrigerants, so manufacturers' advice should be sought.

CHECK YOUR UNDERSTANDING

8.20 Give three examples of sustainable work practice that can minimise or reduce electricity consumption.

8.21 Identify three sustainable work practices that can help to minimise the use of oil and gas.

8.22 List two examples of energy-efficiency technologies.

8.23 Compare similar sustainable work practices for gas and electricity that contribute to reducing energy consumption.

8.13 Renewable energy

Renewable energy is any form of energy that can be used over again and/or is self-replenishing and/or is generated by a near-limitless source such as the sun.

Renewable energy technologies, sometimes referred to as 'green' energy technologies, use natural resources that can be continually replaced to create energy. Table 8.4 shows the main types of renewable energy.

TABLE 8.4 Renewable energy types

Common technologies	Energy-harnessing technologies	Grid-strengthening technologies
Solar energy	Geothermal energy	Battery storage
Wind energy	Marine energy	Smart technology
Hydro energy	Bioenergy	

Data source: arena.gov.au

8.13.1 Solar energy

Photovoltaic panels convert sunlight into electricity, and concentrated solar thermal collectors use energy from the sun to heat up water or oil that can then be used to heat water to make steam. This can be harnessed to drive a turbine attached to a generator.

Photovoltaic panels are found in both small-scale rooftop installations (see Figure 8.9) and on large-scale utility-sized installations (as in Figure 8.10). According to the Clean Energy Council, in 2019 small-scale solar was responsible for 22.3% of Australia's clean energy generation and produced 5.3% of its total electricity.

Table 8.5 shows data comparison estimates from Clean Energy Australia Reports produced by the Clean Energy Council in 2020 and 2021, relating to large-scale renewable energy projects that were either under way or had received the financial commitments necessary to begin. The data is from 1 January 2017 to 30 August 2021. Previously anticipated estimates as at 10 March 2020 are shown in brackets. Significant variations can be seen due to factors including the effects of national bush fires, Covid-19 and investment uncertainty. Job creation figures do not reflect current employment status.

TABLE 8.5 Large-scale renewable energy projects

Large-scale renewable energy projects			
State/territory	**Capacity in megawatts**	**Investment (dollars)**	**Employment (number of jobs created)**
ACT and NSW	3778 (3333)	5841 (5261) million	4324 (4128)
Victoria	2987 (3355)	4518 (5615) million	2821 (3992)
Tasmania	No new projects (260)	(580) million	(350)
South Australia	1517 (1752)	2396 (2998) million	992 (1486)
Western Australia	328 (842)	1854 (2956) million	1661 (2380)
Northern Territory	64 (54)	119 (100) million	272 (190)
Queensland	2661 (1553)	4185 (2901) million	3294 (2153)

Data source: Clean Energy Council, 2019, Large-scale solar PV

FIGURE 8.9 Small-scale solar installation
Federico Rostagno/Shutterstock

FIGURE 8.10 Large-scale solar installation
krasnoyarsk/123RF

8.13.2 Wind energy

Wind power is currently the cheapest source of large-scale renewable energy. Turbines capture energy as the wind moves past the blades, causing them to turn (see Figure 8.11); the blades are coupled to a generator. According to the Clean Energy Council, Australia's wind farms produced 35.9 (35.4)% of the country's clean energy and supplied 9.9 (8.5)% of its overall electricity in 2020 (2019 figures are in brackets). Table 8.6 lists states' recent wind power generation.

Eight wind farms with an overall capacity of 837 megawatts were commissioned in 2019. Another 30 wind farms were under construction or had received the financial commitments necessary for building to begin.

FIGURE 8.11 Wind power turbines
Spaces Images/Blend Images

TABLE 8.6 Wind power generation by state for 2019 and 2020

	NSW		Victoria		Tasmania		South Australia		Western Australia		Queensland	
	2019	2020	2019	2020	2019	2020	2019	2020	2019	2020	2019	2020
% Distribution	22.6	20.4	27.8	29.7	6.3	6.8	29.2	25.9	11	11.2	3.2	6.0
Generation (GWh)	4399	4604	5408	6720	1237	1535	5683	5845	2136	2534	625	1366

8.13.3 Hydro energy

The movement of water causes a turbine to spin, and the turbine is connected to a generator that produces electrical power from the mechanical power of the water. On large-scale systems, the water often flows from a high level (a dam, such as that shown in Figure 8.12) to a lower level (a river). In some cases, other intermittent renewable sources pump water from low areas into higher points to allow the hydrosystem to act as a large renewable battery. This system has been proposed as part of the Snowy Mountains Scheme (see Chapter 2) and will potentially increase its energy-producing capacity by 50%.

8.13.4 Geothermal energy

Geothermal energy uses naturally occurring heat from deep underground to create steam that is used to drive turbines. Iceland and the Philippines generate 25% and 17% respectively of their energy via geothermal means. Countries including El Salvador, Kenya, the Philippines, Iceland, New Zealand and Costa Rica generate more than 15% of their electricity from geothermal sources. Figure 8.13 shows a geothermal energy station.

FIGURE 8.12 Guthega Dam, a part of the Snowy Mountains Scheme
Phillip Minnis/Shutterstock

FIGURE 8.13 Geothermal energy station
Javarman/Shutterstock

Even though Australia does not have any significant volcanic activity, geothermal is a potentially viable renewable energy source. Large-scale geothermal projects are not commercially viable in Australia. However, projects for residential heating and cooling are in early development.

8.13.5 Marine energy

The term 'marine energy' refers to the use of ocean thermal energy, tidal and wave energy to generate electricity.

Ocean thermal energy

Energy is extracted from water using a heat-exchange process to produce electricity. This is possible because water temperature can vary by as much as 20°C in tropical regions. Water temperatures near the surface are warmer than those at deeper levels.

Tidal energy

Tides cause movements in ocean waters, and constrained topology near coastlines can accelerate these movements. Tidal energy generates electricity using the regular local flows of the tidal cycle (see Figure 8.14).

The Kimberley and Pilbara coasts of northern Western Australia see the country's largest tides. Other potential sources of tidal power are the Torres Strait off the coast of Darwin, Broad Sound in Queensland and Bass Strait in Tasmania.

FIGURE 8.14 The generation of tidal power
Shutterstock/Alex Mit

Wave energy

Waves are created by wind passing over the surface of the ocean. Wave-power plants can harvest the energy in the undulating motion of waves and convert it into electricity.

According to the Clean Energy Council, wave energy is strongest where there are trade winds and ocean swells. Australian wave-energy resources are greatest along the southern coastline.

8.13.6 Bioenergy

Bioenergy uses biomass to create energy. ('Biomass' is the name given to all plants and animals—including human beings—on Earth. It includes organic material such as wood, straw and grain crops.) 'Energy from biomass' refers to ways of using plants and animals as energy sources. Biofuel (see Figure 8.15) is fuel that is produced from biomass. Typically, the energy is extracted via either burning the biomass using the heat given off to create steam which drives a turbine, or by anaerobic or aerobic digestion. Anaerobic digestion is a process where organic matter such as animal manure, wastewater and food wastes is broken down by bacteria in the absence of oxygen. Aerobic digestion is a process to break down organic and biological waste which uses bacteria and oxygen. Newer technologies can convert the biomass directly to petrol and diesels.

FIGURE 8.15 Biofuel production
Fransen/Shutterstock

Common sources of bioenergy are:

- sugar cane residues (also known as 'bagasse')
- landfill gas (the methane produced by landfills)
- agricultural crop and livestock waste
- household garbage
- sewage gas
- wood waste
- black liquor (a by-product of the paper-making process).

In Australia, the bioenergy sector currently generates approximately 3314 gigawatt hours of energy per annum. This represents 1.4% of total electricity generation and equates to 6% of total clean energy generation.

Figure 8.16 gives an overview of various renewable generation sources' capacities for electrical generation.

As well as the individual renewable energies, there are also hybrid schemes (hybrid energy combines renewable energies, examples being biomass and wind, and solar and wind). Hybrid energy schemes may also include a traditional power source such as, for example, solar combined with a diesel generator.

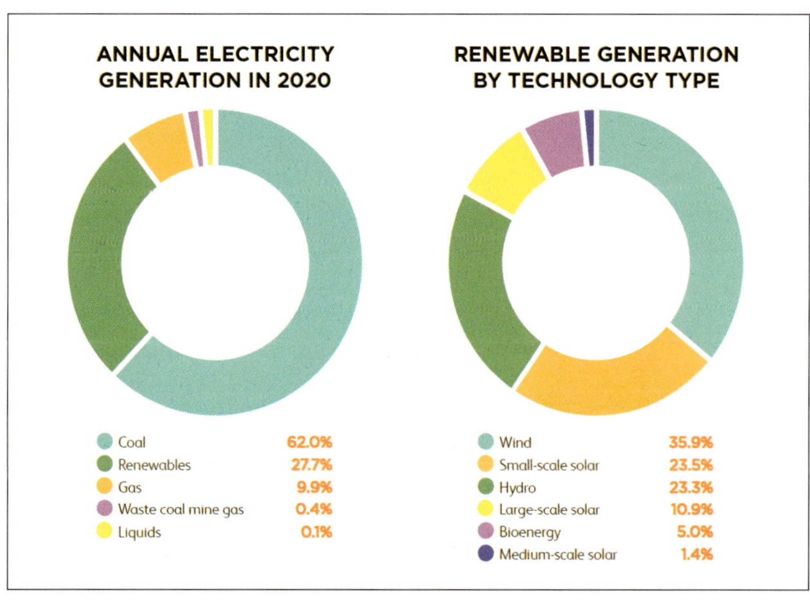

FIGURE 8.16 Electricity generation by energy type
Clean Energy Council

Table 8.7 gives an overview of state and territory electricity generation targets through renewable energy sources. In 2019, the ACT achieved 100% of its electricity supply through renewable energy sources and is aiming for zero net emissions by 2045. According to the Climate Council, the remaining states and territories have all set zero net emission targets for 2050.

TABLE 8.7 State and territory targets for electricity generation from renewable energy

ACT	NSW	Victoria	Tasmania	South Australia	Western Australia	Northern Territory	Queensland
100% by 2019	No target	50% by 2030	200% by 2040	100% by 2030	No target	50% by 2030	50% by 2030

CHECK YOUR UNDERSTANDING

8.24 Define renewable energy.

8.25 Give an example of a grid-strengthening technology.

8.26 Identify three sources of bioenergy.

8.27 Which renewable energy source contributed most to Australia's electricity generation in 2019–20?

INDUSTRY IN FOCUS: BATTERIES

Batteries are a form of energy-storage technology that uses chemicals to absorb and then release energy when it is needed. Types of batteries include lithium-ion and vanadium flow batteries (VFBs). Lithium-ion batteries degrade with cycling, whereas vanadium flow batteries do not—they store energy in a non-flammable liquid electrolyte.

The last few years have seen some exciting developments in battery construction. As of early 2021, Monterey, California was the site of the world's largest battery-energy storage system (BESS), the Moss Landing Energy Storage Facility, which has a capacity of 300 megawatts/1200 megawatt hours. The first part of the Moss Landing project was connected to the U.S. power grid and began operating in December 2020. In Australia, the initial stage of a 700 megawatt/2800 megawatt hour BESS installation at the Eraring Power Station in New South Wales is due for commissioning by late 2022. It will have more than double the capacity of Moss Landing.

An upgrade for Australia's (and the world's) largest lithium-ion battery was begun in 2020. The 50 megawatt/64.5 megawatt hour expansion of the Tesla Hornsdale Power Reserve battery will benefit the grid in South Australia, the National Electricity Market (NEM) and Australian consumers. The original project reportedly saved consumers over $150 million in its first two years of operation.

There are two types of grid-battery supplies—small-scale battery installations and large-scale battery storage. Small-scale battery installations in homes can provide backup power and can also be operated as distributed energy resources (DER), which are collectively referred to as a 'virtual power plant' (VPP). DERs are renewable energy units or systems that are commonly found at domestic and business locations. They include rooftop solar PV units, battery storage, thermal energy storage, electric vehicles and chargers, smart meters and home energy-management technologies.

Coupling (connecting) batteries is becoming more popular in both domestic and business environments. Coupling a large number of batteries installed together is known as 'grid-scale' or 'large-scale battery storage' (LSBS). BESS projects with a capacity of 7 gigawatts have either been proposed or are in the planning process for completion and connection to the grid by 2024. This includes 900 megawatts of committed and significantly progressed stages.

The Australian Renewable Energy Agency (ARENA) has been providing funding for large-scale battery storage projects in Australia since 2012.

8.14 Relevant workplace documentation, policies and procedures

Companies often implement policies requiring employees to minimise their negative environmental impact. Examples include directives for managers to consider sustainability issues as part of the planning process, promote environmental awareness and ensure that employees are aware of their environmental responsibilities.

Electrical licence-holders perform work observing wiring rules and standards, applying environmentally sustainable procedures in the energy sector. The Council of Australian Governments (COAG) Energy Council's work on national energy and resources reform involves overseeing the activities of the:

- Australian Energy Market Commission (AEMC)—the rule-maker and market-development advisory body
- Australian Energy Market Operator (AEMO)—the system operator
- Australian Energy Regulator (AER)—the economic regulator and rule-enforcer.

The responsibilities of workplace employees include following organisations' policies and procedures for sustainable work practices. Most of these will focus on meeting relevant work health and safety (WHS) and environmental legislation/regulations. Examples include:

- carrying out environmentally safe waste disposal
- following energy-efficient work practices
- following recycling practices and procedures
- reusing work resources
- using renewable energy sources
- enhancing work practices to maintain or improve the ecosystem, workplace and community
- completing relevant reports and workplace documentation.

Documents and records can be completed using electronic databases and spreadsheets or paper-based methods. Such documents could include:

- manufacturer information about product features and specifications (for energy efficiency and sustainability)
- manufacturer installation and maintenance instructions
- job sheets and reports
- material checklists
- safety data sheets (SDS)
- hazard inspection reports
- energy usage and meter readings
- wastage reports (to establish the volume of wastage).

8.14.1 Relevant reporting and risk-mitigation processes

An unplanned event or an environmental breach may necessitate internal and external reporting.

Internal reporting

An internal report should be made to one of the following:

- the person listed in the reporting procedure
- the relevant supervisor or manager
- the organisation's CEO.

External reporting

An external report may need to be made to:

- industry associations—these are usually aware of their industry's legislative requirements and will provide advice on how to report the unplanned event or breach
- the Environmental Protection Authority (EPA)—the legal body in each state or territory that ensures that environmental and biodiversity laws are being complied with. EPAs provide guidance on the reporting process.

CHECK YOUR UNDERSTANDING

8.28 Provide two examples of energy-saving improvement strategies and two examples of retrofits for electrical, refrigeration or air-conditioning installations.

8.29 Who should a worker report to within an organisation when responding to an unplanned event or environmental breach?

8.30 Identify three types of compulsory documents or reports that include information necessary to meet legal and organisational requirements for relevant WHS and environmental rules.

8.31 What ongoing checks can be undertaken to ensure all work is completed in accordance with environmentally sustainable work practices and procedures?

SUMMARY

- Human beings need to reduce their consumption of the Earth's resources.
- Sustainable development has been defined by the United Nations as development which meets the needs of the present without compromising the ability of future generations to meet their own needs.
- Methods of being sustainable include using less, reducing waste, reusing materials and recycling materials.
- Not adopting sustainable practices causes harm to the environment and can lead to the collapse of natural systems that support life.
- The greenhouse effect is where the heat from the sun is trapped within the Earth's atmosphere.
- Greenhouse gasses (including water vapour) are causing more heat than is normal to be trapped, thereby heating the planet up.
- Increases in temperature cause damage to the environment.
- International agreements have been reached in Kyoto and Paris to lower the amount of greenhouse gasses being released into the atmosphere.
- The *Environment Protection and Biodiversity Conservation Act (EPBC) 1999* is the Australian Government's environmental legislation. It covers environmental assessment and approvals, protects significant biodiversity and integrates the management of important natural and cultural places.
- Reducing greenhouse gas emissions also reduces the cost of materials and energy.
- A significant way of reducing emissions is to turn off energy-consuming devices, appliances or tools when they are not in use.
- The use of renewable energy sources reduces emissions.

- The most common types of renewable energy sources include:
  - hydro
  - wind
  - solar (PV)
  - solar thermal.
- Energy-harnessing technologies include:
  - biomass
  - wave
  - tidal
  - geothermal.
- Grid-strengthening technologies include small-scale battery installations and large-scale battery storage.
- Renewable energy battery technologies include lithium-ion and vanadium flow batteries.
- Distributed energy resources (DER) include rooftop solar PV units, battery storage, thermal energy storage, electric vehicles and chargers, smart meters and home energy-management technologies.
- Organisations' senior managers and managers should promote environmental awareness and ensure that employees are aware of their environmental responsibilities.

END-OF-CHAPTER QUESTIONS

8.1 Explain the term 'sustainability'.

8.2 Give an example of two uses of copper in the electrical industry.

8.3 Name two pollutants used in the electrotechnology industry.

8.4 Give an example of a sustainable work practice.

8.5 State three ways of reducing the consumption of electricity.

8.6 Give an example of where surplus materials in industry can be reused.

8.7 Give two examples of materials in industry that can be recycled.

8.8 What does 'ABRI' stand for?

8.9 What is a disadvantage of not repairing some relatively cheap products?

8.10 Name three consequences of neglecting sustainable work practices.

8.11 Name four causes of pollution as a result of neglecting sustainable work practices.

8.12 What effect will neglecting sustainable work practices have on future generations?

8.13 Define the 'greenhouse effect'.

8.14 Give one cause of the greenhouse effect for which human beings are responsible.

8.15 Give two examples of 'feedback loop' mechanisms that can increase the greenhouse effect.

8.16 What is the primary greenhouse gas that contributes to the greenhouse effect?

8.17 Name five consequences of the greenhouse effect.

8.18 Give the names of two international initiatives created to limit greenhouse gas emissions.

8.19 What is Australia's carbon emissions target for 2050?

8.20 Which country has the highest penetration rate of solar panels on household roofs?

8.21 There are numerous national initiatives in place to reduce carbon gas emissions. Name seven of them.

8.22 What was the aim of the Kyoto Protocol?

8.23 What is the aim of the Paris Agreement?

8.24 Name four regulatory bodies that were formed to clean up emissions and the climate.

8.25 Name eight methods of reducing greenhouse gas emissions.

8.26 What are the three main areas that the electrotechnology industry can affect to bring about a reduction in greenhouse gas emissions?

8.27 How can we protect our reserves of copper?

8.28 How can we reduce the consumption of electricity?

8.29 How can we reduce the consumption of oil and gas?

8.30 Name five technologies that can be utilised to reduce energy consumption.

8.31 Name five ways to reduce energy consumption.

8.32 What is renewable energy?

8.33 Explain how solar energy works.

8.34 Explain how wind energy works.

8.35 Explain how hydro energy works.

8.36 Explain how geothermal energy works.

8.37 Explain how ocean energy works.

8.38 Explain how bioenergy works.

CHAPTER 9

Alternating current circuits: single phase

LEARNING OBJECTIVES

- Identify alternating current (a.c.) waveforms
- Understand the use of oscilloscopes to measure d.c. and a.c. voltage levels
- Explain how an a.c. EMF is generated
- Understand sinusoidal waveforms and associated terminology
- Use phasor diagrams to solve a.c. circuit problems
- Understand resistance, reactance and impedance in series and parallel a.c. circuits
- Perform calculations involving voltage, current, resistance, reactance, impedance, power factor, phase angle, frequency and power factor correction in a.c. circuits
- Recognise the dangers of resonance in a.c. circuits
- Understand harmonics and the problems they cause in a.c. circuits

PREREQUISITE KNOWLEDGE

Understand d.c. circuit calculations

9.1 Alternating current

9.1.1 Introduction

Alternating current (a.c.) is generated when a *loop conductor*, or coil, is rotated within a uniform magnetic field. It is used in preference to direct current for power transmission, transformers and effects that require self- or mutual inductance.

A coil of wire rotating within a magnetic field produces a waveform with a positive voltage for one half-revolution and a negative voltage for a second half-revolution. The waveform has a constant rate of change, so there are no sharp corners. Such a perfect waveform is called a 'sine wave' (see Figure 9.1).

The laws of physics naturally produce sinusoidal-shaped voltages, and the maths of a.c. electricity is based on sinusoidal waveforms. The term 'sinusoidal' means that form follows the strict mathematical function of the sine function.

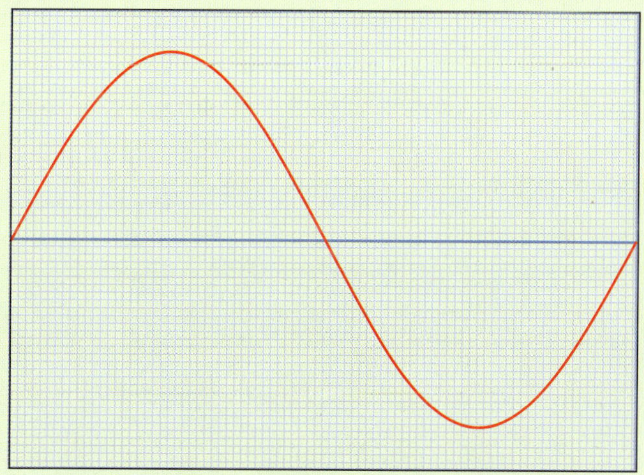

FIGURE 9.1 Graphic representation of a sine wave

9.1.2 Sinusoidal waveforms

When carrying out circuit calculations, it is easiest to work with pure d.c. and pure sinusoidal a.c., as shown in Figure 9.1.

When tables of sine values are plotted as a graph, the resulting characteristic curve is a sine wave—it is sinusoidal in shape. The first half of the curve is a mirror image of the second half and is always positive; the second half is always negative.

The advantages of the sine wave are:

- It is an easy shape to reproduce because it is a naturally occurring waveform based on the trigonometry of a circle.
- It is the only waveform that produces a current flow with the same waveform as the voltage.
- Its shape allows the maximum permissible power per unit size of machine.
- As the sine wave has an alternating voltage, current and magnetic flux, it can operate any equipment that relies on mutual induction, such as transformers.

9.1.3 Other waveforms

In the electrical power industry, any waveform other than sinusoidal is considered to be distorted. There are however other waveforms in use. These are generated for specific purposes but not for the transmission of electrical energy. The examples below show some of the more common waveforms.

Figure 9.2(a) shows the straight line of a d.c. voltage. It attains a steady value and remains constant at that value. Although the voltage value may change over time, it does not change periodically like the other waveforms.

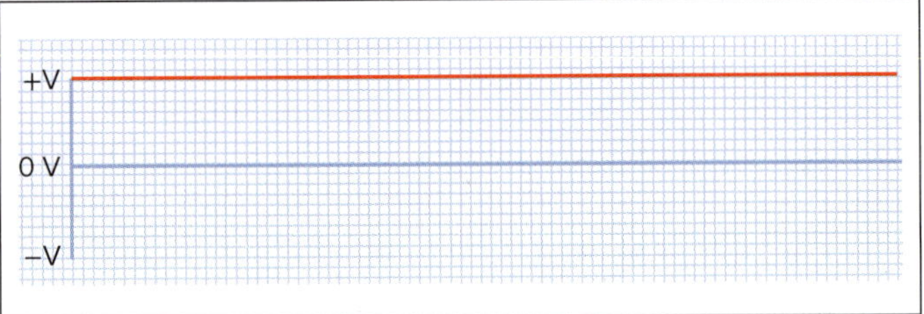

FIGURE 9.2(a) Examples of waveforms: straight line of a d.c. voltage

Figure 9.2(b) shows the sinusoidal waveform. A pure sine wave is continually changing at a constant rate. Its shape is symmetrical around the zero line. It is not, as some may think, a series of semicircles.

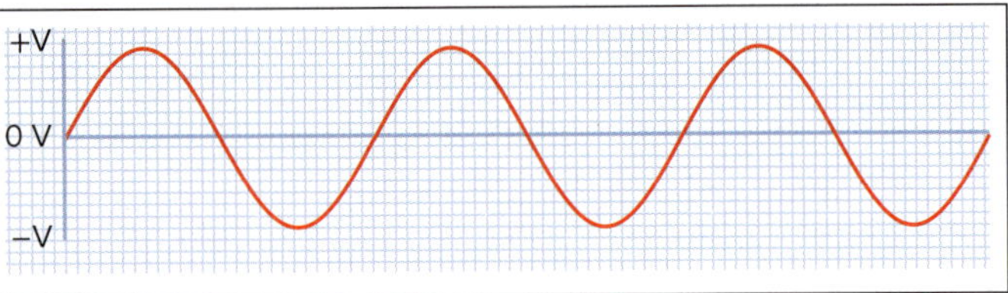

FIGURE 9.2(b) Examples of waveforms: sinusoidal waveform

The next waveform is the sawtooth wave in Figure 9.2(c). It is immediately above the triangular wave in Figure 9.2(d), and the two are often confused. Both rise symmetrically but, while one decreases at the same rate as it increases, the other changes abruptly to its most negative value. Both are used in timing circuits because the voltage changes proportionally to time.

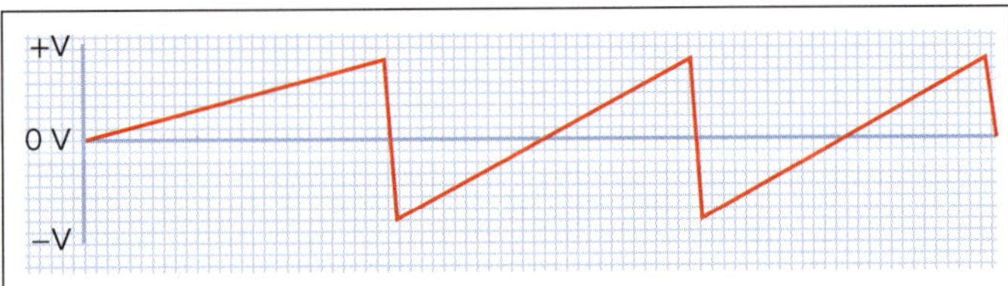

FIGURE 9.2(c) Examples of waveforms: sawtooth wave

Figure 9.2(e) shows the square waveform. When the positive or 'on' time period is equal to the zero or 'off' time period, it is considered square or symmetric. This switches a voltage from positive to zero (or negative) at a steady rate: there is a positive voltage for a period, and then a zero (or negative) value. The transition from positive to zero (or negative) is abrupt and seems to be instantaneous. But in real waveforms, this must take a definite period of time.

FIGURES 9.2(d) Examples of waveforms: triangular wave

When the 'on' period is not equal to the 'off' period, the waveform is described as 'asymmetric' or 'rectangular'. To ensure that an electronic circuit is switched off completely, it is sometimes necessary to make sure there is a negative voltage.

The square waveform is also used in logic circuits and variable frequency motor drives. Electronic components sometimes usc it in comparison-type circuits to indicate that an event has occurred, for example that one voltage has exceeded another voltage.

The waveform in Figure 9.2(f) is a variation of the square wave used to control power circuits. There are two ways of varying the amount of power, and each provides power only when the square wave is positive. One way is to vary the ratio between the time period when the voltage is on and the time period when it is off. This means that the frequency of the power pulses does not change. The other method simply varies the amount of time the power is off, changing the frequency of the power pulses. In both methods, the ratio of on-time to off-time is known as the 'mark/space ratio'.

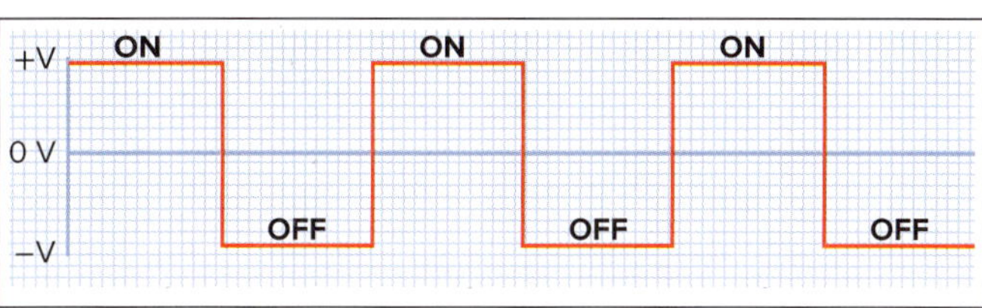

FIGURE 9.2(e) Examples of waveforms: square waveform (symmetrical)

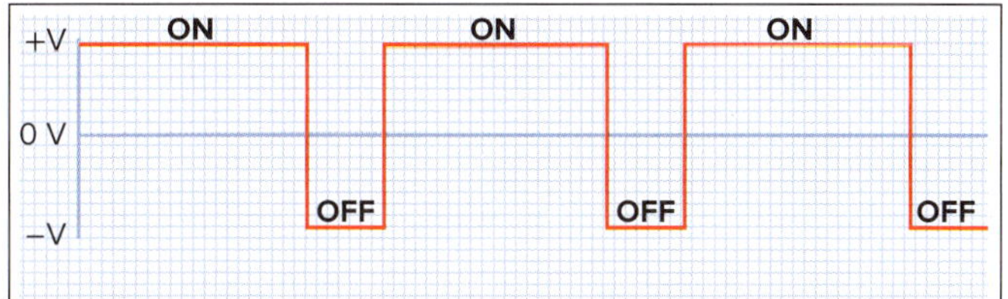

FIGURE 9.2(f) Examples of waveforms: square waveform (asymmetrical)

Representing values

The custom when calculating and representing values of voltage and current is to use upper-case letters for d.c. values and lower-case letters for a.c. values. However, as *root-mean-square (RMS)* values are equivalent to d.c. values, they are also given in capital letters. (RMS values are discussed in section 9.3.4.) In this textbook, lower-case letters usually refer to instantaneous values.

Another custom is to always use RMS a.c. values unless otherwise specified. Modern a.c. instruments are calibrated to read RMS values, and some are what are known as 'true RMS' instruments. This means that they read RMS even if the waveform is not a true sine wave.

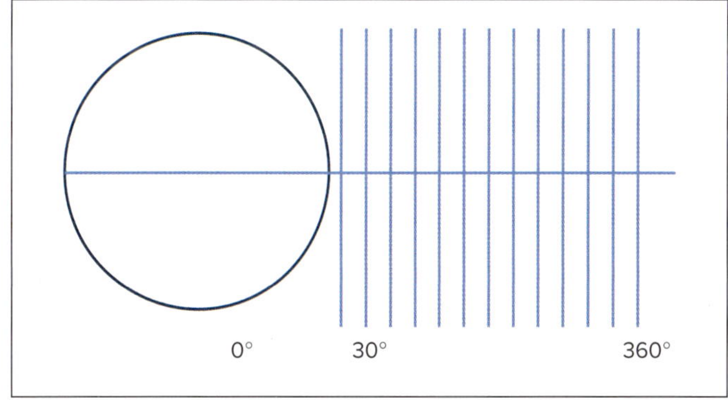

FIGURE 9.3(a) Calculating values along a sine wave, Step 1

9.1.4 Sine, cosine and tangent ratios of a right angle triangle

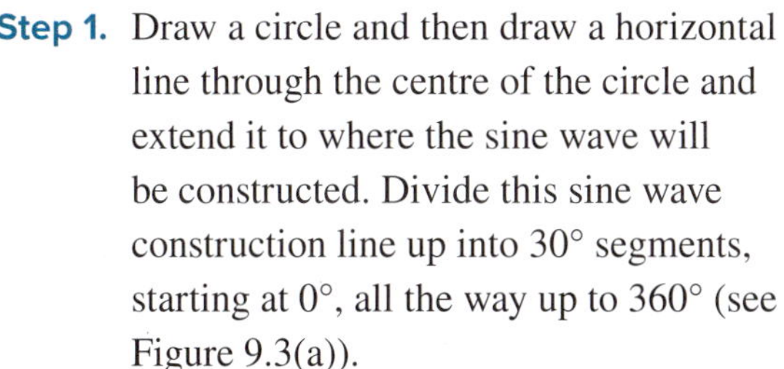

The output generated from an alternator is a sine wave. We can use a circle to help calculate values along it because points on the sine wave correspond to points on the circle. The calculation is done as follows.

Step 1. Draw a circle and then draw a horizontal line through the centre of the circle and extend it to where the sine wave will be constructed. Divide this sine wave construction line up into 30° segments, starting at 0°, all the way up to 360° (see Figure 9.3(a)).

Step 2. Draw lines from the centre of the circle to the perimeter at angles of 30° (Figure 9.3(b)).

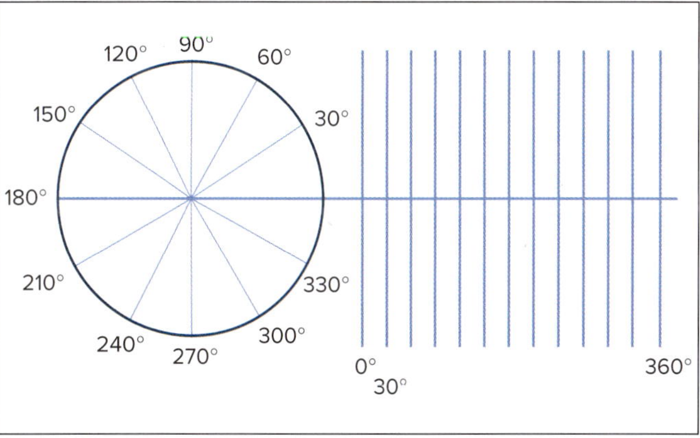

FIGURE 9.3(b) Calculating values along a sine wave, Step 2

Step 3. Now draw a horizontal line between the point at which the angled line radiating from the centre meets the perimeter of the circle and the 30° vertical line (Figure 9.3(c)).

Continue this process for the remaining degrees around the outside of the circle. The intersection points for each angle can then be connected to form the sine wave (Figure 9.3(d)).

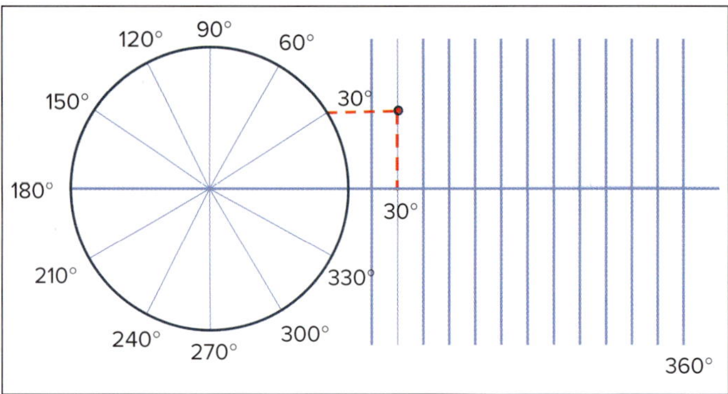

FIGURE 9.3(c) Calculating values along a sine wave, Step 3

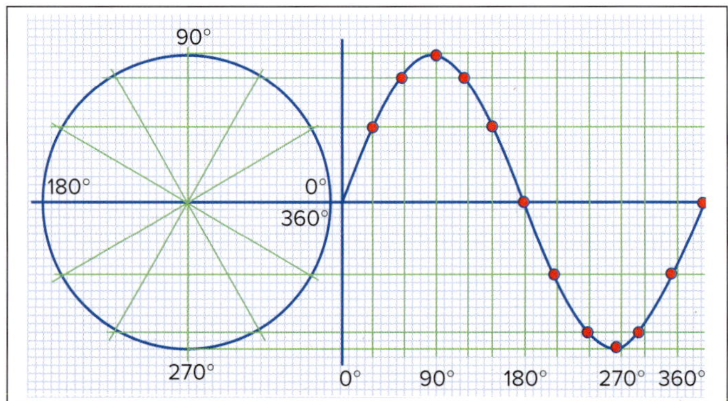

FIGURE 9.3(d) Sine wave construction

9.1.5 Alternative construction of sinusoidal curves

An alternative to the sine wave construction above is to calculate and plot instantaneous values.

The calculated instantaneous values method works as follows.

Step 1. Axes for a graph are constructed as in the previous method, preferably on graph paper. The vertical (y-) axis is scaled for both positive and negative values of voltage (or current). The horizontal (x-) axis is marked out in degrees of rotation.

Step 2. The instantaneous values are then calculated for angles using the equation $v = V_{max} \sin \theta$. Calculate every 10 degrees from 10° to 80°, remembering that the same values are used for 100° to 170°. Converting these values into negative numbers produces the remaining half of the sine wave.

Step 3. The values are then plotted against the corresponding angle, as shown in Figure 9.3(e).

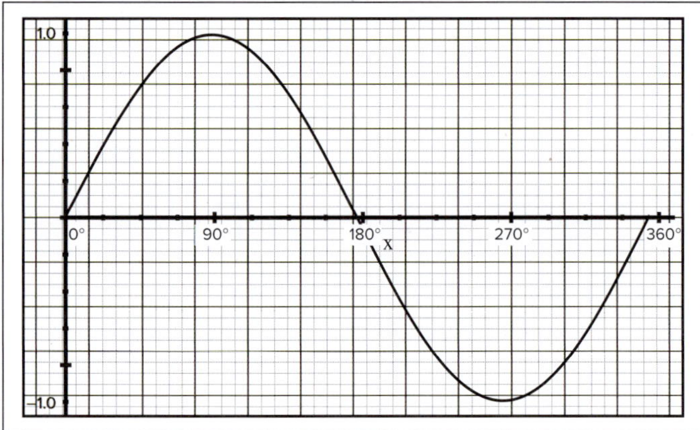

FIGURE 9.3(e) Computer-drawn sine wave

The curve is identical to that produced by the previous method except that it has more plot points. *Some computer programs will plot the sine wave directly, given a Vmax value and the equation.* These programs include GNUPlot, Grapher.app, Graph.exe and DPlot.

Trigonometry can also be used to calculate values of the sine wave.

Consider the triangle in Figure 9.3(f) that was constructed using the 60° line and a line taken down to the horizontal at 0°.

The side going from the centre of the circle to the perimeter is the radius. The magnitude of this line equals the peak value of the sine wave. It is the same value all the way around the circle. This side is called the 'Hypotenuse'.

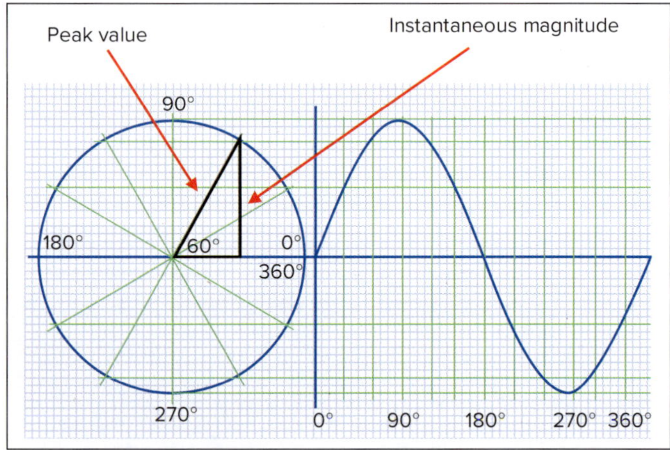

FIGURE 9.3(f) Construction of instantaneous magnitude

The point that corresponds with 60° when extended perpendicular to the horizontal line forms a right angle triangle in the centre of the circle. Here it has a 60° angle. The line on the triangle from this point to the horizontal line is equivalent to the instantaneous magnitude of the sine wave at 60° ('instantaneous magnitude' meaning here the distance between zero and the point that corresponds with 60°). This side is called the 'Opposite'—it is the side opposite to the 60° angle.

The remaining side is called the Adjacent as it is adjacent to the 60° angle.

We can use trigonometry to calculate these values consistently when the magnitude of the maximum value and the desired angular point are known.

The main equation operators are known as Sine (sin), Cosine (cos) and Tangent (tan), and these relate to the ratios of the sides. (Equation operators are mathematical tools that aid calculations and are always the same for each angle.)

The Sine of the angle—in this example sin 60°—is determined by the Opposite divided by the Hypotenuse. This ratio will always be the same, regardless of the size of the triangle, as long as the angle remains the same.

$$\text{Sine} = \frac{\text{Opposite}}{\text{Hypotenuse}}$$

The Cosine of the same angle (in this example cos 60°) is determined by the Adjacent divided by the Hypotenuse. Again, this will always be the same as long as the angle remains the same.

$$\text{Cosine} = \frac{\text{Adjacent}}{\text{Hypotenuse}}$$

The Tangent of the angle, tan 60°, is determined by the Opposite divided by the Adjacent.

$$\text{Tangent} = \frac{\text{Opposite}}{\text{Adjacent}}$$

A simple way of remembering these ratios is to use the mnemonic 'SOH CAH TOA':

1. SOH: <u>S</u>ine $= \frac{\text{Opp}}{\text{Hyp}}$
2. CAH: <u>C</u>osine $= \frac{\text{Adj}}{\text{Hyp}}$
3. TOA: <u>T</u>angent $= \frac{\text{Opp}}{\text{Adj}}$

9.1.5 Applying Pythagoras' theorem to a right angle triangle

Pythagoras' theorem is a useful tool that is associated with right angle triangles. It shows how, if the lengths of two of a triangle's sides are known, the third side can be determined (see Figure 9.4).

The theorem states that the sum of the Opposite squared and the Adjacent squared equals the Hypotenuse squared.

$$\text{Hypotenuse}^2 = \text{Opposite}^2 + \text{Adjacent}^2$$

We can make the Hypotenuse the subject of the equation by taking the square root of each side of the equals sign:

$$\sqrt{\text{Hypotenuse}^2} = \sqrt{(\text{Opposite}^2 + \text{Adjacent}^2)}$$

Since the square root of the Hypotenuse squared is just the Hypotenuse, the equation can be simplified as:

$$\text{Hypotenuse} = \sqrt{(\text{Opposite}^2 + \text{Adjacent}^2)}$$

To determine the Opposite or the Adjacent, the first equation is transposed:

$$\text{Opposite}^2 = \text{Hypotenuse}^2 - \text{Adjacent}^2$$

$$\text{Adjacent}^2 = \text{Hypotenuse}^2 - \text{Opposite}^2$$

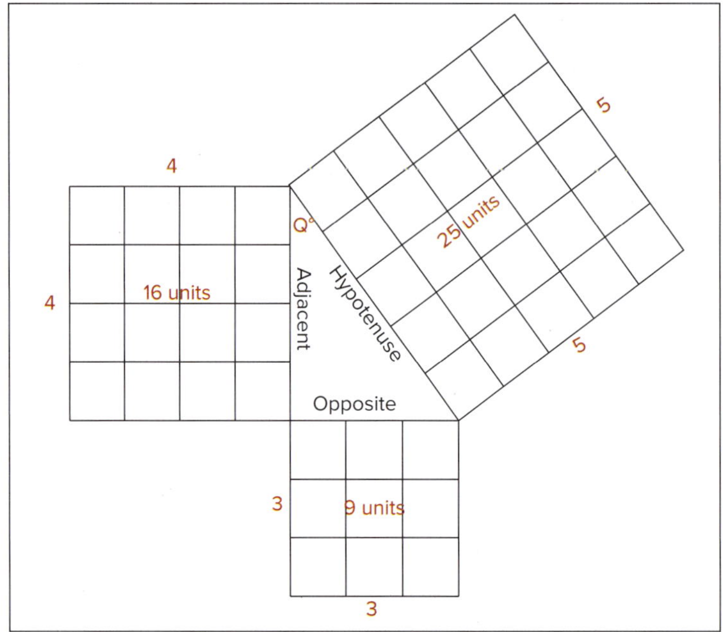

FIGURE 9.4 Applying Pythagoras' theorem to a right angle triangle

Again, these can be simplified in the same way:

$$\sqrt{\text{Opposite}^2} = \sqrt{(\text{Hypotenuse}^2 - \text{Adjacent}^2)}$$

$$\therefore \text{Opposite} = \sqrt{(\text{Hypotenuse}^2 - \text{Adjacent}^2)}$$

And:

$$\sqrt{\text{Adjacent}^2} = \sqrt{(\text{Hypotenuse}^2 - \text{Opposite}^2)}$$

$$\therefore \text{Adjacent} = \sqrt{(\text{Hypotenuse}^2 - \text{Opposite}^2)}$$

CHECK YOUR UNDERSTANDING

9.1 Find the Hypotenuse of a right angle triangle which has an Adjacent side of 40 units and an Opposite side of 50 units.

9.2 Calculate the Adjacent side of a right angle triangle which has a Hypotenuse of 100 units and an Opposite side of 70 units.

9.3 If the Adjacent side of a right angle triangle is 60 units and the Hypotenuse is 90 units, what is the value of the Opposite side?

9.2 EMF generation

9.2.1 Alternators

A machine producing a voltage with an alternating waveform at its terminals is called an 'alternator' or, occasionally, an 'alternating current (a.c.) generator'. Both terms are correct, although 'generator' is usually reserved for d.c. machines. The modern alternator consists of two sets of electromagnetic coils, one set of which rotates to produce an EMF at the terminals. These coils may be connected in either series or parallel to produce a desired voltage.

9.2.2 Loop rotating in a magnetic field

Figure 9.5 shows a single-loop generator in which the loop may be rotated in the field between two fixed, permanent magnets. As this happens, an EMF will be generated within the loop. Two metallic rings called 'slip-rings' are connected to the loop ends to connect the loop to an external circuit, and carbon brushes running over the slip-rings complete the connection.

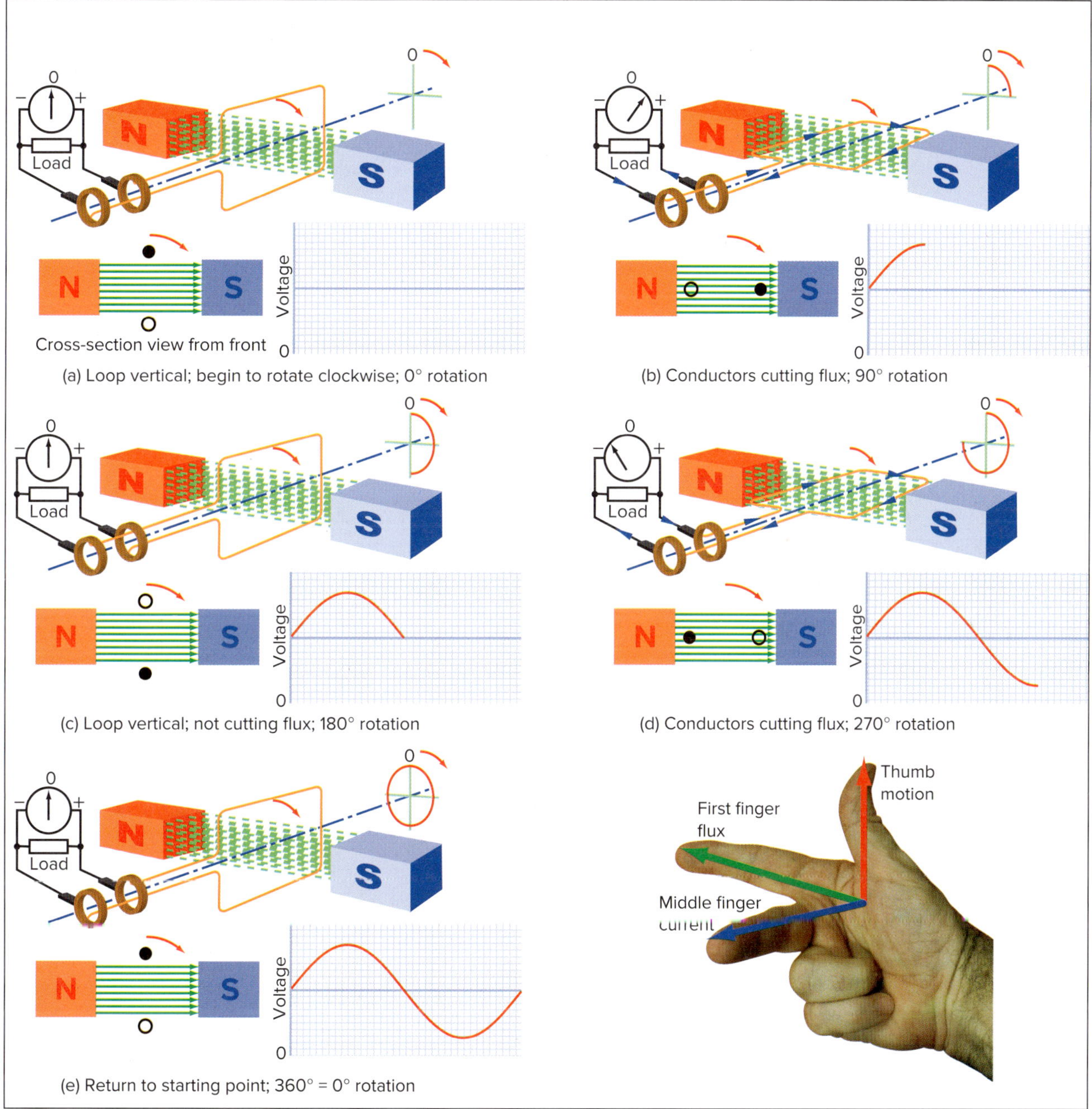

FIGURE 9.5 **EMF in a loop and Fleming's right-hand rule**

9.2.3 Direction of an induced EMF

The direction of a generated voltage depends on the direction of relative motion between a magnetic flux and the conductors that link with it. Fleming's right-hand rule (see Figure 9.5) finds the direction of an induced EMF if the direction of flux and direction of relative motion are known.

9.2.4 Magnitude of a generated EMF

The value of an induced EMF depends on three factors—the magnetic field density, the number of conductors in series (i.e. the total length of conductor in the magnetic field) and the relative rate of motion between the first two factors.

It is easy to determine the rate of rotation of machines and then factor in the comparative direction of the conductors in relation to the magnetic field. By combining these two factors, the relative rate of motion can be found from the expression: v sin (θ). By combining all these factors, an equation can be developed. The generated voltage can be found from:

$$e = Blv \sin \theta$$

where:

e = value of induced voltage
B = flux density in webers
l = length of conductor in metres
v = velocity in metres per second.

This agrees with Fleming's right-hand rule, where the movement of the conductor is at right angles to the magnetic flux. It also takes into account the fact that a conductor might not always be travelling at right angles to the magnetic flux—it could be travelling at some other angle.

9.2.5 Effect on EMF as the loop rotates

The position of the loop shown in Figure 9.5(a) means that the initial movement of its sides is parallel to the magnetic flux. No cutting across the flux will occur, and no EMF will be generated.

When the loop begins rotating, it starts to cut across the flux at an acute angle. This generates a small EMF in each loop side. Further rotation causes a corresponding increase in the angle between the flux and the path of the conductors. This increases the value of generated EMF. In the position shown in Figure 9.5(b), the loop sides cut across the flux at 90°, resulting in the generation of maximum EMF at this position. The direction of the generated EMF is shown by the arrows. It can be confirmed by Fleming's right-hand rule.

Further rotation of the loop causes a gradual decrease in EMF. The angle of cutting becomes increasingly smaller until, at the position shown in Figure 9.5(c), the EMF will fall to zero again.

As the loop rotates from position (c) towards position (d), the direction of relative motion of the coil sides reverses. Since the field direction remains fixed, the direction of the generated EMF must reverse. This is indicated by the arrows in Figure 9.5(d).

In Figure 9.5(d), the EMF will again reach a maximum, but in a direction opposite to that of the previous maximum. Further rotation will cause a gradual decrease in EMF until it again becomes zero, when the loop completes one revolution and returns to its original position. This is shown in Figure 9.5(e).

9.2.6 Alternative alternator construction

The basic construction of an alternator is a loop rotating in a magnetic field, as in Figure 9.6(a). The alternating voltage produced when the ends of the loop are connected to slip-rings causes current to flow in alternate directions.

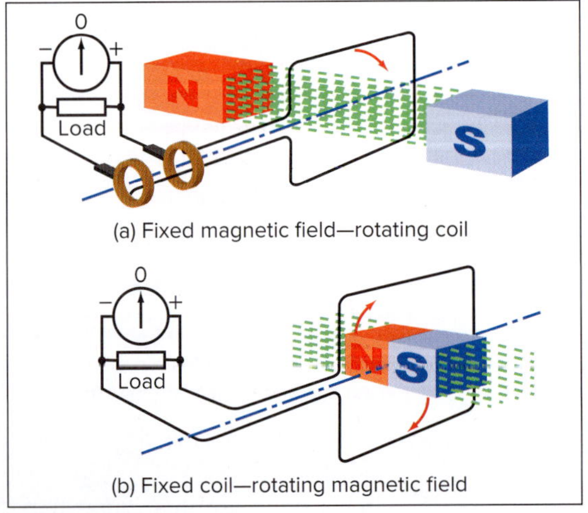

FIGURE 9.6 (a) Rotating coil vs (b) rotating field

This result can also be achieved by keeping the loop stationary and rotating the magnetic field to produce the relative motion. It is often easier to have the generating coils on the outside of the magnetic field, as shown in Figure 9.6(b). This is especially true when there are three phases instead of one.

There are several advantages to this type of construction:

- Slip-rings are not necessary for connection, so solid connections can be made.
- With small alternators, the rotating magnetic field can be obtained from a permanent magnet.
- With large alternators, the field can be driven by a smaller excitation generator (in a similar fashion to a separately excited d.c. motor). This avoids the need for brushes but still allows adjustment of the excitation.

In machines, solid connections allow higher voltages and currents to be handled and connected safely to the external circuit. Sliding connections via slip-rings, brush gear and brushes are removed using permanent magnets (or separate excitation). This reduces friction, wear and other maintenance and safety problems. Voltage may be increased by adding turns to coils or simply adding more coils (see Figure 9.7).

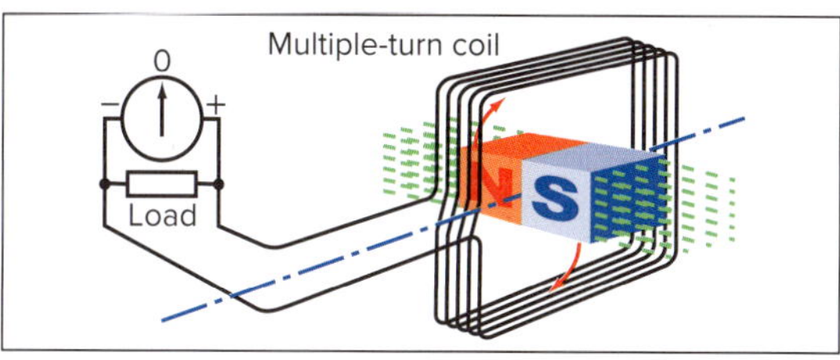

FIGURE 9.7 Increasing voltage with a multiple-turn coil

CHECK YOUR UNDERSTANDING

9.4 You can calculate the output of an alternator to show the positive, negative and zero values throughout the 360 degrees of the waveform using the equation e = Blv sin θ. Given that e = Blv = 325 V, determine the value of output voltage at angles of 0°, 90°, 180°, 270° and 360°. The first calculation has been completed for you below. (If you are not familiar with the use of a scientific calculator, study Example 9.1 and come back to this.)

Output voltage calculation for 0°:

e = Blv sin (0) e = 325 × sin (0) e = 0 V

9.3 Sinusoidal waveform

9.3.1 Instantaneous values

Figure 9.8 shows the computer-drawn sine wave of EMF generated as an alternator coil is rotated through 360°E (electrical degrees). In this example, the maximum voltage generated occurs when θ = 90° (hopefully you would have observed this in the previous calculations as sin 90 = 1). This is usually indicated by the symbol V_{max} or V_m. At this instant, the coil cuts the flux at 90°.

The instantaneous voltage (v) at any point on the waveform can be found from:

$$v = V_{max} \times \sin\theta$$

Since V_{max} occurs when θ = 90°, V_{max} corresponds to a sine ratio of 1. The sine curve, however, varies between 0 and 1, as θ varies between 0° and 90°. The instantaneous voltage (v) must also vary between 0 and a maximum value:

$$v = V_{max} \sin\theta$$

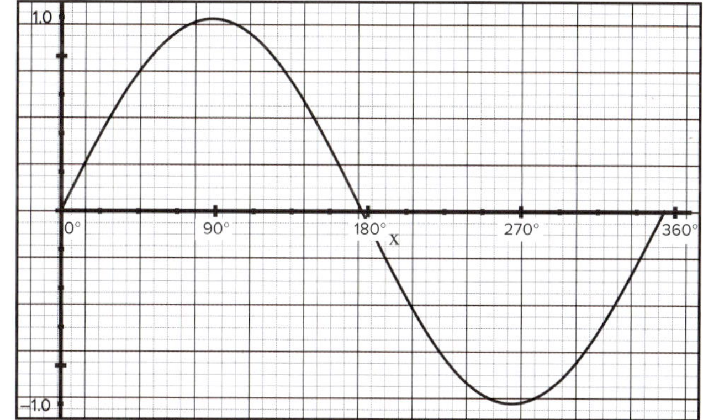

FIGURE 9.8 Computer-drawn sine wave

where:

$$v = \text{instantaneous voltage}$$
$$V_{max} = \text{maximum voltage}$$
$$\theta = \text{angle of rotation.}$$

In general, if an alternating voltage is applied to a load, the current that flows is also alternating. So a similar equation can be used for instantaneous current:

$$i = I_{max} \sin \theta$$

where:

$$i = \text{instantaneous current}$$
$$I_{max} = \text{maximum current}$$
$$\theta = \text{angle of rotation.}$$

Sine values can be derived from a scientific calculator.

EXAMPLE 9.1

The maximum EMF generated in an alternator coil is 200 V. Calculate the instantaneous voltage (v) for the following angles of rotation:

- $\theta = 30°$
- $\theta = 75°$
- $\theta = 150°$
- $\theta = 240°$
- $\theta = 310°$

$$v_a = V_{max} \sin \theta \quad (1)$$
$$= 200 \times \sin (30) \quad (2)$$
$$= 200 \times 0.5 \quad (3)$$
$$= \underline{100 \text{ V}} \quad (4)$$
$$v_b = V_{max} \sin \theta \quad (5)$$
$$= 200 \times \sin (75) \quad (6)$$
$$= 200 \times 0.966 \quad (7)$$
$$= \underline{193 \text{ V}} \quad (8)$$
$$v_c = V_{max} \sin \theta \quad (9)$$
$$= 200 \times \sin (150) \quad (10)$$
$$= 200 \times 0.5 \quad (11)$$
$$= \underline{100 \text{ V}} \quad (12)$$
$$v_d = V_{max} \sin \theta \quad (13)$$
$$= 200 \times \sin (240) \quad (14)$$
$$= 200 \times -0.866 \quad (15)$$

$$= \underline{-173 \text{ V}} \quad (16)$$

$$v_e = V_{max} \sin \theta \quad (17)$$

$$= 200 \times \sin(310) \quad (18)$$

$$= 200 \times -0.766 \quad (19)$$

$$= \underline{-153 \text{ V}} \quad (20)$$

9.3.2 Voltage and current cycles

The term 'cycle' means a recurrent order of events. For example, the seasons of the year follow a definite cycle: summer, autumn, winter, spring.

The output of an alternator also follows a definite cycle. Figure 9.9(a) shows the output EMF of a two-pole alternator for one of these cycles or one complete revolution. During this time, the EMF starts from 0°, rising to a maximum positive voltage as the conductors pass the magnetic north pole at an angle of 90°. The voltage then reduces and falls back to zero at 180°. In the second half of the cycle, the voltage increases to maximum negative voltage at 270°, then reduces again before finally returning to zero voltage at 360°. During the period of one revolution, the value of EMF has made one complete electrical cycle. Rotation of the coil through further revolutions causes the same cycle for each.

Consider the effect of four poles being used in the alternator, as in Figure 9.9(b). During one revolution, a particular side of the coil will rotate from 0° through a north pole to a south pole, then to the next north pole and on to the next south pole. Finally, it returns to the original position. During this time, the output EMF rises to a maximum in each direction. It does the same again before reaching 360° of mechanical rotation. The rotor therefore completes two electrical cycles in one mechanical cycle, as shown in the waveform of Figure 9.9(b).

So, the number of completed cycles of EMF in a given time depends on the number of poles that a rotor passes in that time, and not only on the number of mechanical revolutions that it rotates through in the same time.

In the majority of cases, electrical motors and generators require two poles per cycle—one north pole and one south pole.

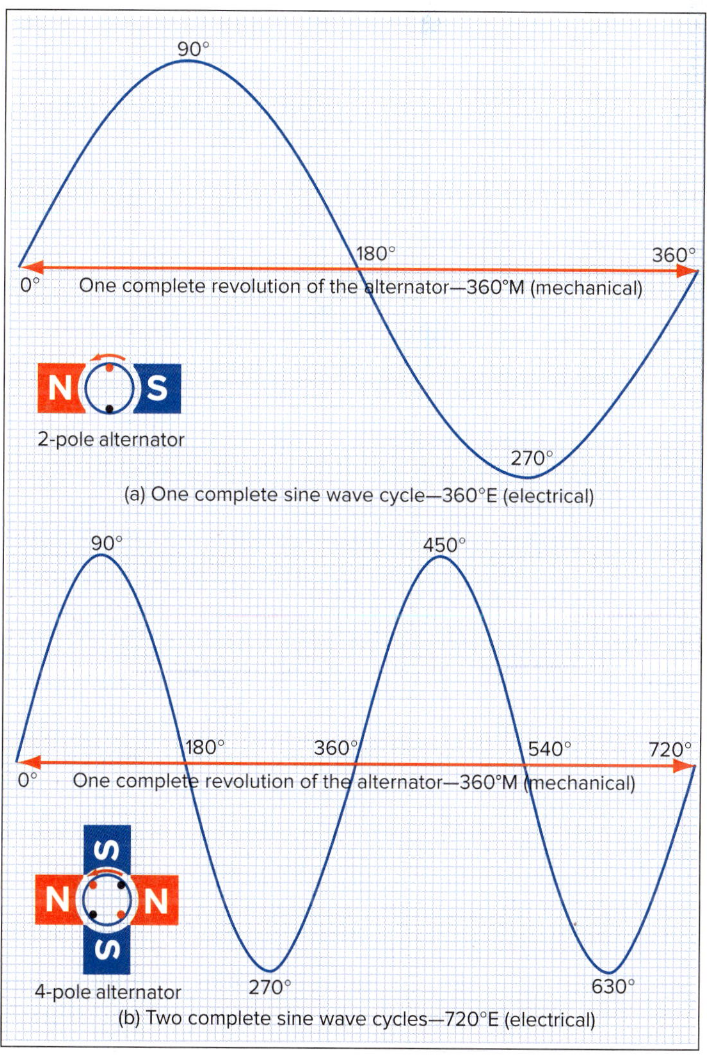

FIGURE 9.9 Electrical vs mechanical degrees

These relationships are shown in the following equation:

$$n = \frac{120\ f}{p}$$

where:

n = speed (rpm)
f = frequency (Hz)
p = poles.

Note that the constant of 120 comes from the conversion of the rpm value from minutes to seconds (× 60) and the fact that there are two poles, making up a pole pair (× 2).

EXAMPLE 9.2

What speed does an alternator need to be driven to produce 50 Hz if it has:

- two poles?
- four poles?

$$n = \frac{120f}{p}$$

$$n_a = \frac{120 \times 50}{2}$$

$$= \underline{3000\ \text{rpm}}$$

$$\text{and } n_b = \frac{120 \times 50}{4}$$

$$= \underline{1500\ \text{rpm}}$$

9.3.3 Electrical frequency

The number of cycles in a given time is a function of mechanical rotation and the number of poles in the machine. Thus, it is necessary to define one electrical cycle as 360°E. Mechanical angles should be referred to as '°M' whenever there could be some confusion.

In Australia, the standard frequency of the electrical supply system is 50 Hz, that is 50 hertz or 50 'cycles per second'. There are 50 sine waves in one second, so the frequency at which the sine waves are generated is 50 times a second.

Another way to look at it is to define how long a waveform is in time. This is known as the 'period' of the waveform. From 0°E through to 360°E is one waveform or period. If there are 50 cycles per second, each period must be $\frac{1}{50}$ seconds long (or 20 ms).

$$f = \frac{1}{T}$$

or

$$T = \frac{1}{f}$$

9.3.4 Sinusoidal wave values

Here is a summary of some of the terms used so far to describe alternating waveforms:

- The **periodic function** is an alternating wave that repeats itself in a cycle.
- The **frequency** (of a wave) is the number of times a wave repeats itself in one second. (This is expressed in cycles per second or hertz.)
- The **period** is the time taken for one cycle to complete itself. It is equal to the reciprocal of the frequency ($\frac{1}{f}$ seconds).
- An instantaneous value of either a voltage or current waveform is a value expressed at only one instant of time.

Below are some other values that are often used to describe a sinusoidal waveform or to do calculations on sinusoidal values.

Average value

As an alternating sinusoidal waveform is continually varying in value, a meaningful standard must apply when describing its value (whether as a voltage or a current). Mathematically, the average value of a sinusoidal function is $\frac{2}{\pi}$ or 0.637.

Therefore, the average voltage or current can be found from:

$$V_{av} = 0.637\ V_{max}$$
$$I_{av} = 0.637\ I_{max}$$

These expressions are accurate for sinusoidal waveforms only.

The average value applies only for individual half-cycles. As the direction of flow alternates, the average value of the second half-cycle in this example is −0.637. Therefore, the average flow for a complete cycle is zero. Average values are more applicable in circuits involving conversion of a.c. to d.c.

Average values are not suitable as a general basis for comparing alternating waveforms with direct current in terms of energy use or power. To compare the real effects of alternating waves, the 'root–mean–square' (RMS) value is used.

Root–mean–square values

The value of a sinusoidal waveform that produces the same heating effect as d.c. can be obtained mathematically by a process called the 'root–mean–square' (RMS) value of that wave. It can be defined as the square root of the arithmetic mean of the squares of a set of values. Calculating this requires an understanding of calculus, which fortunately we do not need to get into. We just need to remember that:

$$V_{RMS} = 0.707\ V_{max}$$
$$I_{RMS} = 0.707\ I_{max}$$

Once again, these expressions are accurate for sinusoidal waveforms only.

The RMS value is also described as the effective voltage that will create an effective current flow to do the same work as a d.c. voltage.

Unless specifically stated otherwise, any given a.c. value should always be considered as the RMS value. The normal domestic a.c. supply is referred to as 230 V, but the maximum value is 325 V.

Throughout this textbook, all a.c. values are RMS values unless otherwise stated.

The maximum value of a 230 V RMS supply can be found from the expression:

$$V_{RMS} = 0.707\ V_{max}$$
$$V_{max} = \frac{V_{RMS}}{0.707}$$
$$= \frac{230}{0.707}$$
$$= 325.3\ V$$

9.3.5 Peak-to-peak values

The value of a wave from its positive peak to its negative peak can be important. With most alternating waveforms, this is equal to twice the maximum value in either direction. For example, as we have just seen, for a 230 V a.c. supply, the maximum value of voltage is found by dividing 230 by 0.707. This value of 325.3 V is the peak value. The peak-to-peak value is twice this, or 650.6 V. This is useful when measuring the values of alternating waveforms on an oscilloscope if there is no zero-voltage line on the trace.

CHECK YOUR UNDERSTANDING

9.5 A sinusoidal a.c. waveform has a period of 50 ms and a maximum value of 300 V. Calculate the:

(a) peak-to-peak value
(b) average value
(c) RMS value
(d) instantaneous value at 45°.

9.4 Use of oscilloscopes to measure d.c. and a.c. voltage levels

9.4.1 Oscilloscopes

An oscilloscope shows an electronic signal as a picture or graph on a screen. It is used to demonstrate how a voltage changes over time. If you need to examine any electrical signal, simply change it into a voltage and you can look at it on the screen for as long as you wish. Oscilloscopes only seem complicated because they are so versatile.

Oscilloscopes are used in most electronic laboratories. They are essential in automotive repair workshops for diagnosing engine faults and tuning fuel injection and engine management systems. They are also used widely in the medical field and in interpreting signals from television stations. Oscilloscopes are vital for maintaining any industrial servicing set-up that has any appreciable amount of electronic equipment.

Although some older-style oscilloscopes called 'cathode ray oscilloscopes' (CROs) may still be in use, modern oscilloscopes such as the one shown in Figure 9.10 have LED screens, are computerised and are referred to as 'DSOs' (digital storage oscilloscopes). They can self-calibrate input signals and retain settings, functions and readings in their memory. Some can even send their stored information to other machines and the internet.

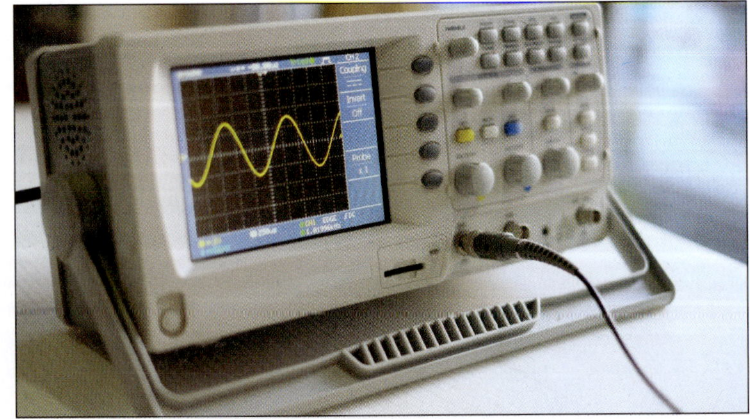

FIGURE 9.10 Modern digital oscilloscope
OhSurat/Shutterstock.com

9.4.2 Interpreting an oscilloscope display

Modern oscilloscopes have a digital grid projected onto the screen, with each square of the grid known as one 'division'. If the oscilloscope is calibrated correctly, each division can be translated into a value dependent on the settings chosen for voltage (vertical divisions) and time (horizontal divisions).

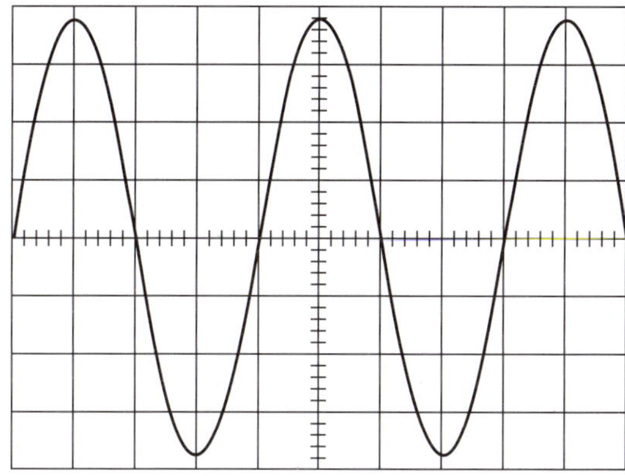

FIGURE 9.11 Sinusoidal waveform cycles

Figure 9.11 shows two cycles of a sinusoidal waveform as they would appear on an oscilloscope screen. The sinusoidal waveform can be evaluated using horizontal and vertical reference values.

As an example, we will use values of 10 V/div vertically and 20 ms/div horizontally.

The peak-to-peak value of the voltage is 3.8 × 2 = 7.6 divisions. Since each division vertically represents 10 V, this can be translated as 7.8 × 10 = 78 V_{P-P} or 38 V_{max} from the centre line to either peak. This can then be converted to an effective value of 34 × 0.707 = 24 V RMS.

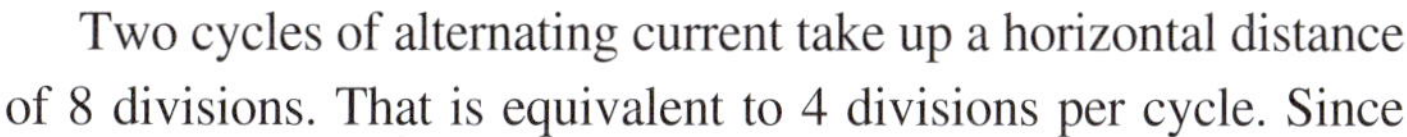

Two cycles of alternating current take up a horizontal distance of 8 divisions. That is equivalent to 4 divisions per cycle. Since the time base is set at 20 ms/div, this represents a time of 4 × 20 = 80 ms. So, one cycle takes up a period of 80 ms, or 0.08 s. Taking the reciprocal of this figure gives the frequency in cycles per second. Here, the frequency is 12.5 Hz.

Frequency is the inverse of the time in seconds. Expressed as an equation it is:

$$f = \frac{1}{T}$$

EXAMPLE 9.3

Find the peak-to-peak voltage and frequency of the sinusoidal waveform in Figure 9.11 if the vertical attenuator is set on 50 V/div and the time base is set on 5 ms/div.

Two cycles take 8 divisions horizontally. This is equivalent to 4 divisions per cycle.

∴ the horizontal time for 1 cycle = 4 × 5 ms = 20 ms (0.02 seconds)

$$\therefore \text{frequency} = \frac{1}{20 \times 10^{-3}} = 50 \text{ Hz}$$

The peak-to-peak value of the wave is 6.8 divisions.

$$\therefore \text{voltage}_{\text{Pk-Pk}} = 6.8 \times 50 = 340 \text{ V}$$

Of course, not all waves are sinusoidal. Figure 9.12 shows a square wave. The same wave appears in each of the diagrams, but with different time bases. The apparent frequency differs considerably, but when the horizontal calibrations are taken into account, all are still the same—the voltage is constant in each.

In Figure 9.12(a), the time base is set at 1 µs/div. One complete cycle takes 10 divisions or 10 µs. The voltage is high for 2.5 µs and low for 7.5 µs. This translates to a frequency of 100 kHz.

In Figure 9.12(b), given a time base of 2 µs/div, one complete cycle takes 5 divisions or 5 × 2 = 10 µs as before. Although the wave looks different, it is the same wave at the same frequency. It is still high for 2.5 µs and low for 7.5 µs.

In Figure 9.12(c), the time base has been altered to 10 µs/div. Each cycle is completed in 1 division and still gives a cycle time of 10 µs and a frequency of 100 kHz. However, this time it is more difficult to determine the high and low periods of each cycle.

Altering the timing of the horizontal time base from one value to another does two things—it gives an overall picture of events and/or enlarges a cycle for analysis.

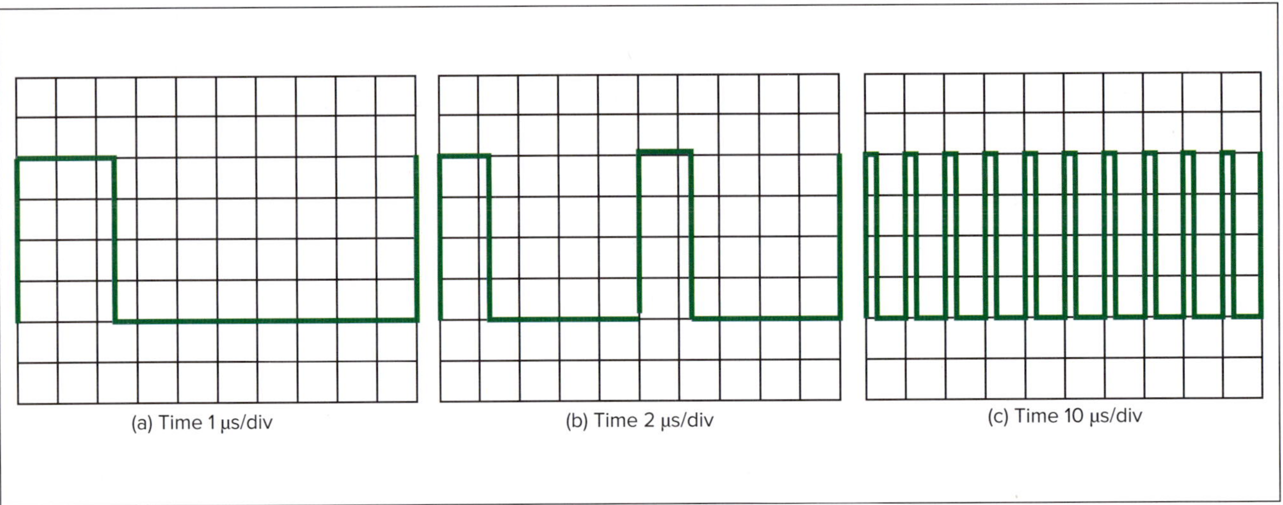

FIGURE 9.12 **Effects of altering the sweep frequency**

9.4.3 Dual trace oscilloscopes

Sometimes it is necessary to compare waveforms. For example, the input waveform to an electronic circuit might need to be compared with its output waveform. Manipulation of the a.c. waveforms may displace them in time from each other. With a single-channel oscilloscope, each waveform can be shown in turn. However, this does not allow a direct comparison between the two, nor does it show any relationship between them. A dual trace oscilloscope allows both to be displayed on the screen at the same time (Figure 9.13(a)). This allows their relationship to be compared.

There are two vertical inputs but still only one time-base circuit. Samples of each input signal are shown on the oscilloscope screen against the same time interval.

Theoretically, there is no limit to the number of traces that can be shown at one time, but in practical terms four appears to be a reasonable limit. A four-channel oscilloscope would be able to display a three-phase supply and show the three waveforms displaced by 120 degrees (Figure 9.13(b)).

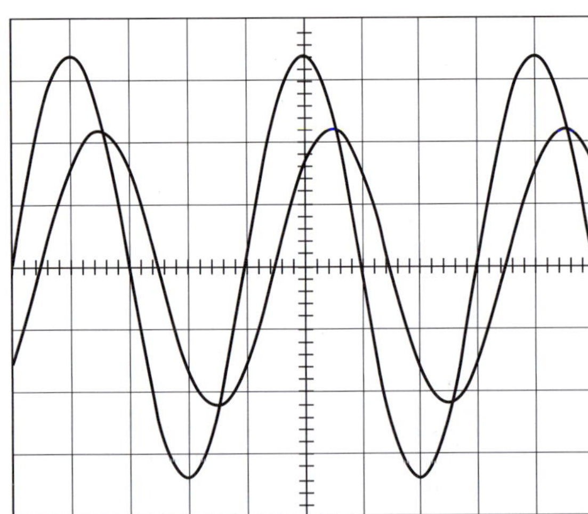

FIGURE 9.13(a) **Dual trace display**

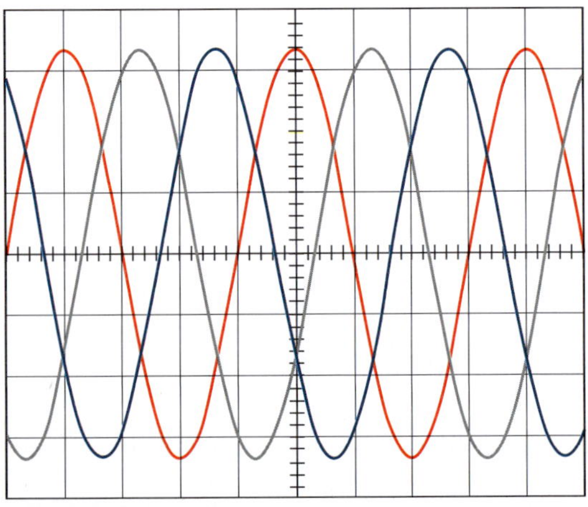

FIGURE 9.13(b) **Three-phase supply shown on a four-channel oscilloscope**

9.4.4 Oscilloscope applications

Oscilloscopes have the following uses.

Electrical:

1. Observing waveforms in electrical circuits
2. Measuring quantities such as voltage, current, power and phase angles
3. Comparing known and unknown frequencies
4. Measuring short time intervals
5. Examining the characteristics of magnetic materials
6. Examining the harmonic content of a.c. wave shapes.

Electronic:

1. Aligning tuned circuits for audio and radio frequencies
2. Television
3. Modulating measurement in transmitters
4. Testing radio components.

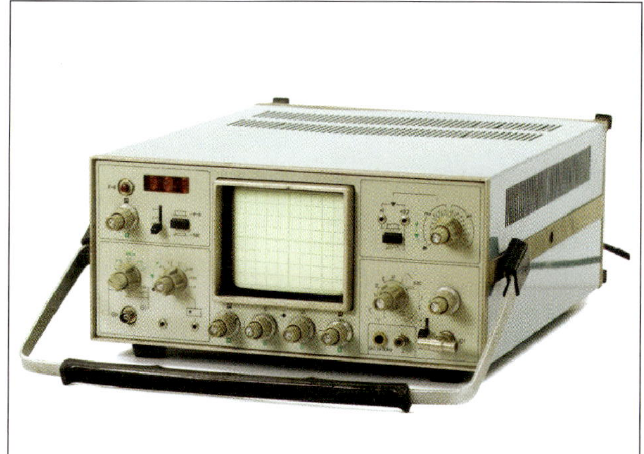

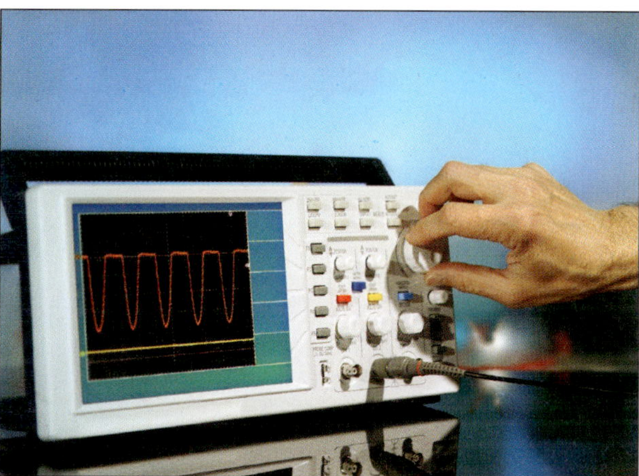

FIGURE 9.14 Size comparison between digital scopes and cathode ray oscilloscopes
(Left) Zoonar GmbH/Alamy Stock Photo; (right) tilialucida/Alamy Stock Photo

Digital storage oscilloscopes with liquid crystal displays have made oscilloscopes far less bulky, with a much smaller footprint. Figure 9.14 illustrates the considerable difference between the bench space required for the digital scope on the left compared with that required for the CRO type on the right.

Even smaller handheld oscilloscopes that are suitable for field work are becoming much more widely available. They have a multitude of functions, much like a multimeter but with a far higher level of sophistication. A handheld power analyser (as in Figure 9.15) is becoming a vital part of an electrotechnology worker's testing equipment.

FIGURE 9.15 Handheld analyser—the Fluke 435 Series II Power Quality and Energy Analyser
Mike Scott

CHECK YOUR UNDERSTANDING

9.6 For the waveform shown in Figure 9.16, the vertical scale of the oscilloscope is set to 5 V/div and the time base (horizontal) is set to 10 ms/div. Determine the:

(a) peak-to-peak voltage
(b) peak voltage
(c) RMS voltage
(d) frequency.

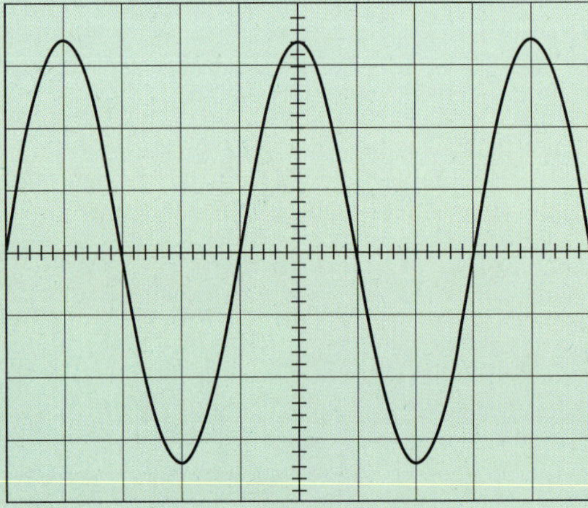

FIGURE 9.16 Waveform displayed on an oscilloscope

9.5 Phasor diagrams

9.5.1 The purpose of phasor diagrams

To understand **phasors**, it is useful to have an understanding of **vectors**. A vector is a line that, by its length and direction, represents the magnitude and angle of application of a force or velocity (or other vector quantity).

A phasor is a rotating vector used in electrical calculations. It has a point of reference, magnitude and rotation. Voltage and current phasors represent the RMS value of the voltage or current by their length and the angle of phase difference between the phasor and the reference plane.

9.5.2 In phase, out of phase, phase angle, lead and lag

Alternating currents and voltages, and their phase relationship, can be represented by sine waves, as in Figures 9.18(a), 9.19(a) and 9.20(a). However, this method is inconvenient; it is simpler to use phasors in a phasor diagram, as in Figures 9.18(b), 9.19(b) and 9.20(b). As the RMS values of a.c. are important, phasor diagrams are nearly always scaled to represent them. Unless otherwise stated, that is the case in this textbook.

Looking at Figure 9.17, the reference phasor is always drawn horizontally and to the right. All phase angles must be measured from it. The reference phasor is usually a quantity that has the same value in all parts of the circuit.

For a series circuit, the current is used as the reference phasor. This is because the current is common to all parts of the circuit.

For a parallel circuit, the voltage is used as the reference phasor. This is because the voltage is common to all parts of the circuit.

When the current and voltage curves pass through the zero position at the same time and increase to their maximum values in the same direction, the waveforms are in phase with each other. This is shown in Figure 9.18(a). The phasor diagram for the in-phase condition is shown in Figure 9.18(b). Note that both phasors start from the same point, called the 'origin'.

This textbook distinguishes between voltage and current phasors by showing the voltage phasor in red and the current phasor in blue.

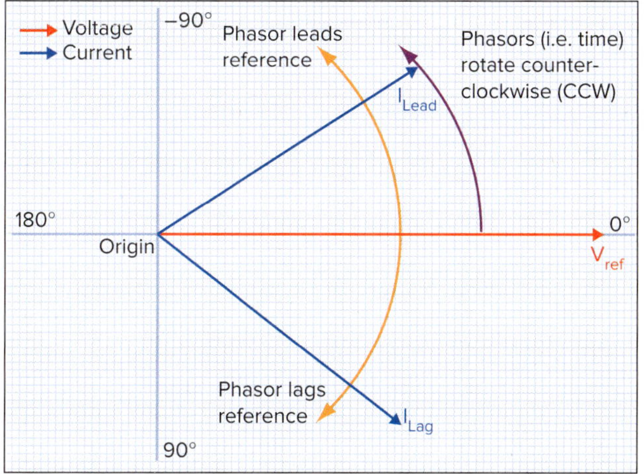

FIGURE 9.17 Phasor diagram

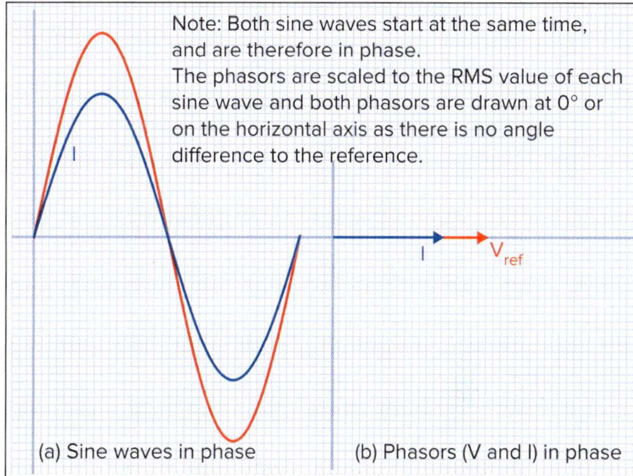

FIGURE 9.18 In-phase phasors

In some circuits, the current and voltage curves do not reach their zero and maximum values simultaneously. In such cases, they are said to be 'out of phase', and the angle of lead or lag (see Figures 9.19(a) and (b), and 9.20(a) and (b)) is called the 'phase angle' (symbol φ—pronounced 'phi', rhymes with fly). This symbol is conventionally used to indicate an angle between two phasors and is not interchangeable with the 'theta' (θ) symbol that indicates an angle of rotation. A second convention is that all phasors are assumed to rotate in an anticlockwise direction, starting at the reference.

Figure 9.19(a) shows the current lagging the voltage by 60° and Figure 9.19(b) shows the same condition using phasors. Figure 9.20(a) shows the current leading the voltage by 60° and Figure 9.20(b) shows the same condition using phasors.

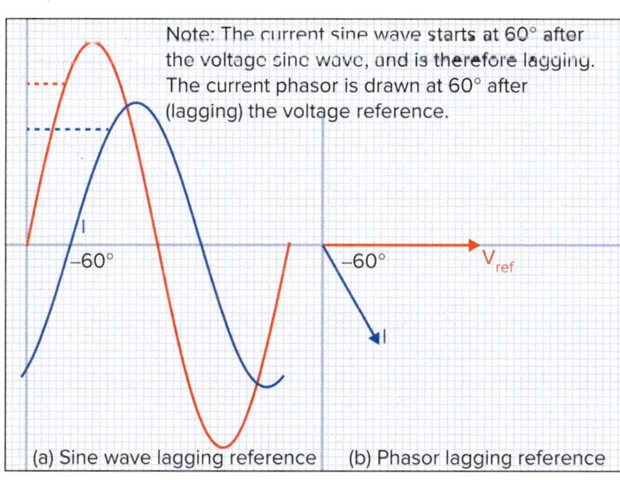

FIGURE 9.19 Lagging phasor

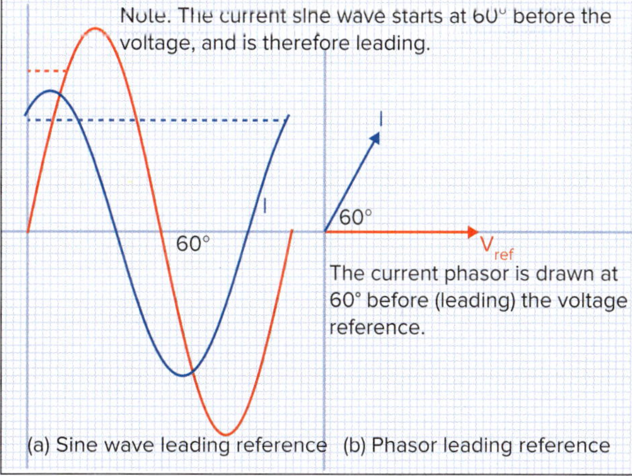

FIGURE 9.20 Leading phasor

9.5.3 Phasor addition by graphical method

Alternating values of current or voltage cannot be added arithmetically unless they are in phase. One way of adding phasors that are out-of-phase is to draw them to scale and angle using RMS values, and then add them together using the *parallelogram method,* as shown in Figure 9.21. This process is explained in detail in Example 9.4.

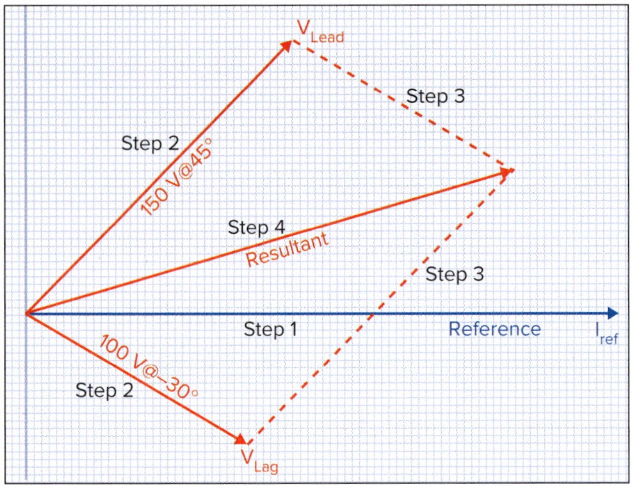

FIGURE 9.21 Parallelogram method

EXAMPLE 9.4

Two voltages A and B are connected in series. Voltage A is 150 V and leads the current by 45°; voltage B is 100 V and lags the current by 30°. Find the total EMF and phase angle.

Since the two voltages are in series, the current is used as the reference phasor. To construct the phasor diagram, follow these steps.

Step 1. Draw the current phasor horizontally to the right as the reference phasor.

Step 2. Draw the phasors for V_A and V_B to scale, measuring the phase angles from the reference phasor with a protractor.

Step 3. Open the compass to the scaled measurement for V_A and then place the compass at the END of V_B. Draw an arc between the two phasors (Figures 9.22 and 9.23).

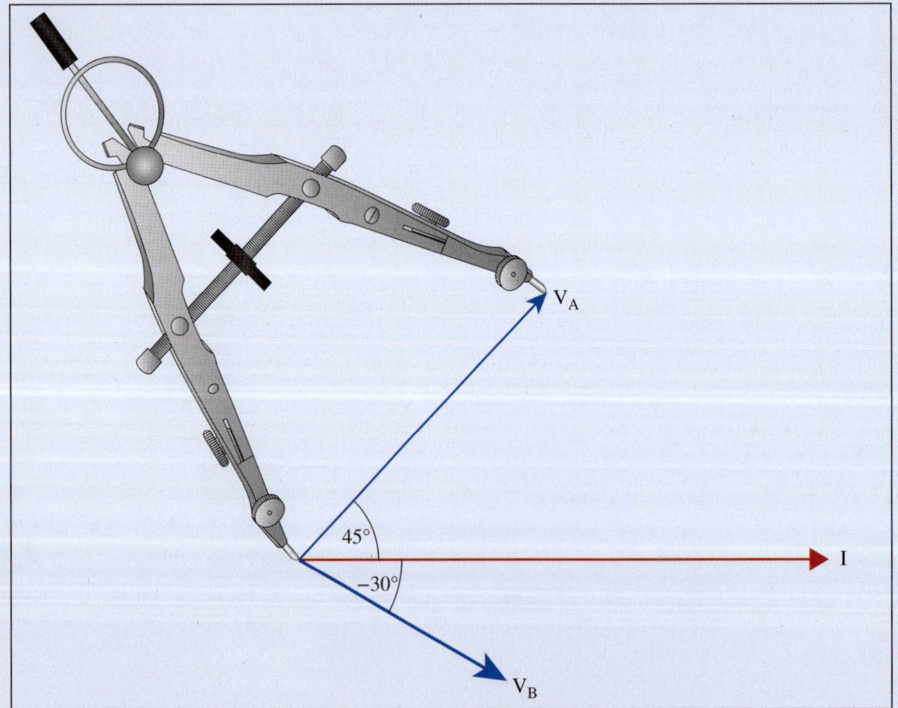

FIGURE 9.22 Step 3(a)

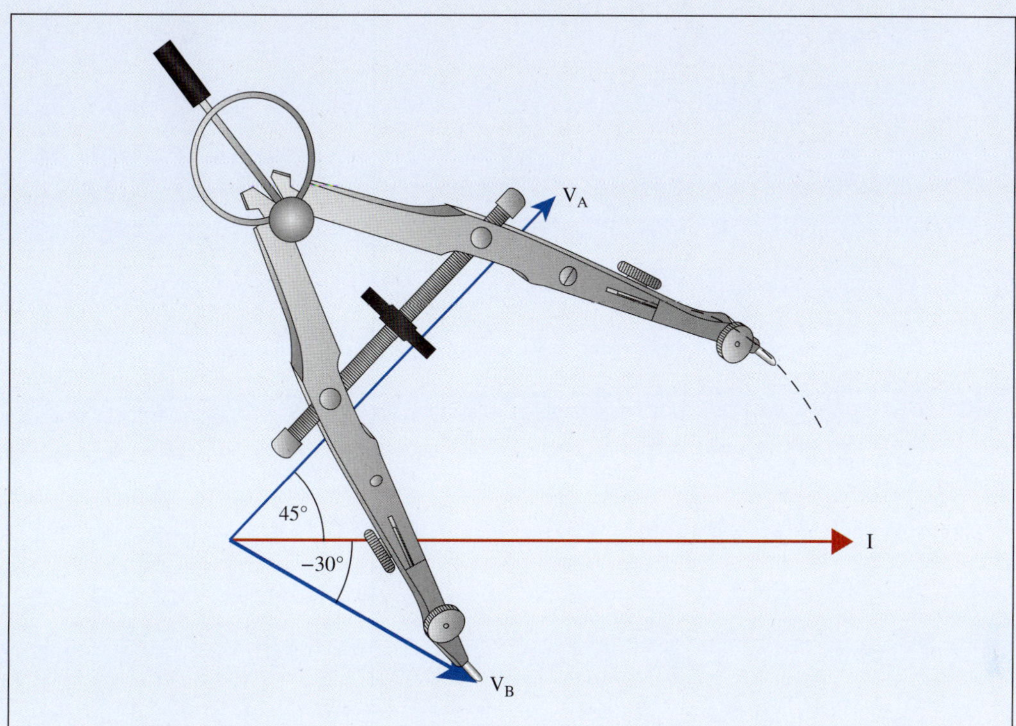

FIGURE 9.23 Step 3(b)

Step 4. Open the compass to the scaled measurement for V_B and then place the compass at the END of V_A. Draw a second arc between the two phasors (Figures 9.24 and 9.25).

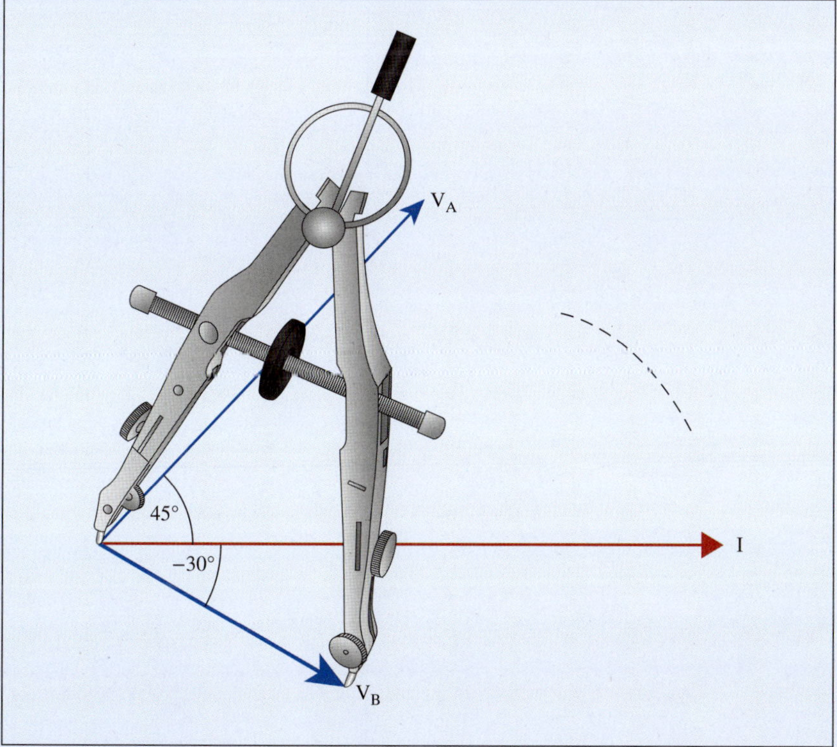

FIGURE 9.24 Step 4(a)

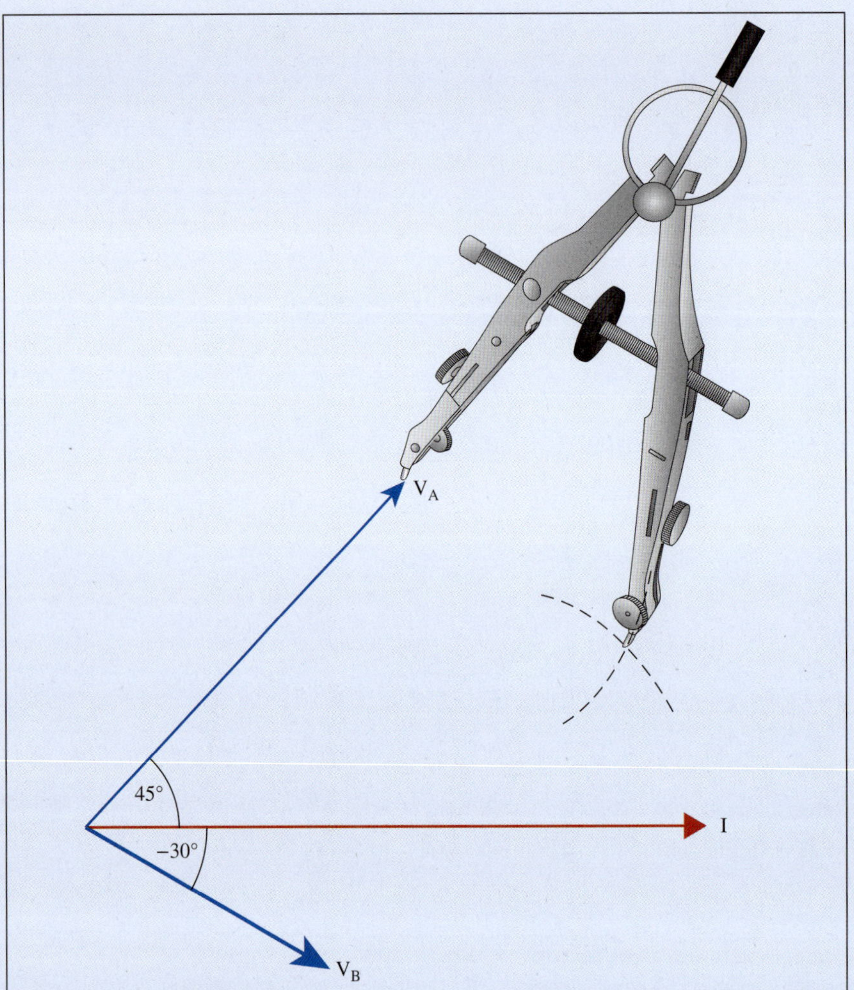

FIGURE 9.25 Step 4(b)

Step 5. Draw a line representing the resultant phasor V_{Total} from the point where all the phasors originate to the intersection of the two arcs (Figure 9.26).

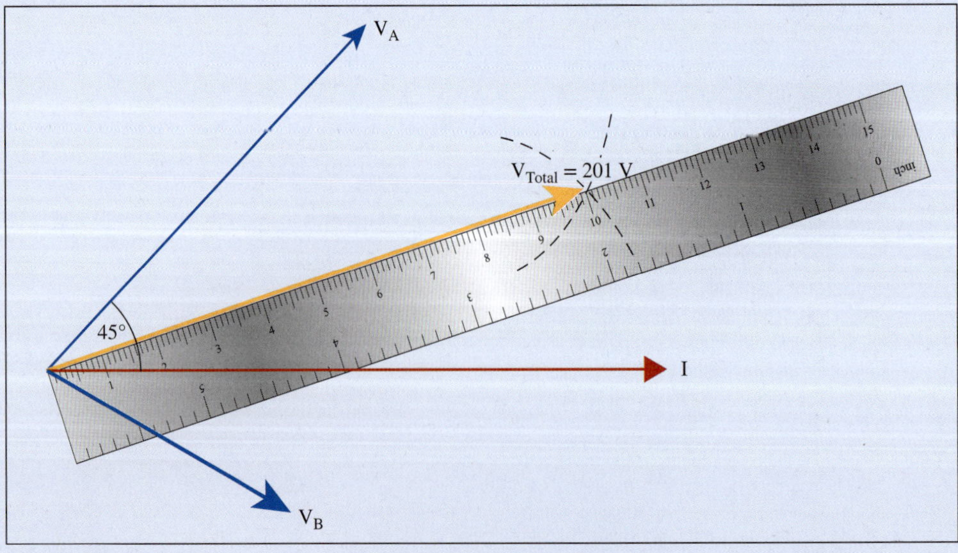

FIGURE 9.26 Step 5

Step 6. Measure the length of V_{Total} and apply the scale to obtain the magnitude of the voltage. In this case it is 201 (Figure 9.27).

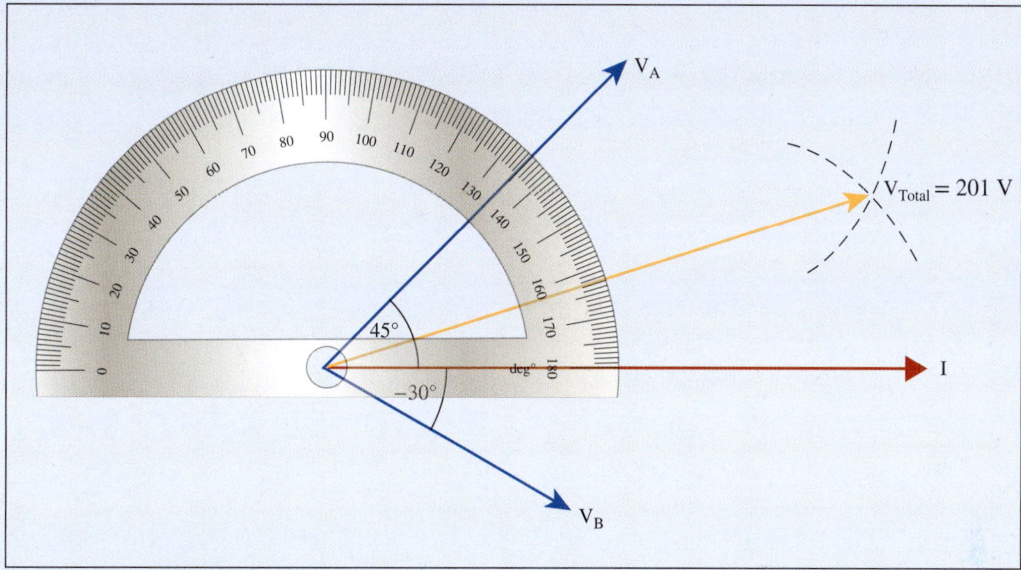

FIGURE 9.27 Step 6(a)

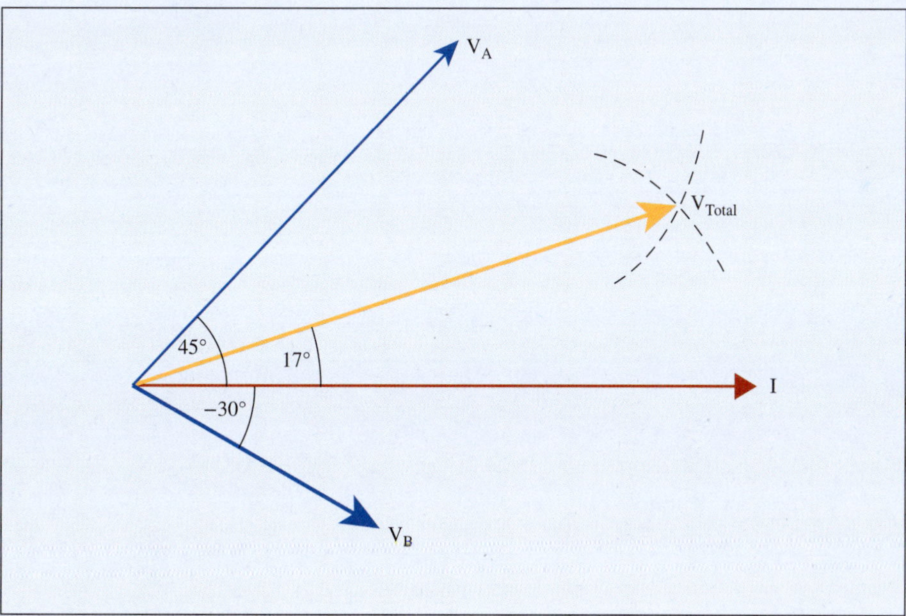

FIGURE 9.28 Step 6(b)

Step 7. Measure the angle between the reference phasor (I) and the resultant phasor (V_{Total}) with a protractor (Figure 9.27) to obtain the phase relationship between the current (I) and the resultant supply voltage (V_{Total}). In this case $\varphi = 17°$ leading (Figure 9.28).

An alternative method of constructing a phasor diagram is to use the 'tip to tail' method. In this, the first phasor is drawn horizontally and then the next phasor's tail is drawn where the tip (meaning the arrowhead end) of the first phasor finished.

Step 1: Draw the voltage V_A at an angle of 45° leading the reference current (Figure 9.29(a)).

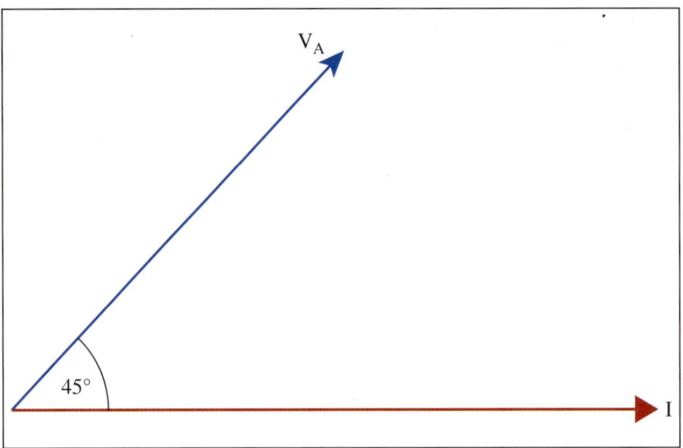

FIGURE 9.29(a) Step 1

Step 2: Now draw the second voltage V_B from the arrowhead (tip) of V_A. Use a second reference line that is parallel to the I reference line to ensure that you draw the voltage at the correct angle. Ensure that you use the same scale and relative direction for the second phasor (Figure 9.29(b)).

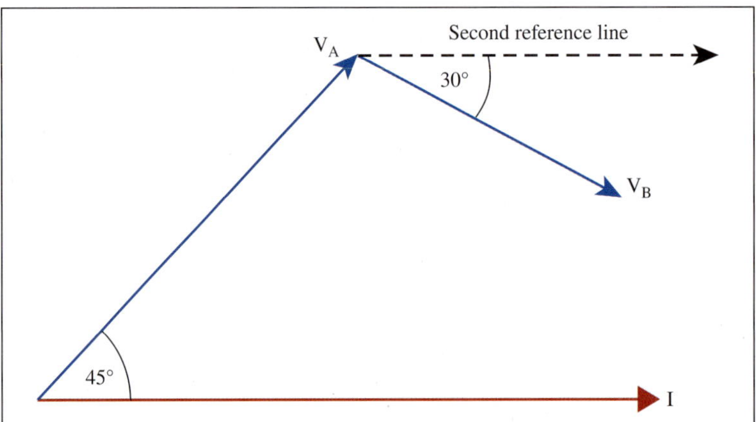

FIGURE 9.29(b) Step 2

Step 3: The resultant phasor is then drawn from the starting point of V_A to the tip of V_B. It gives the same result as before—201 V at 17° leading (Figure 9.29(c)).

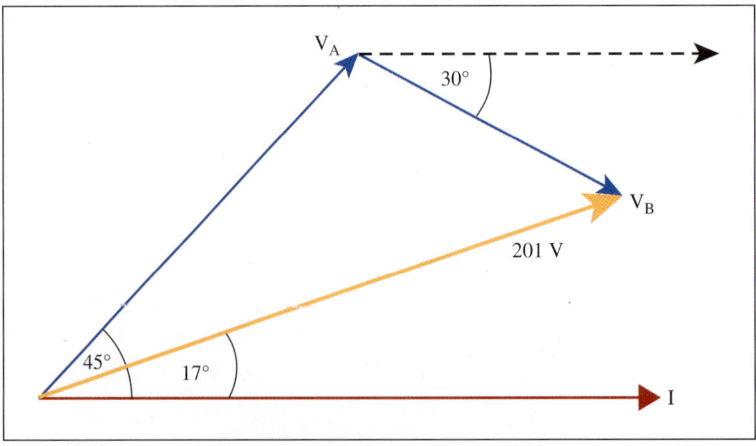

FIGURE 9.29(c) Step 3

Where two phasors are at 90° to each another, with one phasor in phase (at an angle of 0°) and the second at an angle at 90° (leading or lagging), phasor addition can be carried out mathematically using trigonometry to determine the resultant phasor and angle.

EXAMPLE 9.5

Two voltages A and B are connected in series. Voltage A is 200 V and is in phase with the current. Voltage B is 113.58 V and leads the current by 90°. Find the total voltage V_T and the phase angle.

Since the two voltages are in series, the current is used as the reference phasor.

Phasor V_A is drawn in phase with the current and phasor V_B is drawn with its 'tail' on the 'tip' of V_A. This makes two sides of a right angle triangle.

As the resultant V_T is the hypotenuse of the triangle, the magnitude of V_T and the phase angle can be determined using trigonometry.

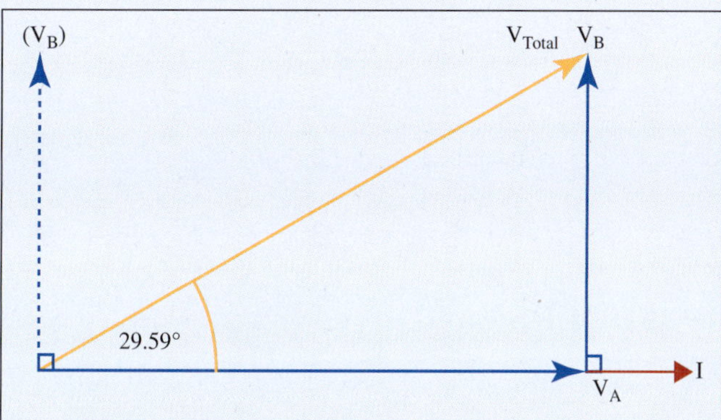

FIGURE 9.30 Determining the V_{Total}

Trigonometry method:

$$\tan \varphi = \frac{\text{Opp}}{\text{Adj}}$$
$$= \frac{V_B}{V_A}$$
$$= \frac{113.58}{200}$$
$$= 0.5679$$

$$\text{Therefore, } \varphi = \tan^{-1} 0.5679$$
$$= 29.59°$$

By using cos (or sin) of 29.59° and the corresponding side, V_{Total} can be determined.

$$\cos \varphi = \frac{\text{Adj}}{\text{Hyp}}$$
$$= \frac{V_A}{V_{Total}}$$

Therefore:

$$V_{Total} = \frac{V_A}{\cos \varphi}$$
$$= \frac{200}{\cos 29.59°}$$
$$= 230 \text{ V leading at } 29.59°$$

While this method takes quite a few steps, it does give an accurate answer for both the magnitude of V_{Total} and the angle.

A quicker method to find just the magnitude is by using Pythagoras' theorem:

$$Hyp^2 = Adj^2 + Opp^2$$

or:

$$V_{T^2} = V_A^{\ 2} + V_B^{\ 2}$$

Therefore:

$$\begin{aligned} V_{Total} &= \sqrt{(V_A^{\ 2} + V_B^{\ 2})} \\ &= \sqrt{(200^2 + 113.58^2)} \\ &= \sqrt{52,900.42} \\ &= 230\ \text{V} \end{aligned}$$

CHECK YOUR UNDERSTANDING

9.7 Two voltages, A and B, are connected in series. Voltage A is 180 V and is in phase with the current. Voltage B is 120 V and leads the current by 90 degrees. Find the:

(a) total voltage (V_T)

(b) resultant phase angle (ϕ).

9.6 Single-element a.c. circuits

9.6.1 Introduction

Alternating current using sin waves is the basis of worldwide electricity generation and distribution, so it is vitally important that electrical workers have an understanding of all aspects of a.c. circuit calculations. This section looks at the effect that inductors and capacitors have on the current and voltage in these circuits. The introduction of frequency will change the way we calculate values and introduces concepts such as reactance, impedance and phase angles. We will also look at power, the consequences of a poor power factor and how we can correct it.

9.6.2 Resistance in a.c. circuits

Ohm's Law enables us to work out the current flowing through a pure resistance at any point in time. It is $I = \frac{V}{R}$ for any part of the cycle. When the resistance remains the same, the current is directly proportional to the applied voltage. Therefore, the current waveform for a purely resistive circuit is in phase with the voltage waveform and $I_{RMS} = \frac{V_{RMS}}{R}$.

Pure or non-inductive resistance

When carrying out calculations, we can think of resistors on power-line frequencies as consisting of 'pure' or 'non-inductive' resistance. Examples include incandescent lamps, radiators and electric kettle elements.

Real resistors are often made from a coil of resistance wire. Theoretically, this wire will exhibit some inductance, but its value is usually so low that it can be ignored. The same goes for capacitance.

At higher frequencies like the ones in modern switch-mode power supplies and motor controllers, inductance can become a problem. It must either be addressed at the design stage or dealt with by using special non-inductive resistors (see Figure 9.31).

The inductive effect can be avoided fairly easily. Half the resistor is wound clockwise on a non-magnetic former, and then the other half is wound on anticlockwise. The magnetic fields produced around the conductors cancel each other out and thus prevent self-induced voltages.

Capacitive effects cannot be overcome easily and so have to be minimised. The main method of achieving this on long-distance transmission lines is to keep them as far apart from each other as is economically possible.

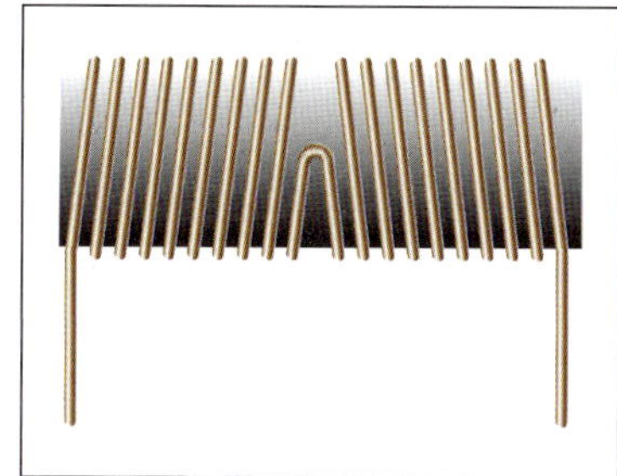

FIGURE 9.31 Non-inductive resistor

EXAMPLE 9.6

A pure resistance of 12 Ω is connected across an a.c. power supply that generates a pure sine wave of 100 V peak voltage. What current will flow and what power will be taken?

$$V_{RMS} = 0.707\,V_{max} \quad (1)$$
$$= 0.707 \times 100 \quad (2)$$
$$= \underline{70.7\ V} \quad (3)$$
$$I_{RMS} = \frac{V_{RMS}}{R} \quad (4)$$
$$= \frac{70.7}{12} \quad (5)$$
$$= \underline{5.89\ A} \quad (6)$$
$$P = V_{RMS} \times I_{RMS} \quad (7)$$
$$= 70.7 \times 5.89 \quad (8)$$
$$= \underline{416.4\ W} \quad (9)$$
$$\text{or} \ldots P = \frac{V_{RMS}{}^2}{R} \quad (10)$$
$$= \frac{70.7^2}{12} \quad (11)$$
$$= \underline{416.5\ W} \quad (12)$$
$$\text{or} \ldots P = I_{RMS}{}^2 R \quad (13)$$
$$= 5.89^2 \times 12 \quad (14)$$
$$= \underline{416.6\ W} \quad (15)$$

Resistors in series and parallel

As the current in resistors is always in phase with the voltage, Ohm's Law still applies, as do Kirchhoff's Voltage Law (KVL) and Kirchhoff's Current Law (KCL). This means that, in a purely resistive circuit, the rules for d.c. circuits continue to work for a.c. This is the case whether it is series, parallel or compound.

9.6.3 Inductors

Inductors play a major role in a.c. electrical circuits, mainly because they can create useful magnetic effects. These include the role they play in electrical machines that can either convert motion to electrical energy or electrical energy to motion. They also include creating mutual induction in transformers and their relay and solenoid applications to either switch circuits or open and close mechanical systems such as valves and air dampeners.

However, inductors have a related property which opposes change in current and current flow. This makes inductive circuits useful in limiting current in devices like fluorescent lighting. When they are used for that purpose, they are sometimes referred to as a 'choke'. They can also be coupled with capacitors to 'tune' radio circuits (to frequencies such as AM or FM). This coupling is also used in mobile telecommunications and to oppose undesirable frequencies in power quality management (in line reactors or filters).

9.6.4 Inductance in a.c. circuits

A change in current flow in an inductive circuit results in an induced EMF. This, according to Lenz's Law, will oppose the change of current flow. In d.c. circuits, this inductive effect causes the current to rise slowly, eventually reaching approximately its maximum value according to the circuit resistance.

In an inductive a.c. circuit, the current is continually changing in value and direction. This generates an induced EMF that must continually oppose the change of current flow. Figure 9.32 shows the relationship between the current, induced EMF and supply voltage for a purely inductive circuit. Using the voltage phasor as the reference, the current phasor lags by 90°E. Consequently, it is drawn at right angles to the voltage reference. Remember that at maximum voltage the current will be zero but rising. At zero voltage the current is at maximum. As strange as this may sound, the effect is a very important one.

On a.c., the change in current flow produces an induced EMF that opposes it. The effect of this current opposition is called 'inductive reactance' (symbol X_L), and is measured in ohms.

Ohm's Law effectively states that the current is equal to the voltage divided by the opposition to current flow. Inductive reactance is a type of opposition to current flow, so for pure inductance $I = \frac{V}{X_L}$.

The value of inductive reactance in a circuit depends on the inductance and the rate of change of current flow, and that depends on the supply frequency. Inductive reactance can be calculated from the equation:

$$X_L = 2\pi fL$$

where:

X_L = inductive reactance in ohms
f = frequency in hertz
L = inductance in henrys.

Once the inductive reactance is known, Ohm's Law works just as if reactance and resistance were the same (they are not, as will be explained in due course).

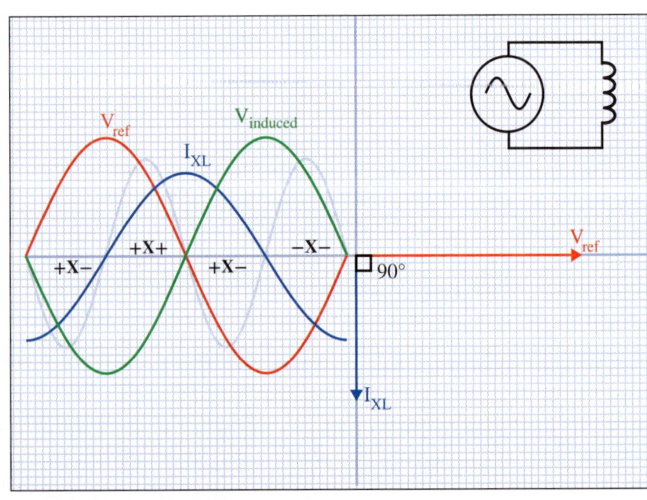

FIGURE 9.32 Relationships in an inductive a.c. circuit

Inductors in series

If two inductors are placed in series, each will produce an induced EMF. This means the total induced EMF will be increased, and that in turn means the opposition to current flow is increased. By placing inductors in series, the total inductive reactance increases in the same way that placing resistors in series increases the total resistance:

$$X_{L\ Total} = X_{L1} + X_{L2} + X_{L3} \ldots + X_{Ln}$$

As the total value of induced EMF is increased, the total inductance is increased. Consequently, the total inductance is found in the same way:

$$L_{Total} = L_1 + L_2 + L_3 \ldots + L_n$$

EXAMPLE 9.7

A coil has an inductance of 0.05 H. What would be the inductive reactance at a frequency of 25 Hz and 50 Hz? At what frequency would it have a reactance of 10 Ω?

$$X_L = 2\pi fL \quad (1)$$
$$= 2 \times 3.142 \times 25 \times 0.05 \quad (2)$$
$$= \underline{7.85\ \Omega} \quad (3)$$
$$X_L = 2\pi fL \quad (4)$$
$$= 2 \times 3.142 \times 50 \times 0.05 \quad (5)$$
$$= \underline{15.7\ \Omega} \quad (6)$$
$$X_L = 2\pi fL \quad (7)$$
$$\therefore f = \frac{X_L}{2\pi L} \quad (8)$$
$$= \frac{10}{2 \times 3.142 \times 0.05} \quad (9)$$
$$= \underline{31.8\ \text{Hz}} \quad (10)$$

EXAMPLE 9.8

A 500 V 50 Hz supply is applied to a coil with an inductance of 0.12 H. Determine the current.

$$X_L = 2\pi fL \quad (1)$$
$$= 2 \times 3.142 \times 50 \times 0.12 \quad (2)$$
$$= \underline{37.7\ \Omega} \quad (3)$$
$$I = \frac{V}{X_L} \quad (4)$$
$$= \frac{500}{37.7} \quad (5)$$
$$= \underline{13.26\ \text{A}} \quad (6)$$

EXAMPLE 9.9

A 230 V 50 Hz supply is applied to a choke coil of negligible resistance and the current through the coil is 2.5 A. Determine the inductance of the coil.

$$X_L = \frac{V}{I} \quad (1)$$

$$= \frac{230}{2.5} \quad (2)$$

$$= \underline{92\ \Omega} \quad (3)$$

$$L = \frac{X_L}{2\pi f} \quad (4)$$

$$= \frac{92}{2 \times 3.142 \times 50} \quad (5)$$

$$= \underline{293\ \text{mH}} \quad (6)$$

EXAMPLE 9.10

Two inductors, one with an inductive reactance of 11 Ω and the other with an inductive reactance of 12 Ω, are connected in series across a 230 V 50 Hz supply.

What is the total inductive reactance? What is the total current?

$$X_{L\ Total} = X_{L1} + X_{L2} \quad (1)$$

$$= 11 + 12 \quad (2)$$

$$= \underline{23\ \Omega} \quad (3)$$

$$I = \frac{V}{X_L} \quad (4)$$

$$= \frac{230}{23} \quad (5)$$

$$= \underline{10\ \text{A}} \quad (6)$$

CHECK YOUR UNDERSTANDING

9.8 An inductor of 0.25 H is connected in series with another inductor of 0.35 H. Calculate the total inductance, inductive reactance and current flowing for each of the following supplies:

(a) 120 V 50 Hz
(b) 200 V 75 Hz
(c) 50 V 100 Hz.

9.7 Capacitors

9.7.1 Capacitors in a.c. circuits

Figure 9.33 shows a capacitor circuit and illustrates a full cycle of alternating voltage and current within it.

Without resistance in the circuit, the capacitor charges according to the rate of change of the applied voltage. So, when the voltage is changing most, the current in the capacitor will be the greatest. When the voltage reaches its maximum value, the current will be zero. But then, as the voltage decreases, the current changes direction.

As the current is already at maximum positive flow when the voltage sine wave crosses zero (going positive), the current seems to come before the voltage. Consequently, it is said that in a capacitive circuit the current 'leads' the voltage.

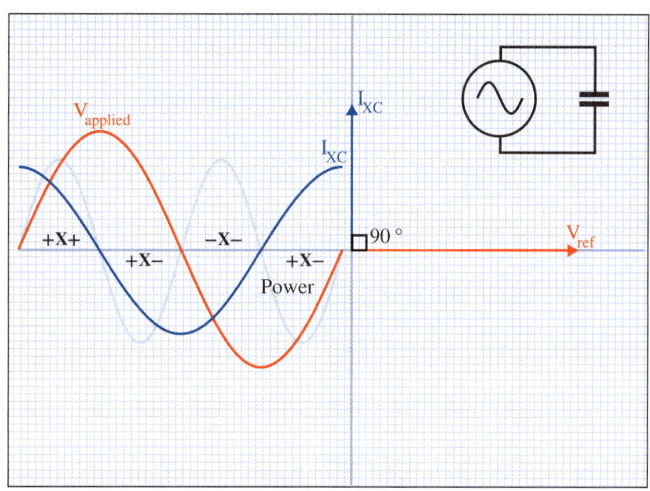

FIGURE 9.33 Capacitive a.c. circuit

For any purely capacitive circuit, the current leads the applied voltage by 90°E, as shown. The phasor diagram in Figure 9.33 shows a current phasor leading the voltage by 90°.

9.7.2 Capacitive reactance

When an a.c. voltage is applied to a capacitor, it is continually being charged and discharged. Current flows in and out of the capacitor at a regular rate, dependent on the supply frequency. An a.c. ammeter connected in the circuit would indicate a current apparently flowing through the capacitor; however, the capacitor has an insulating dielectric between the two plates, so the ammeter is actually recording a displacement current.

The value of displacement current will depend upon the applied voltage, the supply frequency and the capacity of the capacitor.

Since a capacitor reacts when connected to a.c., it is said to have a type of reactance called 'capacitive reactance'. The symbol for this is X_C and the unit is the ohm. It can be calculated using the equation:

$$X_C = \frac{1}{2\pi fC}$$

where:

X_C = capacitive reactance in ohms
f = frequency in hertz
C = capacitance in farads.

9.7.3 Capacitors in series

When two capacitors are placed in series, the effect is as if the distance between the outside plates were increased and the capacity therefore decreased. On an alternating-current supply, this effectively increases the opposition to a current flow in a similar way to that of resistors placed in series, that is:

$$X_{C\ Total} = X_{C1} + X_{C2} + X_{C3} \ \dots + X_{Cn}$$

EXAMPLE 9.11

Find the capacitive reactance of an 8 μF capacitor and the current flowing when it is connected to a 100 V 50 Hz supply. If it is then connected in series with another capacitor of the same capacity, find the new current flowing.

$$X_C = \frac{1}{2\pi fC} \quad (1)$$

$$= \frac{1}{2 \times \pi \times 50 \times 8 \times 10^{-6}} \quad (2)$$

$$= \underline{397.9\ \Omega} \quad (3)$$

$$I = \frac{V}{X_C} \quad (4)$$

$$= \frac{100}{397.9} \quad (5)$$

$$= \underline{251\ \text{mA}} \quad (6)$$

$$I = \frac{V}{X_{C1} + X_{C2}} \quad (7)$$

$$= \frac{100}{397.9 + 397.9} \quad (8)$$

$$= \underline{125.7\ \text{mA}} \quad (9)$$

EXAMPLE 9.12

Calculate the current drawn by a 16 μF capacitor when connected to a 230 V 50 Hz supply.

$$X_C = \frac{1}{2\pi fC} \quad (1)$$

$$= \frac{1}{2 \times \pi \times 50 \times 16 \times 10^{-6}} \quad (2)$$

$$= \underline{198.9\ \Omega} \quad (3)$$

$$I = \frac{V}{X_C} \quad (4)$$

$$= \frac{230}{198.9} \quad (5)$$

$$= \underline{1.156\ \text{A}} \quad (6)$$

9.7.4 Applications of capacitive a.c. circuits

One of the most common uses for capacitive a.c. circuits is for power factor correction. A capacitor is placed in parallel with an inductive load such as a motor or an older fluorescent light fitting in order to counteract the inductive reactance of the load. Capacitors may also be connected at the main switchboard to counteract the inductive reactance of the whole installation. This is covered in greater detail in section 9.12. Capacitive a.c. circuits are also used in the starting circuits of single-phase motors. This is explained in Chapter 12.

As well as 'tuning' resonant circuits, capacitive a.c. circuits are used in TV and mobile communication technology and as a method of switching frequency-sensitive relays for off-peak and other loads. These typically require a 'matched' inductor. This is an inductor with an inductive reactance equal to the capacitive reactance of the capacitive circuit at the desired frequency. Resonance and harmonics are discussed in section 9.13.

The transformerless power supply has enabled a.c. mains-powered electronic devices to become a lot smaller. Specifically designed capacitors are placed in series with the load to substantially drop the applied voltage (for example, from 230 V a.c. to 5 V a.c.) before it is converted to d.c.

9.7.5 AS/NZS 3000:2018 requirements for the installation of capacitors

These requirements were established because capacitors provide energy storage and can have high potentials across their terminals when connected in series a.c. circuits. The only exception to these requirements is when capacitors are an integral part of other electrical equipment such as motors and fluorescent lights. Capacitors installed in electrical equipment are deemed to satisfy these requirements as they have been engineered to meet the current-carrying capacity and other rating requirements. Also, they are designed to provide a discharge path for energised capacitors when the equipment is isolated.

The requirements are outlined in AS/NZS 3000:2018, Clause 4.15 and include the following points:

- Circuit breakers, switches and contactors should be of a type that is suitable for switching the reactive component of the current. A utilisation category of AC-6b (specifically for switching capacitors) is an example of an appropriate type. The utilisation category refers to an international standard for categorising the types of load that can be switched by the respective devices.
- The provision of a circuit breaker for capacitors (rated individually or in banks of more than 100 kVAR) not connected in parallel with individual appliances. For example, power factor correction banks at the main switchboard. In this example, the capacitors are specifically connected for power factor correction and are not associated with any particular circuit or equipment. Hence, they require a circuit breaker for protection and control.
- Regarding the current-carrying capacity of the conductors connecting the capacitors: there are several arrangements, depending on whether the capacitor is connected in parallel with individual appliances or not.
- The provision of a discharge path for capacitors greater than 0.5 μF to allow any charge to be removed. This is typically via an appropriate resistor connected in parallel across the capacitor terminals.

The provision for discharge is to allow qualified people to work on the associated equipment. A warning notice is required and should state something like:

Warning: Ensure that capacitors are completely discharged before working on equipment.

The discharge times for capacitors are one minute for those rated up to 650 V and five minutes for those rated above 650 V. At the end of this time, the capacitor should be not more than 50 V across the terminals.

CHECK YOUR UNDERSTANDING

9.9 A 10 μF capacitor is connected to a 200 V 50 Hz supply. Calculate the:

(a) capacitive reactance
(b) current flowing
(c) current that would flow if another capacitor of 20 μF was connected in series.

9.8 R–C and R–L series a.c. circuits

9.8.1 Series circuits on a.c. current

Pure components are theoretically possible but practically impossible. This means that most a.c. circuits will exhibit a combination of the three types of load described earlier.

As a reminder, the three basic components and their effects are:

- Resistance—no phase shift between voltage and current
- Inductance—90° phase shift; current lagging voltage
- Capacitance—90° phase shift; current leading voltage.

It is necessary to learn how to deal with circuits that contain different types of loads in series and/or parallel.

In an a.c. circuit, the combined opposition to a current flow of both resistance and reactance is called the 'complex impedance'. This is given the symbol Z and is measured in ohms. Ohm's Law is still applicable:

$$Z = \frac{V_{RMS}}{I_{RMS}}$$

9.8.2 Series circuit impedance

In a series a.c. circuit containing resistance and inductance, each voltage drop can be expressed as a factor of current, i.e. $V_Z = IZ$, $V_R = IR$ and $V_L = IX_L$. The voltage phasors in Figure 9.34(a) can also be drawn in triangle form, as in Figure 9.34(b).

The current is common to each component and is the same in each. Therefore, each resistance, reactance or impedance has a voltage developed across it that is proportional to its opposition to current flow.

This means the triangle in Figure 9.35(b) can also represent R, X_L and Z. Since it is a right angle triangle, Pythagoras' theorem applies, i.e. $Z^2 = R^2 + X_L{}^2$ and, therefore, $Z = \sqrt{(R^2 + X_L{}^2)}$

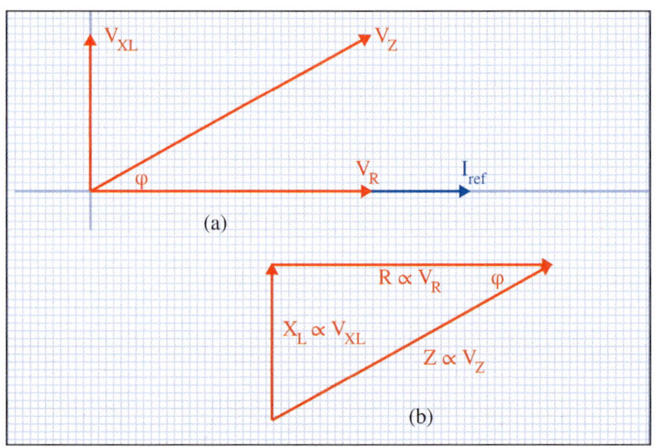

FIGURE 9.34 **Series R–L circuit**

If a series a.c. circuit contains resistance and capacitance, the equation is modified to

$$Z = \sqrt{(R^2 + X_C{}^2)}$$

If it is a series R–L–C circuit, the effective reactance is $(X_L - X_C)$. In this case, the complete equation is:

$$Z = \sqrt{R^2 + (X_L - X_C)^2}$$

where:

Z = impedance
R = resistance
X_L = inductive reactance
X_C = capacitance reactance.

If X_C has a greater value than X_L then the value of $X_L - X_C$ is a negative quantity. This has no effect on the value of Z within the equation because when a negative number is squared it becomes a positive number. However, it is often easier for calculator entry to calculate the difference first by taking the smaller number away from the larger one.

The angle φ between the adjacent sides R and Z has the same value as the angle φ between the voltage phasors V_R and V_Z. This is because the two triangular figures are classed as similar triangles.

The phase angle can be found using trigonometric ratios from the sides of the triangle:

$$\varphi = \tan^{-1}\left(\frac{X}{R}\right) = \cos^{-1}\left(\frac{R}{Z}\right) = \sin^{-1}\left(\frac{X}{Z}\right)$$

Capacitive reactance tends to produce a current leading the voltage by 90°. Inductive reactance tends to produce a current lagging the voltage by 90°. When the values of capacitive and inductive reactance are equal, they cancel each other out. This leaves only resistance as an effective circuit component. When this happens, the circuit is said to be 'resonant'. The circuit then behaves as a purely resistive circuit, at least as far as externally measured. See section 9.13.5 for more information about resonance.

9.8.3 Series R–C circuits

When current flows through a series R–C circuit (Figure 9.35), it causes a voltage drop V_R (due to the resistance), which is in phase with the current. It also causes a voltage drop V_C (due to the capacitive reactance). This lags the current by 90°. The total voltage V is the phasor sum of the two voltage drops V_R and V_C. The current leads the applied voltage by φ.

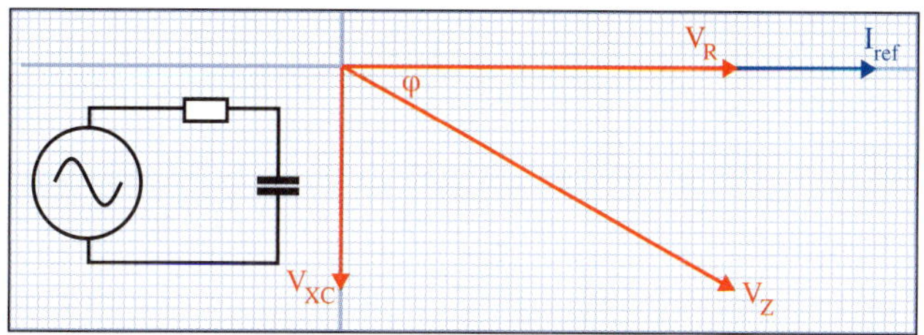

FIGURE 9.35 Series R–C circuit

Commercially made capacitors are considered as pure capacitance, and the problem of losses (as shown in the series R–L circuits) does not usually occur at line frequency (50 Hz).

9.8.4 Series R–L circuits

In a series circuit, the current is common to all parts. The total voltage is the phasor sum of the individual voltage drops. In Figure 9.36, the voltage drop V_R across the resistor is in phase with the current. The voltage drop V_L across the pure inductor is 90° out of phase with the current.

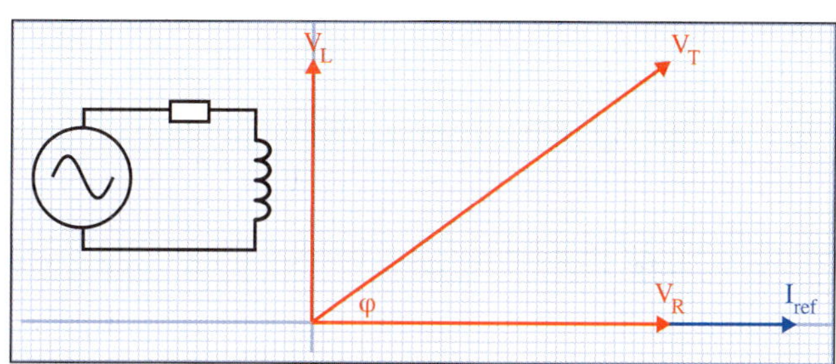

FIGURE 9.36 Series R–L circuit

In an inductive circuit, the current lags the voltage (or the voltage leads the current). The two voltage drops V_R and V_L are out of phase. This means they cannot be added numerically but must be added as phasors.

Since the current in a series circuit is the common value, it is used as the reference phasor. The voltage phasors are drawn as shown in Figure 9.36, with V_L leading I by 90°. The result of the two voltages V_R and V_L represents the total voltage V_T. The angle φ represents the angle of phase difference between the applied voltage and the current (meaning that the applied voltage leads the current by φ).

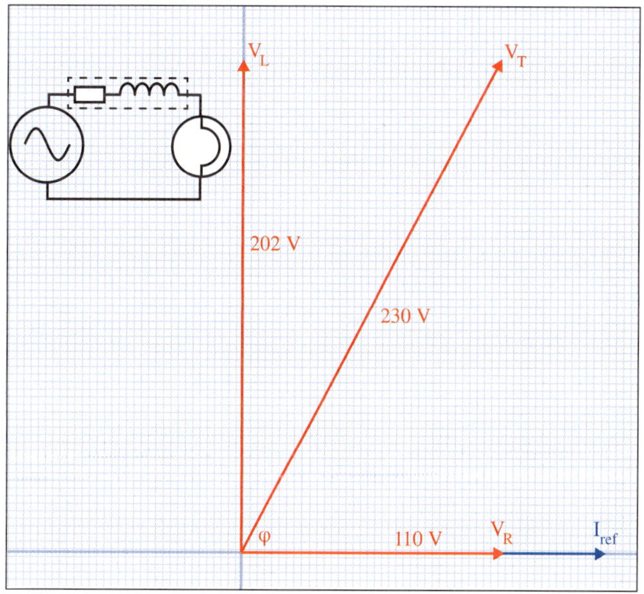

FIGURE 9.37 Series R–L ballasted lamp

With an actual inductor, the winding itself might provide the circuit resistance and there might be no added external resistance. The power losses due to the winding resistance (copper losses) and the core (iron losses) will be real. Accordingly, the lagging phase shift of the current due to the inductance will not be a full 90°—it will be a value corresponding to a combination of both the inductive reactance and the equivalent resistance of the circuit.

Figure 9.37 illustrates one practical use for an inductor in an a.c. circuit. It provides a way of reducing a 230 V supply to 110 V for a projector lamp with minimal power loss and heat generation. The inductor also helps to smooth the current to remove spikes that might damage the lamp.

The voltage V_R is in phase with the current, and the values of V_T, V_R and V_L can be obtained using a voltmeter. Using these values, the phasor diagram can be constructed as shown in Figure 9.37. As the inductor is impure, the voltage does not lead the current by 90° but by an angle less than that (i.e. φL). The voltage V_{RL} is the equivalent voltage drop.

CHECK YOUR UNDERSTANDING

9.10 Read the following statements about series a.c. circuits and decide which of them are true.

(a) In a series R–C circuit the voltage V_Z will lag the current.
(b) In a series R–C circuit the voltages V_R and V_{XC} can be added numerically.
(c) In a series R–L circuit the voltage V_Z will always lead the current.
(d) In a series R–L circuit the voltages V_R and V_{XC} must be added as phasors.

Figure 9.37 shows the relationship between the respective voltages in this series R–L circuit. The current I lags behind the applied voltage V by φ.

9.9 R–L-C series a.c. circuits

9.9.1 Series R–L–C circuits

In Figure 9.38 the voltage V_R is in phase with the current, V_L leads the current by 90° and V_C lags the current by 90°.

The two voltages V_L and V_C are 180° out of phase with each other. Thus, the voltage drop across the total reactance is $V_L - V_C$, as shown in Figure 9.38. The total voltage V is the phasor addition of V_R and $(V_L - V_C)$ and φ is the phase angle. The current is lagging the applied voltage by φ°. If V_C had been greater in value than V_L, the current would have led the voltage by some angle.

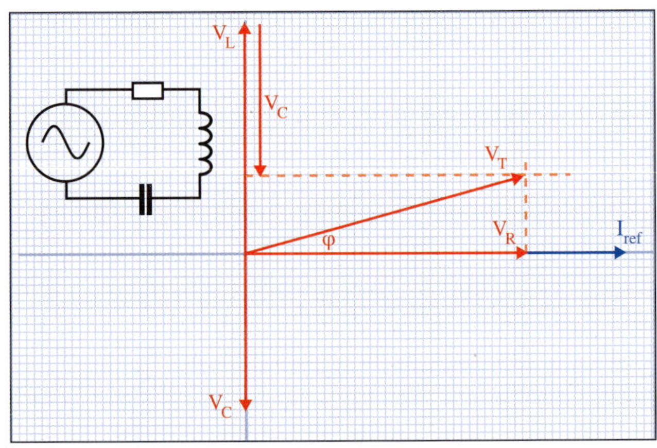

FIGURE 9.38 Series R–L–C circuit

9.9.2 Calculation of total impedance for a series R–L–C circuit

EXAMPLE 9.13

A resistance of 30 Ω is connected in series with an inductive reactance of 60 Ω and a capacitive reactance of 20 Ω. What is the impedance of the circuit?

$$Z = \sqrt{R^2 + (X_L - X_C)^2} \quad (1)$$
$$= \sqrt{30^2 + (60 - 20)^2} \quad (2)$$
$$= \sqrt{30^2 + 40^2} \quad (3)$$
$$= \sqrt{900 + 1600} \quad (4)$$
$$= \sqrt{2500} \quad (5)$$
$$= \underline{50\ \Omega} \quad (6)$$

EXAMPLE 9.14

A circuit that has a 20 Ω resistor in series with a 0.25 H inductor and an 80 μF capacitor is connected to a 230 V 50 Hz supply. Determine the impedance of the circuit, the current and its phase angle.

$$X_L = 2\pi fL \quad (1)$$
$$= 2 \times 3.142 \times 50 \times 0.25 \quad (2)$$
$$= \underline{78.5\ \Omega} \quad (3)$$
$$X_C = \frac{1}{2\pi fC} \quad (4)$$
$$= \frac{1}{2 \times 3.142 \times 50 \times 80 \times 10^{-6}} \quad (5)$$
$$= \underline{39.8\ \Omega} \quad (6)$$
$$Z = \sqrt{(R^2 + (X_L - X_C)^2} \quad (7)$$
$$= \sqrt{(20^2 + (78.5 - 39.8)^2} \quad (8)$$
$$= \sqrt{20^2 + 38.7^2} \quad (9)$$
$$= \sqrt{400 + 1498} \quad (10)$$
$$= \sqrt{1898} \quad (11)$$
$$= 43.6\ \Omega \quad (12)$$
$$\varphi = \cos^{-1}\left(\frac{R}{Z}\right) \quad (13)$$
$$= \cos^{-1}\left(\frac{20}{43.6}\right) \quad (14)$$
$$= \underline{62.7^\circ} \quad (15)$$

9.9.3 Comparison of current-limiting characteristics of inductors and resistors

While a resistor and an inductor will both limit current in a.c. circuits, inductors have several advantages. As discussed in section 9.11.4, resistors consume power. So when they are used to limit current, they reduce the efficiency of a system through the resulting power loss. Inductors are made of a coil of wire and generally have a comparatively low amount of resistance. This means they only consume a small amount of power. They also have a high level of inductive reactance that opposes a.c. current flow.

The other main advantage of inductors over resistors is opposition to change in current. With a high inrush current, for example under short-circuit conditions or when a fluorescent tube is ignited, the back EMF generated from the rapid increase in current will oppose the current flow and reduce its magnitude. This is one of the major reasons for inductors being used as line reactors in power-supply and quality systems. It is also why they are used as **ballasts** in older fluorescent tube light fittings. (A ballast is a device placed in series with the load to limit the amount of current in an electrical circuit.)

9.9.4 Practical examples of R–L–C series circuits

Series R–L–C circuits are generally limited to electronic applications such as tuning circuits in radios. However, a similar circuit can be used as a low-pass filter in electrical supplies. By using a series R–L–C circuit with the appropriate values of reactance, frequencies higher that 50 Hz can be filtered out of the supply.

At one time, series R–L–C circuits were used in car ignition systems. They created the high voltage arc that ignited the fuel via the spark plugs.

CHECK YOUR UNDERSTANDING

9.11 A circuit that has a 30 Ω resistor in series with a 0.3 H inductor (of negligible resistance) and a 100 μF capacitor is connected to a 230 V 50 Hz supply. Determine the:

(a) impedance
(b) current
(c) phase angle.

9.10 Parallel a.c. circuits

9.10.1 Introduction

Most a.c. equipment requires a supply voltage that is fixed (such as 230 V). This means that most work-producing equipment or loads that are connected in electrical installations are connected in parallel. For example, when a lighting or socket-outlet circuit is connected with more than one light or socket-outlet, they are all connected in parallel. This is done by looping all active and neutral connections in the active and neutral terminal at each light or socket-outlet.

Another significant piece of electrical circuitry that is connected in parallel is power factor correction equipment (see section 9.12.3).

9.10.2 Parallel R–L circuits

In a parallel circuit, the voltage is common to all the components. The total current comprises the individual components, as shown in Figure 9.39.

In drawing the phasor diagram, the voltage V is used as the reference phasor. Assuming R and L are both pure values, then:

- I_R is in phase with V
- I_{XL} lags V by 90°
- I_{Total} is equal to the phasor sum of I_R and I_{XL} (as shown in Figure 9.39, where I_{Total} lags V by φ).

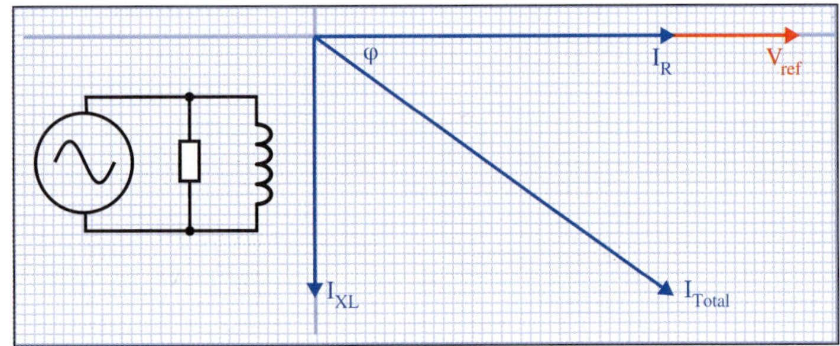

FIGURE 9.39 Parallel R–L circuit

Practical inductors have some series resistance associated with them. A more common circuit would be as shown in Figure 9.40, where the series resistance of the inductor is large enough to affect the phase angle of the inductive phasor (that is no longer 90°).

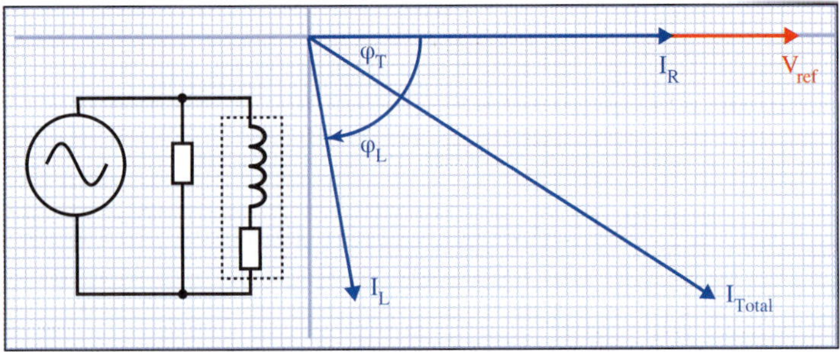

FIGURE 9.40 Parallel R–Z circuit

The current I_R is in phase with V but the current I_{XL}, through the inductor, lags the voltage by φ_L. As the inductor is a series R–L circuit, the angle $\varphi_L = \cos^{-1}\left(\frac{R_L}{Z_L}\right)$.

The phasor addition of I_R and I_L represents the total current I_{Total}, which lags the voltage V by φ_T.

9.10.3 Parallel R–C circuits

An R–C parallel circuit is shown in Figure 9.41. For all practical purposes, the resistor current I_R is in phase with the applied voltage V. The capacitor current leads both by 90°. The total current is the phasor sum of the two individual currents and leads V by φ.

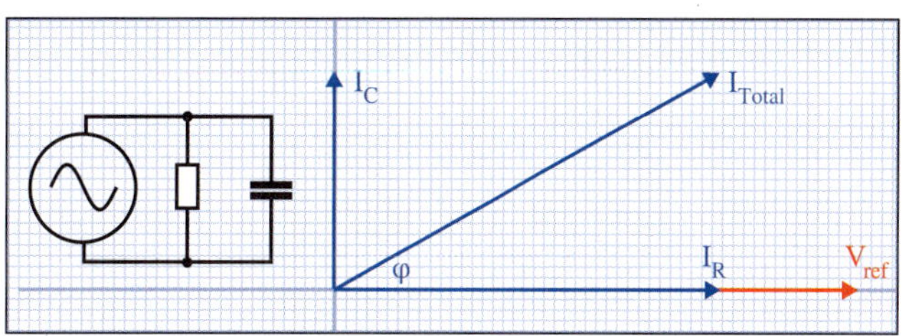

FIGURE 9.41 Parallel R–C circuit

9.10.4 Parallel R–L–C circuits

The total current in this type of circuit is equal to the phasor addition of I_R, I_L and I_C (see Figure 9.42). When L and C are pure quantities, the two currents I_L and I_C are 180° out of phase and can therefore be subtracted ($I_L - I_C$). In this case, I_{Total} lags the voltage by φ. A greater value of I_C could cause the total current to lead the applied voltage.

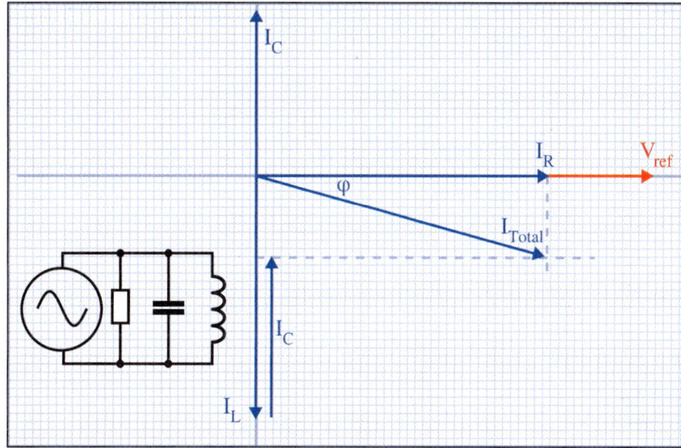

FIGURE 9.42 Parallel R–L–C circuit

9.10.5 Capacitors in parallel

When two capacitors are connected in parallel, the area of the capacitor plates increases which causes the total capacitance to increase. As capacitive reactance (X_c) is inversely proportional to capacitance (C), the overall effect will be an increase in total circuit current. The capacitors in parallel on an alternating current (a.c.) circuit have the same characteristics as resistors in parallel. Each parallel path allows current flow according to its reactance. Two equal-sized capacitors would each draw their usual current, but the total current flow would be double the current flow to a single capacitor.

The total opposition to current for the parallel network is:

$$\frac{1}{X_{C\,Total}} = \frac{1}{X_{C1}} + \frac{1}{X_{C2}} + \frac{1}{X_{C3}} + \dots \frac{1}{X_{Cn}}$$

In the following example, the same capacitor values and supply voltage have been used as in Example 9.15, so that the results can be compared. *The results will differ*.

EXAMPLE 9.15

Two 8 μF capacitors are connected in parallel to a 100 V 50 Hz supply. Find the current flowing through each capacitor and the total current flowing.

$$X_C = \frac{1}{2\pi fC} \quad (1)$$

$$= \frac{1}{2 \times \pi \times 50 \times 8 \times 10^{-6}} \quad (2)$$

$$= \underline{397.9\ \Omega} \quad (3)$$

$$= \frac{V}{X_C} \quad (4)$$

$$= \frac{100}{397.9} \quad (5)$$

$$= 251\ \text{mA} \quad (6)$$

$$X_{C\,Total} = \frac{1}{\frac{1}{X_{C_1}} + \frac{1}{X_{C_2}}} \quad (7)$$

$$= \frac{1}{\frac{1}{397.9} + \frac{1}{397.9}} \quad (8)$$

$$= 198.9\ \Omega \quad (9)$$

$$I = \frac{V}{X_{C\ Total}} \quad (10)$$

$$= \frac{100}{198.9} \quad (11)$$

$$= \underline{502.8\ \text{mA}} \quad (12)$$

9.10.6 Inductors in parallel

If two pure inductors are connected in parallel, each draws its own current from the supply and the line current is the phasor sum of the separate currents. Each current lags the voltage by 90°. Therefore, they are in phase with each other and can be added arithmetically.

The total inductive reactance is therefore reduced in proportion to the increase in current. This allows us to use the same type of equation as for parallel resistors:

$$X_{L\ Total} = \frac{1}{\frac{1}{X_{L1}} + \frac{1}{X_{L2}} + \frac{1}{X_{L3}} + \ldots \frac{1}{X_{LN}}}$$

where:

$X_{L\ Total}$ = total reactance
X_{L1} = reactance 1
X_{L2} = reactance 2
X_{L3} = reactance 3
X_{LN} = more reactances.

The same method is used to find the total inductance of a circuit that has inductors connected in parallel:

$$L_{Total} = \frac{1}{\frac{1}{L_1} + \frac{1}{L_2} + \frac{1}{L_3} + \ldots \frac{1}{L_N}}$$

Where:

L_{Total} = total inductance
L_1 = inductance 1
L_2 = inductance 2
L_3 = inductance 3
L_N = more inductances.

EXAMPLE 9.16

Two inductors, one with an inductive reactance of 10 Ω and the other with an inductive reactance of 8 Ω, are connected in parallel across a 230 V 50 Hz supply.

What is the total inductive reactance? What is the total current?

$$X_{L\ Total} = \frac{1}{\frac{1}{X_{L1}} + \frac{1}{X_{L2}}} \quad (1)$$

$$= \frac{1}{\frac{1}{10} + \frac{1}{8}} \quad (2)$$

$$= \underline{4.44\ \Omega} \quad (3)$$

$$I = \frac{V}{X_L} \quad (4)$$

$$= \frac{230}{4.44} \quad (5)$$

$$= \underline{51.75\ A} \quad (6)$$

9.10.7 Parallel circuit impedance

In parallel circuits with the voltage as reference, the branch currents must be added using a phasor diagram to find the total current. The impedance, reactance and resistance are all proportional to the inverse of the respective current. Therefore, just as parallel resistors in a d.c. circuit cannot be simply added together, the impedance triangle method for series circuits must not be applied to parallel circuit impedances.

The recommended method is to calculate the current flow in each component and add them by use of a phasor diagram.

EXAMPLE 9.17

A resistance of 115 Ω is connected to a 230 V 50 Hz supply in parallel with a pure inductive reactance of 77 Ω (negligible resistance) and a capacitive reactance of 120 Ω. What is the total current and the impedance of the circuit? What is the phase angle of the total current?

$$I_R = \frac{V}{R} \quad (1)$$

$$= \frac{230}{115} \quad (2)$$

$$= \underline{2\ A} \quad (3)$$

$$I_L = \frac{V}{X_L} \quad (4)$$

$$= \frac{230}{77} \quad (5)$$

$$= \underline{2.99\ A} \quad (6)$$

$$I_C = \frac{V}{X_C} \quad (7)$$

$$= \frac{230}{120} \quad (8)$$

$$= \underline{1.92\ A} \quad (9)$$

$$I_X = I_L - I_C \quad (10)$$

$$= 2.99 - 1.92 \quad (11)$$

$$= 1.07\ A \quad (12)$$

$$I_Z = \sqrt{I_R^2 + I_X^2} \quad (13)$$

$$= \sqrt{2^2 + 1.07^2} \quad (14)$$

$$= \underline{2.268\ A} \quad (15)$$

$$Z = \frac{V}{I_Z} \quad (16)$$

$$= \frac{230}{2.268} \quad (17)$$

$$= 101.4\ \Omega \quad (18)$$

$$\Phi = \tan^{-1}\frac{I_X}{I_R} \quad (19)$$

$$= \tan^{-1}\frac{1.07}{2} \quad (20)$$

$$= 28.15^\circ \quad (21)$$

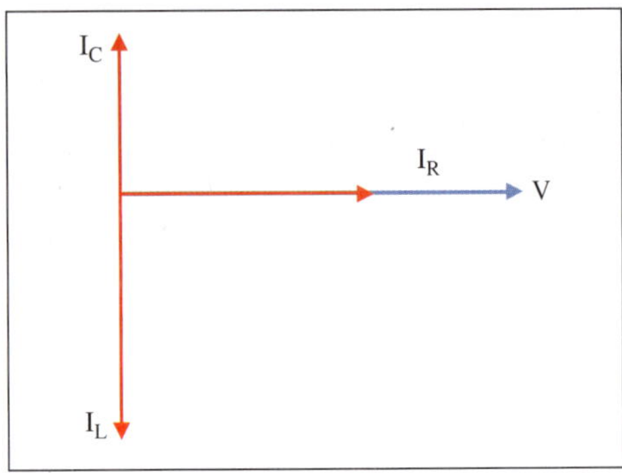

FIGURE 9.43(a) I_R, I_C and I_L phasors

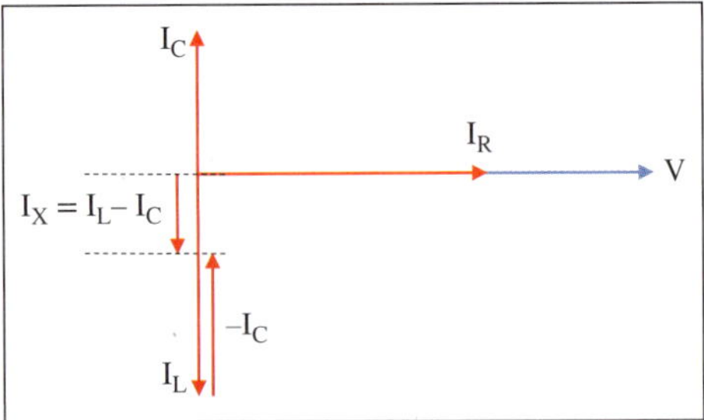

FIGURE 9.43(b) Resultant phasor of I_L and I_C

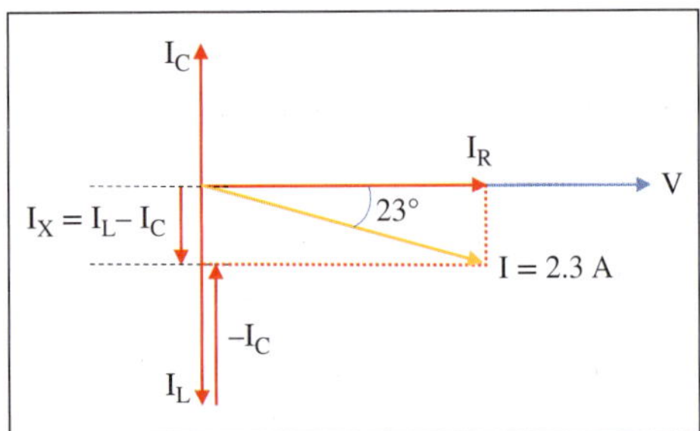

FIGURE 9.43(c) Resultant phasor of I_R, I_C and I_L

We can now use these calculated current values to draw the phasor diagram. The resulting phasor will represent the resulting current.

Using voltage as the reference, draw the three currents to scale (Figure 9.43(a)). The inductive current and the capacitive current are in opposition and so will try to cancel each other out.

The resultant phasor will be IL – $I_{C,}$ which is shown as I_X. It is drawn in the direction of I_L (lagging) because I_L is a greater value than IC (Figure 9.43(b)).

A parallelogram is then drawn, using the value of I_X and I_R (Figure 9.43(c)).

The resultant can then be determined using the scale of the drawing and the angle measured.

The resultant here is approximately 2.3 A at 23° lagging. This can be checked by calculation, which should give a similar—but more accurate—result.

9.10.8 Practical examples of parallel circuits

As stated previously, all equipment that is connected by the end user of a.c. circuits must have the same voltage applied to it. So all electrical equipment connected to final subcircuits is connected in parallel.

Other circuits in the electrotechnology industry that require a parallel connection include:

1. power factor correction capacitors connected in parallel with the equipment load (or with the incoming supply)
2. high-pass filtering that allows 50 Hz a.c. through but opposes low-frequency interference
3. shunting circuits that shunt over-voltage current such as lightning strikes and high-voltage injection (through equipment failure) to earth.

CHECK YOUR UNDERSTANDING

9.12 A resistance of 130 Ω is connected to a 230 V 50 Hz supply, in parallel, with an inductive reactance of 90 Ω and a capacitive reactance of 110 Ω. Determine the:

(a) current in each parallel branch
(b) total current
(c) impedance of the circuit
(d) phase angle.

9.11 Power in an a.c. circuit

9.11.1 Power in a resistive circuit

If the values of voltage (V) and current (I) are taken at a given instant, then the instantaneous power P = V × I. By taking the product of V and I for a number of instantaneous values, we can plot a curve of power for each cycle of a.c. In Figure 9.44, only one cycle is shown as any one cycle represents what any other cycle will do.

The curves for voltage, current and power are shown as generated by a computer program from the maths. The following points are particularly important:

- The power curve is sinusoidal in shape, which agrees with the laws of trigonometry.
- There are no negative values of power—a negative multiplied by a negative becomes positive.
- The power curve completes two cycles for each complete cycle of current or voltage.
- As the power curve is a sine wave, the area underneath it is identical to the area between it and a line drawn to just touch the peak value of power. Therefore, the power used is half what the peak values would have suggested. This proves that the values for RMS must be $\frac{1}{\sqrt{2}V_{max}}$ and $\frac{1}{\sqrt{2}I_{max}}$.

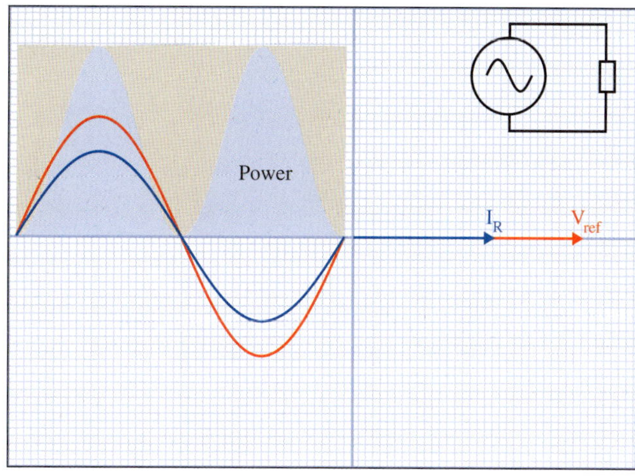

FIGURE 9.44 Resistive a.c. circuit

When calculating power in an a.c. circuit, the RMS values of current and voltage must be used. In order to show this we will look at a circuit with RMS values of 1 A and 1 V and calculate the power.

$$P = V_{RMS} \times I_{RMS} = 1 \times 1 = 1 \text{ W}$$

Using maximum values, which are 1.414 times RMS values, gives an answer of twice the actual power used:

$$P = V_{max} \times I_{max} = 1.414 \times 1.414 = 2 \text{ W}$$

This result confirms the need to use RMS values when working with alternating current.

For purely resistive circuits, Ohm's Law and the power law both work using RMS values of voltage and current:

$$V = I \times R \text{ and } P = V \times I = \frac{V^2}{R} = I^2R$$

But with a.c., V and I must always be measured in RMS values.

For sinusoidal waveforms only, if peak values are used to obtain a power rating, the average power value obtained is always half that of the peak power value. That is, average power equals half maximum power.

Substituting maximum values in the power equation gives:

$$P_{avg} = 0.707 \times V_{max} \times 0.707 \times I_{max} = 0.5P_{max}$$

9.11.2 Power in a capacitive circuit

As with inductors, capacitors charge and discharge. The energy stored in the capacitor in one quarter-cycle is returned in the next quarter-cycle. Therefore, the average power in a pure capacitive circuit is zero.

In Figure 9.45 the shaded power waveform is the result of multiplying the instantaneous values of voltage and current. When both are positive, the capacitor is charged; when both are negative, the capacitor is charged in the opposite polarity. But when one is positive and the other negative, the charge is returned to the power supply. As the charge is the same size as the discharge, no power is consumed.

There is as much of the power curve above the zero line as there is below it. The average power in a purely capacitive circuit is zero.

9.11.3 Power in an inductive circuit

Inductors store energy as a magnetic field, which is returned to the circuit when the field collapses. This happens every half-cycle. As there is no resistance (in theory), there are no losses and all the energy is returned.

Figure 9.45 shows the applied voltage as the red sine wave and the back EMF as the blue sine wave. When the resistance either does not exist or is negligible, the back EMF is equal to the applied voltage and is of opposite polarity (or it is out of phase by 180°).

The back EMF is proportional to the change in current flow. Therefore, it is out of phase with the current by 90°. Similarly, the current is also out of phase with the applied voltage by 90°, with maximum current flow occurring when the applied voltage changes polarity (that is, when it crosses zero).

In Figure 9.45, the power is represented by the shaded sine wave. We can see that it has twice the frequency of the voltage or current and that there are two positive pulses of power. Generally, we would consider this to be energy that has been used; now we must consider it to be energy that has been stored in the inductor.

But the negative pulses of the power curve do not mean that we have discovered negative power. Rather, they mean that we have found energy that has been returned to the circuit.

That means that the sign of the power waveform reverses every quarter of a cycle, showing that power is alternately fed into and returned from the inductor.

As the current rises, energy is used to produce the magnetic field. As the current falls, the magnetic field collapses and the energy is returned to the supply. Over a complete cycle, the positive and negative sections of the power waveform cancel each other. As a result, the average power consumed by a pure inductor is zero.

Figure 9.45 shows that the power wave is sinusoidal if the voltage and current waveforms are sinusoidal. It also shows that the frequency of the power wave is twice that of the line frequency.

9.11.4 Power in a.c. circuits

A purely resistive load consumes power, while a purely inductive load consumes no power. When resistance and inductance are combined in one circuit, there will be a value of power consumed that is dependent on the resistance, or resistive load, in the circuit.

Figure 9.45 shows the waveforms for voltage, current and power in a series R–L circuit where I lags V by 45°.

As there is more positive than negative power, the resultant or average power will be positive. It will be a lesser value than for a purely resistive load, but a greater value than a purely inductive load.

True power (P)

The applied voltage is split into two components at right angles to each other, as shown in the phasor diagram in Figure 9.43. V_R is in phase with the current, while V_L is leading the current by 90° and is shown at 90° to V_R.

Previously it was stated that, in a purely resistive circuit, V and I are in phase and the power consumed is found from P = VI. This is still true as long as the resistive or in-phase component of voltage is used. That is:

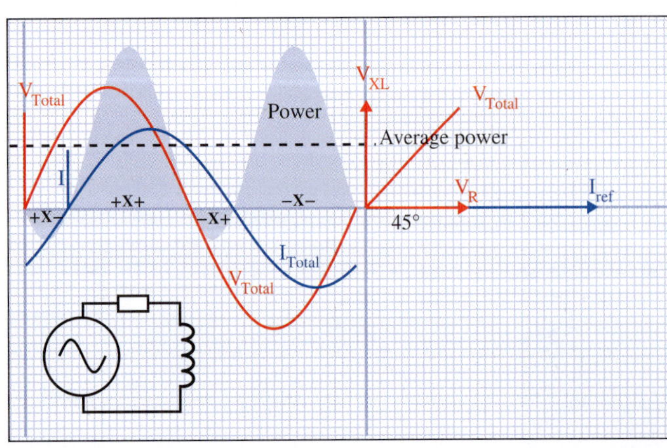

FIGURE 9.45 Power in an a.c. R–L circuit

$$P = V_R I$$

The ratio $\frac{V_R}{V_{Total}}$ is equal to the Cosine of the angle of lag φ and it is usual to express the power consumed in terms of the line voltage V. That is:

$$\frac{V_R}{V} = \cos \varphi$$

$$\therefore V_R = V \cos \varphi$$

Since $P = V_R I$, then also $P = V \cos \varphi\, I$, that is,

$$P = VI \cos \varphi$$

In electrical power work (which mainly deals with sinusoidal waveforms), 'cos φ' usually applies to the power factor (or PF). However, the general expression for power factor for all wave shapes is the symbol lambda (λ). For electrical work, this symbol is synonymous with the expression cos φ (i.e. PF = cos φ = λ):

$$P = VI \cos \varphi = VI\lambda$$

Although you could use lambda (λ) to represent cos φ, using the term cos φ in your equations will help you remember what lambda and the equation mean.

Apparent power (S)

For circuits that have both resistance and reactance, the product of the measured line voltage and line current produces a value greater than the power consumed. Therefore, it cannot be expressed in watts. The value is known as the 'apparent power'. This is useful in dealing with electric machinery, as will be seen.

Apparent power is measured in volt-amperes (VA). Many a.c. machines, particularly alternators and transformers, are rated in VA to give an indication of both core and iron size and the current rating of the windings. *The current rating of the windings is greater than the current required for the power consumed.* For example, an alternator could be rated at 100 kVA yet only deliver 80 kW:

$$S = VI$$

Reactive power (Q)

Reactive power is found from the product of the line voltage and the reactive proportion of the line current that does not consume power.

A capacitor exhibits only reactive power. This is because the current flowing is leading the voltage by 90° and the true power is zero. Reactive power is sometimes called 'wattless power' for this reason. It is measured in volt-amperes reactive (VAR):

$$Q = VI \sin \varphi$$

True power, apparent power and reactive power can be represented by a power triangle, as shown in Figure 9.46. True power equals the apparent power in volt-amperes multiplied by the power factor of the circuit. If cos φ is calculated from 0° to 90°, the power factor will vary from 1 for a purely resistive circuit to 0 for a purely reactive circuit.

Similarly, if sin φ is calculated from 0° to 90°, the reactive power ratio will vary from 0 for a purely resistive circuit to 1 for a purely reactive circuit.

9.11.5 Power losses

An inductor is a coil of wire that has resistance. If an iron core is used, eddy current and hysteresis losses are produced when an a.c. current is applied.

For example, a 40 W fluorescent lamp ballast typically has a resistance of 36 Ω, draws 0.4 A and consumes 10 W. The power loss due to the resistance of the windings is called 'copper loss':

$$P = I^2R = 0.4 \times 0.4 \times 36 = 5.76 \text{ W}$$

The remaining 4.24 W loss is due to eddy currents and hysteresis in the iron core. This is known as an

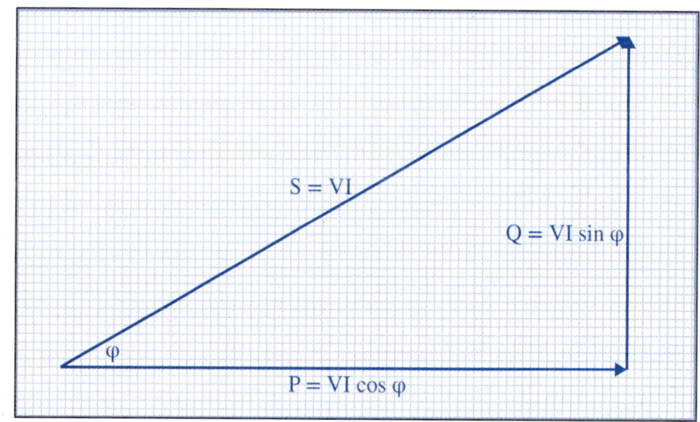

FIGURE 9.46 Power triangle

'iron loss' because the losses occur in the iron core. The total losses in the inductor equal the sum of the copper and iron losses.

Since practical inductors consume some power, they must consist of both pure inductance and pure resistance.

9.11.6 Defining 'power factor' and 'phase angle'

The power factor (PF or λ) is the factor or ratio by which apparent power is multiplied to obtain the true power, or the actual power being consumed.

The phase angle is a value for sinusoidal waveforms which corresponds to the Cosine of the angle between voltage and current.

For all electrical power work with sinusoidal waveforms:

$$\text{Power factor (PF)} = \cos\varphi = \frac{R}{Z}$$

The relationship between true power and apparent power can also lead to obtaining a value for the power factor:

$$\text{Power factor (PF)} = \cos\varphi = \frac{\text{true power}}{\text{apparent power}}$$

$$\text{PF} = \cos\varphi = \frac{P}{S}$$

EXAMPLE 9.18

Determine the power factor and phase angle of a ballast that has 1.8 Ω of resistance and an inductance of 150 mH at 50 Hz.

$$\begin{aligned} X_L &= 2\pi fL \\ &= 2 \times \pi \times 50 \times 0.15 \\ &= 47.12\ \Omega \\ Z_L &= \sqrt{(R^2 + X_L^2)} \\ &= \sqrt{(1.8^2 + 47.12^2)} \\ &= 47.16\ \Omega \end{aligned}$$

$$\begin{aligned} \text{Power factor (PF)} &= \cos\varphi \\ &= \frac{R}{Z} \\ &= \frac{1.8}{47.16} \\ &= 0.0381 \end{aligned}$$

$$\begin{aligned} \text{Phase angle } (\varphi) &= \cos^{-1}(\text{PF}) \\ &= \cos^{-1}(0.0381) \\ &= 87.81^\circ \end{aligned}$$

EXAMPLE 9.19

A single-phase 230 V motor draws 4.5 kW from the supply. If the supply current is 30 A, determine the:

- apparent power
- power factor
- phase angle.

$$S_{motor} = V_{supply} \times I_{motor}$$
$$= 230 \times 30$$
$$= 6.9 \text{ kVA}$$

$$\text{Power factor (PF)} = \cos \varphi$$
$$= \frac{P}{S}$$
$$= \frac{4500}{6900}$$
$$= 0.652$$

$$\text{Phase angle } (\varphi) = \cos^{-1}(\text{PF})$$
$$= \cos^{-1}(0.652)$$
$$= 49.29°$$

9.11.7 Methods for measuring single-phase power, energy and demand

Unlike the case with d.c., a voltmeter and ammeter cannot be used to measure true power in a.c. This is because the product of these two values will give **apparent power** unless the load measured is purely resistive. However, a wattmeter (or dynamometer) can be used, as with d.c. This is due to the current winding measuring the average instantaneous value of current and therefore providing a reading that indicates the true power. If the load is purely reactive, the 90-degree phase shift between the voltage and current will cause the average value of the power waveform to be zero.

The current coil consists of a winding of relatively large cross-sectional area and low amount of turns. The potential coil is a winding of small cross-sectional area and high amount of turns. The terminals on the wattmeter are typically labelled M (mains) and L (load) for the ammeter connection and C (common) and V (voltage) for the voltage connection. Alternative labels are A1 and A2 for the ammeter and V1 and V2 for the voltmeter. Combinations of these two systems also exist.

Figure 9.47 shows the connection for the wattmeter with:

- M connected to the supply active (line side)
- L connected to the load active
- V_1 (C) connected as a common connection to M
- V_2 (V) connected to neutral.

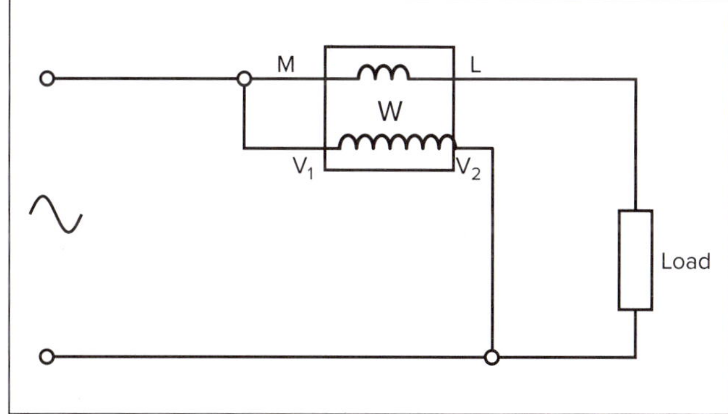

FIGURE 9.47 Connection for the wattmeter

More sophisticated digital wattmeters are available. These have many other functions such as multiple ranges and memory storage. Power analysers have true, apparent and reactive power measurement functions. They typically measure the current component with a tong clamp or flexible current probes fitted around the cables to be tested.

Electrical energy is measured using a kilowatt hour meter, which is a dynamometer with a time-logging function. Older versions of kilowatt hour meters used a rotating disc to turn a set of gears which clocked the energy usage. These systems are generally being replaced by electronic 'smart' meters that can monitor multiple tariffs for different demand periods of the day. They are also able to communicate remotely with electricity retailers. Some power analysers also have time-of-use capabilities, data logging and communication facilities.

CHECK YOUR UNDERSTANDING

9.13 A coil has a resistance of 2.7 Ω and an inductance of 175 mH at 50 Hz. Calculate the:

(a) reactance of the coil
(b) impedance of the coil
(c) power factor
(d) phase angle.

9.12 Power factor improvement

9.12.1 Effects of low power factor

Broadly speaking, the lower the value of the power factor, the greater the current required to supply the same true power. The extra current flowing has the following consequences:

- larger cross-sectional area conductors are required
- larger transformers are required
- higher-rated switch gear
- fuses of a higher current rating
- higher voltage drops along conductors
- extra copper losses
- decreased efficiency such as higher losses and more fuel used
- higher generating and capital costs.

The effects of a low power factor mean that rating electrical equipment by power consumption is not always satisfactory. Some electrical equipment is rated by the volt-ampere (VA) method. The voltage determines the insulation required and the current determines the size of the conductors.

For example, Figure 9.48 describes a transformer designed for 250 V and 40 A. The rating of this transformer is 250 × 40 = 10 000 VA = 10 kVA. This is the maximum VA of the transformer.

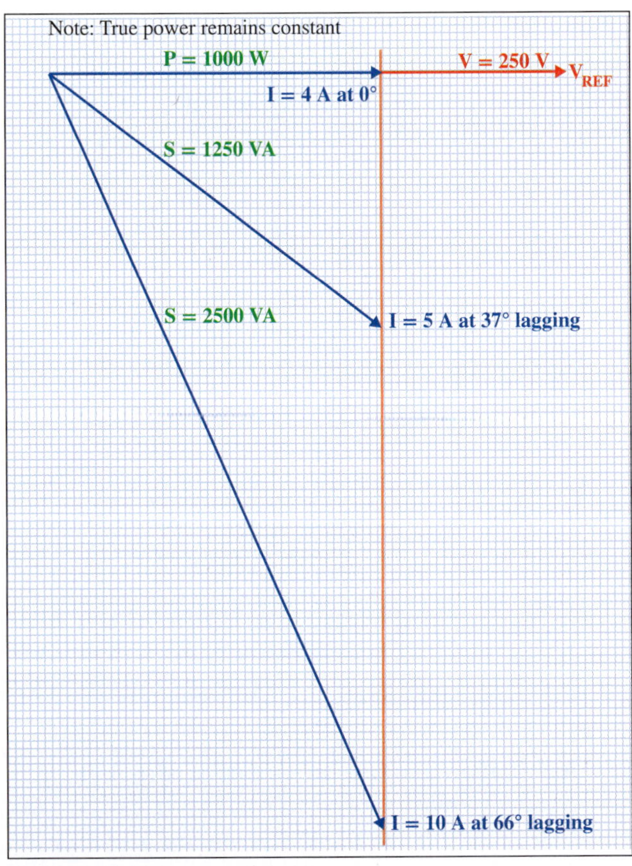

FIGURE 9.48 Power factor effect on current

EXAMPLE 9.20

If a 1 kW load is connected to a 250 V a.c. supply, find the current flowing at:

- unity power factor ($\varphi = 0°$)
- power factor = 0.8 ($\varphi = 37°$)
- power factor = 0.4 ($\varphi = 66°$).

$$P = VI_{\cos}\varphi \quad (1)$$

$$\therefore I = \frac{P}{V_{\cos}\varphi} \quad (2)$$

$$I_A = \frac{1000}{250 \times 1} \quad (3)$$

$$= \underline{4\ A} \quad (4)$$

$$\text{and } I_B = \frac{1000}{250 \times 0.8} \quad (5)$$

$$= \underline{5\ A} \quad (6)$$

$$\text{and } I_C = \frac{1000}{250 \times 0.4} \quad (7)$$

$$= \underline{10\ A} \quad (8)$$

If the transformer was used to supply an electric furnace with a unity power factor (PF = 1), it could supply a maximum of 10 000 × 1 = 10 kW of power. Alternatively, if it was used to supply a load at 0.8 power factor, the maximum power it could supply would be 10 000 × 0.8 = 8 kW.

A transformer must be rated in VA. This value will give its maximum rating, independent of the power factor of the load.

9.12.2 Causes of low power factor

Many circuits are inductive and as a result cause currents to lag. A circuit with a leading current (i.e. a capacitive circuit) is rare, although the effect can occur in long-distance transmission lines.

The major causes of a low power factor are lightly loaded electric motors and transformers, and fluorescent lighting circuits. Motors and transformers should be designed to run at or near full load to improve their power factor. For fluorescent lighting circuits using iron core ballasts, capacitors are added to the circuit to improve the power factor. Modern electronic ballast types are designed to produce a PF nearer to unity.

Electricity generating authorities have conditions-of-supply requirements for the use of equipment at poor power factor values.

9.12.3 Power factor correction

For economic reasons, the recommended value of the power factor is 0.9. Below 0.9, the current increases rapidly; above 0.9, the cost of the equipment necessary to correct the power factor is not worth the overall benefit gained.

Reduced power factor, perhaps due to oversized motors operating at less than full load, causes a sharp rise in current. This is shown in Figure 9.49, where the total current is plotted against the phase angle from the resistive current to ten times the resistive current at around 85° (or a PF of 0.1).

To correct the power factor, it is necessary to reduce the phase angle between the line current and voltage.

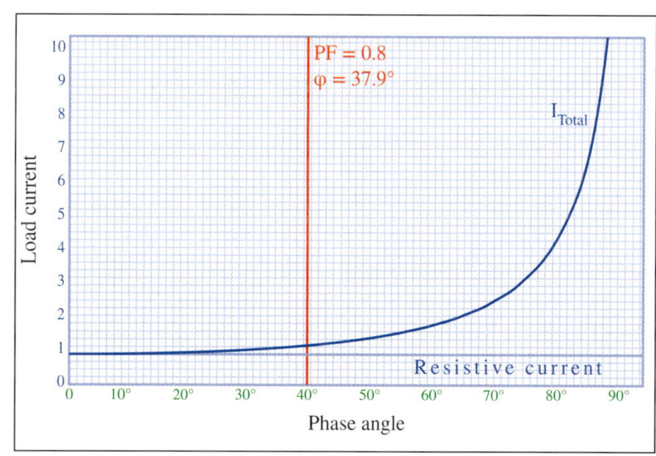

FIGURE 9.49 Power factor vs total current

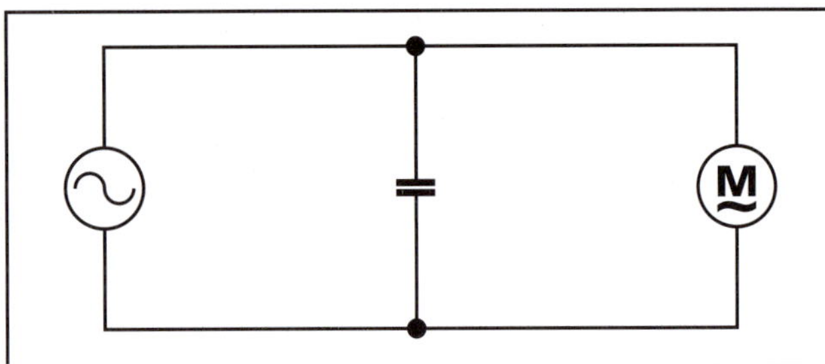

FIGURE 9.50 Motor with PF correction

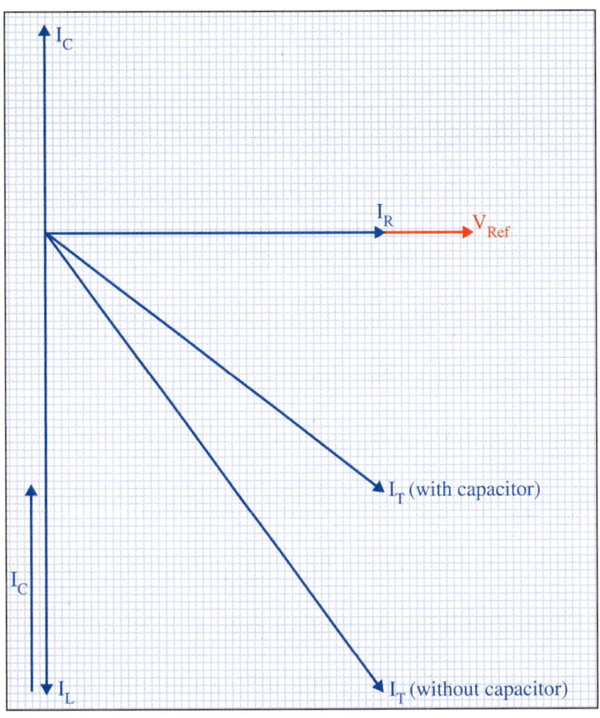

FIGURE 9.51 PF correction phasor diagram

This needs to be done without affecting the values of voltage or load current.

Most low power factor problems are caused by inductive loads, such as induction motors and transformers. One way of improving the power factor in this type of circuit is to connect a capacitor in parallel with the load, as shown in Figure 9.50. The effects of doing that are shown in Figure 9.51.

I_L remains the same and each load operates at its own power factor, but the overall power factor of the combined circuit is improved. A pure capacitor is a load that operates at a leading zero power factor and, when connected across an inductive load, tends to counteract the lagging effect of the inductance, but without consuming any power.

For example, a 230 V single-phase installation supplies a number of motors requiring 10 kW of power. On single-phase, this results in a load current of 43.5 A. If the power factor of the load is 0.6 (cos 53°), the total current in the line will be 72.5 A. Correcting the PF to 0.8 (cos 37°) would reduce the current to 54.5 A, a decrease of approximately 20% for the same true power. This illustrates the need for improving the power factor.

EXAMPLE 9.21

A single-phase 230 V 50 Hz induction motor draws 15 A at 0.6 power factor. Determine the line current and power factor when an 80 μF capacitor is connected across the line. Draw the current phasor diagram showing the condition before and after correction.

$$I_R = I_{Motor} \times PF \quad (1)$$
$$= 15 \times 0.6 \quad (2)$$
$$= \underline{9\ A} \quad (3)$$
$$\varphi_{Motor} = \cos^{-1}(PF) \quad (4)$$
$$= \cos^{-1} 0.6 \quad (5)$$
$$= \underline{53.13^\circ} \quad (6)$$
$$I_{XL} = I_{Motor} \times \sin \varphi \quad (7)$$
$$= 15 \times \sin (53.13) \quad (8)$$
$$= \underline{12\ A} \quad (9)$$
$$X_C = \frac{1}{2\pi fC} \quad (10)$$
$$= \frac{1}{2\pi \times 50 \times 80 \times 10^{-6}} \quad (11)$$
$$= \underline{39.8\ \Omega} \quad (12)$$

$$I_{XC} = \frac{V}{X_C} \quad (13)$$

$$= \frac{230}{39.8} \quad (14)$$

$$= \underline{5.78\ A} \quad (15)$$

$$I_X = I_{XL} - I_{XC} \quad (16)$$

$$= 12 - 5.78 \quad (17)$$

$$= \underline{6.22\ A} \quad (18)$$

$$I_T = \sqrt{I_R{}^2 + I_X{}^2} \quad (19)$$

$$= \sqrt{9^2 + 6.22^2} \quad (20)$$

$$= \underline{10.9\ A} \quad (21)$$

$$PF = \frac{I_R}{I_T} \quad (22)$$

$$= \frac{9}{11} \quad (23)$$

$$= \underline{0.825} \quad (24)$$

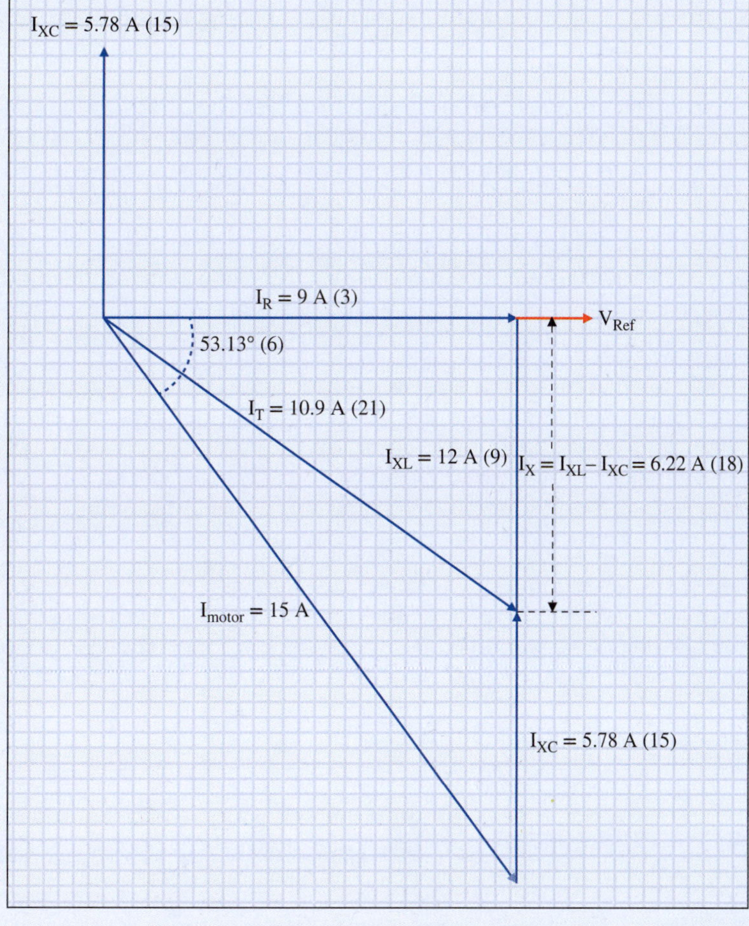

FIGURE 9.52 **Phasor diagram for Example 9.21**

Important points here are that the:

- line current has been reduced, even though an extra component (the capacitor) has been added.
- in-phase or power component of current (I_R) has retained its original value.

- motor current and power factor remain unaltered.
- power factor of the resulting combined load has been improved.
- power consumption remains at its original value.
- volt-ampere rating of the combined circuit has been reduced.
- reactive power rating of the combined circuit has been reduced.

The purpose of using capacitors to improve the power factor is to provide a leading current. This will counteract the lagging current drawn by the load but not increase the value of power consumed. Supply authorities try to keep as high a value of power factor as is economically possible. They do this by regulating the minimum allowable value of power factor for any load connected to the supply. Consequently, there is regular demand for power factor correction capacitors.

Power factor improvement using a capacitor can be calculated as shown in Example 9.21. However, in that example, the power factor using the suggested capacitor still did not improve beyond the 0.9 usually required by supply authorities. A 100 μF capacitor might give an acceptable PF value, but that would just be another guess.

A better method would be to calculate the value of capacitance required to improve the overall power factor to an acceptable specified value. Example 9.22 shows a method of doing that. In this case, a PF of 0.9 has been chosen.

A 44.84 μF capacitor would have to be specially made, which would be very expensive. In this case, a 47 μF capacitor would do the job with the resulting power factor being slightly higher than 0.9. For a single load, a single capacitor is generally used. Stepped capacitors are only used for an entire installation where the total load is varying.

Power factor correction requirements can also be calculated by the reactive power approach. The value of current flowing in the capacitor, multiplied by the voltage across it, gives the reactive power or VAR value.

For example, in Example 9.22 the motor takes 230 V at 10 A or 2.3 kVA. The true power is found from $P = VI \cos \varphi$ or $230 \times 10 \times 0.65 = 1.495$ kW. If the power factor is improved to 0.85, then the true power will remain the same. But the kVA will be reduced, and therefore the line current will be reduced.

If $P = VI \cos \varphi$ for either PF values, it follows that:

$$VI_1 \cos \varphi_1 = VI_2 \cos \varphi_2$$

The voltage is the same each time and so cancels out, giving us an equation for the improved line current:

$$I_2 = I_1\left(\frac{\cos \varphi_1}{\cos \varphi_2}\right) = 10 \times \frac{0.65}{0.9} = 7.22 \text{ A}$$

This tells us how much the current will improve at the specified new power factor.

We can see the effect of the power factor correction by looking at the power triangles before and after correction in Figure 9.53 (the triangles are not to scale).

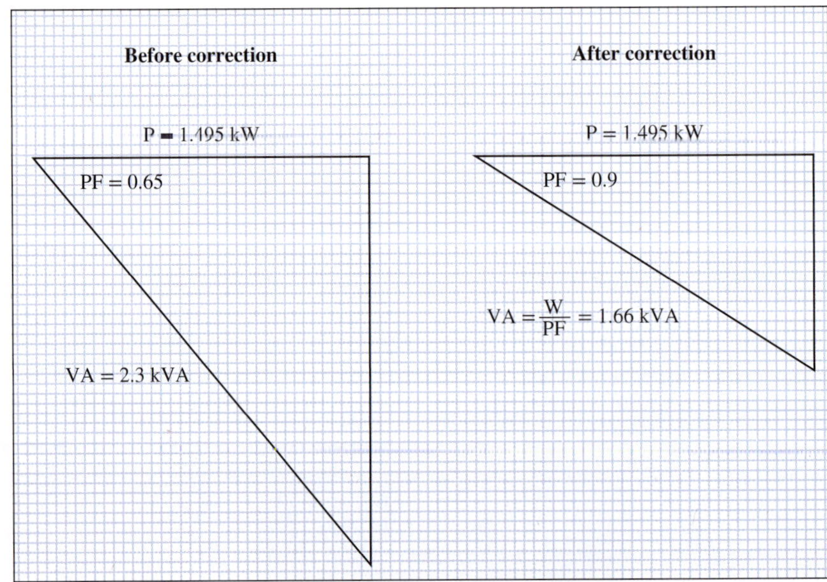

FIGURE 9.53 Power triangles before and after correction

EXAMPLE 9.22

A motor takes a current of 10 A at 0.65 power factor (lagging) from a 230 V 50 Hz supply. What size of capacitor is required to improve the power factor to 0.9 lagging? Draw the current phasor diagram to show the effect of the capacitor.

$$
\begin{aligned}
I_R &= I_Z \times PF && (1)\\
&= 10 \times 0.65 && (2)\\
&= \underline{6.5\text{ A}} && (3)\\
\varphi &= \cos^{-1}(PF) && (4)\\
&= \cos^{-1} 0.65 && (5)\\
&= \underline{49.45^\circ} && (6)\\
I_{XL}\ (@0.65PF) &= I_Z \times \sin(\varphi) && (7)\\
&= 10 \times \sin(49.45) && (8)\\
&= \underline{7.6\text{ A}} && (9)\\
\varphi\ (@0.9PF) &= \cos^{-1}(PF) && (10)\\
&= \cos^{-1} 0.9 && (11)\\
&= \underline{25.84^\circ} && (12)\\
I_X\ (@0.9PF) &= I_Z \times \sin(\varphi) && (13)\\
&= 10 \times \sin(25.84) && (14)\\
&= \underline{4.36\text{ A}} && (15)\\
\therefore I_{XC} &= I_{XL} - I_X && (16)\\
&= 7.6 - 4.36 && (17)\\
&= \underline{324\text{ A}} && (18)\\
X_C &= \frac{V}{I_{XC}} && (19)\\
&= \frac{230}{3.24} && (20)\\
&= 70.99\ \Omega && (21)\\
C &= \frac{1}{2\pi f X_C} && (22)\\
&= \frac{1}{2\pi \times 50 \times 70.99} && (23)\\
&= 31.8 \times 10^{-6} && (24)\\
&= 44.84\ \mu\text{F} && (25)
\end{aligned}
$$

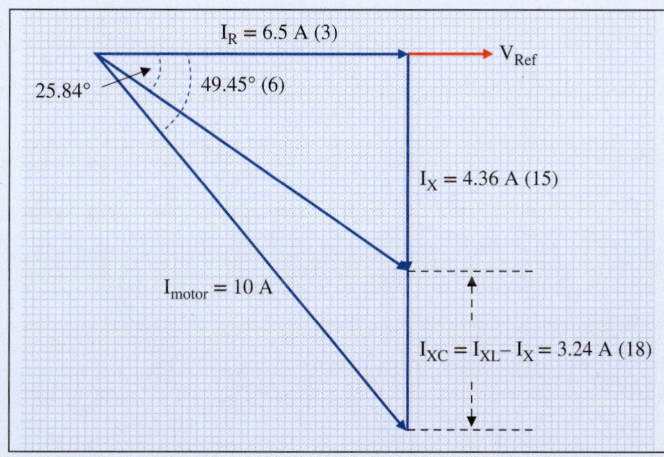

FIGURE 9.54 Phasor diagram showing currents before and after correction

The reactive power is calculated from $Q = VI \sin \varphi$ and $\varphi = \cos^{-1}(PF)$ so reactive power can also be found from

$$Q = VI \sin (\cos^{-1}(PF))$$

EXAMPLE 9.23

A non-standard voltage of 480 V is used to feed a single-phase installation. The installation draws 54 A at a power factor of 0.55. What will the current be if the power factor is improved to 0.8? Determine the kVAR rating of the capacitor required to improve the power factor to 0.8.

$$I_2 = I_1 \times \frac{PF_1}{PF_2} \quad (1)$$

$$= 54 \times \frac{0.55}{0.8} \quad (2)$$

$$= \underline{37.1\ A} \quad (3)$$

$$Q_1 = VI_1 \sin (\cos^{-1}(PF_1)) \quad (4)$$

$$= 480 \times 54 \sin (\cos^{-1}(0.55)) \quad (5)$$

$$= \underline{21.6\ kVAR} \quad (6)$$

$$Q_2 = VI_2 \sin (\cos^{-1}(PF_2)) \quad (7)$$

$$= 480 \times 37.1 \sin (\cos^{-1}(0.8)) \quad (8)$$

$$= \underline{10.7\ kVAR} \quad (9)$$

$$Q_C = Q_1 - Q_2 \quad (10)$$

$$= 21.3 - 10.7 \quad (11)$$

$$= \underline{10.6\ kVAR} \quad (12)$$

9.12.4 Requirements regarding the power factor of an installation and power factor improvement equipment

As we have seen, AS/NZS 3000:2018 outlines the safety requirements for the installation of capacitors. However, as power factor is performance-based (not safety-based), there is little further reference to the subject. The requirements for power factor correction are instead stated in the local supply authority service and installation rules. These are the documents adhered to by supply authorities responsible for the safety, security and maintenance of electricity grids. Some of the organisations that produce these rules are private while others are government run. It varies by jurisdiction.

The requirements that apply to the installation of power factor equipment generally only apply to installations that are charged in kVA hours instead of kW hours. These installations have primarily been non-domestic. This situation is changing, though, with the evolution of the grid and the supply of electricity in Australia.

The performance factors that generally apply to these rules include:

- AS/NZS 3000:2018 should be adhered to for the switching and isolation requirements for capacitor installations.
- The power factor should be maintained between 0.9 lagging and unity and should never become leading.

- Steps in the kVAR rating of the capacitor bank should not be excessive and should be, for example, limited to 50 kVAR increments.
- Series resonance should be avoided.
- Harmonics and spikes in the power supply should be avoided and measures put in place to control their effects on the grid and other customers.

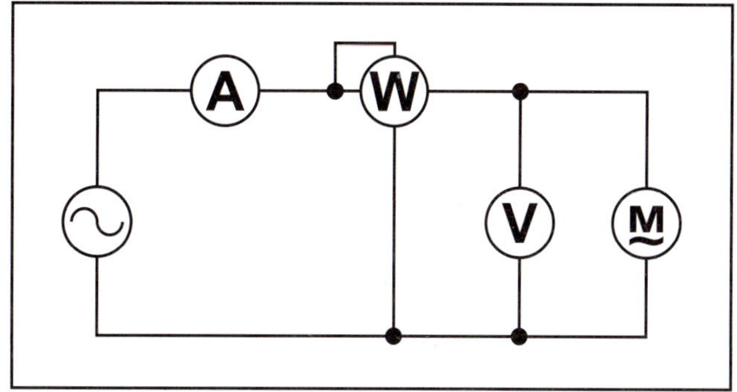

FIGURE 9.55 Power factor measurement

9.12.5 Determining power factor

The power factor of a circuit can be obtained by using a voltmeter, ammeter and wattmeter. Figure 9.55 shows the circuit for measuring the power factor of a single-phase a.c. motor.

The power factor can then be found from:

$$\text{Power factor (PF)} = \frac{P}{S} = \frac{\text{watts}}{\text{volt-amperes}}$$

EXAMPLE 9.24

A single-phase motor draws 2.7 A on 230 V and a wattmeter in the circuit reads 450 W. Find the power factor and phase angle.

$$\text{PF} = \frac{P}{S} \quad (1)$$

$$= \frac{450}{230 \times 2.7} \quad (2)$$

$$= \underline{0.7246} \quad (3)$$

$$\varphi = \cos^{-1}(\text{PF}) \quad (4)$$

$$= \cos^{-1}(0.7246) \quad (5)$$

$$= \underline{43.6^\circ} \quad (6)$$

The power factor can also be measured by using a power factor meter, as shown in Figure 9.56. The meter is connected like a wattmeter and the scale is calibrated in values of power factor from 0 to 1.0.

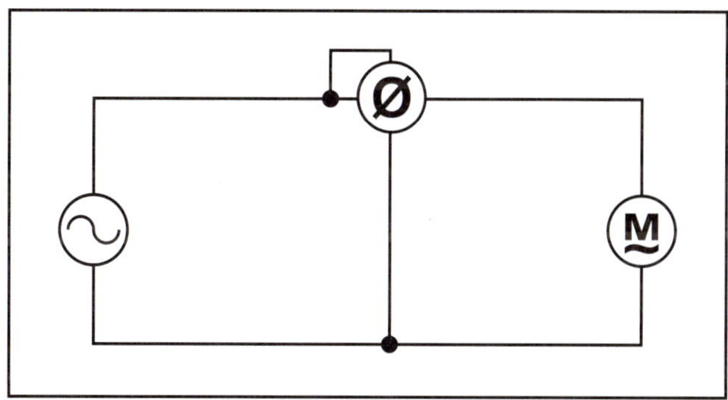

FIGURE 9.56 Power factor meter connected in a circuit

Calculation

In a series circuit, the power factor can be obtained by using the values of resistance and impedance.

EXAMPLE 9.25

An inductor draws 20 A on 230 V d.c. and 10 A on 230 V a.c. Calculate the angle of lag when on a.c.

$$R_{d.c.} = \frac{V}{I} \quad (1)$$

$$= \frac{230}{20} \quad (2)$$

$$= \underline{11.5\ \Omega} \quad (3)$$

$$Z_{a.c.} = \frac{V}{I} \quad (4)$$

$$= \frac{230}{10} \quad (5)$$

$$= \underline{23\ \Omega} \quad (6)$$

$$\varphi = \cos^{-1}\frac{R}{Z} \quad (7)$$

$$= \cos^{-1}\frac{11.5}{23} \quad (8)$$

$$= \cos^{-1}0.5 \quad (9)$$

$$= \underline{60^{\circ}} \quad (10)$$

Industry example

Table 9.1 shows a range of capacitors that are available to improve power factors.

TABLE 9.1 Single-phase capacitors available for power factor correction

Single-phase capacitors—UCW											
	50 Hz		60 Hz								
Rated voltage (V)	Reactive power (kvar)	Rated current I_n (A)	Reactive power (kvar)	Rated current I_n (A)	Capacitance (uF)	Series	Reference	Dimensions ∅ × H (mm)	Discharge resistor		Weight (kg)
208	0.62	3.0	0.74	3.6	45.5	B	UCW0.83V25J4	53 × 85	Not included	270 kΩ/3 W	0.27
	0.62	3.0	0.74	3.6	45.5	B	UCW0.83V25 L6	60 × 105	Not included	270 kΩ/3 W	0.34
	1.24	6.0	1.49	7.2	91.5	B	UCW1.67V25 L6	60 × 105	Not included	150 kΩ/3 W	0.35
	1.86	9.0	2.23	10.7	137.0	B	UCW2.5V25 L10	60 × 156	Not included	82 kΩ/3 W	0.51
	2.48	11.9	2.98	14.3	182.5	B	UCW3.33V25 L10	60 × 156	Not included	56 kΩ/3 W	0.51
	3.72	17.9	4.47	21.5	274.0	C	UCW5V25 N14	75 × 205	Included	41 kΩ/3 W	1.19
	4.97	23.9	5.96	28.7	365.6	C	UCW6.67V25 N14	75 × 205	Included	28 kΩ/3 W	1.22
220	0.69	3.1	0.83	3.8	45.5	B	UCW0.83V25 L4	53 × 85	Not included	270 kΩ/3 W	0.27
	0.69	3.1	0.83	3.8	45.5	B	UCW0.83V25 L6	60 × 105	Not included	270 kΩ/3 W	0.34
	1.39	6.3	1.67	7.6	91.5	B	UCW1.67V25 L6	60 × 105	Not included	150 kΩ/3 W	0.35
	2.08	9.5	2.50	11.4	137.0	B	UCW2.5V25 L10	60 × 156	Not included	82 kΩ/3 W	0.51
	2.78	12.6	3.33	15.1	182.5	B	UCW3.33V25 L10	60 × 156	Not included	56 kΩ/3 W	0.51
	4.17	18.9	5.00	22.7	274.0	C	UCW5V25 N14	75 × 205	Included	41 kΩ /3 W	1.19
	5.56	25.3	6.67	30.3	365.6	C	UCW6.67V25 N14	75 × 205	Included	28 kΩ/3 W	1.22

TABLE 9.1 Single-phase capacitors available for power factor correction (continued)

Single-phase capacitors—UCW											
	50 Hz		60 Hz								
Rated voltage (V)	Reactive power (kvar)	Rated current I_n (A)	Reactive power (kvar)	Rated current I_n (A)	Capacitance (uF)	Series	Reference	Dimensions ∅ × H (mm)	Discharge resistor		Weight (kg)
230	0.83	3.6	1.00	4.3	49.9	B	UCW0.83V34 L6	60 × 105	Not included	180 kΩ/3 W	0.32
	1.67	7.3	2.00	8.7	100.5	B	UCW1.67V34 L8	60 × 141	Not included	120 kΩ/3 W	0.44
	2.50	10.9	3.00	13.0	150.4	B	UCW2.5V34 L10	60 × 156	Not included	56 kΩ/3 W	0.51
	3.33	14.5	4.00	17.4	200.4	B	UCW3.33V34 L10	60 × 156	Not included	56 kΩ/3 W	0.50
	5.00	21.7	6.00	26.1	300.9	C	UCW5V34 N14	75 × 205	Included	28 kΩ/6 W	1.18
240	0.69	2.9	0.83	3.5	38.2	B	UCW0.83V29 L6	60 × 105	Not included	270 kΩ/3 W	0.36
	1.39	5.8	1.67	7.0	76.9	B	UCW1.67V29 L6	60 × 105	Not included	150 kΩ/373 W	0.33
	2.08	8.7	2.50	10.4	115.1	B	UCW2.5V29 L8	60 × 141	Not included	82 kΩ/3 W	0.45
	2.78	11.6	3.33	13.9	153.4	B	UCW3.33V29 L10	60 × 156	Not included	82 kΩ/3 W	0.51
	4.17	17.4	5.00	20.8	230.3	C	UCW5V29 N14	75 × 205	Included	60 kΩ/3 W	1.18
380	0.69	1.8	0.83	2.2	15.2	A	UCW0.83V40 G4	40 × 85	Not included	560 kΩ/3 W	0.19
	0.69	1.8	0.83	2.2	15.2	B	UCW0.83V40 J2	53 × 68	Not included	560 kΩ/3 W	0.23
	0.69	1.8	0.83	2.2	15.2	B	UCW0.83V40 L4	60 × 85	Not included	560 kΩ/3 W	0.27
	1.39	3.7	1.67	4.4	30.7	B	UCW1.67V40 J4	53 × 85	Not included	390 kΩ/3 W	0.28
	1.39	3.7	1.67	4.4	30.7	B	UCW1.67V40 L4	60 × 85	Not included	390 kΩ/3 W	0.28
	2.08	5.5	2.50	6.6	45.9	B	UCW2.5V40 J8	53 × 141	Not included	270 kΩ/3 W	0.43
	2.08	5.5	2.50	6.6	45.9	B	UCW2.5V40 L6	60 × 105	Not included	270 kΩ/3 W	0.37
	2.78	7.3	3.33	8.8	61.2	B	UCW3.33V40 J8	53 × 141	Not included	150 kΩ/3 W	0.43
	2.78	7.3	3.33	8.8	61.2	B	UCW3.33V40 L8	60 × 141	Not included	150 kΩ/3 W	0.52
	4.17	11.0	5.00	13.2	91.8	B	UCW5V40 L10	60 × 156	Not included	120 kΩ/3 W	0.52
	5.56	14.6	6.67	17.6	122.5	B	UCW6.67V40 M10	70 × 156	Not included	82 kΩ /3 W	0.60
	6.25	16.4	7.5	19.7	137.8	C	UCW7.5V40 N14	75 × 205	Included	75 kΩ/6 W	1.19
	6.94	18.3	8.33	21.9	153.0	C	UCW8.33V40 N14	75 × 205	Included	60 kΩ/6 W	1.18
	7.64	20.1	9.17	24.1	168.5	C	UCW9.175V40 N14	75 × 205	Included	60 kΩ/6 W	1.23
	8.33	21.9	10.00	26.3	183.7	C	UCW10V40 N14	75 × 205	Included	60 kΩ/6 W	1.23

Courtesy of TR Industries Ltd

We will use Table 9.1 to specify the capacitor required in a 230 V single-phase installation which has a total apparent power of 4375 VA at a power factor of 0.6 lagging that is required to be corrected to 0.9345 lagging.

Solution

Step 1: First find all sides of the power triangle before correction (the diagrams are not to scale). See Figure 9.57.

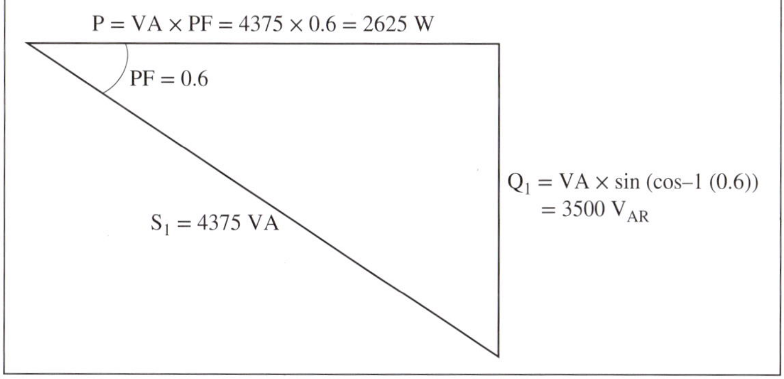

FIGURE 9.57 Step 1

Step 2: After correction, the watts will be the same, but the VAR and VA will be reduced, as shown in Figure 9.58.

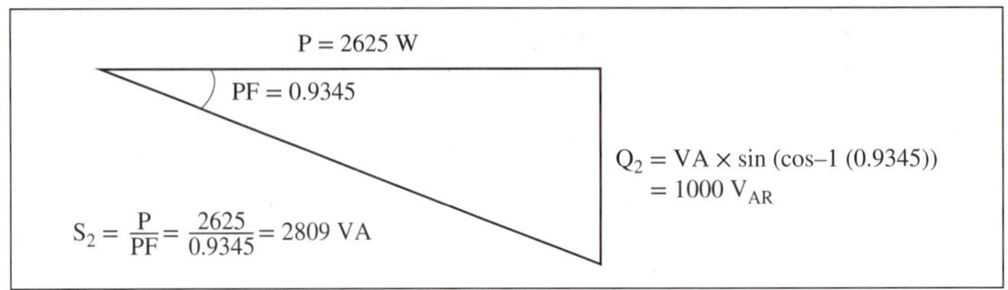

FIGURE 9.58 Step 2

Step 3: The reduction in VAR will be the difference in VAR before and after correction.

VAR required = 3500 – 1000 = 2500 VAR

As can be seen from the extract from Table 9.1, a 230 V 2.5 kVAR capacitor is available. The product code is UCW2.5V34L10.

Single phase capacitors - UCW										
Rated voltage (V)	50 Hz		60 Hz		Capacitance (uF)	Series	Reference	Dimensions ∅ × H (mm)	Discharge resistor		Weight (kg)
	Reactive power (kvar)	Rated current In (A)	Reactive power (kvar)	Rated current In (A)							
230	0.83	3.6	1.00	4.3	49.9	B	UCW0.83V34 L6	60 × 105	Not included	180 kΩ / 3 W	0.32
	1.67	7.3	2.00	8.7	100.5	B	UCW1.67V34 L8	60 × 141	Not included	120 kΩ / 3 W	0.44
	2.50	10.9	3.00	13.0	150.4	B	UCW2.5V34 L10	60 × 156	Not included	56 kΩ / 3 W	0.51
	3.33	14.5	4.00	17.4	200.4	B	UCW3.33V34 L10	60 × 156	Not included	56 kΩ / 3 W	0.50
	5.00	21.7	6.00	26.1	300.9	C	UCW5V34 N14	75 × 205	Included	28 kΩ / 6 W	1.18

9.13 Harmonics and resonance effect in a.c. systems

9.13.1 Harmonics

All waveforms, with the exception of pure d.c., comprise sinusoidal waveforms. A single a.c. wave has only one sine wave, the *fundamental wave*. All other waveforms, such as triangle and square, are a combination of multiple sine waves of differing frequencies and amplitudes. This perhaps surprising fact can be proven both experimentally and mathematically.

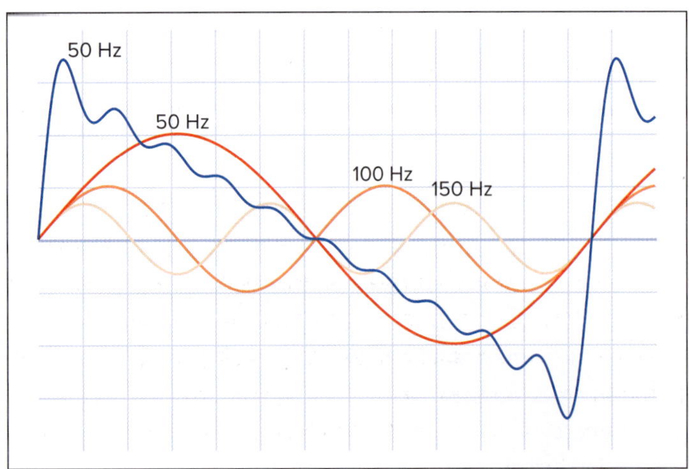

FIGURE 9.59 Sine waves added to form a sawtooth wave

Figures 9.59 and 9.60 show the fundamental wave and two harmonically related waveforms. We will take the fundamental wave to be 50 Hz. In Figure 9.60, the second harmonic is 2 × 50 or 100 Hz at half the amplitude in this case. The third harmonic is at 150 Hz and one-third the amplitude. The contributing waveforms are shown in shades of red.

Showing more harmonics would make things unnecessarily complicated.

The combined waveform (coloured blue) shows the result of adding the fundamental wave to nine harmonics to produce what becomes a sawtooth waveform. A perfect sawtooth waveform would consist of an infinite number of harmonics.

In Figure 9.60, the fundamental wave is added to the 3rd harmonic at 150 Hz at one-third the amplitude; it is added to the 5th harmonic at 250 Hz and one-fifth the amplitude. Here too the contributing waveforms are shown in shades of red.

The combined waveform (coloured blue) shows the result of adding the fundamental wave to nine odd harmonics from the 3rd to the 19th, to produce what becomes a square waveform. A perfect square wave would consist of an infinite number of odd harmonics only.

Finally, in Figure 9.61, a 50 Hz waveform is shown experiencing interference from another sine wave 20 times the fundamental frequency or 1000 Hz, at $\frac{1}{20}$ the amplitude.

We will see later that harmonics can be useful or harmful, depending on how they are applied. Radio and TV would be impossible without them; but their presence in transformers or in the distribution system can cause serious trouble.

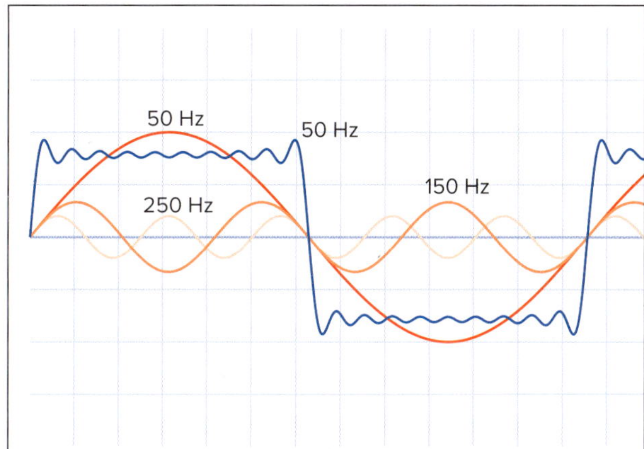

FIGURE 9.60 **Sine waves added to form a square wave**

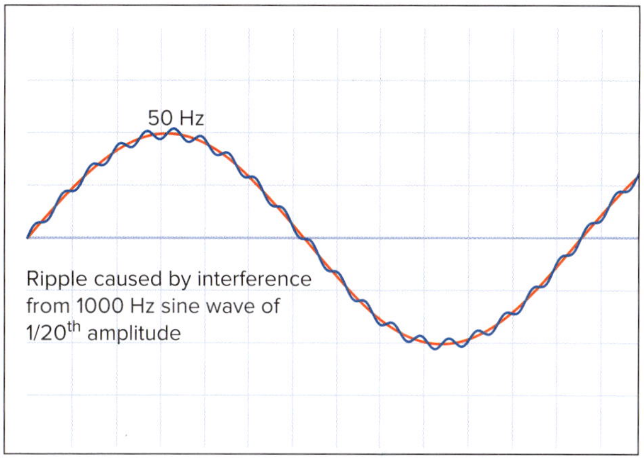

FIGURE 9.61 **50 Hz sine wave modified by a 1 kHz sine wave**

9.13.2 Sources in a.c. systems that produce harmonics

Power-consuming loads connected to the electrical grid can be broadly separated into two groups: linear and non-linear loads. Linear loads are generally the older technology that is resistive and/or reactive. This includes heaters, induction motors, halogen lamps and transformers, and these generally have little effect on the shape of the sinusoidal a.c. waveform. Non-linear loads are usually electronic and involve a degree of 'cutting up', a term that describes high-frequency switching of the a.c. waveform to achieve different voltage, current and frequency levels. Cutting up the fundamental 50 Hz sine wave results in other frequencies being generated in the supply. Equipment that uses non-linear technology is more modern, efficient and compact. Examples include switched-mode power supplies (in computers and other small electronic appliances), electronic ballasts, electronic low-voltage transformers, variable frequency drives (in motor speed control circuits) and inverters (including solar photovoltaic and battery types).

9.13.3 Problems that may arise in a.c. circuits as a result of harmonics

Harmonics can cause problems that are confusing and hard to diagnose and rectify. These problems can be separated into two main groups: those associated with the harmonic order and those associated with the harmonic sequence. The harmonic sequence group can be split into three smaller groups: positive, negative and zero-sequence harmonics.

Positive sequence harmonics include the fundamental, the 4th, 7th, 10th etc. In motors they produce magnetic and current fields that rotate in the *same* direction as the fundamental frequency.

Negative sequence harmonics include the 2nd, 5th, 8th, 11th etc. In motors they produce magnetic and current fields that rotate in the *opposite* direction to the fundamental frequency.

Zero-sequence harmonics include the 3rd, 6th, 9th, 12th etc. They have no effect on the rotating magnetic and current field but can cause additional I^2R power losses in the machine.

The harmonic order can simply be divided into two groups: even and odd.

Even harmonics tend to cancel each other and consequently have a less alarming effect than odd harmonics. Three-phase non-linear loads produce odd harmonics that are not multiples of three. Single-phase non-linear loads produce zero-sequence harmonics (3rd, 6th, 9th etc.) accompanied by other odd-order harmonics. The odd harmonics that are multiples of three (3rd, 9th, 15th etc.) in three-phase systems can combine in the neutral and cause it to heat up. Other effects of these third-order harmonics include overheating the supply transformer and nuisance-tripping circuit protection devices.

The main ways of overcoming the effects of harmonics on an electrical installation are: removal of the source; increasing the neutral conductor size in three-phase systems or supplying each phase with a separate neutral; and the use of inductor/capacitor filters that can be tuned to reduce harmonic current flow at desired frequencies. The inductors that are used for this are called 'line reactors'.

9.13.4 Methods and test equipment used to test for harmonics

Harmonics can be measured with oscilloscopes, power analysers or clamp meters. Using an oscilloscope or power analyser, harmonics can be detected either by looking at the wave shape or by analysing the harmonic order. The clamp meter option involves comparing the difference in the neutral current and requires the use of two meters, a standard RMS clamp meter and a clamp meter with multiple frequency and true RMS capabilities. The difference between the readings from each device will indicate if there is any harmonic, as compared with out-of-balance current.

9.13.5 Resonance

When an electrical circuit has its power factor corrected to unity, the current is brought into phase with the voltage. Although it could still contain both capacitive and inductive reactances, it behaves as a purely resistive circuit.

Capacitive reactance and inductive reactance are both determined by the frequency. When they are exactly equal, the circuit is said to be 'resonant'. Resonance is the frequency that conserves energy. At resonance, energy passes back and forth between the reactances at the resonant frequency.

In both series and parallel resonant circuits, capacitive reactance is equal to inductive reactance. The energy stored in the magnetic field of the inductor for one part of a cycle is transferred to the electrostatic field of the capacitor in the next part. This energy transfer produces different effects in series and parallel circuits.

9.13.6 Series-circuit resonance

Since at resonance $X_L = X_C$, then $X_L - X_C = 0$

That is:

$$Z = \sqrt{R^2} \text{ or } Z = R$$

As the impedance is equal to the resistance, the current flow from the supply will be calculated by Ohm's Law as $I = \frac{V}{R}$.

However, as the same current will flow between the inductor and the capacitor, the reactive components will generate voltages according to Ohm's Law and the reactance of each component: $V = IX_L = IX_C$. The voltage across the reactive components will cancel each other out as they will be equal and opposite. However, the voltage may be many times the circuit terminal voltage and may destroy insulation and become a shock and fire hazard.

The major characteristics of the series resonant circuit are a power factor of unity and a minimum impedance:

$$Z = \sqrt{R^2 + (X_L - X_C)^2}$$

The series resonant circuit should be avoided for general electrical work, unless special precautions are taken. One precaution is ensuring that adequate series resistance is in the circuit. If necessary, extra resistance should be added to ensure that the current (and therefore the voltage across the reactors) is limited.

Where circuits are operated such that voltages higher than the supply voltage are generated, the circuit must be insulated for those higher voltages and suitably rated components must be selected.

EXAMPLE 9.26

A 10 Ω resistor, 460 mH inductor and a 22 μF capacitor are connected in series across a 230 V 50 Hz supply. Calculate the supply current and the voltage drop across each component.

$$R = 10\ \Omega \quad (1)$$

$$X_L = 2\pi fL \quad (2)$$

$$= 2 \times \pi \times 50 \times 0.46 \quad (3)$$

$$= \frac{\underline{145\ \Omega}}{1} \quad (4)$$

$$X_C = \frac{1}{2\pi fC} \quad (5)$$

$$= \frac{1}{2 \times \pi \times 50 \times 22 \times 10^{-6}} \quad (6)$$

$$= \underline{145\ \Omega} \quad (7)$$

$$Z = \sqrt{R^2 + (X_L - X_C)^2} \quad (8)$$

$$= \sqrt{10^2 + (145 - 145)^2} \quad (9)$$

$$= \underline{10\ \Omega} \quad (10)$$

$$I = \frac{V}{Z} \quad (11)$$

$$= \frac{230}{10} \quad (12)$$

$$= \underline{23\ A} \quad (13)$$

$$V_R = IR \quad (14)$$

$$= 23 \times 10 \quad (15)$$

$$= \underline{230\ V} \quad (16)$$

$$V_{XL} = IX_L \quad (17)$$

$$= 23 \times 145 \quad (18)$$

$$= \underline{3335\ V} \quad (19)$$

$$V_{XC} = IX_C \quad (20)$$

$$= 23 \times 145 \quad (21)$$

$$= \underline{3335\ V} \quad (22)$$

Note the very high voltages across the inductor and the capacitor despite the supply voltage only being 230 V. The voltages are 180° out of phase so cancel each other out, but each component must still be suitably rated for these voltages. Care should also be taken when testing with a voltmeter due to the high voltages present.

A useful characteristic of the series resonant circuit is that the capacitive reactance is equal to the inductive reactance at only one frequency. Other frequencies above or below resonance experience higher impedances due to the reactance no longer cancelling out. They will therefore be attenuated.

Knowing that at resonant frequency (f_0), X_C and X_L are equal, the equation for resonant frequency can be developed, as:

$$X_L = 2\pi fL$$

$$X_C = \frac{1}{2\pi fC}$$

$$\text{at resonance } X_L = X_C$$

$$\therefore 2\pi fL = \frac{1}{2\pi fC}$$

$$\therefore f \times f = \frac{1}{2\pi L \times 2\pi C}$$

$$\therefore f_r = \sqrt{\frac{1}{2\pi L 2\pi C}}$$

$$\therefore f_r = \frac{1}{2\pi \sqrt{LC}}$$

where:

f_r = resonant frequency
X_L = inductive reactance
X_C = capacitance reactance.

9.13.7 Parallel circuit resonance

When an inductor and capacitor are connected in parallel and their respective reactances are equal, the reactive currents are equal but are 180° out of phase with each other.

Theoretically, the current drawn from the supply would be zero, while a comparatively high circulating current would oscillate between the inductor and capacitor at the supply frequency. In practice, an inductor has some finite value of resistance and so the parallel resonant circuit appears as a pure resistance, and some current in phase with the voltage would flow from the supply.

The approximate resonant frequency in a parallel circuit with zero resistance is found using the same equation as for series resonant frequency (see Figure 9.62). For a practical circuit, inductive reactance is actually a series circuit of resistance and inductive reactance. Although the resonant frequency of a parallel resonant circuit is not exactly as calculated from the series equation, the difference may be less that any error in calculation.

For parallel resonant circuits operating on very high frequencies, the inductive reactance can be achieved with very short lengths of conductors. The resistance can then be ignored.

When the supply frequency to a parallel resonant circuit is varied, the resistance in the circuit is unchanged but the impedance will be a maximum only at the resonant frequency.

At resonance, energy is being transferred from the electromagnetic field to the electrostatic field and back again via the circulating current. The circulating current cannot be related to the input current from the supply. The supply current at resonance is at a minimum because the impedance is at its maximum. The current is sufficient only to make up the losses in the circuit. The lower the losses, the lower the value of input current.

In electronic circuits, a capacitance in parallel with an inductance can be used to get the highest possible impedance for a particular frequency across that parallel section. Making either the capacitor or the inductor adjustable means a particular resonant frequency can be chosen.

When a variety of frequencies (such as different radio stations) is presented to a parallel circuit, only the resonant frequency is passed to the next part of the circuit. This is the method used in tuning a radio to a set frequency. In other words, the non-resonant frequencies are presented with a low-impedance path to earth and are thus removed from the circuit.

Test equipment sometimes uses the principle of parallel resonance for measuring capacitance. Frequency-dependent transducers use the phenomenon as a mcans of monitoring and regulating frequency. Other circuits use the property of resonant circuits as a metal detector.

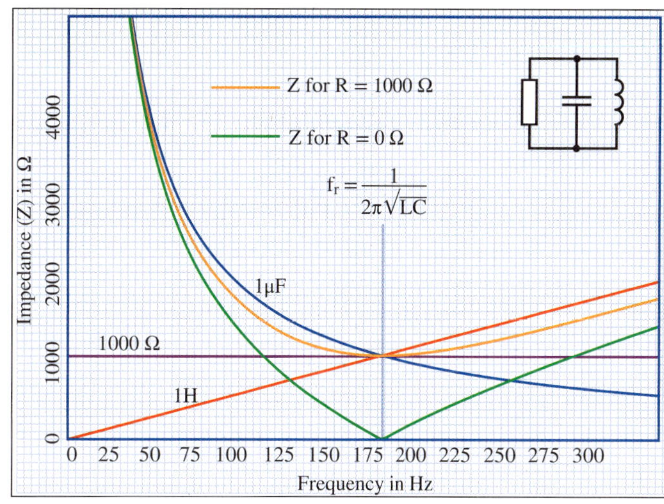

FIGURE 9.62 Graphical representation of resonant frequency

9.13.8 Standards relating to harmonics

AS/NZS 3000:2018 and local Service and Installation Rules have strict guidelines about equipment that might generate harmonics.

In AS/NZS 3000:2018, the main guidelines relate to the current rating and size of conductors. This is especially important in the neutral conductor, where harmonic currents can be cumulative. Clause 3.5.2 outlines these requirements. It notes that a harmonic current bigger than 40% of the phase current is enough to warrant a change in the cross-sectional area of the conductor(s). It also explains that the 3rd harmonic can make it necessary for a neutral that is larger than the associated actives to carry the harmonic current.

Local Service and Installation Rules focus more on the effect that harmonics have on the network and other customers. These include switched-mode power supplies, and there are requirements for inverters to be limited. This is to prevent harmonic distortion (which changes the RMS values) and other unwanted effects such as data corruption and frequency relay switching.

9.13.9 Standards relating to resonance

AS/NZS 3000:2018 and the Service and Installation Rules set requirements about the effects of resonance in an installation. Series resonance can cause overvoltage within the circuit, which would be a safety concern. AS/NZS 3000:2018, Clauses 1.5.11 and 2.7 state that, where danger to persons or property can occur, the installation should be protected to prevent harm from overvoltage. This requirement can be satisfied by either insulation/separation or by protective devices.

Parallel resonance can cause excessive currents and conductors carrying them need to be sized correctly. Local Service and Installation Rules generally require some measures to be in place to avoid resonance with the network that can cause high inrush currents. Effective measures include de-tuning reactors and resistors.

CHECK YOUR UNDERSTANDING

9.14 A 20 Ω resistor, 253 mH inductor and a 40 µF capacitor are connected in series across a 230 V 50 Hz supply. Calculate the:

(a) reactance of the inductor and the capacitor
(b) impedance of the circuit
(c) current flowing in the circuit
(d) voltage drop across the inductor
(e) voltage drop across the capacitor.

9.15 What can we say about this circuit?

SUMMARY

- A conductive loop rotating in a magnetic field produces an alternating voltage waveform.
- When the field is uniform and the speed of rotation is constant, the shape of the waveform is sinusoidal.
- An alternator is a generator that produces a voltage of alternating voltage polarity and current flow.
- There are two types of alternators: conductors rotated inside a stationary magnetic field and a magnetic field rotated through stationary conductors.
- The direction of the induced voltage can be determined from Fleming's right-hand rule.
- The magnitude of the voltage can be found from $e = Blv \sin \theta$
- The output voltage of an alternator can be increased by increasing the number of conductors, increasing the magnetic flux or adding an iron core to the alternator.
- The iron core rotates within a magnetic field that generates a current within the metal, giving rise to iron losses (eddy currents and hysteresis).
- One cycle of alternating current occupies 360°E (electrical degrees). It is not always equal to mechanical degrees.
- Phasors are rotating vectors.
- Phasors that are 'in phase' have a zero angle between them. Phasors that have a difference of angle between them are said to be 'out of phase'.
- The 'reference phasor' is one that is common to all parts of the circuit.
- Phasors are assumed to rotate in an anticlockwise direction.
- Phasors ahead of the reference phasor are 'leading' the reference value. Phasors behind the reference phasor are 'lagging' the reference value.
- Phasors cannot be added arithmetically unless they are in phase.
- Phasors that are out of phase must be added vectorially, using either the graphical method or trigonometry.
- Every repeating regular waveform is made up of sine waves of varying amplitudes and frequencies.
- Harmonics are sine waves at multiples of the original (fundamental) frequency.
- Harmonics may be detrimental to power-supply systems and electrical machines.
- With alternating current, the current might not be in step with the voltage, in which case it is said to be 'out of phase'.
- For resistive circuits, current is 'in phase' with the voltage.
- For purely inductive circuits, the current lags the voltage by 90°.
- For purely capacitive circuits, the current leads the voltage by 90°.
- For circuits containing resistance and reactance, the phase displacement of the current will be less than 90° and will depend on the proportions of resistance and reactance in the circuit.
- The angle of phase displacement is the 'phase angle' (symbol φ).
- A power triangle can be used for an a.c. circuit. A power triangle consists consisting of the components 'true power' (P), 'apparent power' (S) and 'reactive power' (Q).
- The power factor is equal to true power divided by apparent power: $PF = \frac{P}{S}$
- A low power factor leads to 'wattless' components of current flowing in distribution lines.
- A low power factor causes poor voltage regulation, excessive line losses, excess fuel consumption and unnecessary loading of alternators.
- The power factor can be obtained by measuring with a voltmeter, ammeter and wattmeter and then calculating the value from $PF = \frac{P}{VI}$

- The power factor can be measured directly with a power factor meter.
- The lowest acceptable power factors could vary across states and territories. In NSW, for example, it is considered to be 0.9.
- For individual loads, capacitors are connected in parallel to improve the power factor.
- For larger loads, power factor correction can be made using leading current synchronous motors or a power factor correction unit installed at the main switchboard.
- Power factor correction VAR can be calculated by subtracting the maximum acceptable VAR (e.g. at a PF of 0.9) from the maximum VAR without correction.
- If inductive reactance is equal to capacitive reactance, the circuit behaves as if it were purely resistive and is said to be 'resonant' at that frequency.
- The resonant frequency can be calculated from $f_r = \frac{1}{(2\pi\sqrt{(LC)})}$
- Resonance can cause dangerously high voltages in series circuits and dangerously high circulating currents in parallel circuits.

END-OF-CHAPTER QUESTIONS

9.1 Define the terms 'periodic time', 'maximum value', 'peak-to-peak value', 'instantaneous value', 'average value' and 'root-mean-square (RMS) value' as they relate to a sinusoidal waveform.

9.2 Explain how the mnemonic "SOHCAHTOA" can be used to remember the relationships in a right-angle triangle.

9.3 What is meant by the terms 'in phase' and 'out of phase'?

9.4 Compare the relationships between voltage and current if a.c. is applied to pure resistance, inductance or capacitance.

9.5 Draw a phasor diagram of a resistor drawing 1 A from a 230 V supply.

9.6 Draw a phasor diagram of a pure inductor drawing 1 A from a 230 V supply.

9.7 Draw a phasor diagram of a pure capacitor drawing 1 A from a 230 V supply.

9.8 Why does a practical inductor never cause the current to lag by 90°? Draw a circuit to represent a practical inductor.

9.9 Calculate the current and power drawn from a circuit consisting of three resistors (100 Ω, 47 Ω and 22 Ω) connected in series across a 100 V 50 Hz supply.

9.10 Describe the dangers of a series resonant circuit when it is connected to a 230 V 50 Hz supply.

9.11 A 30 Ω resistor in series with a 0.1 H inductor is placed across a 230 V 50 Hz supply. Calculate the total current, total power and phase angle between the current and supply voltage?

9.12 In a circuit containing a 0.5 H inductor in parallel with a 100 Ω resistance supplied by 100 V at 50 Hz, calculate the current in each component and the total current.

9.13 When does a parallel circuit draw minimum current from the supply?

9.14 Explain why a pure inductor does not consume power.

9.15 What is 'power factor'? Give its maximum and minimum values.

9.16 Explain the differences between true power, apparent power and reactive power.

9.17 What are the effects of poor power factor on a distribution system?

9.18 Explain how a capacitor can correct poor power factor.

9.19 Calculate the total current, phase angle and power factor for a 230 V 50 Hz series circuit consisting of a 33 Ω resistive element, a 0.3 H inductive coil and a 100 μF capacitance.

9.20 A 230 V 50 Hz circuit passes 1.6 A at a power factor of 0.8. Calculate the phase angle, true power, apparent power and reactive or wattless power.

9.21 At what frequency will a series circuit resonate if it has an inductance of 25 mH, 100 μF and a resistance of 10 Ω? What is the impedance at resonance?

9.22 A single-phase motor draws 1150 W from a supply of 230 V 50 Hz. A power factor meter reading in the circuit is 0.54. Determine the current taken from the supply.

9.23 A 400 V single-phase alternator is rated at 32 kVA. What would be the maximum safe power output for this machine at (a) unity power factor (b) a power factor of 0.8?

9.24 What is the 3rd harmonic for a 50 Hz a.c. sine wave?

9.25 A single-phase 230 V installation has a total loading of 20 kW at a power factor of 0.6. Calculate the size of the capacitor (in kVAR) to correct the installation so that it has a power factor of 0.9. What is the current drawn from the supply with the new power factor, and what is the current rating of the capacitor used for power factor correction?

CHAPTER 10

Alternating current circuits: three phase

LEARNING OBJECTIVES

- Understand that three-phase power is more efficient and economical than single-phase power
- Describe two-phase systems
- Describe how three-phase power is generated and state its advantages
- Construct three-phase sine waves in sequence
- Describe three-phase star and delta connections and the relationship between their phase and line values
- Explain the concept of infinite grid
- Describe typical transmission voltages
- Understand a SWER system
- Differentiate between balanced and unbalanced loads
- Calculate line values from phase values, and phase values from line values
- Calculate the neutral current of an unbalanced load
- Calculate power in three-phase systems
- Describe methods of power measurement in three-phase loads
- Understand and comply with the requirements of AS/NZS 3000:2018 and AS/NZS 3008.1:2017 for single- and three-phase voltage drop
- Understand and comply with the requirements of AS/NZS 3000:2018 and AS/NZS 3008.1:2017 for fault-loop impedance

PREREQUISITE KNOWLEDGE

Understand single-phase a.c. circuits

INTRODUCTION

Three-phase electricity is the standard for power generation, distribution and industrial use around the world. Although d.c. is still used, and single-phase a.c. is common in domestic and many commercial situations, two-phase electricity has mostly disappeared (with the exception of its use in centre-tapped SWER transformers).

Many industrial electricians will work on three-phase plant, motors and transformers for most of their working day. Others will work on three-phase refrigeration and air-conditioning units.

This chapter examines why using three phases is the preferred method of producing electricity and why having more than three phases is uneconomical.

10.1 Three-phase systems

10.1.1 Efficiency in generation and distribution

Although the alternating current (a.c.) electrical system has some major advantages over direct current (d.c.), it has one considerable disadvantage: its low efficiency in generation, distribution and machine loads. This is a result of the voltage and current passing through zero twice each cycle. When no voltage is generated, no power is delivered—that equals wasted time. Power taken from the driving machinery also decreases to zero one hundred times per second, creating 100 Hz mechanical vibrations in the machine (see Figure 10.1).

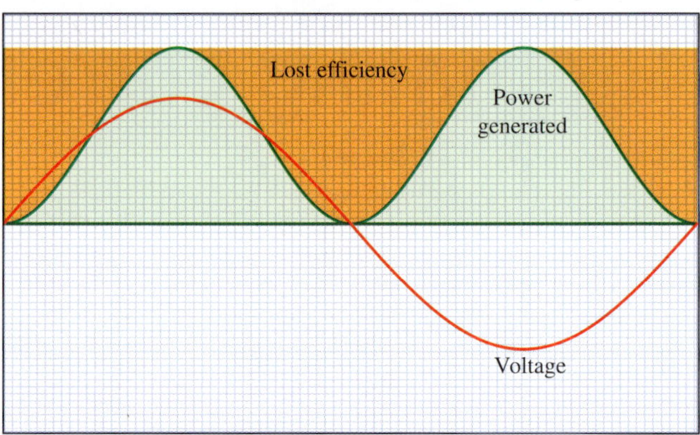

FIGURE 10.1 a.c. sine wave showing power lost

One way of improving a.c. efficiency is by generating a second a.c. voltage so that the peak of the second waveform occurs at the same time as the zero crossing of the first.

This is called a 'two-phase' system. Generating two a.c. voltages at 90° phase angle to one another improves the efficiency to 90% (see Figure 10.2).

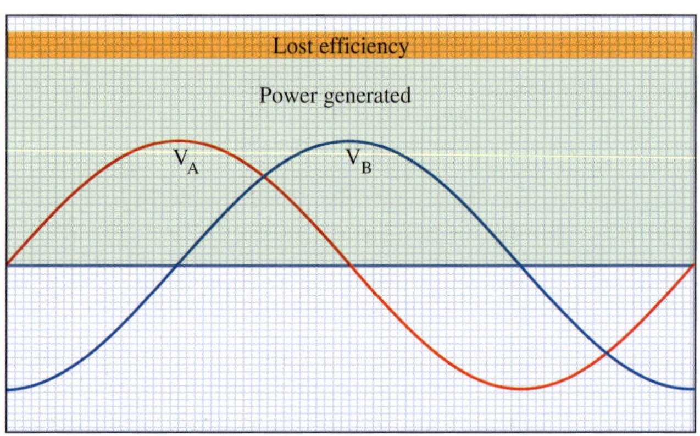

FIGURE 10.2 Total power with a second phase

If a third waveform was added, instead of a phase angle of 90°, three phases would need to be balanced across 360° (360° divided by 3 is 120°, which is mathematically the best phase-angle difference for three-phase power).

The three-phase (or 3ϕ) system has an efficiency of around 96% (see Figure 10.3). There is no disadvantage to using three phases instead of two as a two-phase system requires at least three wires, and usually uses four. The same applies to a three-phase system; it requires three or four wires. That is why two-phase systems are so rare today.

Systems using more than three phases require an extra wire for each extra phase. So, although they slightly increase efficiency, they also increase the costs of transmission and distribution. The three-phase system is the most efficient and economical polyphase system available. (Systems with more than one phase are called 'polyphase', or sometimes 'multiphase'.)

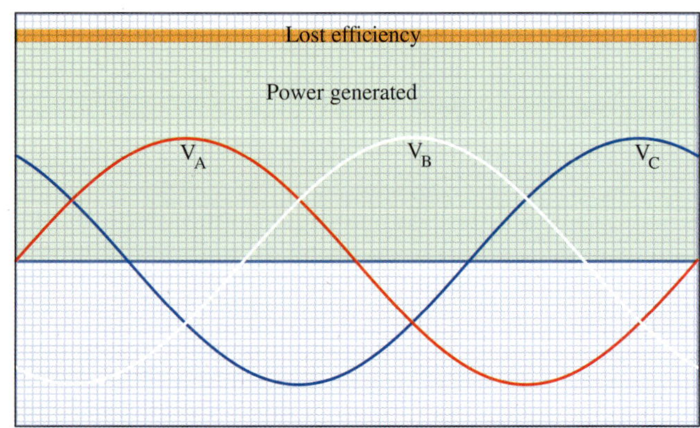

FIGURE 10.3 Total power with a third phase

10.1.2 Two-phase systems

In a two-phase system, two coils are physically displaced by 90° mechanical (written as '90°M'), as shown in Figure 10.4. The result is that the two output voltages have a 90° electrical phase difference (written as '90°E').

As discussed previously, three-phase systems are more efficient and have almost completely replaced two-phase systems in Australia. There may still be some SWER systems in use to supply rural properties on the fringe of the electrical grid.

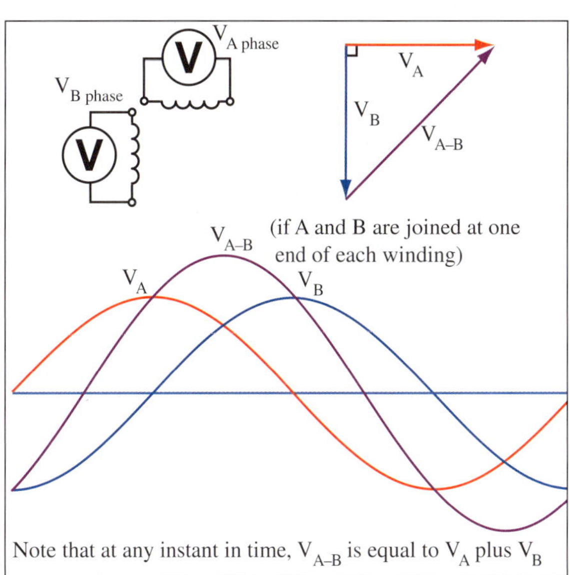

FIGURE 10.4 Two-phase system

10.1.3 Three-phase systems

A three-phase system has three generating windings, each out of phase with the other two by 120°E. Figure 10.5(a) illustrates a simplified three-phase alternator with three windings A, B and C at 120°E intervals.

Figure 10.5(b) shows the waveforms generated by the three windings, and illustrates the phase shift between waveforms.

Figure 10.5(c) shows the phasor diagram of the three phasors.

The phasor of the A-phase voltage V_A is shown as the reference phasor. This is drawn horizontally to the right, i.e. at 0°. The phasor of the B-phase voltage V_B is drawn 120° after V_A, putting it at the bottom left of the diagram. *The phasors rotate in a counterclockwise direction, so lagging phasors are clockwise from the reference phasor.* The phasor of the C-phase voltage V_C is drawn 240° after V_A. This puts it at the top left of the diagram. It is also 120° counterclockwise from V_A, which completes the cycle.

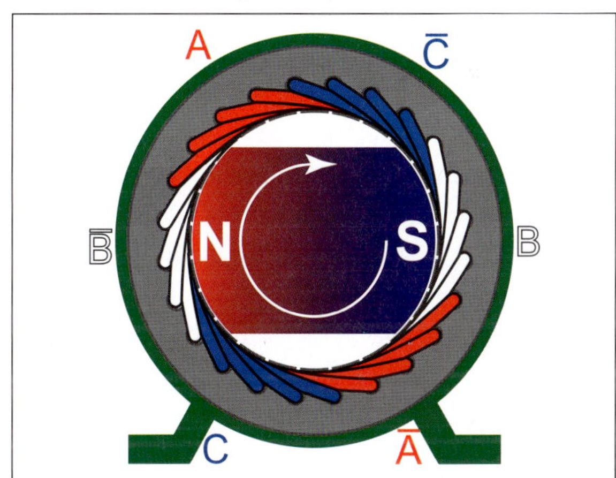

FIGURE 10.5(a) A three-phase machine

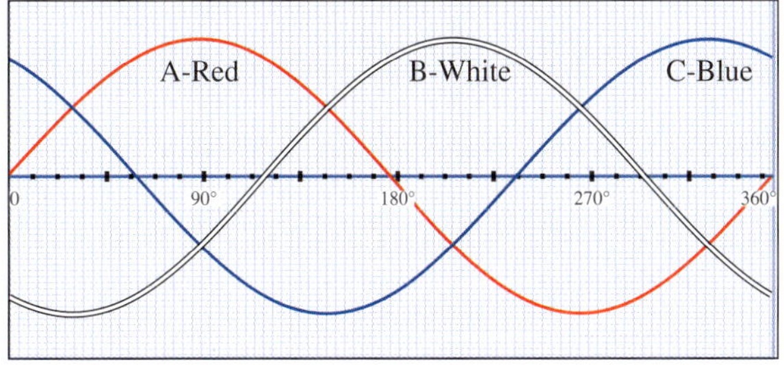

FIGURE 10.5(b) Phase shift between waveforms

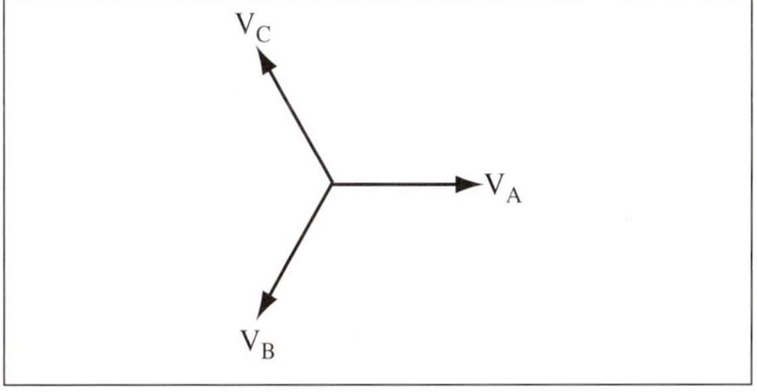

FIGURE 10.5(c) Phasor diagram of three phasors

The advantages of a three-phase system are:

- For the same size or weight, a three-phase machine can produce higher outputs than a single-phase machine.
- A three-phase machine can be smaller than a single-phase machine for the same power output.
- The power delivered to (or taken from) a three-phase system has a more constant value. In a single-phase system, the power pulses at twice the line frequency. With three phases, the power pulses are six times the line frequency, with far less amplitude than single-phase power. Since the power is more constant, the torque of a rotating machine is more constant. This results in much less vibration from the machine.
- With one type of three-phase connection, there are two voltages available: 230 V and 400 V.
- In a distribution system, the total quantity of material needed for three conductors is less than that required for the equivalent single-phase system. This is due to higher efficiency.

10.1.4 Generating a three-phase supply

A three-phase supply is produced by an electric machine that contains three windings. These windings are separated by 120°E, as seen in Figure 10.5(a). This causes them to naturally produce three voltage waveforms 120° apart.

The three phases follow in a fixed sequence known as the 'phase sequence' or 'phase rotation'. First, the three phases must be identified to allow the sequence to be stated and loads to be balanced across them. In some applications, the letters A, B and C are used as A-phase, B-phase and C-phase. For general-purpose supply and distribution identification in Australia, the phases are given the colour coding red, white and blue (prior to 1981, the colours were red, yellow and blue).

As the phasors rotate, they pass through the reference position in the order red, white, blue. This sequence must be followed in connecting equipment on three-phase circuits so that motors rotate in the expected direction.

When three-phase phasors are drawn, the usual practice is to draw the red phasor (A) in the reference position; if it helps, either the white (B) or the blue (C) phasor may be used as the reference as long as the sequence is still correct.

Phases may also be marked as L_1, L_2 and L_3, meaning Line 1, Line 2 and Line 3.

Three-phase winding arrangements

Unlike d.c. machines, the poles in a three-phase machine generally overlap. This is a factor in balancing the current and power of a three-phase machine.

Figure 10.6 shows a typical 24-slot stator lamination set. To use this for a three-phase four-pole electric motor, there must be three sets of coils for four poles fitted into 24 slots. ('Four-pole' here means that there are four sets of coils (poles) for each phase.) This would usually result in a winding of 3 × 4 × 2 coils, meaning that two coils form one pole for one phase.

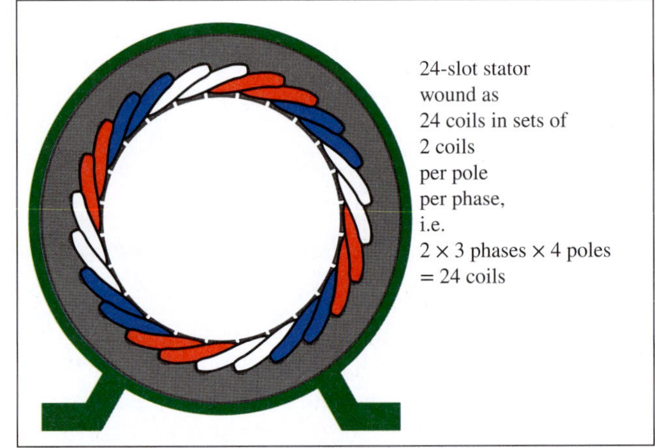

FIGURE 10.6 A three-phase induction motor winding

Phase A is drawn in red, with one side of the coils on the outside and the other on the inside of the laminated stator. Phase B is white and phase C is blue. The motor is four pole. Each phase occupies one-third of the total number of slots. A four-pole machine has 720°E in one complete rotation (360°M).

10.1.5 Three-phase machine alternator construction

An alternator basically consists of coils rotating in a magnetic field. There is a more advantageous form where the a.c. windings are stationary and the magnetic field system rotates.

The same principles apply to both single- and three-phase alternators. The only real difference is whether there is one winding or three identical windings.

10.1.6 Three-phase sine wave construction

The construction of the three sine wave curves for a three-phase system follows the same method as that of a single sine wave (as discussed in Chapter 9). The phase displacement of 120°E between the phases must be factored in. The conventional sine wave set for three-phase is illustrated in Figure 10.7. The sine waves for B and C phase are the same sine wave drawn 120° and 240° after A phase.

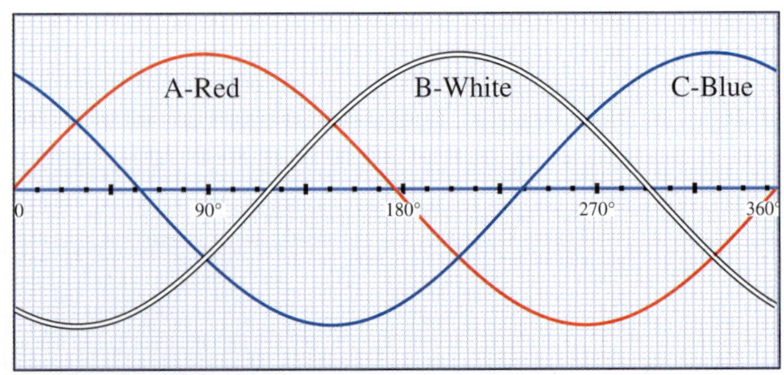

FIGURE 10.7 Three-phase waveforms

An alternative method, as used in this textbook, is to use a computer program to draw the three phases. Grapher.app or Plot are suitable programs, but an internet search should locate free software.

10.1.7 Three-phase connections

Three-phase voltages are produced by mechanically interconnecting three sets of windings to form a three-phase a.c. source. (As there are three separate voltages, each could be used as a single-phase source.) Connection methods are discussed in section 10.2.

10.1.8 Phase sequence

If a magnet is rotated anticlockwise within the coils in Figure 10.6, the voltage generated in each set of coils will reach a positive peak value in the order red, white, blue (or A, B, C, in which case the alternator is said to have a phase sequence of ABC). If the magnet were driven clockwise, the phase sequence would be ACB. In a three-phase system, where the three voltage sources are connected to feed a three-phase load, the phase sequence can be important. This is the case in rotating machinery, as the sequence can affect the direction of rotation.

10.1.9 Determining the phase sequence of a three-phase supply

By convention, the phase rotation produced by a set of windings will produce a phase sequence ABC, as just described. But in a complex system such as an electrical grid, this sequence can become mixed up. So to maintain a safe operating system and prevent short-circuits, voltage checks must be made. These should also be carried out to ensure that all connections (both those within the grid and those to installations and loads) have no potential difference between them.

For example, if a conductor labelled as ‘A’ or ‘red phase’ is to be connected to a parallel conductor that is also labelled as ‘A’ or ‘red phase’, the voltage between the two conductors should be zero. This is because both are at the same potential. If the marking of one of the cables was incorrect, a voltage of 400 V would be present across the two cables.

It is also important to maintain *phase sequence* within an installation where three-phase machinery is to be connected in multiple places. This is vital when the equipment is connected via a plug and socket: if the phase sequence is the same throughout the entire installation, when equipment is connected to different sockets the machine will still rotate in the same direction.

A phase rotation meter can be used to determine the phase sequence of an installation. The analogue version of this device is a three-phase motor that simply rotates clockwise or anticlockwise, depending on the sequence of the socket it is connected to. Ensuring that the meter rotates in the same direction when connected to the same conductors in each socket means that each will have the same phase rotation.

CHECK YOUR UNDERSTANDING

10.1 What is the:

(a) angle between phases on a three-phase supply system?

(b) percentage efficiency of a three-phase supply system?

10.2 Would a four-phase supply be more efficient than a three-phase supply? If so, why is it not used?

10.3 Give three advantages of a three-phase supply system.

10.4 Why is the phase sequence important in rotating machinery?

10.5 What can be used to determine the phase sequence of an installation?

10.2 Three-phase star connections

10.2.1 Three-phase star connections

One method of forming a three-phase system is to connect the three similar ends of the windings together, as shown in Figure 10.8(a). Either the start or finish ends of the windings can be connected. The three phases are said to be 'star connected', and the common connecting point is called the 'star point'.

An alternative method of drawing and labelling the windings is shown in Figure 10.8(b). Once again, similar ends are connected to the star point. This diagram resembles a 'star' which is where the connection gets its name.

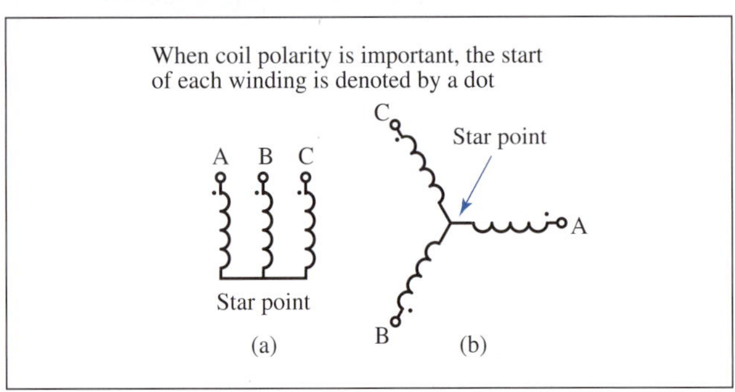

FIGURE 10.8 Three-phase star (wye) connection

The three supply actives (or lines) are connected to the other end of each phase winding labelled A, B, C. The voltage across these lines is called the 'line voltage' (V_{line} or V_L). The current flowing through the lines is called the 'line current' (I_{line} or I_L). The neutral, which is common to the three actives, is connected to the star point. (The neutral is usually earthed at the main switchboard and is not considered to be a line.)

The voltage across a single phase winding is called the 'phase voltage' (V_P). This distinguishes it from the line voltage. The line voltage is not equal to the phase voltage as two phase windings are connected across each pair of lines. The current flowing through the phase winding is called the 'phase current' (I_P).

The three phases are shown in Figure 10.9 as red, white and blue (or A, B and C). The line voltages are shown as A–B, B–C and C–A, with appropriate amplitude and phase angle to the phase voltages. The line voltage at any point in time is the voltage between the two phase values. So, $V_{A\ B}$ is equal to $V_A - V_B$. A rule or a set of dividers can be used to check this.

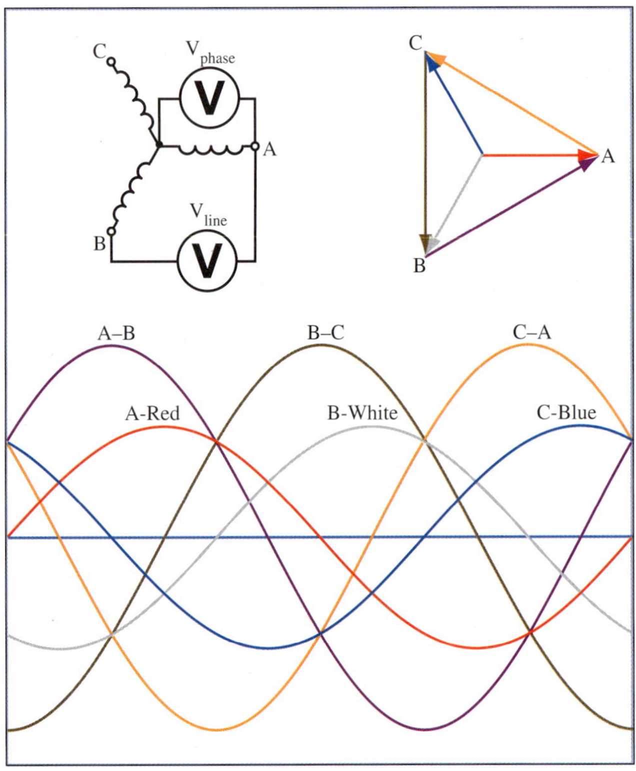

FIGURE 10.9 Phase vs line voltage

This is the conventional sequence, but if the line voltages are given as A–C, C–B and B–A, the line phasors will be 180° to those shown. Nothing will be changed electrically as this is simply a naming issue.

Line A is in series with winding A, so the current is the same. There is only one path for the current. As this applies to all three phases, in a star-connected system the line current equals the phase current:

$$\text{For a star system } I_L = I_P$$

V_{line} is equal to the phasor difference between the two phase voltages. This can be estimated graphically or calculated accurately by phasor addition (see Figure 10.10).

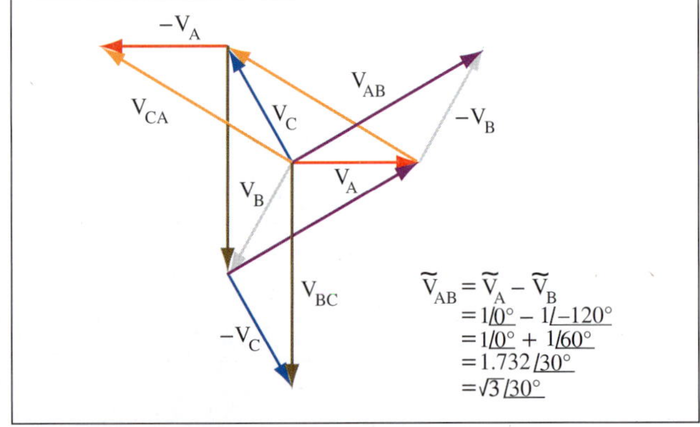

FIGURE 10.10 V_{line} to V_{phase} ratio

Using the methods shown in Chapter 9, section 9.5.3, the two phase voltages are subtracted to show that the line voltage is 1.732 ($\sqrt{3}$) of the phase voltage and leads the phase voltage by 30°.

For a star system $V_L = \sqrt{3}\,V_P$

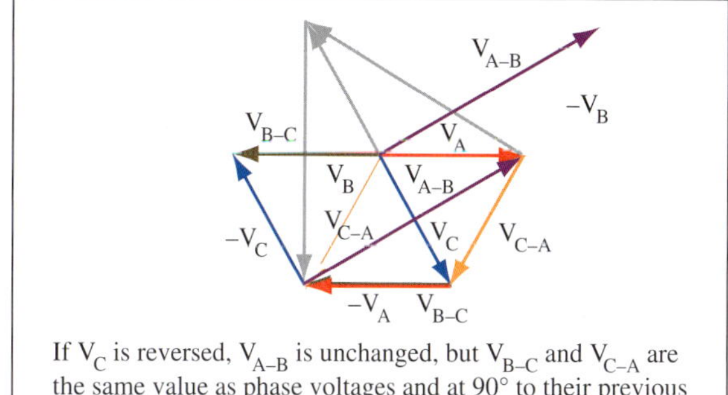

If V_C is reversed, V_{A-B} is unchanged, but V_{B-C} and V_{C-A} are the same value as phase voltages and at 90° to their previous phase angles. (Original phasors are shown in grey.)

FIGURE 10.11 One phase reversed

10.2.2 Effect of phase reversal on a star system

If one phase winding is reversed, the start ends of the three windings are now no longer 120° apart. If phase winding C is reversed, for example, the phase voltage of C has a phase shift of 180°. The waveforms of the three phase voltages then have a displacement of 120° between A and B and 60° between A and C and C and B, as shown in Figure 10.11.

Although the phase voltages are equal, they are not in the correct phase sequence. Two of the line voltages are reduced to the same value as the phase voltages. The load is no longer balanced, and motors will be likely to run backwards (if they run at all).

CHECK YOUR UNDERSTANDING

10.6 Draw a star-connected three-phase system and indicate the star point.

10.7 What is the relationship between phase current and line current in a star-connected system?

10.8 What is the relationship between phase voltage and line voltage in a star-connected system?

10.9 (a) A star-connected three-phase system has the following values:

$I_P = 10$ A

$V_P = 230$ V

Determine the values of I_L and V_L.

(b) A star-connected three-phase system has the following values:

$I_L = 10$ A

$V_L = 500$ V

Determine the values of I_P and V_P.

10.3 Three-phase four-wire systems

10.3.1 Purpose of the neutral conductor in a three-phase four-wire system

In a balanced three-phase four-wire system there is no need for a neutral. This is because these systems have equal current flowing in and out of each of the phases. As a result, the phase voltage of each phase will remain equal across each phase. If the single-phase load that is connected to each phase is identical, to meet the voltage requirement of these loads they simply need to be connected to each phase and joined together at their tails.

However, if each of the single-phase loads is different, a different phase voltage will be present on each phase. For this reason, the star point of the loads is connected to the star point of the supply via the neutral conductor. This provides a zero-volt point for each phase and maintains the voltages, regardless of the loads.

The neutral conductor also provides a current path for the unbalanced load current where the phasor difference of the three phases is the magnitude of current flowing in the neutral.

10.3.2 Effects of a broken neutral

With single-phase loads on three-phase four-wire systems, the neutral must not be open-circuited. One major reason for this is personal safety. With equipment operating on one phase at 230 V, an open-circuit neutral can prevent it from working. Although the equipment appears to be without power, it still has the active conductor connected. A person could contact the active conductor, complete the circuit via the body to earth and receive an electric shock.

With the Multiple-Earthed Neutral (MEN) system, a break in the supply neutral can create a hazard as the potential of earthed appliances can sometimes rise to levels approaching that of the mains voltage.

With three-phase unbalanced loads, a broken neutral has several effects. The voltage across the lightest load increases; the voltage across the heaviest load decreases; and the voltage across the third phase may shift either up or down.

At the same time, the power factors of the individual line currents may also be affected. Although any load located before the neutral break is not affected, single-phase loads and unbalanced three-phase loads after the break will be affected. These are illustrated in Figure 10.12.

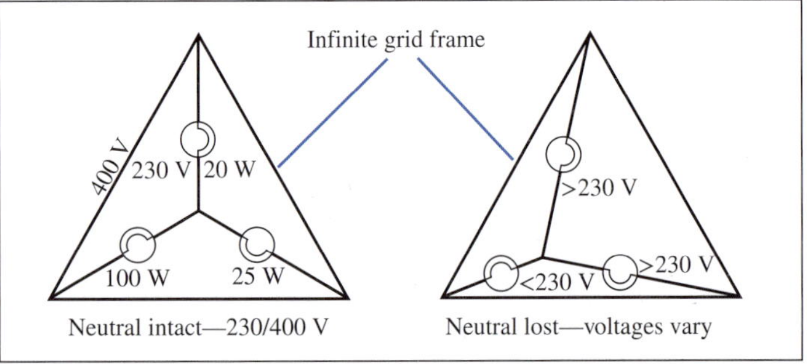

FIGURE 10.12 Lost neutral effects

To prevent the neutral from being open-circuited, neutral switches and links in substations are often locked or bolted. Neutrals are never fitted with fuses under normal installation conditions.

10.3.3 Balanced loads

With balanced loads, the line currents are 120° out of phase with one another. Figure 10.13 shows the waveform diagram for a balanced load with three phase currents, I_A, I_B and I_C.

At the point K, the current I_A is a maximum at +1 A and the currents I_B and I_C are both −0.5 A. At point L, the current I_A is zero, I_B is +0.866 A and I_C is −0.866 A.

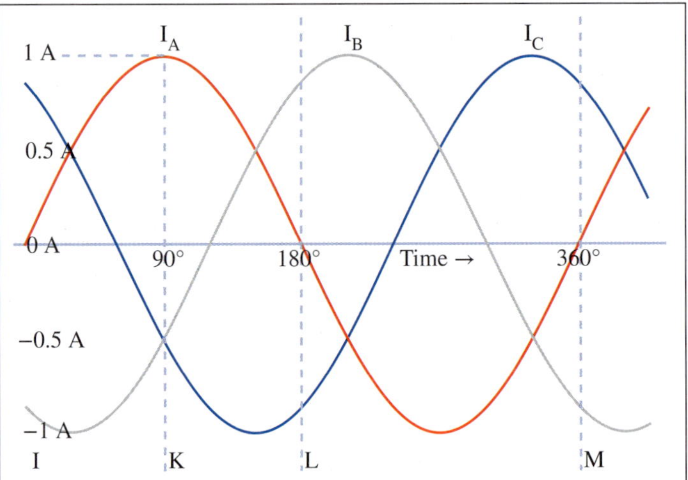

FIGURE 10.13 Current in a balanced system

Although the three currents are all changing in value and direction, the *phasor sum* of the instantaneous currents is zero ($I_A + I_B + I_C = 0$). Therefore, the current flowing through the neutral in a balanced circuit is also zero.

The same result can be obtained from the phasor sum of the line currents using RMS values.

Two types of load can exist in a three-phase system—balanced and unbalanced. If the current and phase angle in all three phases are equal ('balanced load'), the system is said to be 'balanced'. These are typically individual three-phase loads where the impedance of each phase is the same. Examples include three-phase motors and heaters (see Figure 10.14).

If the current and/or phase angle of any of the single phases are not equal, the system is called 'unbalanced' or an 'unbalanced load'. Examples include a three-phase main switchboard and three-phase distribution for houses (see Figure 10.14). This is more common in three-phase systems, where multiple combinations of single-phase equipment are connected and a complete balance is unlikely. While a balanced three-phase system should always be the ultimate aim, local service rules have a degree of tolerance for unbalanced loads.

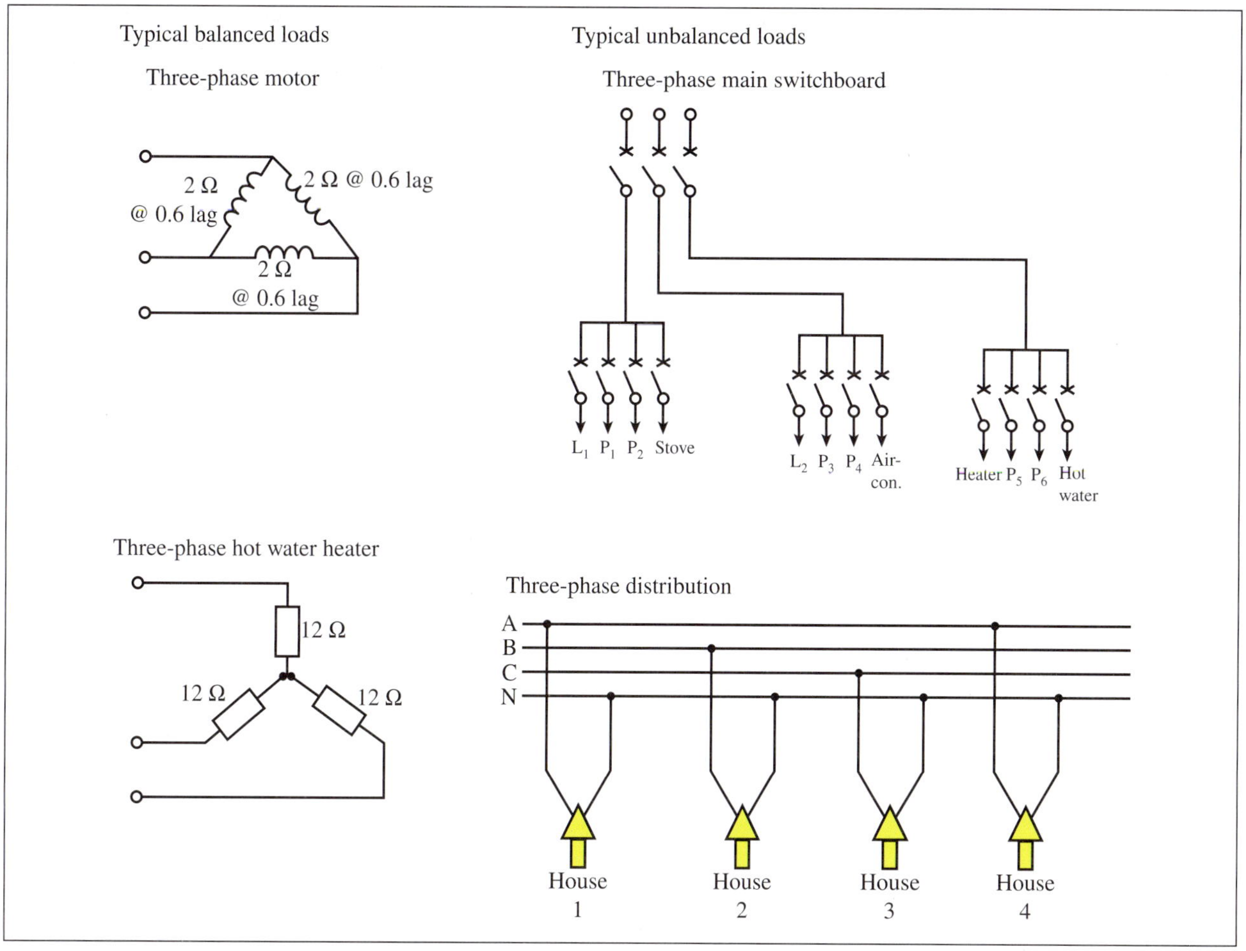

Typical balanced loads
Three-phase motor
2 Ω @ 0.6 lag
2 Ω @ 0.6 lag
2 Ω @ 0.6 lag
Three-phase hot water heater
12 Ω
12 Ω
12 Ω
Typical unbalanced loads
Three-phase main switchboard
L1 P1 P2 Stove
L2 P3 P4 Air-con.
Heater P5 P6 Hot water
Three-phase distribution
A
B
C
N
House 1
House 2
House 3
House 4

FIGURE 10.14 **Examples of a balanced and unbalanced loads**

For a balanced load (e.g. a three-phase motor) the neutral current is zero. This means the neutral wire is unnecessary and is usually omitted unless it is needed for control devices.

If a distribution system is subject to load changes on one phase, this puts the system out of balance and so a neutral conductor must be installed.

The two requirements for a balanced load are: there must be the same current loading on each phase; and each current must have the same power factor. When both requirements are met, the neutral current is zero.

When the two conditions for balance are not met, a system is said to be 'unbalanced' and a current may flow in the neutral (see Figure 10.15). By convention, the generated voltage is assumed to be positive when acting from the star point to the line terminal. The phase current is also regarded as positive when flowing in the same direction.

VC
IC
θC
IA+IB+IC
180°
VA
IA
IN
θB
IB
IA+IB
VB

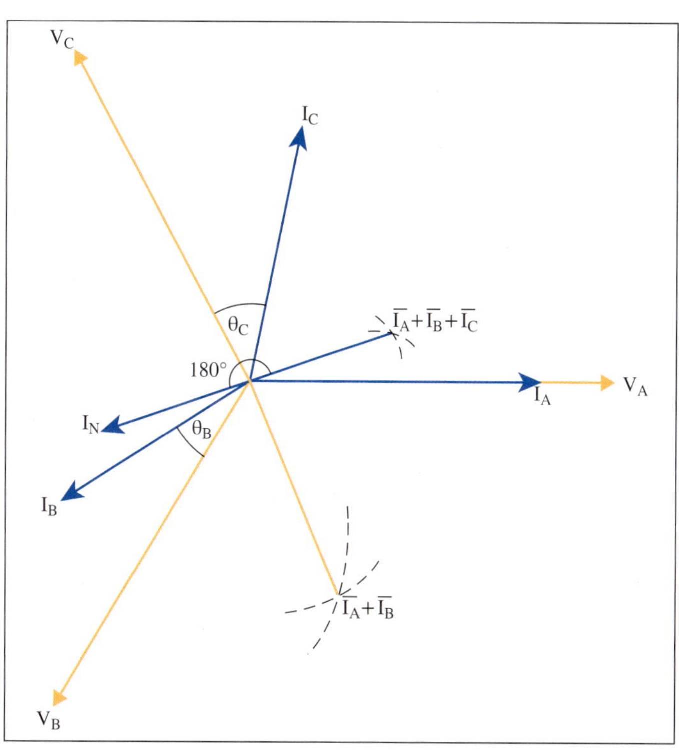

FIGURE 10.15 **Neutral current in an unbalanced three-phase system**

The value of I_N is equal to the phasor sum of I_A, I_B and I_C but the direction of current flow in the neutral wire is reversed.

The current in a neutral conductor of any three-phase four-wire system is equal to the negative of the phasor sum of the line currents.

$$\overline{I_N} = -(\overline{I_A} + \overline{I_B} + \overline{I_C})$$

The overbar '—' in the equation above represents phasor quantities.

10.3.4 AS/NZS 3000 requirements for neutral conductors

Clause 3.5.2 of AS/NZS 3000:2018 states that the current-carrying capacity and associated cable size of the neutral conductor should be the same as the active conductor for single-phase circuits, and the same as the largest associated active conductor for multiphase circuits. The harmonic content of the load supplied by multiphase mains, submains and final subcircuits must also be considered, especially the 3rd harmonic and the odd multiples of it, which can result in higher current in the neutral. In the instance of these harmonic currents, the neutral conductor must be sized to accommodate them.

EXAMPLE 10.1

Find the value of current flowing in the neutral conductor in a three-phase star-connected distribution system where the phase currents are:

- A phase 125 A at a power factor of 0.79 lagging
- B phase 147 A at a power factor of 0.85 lagging
- C phase 215 A at a power factor of 0.80 lagging.

Step 1. Convert power factors to phase angles (note that lagging is negative).

$$\varphi_A = \cos^{-1}0.79 \quad (1)$$
$$= -37.8° \quad (2)$$
$$\varphi_B = \cos^{-1}0.85 \quad (3)$$
$$= -31.8° \quad (4)$$
$$\varphi_B = \cos^{-1}0.80 \quad (5)$$
$$= -36.87° \quad (6)$$

Step 2. Select an appropriate scale to draw the three-phase phasor diagram (e.g. 10 mm = 25 A).

Step 3. Using V_A as a reference, plot all three-phase voltages with a phase displacement of 120°.

Step 4. Plot I_A lagging V_A by 37.8°.

Step 5. Plot I_B lagging V_B by 31.8°.

Step 6. Plot I_C lagging V_C by 36.87°.

Step 7. Add the two phasors I_A and I_B.

Step 8. Add the phasor I_C to the resultant of the addition of I_A and I_B.

Figure 10.16 shows the type of three-phase phasor diagram that should be generated in the course of answering the question.

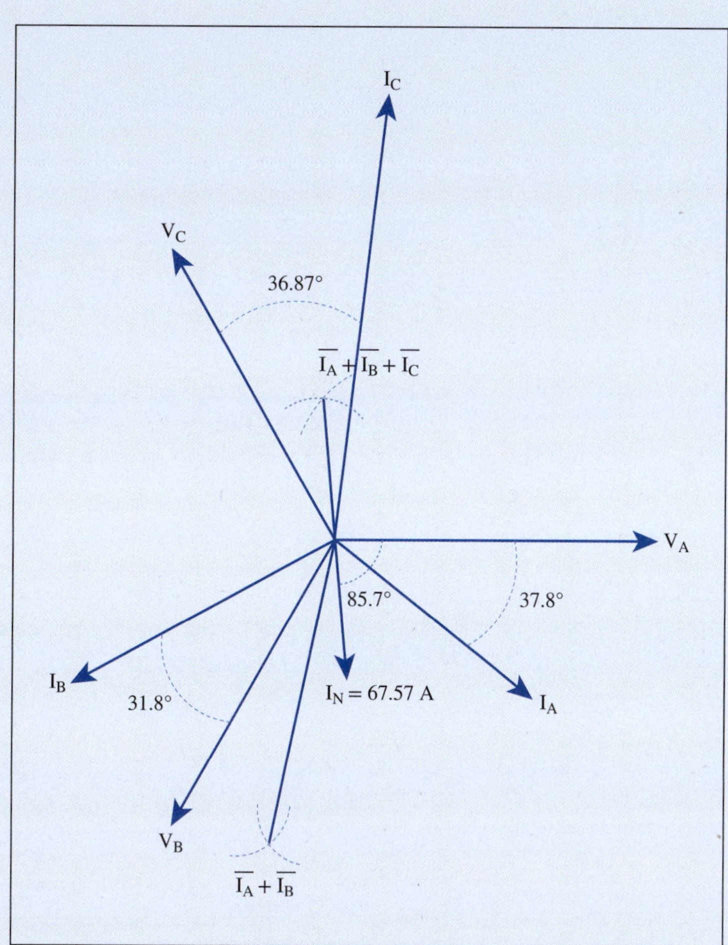

FIGURE 10.16 Three-phase phasor diagram for Example 10.1

Step 9. Plot in 180° in the opposite direction of the resultant of the sum of I_A, I_B and I_C. This is the current phasor for I_N.

Step 10. Measure the angle between I_N and V_A (the reference) and scale the measured value of I_N to get the resultant of $I_N = 67.57\text{A} \angle 85.7°$ lagging.

Alternative method of determining the neutral current

Another method for determining the neutral current is to use the 'tip to tail' method. This involves drawing the second current, I_B, on the end of I_A and then drawing I_C on the end of I_B. You will see in Figure 10.17 that the tip of I_C finishes at the same point as in the previous method. (The reference voltages of V_B and V_C have been removed from these diagrams to make the process clearer, but the current phasors have been drawn to the appropriate leading/lagging angle.)

Reduction of the neutral current

The neutral conductor of a multiphase circuit can also be reduced in size under either of the following conditions: there is a detection device connected and arranged to prevent the neutral current from exceeding the current-carrying capacity; or the multiphase circuit is predominantly supplying multiphase equipment and the current-carrying capacity of the neutral is not exceeded by out-of-balance and harmonic currents.

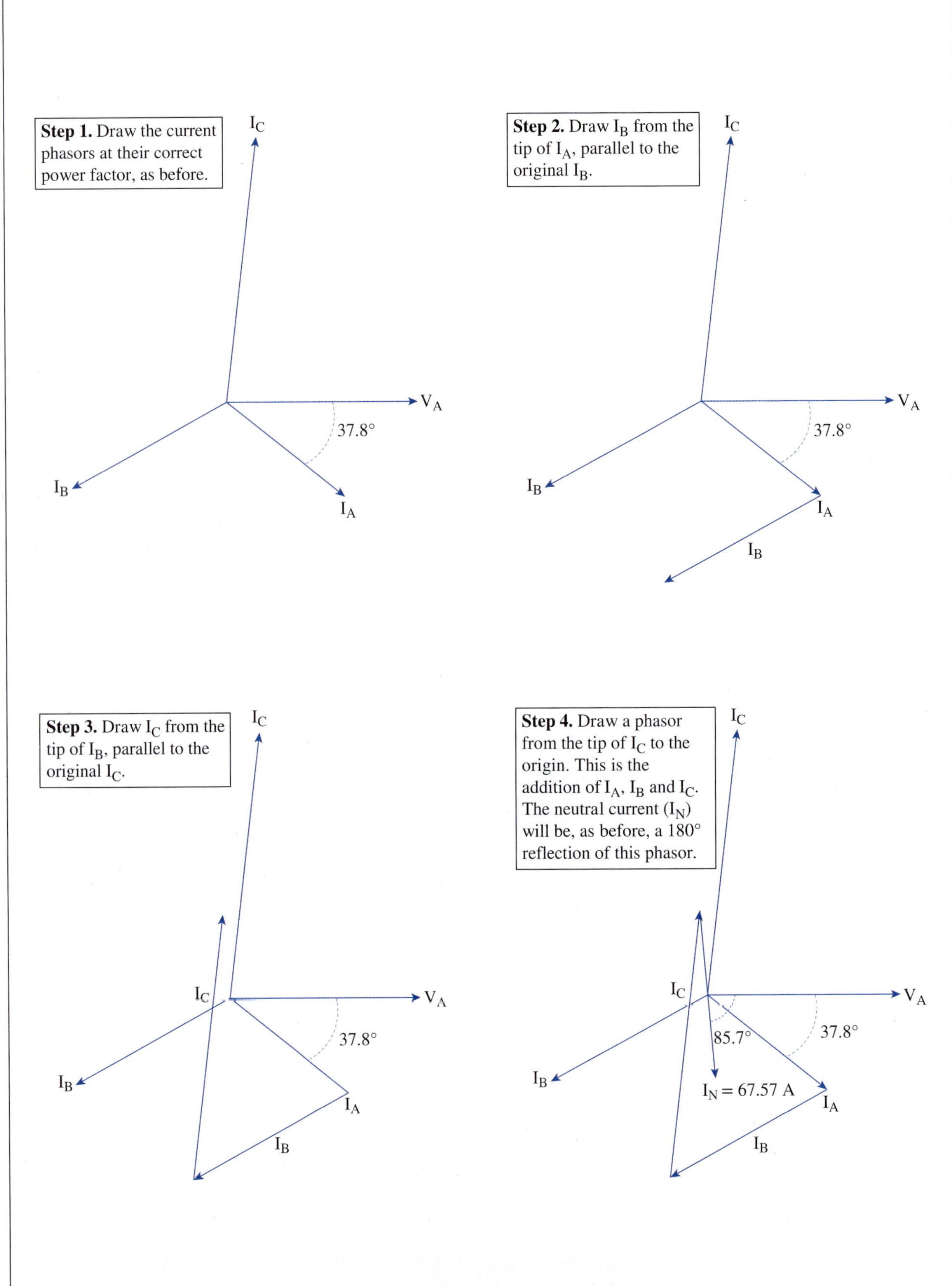

FIGURE 10.17 Alternative method of determining the neutral current

CHECK YOUR UNDERSTANDING

10.10 Why does the neutral of the load need to be connected to the star point of a three-phase four-wire supply system?

10.11 What is the danger of having an open circuit neutral on a three-phase four-wire supply system?

10.12 Under normal installation conditions, should a fuse be placed in the neutral? If not, why not?

10.13 Explain the difference between a balanced and an unbalanced load.

10.14 What is the value of neutral current in a balanced three-phase four-wire supply system?

10.4 Three-phase delta connections and interconnected systems

10.4.1 Three-phase delta connections

The windings of a three-phase alternator can also be connected as shown in Figure 10.18(a). This forms a closed loop with dissimilar ends joined and the lines are connected to the junctions. An alternative method of drawing the windings and connections is shown in Figure 10.18(b). This system is called a 'delta' connection because of the similarity in shape with the Greek letter Δ (delta).

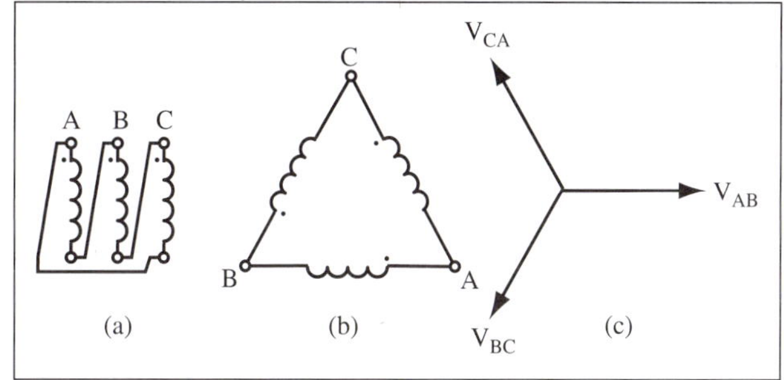

FIGURE 10.18 Three-phase delta connections

Each phase winding is now connected across a pair of lines, so the phase voltage is equal to the line voltage.

For a delta system $V_L = V_P$

Each phase winding is also connected in parallel across the other two phase windings, which are in series with one another. A quick check should confirm, however, that the phasor sum of the other two windings is equal in voltage and phase to the first winding, noting winding polarities.

The phase currents in Figure 10.19 are I_A, I_B and I_C and the line currents are I_1, I_2 and I_3. If the phase currents are of equal value, then by phasor addition the line current is equal to $\sqrt{3}$ times the phase current.

For a delta system $I_L = \sqrt{3}\, I_P$

Also $I_{A\text{-}B}$ leads I A by 30°, so I_L leads I P by 30°.

Where a *star* connection gives the *line voltage* a phase shift of 30° causing it to lead one of the phase voltages, a *delta* connection causes the *line current* to lead one of the phase currents by 30° (assuming a balanced three-phase load). This effect is put into use when larger numbers of phases are required (as with large industrial rectifiers, where filtering costs become prohibitive).

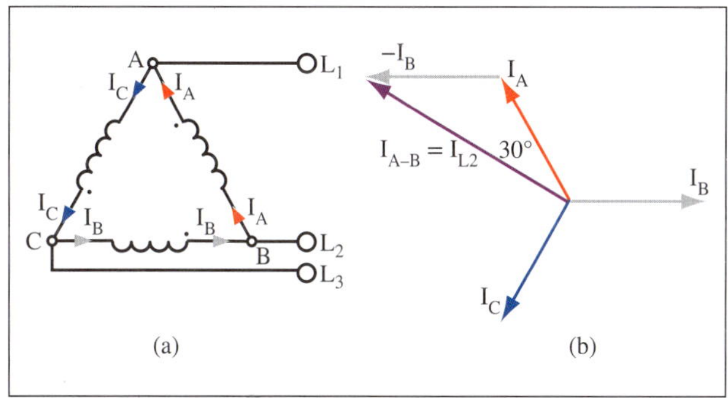

FIGURE 10.19 Line current vs phase current in delta

10.4.2 Limitations and uses of open delta connections

Three-phase systems can be connected by using two windings. The start of one winding is connected to the end of the other. The three phases are connected to each free end of the windings and this junction. This form of connection is referred to as 'open delta' or 'v' connection (See Figure 10.20). It is relatively cheap as only two windings are required.

The power in a closed three-phase system is:

$$P_{3phase} = \sqrt{3} \times V_L \times I_L \times PF$$

or:

$$P_{3phase} = 3 \times V_L \times I_P \times PF$$

(Note: $V_L = V_P$ in a closed delta system.)

The path for current in an open delta system is limited to $\sqrt{3}\ I_P$ with one of the windings missing. This equates to a power of:

$$P_{open\ delta} = \sqrt{3} \times V_L \times I_P \times PF$$

So, the ratio of open to closed delta equals:

$$\frac{\sqrt{3} \times V_L \times I_P \times PF}{3 \times V_L \times I_P \times PF}$$

and as $V_L \times I_P \times PF$ cancels out on both sides the ratio is:

$$\frac{\sqrt{3}}{3}$$

which equals 57.7%.

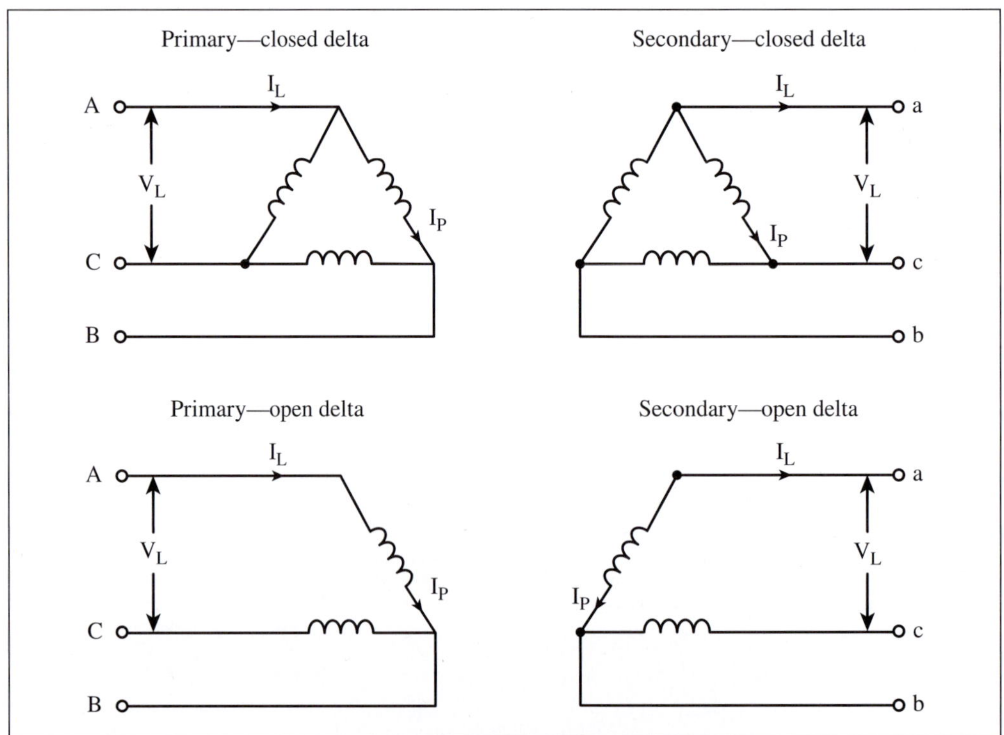

FIGURE 10.20 Open delta connections of a transformer

So, the major limitation of an open delta system is that it is only able to deliver 57.7% of the power of a closed delta system. Although this limits the uses of the system, it does provide a cheaper option when initially installed. It can be upgraded to a closed delta system later.

10.4.3 Effect of phase reversal on a delta system

If the ends of one phase winding are reversed, a phase shift of 180°E occurs. This causes the voltage on that phase to be added to the sum of the other two phases. This in turn produces a voltage that is twice the phase voltage. In a 400 V delta system, the resulting voltage is not zero, which would not allow the delta loop to be closed safely; it is in fact 800 V which, if closed, would cause a circulating short-circuit and a very large current to flow.

The phasor diagrams for line currents I_2 and I_3 are identical but 120° phase shifted.

Because this higher voltage is generated within the closed circuit of the windings, heavy circulating currents will flow in them, causing them to burn out quickly. Great care must be taken when the phase windings are connected in delta to ensure that dissimilar ends are joined, e.g. connect a2 to b1 and not a2 to b2.

A simple method of testing the connections is to leave one junction open, as shown in Figure 10.21, and connect a voltmeter across the open winding ends. If the connections are correct, the voltmeter will read zero. If one phase winding is reversed, the meter will register a voltage equal to twice the phase voltage.

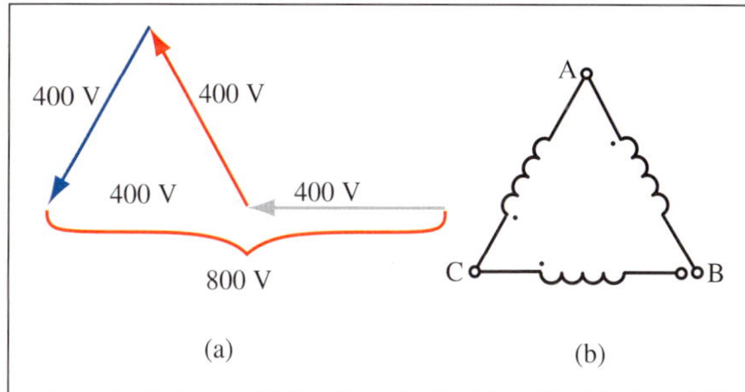

FIGURE 10.21 One phase winding reversed

Figure 10.22(a) shows the healthy delta system with all three supply lines connected. If one supply line feeding a three-phase delta-connected load was lost (through a disconnection or broken conductor), the system would become a single-phase 400 V series/parallel system. This can be seen in Figure 10.22(b).

With the loss of supply line B, the system is only being fed by supply lines A and C and so becomes a single-phase supply. The voltage between the supply lines (A and C) remains 400 V.

Winding 1 is now in parallel on its own across A and C. Windings 2 and 3 are in series with each other (and so their impedances would be added) but they are also in parallel across A and C.

There would be two parallel paths of current, one through winding 1 and the other through the series combination of windings 2 and 3.

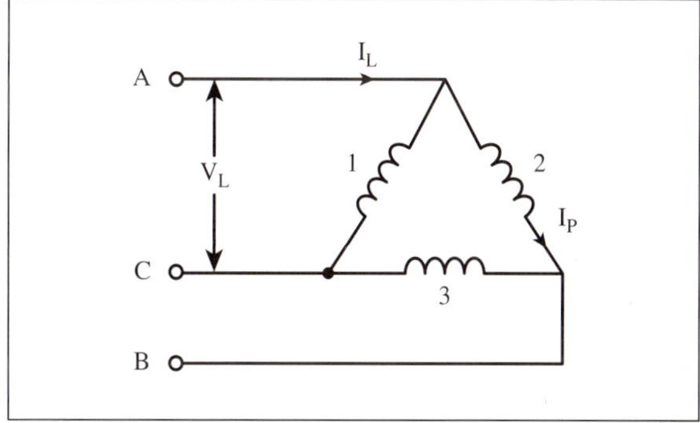

FIGURE 10.22(a) Healthy delta connected system

10.4.4 Loads in typical power systems

Three-phase power systems are connected in star or delta, depending on circumstances and requirements.

Where the load current in each phase has minimal difference, delta systems are used as they require one fewer conductor. Examples include generators, motors and high-voltage transmission and distribution systems.

Alternatively, star-connected systems are used where there is a higher difference between phase currents and where single-phase loads are required. For this reason, all low-voltage consumer supplies and associated distribution

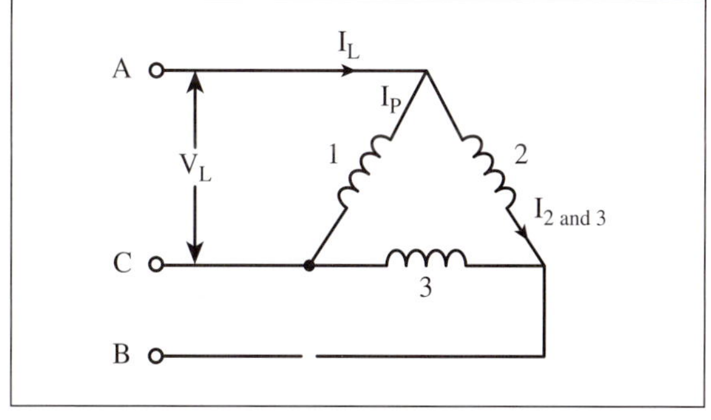

FIGURE 10.22(b) Delta system with loss of B phase

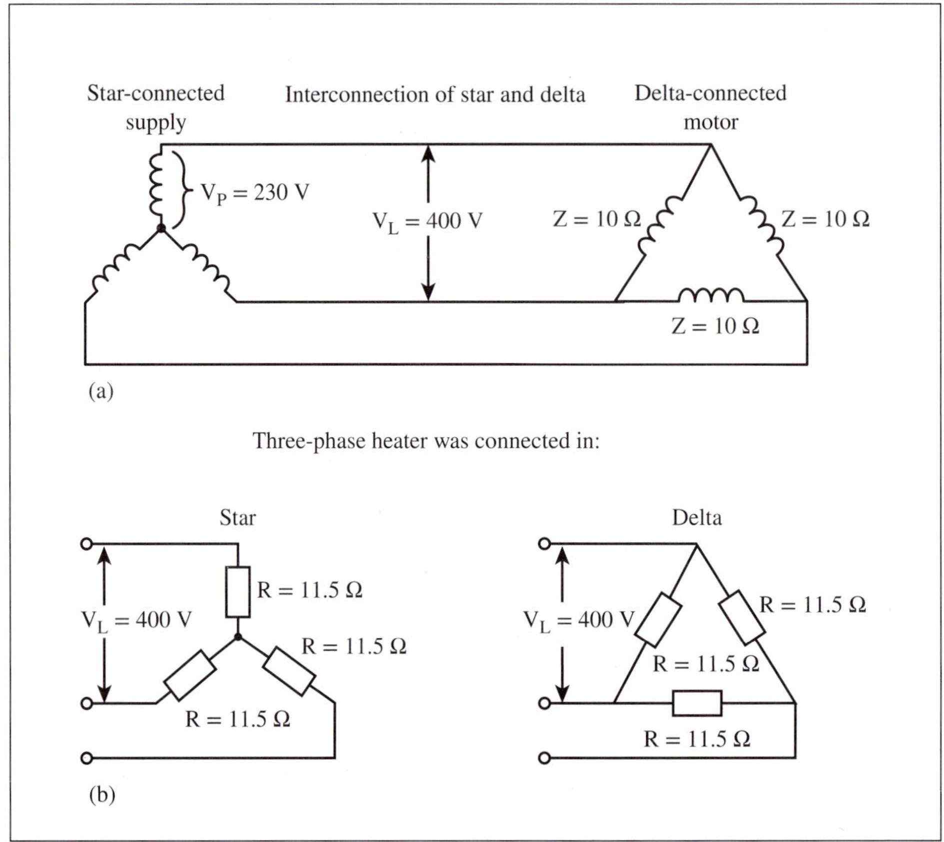

FIGURE 10.23 Typical combinations of three-phase interconnected systems

systems are connected in either three-phase four-wire or single-phase two-wire. Some older low-voltage supplies were also connected with two phases using three-wire (two phases and neutral).

Star- or delta-connected systems are also used to vary the output of a three-phase load or restrict start-up and inrush currents in motor starter circuits.

Figure 10.23(a) is a diagram of a three-phase delta-connected motor supplied from a star-connected three-phase 230/400 V supply. As the impedance of each winding of the motor is 10 Ω, the current can be determined in each winding of the motor:

$$
\begin{aligned}
I_{\text{motor phase}} &= \frac{V_{\text{supply line}}}{Z_{\text{motor phase}}} \\
&= \frac{400\text{ V}}{10\ \Omega} \\
&= 40\text{ A}
\end{aligned}
$$

That is to say that there is 40 A flowing in each of the windings in the motor. The line current and phase current are not the same in delta-connected loads and so we need to calculate I_{line}. The line current to the motor will be:

$$
\begin{aligned}
I_{\text{motor line}} &= \sqrt{3} \times I_{\text{motor phase}} \\
&= \sqrt{3} \times 40 \\
&= 69.3\text{ A}
\end{aligned}
$$

And as the phase and line currents are equal in star-connected systems, the supply will have a line current and a phase current of 69.3 A.

In Figure 10.23(b), if we consider a three-phase heater connected in either star or delta, we can see the difference in power output. As the phase voltage in a star-connected 400 V system is 230 V, the current in each element of the star-connected heater is:

$$\begin{aligned} I_{phase} &= \frac{V_{phase}}{R} \\ &= \frac{230\text{ V}}{11.5\ \Omega} \\ &= 20\text{ A} \end{aligned}$$

And as the line current is equal to the phase current, the power is (assuming a unity power factor):

$$\begin{aligned} P_{star} &= \sqrt{3} \times V_{line} \times I_{line} \times \text{power factor} \\ &= \sqrt{3} \times 400\text{ V} \times 20\text{ A} \times 1 \\ &= 13.9\text{ kW} \end{aligned}$$

The same load connected in delta will have a phase current of:

$$\begin{aligned} I_{phase} &= \frac{V_{phase}}{Z_{phase}} \\ &= \frac{400\text{ V}}{11.5\ \Omega} \\ &= 34.78\text{ A} \end{aligned}$$

And a line current of:

$$\begin{aligned} I_{line} &= \sqrt{3} \times I_{phase} \\ &= \sqrt{3} \times 34.78 \\ &= 60.25\text{ A} \end{aligned}$$

So, the three-phase power for the delta-connected heater is:

$$\begin{aligned} I_{line} &= \sqrt{3} \times V_{line} \times I_{line} \times \text{power factor} \\ &= \sqrt{3} \times 400\text{ V} \times 60.25\text{ A} \times 1 \\ &= 41.7\text{ kW} \end{aligned}$$

From this, we can see that the same heating elements in the two different configurations can produce two different levels of heat. It is important to note that the power in delta is *three times* the power in star (41.7 kW is approximately three times 13.9 kW).

CHECK YOUR UNDERSTANDING

10.15 Draw a delta-connected three-phase system.

10.16 In a delta-connected system, what is the relationship between (a) phase current and line current and (b) phase voltage and line voltage?

10.17 A delta-connected three-phase system has the following values:

$I_p = 10$ A

$V_p = 230$ V

Determine the values of I_L and V_L.

10.18 How much greater is the power produced by a delta system than the power produced by the same system connected in star?

10.19 A system produces a power of 3 kW when connected in star. Determine the power produced if the system was delta connected.

10.5 Energy and power requirements of a.c. systems

10.5.1 Power transmission

Infinite grid

The distribution grid is so large in comparison with most loads that it appears to be infinite, not only visually but in most calculations as well. If a load took 100 A on each phase in a 230/400 V three-phase system, this might seem like a large load. However, a relatively small 500 MW power station can supply over 700 000 A per phase at 230 V, so the 100 A is relatively insignificant.

The entire grid has a much greater capacity than that.

What this means when carrying out calculations for loads on the grid is that the voltage, frequency and phase difference between lines are fixed. They cannot be changed by any relatively small load that we apply.

This makes calculations much easier for electricians, although engineers may still have to consider supply impedances in their calculations.

The infinite grid means that we can always assume a fixed voltage of 230/400 V, a frequency of 50 Hz and a phase separation of 120°.

10.5.2 Transmission

Electrical power could be transmitted using low voltage and high current. But higher current results in higher transmission losses, according to the equation $P = I^2R$. So, to reduce what are known as 'I^2R losses', transmission lines generally use high voltage and low current.

Energy is consumed in a power grid relative to the load (or current used). For the same power, the line current can be reduced by increasing the transmission voltage. This also allows for a reduction in conductor size for transmission lines, which still produces a lower power loss in the line.

In economic terms, the higher the voltage used for transmission, the lower the cost of installing and maintaining the transmission lines. For long distances between the source of supply and the consumer, voltages up to 576 kV are used for power transmission and distribution. (Typical voltage values vary between states and territories, and even within a state or territory.)

Transmission system voltages are far higher than the voltages required by the average consumer; therefore, the local distribution voltage is transformed in stages at substations and sometimes in pole transformers. Figure 10.24 shows a simplified distribution network.

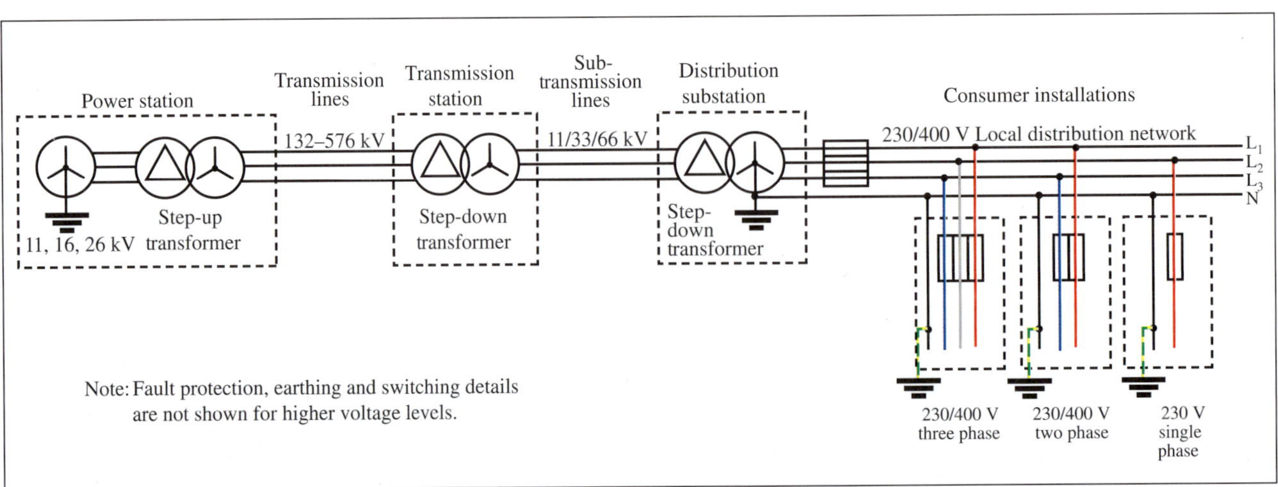

FIGURE 10.24 Electrical power distribution

Most main transmission and sub-transmission lines are duplicated, with alternative routes provided so that different localities can be fed by other lines and substations in the event of essential maintenance or breakdowns.

Some systems are ring fed, which means that the distribution lines form a complete circle. The reason for this is that, in the event of a fault, a small section of the ring can be shut down while the remaining sides remain online; a smaller number of consumers are affected. Electricians must be aware of ring feeds, as two disconnections must be made to render the section safe.

10.5.3 SWER distribution

Rural areas are generally a special case, where load units are comparatively small and their points of application are widely dispersed. One method of providing electrical power in these areas is the Single-Wire Earth-Return (SWER) system. Figure 10.25 shows an isolating transformer connected to the sub-transmission lines. The secondary voltage of the transformer is connected to a single conductor that traverses the countryside. The voltage of this line can vary between states or localities.

If the isolating transformer is omitted and the SWER line is fed directly from one conductor of a three-phase 33 kV line, the phase voltage to earth is 19 kV. In some localities, the other phases are led out in different directions. This provides a balancing effect on the line.

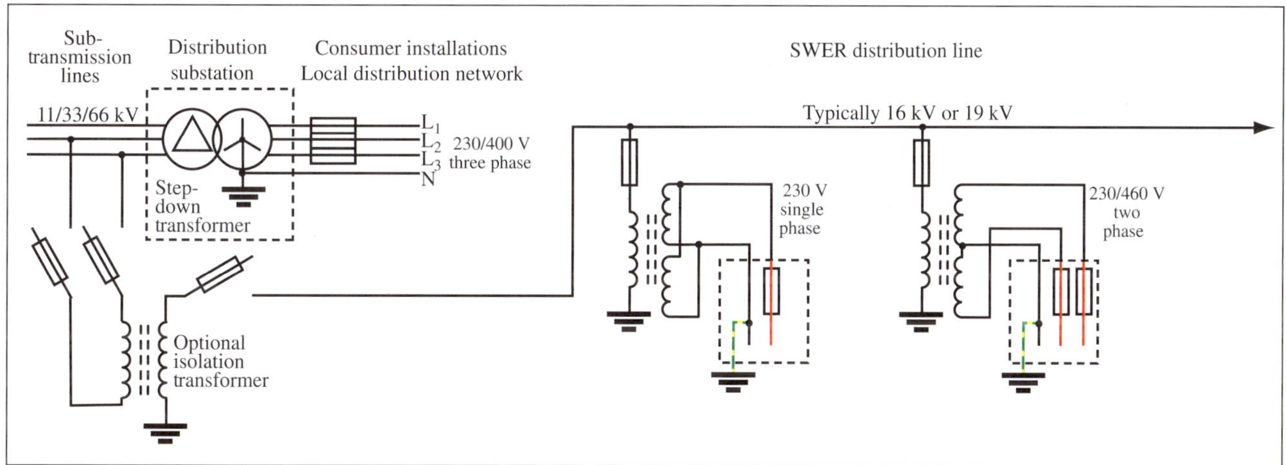

FIGURE 10.25 Single-Wire Earth-Return (SWER) distribution

Due to the high voltage, both the line currents and the earth-return currents are small. These ground currents effectively prevent the system from being used in highly populated areas. This is because, in areas where large amounts of metal are buried in the ground, electrolytic corrosion is a problem. Special bonding arrangements might have to be set up to limit the amount of corrosion, and excessive metal would therefore make a SWER line uneconomical.

Transformers are provided at the point of distribution. There are two secondaries on a SWER transformer which can be connected in parallel to supply 230 V or in series for 460 V at half the current (see Figure 10.26).

FIGURE 10.26 A SWER consumer transformer
Roman Tiraspolsky/Shutterstock

10.5.4 Three-phase distribution

For general purposes, three-phase power may be supplied using either a three-wire or a four-wire system.

A three-wire system is one that uses only the three line conductors, as shown in Figure 10.27(a). The phase windings are shown connected in delta, but they can also be connected in star, with or without the star point earthed.

A four-wire distribution system is one that uses the three line conductors plus the neutral. The neutral is connected to the star point of the phase windings and earthed, as shown in Figure 10.27(b).

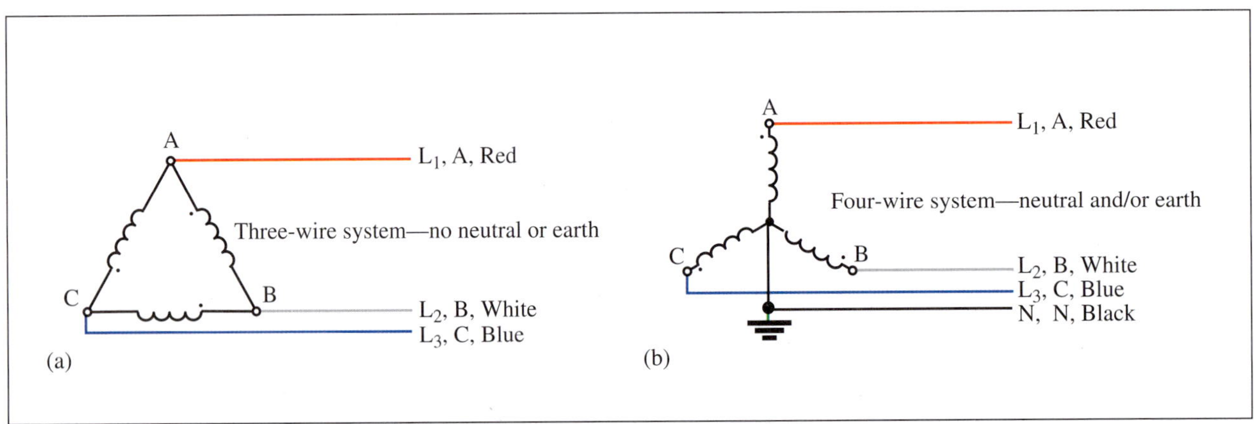

FIGURE 10.27 (a)Three-wire vs (b) four-wire

For the supply authority, it is no great problem to supply power with star or delta systems. For the consumer, the star system is safer and more versatile because it provides a choice of voltages and an earth reference point. Single-phase loads present a problem to an otherwise balanced three-phase system but are more convenient and safer for smaller consumers.

Voltage drop is a major problem in power distribution because it causes consumers furthest from the supply to receive power at a reduced voltage. Various techniques are used to try to overcome this. One method that works for small groups of consumers is to supply them from a centrally located transformer fed by a high-voltage primary feeder (a feeder is a grid supply line). Radiating from the central point are the secondary or supply mains, consisting of three lines and a neutral. Characteristically, the mains become smaller in size as the distance from the transformer increases. This method is typical for very small country towns.

For higher-density distribution, as in the suburbs of a larger town, it is more common to have each block of consumers supplied from a transformer and have the mains from various blocks interconnected. This means that any one consumer can be supplied from several sources. This method has the disadvantage of needing more protection devices for the transformers, and it becomes more difficult to isolate sections of mains for maintenance purposes.

10.5.5 Three-phase power

Star connection

The power consumed by a load connected to a single-phase a.c. supply is found by using the equation $P = V\,I \cos\phi$

If three identical loads are connected in parallel to one supply, the total power used is $3 \times V\,I \cos\phi$. This means that three loads of the same size use three times the power of one load in three-phase systems. Three loads on the three phases use three times the power of one load, at least when working with phase voltage and phase current.

In general, however, it is easier to measure and use line values rather than phase values.

In a star-connected system, $I_L = I_P$ and $V_L = \sqrt{3}.V_P$. Substituting line values for phase values:

$$P = 3 \times \left(\frac{V_L}{\sqrt{3}}\right) I_L \times \cos\phi = \sqrt{3}\, V_L\, I_L \cos\phi$$

This is usually expressed as:

$$P_T = \sqrt{3}\, V_L\, I_L \cos \phi$$

where V and I are line values.

Delta connection

If the three individual loads are connected in delta, the power consumed is still:

$$P = 3 \times V_P I_P \times \cos \phi \text{ (for a balanced condition)}$$

In a delta system, $V_L = V_P$ and $I_L = \sqrt{3}\, I_P$
Substituting these:

$$\begin{aligned} \text{total power } P &= 3 \times \left(\frac{V_L}{\sqrt{3}}\right) \times I_L \times \cos \phi \\ &= \sqrt{3}\, V_L\, I_L \cos \phi \end{aligned}$$

In general terms:

$$P_T = \sqrt{3}\, V_L\, I_L \cos \phi$$

where V and I are line values: $\cos \phi$ = power factor.

This equation is the same as for a balanced star-connected load. It is therefore applicable for both star- and delta-connected balanced three-phase loads. *It is not to be used for unbalanced loads.*

EXAMPLE 10.2

A three-phase 400 V motor draws 12 A at 0.85 lagging power factor. How much power is consumed?

$$\begin{aligned} P_T &= \sqrt{3}\, V_L\, I_L \cos \phi && (1) \\ &= \sqrt{3} \times 400 \times 12 \times 0.85 && (2) \\ &= \underline{7067 \text{ W } (7.07 \text{ kW})} && (3) \end{aligned}$$

10.5.6 Power and energy meters

In the days of mechanical instruments, power meters required two coils: one for voltage connected in parallel across the circuit and one for current to be connected in series in the circuit. Their combined magnetic field would cause the meter to register power. The instrument was reasonably reliable and accurate. Unfortunately, it was also fragile, sensitive to temperature changes and required occasional recalibration. Energy meters were similar, although generally fixed in use, and were almost free of recalibration.

Power being consumed in a circuit is measured with a wattmeter, and these are often constructed with a dynamometer movement. This type of movement usually has two internal electrical circuits (see Figure 10.28).

Although electronic instruments are good at addition and subtraction (they can even emulate square and square-root functions), multiplication of two rapidly varying values is difficult and only accurate over a narrow band of values. It took digital electronics and computerised instrumentation to really replace analogue instruments.

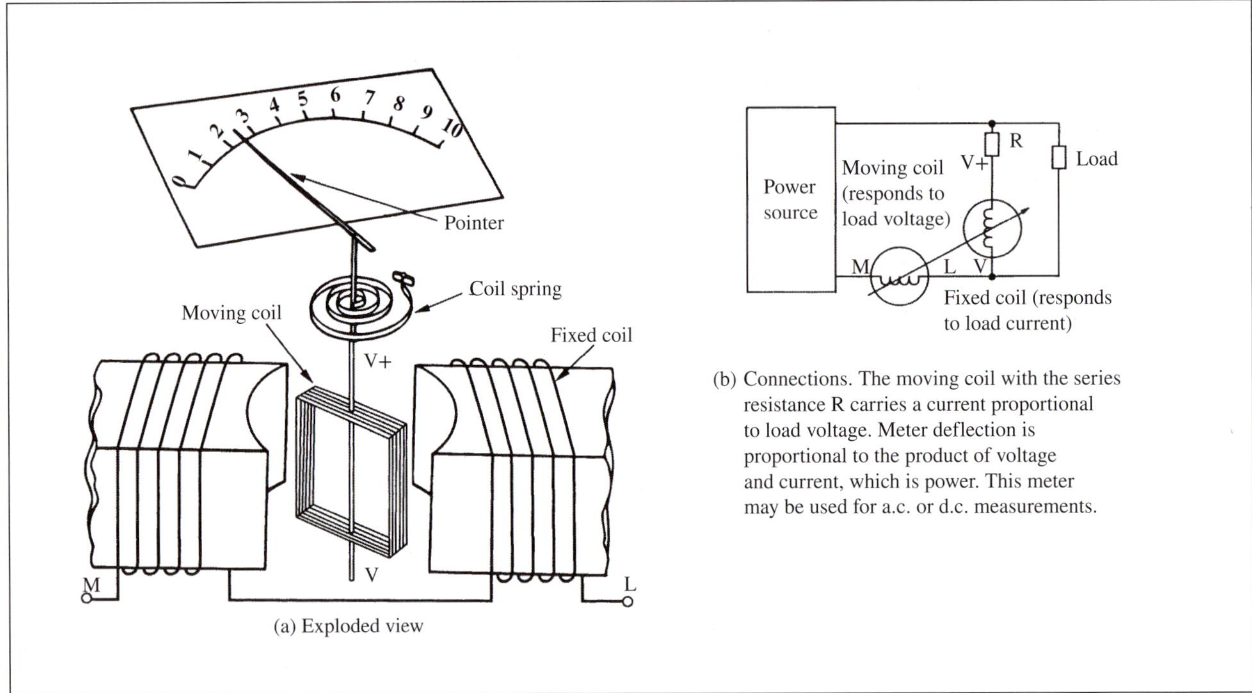

FIGURE 10.28 A dynamometer movement

10.5.7 Methods of three-phase power measurement

One wattmeter four-wire system

The single wattmeter is connected between one line and neutral in a three-phase four-wire system (see Figure 10.29). The total power drawn from a three-phase supply is found by adding the separate values of power consumed by each phase.

In the case of a balanced load:

$$P_{Total} = 3P_A = 3P_B = 3P_C = \text{three times the wattmeter reading in any of the three lines.}$$

For an unbalanced load, the wattmeter must be connected or switched into each phase in turn and the individual power readings added:

$$P_{Total} = P_A + P_B + P_C$$

Advantages

1. One wattmeter only is required.
2. It is suitable for both balanced and unbalanced loads.

Disadvantages

1. A neutral connection is required for the wattmeter.
2. It is not accurate for unbalanced fluctuating loads.
3. The wattmeter must be connected or switched into each phase in turn for unbalanced loads. The switch must not break the line when switching.

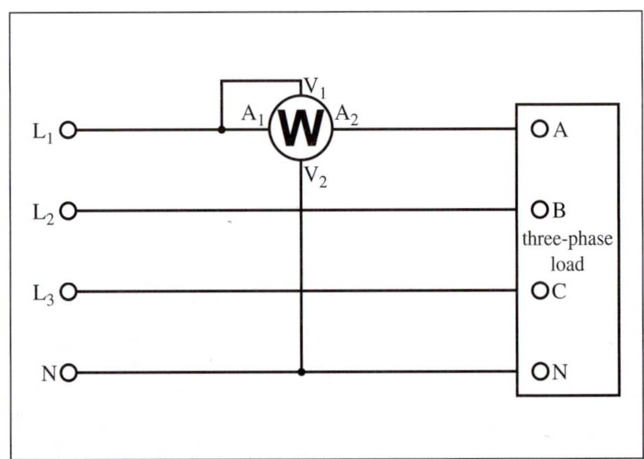

FIGURE 10.29 One wattmeter four-wire system

One wattmeter three-wire system

With a three-wire system, no neutral is available. As there is a 30° phase shift between line and phase voltages, it is necessary to provide an artificial star point so that the correct voltage at the correct phase angle is applied to the wattmeter. Two impedances, each matching the impedance of the voltage circuit in the wattmeter, must be connected in star with the meter voltage circuit (see Figure 10.30) and to the other two lines. For this reason, resistors matching the resistance of the voltmeter circuit must be very close approximations.

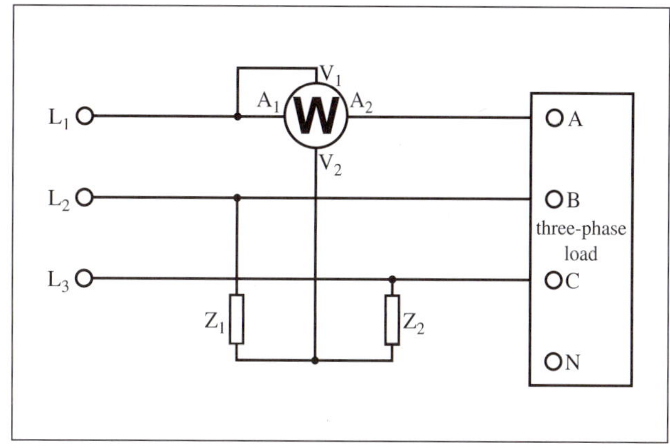

FIGURE 10.30 One wattmeter three-wire system

For balanced loads only:

$$P_{Total} = 3P_A = \text{three times the wattmeter reading}$$

For balanced and unbalanced loads:

$$P_{Total} = P_A + P_B + P_C$$

Advantages

1. One wattmeter only is required.
2. It is suitable for both balanced and unbalanced loads.

Disadvantages

1. Two matching impedances are required to provide an artificial neutral.
2. It is not accurate for unbalanced fluctuating loads.
3. The wattmeter must be connected or switched into each phase in turn for unbalanced loads. The switch must not break the line when switching.

Two wattmeters three-wire system

A method for measuring the power consumed in a three-phase three-wire circuit is shown in Figure 10.31. The two meters have their current windings in any two lines and both voltage windings are connected to the third line. Neither meter alone indicates the total power in the circuit. But the two meters together, by their algebraic sum, indicate the power consumed. That is:

$$P_{Total} = W_1 + W_2$$

For a balanced load with unity power factor, both meter readings will be equal. For all other conditions, the meters will show different readings.

If the lower-value meter indication is W_1 and the higher-value one W_2, as the power factor decreases, W_1 registers less and less of the total power. When the power factor is 0.5 with a balanced load, W_1 will read zero and W_2 will read the total power. Should the power factor fall further (to, say, 0.3), W_1 will read even less, meaning that W_1 will attempt to read a negative value of power consumption. If the current or voltage

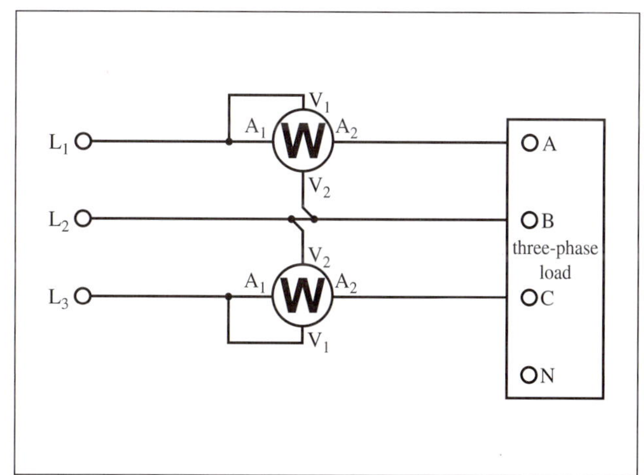

FIGURE 10.31 Two wattmeters three-wire system

connections to the meter are reversed, the numerical value of W_1 can be obtained. *But it is a negative value.* The total power in this case is still the algebraic sum of W_1 and W_2. That is:

$$\begin{aligned} P_{Total} &= (-W_1 + W_2) \\ &= W_2 - W_1 \end{aligned}$$

Should the power be reduced to zero with a balanced load (pure capacitance or inductance), W_1 would read a negative value numerically equal to W_2 and the algebraic sum would be zero:

$$P_{Total} = -W_1 + W_2 = 0$$

This also agrees with the concept of a purely reactive circuit.

The two-wattmeter method may be used on three-phase three-wire systems to obtain load power values whether the load is balanced or unbalanced. The same applies whether it is star or delta connected. However, it cannot be used on a four-wire, star-connected system because a single-phase component of current might be flowing in the line (and neutral) with no wattmeter current-coil connection. The power being consumed would not be correctly recorded.

Only when the three-phase load is balanced is it possible to find the power factor of the load from the wattmeter readings. The angle of lag or lead is found from:

$$\varphi = \tan^{-1}\sqrt{3}\left(\frac{W_1 - W_2}{W_1 + W_2}\right)$$

(This applies to balanced loads and sinusoidal waveforms only.)

The Cosine of this angle gives the power factor of the load.

EXAMPLE 10.3

When connected to a three-phase motor, two wattmeters gave readings of 5 kW and 1 kW.

Find the total power being consumed and the power factor of the motor.

$$\begin{aligned} P_{Total} &= W_1 + W_2 && (1) \\ &= 5 + (-1) && (2) \\ &= \underline{4\ \text{kW}} && (3) \\ \varphi &= \tan^{-1}\sqrt{3}\left(\frac{W_1 - W_2}{W_1 + W_2}\right) && (4) \\ \varphi &= \tan^{-1}\sqrt{3}\left(\frac{5-(-1)}{5+(-1)}\right) && (5) \\ \varphi &= \tan^{-1}\sqrt{3}\left(\frac{5+1}{5-1}\right) && (6) \\ \varphi &= \tan^{-1}\sqrt{3}\left(\frac{6}{4}\right) && (7) \\ &= \tan^{-1} 2.598 && (8) \\ &= 68.9^\circ && (9) \\ \therefore \text{PF} &= \underline{\cos 68.9^\circ = 0.3593} && (10) \end{aligned}$$

Advantages

1. Only two wattmeters are required.
2. It is useful for both balanced and unbalanced three-phase three-wire loads.
3. The power factor can be obtained for balanced loads.
4. No neutral connection is required.

Disadvantages

1. It is suitable only for three-phase three-wire loads.
2. Care must be used in determining the polarity of W_1.
3. The power factor cannot be obtained for unbalanced loads.
4. It is not suitable for power or power factor readings with three-phase four-wire systems.

Three wattmeters three-wire system

With a three-wire system, no neutral is available; an artificial neutral must be provided. However, if identical wattmeters are used, the three voltage circuits can be connected to provide a star point, as shown in Figure 10.32:

$$P_{Total} = W_1 + W_2 + W_3$$

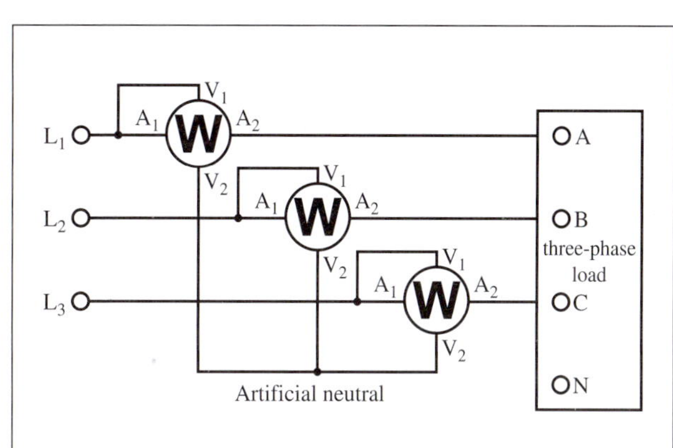

FIGURE 10.32 Three wattmeters three-wire system

Advantages

1. It is suitable for both balanced and unbalanced loads.
2. It is convenient for obtaining total power.
3. It is more accurate than one wattmeter for fluctuating loads.

Disadvantage

1. Three wattmeters are needed.

Three wattmeters four-wire system

The three-phase four-wire system is basically three separate supplies with only a common neutral. The total power is obtained by connecting three wattmeters, as shown in Figure 10.33:

$$P_{Total} = W_1 + W_2 + W_3$$

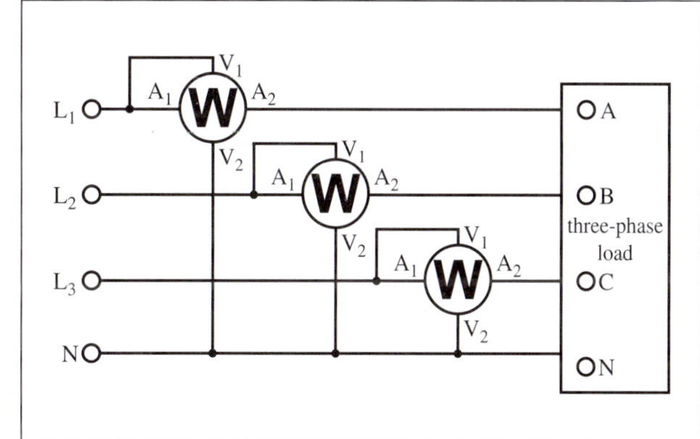

FIGURE 10.33 Three wattmeters four-wire system

Advantages

1. It is suitable for both balanced and unbalanced loads.
2. It is convenient for obtaining total power.
3. It is more accurate than one wattmeter for fluctuating loads.

Disadvantage

1. Three wattmeters are needed.

Electronic three-phase power meter

A typical electronic power meter measures all three line voltages (V_{AB}, V_{BC} and V_{CA}) and line currents to determine the power. For higher current measurements, current transformers (CTs) are used (see Figures 10.34 and 10.35).

Total power is measured by linear electronics in some instruments. They use operational amplifiers to calculate power continuously before displaying data on an analogue meter or digital readout.

Digital instruments may use analogue-to-digital converters to measure voltage and current, and zero-crossing comparators to measure phase angle. From those values, digital electronics calculate true power, reactive and apparent power, PF and phase angle.

The most recent instruments use computer technology to improve on digital instruments. This gives greater accuracy and more functions and features. These include frequency, total harmonic distortion (THD), true RMS measurement and possibly even the ability to display the waveforms or phasors. Figure 10.36 shows a technician using a power analyser.

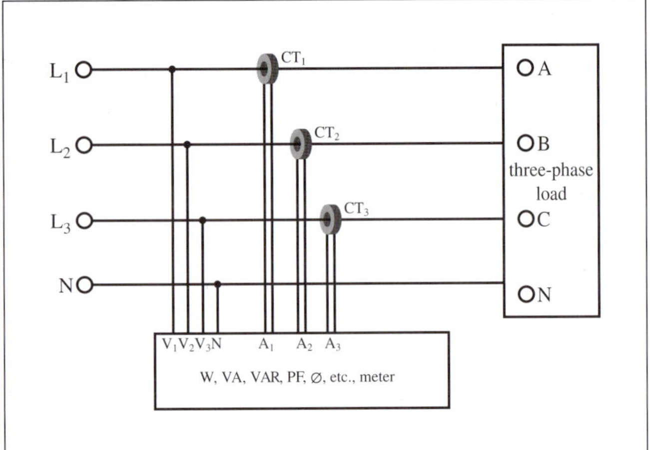

FIGURE 10.34 Electronic three-phase power meter

FIGURE 10.35 Panel power meter
CREATE STUDIO/Shutterstock.com

Advantages

1. It suits any type of load or system (three- or four-wire).
2. It can measure V, I, W, VA, VAR, PF, ϕ, f, THD, etc.
3. No calculations are required.
4. Some units show waveforms and/or phasors.
5. It has robust construction.
6. It offers good environmental protection, often IP56 or better.
7. Once set up, measurements are selected by push button.

Disadvantage

1. None, other than it may require a battery!

10.5.8 Volt-ampere reactive (VAR) measurement

In power stations operating under normal conditions, the values of voltage and current are generally too high and unwieldy to be used directly with portable

FIGURE 10.36 Fluke power analyser in use
Mike Scott

instruments. Potential transformers (PTs) and current transformers (CTs) must be used to avoid these high-energy circuits.

The constant measurement of voltage, current, power, VA and VAR, as well as PF, phase angle and frequency, means that the CTs and PTs must be permanently installed in the conductors. The instruments must be installed in a suitable location such as a power station control room.

Reactive power measurement is often required when a power station supplies a grid system. There may also be more than one generating station supplying power to the same grid. It can be desirable to know the amount of reactive power that may be circulating between power stations and the grid. To keep track of this, a readout is provided in a power station's control room.

Under 'Conditions of supply to consumers', suppliers try to ensure that the overall power factor is no worse than 0.9 lagging. This reduces the amount of VAR circulating in their supply mains. Somewhat ironically, at the same time, due to the capacitive effect between long-distance transmission lines, the power station also has trouble with leading power factors. The VAR load travels up the transmission lines with a circulating current and can cause unwanted rises in voltage at the end that is remote from the power station. The problem can be even worse with underground transmission lines.

10.5.9 Handheld wattmeters

Handheld wattmeters are similar in size and shape to a multimeter. They are battery-powered electronic devices that provide a digital readout. Range selection is by a rotary switch. The meter uses RMS values of current and voltage, irrespective of the actual waveform. It has an accuracy of around 5% of the readout, a figure which is sufficiently accurate for a portable instrument.

The rotary switch has three sections: voltage, current and power. Maximum ranges are up to around 750 V and 20 A, giving a power range from 400 W to 15 kW. The instrument has the added advantage of being comparatively accurate from 15 Hz to 1 kHz.

Because there are three groups of readings, individual readings of voltage, current and power can be obtained. Some models also indicate the displacement, if any, between voltage and current.

10.5.10 Bench-type wattmeters

Bench-type wattmeters have a far higher degree of accuracy than the portable version. They are also far more expensive and so are rarely taken into the field—they are generally kept in the protective environment of the workshop.

They are usually 230 V mains powered, but later models are electronically operated. With an analogue readout, the operating frequency is usually from d.c. ($f = 0$) to around 15 kHz. Current ranges are up to 10 A, with a maximum voltage of 1000 V. This gives a maximum power range from 250 mW to 10 kW.

10.5.11 VA and VAR meters

VA is of course simply the product of V and I, which digital electronics with computerised instrumentation can easily calculate. VAR is calculated using Pythagoras' theorem, or $(VA)^2 = P^2 + VAR^2$, i.e. $VAR = \sqrt{(VA^2 - P^2)}$. This means that once P has been measured, VA and VAR are relatively easy to calculate inside the instrument if a microcomputer is used.

10.5.12 Power factor meters

The same method as above can be used to calculate power factor. However, another method in analogue electronics uses the zero crossing point of sine waves. That is, when the voltage waveform crosses zero in the positive direction, a circuit called a 'flip-flop' goes positive. Then, when the current waveform crosses zero and goes negative, the flip-flop goes negative again. That should happen 180° later. The time that the gate is on for is smoothed to make a voltage that is proportional to the power factor. Half the voltage is a power factor of unity, while less than half is leading and more than half is lagging.

10.5.13 Energy meters (kWh meters)

Energy meters are power meters that integrate the power over time. In other words, the energy meter adds the power to increase a value that is stored in memory. Most energy meters are much more complex than that, being able to record a vast amount of data for later harvesting and analysis by technicians.

10.5.14 Frequency meters

Digital electronics are perfect for frequency measurement. Of course, the circuit could count the cycles every second and give a readout each second, but they are too fast for that. Digital instruments can count the period of the cycle and invert the result to give an answer in hundredths of a hertz, 50 times a second.

10.5.15 High-frequency wattmeters

Wattmeters intended for use on frequencies well above powerline frequencies use different principles of operation. Most rely on the heating effect of the current flowing in the circuit. The heat produced generates a voltage proportional to the temperature of a thermocouple. The voltage is then processed and indicated on a meter, whether analogue or digital.

10.5.16 Ultra-high-frequency wattmeters

For frequencies in excess of 300 MHz, parallel-line meters are used. One of the parallel lines has the load current flowing through it. The other line has a voltage induced in it. This voltage is rectified and read against the scale of a meter calibrated for that frequency.

10.5.17 Total harmonic distortion meters

While not something that is typically shown on power meters, THD can be readily monitored with modern technology to display harmonics, spikes and other issues with power supplies. Harmonics can have disastrous effects within an electrical installation, and technicians need to be familiar with the use and operation of THD meters.

10.5.18 Power factor improvement

Power factor improvement is theoretically the same for three phase as it is for single phase. The major difference is in the location of the power factor improvement equipment. For single phase, this is more often in the equipment connected at the supply terminals. For three-phase installations, the power factor equipment can be a separate component that is usually located close to the point of supply. Three-phase power factor correction equipment is also generally three- or four-wire when incorporated into three-phase components (see Figure 10.37).

FIGURE 10.37 Types of power factor correction capacitors
© KYOCERA AVX Components Corporation. Power Factor Correction (PFC) Capacitors. Dec. 2021, www.kyocera-avx.com

CHECK YOUR UNDERSTANDING

10.20 Why is electrical power generally transmitted at a high voltage?

10.21 Why are transmission systems generally 'ring fed'?

10.22 A three-phase 400 V motor draws 15 A at a lagging power factor of 0.89. How much power is consumed?

10.23 A single-phase wattmeter connected to a balanced four-wire load is showing 50 W. What is the total power taken by the system?

10.24 Identify five things that a power analyser can measure.

10.6 Fault-loop impedance

10.6.1 Fault-loop impedance of an a.c. power system

When a fault occurs in an installation, there is the risk of the exposed conductive parts of the installation becoming live. That would be dangerous. The provision of an earth connected to all the exposed conductive parts of an installation creates a low-impedance conductive path to conduct this hazardous current. Under fault conditions, this ensures that sufficient fault current flows and the circuit protection operates (open circuits) and de-energises the fault. This path is called the 'fault-loop' and its impedance is called the 'fault-loop impedance'. It consists of the:

- impedance of the windings of the electrical supply transformer
- impedance (resistance and reactance) of the supply authority service line active
- impedance (resistance and reactance) of the consumer mains active
- impedance (resistance and reactance) of the submain active (if any)
- impedance (resistance and reactance) of the final subcircuit active
- impedance (resistance and reactance) of the fault, which is generally negligible for the determination of the fault-loop impedance
- impedance (resistance and reactance) of the final subcircuit protective earth
- impedance (resistance and reactance) of the submain earth (if any)
- impedance (resistance and reactance) of the M.E.N. connection, which is generally negligible for the determination of the fault-loop impedance
- impedance (resistance and reactance) of the consumer mains neutral
- impedance (resistance and reactance) of the supply authority service line neutral.

The dashed red line of Figure 10.38 highlights the fault-loop impedance of an installation under fault conditions.

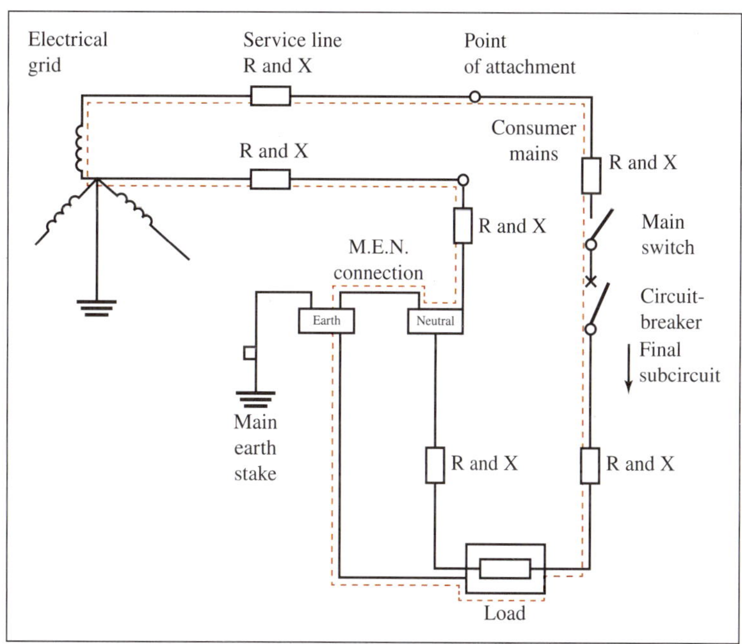

FIGURE 10.38 Fault-loop impedance under fault conditions

AS/NZS 3000:2018 has guidance on (and examples of) general arrangements for installations. These determine appropriate conductor size and associated protection ratings. The particular requirements for fault-loop impedance are covered in paragraphs B4 and B5.

Paragraph B4.4 outlines the components of the fault-loop and states that it is determined by the sum of the impedances of all these components (calculation B3). B4.4 also states a demarcation point between the 'easily known' and the 'not so easily known' parts of the fault-loop. The 'easily known' part is referred to as the 'internal fault-loop impedance' (Z_{int}); the 'not so easily known' part is referred to as the 'external fault-loop impedance' (Z_{ext}). The boundary between these points is the protective device that the final subcircuit originates from. Z_{ext} is upstream from the protective device and includes the components that may be known. These include the submain and consumer mains impedance as well as perhaps less-well-known components such as service line, distribution, supply transformer and transmission impedance. These are the responsibility of the supply authority as they are beyond the point of supply (or point of attachment). However, the acceptable internal fault-loop impedance (Z_{int}) directly relates to the circuit under scrutiny and the impedance is in the control of electrical workers. A simple diagrammatic representation of the fault-loop impedance is shown in figure B5 of AS/NZS 3000:2018.

Under fault conditions, it is assumed that there will be at least 80% of the supply voltage at the protective device. This is the case even with an excessive voltage drop due to the high fault current. This means the Z_{int} can be calculated when designing a circuit by the equation in paragraph B5.2.1 in AS/NZS 3000:2018:

$$Z_{int} = \frac{0.8 \times U_o}{I_a}$$

where:

U_o is the nominal phase voltage (230 V) and I_a is the value of current required to trip the protective device.

10.6.2 Determining fault-loop impedance using resistance and reactance values

Although outside the scope of this textbook, the impedance of cables for voltage drop can be determined by Tables 30 to 39 in AS/NZS 3008.1:2017. The same method can be applied to determine the fault-loop impedance for the final subcircuit. The difference is that the protective earth conductor for the final subcircuit must be considered instead of the neutral. This is because the neutral is excluded from the fault circuit under the worst-case-scenario fault condition.

10.6.3 Measuring fault-loop impedance of typical circuits

There are two methods for measuring fault-loop impedance:

1. Using a fault-loop impedance tester. This type of tester is connected to a live circuit under normal conditions. It measures the complete fault-loop impedance (Z_s). Values obtained using a fault-loop impedance tester should be compared to Table 8.1 in AS/NZS 3000:2018 for verification of compliance. These are known as 'hot' values as the circuit is energised and can be up to the temperature limits stated in AS/NZS 3000:2018, Table 3.2.
2. Using an ohmmeter. This is the method outlined in AS/NZS 3017 Electrical Installations – Verification Guidelines. It gives a value of resistance for the internal part of the fault-loop (Z_{int}), and the values obtained should be compared with AS/NZS 3000:2018, Table 8.2. The values in Table 8.2 are 64% of the values in Table 8.1. This is due not only to the reduction in voltage at the protective device (80%) but, since the conductors are de-energised and at ambient temperature, their resistance is reduced to 80% of their hot value (i.e. $0.8 \times 0.8 = 0.64 = 64\%$). They are therefore known as 'cold' values.

10.6.4 Fault-loop impedance testers

Under existing self-regulatory conditions for electrical workers, the worker is expected to check the fault-loop conditions for an installation. The fault-loop is the path from the fault to the installation earth bar, back to the transformer star point, then through the transformer winding, along the supply active and into the installation to the point of the fault. Fault-loop impedance testers are designed to give an indication of the impedance of that path.

The tester itself may be a combination of insulation resistance tester and continuity tester. It may also perform other functions such as testing the operation and effectiveness of any residual current device in the installation. The loop test places a known resistor between one phase of the installation and measures both the no-fault voltage and the voltage under simulated fault conditions. This is then processed in the instrument, providing a reading in ohms impedance for the loop back to the supply via the earth return.

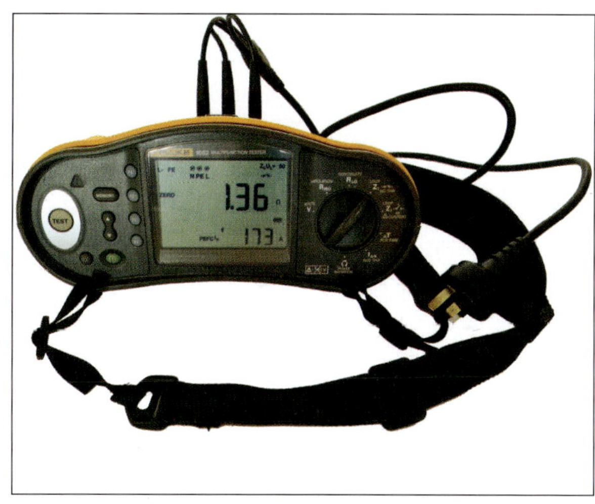

FIGURE 10.39 **Fault-loop impedance tester**
Mike Scott

For safety, and also for the operation of residual current devices, a good earth return is generally considered a must. With modern electronic devices, the testing of the loop impedance can cause tripping and a possibility of damage to the device itself. In either case, the effects of an interruption to the supply service can vary from a minor nuisance in domestic cases to a major problem involving safety in an industrial installation.

Figure 10.39 shows a typical fault-loop impedance tester. This instrument can also test for prospective fault levels within an installation. Typically, these testers have two types of leads: one with a male plug for testing socket-outlets and one with alligator-type clamps for testing electrical equipment and apparatus.

10.6.5 Procedures for testing fault-loop impedance

When using a fault-loop impedance tester, the tester is connected to an energised circuit and then a reading is taken. WHS regulations stipulate de-energised testing in most circumstances. AS/NZS 3017 provides guidance and procedures for carrying out this kind of testing safely.

Testing the fault-loop impedance of circuits supplying socket-outlets that are NOT protected by RCDs is mandatory in AS/NZS 3000:2018. It must be done prior to connecting the circuit to the supply. AS/NZS 3017:2018 outlines a detailed step-by-step process for doing this, as follows:

1. Verify that the circuit to be tested is isolated from the supply.
2. Bridge the active and protective earth conductor of the circuit to be tested *at the origin of the circuit* (i.e. in the switchboard that the circuit originates from).
3. Measure the resistance between the active and protective earth contacts on the furthest point on the circuit.
4. Compare the reading with the values in AS/ZNS 3000:2018, Table 8.2.

Table 8.2 also has values for the individual active and protective earth conductors (R_{ph} and R_e). The time taken for testing can be reduced by incorporating the test to verify the resistance of the protective earth conductor's resistance into the test to verify fault-loop impedance.

CHECK YOUR UNDERSTANDING

10.25 Why is a star connection better than a delta connection for the consumer side of a supply?

10.26 A three-phase 400 V motor draws 15 A at a power factor of 0.87. Calculate the power consumed.

10.27 When connected to a three-phase motor, two wattmeters gave readings of $W_1 = 4$ kW and $W_2 = 2$ kW. Find the total power consumed and the power factor of the motor.

10.28 State five advantages of a digital power analyser.

10.29 Draw a diagram showing the fault-loop impedance under fault conditions.

SUMMARY

- As single-phase alternators do not use the whole cycle of the driving machine, the output is only 63% efficient.
- Two-phase systems have two voltages locked together at 90°E and efficiency increases to 90%.
- Three-phase systems have three voltages locked together 120°E apart and efficiency increases to 94.5%.
- Four-phase systems have too small an advantage over three-phase systems to be of any economic interest.
- Three-phase systems are more cost-efficient than any other multiphase system.
- Three-phase systems are more mechanically efficient than two-phase or single-phase systems.
- The phase sequence of a three-phase system determines whether systems can be interconnected. Phase sequence also determines the direction of rotation of induction motors and similar machines.
- Three-phase motors can start without additional circuitry as three-phase systems have a phase sequence.
- Phases are designated A, B and C and identified by the colours red, white and blue.
- Lines are generally designated L_1, L_2 and L_3.
- Three-phase systems have two possible connections—star and delta. Both have their advantages and have different current and voltage ratios between line and phase values.
- Both star and delta systems can be either balanced or unbalanced.
- An unbalanced delta system tends to be unstable.
- Windings must be connected correctly in star to prevent reduced voltages.
- Windings must be connected correctly in delta, otherwise high voltages occur at open terminals or high circulating currents occur if the triangle is closed.
- Three-phase balanced system power is found from

  $P_T = \sqrt{3}\, V_L\, I_L \cos \phi$ where V and I are line values and cos ϕ is the power factor.
- In unbalanced systems, the power in each load must be added to find total power.
- With a star-connected system in balance, there is no neutral current.
- If a broken neutral occurs in a star-connected system, voltages and currents change, with the greatest load reducing in voltage and the smallest load increasing in voltage.
- Local (consumer) power is usually distributed with star-connected systems. Delta-connected systems are more usual in high power transmission. Loads such as electric motors can be connected in star or delta configurations at appropriate voltage levels.
- In remote areas, SWER systems are sometimes used to save money. Only one active wire is used, and the earth is used as a return conductor. There are usually two voltages available to the consumer: 230 V and 460 V.
- There are several methods by which three-phase power can be measured. The same methods can apply to the measurement of VAR, as long as due attention is paid to the phase relationship of the voltage used.
- Modern measuring instruments have a far greater selection of available measurements, including V, I, P, Q, S, PF, ϕ, f, THD, etc.
- High-energy-level circuits reduce the risk of shock or dangerous arcs by using CTs and PTs.

END-OF-CHAPTER QUESTIONS

10.1 Explain why the three-phase system is better than a system with one, two, four or more phases.

10.2 Explain how three phases are produced in an electric machine to produce the 230 V 50 Hz three-phase power supply.

10.3 Describe the differences between star and delta connections. Show the voltages in each system for current Australian and New Zealand practices.

10.4 Explain the purpose of the neutral conductor in: (a) a balanced load; and (b) an unbalanced load.

10.5 Draw a circuit showing methods of measuring power in a balanced three-phase system.

10.6 Draw a circuit showing methods of measuring power in an unbalanced star-connected, three-phase distribution system.

10.7 A three-phase alternator has three separate 230 V windings. Calculate the line and phase voltages for: (a) star connection; and (b) delta connection.

10.8 A three-phase alternator has a line voltage of 400 V. Find the phase voltage and current for a balanced load consisting of 3 × 100 Ω resistors connected in: (a) star connection; and (b) delta connection.

10.9 A three-phase star-connected alternator supplies an induction motor at a line voltage of 400 V. The current in each delta-connected motor coil set is 40 A. Find:

(a) the line and phase voltages and line current of the motor
(b) the phase voltage and current in each phase of the alternator.

10.10 A three-phase delta-connected alternator supplies a star-connected induction motor at a phase voltage of 230 V. The current in each coil set is 25 A. Find the:

(a) line and phase voltages and phase current of the alternator
(b) line voltage and current in each phase of the motor.

10.11 A 220 V three-phase motor can be connected in star or delta (with 220 V on each coil set). What are the power-supply options for this motor?

10.12 A three-phase machine is connected in delta. If the line currents are 5 A each, find the phase current in each winding.

10.13 Three heating elements can be connected in star or delta to a three-phase 230/400 V system. If they are in star and 10 A flows in each resistor, determine:

(a) the resistance of each heating element
(b) the power drawn by the three elements connected in star
(c) the power drawn by the three elements connected in delta.

10.14 In a three-phase four-wire system, the line currents are each 100 A. The power factors in the three lines are (a) 0.9 (b) 0.8 (c) –0.9 (leading). Determine what (if any) current flows in the neutral conductor.

10.15 The power input to a 230/400 V three-phase induction motor is measured by the two-wattmeter method. The wattmeters show readings of W_1 = 13.5 kW and W_2 = 7.5 kW—both positive. Calculate the line current, power input and power factor of the motor.

10.16 A factory supplied by a 230/400 V three-phase system has three lighting circuits, each consisting of 120 x 40 W fluorescent lamps with a power factor of 0.99 on a 20 A breaker.

(a) Is that too many lamps on a 20 A circuit?
(b) What is the actual power supplied to the lighting circuits?
(c) If 20 lamps fail to operate on B-phase and 30 lamps fail to operate on C-phase, what are the circuit currents and neutral current?

10.17 Three resistors of 10, 20 and 30 Ω respectively are connected to a 230/400 V three-phase supply in both star and delta configuration. Find the line current and total power drawn from the supply in each case.

10.18 List the circuit components that make up the fault-loop impedance of a final subcircuit that has a short of negligible impedance from the active to the earthed frame of an appliance.

10.19 When designing and/or installing a final subcircuit, the installer must consider the internal fault-loop impedance. What device is the demarcation point from which the internal fault-loop can be determined?

CHAPTER 11

Single- and three-phase transformers

LEARNING OBJECTIVES

- Define the terms 'transformer', 'primary winding', 'secondary winding' and 'magnetic core'
- Describe typical single-phase and three-phase transformer construction
- Understand the operating principle of a two-coil transformer
- Explain turns ratio, voltage ratio and current ratio
- Describe typical failure modes and testing transformers for fault finding and commissioning
- Define and calculate transformer losses, including iron losses and copper losses
- Understand how to sketch the phasor diagram of a transformer under no-load and load conditions
- Define and calculate transformer efficiency, voltage regulation and percentage impedance
- Describe transformer connections, including winding polarity, paralleling and three-phase configurations
- Identify power transformer components
- Describe cooling methods employed in power transformers
- Understand other special-use transformers such as multiple secondaries, tapped windings, autotransformers (e.g. variacs), isolation transformers and high-reactance and flux leakage types
- Describe instrument transformers and associated safety considerations

PREREQUISITE KNOWLEDGE

- Electrical energy transmission and distribution
- Measuring electricity
- Electromagnetism, including the right-hand grip rule
- Electromagnetic induction, including Lenz's Law and mutual induction
- Inductors and inductance
- Three-phase star and delta connections

INTRODUCTION

The **transformer** is one of the most important electrical devices ever invented. In fact, without them, our present electrical transmission and distribution system would not exist.

A simple double wound transformer consists of two separate windings (coils) wound with insulated copper wire. The *primary winding* is connected to a source of electrical energy and the *secondary winding* to a load. Typically, an iron core is used to increase the magnetic field density and efficiency of the transformer. A small, single-phase, double wound transformer is shown in Figure 11.1(a); note the two windings (coloured red and blue).

Electrical energy is transferred from the primary to the secondary windings by a phenomenon called 'mutual induction' (also known as 'transformer action'). Many transformers are fully reversible in operation, so the winding connected to the source of supply is *always* referred to as the 'primary winding'.

Two very different transformer symbols are shown in Figure 11.1. Figure 11.1(b) shows the circuit diagram symbol for a double wound transformer with an iron core, and Figure 11.1(c) presents the 'general' transformer symbol that is found in single-line diagrams. Other transformer types have different symbols.

Transformers are primarily used to step between different magnitudes of voltage or current. The voltage of the secondary winding can be made to be higher, lower or the same as the primary supply voltage. If higher, it is called a 'step-up transformer'. If lower, it is called a 'step-down transformer'. If it is the same voltage, it is referred to as a 'one-to-one' or an 'isolation transformer'. The transformer shown in Figure 11.1(a) *steps down* the primary voltage from 230 V to 12 V.

Transformers range in power rating from volt-ampere (VA) size to hundreds of megavolt-ampere (MVA), with operational efficiency over 99% in **transmission** and **distribution transformers** (collectively known as 'power transformers'). This level of efficiency basically means that power out equals power in, and is far higher than any other electrical or mechanical machine.

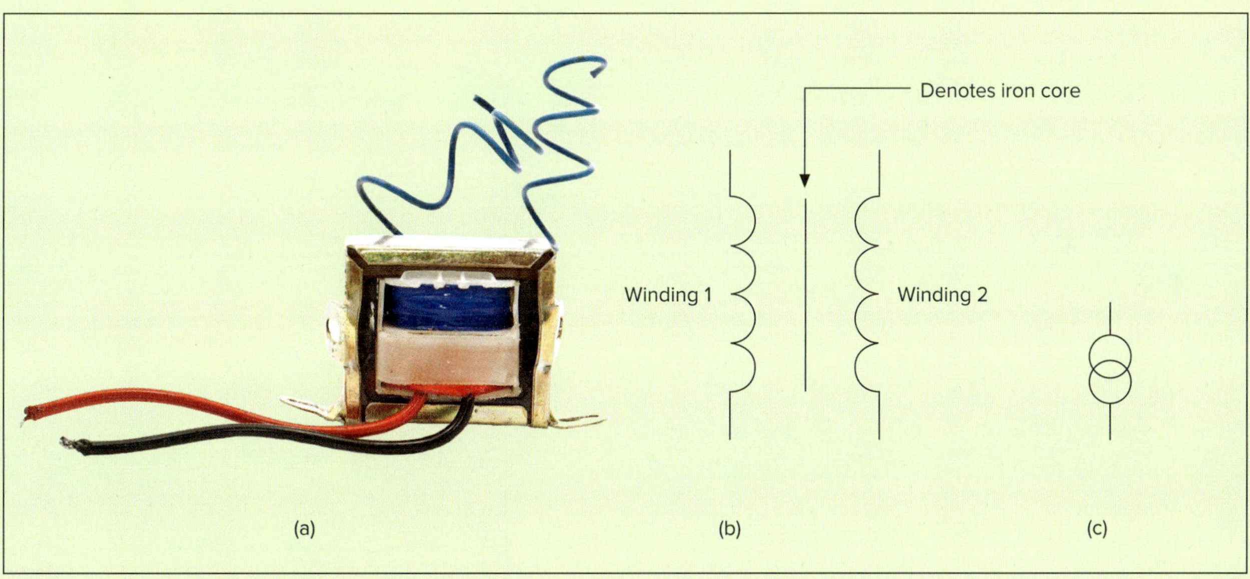

FIGURE 11.1 Transformer and drawing symbols
(a) Dmitriy Sechin/Alamy Stock Photo

11.1 Transformer construction

11.1.1 Types of laminations and core construction

The design and construction of transformers is governed by a number of Australian/New Zealand Standards. For example, AS/NZS 60076 provides general requirements for power transformers.

A transformer consists of a common magnetic circuit linking the primary and secondary windings. The form of core construction is determined by the arrangement of the laminations and the way they are stacked together. The laminations are usually made from high-grade silicon steel of a definite thickness.

Figure 11.2 shows two methods for making up the stack of a transformer core. Figure 11.2(a) shows U–I-shaped laminations, which are stacked in alternate directions to make a *core-type magnetic circuit*. Figure 11.2(b) shows E–I-shaped laminations, also stacked in alternate directions, to make a *shell-type magnetic circuit*.

Although there are several variations of these types of construction, transformers are generally classified as either core or shell. There is also a **toroidal**-type construction (see Figure 11.2(c)).

FIGURE 11.2 **Cores—core, shell and toroidal**

Core

With the core-type transformer, the windings surround the laminated core (see Figure 11.3(a)). To provide a uniform flux density throughout the magnetic core, the cross-sectional area (CSA) of the core is uniform.

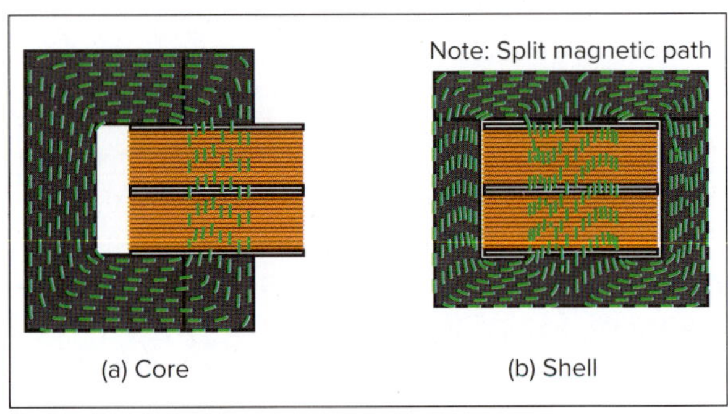

FIGURE 11.3 **Windings—core and shell**

Shell

For single-phase operation, the majority of transformer cores use shell-type construction. This has the magnetic core surrounding the windings (see Figure 11.3(b)). Because the core provides a parallel magnetic path for the flux, the centre limb is twice the CSA of the outer limbs, maintaining uniform flux density throughout the iron core.

By comparison, the core-type construction has a lighter core of smaller CSA but a greater length of magnetic circuit. It also has a relatively greater number of turns but these have shorter mean length. The core type, with its larger window space, is more suitable for higher voltages (those above 1000 V a.c.) that require many turns and a bigger space for insulation. The shell type is particularly suited for moderate voltages (those between 50 and 1000 V a.c.) that require fewer turns, less insulation, larger currents and lower frequencies.

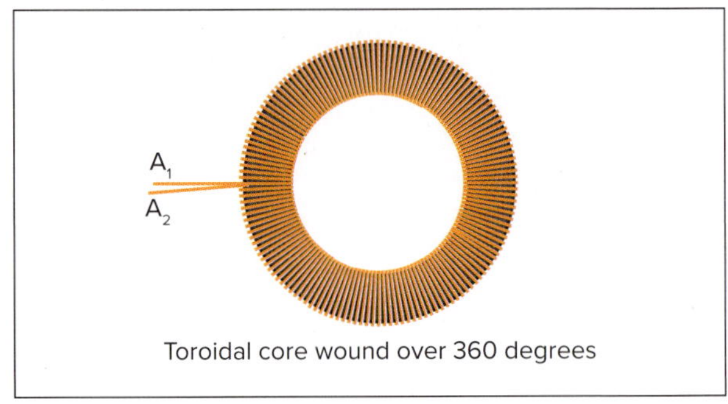

FIGURE 11.4 **Toroidal winding**

Toroidal

The toroidal core is formed from a continuous ribbon of thin metal tape made from a special alloy. It is wound tightly around a former by a special machine that passes the insulated coil wire through the centre of the toroid many times. It is then consolidated under pressure into a solid mass.

One advantage of the toroidal type is that the windings are spaced around the whole core, resulting in a shorter, constant cross-section magnetic path, with very low leakage flux (see Figure 11.4).

A toroidal core may alternatively be sliced into two C-shaped pieces—the finished article is sometimes referred to as a 'C-core'. The cut faces are ground to ensure good surface contact between the two halves. These halves are placed around the transformer windings and clamped with a metallic band under moderate pressure to counter the effect of an air gap.

11.1.2 Three-phase core construction

A three-phase transformer can be obtained using three identical single-phase transformers. Usually, though, a common three-phase magnetic core is used with three identical sets of primary and secondary windings mounted on it. Although the same variations in single-phase cores apply to three-phase cores, the majority of three-phase transformers are of a core-type construction.

Three-phase core type

The shape shown in Figure 11.5(a) is usually employed in smaller distribution-type transformers. The core-type construction has a shorter length per turn of winding than the shell type but has a longer magnetic path. While similar in appearance to the single-phase shell type, each leg of the core has an equal CSA.

Three-phase shell type

This shape of core overcomes the tendency of the core type to have unequal flux densities (see Figure 11.5(b)).

Three-phase cruciform or stepped core

With conductors of large CSA, it becomes difficult to construct windings that have 90° bends in the conductors. The core is shown in cross-section in Figure 11.5(c), and it is evident that a great number of different-sized laminations are required. This form of construction is expensive and is generally used only on large power transformers.

Three-phase toroidal

Generally speaking, toroidal cores can be obtained in most shapes for three-phase transformers (see Figure 11.6).

11.1.3 Arrangement of windings

The actual placement of the windings on the transformer core depends on the type of core and the intended application. Other factors that influence this arrangement are the operating frequency and the size or power rating of the transformer.

Some typical winding layouts are shown in Figure 11.7. While the core-type transformer construction is illustrated in the diagrams, the winding arrangement applies equally to the shell-type construction.

Fig 11.7(a) shows a simple method of placing the primary and secondary windings *side by side*. Although this method would reduce the magnetic leakage more than placing windings on separate

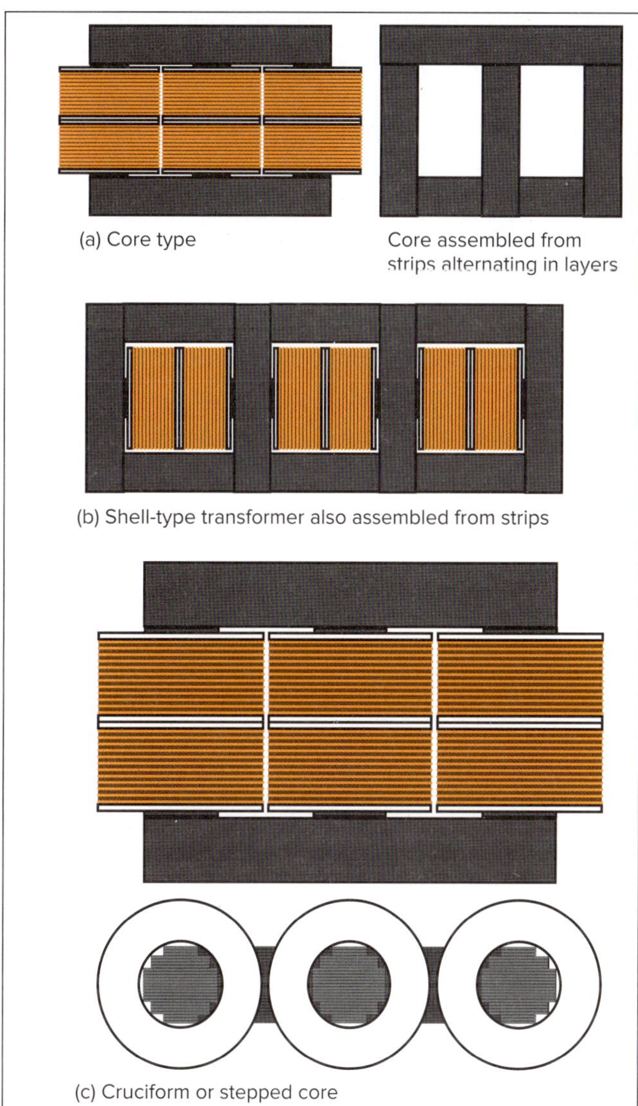

FIGURE 11.5 Three-phase core types

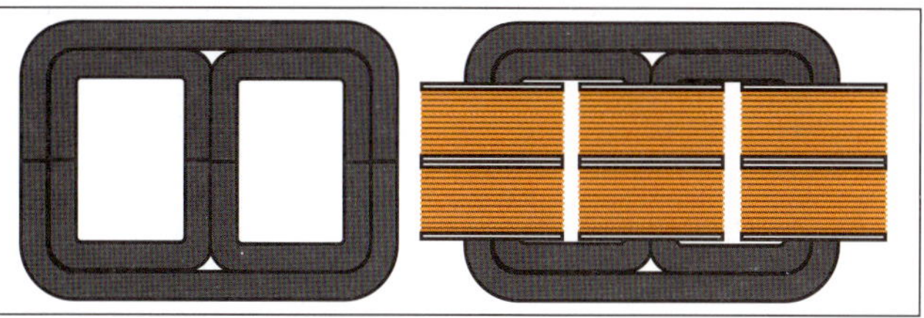

FIGURE 11.6 Three-phase C-core formed from rolled and bonded transformer steel strip

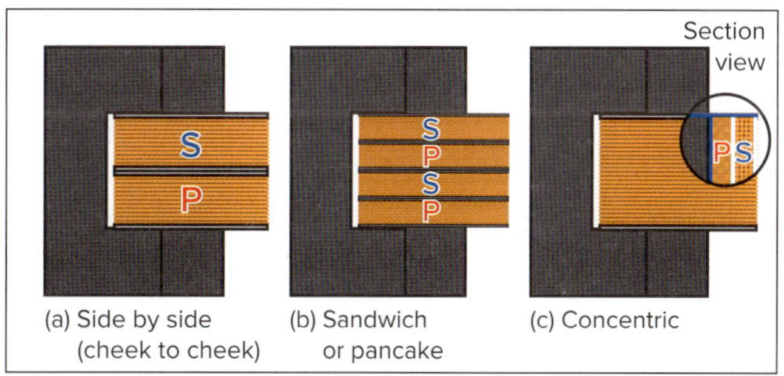

FIGURE 11.7 Winding arrangements

limbs, a better configuration is the *sandwich-* or *pancake*-type (see Figure 11.7(b)). This configuration is used where closely coupled windings are required, so that the magnetic leakage can be reduced to a minimum. The same factors that affect windings and cores for single-phase transformers apply to three-phase transformers. Some distribution transformers are wound in the sandwich or pancake style as the method lends itself to ease of winding handling, construction and repair. The sandwich method is also used in smaller transformers operating at higher audio frequencies.

With the *concentric* method (see Figure 11.7(c)), one winding is wound on the top of the other (primary or secondary) and suitable insulation is installed between the two. This type of winding arrangement is becoming more common for distribution transformers, due in part to Australian Standards recommendations for insulation requirements between primary and secondary windings. The example shown is a step-up transformer as it has the primary, low-voltage winding next to the core and the secondary, high-voltage winding on the outside. The low-voltage winding is always next to the core as it is easier to insulate it from the earthed core than the high-voltage winding. A step-down transformer would have the windings reversed with the secondary (LV) next to the core and the primary (HV) away from the core.

Coupling between windings

The degree to which the primary and secondary windings are magnetically coupled depends on the intended purpose of the transformer. A transformer is said to be 'close coupled' when all the primary flux passes through the secondary turns. If a large proportion bypasses the secondary windings, the transformer is said to be 'loosely coupled'.

There are also intermediate degrees of coupling. For example, a distribution transformer is less than close coupled as a form of current limitation, to allow for the event of damage to overhead lines connected to its secondary. However, the degree of coupling for the transformer of an illuminated sign is far less than that for a distribution transformer. In this case, the on-load voltage must be considerably less than the open-circuit voltage.

11.1.4 Transformer insulation

Insulation oil plays an important role in most power transformer insulation systems, acting as a liquid dielectric and coolant. However, in non-power transformers there is usually no need for insulating oil as heat dissipation is very low.

Several materials are used to provide insulation between the windings, between the windings and the core/frame and for the winding wire itself. These materials include:

- presspaper and pressboard
- lacquer, enamel, polymer or varnish film
- flexible multi-layer insulation materials
- mica.

11.1.5 Transformer nameplate

The *transformer nameplate* (see Figure 11.8) contains important specifications, including power rating and primary and secondary

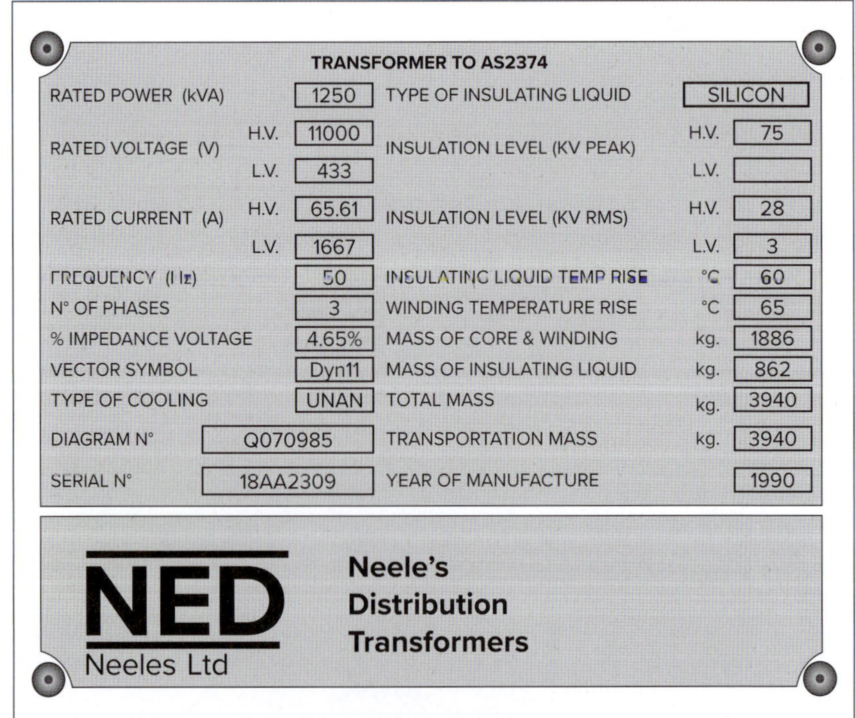

FIGURE 11.8 Distribution transformer nameplate

voltage and current ratings. The following is a list of the standard information found on the nameplate of an oil-cooled distribution transformer:

- manufacturer and serial number
- year of manufacture
- number of phases
- VA rating (kVA, MVA etc.)
- frequency
- voltage rating (primary and secondary)
- current rating (primary and secondary)
- tap voltages
- connection diagram
- cooling class/type
- winding and oil temperature rise
- vector group and diagram
- percent impedance
- approximate weight
- type of insulating liquid
- oil volume.

11.2 Transformer operation

11.2.1 Principle of mutual induction of a transformer

Transformer operation is based on the principle of *mutual induction*. The primary and secondary windings and the magnetic core of the transformer are all stationary in relation to each other.

The primary winding is connected to an alternating supply, causing an alternating magnetic flux to be produced in the magnetic core, with the magnitude of the flux changing over time.

The changing current in the primary winding produces changing flux in both windings, causing a back EMF in the primary winding and an induced voltage in the secondary winding. The general concept is illustrated in Figure 11.9.

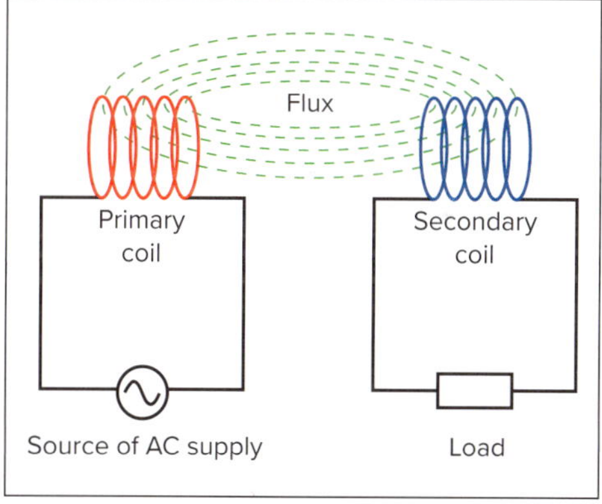

FIGURE 11.9 Mutual induction

11.2.2 Value of induced voltage

The value of an *induced voltage* in a transformer depends on three factors: frequency (f), number of turns (N) and the maximum instantaneous flux (Φ_{max}). Provided that the current waveform, and consequently the flux distribution, is sinusoidal, the equation for the RMS value of induced voltage is given by:

$$V' = 4.44\Phi_{max}\, fN$$

where:

$\frac{2\pi}{\sqrt{2}} = 4.44$ (adjustment for an RMS sine wave)

Φ_{max} = maximum instantaneous flux

f = frequency

N = number of turns.

Since transformer cores are usually designed on the basis of permissible flux density, the equation may also be expressed as:

$$V' = 4.44\ B_{max}\ AfN$$

where:

B_{max} = maximum permissible flux density in weber
A = cross-sectional area of core in square metres.

A point to note is that $\Phi = BA$

11.2.3 Transformation ratios

Turns and voltage ratio

The mutual flux is common to each winding. Therefore, it must induce the same voltage per turn in each winding. If V_1' is the total induced voltage in the primary winding having N_1 turns, then the induced voltage per turn is $\frac{V_1'}{N_1}$. Similarly, the induced voltage per turn in the secondary winding is $\frac{V_2'}{N_2}$.

Under no load, the applied voltage V_1 and the self-induced voltage V_1' are almost equal and $V_2 = V_2'$, so the above ratios are transposed and usually expressed as:

$$\frac{V_1}{V_2} = \frac{N_1}{N_2}$$

That is, on no load, the ratio of the voltages is equal to the ratio of the turns.

EXAMPLE 11.1

A transformer has 1000 turns on the primary winding and 200 on the secondary. If the applied voltage is 250 V, calculate the output voltage of the transformer.

$$\frac{V_1}{V_2} = \frac{N_1}{N_2}$$

$$V_2 = V_1 \times \frac{N_2}{N_1}$$

$$= 250 \times \frac{200}{1000}$$

$$= 50\ \text{V}$$

Current ratio

When the transformer is connected to a load, the secondary current I_2 produces a demagnetising flux proportional to the secondary ampere-turns I_2N_2. The primary current increases, providing an increase in the primary ampere-turns I_1N_1 to balance the effect of the secondary ampere-turns. Therefore, the primary ampere-turns equal the secondary ampere-turns:

$$I_1N_1 = I_2N_2$$

Transformation ratio

Together, the turns ratio, voltage ratio and current ratio can be combined as the overall transformation ratio:

$$\frac{V_1}{V_2} = \frac{N_1}{N_2} = \frac{I_2}{I_1}$$

EXAMPLE 11.2

A transformer has 1000 turns on the primary winding and 500 on the secondary. If the applied voltage is 220 V @ 50 Hz and a purely resistive load of 40 Ω is connected to the secondary, calculate the:

- output voltage of the transformer
- secondary winding current
- primary winding current.

$$\frac{V_1}{V_2} = \frac{N_1}{N_2}$$

$$V_2 = V_1 \times \frac{N_2}{N_1}$$

$$= 220 \times \frac{500}{1000}$$

$$= 110\text{ V}$$

$$I_2 = \frac{V_2}{R}$$

$$= \frac{110}{40}$$

$$= 2.75\text{ A}$$

$$\frac{N_1}{N_2} = \frac{I_2}{I_1}$$

$$I_1 = \frac{N_2 I_2}{N_1}$$

$$\frac{500 \times 2.75}{1000} = 1.375\text{ A}$$

As can be seen in Example 11.2, the turns ratio is 2:1 (1000:500), the voltage ratio is 2:1 (220:110) and the current ratio is 1:2 (1.375:2.75).

11.2.4 Phasor diagram of transformer on no load

Under no-load conditions, i.e. with an open circuit on secondary, the supply voltage is applied to the *highly inductive* primary winding. Because of the relatively low resistance, a pure d.c. would cause a large current to flow, probably burning out the transformer in a very short time. The a.c. current, however, produces a self-induced voltage V_1', only slightly less than the applied voltage and in opposition to the applied voltage.

The only losses are the one required to produce the magnetic field and the current flowing through the resistance of the primary winding. This no-load current (or **excitation** current) is typically very small compared with the full-load current—as low as 1% to 3%, for example.

The excitation current causes an alternating flux, called the 'mutual flux', to be set up in the core linking both primary and secondary windings. The mutual flux causes a voltage to be induced in the secondary winding—the secondary voltage V_2'—but no current can flow until a load is connected (see Figure 11.10).

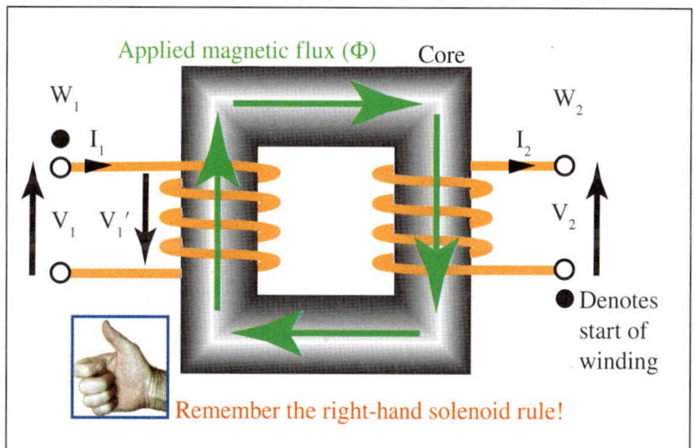

FIGURE 11.10 Self-induced voltage and mutual flux in non-loaded transformer

Parallel circuits use the voltage as the reference phasor and series circuits use the current; in each case the reference phasor is common to all of the components in the circuit. For transformers, the mutual flux Φ produced by the magnetising component is common to both windings and is used as the reference phasor when drawing phasor diagrams.

The excitation current can be resolved into two rectangular components called the 'energy' and 'magnetising' components, as shown in the phasor diagram of a non-loaded transformer in Figure 11.11. The magnetising component of the excitation current is *in phase* with the mutual flux. Both Φ and I_m represent the purely inductive part of the circuit, and thus will *lag* 90°E behind the applied voltage V_1.

This means that with flux as the reference phasor, the voltage will be leading the flux by 90°E. The energy component of current I_e that represents the losses in the iron circuit and the small copper losses (described later) is resistive and will be represented by a phasor *in phase* with the voltage. A wattmeter connected in the primary circuit would show power being used to cover these losses.

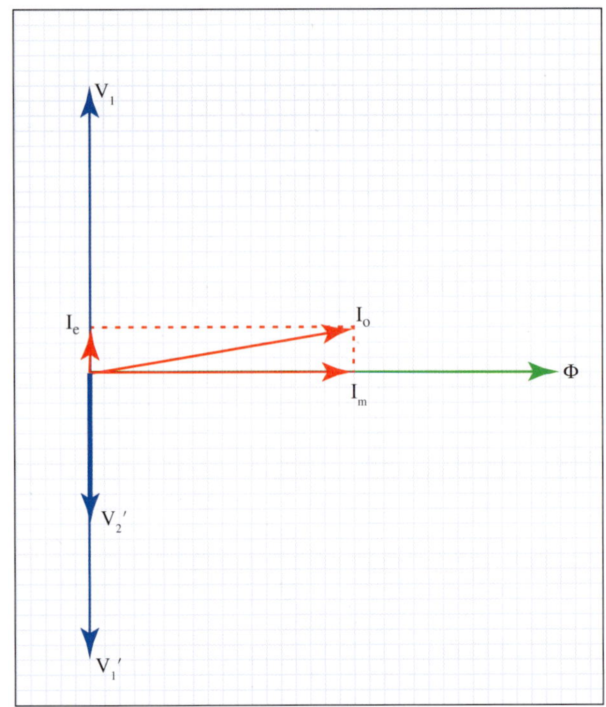

FIGURE 11.11 Phasor diagram for non-loaded transformer

The phasor sum of I_m and I_e add up to the no-load current I_o, and the large angle (perhaps approaching 90°) between V_1 and I_o indicates a very poor power factor for a transformer on no load.

The self-induced voltage V_1' in the primary winding is 180°E out of phase with V_1 since it opposes the applied voltage.

11.2.5 Phasor diagram of transformer under load

When a load is applied to the secondary terminals, a secondary current I_2 will flow. Its magnitude and phase relationship with the secondary terminal voltage V_2 is determined by the type of load.

Lenz's Law tells us that the direction of this secondary current I_2 will always oppose any change in the flux Φ. In Figure 11.12, W_1 is the primary winding with the start of the winding marked by a solid dot '•'.

Assume that at a particular instant in time the primary current I_1 flows from the start to the end of the winding, establishing a flux with a magnetic polarity in a clockwise direction around the iron core as shown. This flux is mutual to both coils.

The mutual flux causes a reaction current in both coils, which has the effect of opposing the establishment of the mutual flux. This can be seen as an opposing reactive flux, but the total effect is to reduce the mutual flux, thus reducing the self-induced voltage V_1' in the primary and allowing more current to flow in both the primary and secondary.

All of these events happen together. The application of a load draws a current in the secondary winding, causing a demagnetising flux and reducing the mutual flux. The self-induced voltage in the primary decreases; the primary current increases; the mutual flux rises to its original value. In practice, the mutual flux in the iron core of a transformer effectively stays at a constant value for all loads.

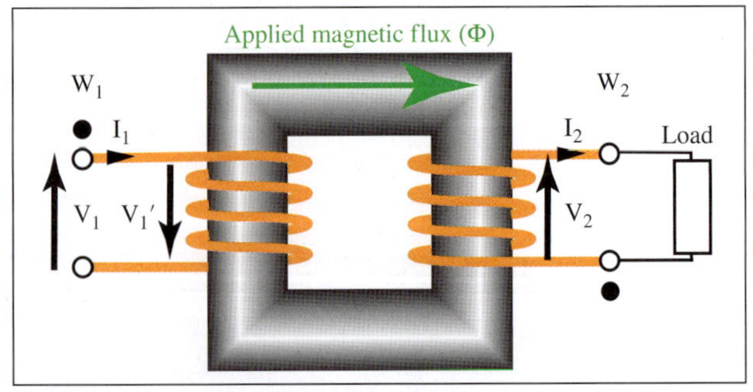

FIGURE 11.12 Loaded transformer

It is important to note that an increase in secondary load current causes an increase in primary line current.

The phasor diagram in Figure 11.13 shows the general case for a transformer on load. Assume for the purposes of the diagram that the transformer has a turns ratio of 2:1 and that the connected load is inductive, so that the secondary current I_2 lags behind the induced voltage V_2' by the phase angle Φ_2.

The equivalent current to supply this load will be the value I_1'. If the transformer were 100% efficient, this value of primary current would be the actual current flowing into the transformer from the supply. Since the excitation current I_o is already flowing in the primary windings to cover core losses, the total primary current will be the phasor sum of these two currents ($I_1' + I_o$). The phasor sum of I_1' and I_o gives the actual primary current of I_1 flowing at a lagging phase angle of Φ_1. It should be noted that the excitation current has been enlarged for the sake of clarity and copper losses in the windings are considered negligible.

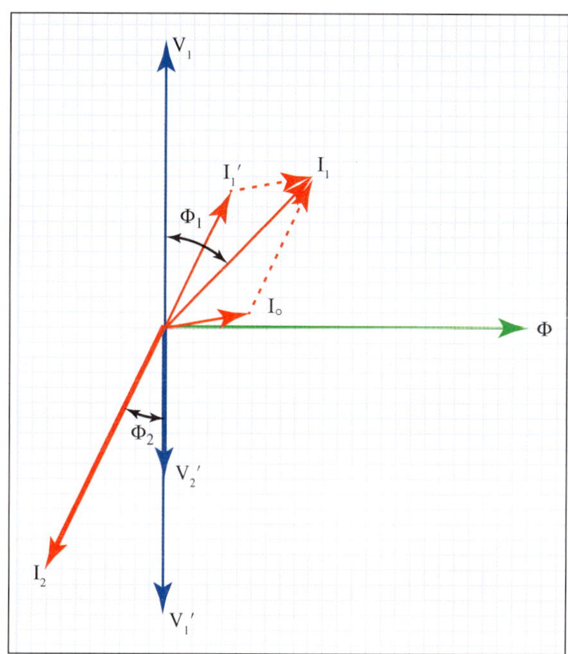

FIGURE 11.13 Phasor diagram for loaded transformer

11.2.6 Failure modes of transformers

Common failure modes of transformers include:

- *earth faults*—insulation failure leading to a winding or connection shorting to earth
- *interwinding faults*—insulation failure leading to shorting between windings
- *shorted turns*—adjacent turns of a winding short together, leading to a change in resistance and operational characteristics
- *open circuit*—faulty connection or break in the winding.

Conditions leading to failures include normal ageing, environmental factors such as high ambient temperature and moisture ingress, internal overtemperature, overvoltage transients, overload and poor workmanship.

11.2.7 Performing basic electrical tests

Several basic tests need to be performed on a transformer when fault finding or prior to putting it into service. These are carried out to ensure that there are no open/short-circuits or other problems. Basic tests include:

- *a visual inspection* is necessary to identify any signs of overheating, mechanical damage or other potential problems.
- *an insulation resistance test* (using an insulation resistance tester) should be performed between each winding and earth and between each of the windings to ensure the insulation resistance meets minimum requirements.
- *a continuity test* (using an ohmmeter) should be performed to ensure that there are no open circuits in the windings. These results may also assist in winding identification.

For power and distribution transformers, it will be necessary to perform a range of specialised tests, including winding resistance using a high-resolution ohmmeter or other suitable instrument.

11.2.8 AS/NZS 3000:2018 requirements for transformers

For the installation of transformers, AS/NZS 3000:2018 predominantly references transformers in Clause 4.14 and Clause 7.4. Clause 4.14 (Transformers) details the installation requirements for equipment connected to secondary circuits, control and protection, isolating transformers and autotransformers. Clause 7.4 of AS/NZS 3000:2018 relates to protection of electrical circuits by electrical separation (isolated supply), including necessary testing and verification (7.4.8). Clause 2.7.2 of AS/NZS 3000:2018 refers to the requirements regarding protection by insulation or separation—specifically the provision of adequate insulation, screening or separation of windings.

WORKPLACE SCENARIO

Transformer inrush (or start-up) current can be as high as 20 times its nominal current (I_N). This means specific circuit-breakers and fuses must be chosen to protect transformers against overload and short-circuit currents.

An electrical contractor was asked to determine why a new motor starter circuit that incorporated an autotransformer was tripping the circuit-breaker upon energisation. It was discovered that the original circuit-breaker was not designed for the large magnetising current produced by inductive loads such as autotransformers. The circuit-breaker was replaced with one of the same nominal current value but with a 'D'-type instantaneous inverse time-current curve.

CHECK YOUR UNDERSTANDING

11.1 Draw the three types of basic transformer construction: core shell and toroidal.

11.2 Name and describe the most common type of winding arrangement used in three-phase distribution transformers.

11.3 Transformer characteristics

11.3.1 Transformer power and current ratings

Engineers design transformers for a specified voltage ratio and current (power) capacity, but once a transformer is placed in service, the load placed on it is beyond the immediate control of both the designer and the electricity network operator.

The actual load depends on the loading of the total number of connected circuits. Transformer design may assume that the individual circuits will not all reach maximum load at the same time, and so the transformer power rating will be less than the total potential load. Further, the load circuits may have a **power factor** that is different from the design expectations and therefore may cause a higher current than a unity power factor. *So, transformers are not rated by power but by their secondary winding voltage and current, which is expressed in terms of apparent power (VA).*

For example, a single-phase transformer capable of delivering 100 A at 500 V would be rated at 500 × 100 = 50 000 VA or 50 kVA. If the power factor of any given load is 0.5, then the maximum safe power output would be 25 kW (now calculated as true power as power factor can be applied). Similarly, at a power factor of 0.8, the safe power output would be 40 kW. In both cases, the full-load current would be 100 A.

EXAMPLE 11.3

Calculate the necessary kVA rating of a transformer for the following load: 20 × 230 V 50 Hz metal halide luminaires, with each fitting drawing 3.2 A at a power factor of 0.87 lagging. The luminaires will be switched on using a time delay between each bank of four.

$$P = VI\cos\phi$$
$$\text{True power} = 230 \times (20 \times 3.2) \times 0.87$$
$$= 12.8\ \text{kW}$$
$$\text{kVA rating} = \frac{12\,806}{0.87} = 14.7\ \text{kVA}$$

Using the above results, a 15 kVA transformer will satisfy the load requirements.

The current rating of the conductors in the windings is dependent on the rate at which the total heat generated in the transformer can be dissipated. The rating limitation of the transformer is a factor of the temperature rise of the unit on load and the ambient temperature. High ambient temperatures result in a lower rating and a low ambient temperature allows a higher rating.

11.3.2 Power losses in a transformer

The power absorbed by the core of a transformer is due to eddy currents and hysteresis, and is referred to as 'iron losses'.

Iron losses—eddy currents

When the alternating flux cuts the steel core, an EMF is induced in each lamination, causing an eddy current to flow in the closed electrical circuit of the lamination. This eddy current flows through the resistance in each lamination, causing heat to be generated in them, and therefore in the core as a whole. Although eddy-current losses are effectively reduced by using laminations for the core, they are never entirely eliminated.

Iron losses—hysteresis

The alternating flux also causes changes in the alignment of the magnetic domains in the magnetic core, with the magnetic polarity reversing one hundred times per second. This change consumes energy and heat is produced within the core. The energy loss is referred to as 'hysteresis loss', the degree of loss being dependent on the nature of the material used for the laminations.

Silicon steel has low hysteresis losses, making it suitable for electrical laminations. Figure 11.14 shows a comparison of two hysteresis curves for different materials. The silicon steel curve has a smaller area, representing a lower energy loss and reduced heat production.

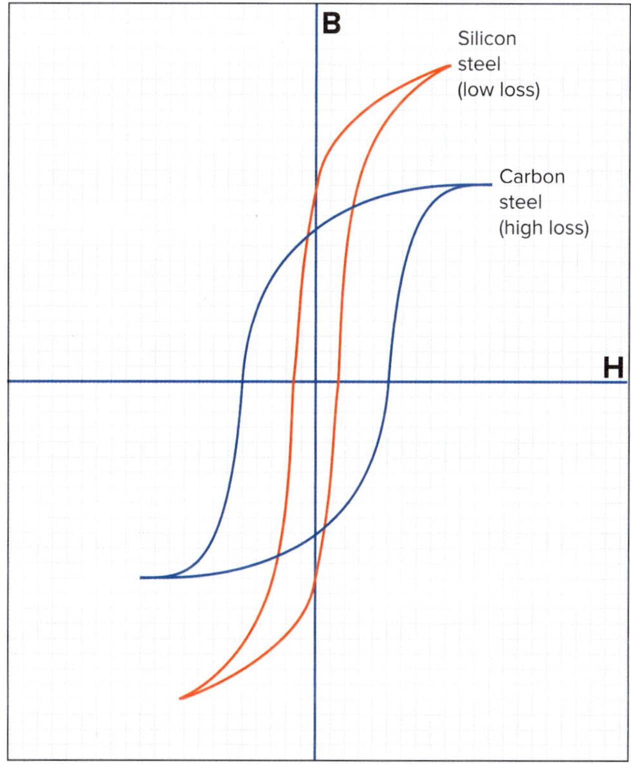

FIGURE 11.14 Hysteresis curves

Copper losses

Another loss that occurs in a transformer is **copper loss**. This is the energy lost in the windings when the transformer is loaded. The resistance of each winding is relatively low, but since the power dissipated in each winding is proportional to the square of the current flowing through that winding (I^2R), it follows that the copper loss is significant when the load current is high.

11.3.3 Determining power losses

To calculate the efficiency of a transformer, it is necessary first to calculate its total losses using two different tests. The first test will determine the iron losses by measuring the power consumed on no load in what is known as a 'no-load test' or 'open-circuit test'. The second test determines the total copper losses and is known as a 'short-circuit test'.

No-load test

The transformer is connected as in Figure 11.15 to a supply at the rated voltage and frequency. The primary current on no load is usually less than 3% of the full-load current, so the primary I^2R loss on no load is negligible compared with the iron loss. The wattmeter reading can then be taken as being the total iron loss of the transformer.

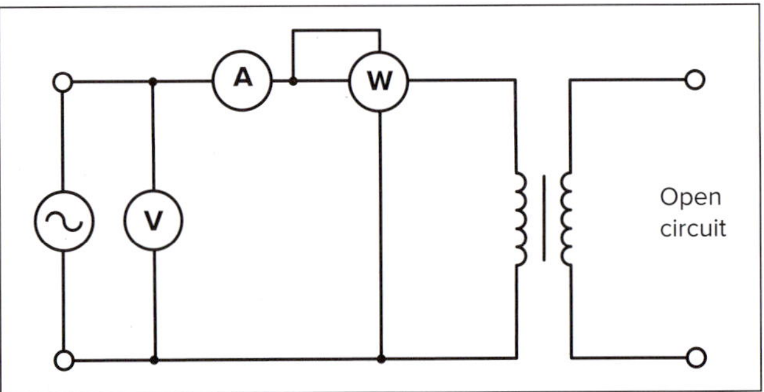

FIGURE 11.15 No-load or open-circuit test

Short-circuit test

The total copper loss is $P_{cu} = I_1^2R_1 + I_2^2R_2$, where R_1 and R_2 are the resistance values of the primary and secondary windings respectively. The copper losses are not constant but change according to the square of the load current. The value of the losses can be obtained by performing the short-circuit test, as shown in Figure 11.16. The typically shaped curve of copper losses can be seen in Figure 11.17.

As Figure 11.16 shows, the secondary winding of the transformer under test is shorted through the ammeter A_2. A **variac** (see section 11.6.3) is used to provide a low-voltage supply to the primary winding of the transformer on test. The output of the variac is increased until the full rated current flows in the primary and secondary circuits.

The supply voltage to the transformer is low, and the flux in the iron core is also low so the iron losses are negligible. The power registered on the wattmeter can be taken as the total copper losses in the transformer on full load.

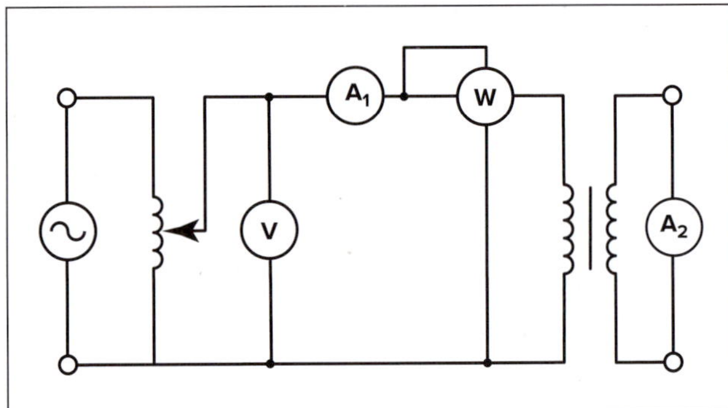

FIGURE 11.16 Transformer short-circuit test

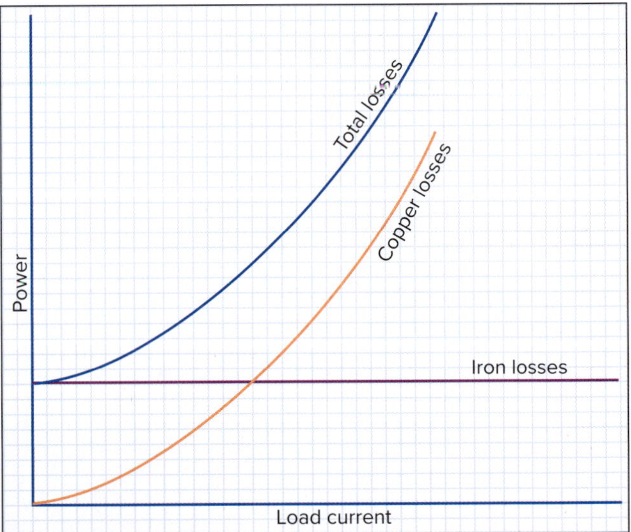

FIGURE 11.17 Transformer losses

11.3.4 Determining transformer efficiency

The efficiency of any machine is expressed as:

$$\eta = \frac{\text{output}}{\text{input}}$$

A transformer normally has a high efficiency, so the difference between the output and input readings is very small (typically 1% to 3%). The efficiency is usually determined from the losses.

$$\eta = \frac{\text{output}}{\text{input}} = \frac{\text{output}}{\text{output} + \text{losses}}$$

That is:

$$\frac{V_2 I_2 \cos \phi_2}{(V_2 I_2 \cos \phi_2 + P_{Cu} + P_{Fe})}$$

where:

P_{Cu} = copper losses
P_{Fe} = iron losses.

Assuming the output voltage V_2 remains constant, the only variables affecting the efficiency of a transformer are load current and power factor.

EXAMPLE 11.4

A single-phase 11 kV/230 V transformer supplies an inductive load current of 30 A at a power factor of 0.9 lagging. A short-circuit test is performed and the wattmeter indicates that the copper losses at full load are 330 W. An open-circuit test reveals the iron losses to be 230 W. Calculate the efficiency of the transformer.

$$\text{Power out} = 230 \times 30 \times 0.9 = 6210$$
$$\text{Power in} = \text{Power out} + \text{losses} = 6210 + 330 + 230 = 6770$$
$$\eta\,\% = \frac{\text{output}}{\text{output} + \text{losses}} \times 100 = \frac{6210}{6770} \times 100 = 91.73\%$$

The loading of a transformer will have an impact on its efficiency as it will cause a change in copper losses. The copper losses will change by a multiple of the square of the load. For example, at half load, the copper losses would be 0.5^{2}, which is 0.25, a quarter of the full-load losses.

It can be shown that the maximum efficiency occurs when the copper losses are equal to the fixed iron loss value—in this case, 230 W. This occurs at a loading of approximately 83.49% (0.8349) of full load. We can show this by calculating the copper losses.

$$\text{Copper losses} = \text{loading}^2 \times \text{copper losses at full load}$$
$$\text{Copper losses} = 0.8348^2 \times 330 = 230 \text{ W}$$

We can carry out the same calculation as in Example 11.4 and show the increase in efficiency at this loading.

$$\text{Power out} = 230 \times 30 \times 0.9 = 6210 \times 0.8349 \text{ (loading)} = 5184.729$$
$$\text{Power in} = \text{Power out} + \text{losses} = 5184.729 + 230 + 230 = 5644.729$$
$$\eta = \frac{\text{output}}{\text{output} + \text{losses}} \times 100 = \frac{5184.729}{5644.729} \times 100 = 91.851\%$$

Loadings either side of 83.48% would have a slightly lower efficiency. You can prove this for yourself by calculating the efficiency for 0.82 and 0.85. The difference will be minimal but noticeable if you calculate to three decimal places. This can be seen in Figure 11.18. The maximum efficiency occurs when the copper losses rise to the fixed-loss value of the iron losses (230 W).

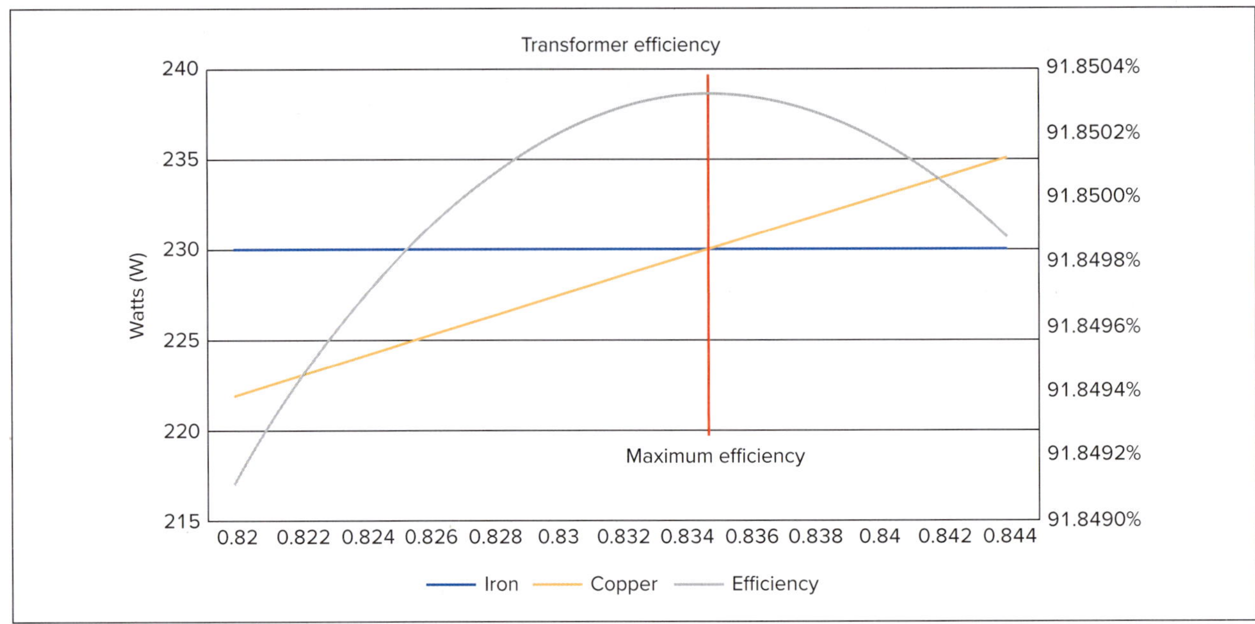

FIGURE 11.18 **Graph showing maximum efficiency**

11.3.5 Relationship between transformer cooling and rating

As with any device, a transformer on load generates heat in both the core and the windings. Over time, heat causes deterioration of the winding insulation, leading to winding faults.

So, the *cooling system* of a transformer affects the life of the transformer and also its kVA rating. An increase in cooling efficiency will increase the kVA rating. For example, a transformer with an 80°C rise in temperature uses between 13% and 23% less energy than the same transformer operating at 150°C. Even a small increase in the operating temperature will have a negative impact on the efficiency of the transformer.

For smaller transformers, the surface area is great enough to remove the generated heat by convection and radiation. As transformer size increases, the surface area becomes proportionately smaller than the volume and eventually the heat being generated cannot be dissipated quickly enough. As a result, the temperature of the transformer begins to rise and additional cooling methods must be used.

11.3.6 Transformer voltage regulation

A transformer is expected to deliver a pre-determined voltage at full load. Because of changes in the connected load impedance and the two major losses introduced previously, the full-load voltage will tend to be less than the no-load voltage.

Voltage regulation of a transformer can be expressed as a percentage of its full-load secondary voltage and can be calculated using the following equation (which is really only accurate for single-phase transformers):

$$\text{voltage regulation} = \left[\frac{(V_{NL} - V_{FL})}{V_{FL}}\right] \times 100\%$$

To obtain a regulation value, the primary input voltage should be maintained at its rated value and the power factor of the load must be known—the regulation value obtained is only relevant for a particular transformer at this value of power factor, voltage ratio and load current.

11.3.7 Transformer percentage impedance

Transformer *percentage impedance* can be defined as 'the percentage of primary voltage (V_P) necessary to cause the rated full-load current to flow in the secondary when the secondary terminals are short-circuited (V_{PS})'.

To measure V_P and V_{PS}, a temporary short-circuit is placed on the secondary side of the transformer and a variac used to increase the primary voltage until the rated secondary current is drawn. The circuit is the same as the short-circuit test shown in Figure 11.15, but without the wattmeter. Once the measurements are made, percentage impedance can be calculated using:

$$Z\% = \frac{V_{PS}}{V_P} \times 100$$

where:

$Z\%$ = transformer impedance percentage
V_{PS} = test voltage applied to the primary winding with the secondary winding short-circuited and the full rated current flowing in the secondary winding
V_P = rated primary voltage.

A typical value of impedance percentage for a distribution transformer is between 1% and 5%. The transformer percentage impedance has an effect on system fault levels and is used to determine the maximum *prospective short-circuit current (PSC)* of the system while under fault conditions.

11.4 Transformer connections

11.4.1 Parallel operation of transformers

Parallel operation involves two or more transformers connected to a common source of supply and their secondaries connected to a common load. Transformers can be connected in parallel to supply a load in excess of the rating of the existing transformer, thus reducing the expense of having to replace the original transformer with a larger-capacity transformer. Further, having a number of transformers in parallel allows maintenance to be carried out on a single transformer without interruption to the supply.

Conditions/restrictions required before two single-phase transformers can be connected in parallel

To connect single-phase transformers in parallel, not only must the output voltages and internal impedance be equal but the instantaneous polarities must be the same.

Equal output voltages

If two unequal voltage sources are connected in parallel, a circulating current caused by the phasor difference is set up between them. One transformer becomes a burden on the other and they are unable to supply full power to the connected load.

The current flow is limited only by the impedances of the windings and will flow despite all other conditions for parallel operation being met. Large quantities of heat are generated, and the circulating current effectively renders both sources of power useless. Essentially, the turns ratios (and hence voltage ratios) must be the same for the same style of connection.

Compatible internal impedance

The two transformers must have a compatible range of internal impedance to allow effective load-sharing across the load range. Only when the two transformers match in all important characteristics can they be expected to share the load evenly, or according to design.

Instantaneous polarities

The two transformers shown in Figure 11.19 have their primary windings wound in the *same direction* around the iron core. When the instantaneous polarity of line A is positive indicated by the dot (•), the mutual flux Φ in each transformer acts in the same direction.

The secondary windings in Figure 11.19 are shown wound in opposite directions to each other. The induced voltage V_2 acts in an upwards direction in (a), while in (b) V_2 acts downwards. In both cases, the secondary flux Φ_2 must oppose the mutual flux (Lenz's Law). This condition is met by the induced voltage acting downwards in (b) and producing an instantaneous current flow, as indicated by the arrows in both figures. That is, when an instantaneously positive voltage is applied to the primary terminals (indicated by dots), there will be an instantaneously positive voltage produced at the secondary terminals (also indicated by dots).

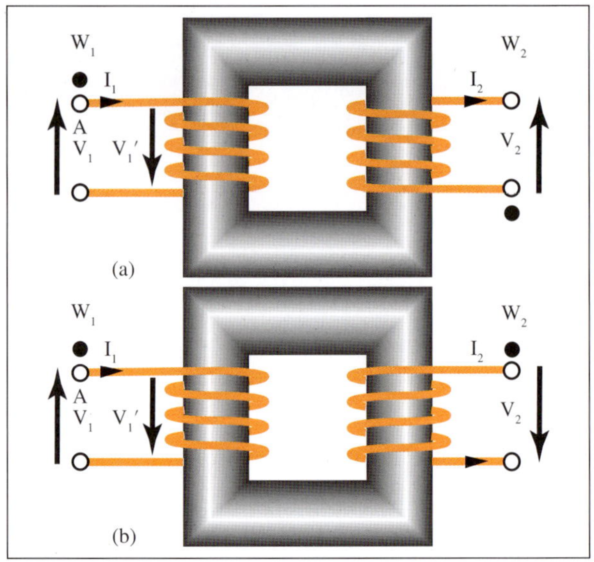

FIGURE 11.19 Winding polarity

In general terms, the positioning of dots on winding ends is used to indicate the similar instantaneous polarities. For single-phase transformers to operate in parallel, their instantaneous polarities must be identical. The correct connections for two transformers in parallel are shown in Figure 11.20.

If terminals of the wrong polarity are connected, a high circulating current is set up in both primary and secondary windings. Effectively, the two secondary windings are connected in series and then short-circuited. The path for this dangerous circulating current is highlighted yellow in Figure 11.21.

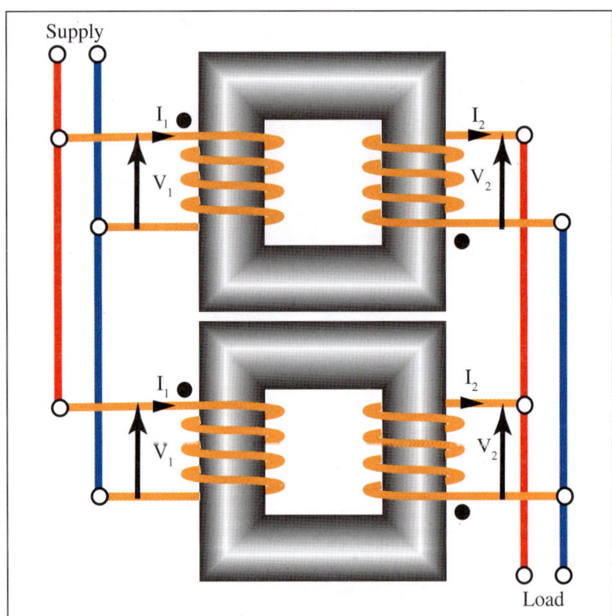

FIGURE 11.20 Parallel transformers correctly connected

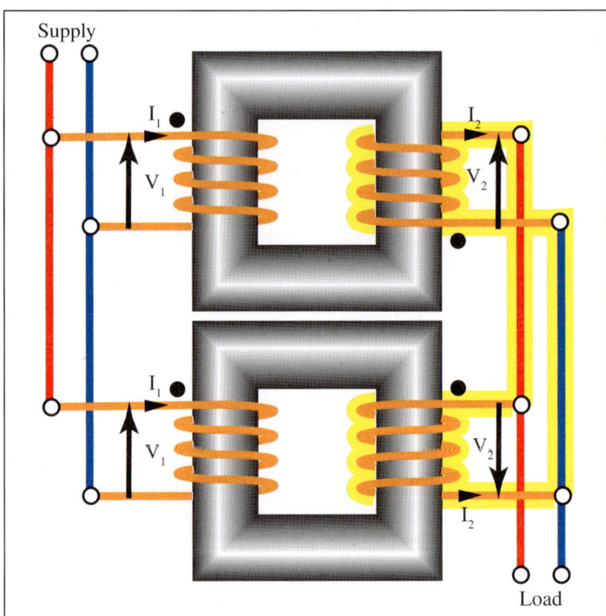

FIGURE 11.21 Path for circulating current

Determining instantaneous polarities

To determine the polarities of transformer windings, one terminal from each winding is connected together and the transformer primary is connected to a voltage source (usually a variac, to keep the test voltage low). A voltmeter is placed across the other terminals. This is illustrated in Figure 11.22.

If the voltmeter reading is greater than the supply voltage, the transformer is said to have *additive polarity* and *dissimilar ends* are bridged together. Therefore, the 'open end' of the primary is marked as the 'start' of the primary and the bridged end of the secondary is also marked as a 'start' end.

If the voltmeter reads less than the supply voltage, then the windings have *subtractive polarity* and must have *similar ends* bridged; both open ends should be marked as starts, shown by a dot (•).

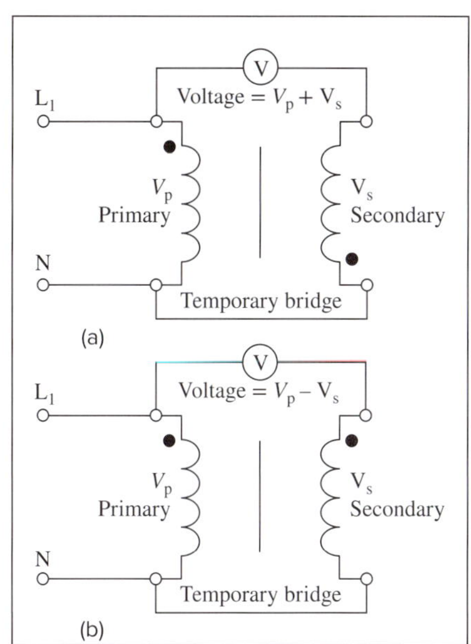

FIGURE 11.22 Checking winding polarity: (a) additive and (b) subtractive

Conditions/restrictions required before two three-phase transformers can be connected in parallel

For three-phase transformers, other conditions must also be met, predominantly phase sequence and phase shift.

Same phase sequence

For three phase transformers, if different phase sequences are connected in parallel, the least that can occur is a short-circuit between the lines. Heavy circulating currents flow, and damage will almost certainly occur to both transformers (and perhaps to the installation).

Same phase-angle shift (vector group)

Further, the change in phase angle from primary to secondary on both transformers must be identical, otherwise dangerous circulating currents will be generated. Satisfactory parallel operation can occur only when the two transformers belong to the same vector grouping and have the same phase shift.

Transformer connections should always be checked before energising and subsequent loading. Serious damage can be caused by an improperly connected transformer, and operators risk injury or electric shock.

11.4.2 Terminal identification

When drawing sketches of transformers, use of the dot or a similar system of identification for winding ends is satisfactory. In practice, it is more usual to be confronted with a transformer and a row of terminals, which makes some general system of identification necessary. AS/NZS 2374 sets out such a system for power transformers.

In brief, all terminals are given an identifying letter and a subscript number; upper-case letters are used for the higher voltage winding and lower-case letters for the lower-voltage winding, e.g. A_1 and a_1 respectively. Where more than one end of a winding is brought out to a terminal, the higher number is the line terminal unless a specific phase shift is required.

An example for a single-phase transformer is shown in Figure 11.23, and Figure 11.24 shows an example for a three-phase transformer. The Standard specifies that the identification

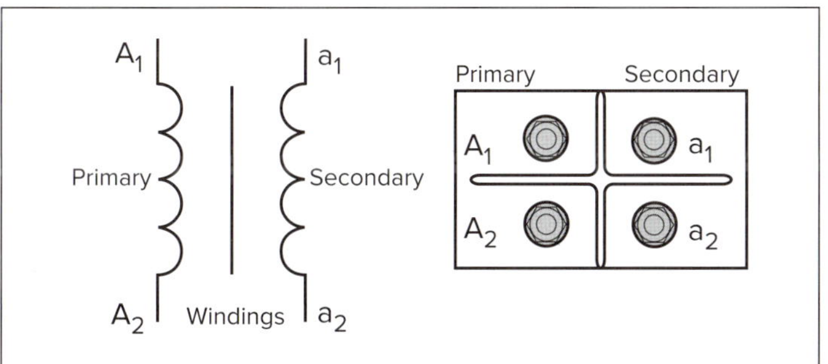

FIGURE 11.23 Typical terminal arrangements for a single-phase transformer

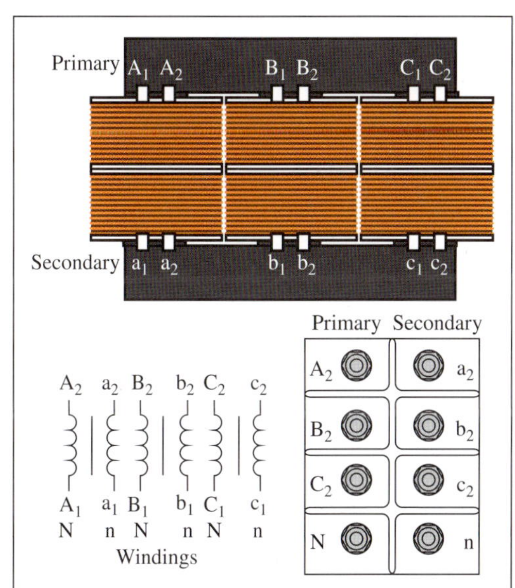

FIGURE 11.24 Typical terminal arrangements for a three-phase transformer

is permanently marked on (or adjacent to) the terminals. Invariably this means stamping the identification into the metal of the terminal or the case adjacent to the terminal.

Note that network operators may not follow this standard and may require further markings on the transformer to assist them in installation or to match phase sequence.

11.4.3 Three-phase transformer connections

Three phase transformer windings (both primary and secondary) can be connected in a number of ways, with the four most common being shown in Figure 11.25. For power and distribution transformers, these connections occur inside the transformer. Delta–star and star–delta produce a phase shift, which, as described earlier, must be considered when paralleling.

Each configuration lends itself to particular applications. For example, star-star-connected transformers are used for HV transmission systems for economic reasons. The voltage across each phase winding is a factor of $\sqrt{3}$ (1.73) *less* than the voltage between lines, meaning the windings do not need to be insulated for the higher line voltage.

The combination of *delta–star step-up* and *star–delta step-down* is undoubtedly the best for long-distance systems with very high voltage.

For the MEN system of earthing, the *delta–star* step-down is used for the final distribution to consumers, as the star point provides both the neutral point and a 'stable' earth connection. The delta–star connection is also used to step up voltage at the beginning of an HV transmission system.

Like single-phase transformers, the voltages across the primary and secondary windings are proportional to the turns ratio of the phase windings. The line voltages will then be a function of the phase connections, as described in Chapter 10.

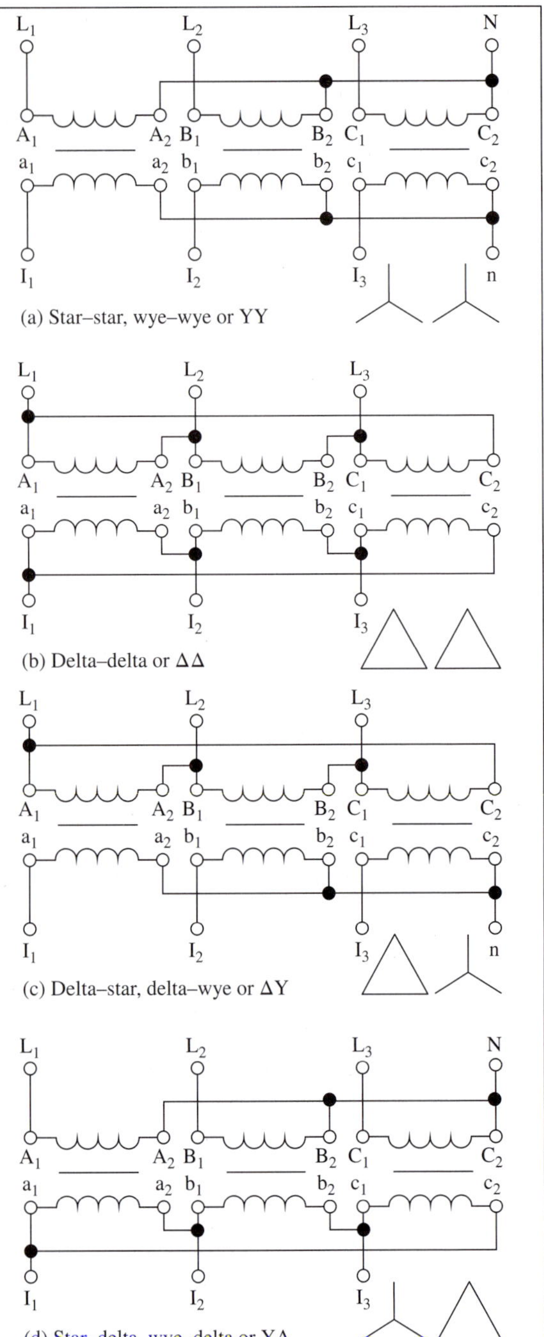

FIGURE 11.25 Three-phase transformer connections

EXAMPLE 11.5

If a 400 V three-phase step-down transformer has 200 turns per phase on the primary windings and 40 turns on the secondaries, find the output line voltage (V_L) for each of the four main connection types.

(a) *Star–star*

$$(star_1)\ V_L = 400\ V$$

$$\therefore V_P = \frac{400}{\sqrt{3}} = 230\ V\ (= V_1)$$

$$\frac{V_1}{V_2} = \frac{N_1}{N_2} = \frac{200}{40}$$
$$V_2 = \frac{V_1 N_2}{N_1} = \frac{230 \times 40}{200}$$
$$= 46\ V\ (= V_P)$$
$$(star_2)\ V_L = \sqrt{3} V_P$$
$$= \sqrt{3} \times 46 = 80\ V$$

(b) *Delta–delta*

$$(delta_1)\ V_L = V_P = 400\ V\ (= V_1)$$
$$\frac{V_1}{V_2} = \frac{N_1}{N_2} = \frac{200}{40}$$
$$V_2 = \frac{V_1 N_2}{N_1} = \frac{400 \times 40}{200}$$
$$= 80\ V\ (= V_P)$$
$$(delta_2)\ V_L = V_P$$
$$= 80\ V$$

(c) *Delta–star*

$$(delta_1)\ V_L = V_P = 400\ V\ (= V_1)$$
$$\frac{V_1}{V_2} = \frac{N_1}{N_2} = \frac{200}{40}$$
$$V_2 = \frac{V_1 N_2}{N_1} = \frac{400 \times 40}{200}$$
$$= 80\ V\ (= V_P)$$
$$(star_2)\ V_L = \sqrt{3} V_P$$
$$= \sqrt{3} \times 80 = 139\ V$$

(b) *Star–delta*

$$(star_1)\ V_L = 400\ V$$
$$\therefore V_P = \frac{400}{\sqrt{3}} = 230\ V\ (= V_1)$$
$$\frac{V_1}{V_2} = \frac{N_1}{N_2} = \frac{200}{40}$$
$$V_2 = \frac{V_1 N_2}{N_1} = \frac{230 \times 40}{200}$$
$$= 46\ V\ (= V_P)$$
$$(delta_2)\ V_L = V_P = 46\ V$$

11.5 Power transformers

Power transformers differ from other transformer types as they are designed to work at transmission and distribution voltages and relatively high currents. Like smaller transformers, they can be classified as single or three phase and core or shell, but they can also be classified by their insulating/cooling fluid, e.g. liquid filled, gas filled or dry type.

11.5.1 Transformer components

Power transformers incorporate a number of additional components for normal operation, control, protection and monitoring. Figure 11.26 shows the typical auxiliary equipment found on a large distribution transformer.

- *Tank*—houses the windings and cooling oil.
- *Conservator tank*—used to fill the tank and also stores oil that has expanded due to temperature increases.
- *Oil level indicator*—located at end of conservator tank and provides quick indication of oil level.
- *Gas-actuated relay (Buchholz relay)*—for detecting gas bubbles caused by an internal fault.
- *Tap changer*—for regulating output voltage.
- *Tap changer motor drive*—usually controlled by an automatic voltage regulator (AVR).
- *Bushing*—Connections to grids.
- *Bushing voltage transformer (VT)*—for metering the current through the passing bushing.
- *Core and windings*—transformer construction.
- *Oil release valve*—allows draining of oil for maintenance.
- *Vacuum valve*—allows air and moisture to be removed from the oil.

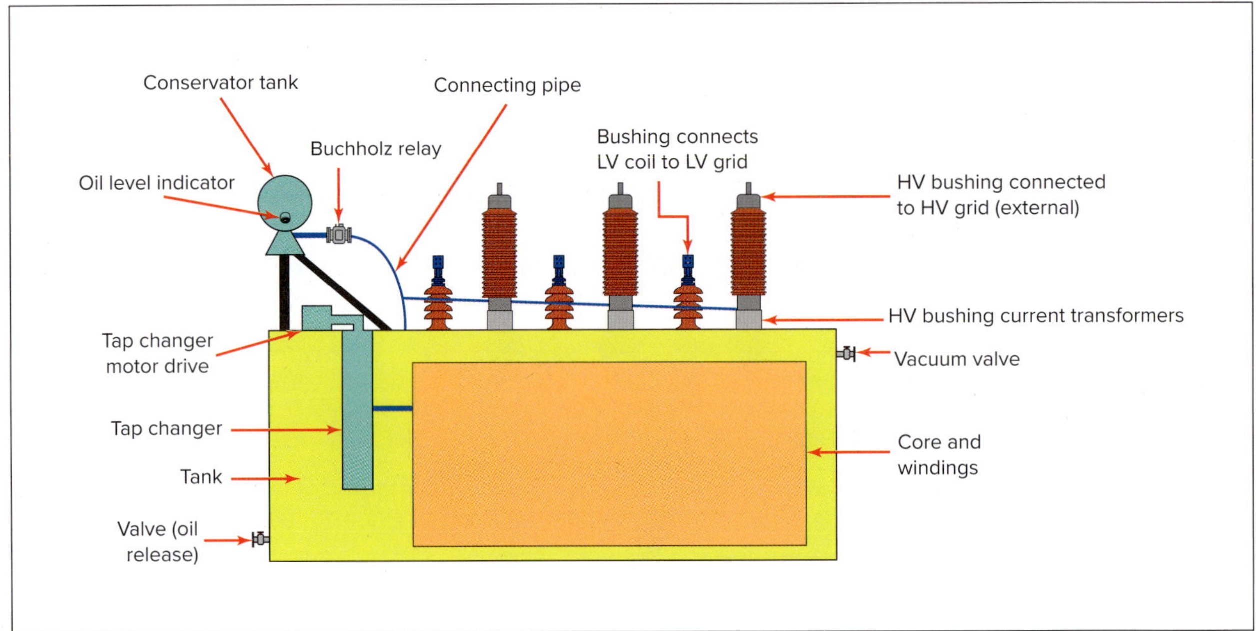

FIGURE 11.26 **Transformer auxiliary equipment**

Transformer tank

The *transformer tank* houses the insulating oil and provides physical protection for the windings, accessories and controls. Changes in the loading can result in alternating tank pressure and, at times, vacuum. These need to be allowed for to protect against deformations and stress fractures.

The transformer tank is usually corrugated to increase the surface area and provide additional heat dissipation. The type of tank surface and colour also play a part in cooling. It has been found that colours such as low-sheen variations of black, green or grey enable the oil to run at lower temperatures than would otherwise be the case.

Tap changing

The majority of power transformers incorporate some method to compensate for variations in voltage caused by load changes or other factors. The adjustment of the turns ratio, and hence the voltage ratio, is made using two different methods—*off-load tap changers* or *on-load tap changers*. The choice of system depends on the importance of the transformer and how often regulation is required.

As the name suggests, on-load tap changing is carried out while the transformer is still energised. This method is more convenient to consumers and lends itself to automatic operation so that the line voltages can be regulated without manual intervention.

11.5.2 Insulating/cooling media

In general terms, there are two commonly used media for transformer cooling—air (A) and oil (O)—and the methods for doing so are many and varied, as are the ways in which the media are combined. Abbreviations stamped on transformer nameplates are from IEC standards, and they identify the medium and how it is circulated. For example, an ONAF transformer refers to 'Oil Natural–Air Forced'.

SF_6 gas transformers are also found in locations where oil may pose a fire, explosion or leak hazard (e.g. in underground and offshore substations).

Air cooling

For air cooling, the transformer must be provided with ducts between the coils, and between the core and the housing, so that air can be blown through them to remove the heat. The air must be filtered so that dust cannot build up in the ducts, become wet and lead to faults.

Air-blast cooling is used on transformers where economy of space and weight is required or where oil cooling may be a fire hazard. It is seldom used in very large transformers or for voltages above 20 kV.

Oil cooling

One of the most common cooling methods is to immerse the transformer in a tank of special transformer oil, providing as large a cooling surface area of the tank as possible by using external tubes, as shown in Figure 11.27.

The oil serves the dual purpose of cooling and insulating. It conducts the heat from the core and the windings to the surface of the tank and the external tubes. The heat is then dissipated into the surrounding air, cooling the oil that circulates through the tank by means of natural convection.

Forced circulation

For very large transformers, convection within the oil does not remove the heat quickly enough, so forced circulation methods are needed. The oil is drawn off at the top of the tank, pumped through a water- or air-cooled heat exchanger and then returned to the bottom of the transformer tank. Figure 11.28 shows a transformer heat exchanger.

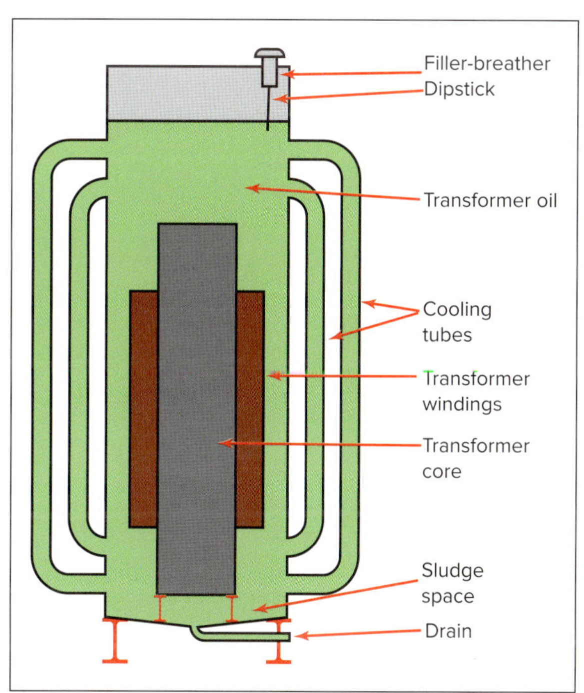

FIGURE 11.27 Example of transformer cooling

11.5.3 Transformer oil

Although the basic function of transformer oil is to provide electrical insulation and heat conduction, its other properties should be taken into consideration. The viscosity of the oil determines its flow rate. Its purity will reduce the risk of oxidisation and sludge build-up. The *flash point* of the oil is the temperature at which it will ignite spontaneously and is typically greater than 140°C. The *fire point* is the temperature at which the oil will burn and is typically greater than 170°C.

FIGURE 11.28 **Transformer heat exchanger**
RachenStocker/Shutterstock.com

Tests conducted on transformer oil

The most common transformer globally is the hermetically sealed oil transformer. Although the oil does not come into contact with air, periodic tests will ensure the oil in the transformer is fit to provide the cooling and insulating functions. A sample of oil is drawn from the transformer tank *while de-energised,* and is usually sent to a specialised, accredited laboratory for testing.

Visual inspections are conducted to check oil colour, level and evidence of sludge, which can bake itself to the insulation and mechanical structures. A *dielectric test* will check the breakdown voltage of the oil, which can be reduced with the introduction of contaminants such as water, dirt or other conductive particles. Other tests include:

- dissolved gas analysis (DGA)
- interfacial tension (IFT)
- acid number test
- power factor test.

WORKPLACE SCENARIO

On-site commissioning of a three-phase power transformer is a specialised task. Verification will usually include (but not be limited to):

- voltage ratios on all three phases for each tap position
- insulation resistance (performed at 5 kV) primary to tank, secondary to tank and primary to secondary
- condition and insulation resistance of control and power cabling (performed at 1 kV)
- vector groupings
- nameplate impedance
- bushing condition
- functional test for all alarm and trip contacts, e.g. temperature and pressure
- winding and oil-temperature indicator accuracy
- fan and oil pump starting operation, direction of rotation and overload protection
- tap changer operation and indication
- oil levels (main tank and tap change)
- oil circuit valves' position and operation
- condition of desiccant (silica gel).

CHECK YOUR UNDERSTANDING

11.3 State the two reasons why oil is used in large power transformers.

11.4 Describe mutual induction in a transformer.

11.5 A transformer has 2000 turns on the primary winding and 1000 on the secondary. If the applied voltage is 230 V @ 50 Hz and a purely resistive load of 80 Ω is connected to the secondary, calculate the:

(a) output voltage of the transformer
(b) secondary winding current
(c) primary winding current.

11.6 Specialised transformers

11.6.1 Transformers with multiple secondaries

Sometimes more than one secondary voltage is desired. The choice is then one of having two or more transformers to obtain the voltages or having one transformer with one primary winding and more than one secondary. It could occasionally be mandatory from a safety point of view to have separate transformers, but it is common (and cheaper) to have one transformer with a slightly larger core and as many secondaries as required. This is illustrated in Figure 11.29, where a transformer is shown with three secondary windings, each having different voltages.

The current and voltage ratios discussed previously still hold true for the individual windings, but it should be remembered that the volt-ampere rating of the transformer will be the sum of the individual ratings of each winding.

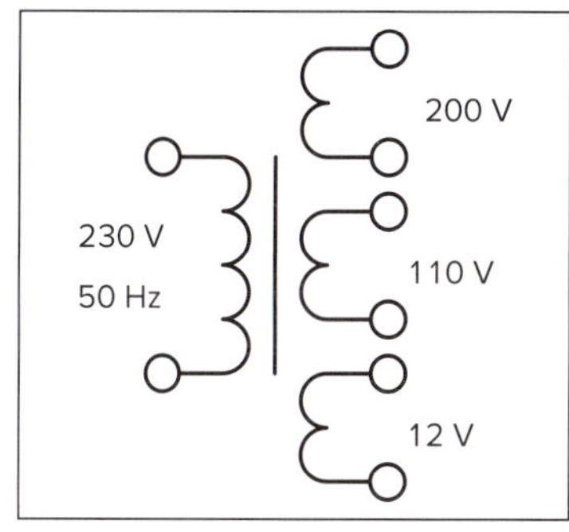

FIGURE 11.29 Transformer with multiple secondaries

For example, to use the windings at ratings of 50 VA, 55 VA and 40 VA, the transformer as a whole would have to be rated at the sum of these figures (145 VA). If due regard is given to phasing the windings (this is described later), it is only one step further to connect all three windings in series (called a 'series-aiding' connection) and have a total voltage of 322 V.

11.6.2 Tapped windings

For small transformers, it is common to see *secondary windings* with tappings brought out to give a range of voltages from a common point. This method of obtaining various voltages from the secondary winding is shown in Figure 11.30. It is also a variation of the autotransformer method (see section 11.6.3).

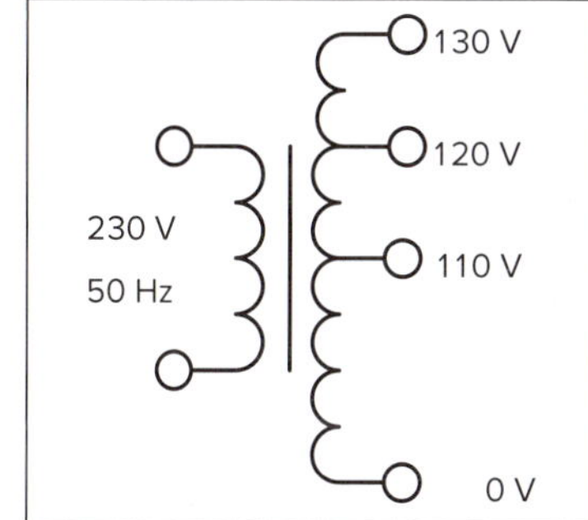

FIGURE 11.30 Tapped secondary winding

While the taps are shown here on the secondary winding, the method is not restricted to secondary windings. It is also quite common for tapped primary windings to accept a range of primary supply voltages such as 220 V, 230 V and 240 V. This is also the case for most distribution transformers, though the primary voltage is much higher (refer back to Figure 11.8).

Once the 'turns per volt' ratio of a transformer is established, it is a straightforward matter of calculating the required number of turns to determine the location of a voltage tap on the winding. As an example, a *centre-tapped transformer* (see Figure 11.31) has a tapping point from the centre of the secondary winding.

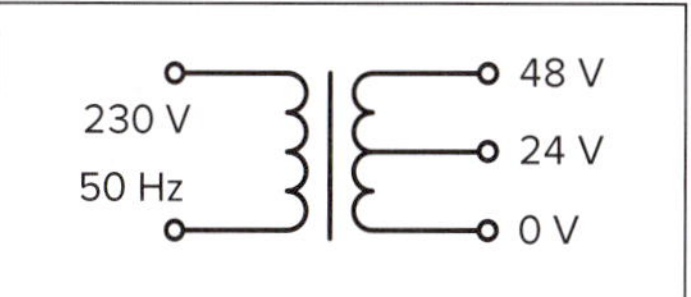

FIGURE 11.31 Centre-tapped transformer

The volt-ampere rating of the transformer always has to be observed, so converting a 24 V transformer to a 12 V transformer will allow for twice the current output, assuming that the winding conductor is an appropriate CSA.

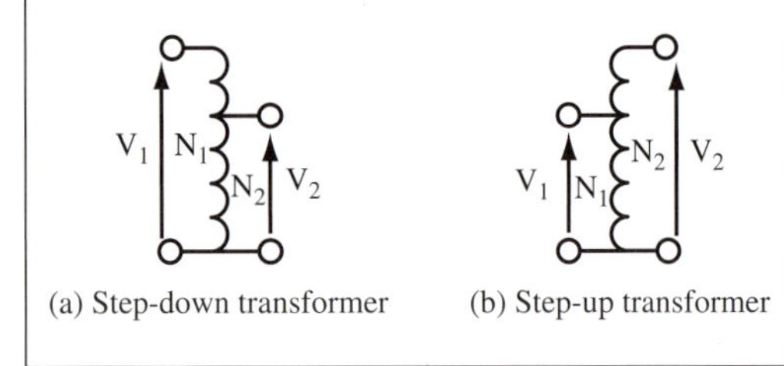

FIGURE 11.32 Autotransformers

11.6.3 Autotransformers

An autotransformer has only one winding, with part of the winding common to both the primary and the secondary circuits. Autotransformers, like most other transformers, may be step up or step down, meaning the output may be a higher or lower voltage than the primary voltage (see Figure 11.32). Note that although autotransformers are usually considered not to have a secondary winding, the output may be referred to as 'secondary'.

The voltage across any number of turns is proportional to the turns per volt established into the primary winding. Therefore, if a certain transformer has one volt per turn and is connected across a 240 V supply, it would require 240 turns. If, however, it was intended to also provide the correct output voltage at either 220 V or 230 V, there might be a tapping at 220 turns and another at 230 turns.

If a voltmeter were placed between the common (neutral) terminal and each of the tappings in turn, the meter would read 220 V, 230 V and 240 V respectively. A load connected to one of the terminals would be provided with that voltage and a maximum current allowed by the volt-ampere rating of the transformer. In this way, a device designed to operate on 230 V could be supplied from a 230, 220 or 240 V supply simply by selecting the appropriate tapping of the primary winding.

EXAMPLE 11.6

Figure 11.33 shows an autotransformer supplied with 230 V and output voltage of 180 V. If a resistive load of 45 Ω is connected across the output, the current is calculated as follows.

$$I = \frac{V}{R}$$

$$I_2 = \frac{180}{45} = 4\text{ A}$$

$$I_1 = \frac{4 \times 180}{230} = 3\text{ A}$$

This means the common winding only carries the difference between the primary and secondary currents, i.e. a current of one ampere (4 – 3 = 1 *A*).

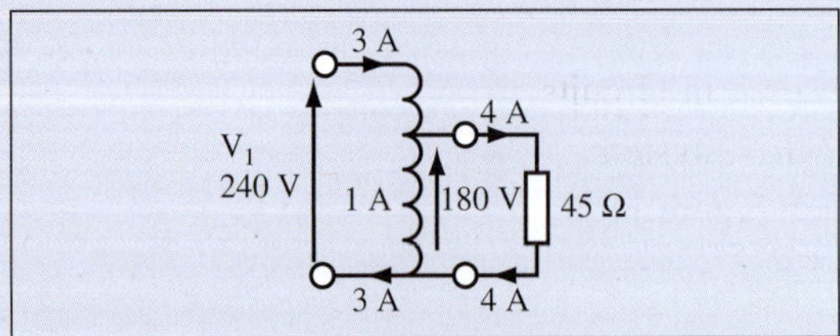

FIGURE 11.33 Autotransformer winding current

Although the autotransformer has advantages, there is one key disadvantage. If the shared portion of the winding is open-circuited during operation, the full *line voltage* is across the open circuit. This can lead to a hazardous situation. For this reason, AS/NZS 3000:2018 places limitations on autotransformers for general use.

Any equipment supplied via an autotransformer cannot have a voltage rating less than the highest input or output voltage of the autotransformer (see AS/NZS 3000:2018, Clause 4.14.4).

Autotransformers find such uses as in three-phase induction motor starters, power supplies, voltage regulation of transmission lines and laboratory experiments.

Variac transformers

An extension of the concept of an autotransformer is that a transformer could have a tapping on every turn, making the range of output voltage almost continuous.

Originally the trade name of an autotransformer, a 'variac' has become the generic term for a toroidal winding with an exposed side cleaned of insulation. A moving 'wiper' is adjusted (see Figure 11.34(a)) along the winding, making an apparently continuously variable voltage. In fact, the voltage would only vary in steps according to the volts per turn if the wiper did not cover more than one turn. Variacs are widely used in laboratories for experiments. A variac is shown in Figure 11.34(b).

(a) Variable autotransformer—variac

(b) A variac

FIGURE 11.34 Variable voltage autotransformer (variac)
(b) David J. Green - electrical/Alamy Stock Photo

11.6.4 Isolation transformers

Almost all transformers except autotransformers provide isolation. However, a specialised transformer called an 'isolation transformer' has an equal number of turns on the primary and secondary windings. This means the output voltage is equal to the supply voltage.

Secondly, there is no connection to earth on the secondary side. Because the output is unrelated to earth, a person would need to make contact with both sides of the transformer to cause a shock situation. This does not eliminate the risk of electric shock, but greatly reduces it in circuits using the multiple-earth neutral (MEN) system of earthing. The concept is illustrated in Figure 11.35.

Isolation transformers have a number of applications, including:

- to provide protection by electrical separation, e.g. for damp situations and medical installations
- to prevent interference to sensitive communication/instrumentation circuits as they block d.c. signals from one circuit to the other but allow a.c. signals to pass

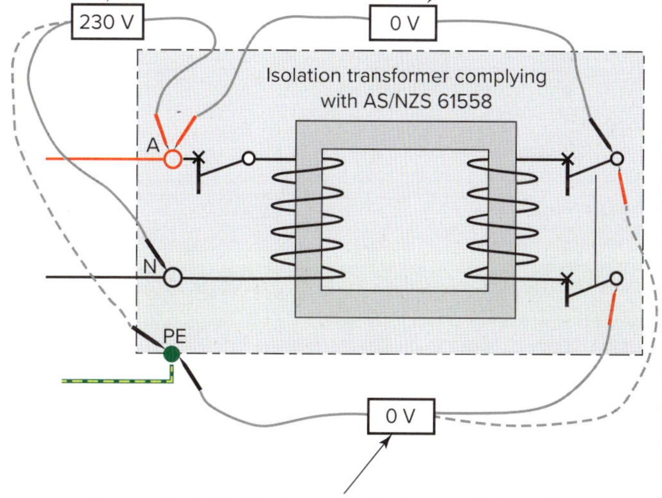

The normal 230 V between active and neutral and active and earth
There is no potential between either of the two conductors and the supply active or neutral
230 V
0 V
Isolation transformer complying with AS/NZS 61558
A
N
PE
0 V
Zero potential exists between the live conductors from the output side of the isolation transformer and earth

FIGURE 11.35 Primary and secondary winding voltages of an isolation transformer

- as the supply transformer in the SWER distribution system
- for fault finding and laboratory work, e.g. testing RCDs for compliance to standards.

11.6.5 High-reactance or flux leakage transformers

When designing transformers, the highest-possible efficiency is usually desired and design features are incorporated to reduce the leakage flux. In some applications, however, transformers with poor efficiency may be deliberately designed to meet particular requirements.

One such transformer, called a 'high-reactance' or 'flux leakage' transformer, produces a very high no-load voltage and a comparatively small short-circuit current. Transformers using this principle are found in such applications as furnace ignition, gaseous discharge lighting and welding machines.

The design permits a low flux leakage on no load but a high flux leakage on increasing load. This is achieved by spacing the primary and secondary windings some distance apart on the core and by using either fixed or variable magnetic shunts. Figure 11.36 shows the transformer of a typical home welder, including the adjustable flux shunt for varying the welding current.

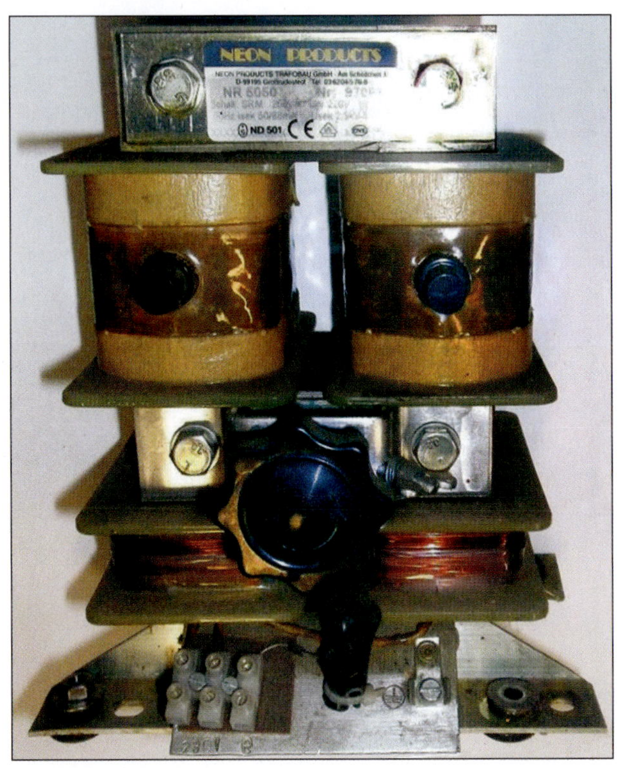

FIGURE 11.36 Flux shunt welding transformer
From Ulf Šustek Seifert/wikimedia.com

On no load, the primary winding produces a flux in the core, which cuts the secondary winding, inducing a voltage in it. The leakage flux is reasonably low because the air gaps in the magnetic shunt circuits produce a reasonably high reluctance in the shunt circuits. The secondary voltage can be calculated using the usual transformation ratio.

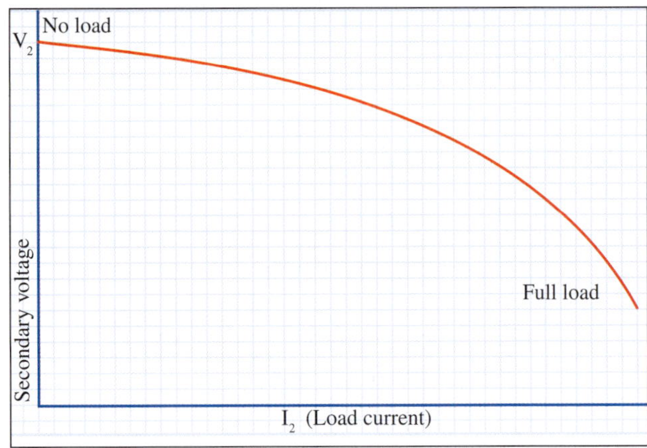

FIGURE 11.37 Flux leakage transformer load curves

When the transformer is loaded, however, the secondary current produces a flux that opposes the primary flux, causing some of the primary flux to be diverted through the magnetic shunts, which reduces the value of flux cutting the secondary turns. This in turn reduces the value of secondary voltage at an increasing rate. Figure 11.37 shows the secondary voltage decreasing as the load current increases.

11.7 Instrument transformers

It is unsafe to directly connect instruments, protection relays and other equipment to high-voltage and high-current circuits, so instrument transformers are used to reduce these voltages and currents to safer values. These specialised transformers include the voltage transformer (also known as a 'potential transformer' (PT) when used in metering circuits) and the current transformer (CT). Figure 11.38 shows how a voltmeter, ammeter and wattmeter are connected to high-voltage lines by means of a PT and CT. The circuit diagram symbols for both PTs and CTs are shown in Figure 11.39.

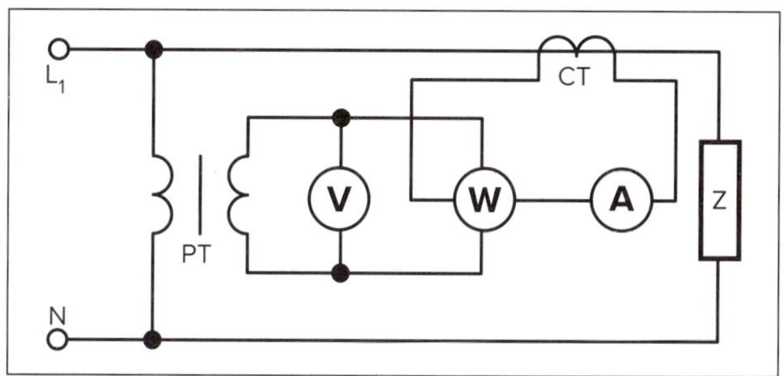

FIGURE 11.38 PT/CT power measurement

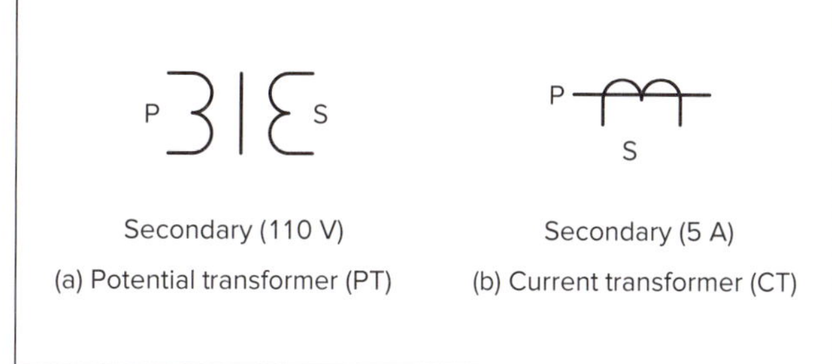

FIGURE 11.39 PT and CT symbols

11.7.1 Voltage transformers

A voltage transformer (VT) is an instrument transformer in which the secondary voltage is substantially lower than, but proportional to, the primary voltage. It is usually referred to as a 'potential transformer' (PT) and is used to provide an accurate voltage for metering purposes or protection. A typical PT suitable for use in a high-voltage substation is shown in Figure 11.40.

FIGURE 11.40 Potential transformer
Wichien Tepsuttinun/Shutterstock.com

Theory of operation

The primary of a potential transformer (PT) has to be insulated for a high voltage and is often immersed in oil for extra protection. As a general guide, a PT usually operates at low flux densities in iron cores of relatively large CSA. The copper conductors have few turns and are large in cross-section for the small current taken.

The PT operates on the same principle as the power transformer, where the ratio of the primary and secondary voltages is proportional to the turns ratio of the primary and secondary windings.

Potential transformers are designed to have a standard output voltage when the full rated voltage is applied to the primary winding. For single-phase work, AS/NZS 1243 specifies a secondary voltage of 110 V; where transformers are used in the star connection for control work in substations, the standard output voltage for each transformer is 63.5 V, giving a line-to-line voltage of 110 V.

The important factors with potential transformers are the accuracy of the voltage ratio and the elimination of phase-angle errors. The energy levels being measured are generally large, and therefore losses in the transformer are inconsequential compared with the power levels. A phase angle of 0° is desirable (180° is also acceptable), particularly where a PT has to supply an instrument such as a wattmeter, which has more than one operating coil.

Potential transformer burden

A PT is also assigned a load or burden rating. This gives an indication of the available full-load secondary current and also the load placed on the supply source. The VA ratings of potential transformers are quite small as these instruments require little energy to operate. For example, a 200 VA potential transformer at 110 V would make available approximately 1.8 A of secondary current for meters and relays:

$$110 \text{ V} \times 1.8 \text{ A} = 200 \text{ VA}$$

Safe working procedures

Potential transformers are designed to restrict the high voltage to a designated area and conduct a safe lower voltage to a monitoring point where it can be connected to instruments or relays. The four terminals of a PT are always designated, and care must be taken to see they are correctly connected and the two voltage systems appropriately isolated. One of the secondary terminals is usually earthed as an added safety measure in the event of a breakdown in the insulation between primary and secondary windings (see Figure 11.41).

Care must be taken in the choice of instruments and their handling to ensure that an additional earth is not introduced at some other point within the circuit. In some cases, a non-magnetic, electrostatic shield is installed between the primary and secondary windings during manufacture. This shield is earthed as a means of protection for the operator as well as to remove electrostatic interference and noise.

When connecting meters to a PT, a short-circuit across the terminals must be avoided, as with any high-impedance device.

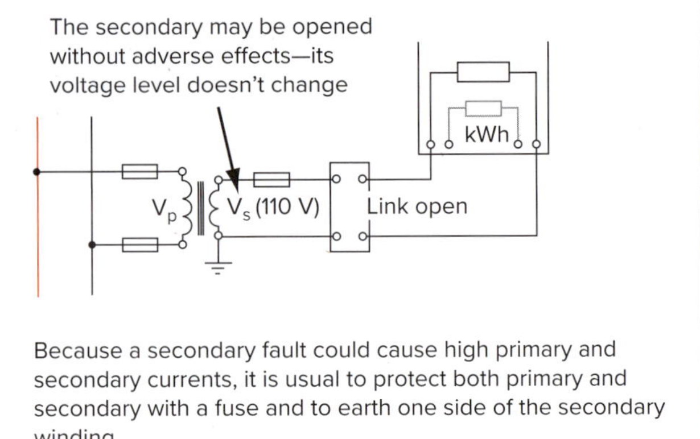

FIGURE 11.41 Safe installation of a PT

11.7.2 Current transformers

A current transformer (CT) is a type of instrument transformer designed to provide a current in its secondary winding that is proportional to the alternating current flowing in its primary. This allows safe measurement of large currents while effectively isolating measurement and control circuitry from the high voltages that are typically present in the circuit being measured.

Current transformers are made in a number of forms, depending on requirements and current ratios. Some types have primary windings that have more than one turn, while others are variations of the one turn primary stage. One variation of the single-turn primary type has a short, straight conductor passing through a hole in the iron core that forms part of the transformer's construction. Another type has an opening in the iron core and the transformer is slipped over the busbar or cable. Some current transformers have a secondary winding wound on a circular iron core, which is also slipped over the busbar adjacent to a circuit-breaker or power transformer.

The CT in Figure 11.42(a) shows a typical toroidal-type CT. The circular winding is the secondary. The primary is the conductor placed through its centre. The diagram of this is shown in 11.42(b). An ammeter would

FIGURE 11.42(a) Toroidal-type CT
Mike Scott

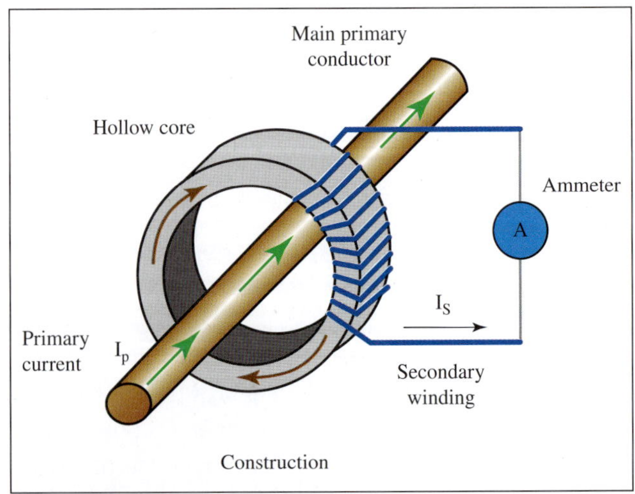

FIGURE 11.42(b) Toroidal-type CT

be connected to the two connections shown in Figure 11.42(a). This can also be seen in Figure 11.42(b).

Figure 11.42(c) shows six CTs (two per phase) monitoring the current flow in a three-phase distribution switchboard for a large engineering workshop. The low current outputs of the CTs feed a revenue meter and control gear.

FIGURE 11.42(c) Current transformers in a three-phase system
Mike Scott

Theory of operation

In a power transformer, the flux density in the core is high, the primary current depends largely on the secondary current and the voltage ratio is the main consideration. However, for the current transformer, the core flux density is very low, the secondary current is connected across a low-impedance load (and therefore depends directly on the primary current) and the current ratio is the main consideration.

The primary winding of a current transformer is connected in series with the load and generally consists of a single conductor (often a copper busbar), which in transformer terms equates to one turn. This arrangement is shown in Figure 11.43(a). In some arrangements this may also consist of a few turns of heavy-gauge conductor. The impedance of the primary winding is therefore so low that the primary current I_P is not affected by the secondary load but depends on the external load connected in the primary circuit.

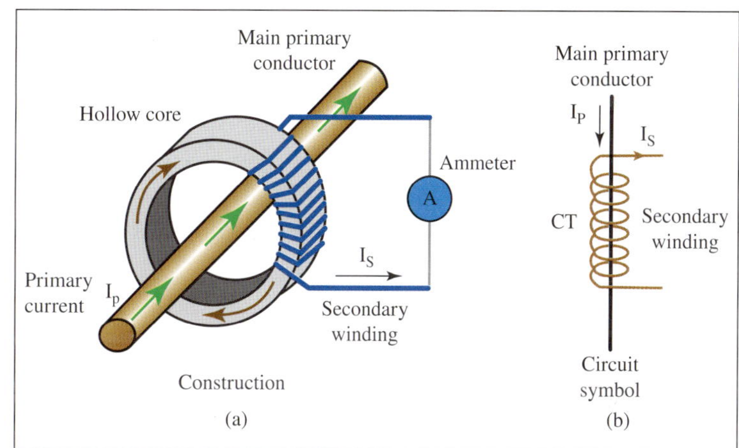

FIGURE 11.43 (a) A CT consisting of a single conductor primary with more turns on the secondary (b) Current transformer circuit symbol

The secondary circuit consists of the current coils of ammeters, wattmeters and protective relays that are all low-impedance loads. It is always closed, so the secondary current produces a flux, which opposes the primary flux. This greatly limits the flux density of the core and therefore any load on the primary. The current ratio is equal to the inverse of the turns ratio, and so there are more turns on the secondary winding.

If the secondary becomes open-circuited, there will be no secondary flux to oppose the primary flux and so the core flux density will increase. Owing to the large ratio of secondary to primary turns, and the excessive core flux, the induced voltage at the secondary terminals increases greatly, producing a safety hazard and the possibility of insulation breakdown. The greater core flux might also cause excessive heat losses and saturation in the core. *Consequently, the secondary of a current transformer must never be open-circuited under any circumstances.*

With most CTs and associated instruments, a shorting link is provided (see Figure 11.44(a)). Any connecting or disconnecting in CT circuits must follow a standard operating procedure that ensures the CT is not open-circuited at any stage.

Figure 11.44(b) shows the shorting links in the three-phase supply in Figure 11.42(c). In this position they are disconnected, but when they are connected to the bar to their left, they short out the CT output (S_1 and S_2 in this case).

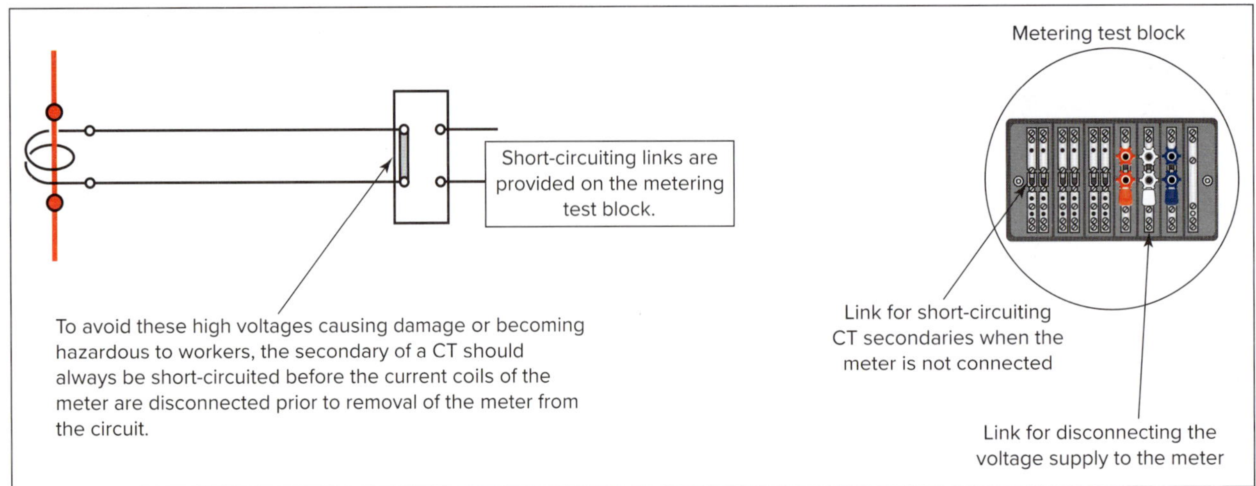

FIGURE 11.44(a) Metering shorting links

IEC 60044, the standard for current transformers, specifies two values for secondary current range—1 A and 5 A. The higher value is still used in many instances, but increasing use is being made of the lower value, especially in substation work where the instrumentation and control relays are some distance away from the transformers. At the lower value of current, the resistance and impedance of the cable are of relatively less importance.

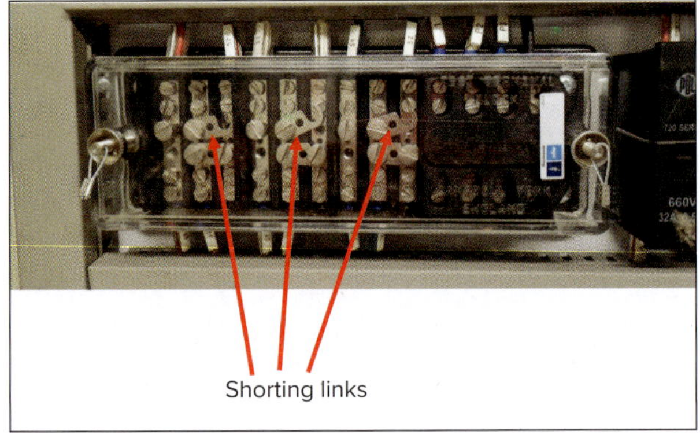

FIGURE 11.44(b) Shorting links
Mike Scott

Current transformer burden

Current transformers are designed to measure line currents without affecting the load in any way. Consequently, while a CT might be able to handle high or low currents, the burden it imposes must be as low as possible (e.g. 5 VA). The CT must have a load on the secondary at all times; therefore, the burden will exist at all times, causing a small voltage drop in the primary conductor at the point of the CT placement.

Safe working procedures

EXAMPLE 11.7

A current transformer has one turn on its primary and 160 turns on its secondary. Determine the voltage across the secondary if the ammeter was inadvertently removed without short-circuiting the secondary and the circuit was energised at 400 V.

$$V_S = V_P\left(\frac{N_1}{N_2}\right)$$

$$= 400 \times \left(\frac{160}{1}\right)$$

$$= 64\,000 \text{ V or } 64 \text{ kV}$$

An additional factor that should be taken into account is that an open circuit can also lead to magnetic saturation occurring within the core that can affect the future accuracy of the CT (this is separate from the high secondary voltage problem).

Current transformers must have all the windings held firmly in place to withstand the magnetic forces created during overloads, current surges and fault conditions. The secondaries often have one side of the winding earthed for protection and, in some cases, a non-magnetic screen between the windings.

CHECK YOUR UNDERSTANDING

11.6 State the three main requirements for connecting transformers in parallel.

11.7 Draw and label the windings of a typical single-phase transformer.

11.8 Draw the diagrams for the following types of specialised transformers:

(a) multiple secondaries
(b) tapped secondary
(c) centre tapped
(d) a step-up autotransformer.

11.9 Give three uses of an isolating transformer.

11.10 What happens if the secondary of a connected CT becomes open circuited?

SUMMARY

- Transformers operate on the principle of mutual induction. Alternating current (a.c.) creates an alternating magnetic flux that cuts both windings and generates a self-induced voltage in the primary winding and a mutually induced voltage in the secondary winding.
- The secondary voltage can be greater or less than the applied voltage, depending on the number of turns on the two windings.
- Transformer rated power is specified in volt-amperes (VA) and not in watts (W) as the load power factor is not known.
- The transformation ratios are:

$$\frac{V_1}{V_2} = \frac{N_1}{N_2} = \frac{I_2}{I_1}$$

- Transformer losses are due to copper and iron losses and affect the transformation ratios in practical situations.
- Transformer efficiency $= \frac{\text{output}}{\text{input}} = \frac{\text{output}}{\text{output} + \text{losses}}$
- Voltage regulation $= \left[\frac{(V_{NL} - V_{FL})}{V_{FL}}\right] \times 100\%$
- Basic transformer tests include visual inspection, insulation resistance and winding continuity tests.
- Transformer cores can be shell type, core type or toroidal for both single-phase and three-phase transformers.
- Windings may be wound as side-by-side, pancake or concentric.
- Cores are usually laminated with special grades of steel to reduce iron losses. Some smaller transformer cores may be made from powdered iron cores set in a medium to hold their shape.
- Toroidal cores are a highly efficient alternative to C-cores.
- Coil winding arrangements depend on the transformer application. Windings may be tightly or loosely coupled.

- Transformer cooling is essential on larger transformers. The method may be air, insulating gas or oil cooling—there are many variations and combinations. Even the colour of the tank holding the transformer has an effect on cooling.
- To connect transformer windings for paralleling purposes it is necessary to know about winding polarity.
- Three-phase transformer connections can cause phase shifts in secondary voltages.
- Factors affecting parallel operation of transformers are: voltages, frequencies, instantaneous polarities and phase relationships affected by connections.
- Commercial transformers have to conform to Australian Standards for terminal plate layouts and identification.
- In long-distance transmission lines, it might become necessary to change transformer ratios to maintain voltage levels. Two methods for changing ratios are off-line and on-line tap changing.
- Special-purpose transformers comprise:
  - power transformers (distribution and transmission)
  - transformers with multiple secondaries
  - transformers with tapped windings
  - autotransformers
  - isolation transformers
  - high-reactance transformers.
- Instrument transformers (potential and current) are used to step down high voltages and currents for connection to metering circuits.
- Potential transformers operate like a standard power transformer, i.e. the voltage ratio is the important transformation ratio. For current transformers, the current ratio is the important transformation ratio.
- There are safe working procedures for CTs and PTs. They must be strictly adhered to.

END-OF-CHAPTER QUESTIONS

11.1 What is meant by the terms 'primary' and 'secondary' windings?

11.2 Name two methods of transformer core construction.

11.3 List four manufacturer specifications that are typically found on a transformer nameplate.

11.4 Name a suitable insulating material for low-voltage (LV) transformers.

11.5 Explain the relationship between the voltages and number of turns of the two windings of a transformer.

11.6 The primary winding of a 400/24 V transformer has 400 turns. How many turns are there on the secondary winding?

11.7 A 230/110 V transformer is connected to a 22 Ω resistive circuit. Calculate the primary current.

11.8 230 V is applied to the primary winding of a transformer with 100 turns. If the secondary has 900 turns, calculate the secondary voltage.

11.9 Name two tests that should be performed on a transformer prior to putting it into service. Identify the relevant test instruments for both.

11.10 What are the major losses in a transformer and how are they affected by the load?

11.11 How can eddy-current losses from the core be effectively reduced?

11.12 What kind of steel has low hysteresis losses, making it suitable for electrical laminations?

11.13 What value is the transformer's percentage impedance used to calculate?

11.14 How can transformer percentage impedance be determined?

11.15 Why might it be necessary to parallel transformers?

11.16 What does a solid dot on a transformer symbol represent?

11.17 What is the effect when paralleling transformers if terminals of the wrong polarity are connected?

11.18 Draw the AS/NZS 2374 terminal arrangements for single- and three-phase transformers.

11.19 Name four methods of connecting the windings of a three-phase distribution transformer.

11.20 List two methods of changing transformer ratios.

11.21 What three factors affect the magnitude of induced voltage in a transformer winding?

11.22 Is the phasor diagram in Figure 11.45 of a transformer on no load or while under load?

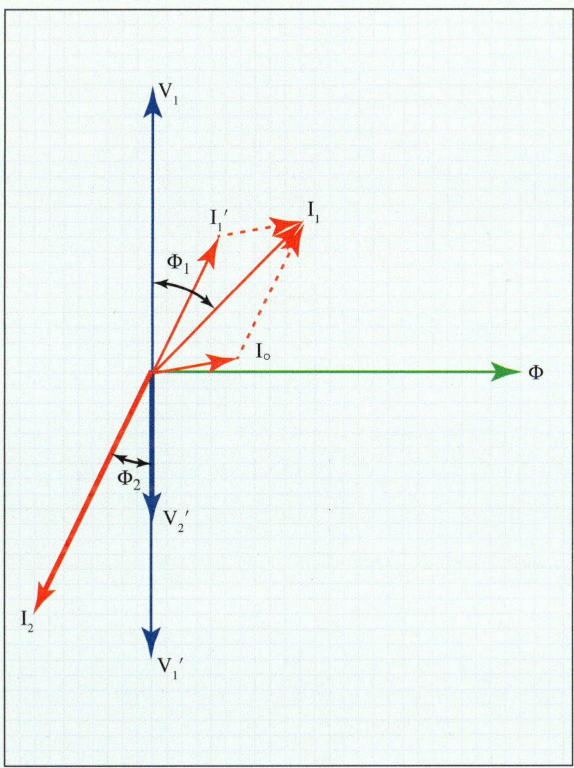

FIGURE 11.45 Phasor diagram

11.23 Tests on a 19 kV/480 V 50 Hz single-phase transformer produced the following results:

(a) open-circuit test—iron losses = 526 W

(b) short-circuit test—copper losses = 570 W.

If the transformer supplies a resistive load of 9.6 Ω at 480 V, calculate the efficiency of the transformer.

11.24 A 5 kVA single-phase 50 Hz transformer has a 240 V primary and a 120 V secondary. Calculate the:

(a) primary current

(b) secondary current.

11.25 State the purpose of a Buchholz relay on a distribution transformer.

11.26 How is the kVA rating of a transformer affected by an increase in cooling efficiency?

11.27 How is transformer voltage regulation expressed?

11.28 Describe an advantage and a disadvantage of an autotransformer.

11.29 Name an application for an autotransformer.

11.30 A step-down autotransformer has 500 turns and is tapped at 300 turns. The primary current is 30 amperes, and the secondary voltage is 330 volts. Determine the:

(a) secondary current

(b) primary voltage

(c) current in the common section of the winding.

11.31 What is an isolation transformer?

11.32 Describe a safety application of an isolation transformer.

11.33 According to AS/NZS 3000:2018, can electrical equipment be supplied from an isolation transformer? State the relevant reference.

11.34 Using a diagram, show the method of connecting the instruments required to measure power in a high-voltage, high-current a.c. circuit.

11.35 Compare the construction of a current transformer with that of a potential transformer.

CHAPTER 12

Alternating current rotating machines and motor protection

LEARNING OBJECTIVES

- Understand how an a.c. machine is constructed
- Identify all the parts of an a.c. machine
- Understand the differences in operation between an a.c. generator and an a.c. motor
- Apply magnetic principles to a.c. machines
- Identify several different types of single-phase a.c. motors
- Explain how three-phase a.c. motors produce torque
- Understand the differences between synchronous and asynchronous machines
- Identify and test for various faults in a.c. machines
- Identify various methods of motor protection

PREREQUISITE KNOWLEDGE

- Understand electromagnetism
- Know about d.c. machines
- Be aware of a.c. theory
- Possess basic fault-finding skills
- Know about electrical test equipment
- Understand basic electrical circuit protection equipment

12.1 Three-phase induction motor construction

12.1.1 Introduction

Most electric motors used in industry are of the a.c. induction type, usually running on a three-phase supply (see Figure 12.1). The induction motor obtains its name from the fact that the currents flowing in the rotor are a result of a voltage induced into the rotor and not drawn directly from the supply. This process is called 'electromagnetic induction'. The basic function of a motor of this type is to convert electrical energy into mechanical energy. Three-phase induction motors are simple, efficient, rugged, auto-starting and have a high degree of reliability. Due to their widespread use throughout industry and their design for purpose, they are arguably one of the greatest inventions of the modern age.

12.1.2 Construction

A three-phase induction motor, as shown in Figure 12.2, consists of the following basic components:

- the body or frame of the machine—the large, stationary component that houses the stator and the stator's laminated core
- the stator windings and the laminated stator core
- a revolving part (the rotor)
- end-shields or end caps which house the bearings that support the rotor shaft and allow the rotor to rotate
- a fan and fan shroud (fitted to the non-drive end of the motor's shaft) which is used to help dissipate heat created by the motor as it runs
- an electrical terminal box, to facilitate connection of the stator to the electrical supply conductors.

FIGURE 12.1 Typical three-phase induction motor
iStock/Getty Images Plus/surasak petchang

12.1.3 Motor nameplate information

All the information needed to identify a motor and its important operating characteristics can be found on the motor's nameplate. Correct motor selection for a given application is vital for efficient, safe and reliable operation; the motor details contained on the motor nameplate are crucial to this. Those details include:

- motor model number or type
- motor serial number
- date of manufacture
- rated voltage and frequency
- full load rated current
- power factor
- enclosure type and IP rating
- winding connection configuration, i.e. star or delta
- rated speed in rpm
- motor insulation class
- efficiency.

FIGURE 12.2 Side section of a three-phase induction motor
iStock/Getty Images Plus/surasak petchang

12.1.4 The stator

Figure 12.3 shows stator laminations. The laminated stator core is made from stamped sheet steel with slots on the inner surface, and three identical windings are then laid out in a series of overlapping coils. In motors of higher power ratings, the stator slots are of the open type to allow the insertion of the large pre-shaped and insulated coils; in lower-power motors, the slots are partially closed to reduce the air gap as much as possible.

The stator core is fixed into the motor frame, which, via the end shields or end caps, carries the bearings that support the rotor. The

FIGURE 12.3 Laminated stator
Simon Dand

frame protects the coils and provides a means for mounting the motor to the machinery it is to drive.

The motor frame takes various forms, depending on the conditions under which the motor will operate. These forms range from *open frame* through to submerged pump motors. It is common for the frame of the motor to have fins cast into it to help move the heat created by the stator out and away from the stator, and thus provide a cooling effect for the motor.

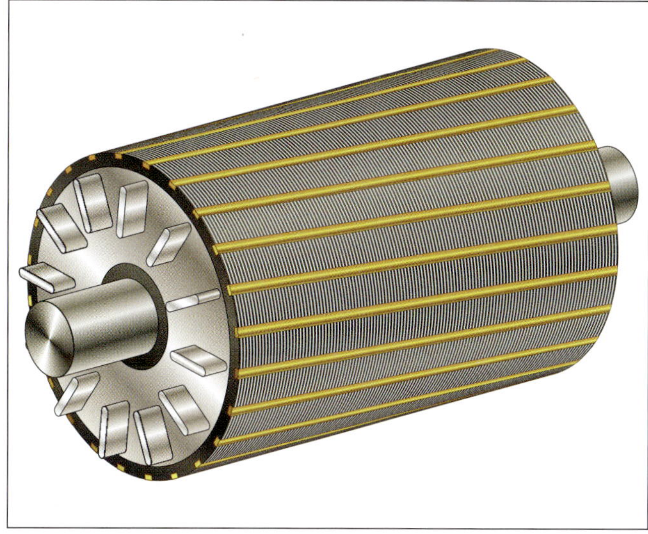

FIGURE 12.4 Squirrel cage rotor
Fouad A. Saad/Shutterstock.com

12.1.5 The rotor

Squirrel cage rotor

The rotor of an induction motor consists of a shaft (or spindle) with bearings, laminated iron core and rotor conductors. The most common type of construction is one with rotor bars in the lamination slots rather than a winding. The rotor bars pass through the rotor via holes punched in the laminations when they are pressed. The ends of the rotor bars are short-circuited at each end by a solid conducting ring, which is often a copper ring that is brazed or fusion-welded to copper rotor bars. For small-to-medium motors, they may be cast in one piece from aluminium (usually including cast fins for creating air movement). The rotor windings, if shown without the laminations, would resemble a metal cage, giving rise to the frequently used name 'squirrel cage' rotors (although the Standards refer to them simply as 'cage' rotors).

Figure 12.4 shows a cast aluminium squirrel cage rotor with end fins. It also shows skewed conductors in the rotor, which means that the conductors are at an angle to the rotor axis. Skewed conductors cause a smoother power transfer to the rotor, resulting in less vibration and a steady acceleration during starting.

Varying the physical design features of the bars affects motor performance. Embedding them deeper into the rotor, for example, increases the inductance and gives a lower starting current but creates a lower pull-out torque. This type of rotor is then restricted to loads requiring low starting torques, such as centrifugal pumps.

Wound rotor

The wound rotor is fitted with insulated windings, like the stator winding and with the same number of poles. The three-phase winding is connected in star configuration with the 'Y' point connected on the rotor and the three phases terminated at one end of the rotor at three slip-rings. Figure 12.5 shows a typical wound rotor.

The slip-rings are connected by brushes to an external star-connected variable resistance, as in Figure 12.6.

This rotor rheostat provides the means of increasing the resistance of the rotor circuit during starting, thereby producing a higher starting torque at a lower starting current. As the speed increases, the external resistance is gradually reduced, lowering the rotor circuit resistance as the rotor reactance decreases.

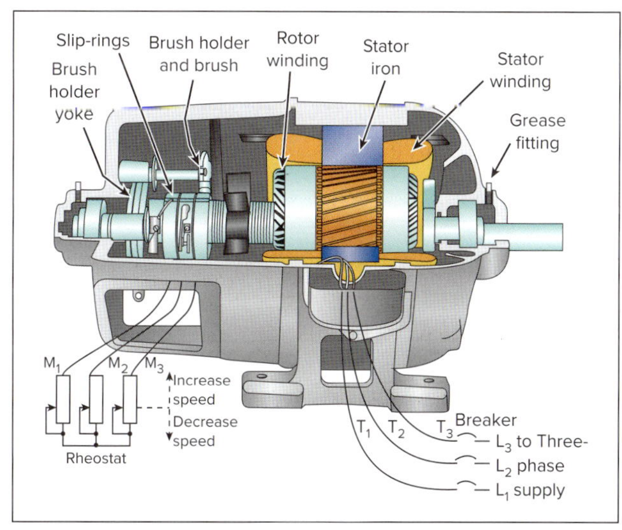

FIGURE 12.5 Typical wound rotor

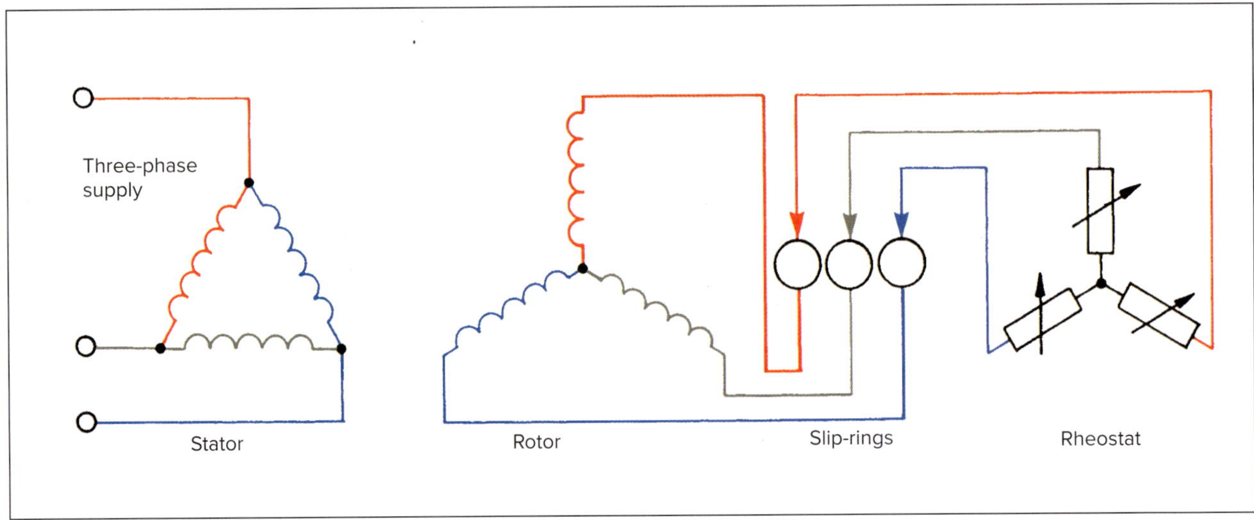

FIGURE 12.6 **Wound rotor with starting resistance**

Under operating conditions, the variations in rotor circuit resistance provide a means of controlling the speed of the motor over a small range. An increase in resistance results in a reduction in speed and torque, also producing a loss in efficiency due to the I^2R losses in the rheostat.

The wound-rotor motor is more expensive than the squirrel cage motor because of its manufacturing cost. It also has a higher starting torque and lower starting current (but poorer running characteristics) than the squirrel cage motor, as well as higher maintenance needs.

12.1.6 Motor enclosures

The conditions governing the actual installation of an induction motor are usually beyond the control of the motor manufacturer. As a result, the motor is manufactured in various enclosures. A motor driving a compressor for a refrigerated display cabinet, for example, may operate under such clean and dry conditions that the motor enclosure need only provide a mounting for the bearings and a means for fixing the motor in a horizontal plane. At the same time, the enclosure provides mechanical protection against accidental spillage and enables cooling air to circulate freely through the motor windings.

Compare this situation with a water turbine pump used for irrigation. In most cases, the motor is mounted vertically at the bore head and is given no protection from the weather. It needs to be totally enclosed to prevent the entry of water, and cooling is by means of heat transfer through the motor housing. The air sealed within the motor housing is circulated by an internal fan, thus transferring the heat generated by the windings to the housing. This heat is then transferred to the atmosphere by a second fan circulating free air across the outside of the motor housing.

AS/NZS 1359 provides detailed information about electric motor standards, including requirements for electrical rotating machines. Electrical rotating machinery is now classified by two letters followed by four numerals. This classification number is different for such categories as cooling, mounting and protection.

12.1.7 Terminal block arrangements

Squirrel cage motors

Most three-phase induction motors have both ends of the three windings brought out to a six-stud terminal block. The only exceptions are small, dedicated-purpose motors that can be started DOL (direct on line) in either star or delta configuration at all times. These motors have only three terminals (not including the earth terminal).

By custom (and some attempts at standardisation), the phase ends are brought out to terminals in one of two possible arrangements. Figure 12.7(a) shows the more usual arrangement, while Figure 12.7(b) illustrates an alternative system of connections.

The idea is to standardise connecting arrangements for people installing the motor or reconnecting it after maintenance. To connect a motor in either star or delta configuration, winding ends are bridged (see Figure 12.8).

The points to note are:

- Individual phase ends are not connected to terminal pairs but are offset (see Figure 12.8).
- Connecting a motor in star configuration means that the windings have only 58% of the line voltage applied to them.
- In star configuration, two terminal links connect all three corresponding ends of each winding.
- The bridge can be placed across the beginnings of the phase windings instead of the ends as shown. The lines are connected to the opposite ends to the bridge.
- There is no alternative option for the bridges when connecting a motor in delta configuration. Each bridge connects to line voltage.
- While bridges or shorting strips are shown in Figure 12.8, it might be necessary to remove the bridges so that all six ends of the windings can be connected to a starter—for example, a star–delta starter.

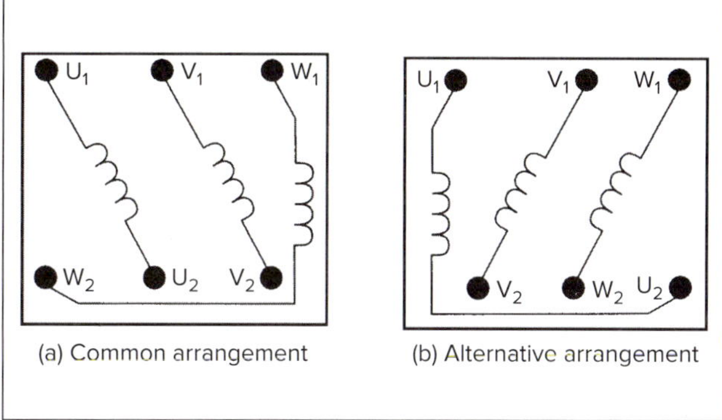

FIGURE 12.7 Three-phase motor terminals
Simon Dand

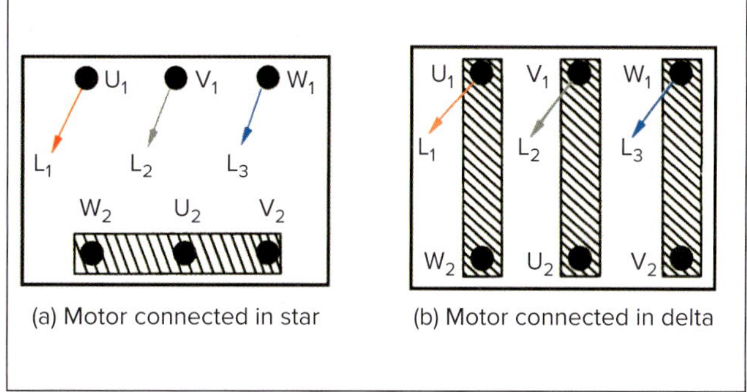

FIGURE 12.8 Bridging connections in a three-phase motor terminal block
Simon Dand

Wound-rotor motors

With wound-rotor motors, it is common to connect the stator windings internally in either star or delta configuration as required—usually delta. The terminal block will still have six terminals because the three appropriate ends of the rotor windings are connected in star configuration internally and the other ends brought out via slip-rings to the terminal block.

It is usual to identify them in some way, such as by separating them from the line terminals of the stator windings or using a different type of terminal connector. Care should be taken to observe that line voltages are not connected to the rotor terminals.

CHECK YOUR UNDERSTANDING

12.1 What are the basic components in a rotating machine's construction?

12.2 List some of the information that is found on a machine's nameplate.

12.3 What is the difference between the stator and the rotor?

12.4 What is the difference between a squirrel cage rotor and a wound rotor?

12.2 Operating principles of three-phase induction motors

12.2.1 Rotating magnetic fields

A three-phase induction motor depends for its operation on a rotating magnetic field being established by the a.c. windings. The three separate windings are installed in the stator at 120°E intervals to one another and provide a fixed number of poles for each phase. This is shown in Figure 12.9(a) for one phase of a two-pole machine. Figure 12.9(b) shows the three phases in relation to one another, giving a total of six poles. Winding U is drawn as a solid line, winding V as a dotted line and winding W as a dashed line. This drawing convention applies to all of the current waveforms, magnetic fields and vectors below.

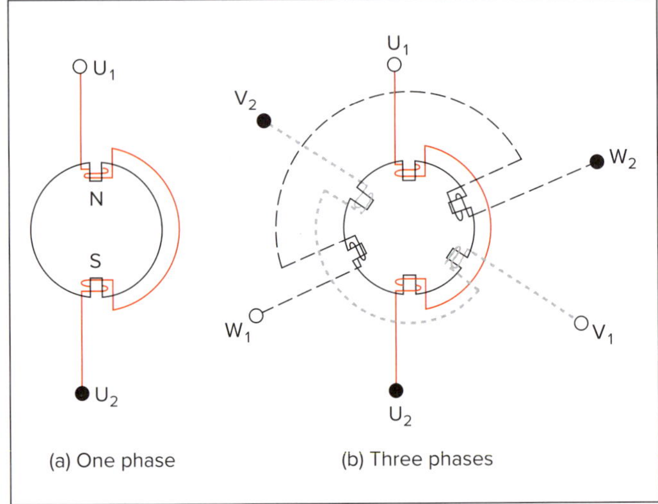

FIGURE 12.9 **Polarities and connections in a two-pole three-phase motor**

Assumption

In the following explanation of how to produce a rotating field, one assumption has been made—that winding ends U_1, V_1 and W_1, when connected to a positive source of voltage, make the adjacent iron core a north magnetic pole. From this, it follows that the opposite pole becomes a south magnetic pole. These details are also shown in Figure 12.9(a). If the current flow is reversed, the magnetic poles are also reversed.

With the three windings connected in star configuration by joining ends U_2, V_2 and W_2 together, and the ends U_1, V_1 and W_1 connected to a three-phase supply, the phase currents I_U, I_V and I_W are 120°E out of phase with one another. This is shown in Figure 12.10.

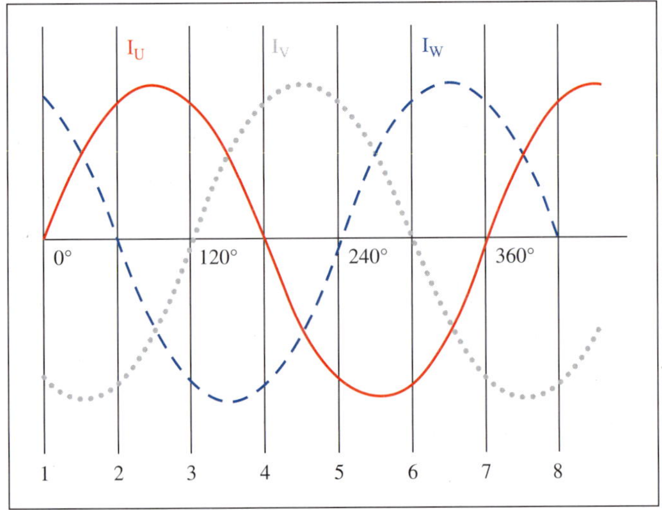

FIGURE 12.10 **Waveform diagram showing three-phase currents at 120°E**

As each current is alternating, each pair of poles sets up a magnetic flux that continually changes from one polarity to the other. Although the flux set up by winding A in Figure 12.9(b) alternates in direction, it does not rotate in any way. It simply varies in strength and direction in the vertical plane. Similarly, a pulsating flux is established by the other two windings, giving a total of three magnetic fluxes that combine into one resultant flux. This flux rotates at **synchronous speed**.

At reference position 1 in Figure 12.10, the current I_U is zero and no flux is produced by the winding U–U_1. Current I_V is negative and so will produce a south pole at V_1 and a north pole at V_2. Current I_W is positive so will produce a north pole at W_1 and a south pole at W_2. As currents I_V and I_W are equal, the two magnetic fields are equal in strength. The directions of these fields are shown in Figure 12.11(a). In the accompanying vector diagram, the addition of these two fields is shown, giving a resultant instantaneous field ϕ_R.

At position 2 in Figure 12.10, I_U is positive, I_V is still negative and I_W is zero. This produces a north pole at U, a south pole at V and nothing at W. This is illustrated in Figure 12.11(b), together with the vector diagram showing the addition of the flux vectors and the resultant instantaneous magnetic field. Since all coils have an

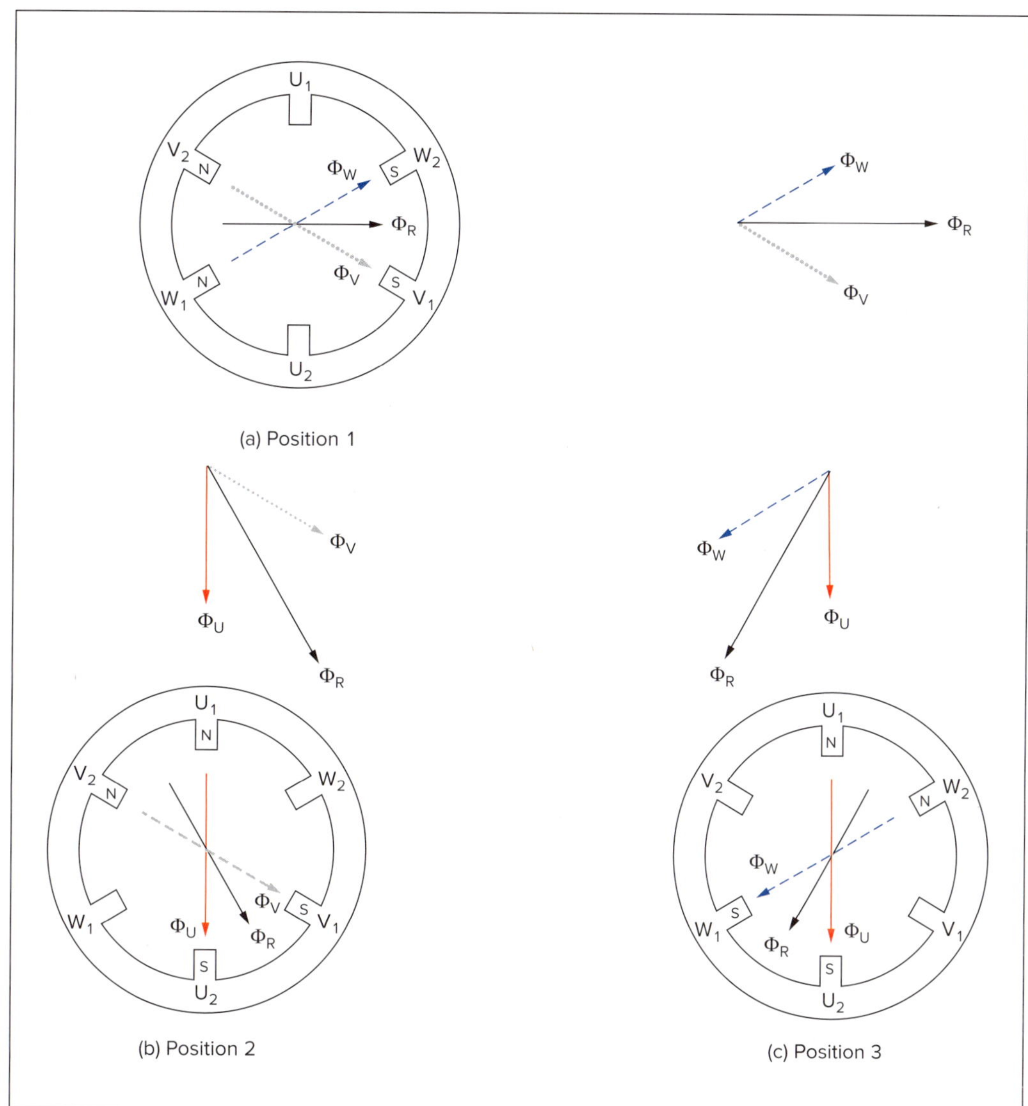

FIGURE 12.11 **Resultant flux produced at positions indicated from Figure 12.10**

equal number of turns, the relative strengths of the magnetic fields can be gauged by measuring the vertical heights of the current waveforms at the positions indicated by the reference number. Here, the direction of the resultant magnetic field has shifted 60°E clockwise from that in position 1. If drawn to scale, it can also be shown that the length of the resultant has remained constant, indicating that the field strength has remained constant.

At position 3 in Figure 12.10, I_U is positive, producing a north pole at U_1 and a south pole at U_2; I_V is zero and I_W is negative, producing a south pole at W_1 and a north pole at W_2. These fields are drawn in Figure 12.11(c), together with their vectors. The resultant field has rotated a further 60°E in a clockwise direction. (There is a 60°E difference between all the numbered positions in Figure 12.10.) For each of the numbered positions, the resultant field rotates a further 60°E in a clockwise direction. For one complete cycle of current (360°E), the resultant magnetic field rotates 360°E.

12.2.2 Rate of rotation and factors affecting it

Comparing Figures 12.10 and 12.11 shows that, for the time intervals of 60°E between positions 1, 2 and 3, the resultant field rotates an equal amount around the stator. For a complete cycle of a.c., the resultant field of a two-pole three-phase winding rotates one complete revolution around the stator. The strength of the resultant magnetic field remains constant and is equal to 1.5 times the maximum flux produced by one phase alone. The speed of the rotating magnetic field is given a specific name, the synchronous speed. The synchronous speed of the rotating magnetic field (RMF) is dependent on two properties of the motor:

1. the supply frequency
2. the number of stator poles.

The synchronous speed in revolutions per minute can be determined from the following equation:

$$n_{syn} = \frac{120f}{p}$$

where:

n_{syn} = synchronous speed in revolutions per minute
f = frequency in Hz
p = number of poles per phase.

EXAMPLE 12.1

A two-pole three-phase motor is connected to a 50 Hz supply. Find the speed at which the magnetic field rotates around the stator.

$$n_{syn} = \frac{120f}{p} = \frac{120 \times 50}{2} = 3000 \text{ rpm}$$

With a four-pole machine, 360°Electrical represents one-half of a full revolution of the stator field, so the speed of rotation of the field is one-half of that of a two-pole machine. Similarly, the speed of field rotation for a six-pole machine is reduced to one-third of that of a two-pole machine. In each case, the speed is usually expressed in revolutions per minute, whereas the frequency is in hertz (cycles per second). If the number of poles needs to be calculated, we can use the following equation:

$$p = \frac{120f}{n_{syn}}$$

where:

n_{syn} = number of revolutions per minute
f = frequency in Hz
p = number of poles.

The speed n_{syn} of the rotating magnetic field is the synchronous speed of the motor. The synchronous speeds of common pole configurations of motors at a frequency of 50 Hz are given in Table 12.1.

TABLE 12.1 Speed of the rotating field in an induction motor for various numbers of poles

Poles	2	4	6	8	10	12
Synchronous speed (rpm)	3000	1500	1000	750	600	500

12.2.3 Direction of rotation and reversal

The direction of rotation of a rotating field depends on the phase sequence of the three currents flowing through the windings. In Figure 12.12(a), the three supply lines R, W and B are connected to terminals U, V and W of the motor. The resultant magnetic field rotates clockwise.

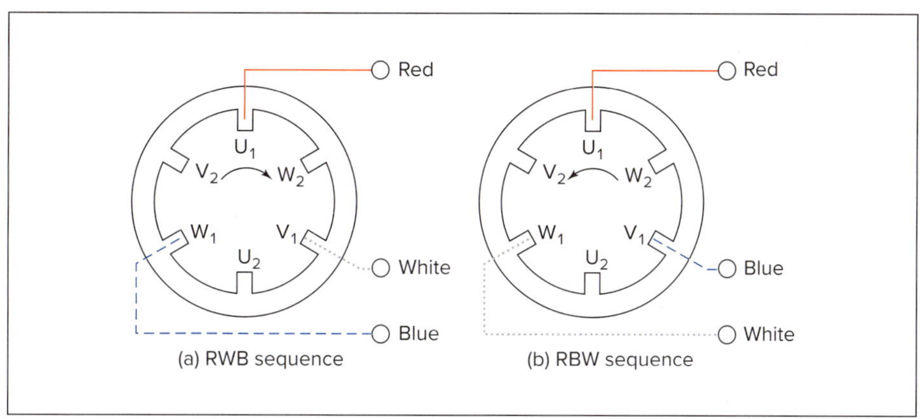

FIGURE 12.12 Phase sequence and field rotation

In Figure 12.12(b), the supply lines to windings V and W have been changed over and, using the procedure from the previous section, it can be shown that the rotation of the magnetic field is reversed. That is, the direction of rotation of the field can be controlled by interchanging any two supply lines to the motor. The part of section 12.2.4 that deals with torque states that the rotation of a three-phase induction motor is in the same direction as that of the rotating field.

12.2.4 Induction and its effects

When the stator windings of a three-phase induction motor are energised from a three-phase supply, a magnetic field is produced, rotating at synchronous speed. This rotating magnetic field crosses the air gap and cuts the rotor conductors, inducing a voltage in them (a magnetic field, the conductors and the relative motion between them). When the rotor circuit is complete (through end rings in the case of the squirrel cage rotor or through external resistance in the case of the wound rotor), the induced voltages cause high currents to flow in the rotor conductors.

Torque

Figure 12.13(a) shows a part of the stator and air gap of an induction motor, with the stator flux rotating clockwise as indicated. When these lines of force cut the rotor conductors from left to right, the relative movement between the stator flux and the rotor conductor is from right to left. By applying Fleming's right-hand rule for generators, the direction of induced current flow in the conductor is towards the reader. Owing to the comparatively high rotor currents flowing, a large flux is established around the conductor (see Figure 12.13(b)). The stator and rotor fluxes react with each other as shown in Figure 12.13(c), forming a resultant field. This resultant field tends to straighten itself out and, in the process, causes a force to be exerted on the rotor conductor, trying to force it to

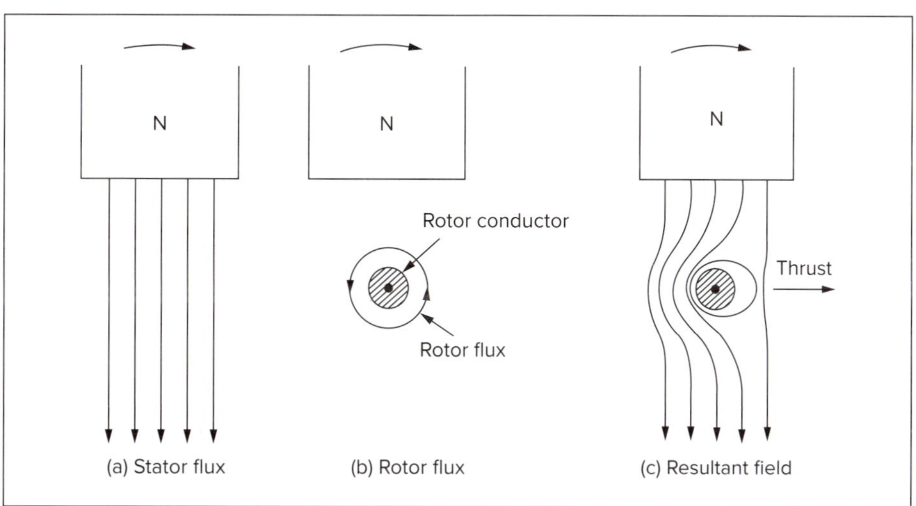

FIGURE 12.13 Production of torque in an induction motor

the right and out of the stator magnetic field. The direction of the conductor is in accordance with Fleming's left-hand rule for motors. A similar force is exerted on all the rotor conductors as the field rotates. If sufficient force is created, the rotor will start rotating in the same direction as the rotating magnetic field. Provided it is free to rotate, the rotor will accelerate until it approaches synchronous speed.

This rotating force is called the 'torque' of the motor. It is the result of the interaction of the two fluxes. The stator flux remains constant, but the rotor flux varies with the rotor current, which is determined by such factors as the impedance, the induced voltage and the relative speed of the rotor conductors.

Slip

To produce torque, there must be a rotor flux caused by current flowing through the rotor conductors. If the rotor could run at synchronous speed, there would be no relative motion between the stator flux and rotor conductors. Consequently, there would be no induced voltage, no rotor current, no rotor flux and no torque; the rotor would slow down. An induction motor therefore cannot run at synchronous speed.

With the rotor running just below synchronous speed, relative motion exists and sufficient torque is developed to keep the rotor turning. The difference between the synchronous speed of the rotating field and the actual speed of the rotor is called the 'slip speed'. It is commonly expressed as a percentage of the synchronous speed.

Generator action within a motor

Once torque is produced within a motor and the rotor begins to rotate, the rotor bars of a squirrel cage motor or the rotor winding in a wound-rotor induction motor begin to cut the magnetic field. This relative motion between a moving conductor and a magnetic field induces a voltage within the rotor conductors and thus a rotor current. Looking at Figure 12.13(c) in the context of Fleming's right-hand rule for generators, it can be seen that the generated current will flow *away* from the reader (or 'into' the page). This is in opposition to the induced current that produced the torque, thus demonstrating Lenz's Law. The generated voltage is known as a 'back EMF'. The magnitude of back EMF is proportional to the speed of the rotor. Therefore, at start-up, when the rotor is stationary, there is no back EMF generated so maximum current can flow within the rotor. As the rotor accelerates to near-synchronous speed, the back EMF increases to a maximum, resulting in a decrease in rotor current. The remaining level of rotor current flowing is just enough to produce the continuous torque required to keep the load turning. An increase in load on the shaft of the motor will initially decrease the speed of the rotor, decreasing the back EMF and allowing more rotor current to flow, thus allowing the motor to produce more torque to meet the load.

As for the stator, when the flux produced in the rotor cuts the stator conductors, a back EMF is induced within it. When the motor is on no load, the back EMF generated within the stator will almost equal the supply voltage, and the resulting voltage within the stator will be sufficient for a low value of stator current to flow from the supply to produce the magnetic flux in the rotational magnetic field. As load is applied to the rotor and its speed decreases, the generated back EMF in the stator also decreases. If the supply voltage remains constant, then the resultant voltage within the stator increases and thus more current will flow into the stator from the supply. This is how more power is drawn from the supply to meet the requirements of additional mechanical load on the motor.

EXAMPLE 12.2

Determine the slip of a four-pole induction motor running at 1440 rpm when connected to a 50 Hz supply.

$$n_{syn} = \frac{120f}{p}$$

$$= \frac{120 \times 50}{4} = 1500 \text{ rpm}$$

$$\text{slip speed} = 1500 - 1440 = 60 \text{ rpm}$$

$$\text{percentage slip} = \frac{60}{1500} \times 100 = 4\%$$

The equation for determining percentage slip is:

$$\text{slip\%} = \frac{n_{syn} - n}{n_{syn}} \times 100$$

where:

slip% = percentage slip

n_{syn} = synchronous speed

n = actual rotor speed.

At standstill (i.e. when starting), the slip is 100%. If the motor could run at synchronous speed, the slip would be zero.

Rotor frequency

When the rotor of a two-pole motor is at standstill and the stator is connected to a 50 Hz supply, each rotor conductor is cut by the rotating magnetic field at a rate of 50 times per second. At standstill, the frequency of the rotor voltage (rotor frequency) is the same as the frequency of the supply (stator frequency).

As the rotor speeds up to half the synchronous speed (1500 rpm), the relative speed between the rotating magnetic field and the rotor is half what it was at standstill. So, the frequency of the voltage induced in the rotor is one-half the supply frequency (i.e. 25 Hz). If the rotor were to revolve at synchronous speed, the rotor frequency would be zero. The rotor frequency depends on the differences in the speeds of the stator flux and the rotor (i.e. the slip of the motor), as shown in Figure 12.14.

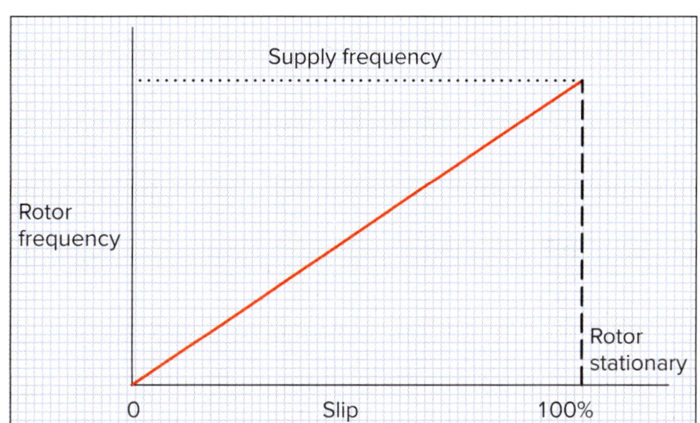

FIGURE 12.14 Relationship between rotor frequency and slip

The rotor frequency can be calculated using the following equation:

$$f_r = \frac{s\%}{100} \times f$$

where:

f_r = rotor frequency in hertz

s% = slip percentage

f = supply frequency in hertz.

EXAMPLE 12.3

Determine the rotor frequency of a two-pole, 50 Hz induction motor if the rotor speed is 2850 rpm.

$$s = \frac{n_{syn} - n}{n_{syn}} \times 100$$

$$= \frac{3000 - 2850}{3000} \times 100 = 5\%$$

$$f_r = \frac{s\%}{100} \times f = \frac{5}{100} \times 50 = 2.5 \text{ Hz}$$

As the rotor frequency varies, so does the rotor inductive reactance. This affects the starting and running characteristics of the motor.

Figure 12.14 shows that at motor start-up, when the rotor is stationary, the frequency within the rotor will equal the supply frequency. In Chapter 9 we saw that inductive reactance within a conductor under the effect of a.c. is directly proportional to the product of the conductor's inductance and frequency, using the following equation:

$$X_L = 2\pi fL$$

Therefore, when the rotor is initially stationary, the inductive reactance within the rotor conductors will be at a maximum. The addition of the rotor conductor's resistance and inductive reactance produces the rotor's impedance. At start-up, the impedance of the rotor's conductors are of a high value and more inductive than they are resistive. The result of this is that the induced currents within the rotor lag the induced voltage within the rotor by almost 90°. This has the effect of producing low torque. As the rotor speed increases, the impedance of the rotor conductors will decrease due the rotor frequency decreasing and thus the inductive reactance decreasing. This allows more rotor current to flow. However, the impedance becoming more resistive also brings the induced rotor current in phase with the induced rotor voltage. The result of more rotor current flowing and the rotor current and voltage being in phase increases the torque produced by the motor. Motor torque will be at a maximum when the rotor resistance *equals* the rotor inductive reactance. The speed of the motor and thus the amount of slip speed will stabilise when the motor torque equals the load torque. A typical motor torque/speed characteristic is shown in Figure 12.15(b).

The maximum torque produced by a motor is also referred to as 'breakdown torque'. If the load torque required by the load exceeds the breakdown torque that a motor is capable of producing, the motor will stall and will thus be unable to rotate. When this happens, the motor will continue to draw its locked rotor current from the supply, which is typically four-to-eight times the full load current rating of the motor. This would result in the tripping of the motor protection, i.e. a fuse or circuit-breaker.

CHECK YOUR UNDERSTANDING

12.5 Explain what a rotating magnetic field is and how it is created.

12.6 How is the direction of rotation determined?

12.7 How is direction of rotation reversed?

12.8 How is torque produced in a motor?

12.3 Three-phase induction motor characteristics

12.3.1 Squirrel cage motors

When power is first applied to a stationary motor, the stator windings act as transformer primary windings, with the resultant magnetic field rotating at synchronous speed. The rotor then behaves as a shorted secondary winding, causing a high circulating current in the rotor bars and a high starting current in the stator windings. As the rotor accelerates in the direction of the rotating field, the difference between its speed and the rotating magnetic field becomes less, as does the generated voltage causing the rotor circulating currents. This in turn reduces the stator current.

The typical relationship between the stator current and the rotor speed is shown in Figure 12.15(a). The initial circulating current in the rotor is affected by the frequency of the supply, the resistance of the rotor bars and the inductance of the rotor circuit—that is, the current-limiting factor is the impedance of the rotor circuit. With the standard type of power transformer, the frequency in both the primary and secondary windings are at line frequency.

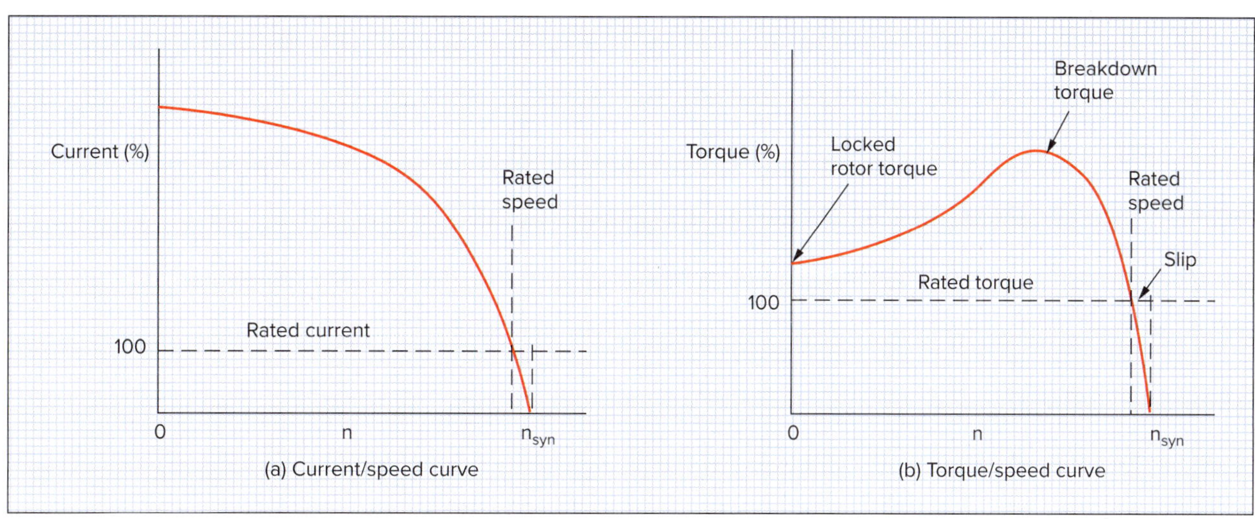

FIGURE 12.15 Relationships between motor speed, line current and torque

However, with a motor the initial frequency of the induced voltage and currents within the rotor are at line frequency and steadily decrease as the motor speed increases. When the rotor voltage and current is at line frequency, the inductive reactance is significant and causes the rotor current to lag the rotor voltage by almost 90°. The result of this is to also place the magnetic flux generated by the rotor almost 90° behind the flux from the stator. This causes low torque. As the motor accelerates and the frequency of the induced voltage and the current in the rotor decreases, the rotor inductive reactance also decreases and the voltage and current move more in phase with each other. The rotor and stator magnetic flux also move more in phase, and this strengthens the torque produced by the motor. Consequently, the torque created can change as the speed changes. Figure 12.15(b) shows the typical relationships between speed and torque.

For small values of slip, the torque is assumed to be proportional to the slip. As the motor load increases, the torque increases and the speed decreases, until the torque reaches a maximum value (the breakdown torque). If the motor is loaded beyond this point, the torque and the speed both decrease and the motor quickly comes to a standstill. For a general purpose motor, an overall figure for starting torque is around 1.5 times the rated torque; the breakdown torque is usually about twice the rated torque. AS/NZS 1359.41 sets out minimum requirements for these torque values and provides a table for a range of motor sizes.

For a squirrel cage motor, the resistance of the rotor conductors remains constant at power-line frequencies for all practical purposes, while the inductive reactance decreases as the rotor speed increases. As a guide, torque reaches a maximum when the rotor resistance in ohms is equal to the rotor reactance in ohms. Since the resistance is fixed, the breakdown torque can be altered only in relationship to the motor speed by altering the inductance of the rotor. This in turn affects the starting torque. AS/NZS 1359.41 allows for only two basic types of rotor—normal and high torque. Any other types necessarily require prior arrangement with the manufacturer. For the high-torque motor, the required breakdown torque remains around twice the rated torque, while the starting torque is increased to approximately 2.5 times the rated torque.

12.3.2 Squirrel cage motor operating characteristics—conditions necessary for an induction motor to produce maximum torque

The rotor bars in a squirrel cage rotor can be designed and manufactured to suit different operating conditions. In general, for any one stator, only the impedance of the rotor can affect the current flowing in the rotor bars. Impedance can in turn be broken down into resistance and inductive reactance. Altering either affects the current flow.

Resistance can be altered by changing the cross-sectional area of the bars, while the inductive reactance can be altered by varying the depth and shape of the bars in the iron core of the rotor.

Figure 12.16 shows typically shaped rotor bars where the starting current is around six to seven times the rated current and the motor has a starting torque of approximately 150% of the rated running torque. Its use is restricted to extremely low starting torque requirements.

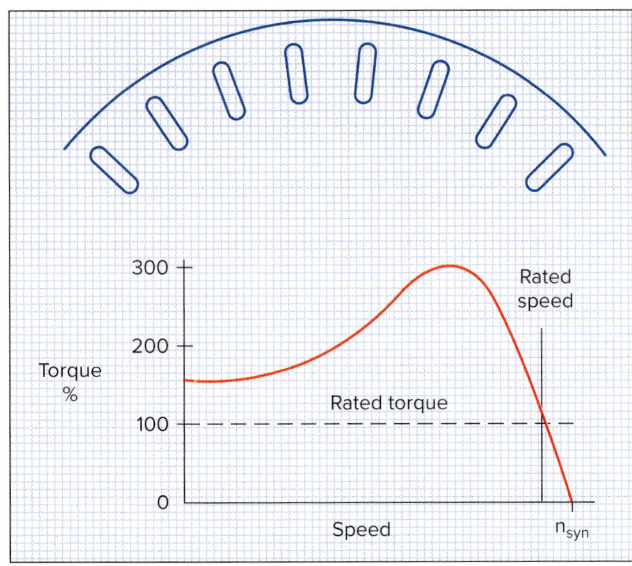

FIGURE 12.16 **Low starting torque rotor**

Figure 12.17 shows rotor bars of greater cross-section, where part is embedded deeper into the rotor magnetic circuit. Starting torque is still about 150% of running torque, but the starting current is reduced to about five times the running current. It is suitable for use with equipment of low starting inertia such as fans, blowers and some types of machinery.

Figure 12.18 gives one example of a rotor with two sets of rotor bars. The inner set has half as many bars as the outer set and includes an optional air gap.

Depending on performance requirements, there may be different-shaped bars, no air gap or a full set of bars in the cage. Starting torque is high—here it is 225% of rated torque—and starting current is about five times the rated running current. Applications are in air compressors, crushers, refrigerator compressor motors or reciprocating force pumps.

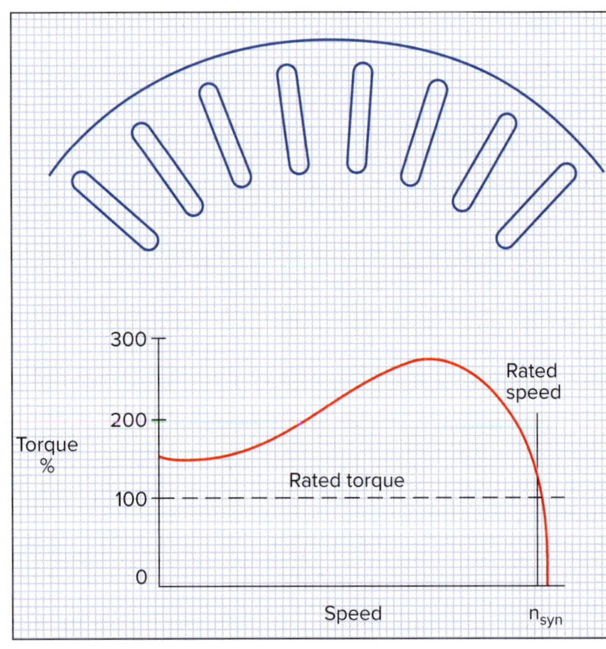

FIGURE 12.17 **Standard rotor bars**

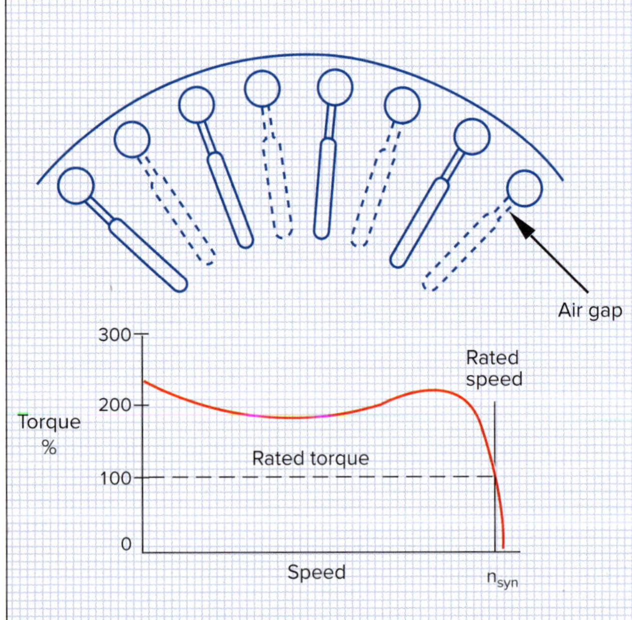

FIGURE 12.18 **Double cage rotor for high starting torque**

A typical example of high-resistance rotor bars with low starting current requirements is shown in Figure 12.19. With this construction, the starting torque can be increased to about 275% with low starting currents. This comes at the cost of a lower-rated speed (i.e. increased slip). Typical uses are flywheel-mounted machinery such as presses and punches. This type of construction is excellent with hoists, where the maximum load occurs at the start of the lift. The above details apply particularly to copper rotor bars.

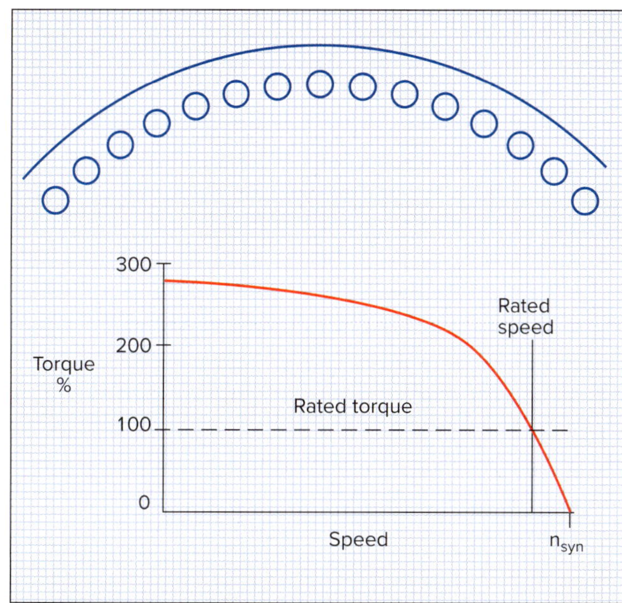

FIGURE 12.19 **High-resistance rotor bars**

If aluminium is used for the rotor bars, their cross-sectional area must be increased to allow for the metal's higher resistivity. The shape may also be changed to incorporate desired starting and running characteristics. Figure 12.20 shows a teardrop-shaped cast aluminium bar. In practice, the shape may also be inverted to alter the characteristics to suit a particular purpose.

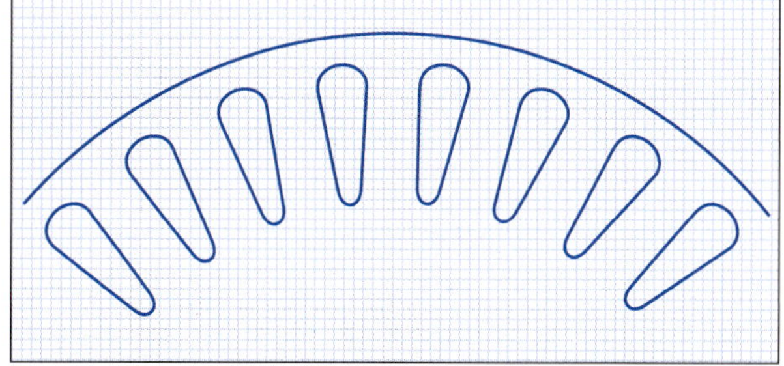

FIGURE 12.20 **Cast aluminium rotor bars**

12.3.3 Wound-rotor motors

The introduction of resistance into the rotor circuit of an induction motor produces three effects:

1. The rotor current is reduced, resulting in less stator current.
2. The starting torque is increased because rotor and stator magnetic fields are more in phase with each other.
3. The slip speed is increased.

An adjustable resistor is used external to the rotor, which is wound with comparatively low-resistance windings. The value of the external resistance can be adjusted as required and, as the motor accelerates, the value is gradually reduced until all the resistance is out of the rotor circuit and the motor behaves as an ordinary induction motor.

The torque-speed characteristic of a typical three-stage wound-rotor motor is shown in Figure 12.21. When all the resistance is in the rotor circuit, the starting current is low and the starting torque is high, as shown by curve (a). If this resistance is left in, the full-load torque would occur at approximately 25% slip, resulting in extremely poor speed regulation.

If one stage of the resistance in the rotor circuit is shorted out, the operating characteristics are modified as shown by curve (b). If all the external resistance in the rotor circuit is shorted out, the operating characteristic is shown by curve (c).

The normal starting procedure is to start the motor with all the resistance in the rotor circuit. As the motor speeds up, the resistance is reduced. The motor increases in speed but maintains a high torque. During the starting procedure, the torque-speed curve is as shown by the thicker curve (d).

By comparing Figures 12.16 and 12.21, it can be seen that full-load torque occurs at a greater slip in a wound-rotor motor than a squirrel cage motor. This is due to the extra resistance of the windings in the wound rotor.

The main use for wound-rotor motors is the starting of high-inertia loads that may take several minutes to attain operating speed. The high starting torque and reduced starting current of this type of motor ensure that the motor windings are not subjected to excessive starting currents for any length of time.

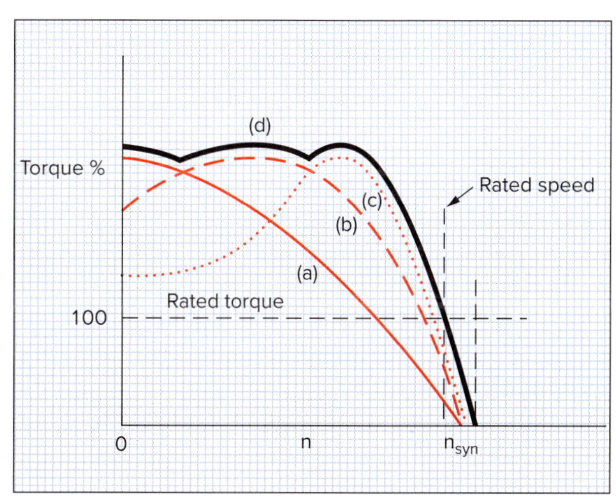

FIGURE 12.21 **Operating characteristics for a wound-rotor motor**

Applications for wound-rotor motors invariably involve the need to get pairs of large flywheels up to speed. Once they are, they can absorb the impact shock of sudden loads and so enable the machinery to continue operating. Applications include large air compressors, metal presses and stone-crusher heads that are used in quarries to reduce excavated material to suitable sizes before grading.

12.3.4 Operational parameters for induction motors

On no load, the stator current of any induction motor is largely a magnetising current, with a small energy component required to supply the no-load losses. Accordingly, the power factor of an induction motor on no load is extremely low. The no-load current is relatively high when compared with a transformer because of the high reluctance of the magnetic circuit resulting from the air gap between the stator and rotor.

The stator flux remains constant from no load to full load, so the magnetising current is also constant. In Figure 12.22(a) the no-load stator current I_0 lags the supply voltage by Φ degrees.

As a load is applied to the motor, a load current I'_1 is required to accommodate it. This load current I'_1 lags the supply voltage slightly, owing to the effect of the stator and rotor reactance. The two current components I_0 and I'_1 combine to give the total stator current I_1 at that load. The phase-angle decreases from Φ to Φ_1, and the power factor of the induction motor increases as the load on the motor increases.

Figure 12.23 gives representative characteristic shapes for some parameters of a three-phase induction motor. It shows the speed decreasing and the slip increasing as the load on the motor is increased. It also shows the line current increasing and the power factor improving simultaneously.

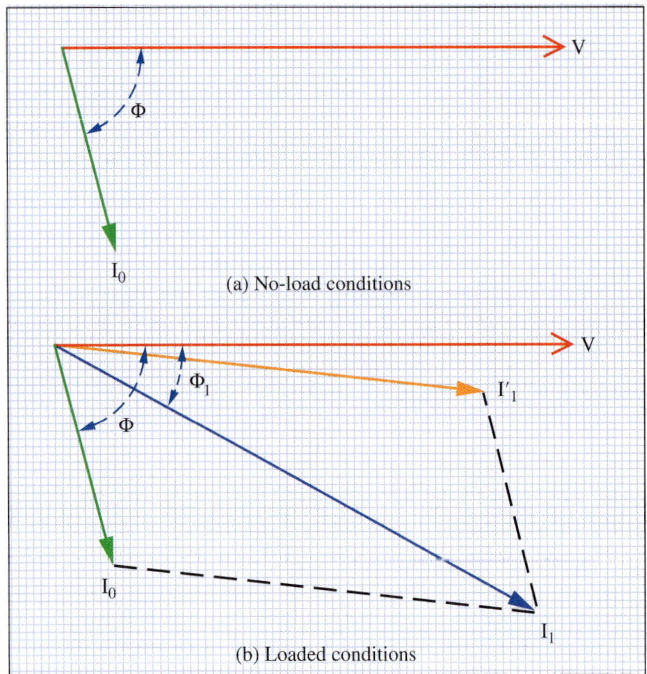

FIGURE 12.22 Current phasors for an induction motor

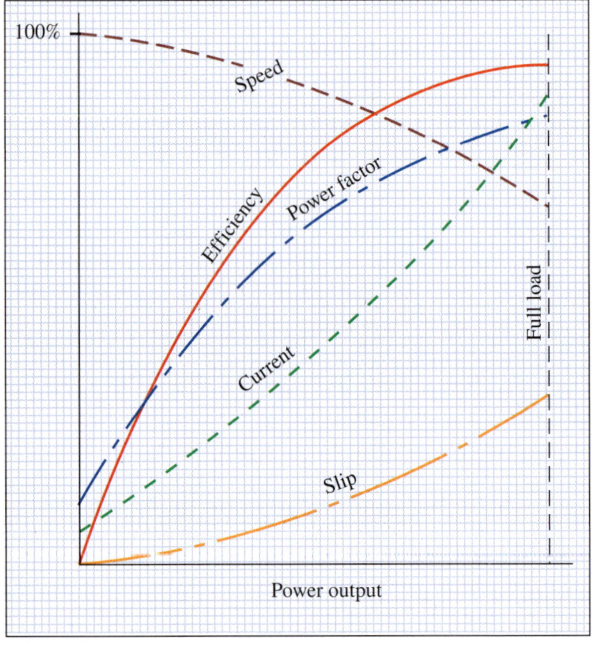

FIGURE 12.23 Typical parameters for a three-phase induction motor

12.3.5 Abnormal operating conditions for three-phase motors

The satisfactory operation of three-phase motors on a three-phase supply depends on the factors described below.

Three equal voltages at the correct phase displacement

Under normal operating conditions, the phase displacement is a function of the generating equipment and stays relatively fixed. However, the line voltages can vary, depending on the individual loads connected at that time. For balanced loads, such as three-phase motors, unbalanced phase voltages lead to unbalanced currents flowing in the motor windings. Consequently, circulating currents are set up, heating is increased and torque is reduced.

Stator windings being correctly connected in either star or delta configuration

If windings are incorrectly connected in star or delta, phase currents may become unbalanced. This will result in additional heat being generated in the windings and torque will be greatly reduced. See section 12.3.6 for a typical example.

All three line voltages being connected to the motor windings

When any one supply line is unable to supply current to the winding to which it is connected, the condition known as 'single phasing' occurs. See section 12.3.7 for more details.

The condition of all four windings in the motor

The three stator windings connected to the supply are prominent and therefore can be obvious areas of concern. Problems with the rotor winding, however, may be less obvious. Noisy operation and reduced torque of a three-phase motor can mean that the bars of the rotor might need attention. Many cages consist of aluminium cast into shape in the laminations, and little can be done in the way of maintenance; however, many of the larger motors have prefabricated bars and rings of copper which are welded into place. It is possible to repair or replace items, whether they are broken or simply loose in the rotor.

12.3.6 Phase reversal

So far in this chapter, the operation of the induction motor has been described based on the assumption that the motor has three identical windings and three equal currents flowing in them, all spaced at 120°E to each other. If one phase is reversed (as shown in Figure 12.24(a) for a star connection), these conditions no longer hold true. Two of the three currents that flow are at 60°E to each other and the load system is unbalanced (Figure 12.24(b)). The same condition applies to delta-connected loads.

As a result of this incorrect connection, the motor loses most of its torque and is often unable to start against even a light load. If it can start at all, it usually rotates very slowly and has unequal values of current in the phase windings. The values of current approach those drawn during normal starting but remain high. The motor usually emits a 'growling' noise and has an associated vibration due to the sustained high-current values.

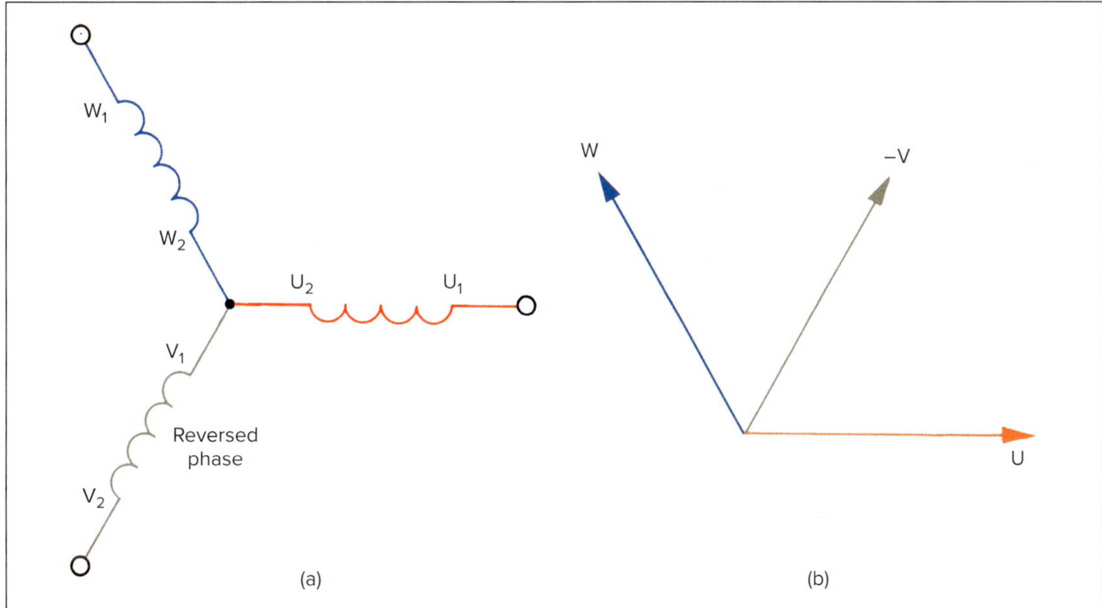

FIGURE 12.24 Reversal of one phase for a star-connected motor

12.3.7 Single phasing

Single phasing is a condition that occurs when one line of a three-phase supply is open-circuited and is not able to supply current to a three-phase load. The term is also used when one of the three-phase windings in a load is open-circuited.

The condition for single phasing in a star-connected load is shown in Figure 12.25(a). A break in either the line or the phase winding reduces the circuit to a single current path.

Figure 12.25(b) shows an open-circuited line for a delta-connected motor. There is one current path from L_1 to L_3 through winding U, and another path from L_1 to L_3 through windings V and W in series. Both currents are in parallel with each other, although not necessarily in phase with each other. In Figure 12.25(c), a delta-connected induction motor is shown with winding W open-circuited. There are two current paths—L_1 through winding U to L_3 and L_2 through winding V to L_3.

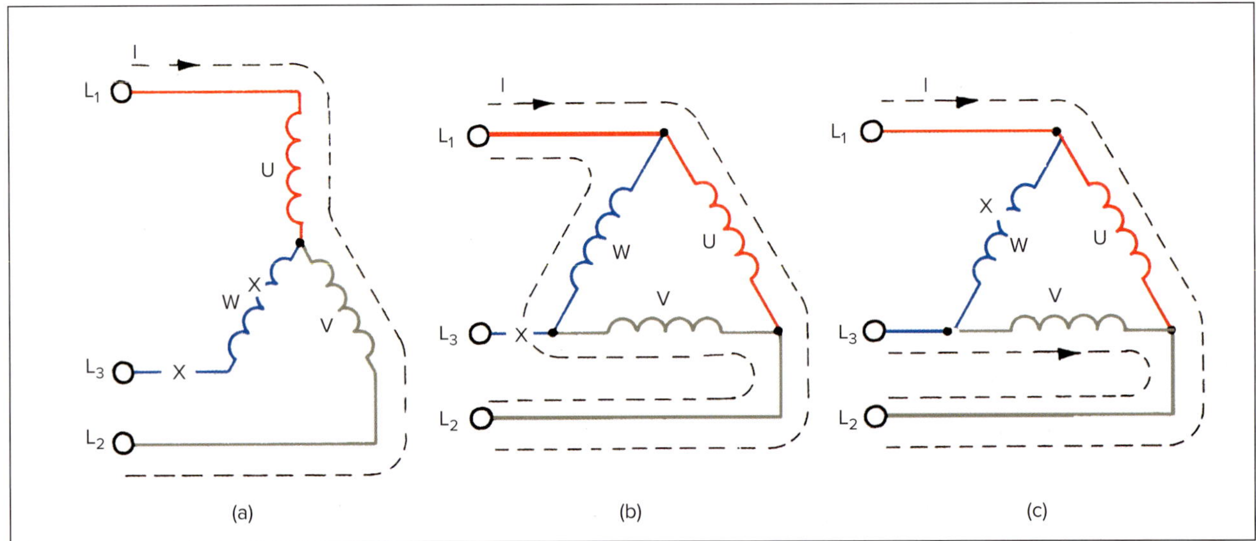

FIGURE 12.25 **Circuit conditions causing single phasing with a three-phase motor**

In each of the cases shown in Figure 12.25, the rotating magnetic field is either destroyed or unbalanced and causes unsatisfactory operation of the motor. The motor rotates at slower speeds (if it is able to start at all) because of a much reduced starting torque. It usually draws higher-than-normal currents in the parts of the circuit still operating, with values approaching starting current values in some circumstances. It can also emit a low-pitched 'growling' noise similar to that which occurs during a phase reversal. If single phasing occurs while the motor is operating at normal speeds, the normal humming sound often changes to a higher-pitched whine. For any of the conditions for single phasing that have been outlined, the three ratios between line and phase values are no longer valid.

12.3.8 Overloading

As discussed in section 12.3.1, standards are laid down for the operation and performance of induction motors, particularly for their starting and running torques. There are also stated limits for the amount of torque a motor can exert when it is unable to start or is stalled by the connected load. The limits are variable, depending on the size or rating of the motor.

These requirements are built into the motor design and are beyond the control of anyone attending to the maintenance or installation of a motor. Overloading is a condition that is usually brought to the attention of a technician only when a motor is behaving in an abnormal manner.

As a general guide, the motor may run at a slower speed and higher temperature than normal. The varnish used on the windings might start to smell, and in extreme cases smoke might start to issue from the windings.

Once a motor is unable to meet the requirements of the applied load, it will come to a rapid stop and *locked rotor* conditions then apply (see Figure 12.15(b), where the 'breakdown torque' point is labelled). Current more than normal full load will then be drawn, and the installed motor protection must rapidly disconnect the motor from the supply to prevent damage.

CHECK YOUR UNDERSTANDING

12.9 How do rotor frequency and rotor resistance affect a motor's torque during motor start-up?

12.10 What are some of the abnormal operating conditions for a three-phase motor?

12.4 Single-phase motors—split phase

12.4.1 Introduction

The induction motor is highly regarded because of its simplicity, ruggedness and reliability. The three-phase induction motors discussed earlier had inherent starting torque due to a rotating field; the single-phase induction motor, by contrast, initially has no rotating field and therefore no starting torque.

In the three-phase motor, the supply consists of three identical currents being supplied to three identical windings in the motor. The resultant magnetic field rotates at a constant speed and strength. Similarly, if two identical currents at 90°E could be supplied to two identical windings displaced by 90°E in a two-phase motor, the magnetic field would rotate at a synchronous speed determined by the supply frequency and the number of poles in the windings.

12.4.2 Single-phase induction motors

A single-phase motor can easily be wound with two identical windings but, on connecting them to a single-phase supply, the individual winding currents may well be closely in phase with each other and not produce a rotating magnetic field. The two currents must be displaced electrically from each other by some means to produce a rotating field. This can be achieved by having windings of different inductances and resistances. Sometimes a capacitor is also added in series with one winding to enhance the phase displacement. A single-phase motor is shown in Figure 12.26.

Once the motor is rotating at sufficient speed, one of the windings can be disconnected and it will continue to rotate. Due to the more uneven magnetisation of the iron core, the motor develops a pronounced vibration that is a characteristic of the single-phase motor. This vibration is at twice the supply-line frequency and tends to make the motor noisier than a three-phase motor in operation.

The motor also develops a small amount of negative torque, which is a function of slip speed. It results in a rather high no-load current at low power factor. When a load is applied, the current increases only marginally but the power factor improves in a similar fashion to that of the three-phase motor.

The above considerations mean that special techniques must be adopted to ensure the starting of the single-phase motor. Due to the various starting methods employed, there are several versions of the single-phase motor.

FIGURE 12.26 Single-phase motor

12.4.3 Split-phase motors

The standard split-phase induction motor has two separate windings (start and run) connected across the supply during the starting process (see Figure 12.27(a)). For normal running, however, only the run winding is used.

The run winding consists of a number of coils connected in series to form a set number of poles. A four-pole machine, for example, has four groups of coils, all in series and physically displaced around the stator to form four separate poles (see Figure 12.27(b)).

The run winding is wound with a heavier-gauge wire to reduce its resistance. To increase the inductance of the run winding, the coils are embedded deep into the slots of the iron core and usually have more turns than the start winding. The current I_R flowing through the run winding is highly reactive, and so it lags the applied voltage V by a considerable angle ϕ_R (see Figure 12.27(c)).

The start winding also consists of a number of coils connected in series to form a set number of poles. If the machine has four poles in the run winding, it will also have four poles in the start winding. However, the start winding is physically displaced by 90°E around the stator core (see Figure 12.27(b)).

The start winding of the standard split-phase motor is wound with finer-gauge wire, increasing the winding's resistance. When compared with the run winding, the start winding has fewer turns and the coils are placed nearer the surface of the slots in the stator core, reducing the inductance of the winding. The net result is that the current I_S flowing in the start winding is more in phase with the voltage V than I_R in the run winding. In Figure 12.27(c), I_R lags V by ϕ_R and I_S lags V by ϕ_S. This produces a phase displacement of ϕ between the two currents, resulting in a phase displacement between the respective fluxes of the two windings.

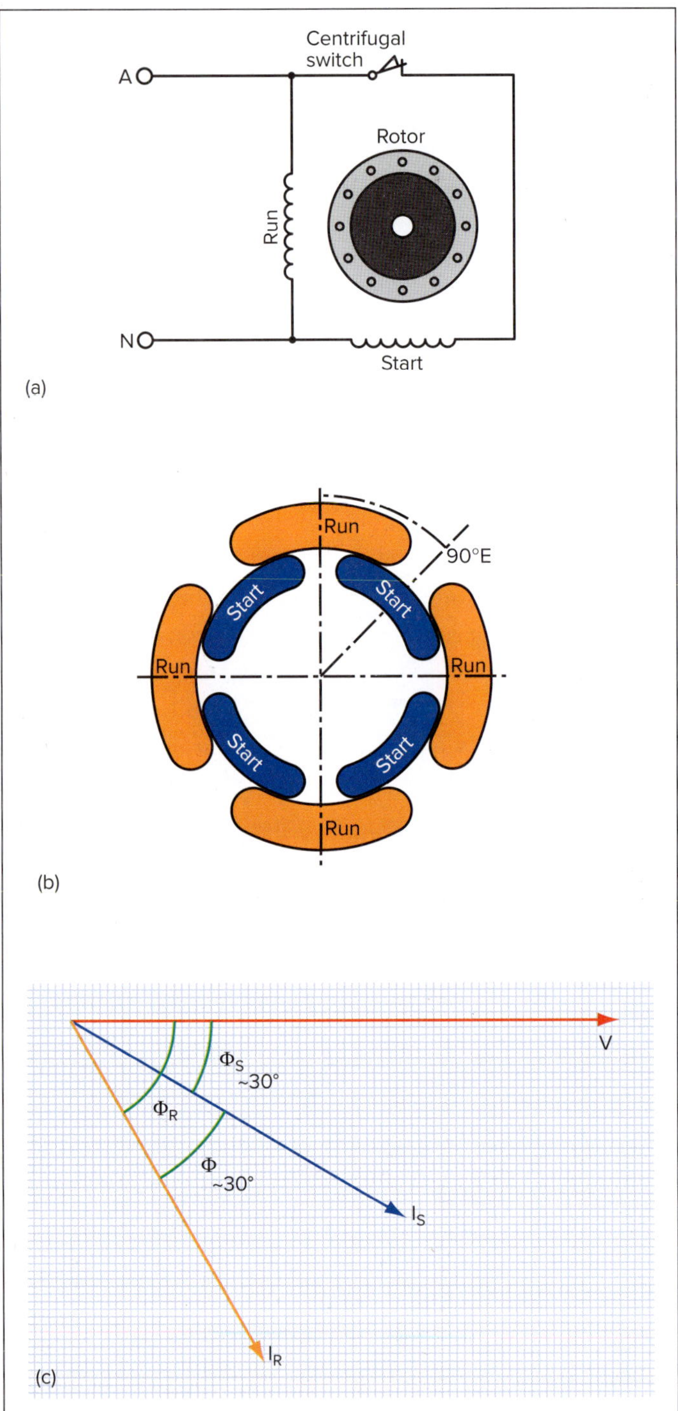

FIGURE 12.27 Split-phase motor: (a) split-phase induction motor circuit (b) four-pole machine (c) phase relationships between start and run winding currents

Starting

For a two-pole machine, the windings are also physically displaced by 90°E (see Figure 12.28(a)). Assuming a 30° phase displacement between the two fluxes, at position 'a' in Figure 12.28(b) the run flux ϕ_R is zero and the start flux ϕ_S is 50% of the maximum value in a positive direction. The resultant stator flux is shown at position 'a' in Figure 12.28(c).

At position 'b' in Figure 12.28(b), ϕ_R is 50% of its maximum value and ϕ_S is 86.6% of its maximum value. These combine to form the resultant stator flux 'b' in Figure 12.28(c). At position 'c' in Figure 12.28(b), ϕ_R is 86.6% of its maximum value and ϕ_S is 100%. The resultant stator flux 'c' is shown in Figure 12.28(c). By taking each position from 'a' to 'l' in Figure 12.28(b), it can be seen that the stator flux rotates one full revolution for one full cycle. The direction of rotation is the same direction as that of the resultant magnetic field.

The stator flux rotates at a speed governed by the supply frequency and the number of poles in a winding. That is:

$$n = \frac{120f}{p}$$

where:

f = frequency
p = number of poles
n = speed in rpm.

For a two-pole machine on a 50 Hz supply:

$$n = 3000 \text{ rpm}$$

For a four-pole machine:

$$n = 1500 \text{ rpm}$$

If the direction of current flow through one winding is reversed, the resultant magnetic field rotates in the reverse direction. That is, the direction of rotation of the motor is reversed by changing the direction of current flow through one winding. This is done by exchanging the two end connections of any one winding.

As seen in Figure 12.28(c), the rotating stator flux is not of uniform value and an elliptical field pattern results. This produces considerable vibrations and humming noise during starting.

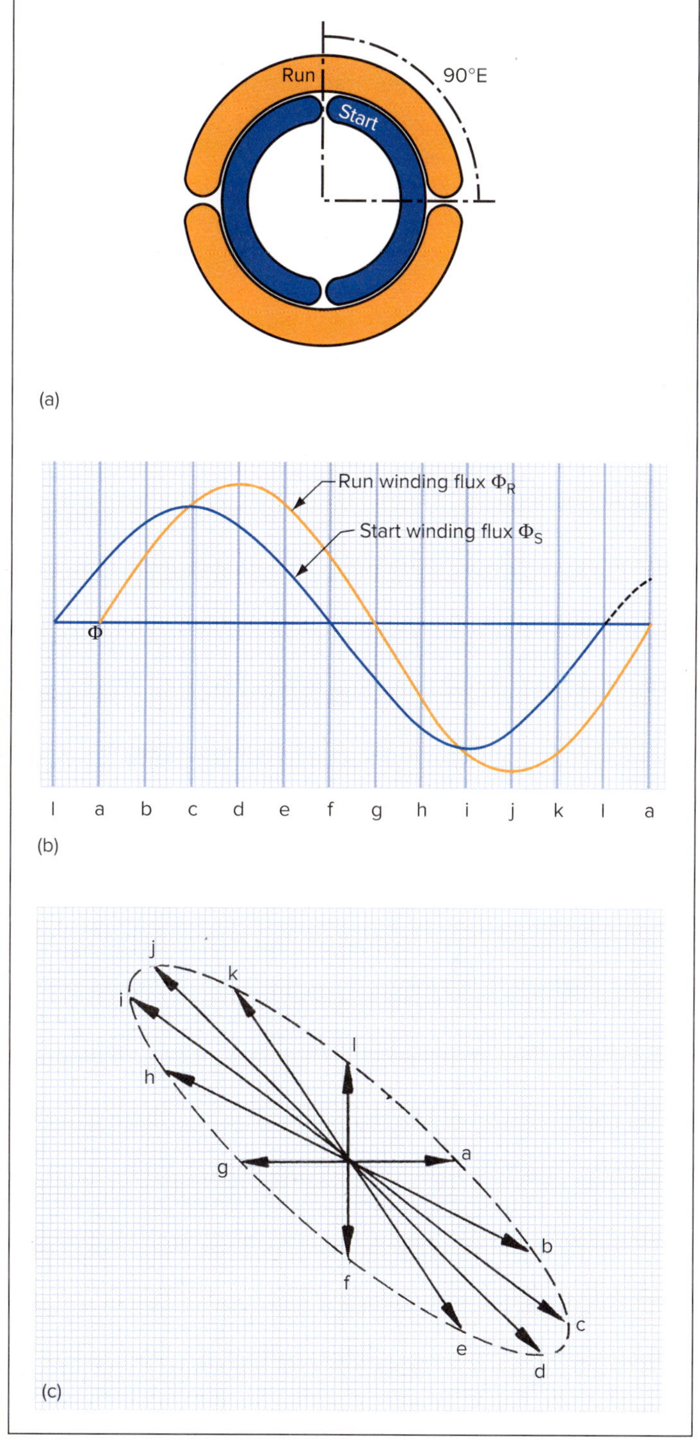

FIGURE 12.28 Rotating field of a split-phase motor: (a) two-pole machine (b) start and run winding flux (c) resultant stator field

The rotating stator field cuts the rotor bars and induces a voltage in them. As the rotor bars are shorted out, a current flows through them and produces a rotor flux. The stator flux and the rotor flux interact to produce a force on the rotor bars, causing the rotor to turn in the direction in which the stator flux is rotating. This rotating force is called the 'starting torque' and largely depends on the relative strengths of the start and run fluxes, as well as on the phase displacement between the currents flowing through both windings.

The start and run windings are connected in parallel across the supply voltage. When the rotor has reached sufficient speed to provide a strong cross-flux, the start winding can be open-circuited. This is usually done by connecting a centrifugally operated switch in series with the start winding.

The centrifugal switch is usually set to open when the rotor speed reaches approximately 75% of the rated speed of the motor. When the motor is switched off, the rotor slows down and the centrifugal mechanism operates, closing the switch contacts again in readiness for the next starting operation. As the start winding is only connected during the starting procedure, it is designed for a very short duty cycle. If the centrifugal switch fails to operate, the start winding will quickly overheat and burn out.

Running

When the rotor speed of the standard split-phase motor reaches approximately 75% of the synchronous speed, the centrifugal switch open-circuits the start winding and only the run winding is connected to the supply.

For a two-pole motor, when the stator current flows in one direction for one half-cycle, a magnetic field is produced in the direction C–A, as in Figure 12.29(a). During the next half-cycle, when the stator current is reversed, the magnetic field also reverses and is in the direction A–C, as in Figure 12.29(b).

This stator field (which is produced by the run windings) varies in strength and direction according to the supply, but it does not rotate. It is a stationary pulsating field. This is why some form of starting (i.e. start winding) is required for split-phase motors.

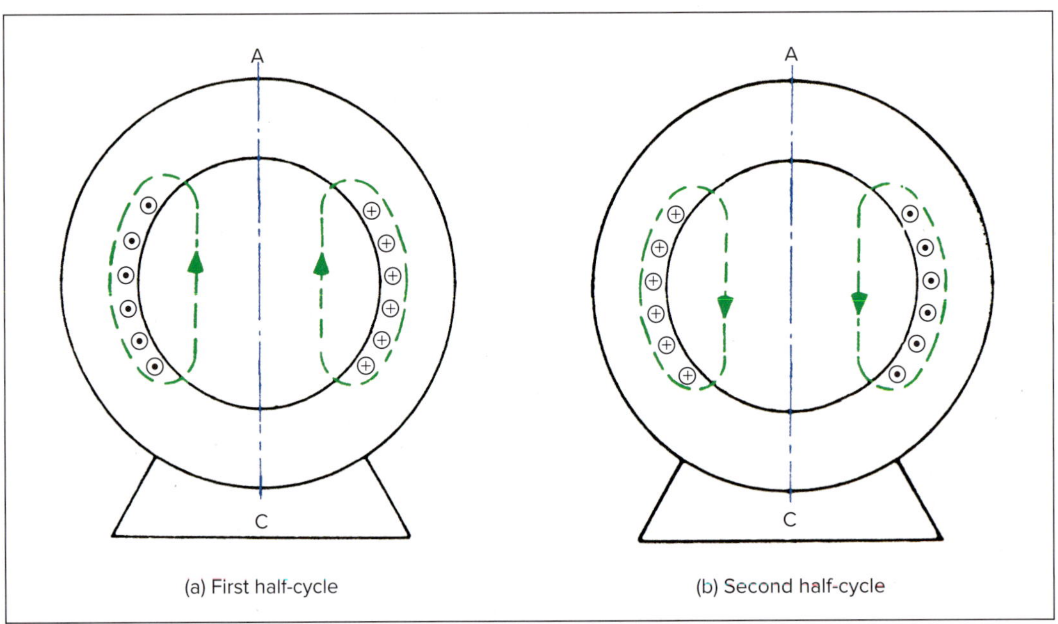

FIGURE 12.29 Pulsating stator field

When the stator winding is connected to the a.c. supply and the rotor is turning, the rotor bars cut the stator flux, causing an EMF to be generated in them. In Figure 12.30(a) the rotor is revolving clockwise, and the stator field is acting in the direction C–A. By Fleming's right-hand rule, the generated EMF in the rotor acts in the direction shown ('out' of the page) in all rotor bars above the axis D–B (indicated by the dots) and 'into' the page in all rotor bars below the axis D–B (indicated by +).

The induced rotor voltages are in phase with the stator flux and cause a rotor current to flow. Because of the low resistance and high inductance of the rotor bars, these currents lag the induced rotor voltage by nearly 90°. Consequently, the rotor currents produce a rotor flux lagging almost 90° behind the stator flux and acting in the direction D–B (see Figure 12.30(b)).

Because the rotor flux is at right angles to the stator flux, it is often referred to as the 'cross-field'. The two fields effectively combine to form a rotating field, which tends to force the rotor bars in the direction in which the field rotates.

For one full cycle of the a.c. supply, the resultant field rotates 360°E. For the two-pole machine described, this constitutes one full revolution. For a four-pole machine, it will rotate a half-revolution for each full cycle of the a.c. supply.

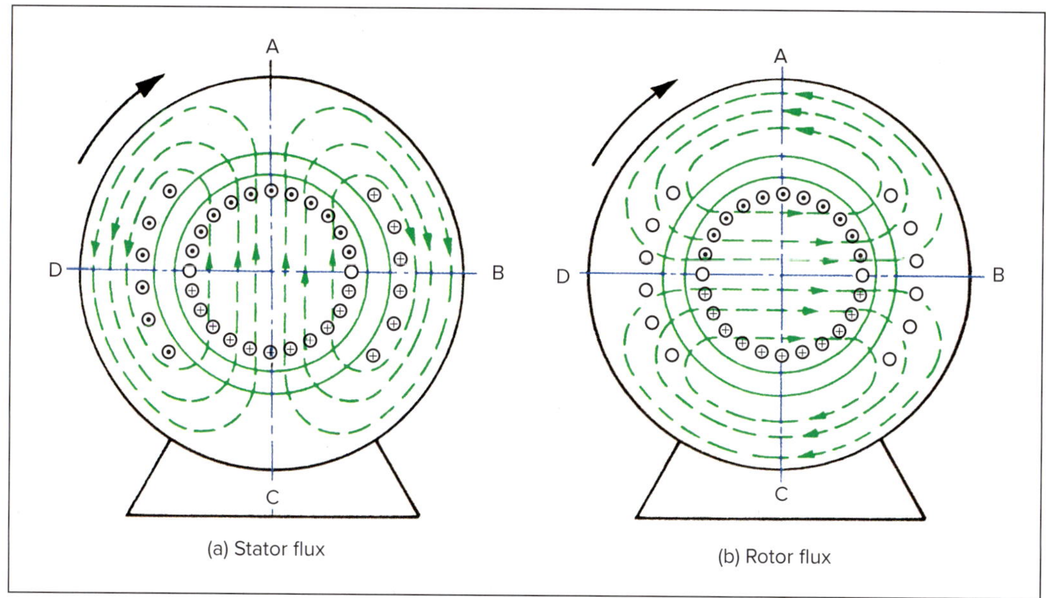

FIGURE 12.30 **Magnetic fields in a split-phase motor**

Owing to the internal losses within the rotor, however, the rotor itself will not rotate at synchronous speed but at a slightly slower speed.

Figure 12.31 shows a typical speed/torque characteristic curve for a split-phase motor. The break in the curve is caused by the switch operating to disconnect the starting winding. This is necessary to limit the losses in the motor and protect the start winding. The torque curves between the running and starting sequences normally do not coincide, so the speed and torque values must adjust when the switch operates. The values shown on the curve must be considered only as representative. They will vary from one make to another, and even within the same make (due to design changes).

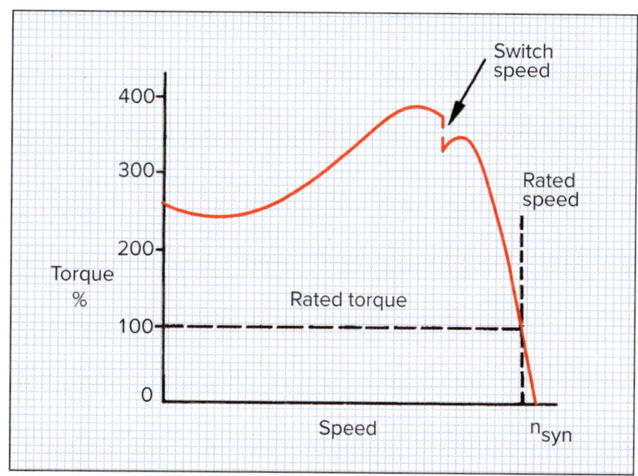

FIGURE 12.31 **Split-phase motor speed/torque curve**

Reversal of a split-phase motor is carried out by swapping the connections of either the auxiliary winding or the main winding—not both.

Split-phase motors have only moderate starting torque so they are limited to uses such as washing machines, blowers, buffing machines, grinders and machine tools.

12.4.4 Abnormal operating conditions—voltage fluctuation

Voltage fluctuation can be of two types—voltage rise or fall, where the voltages remain symmetrical or variations in individual phase voltages might occur (the latter is especially detrimental to the performance of three-phase motors).

The torque produced is proportional to the square of the voltage. That is, if the voltage drops to 90% of its nominal value, the torque reduces to 81% of its rated value. Similarly, if the voltage rises to 110%, then the torque increases to 121% of its rated value. For example, a 10 kW motor with a 10% voltage variation should now be rated at 8.1 kW or 12.1 kW.

Under normal operating conditions, a voltage variation of this magnitude should make only minor differences to the motor's characteristics. With a voltage increase, for example, the increased torque reduces the slip only slightly, so a motor rotating at 1450 rpm on 50 Hz would increase its speed to approximately 1455 rpm. If advantage is taken of the increased torque, the operating temperature could be expected to increase. With voltage variations greater than 10%, the motor must be de-rated to prevent excessive temperature rise. Motors are normally given a full-time rating for a specified temperature rise, and under these conditions might have to be switched off after a duty period and be allowed to cool down.

Starting and breakdown torque values are also affected. A voltage rise increases torque, while a voltage reduction decreases it. In the latter case, care must be taken to see that the motor does not stall while under load.

For three-phase motors, a more serious problem occurs when only one phase shifts in value. The phase current is affected to a greater degree, which affects the rotating magnetic field in the motor. This results in reduced starting, running and breakdown torques, increased running noise and vibration, full-load speed reduction and a higher operating temperature.

12.4.5 Abnormal operating conditions—higher operating temperatures

Common causes of overheating in motors are inadequate or restricted ventilation and overloading. Apart from accelerated deterioration of lubricants, possibly the most serious effect is on the insulation—at increased temperatures, there is a marked reduction in the life of insulation. For example, insulation designed to work at 90°C may have an expected life of 25 years. If the operating temperature is doubled to 180°C, the life expectancy is reduced to about 1.25 years. The result is increased efficiency of the cooling system and a decrease in the load applied to the motor.

12.4.6 Abnormal operating conditions—frequency variation

Motor speed can be regulated by the frequency of the supply. Allowances are made when selecting a motor for this purpose, but a variation in supply frequency under other circumstances can affect motor operation. The obvious effect is a change in speed, but there are also changes in power factor, efficiency and torque. A higher frequency causes an increase in power factor, a slight increase in efficiency and a decrease in torque. With a decreased frequency, the opposite occurs. A variation in the frequency of supply usually occurs when there are comparatively few large loads connected to a smaller supply.

12.4.7 Abnormal operating conditions—overloading

Manufacturers build into their motors the capability to handle short-duration overloads, as specified in AS/NZS 1359.41. In general terms, the Standard requires the motor to be able to withstand 1.5 times full load for a period of 15 seconds without appreciable change in speed or excessive heating. For a motor to operate under these conditions, it must also have a breakdown torque greater than the overload test figure.

The heating effect in a machine winding is related to the square of the current and the time it is flowing, so any excess current must result in a temperature rise. With short-duration overloads, the amount of heat generated is small and can be expected to be dissipated by the normal cooling process. Long periods of overload, however, can lead to excessive increases in temperature, which in turn lead to a shortened motor life. Other effects are a slight decrease in speed, decreased efficiency, decreased power factor and an increased possibility of stalling because the working torque is closer to the breakdown torque. When the motor stalls, it draws starting current at full line voltage until its protection system operates. Large amounts of heat can be generated in short periods of time under these conditions. Special types of motors are given restricted duty cycles. In effect, they are overloaded for short periods of time, after which they must be switched off and allowed to cool down to room temperature. If this type of motor is required to run on a continuous duty cycle, it must be de-rated to a lower power value.

12.4.8 Abnormal operating conditions—frequent starting

Motors in general are mechanically strong enough to handle normal loads with a fair safety factor. However, the number of times a motor is started cannot be provided for by the manufacturer unless this is specifically requested at the time of purchase or manufacture.

When a motor is started, there is a high current flow that decreases as the motor accelerates up to its operational speed. While this current is flowing, heat is being generated within the windings at a rate more than the usual heat dissipation rate. Under normal conditions, this excess heat is removed by the cooling system while the motor is in operation. With repetitive starting, however, the heat generated does not have sufficient time to be removed and the temperature of the motor rises.

The circumstances are similar for repeated reversing and plug braking. Starting current values repeatedly stress the windings, whether they are being used for starting, reversing or braking. Unless the coils are firmly braced, they rub against one another, eventually rubbing through the insulation and causing short-circuits within the windings of the motor.

12.4.9 Abnormal operating conditions—other factors

Other factors can cause—or constitute—abnormal operating conditions. These include exposure to corrosive fumes, explosive vapours, dust, steam, salt air, high humidity and operation in ambient temperatures of below approximately 10°C or above 40°C, or operation at altitudes in excess of 1000 metres. Generally, all these factors can subject a motor to damage of some kind; initially, though, the damage caused by them is due to the selection of the wrong type of stator housing at the time of purchase. Even motors selected for operation at elevated altitudes are subject to the same restrictions and must be rated by the manufacturer at the time of purchase for the conditions under which they will work.

Probably the most common abnormal operating condition for single-phase motors is encountered with centrifugal switch failure, which usually shows itself in the following two conditions.

1. Starting switch contacts remain closed after motor starts

A starting switch that remains closed effectively prevents the isolation of the starting winding from the power after the starting process has been achieved. Since most (if not all) starting windings are designed with a limited duty cycle, there is very little tolerance in terms of the amount of time they can remain energised. After a period of about 20 to 30 seconds, they overheat and can be permanently damaged.

2. Starting switch contacts open-circuit

A fault of this nature usually occurs when the switch becomes jammed in its running position. The next time the motor is required to start, the starting windings cannot function. The motor is unable to start its rotation and remains stalled, with full line voltage applied. A high starting current is maintained and is not able to reduce to a normal running value. The effect is that the excessive current continues and overheats the running winding, leading to possible burnout.

In both conditions, there is also the real possibility that the heat generated in one winding can be transferred to the other, causing damage to that winding as well. Without suitable motor protection, only prompt action on the part of the operator will prevent further damage.

12.4.10 Abnormal operating conditions—driven load

There are various mechanical load factors that can cause abnormal operation of motors. When installing or servicing a motor or testing it for faults, it is not only the electrical characteristics of the motor that need to be considered, but the mechanical factors of the load as well. These include:

- incorrect tension or misalignment on drive belts, resulting in excess wear and/or belt slippage
- poor alignment between the motor drive shaft and the mechanical load (shims inserted under the motor can correct misalignment)
- excess vibration due to loose mechanical components, misalignment, inadequate motor mountings, etc
- failure of load bearings due to any of the factors already mentioned
- loss of oil, grease or other lubricants required in the load, resulting in mechanical seizure
- the load being too great for the motor, resulting in excessive overloads, longer start-up time or stalling of the motor.

CHECK YOUR UNDERSTANDING

12.11 How does a magnetic field behave in a single-phase motor?

12.12 What conditions need to be met to allow a single-phase motor to start up and continue running?

12.13 List some of the abnormal operating conditions for a single-phase motor.

12.5 Single-phase motors—capacitor and shaded pole types

12.5.1 Capacitor-start motor

Design limitations restrict the split-phase motor to a maximum of about 30°E between the starting and running winding currents. To increase this angle and produce improved starting characteristics, a capacitor is connected in series with the starting winding (see Figure 12.32).

A capacitor-start motor is identical to a split-phase motor, with the addition of a capacitor which is usually mounted on top of the motor (see Figure 12.33). If the correct size capacitor is selected, the two currents will be at 90°E to each other and improved starting torque will be obtained.

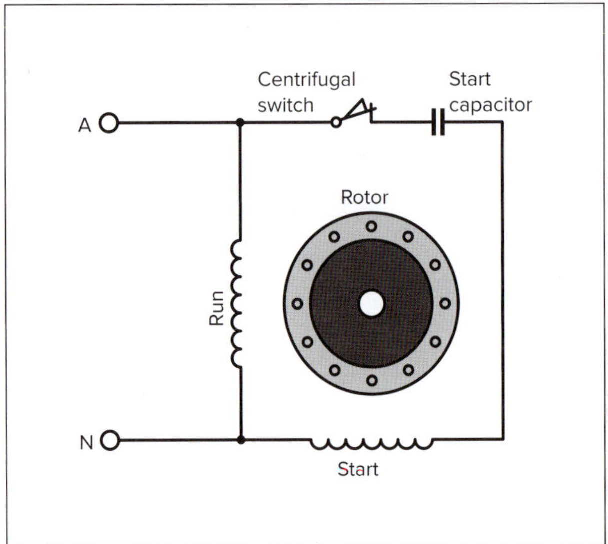

FIGURE 12.32 **Capacitor-start motor circuit**

FIGURE 12.33 **Capacitor-start motor**

On start, any value capacitor will increase the angle, but values that enable the starting winding to approach resonance must be avoided. For this reason it is advisable to use a capacitor much larger than necessary to ensure that resonance does not occur. The phasors in Figure 12.34 indicate the ideal phase displacement of 90°E. This angle of phase displacement between I_R and I_S provides a more uniform strength of stator flux during starting—a comparison of Figures 12.34 and Figure 12.35 is useful here. Owing to this more even field strength, the starting torque is higher than for the equivalent size split-phase motor.

Figure 12.36 shows the speed/torque curve for a capacitor-start motor (it is drawn to the same scale as Figure 12.31 to facilitate comparison). It demonstrates a large increase in starting torque due to the addition of the capacitor, while the torque is the same as for the split-phase motor after the switch has operated. In a similar fashion

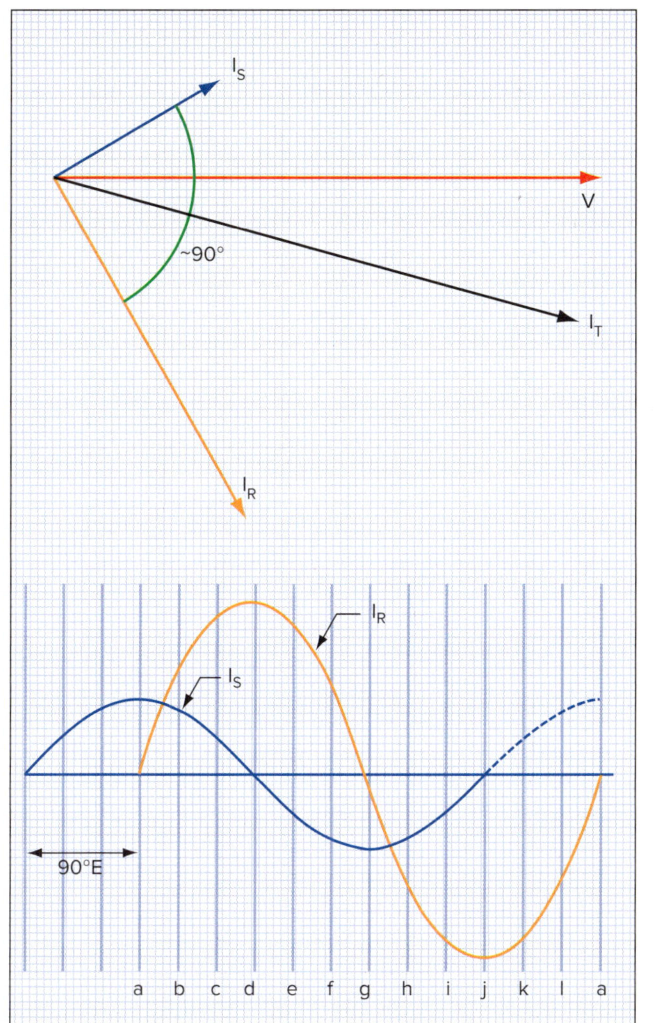

FIGURE 12.34 Phasor and waveform diagram

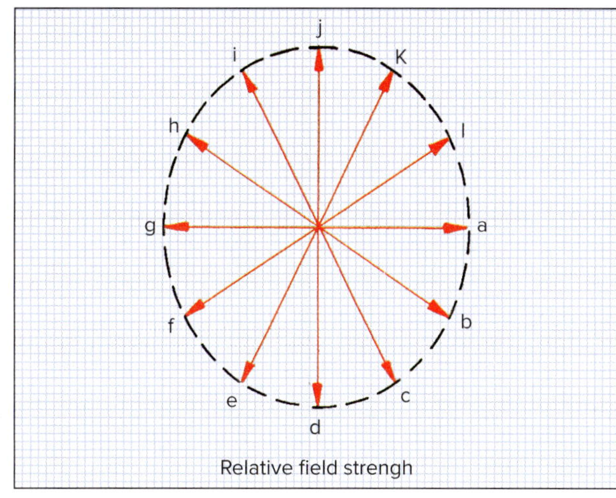

FIGURE 12.35 Resultant stator field

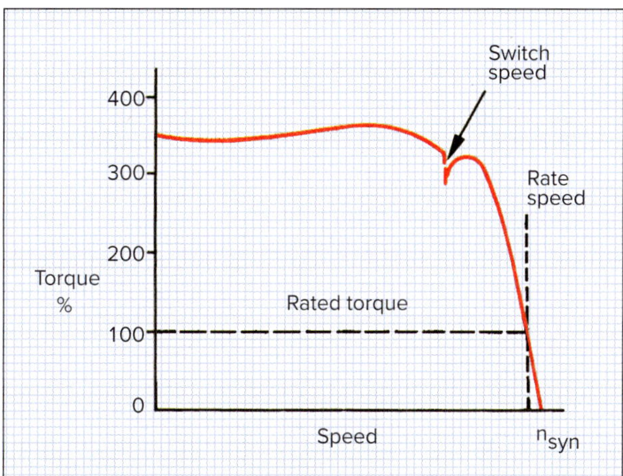

FIGURE 12.36 Capacitor-start motor speed/torque curve

to the split-phase motor, the switch operates at approximately 75% of full-load speed. The actual start windings for the two types of motor may consist of different data. The reversal of rotation is achieved by the same principles that apply to the split-phase motor. That is, the motor can be reversed by changing over the two leads of either start or run winding (but not both).

Capacitor-start motors are used in general-purpose, heavy-duty applications requiring high locked-rotor starting torque, such as that used in starting refrigerators and air compressors.

12.5.2 Capacitor-start/capacitor-run motor

This type of motor has both windings permanently connected across the supply, and these are referred to as the 'main' and 'auxiliary' windings. During starting, additional capacitance is connected in series with the auxiliary winding to provide the necessary phase displacement between the winding currents for maximum torque. The starting capacitor is therefore connected in parallel with the running capacitor. When the rotor speed reaches about 75% of the rated speed, the centrifugal switch disconnects the starting capacitor from the circuit (see Figure 12.35).

Figure 12.37 shows the circuit diagram from a capacitor-start/capacitor-run motor and Figure 12.38 shows a capacitor-start/capacitor-run motor. During operating conditions, the running capacitor ensures the correct phase displacement between the two currents in the windings, thus providing a constant-strength rotating magnetic field.

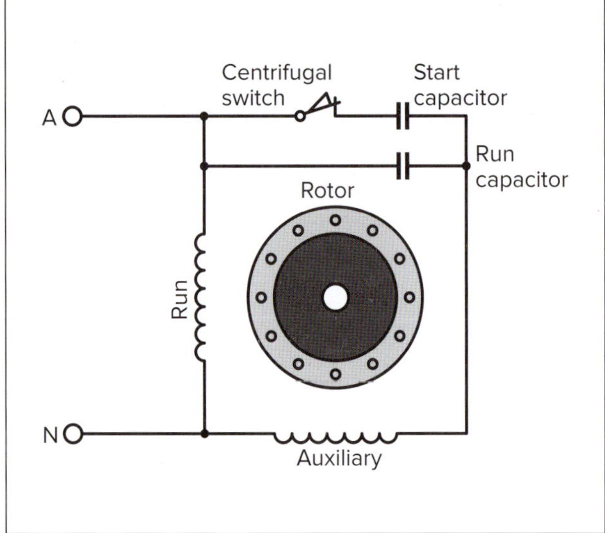

FIGURE 12.37 Capacitor-start/capacitor-run motor circuit

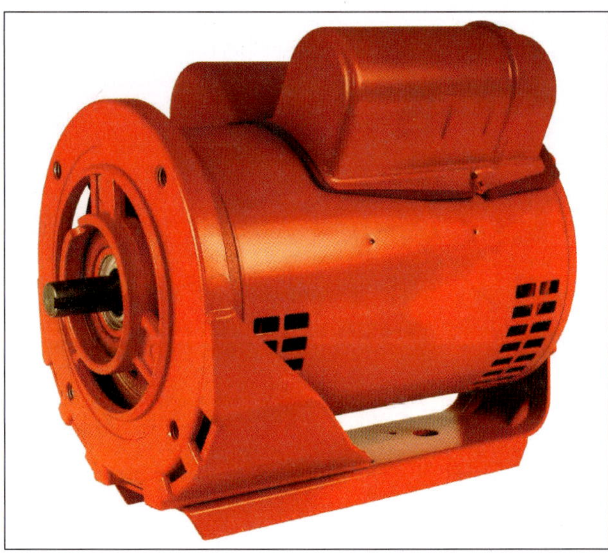

FIGURE 12.38 Capacitor-start/capacitor-run motor
I. Pilon/Shutterstock.com

The starting capacitor can be rated for intermittent duty, but the running capacitor must be of a construction suitable for continuous rating, such as the paper-spaced oil-filled type. The two-capacitor motor provides substantially the same running torque as the capacitor-start, induction-run type, but has beneficial effects. Adding the second capacitor:

- increases the breakdown torque
- improves full-load efficiency and power factor
- reduces operational noise and vibration
- increases locked-rotor torque.

Again, the direction of rotation can be reversed by changing over the two leads of either winding, but not of both. This changes the direction of rotation of the magnetic field in the stator.

The capacitor-start/capacitor-run motor is suitable for heavy-duty loads where quietness is a consideration and substantial starting torque is necessary (e.g. wall-mounted air-conditioning units where high head pressures are encountered in hot weather).

12.5.3 Permanent split capacitor (PSC) motor

The permanent split capacitor (PSC) motor also has both windings permanently connected across the supply, with a capacitor in series with one of them (see Figure 12.39).

For this type of motor, both windings are identical in wire size and number of turns, and they are also referred to as the main and auxiliary windings. Because the capacitor is in series with one winding, the current in that winding leads the current in the other, providing the phase displacement necessary to produce a rotating stator field. However, the phase displacement between the two fluxes is relatively small and so the starting torque is low.

By changing the line connection to the 'reverse' position shown in Figure 12.39, the capacitor is then in series with the main winding instead of the auxiliary winding. The current in the main winding leads that in the auxiliary winding and the rotor runs in the reverse direction.

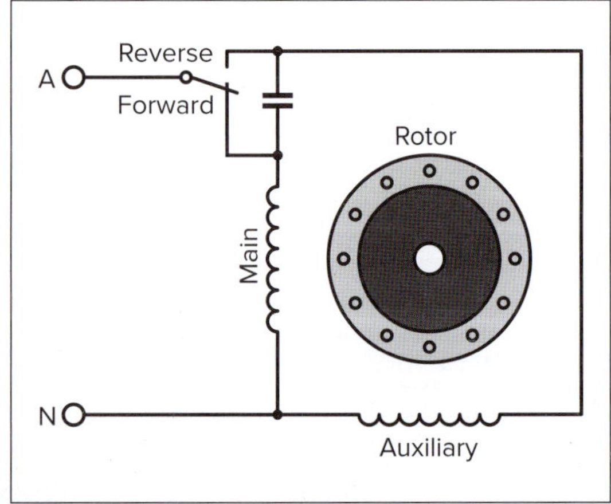

FIGURE 12.39 Permanent split capacitor motor circuit

The permanent split capacitor motor is suitable for light applications with low starting torque (e.g. fans and blowers) which might need to be reversed frequently. It is also suitable for remote control of induction regulators and dampers for regulating air flow in air-conditioning systems. The permanent split capacitor motor is suitable for unit heaters and fans because its speed can be varied easily with series inductances.

12.5.4 Shaded pole motor

The shaded pole motor has a cage rotor with salient poles in the stator. On one side of each pole a slot is cut, and a shading ring is embedded into it (see Figures 12.40 and 12.41). The shading rings are made of copper bar formed into a closed loop, providing a low-resistance path through the ring.

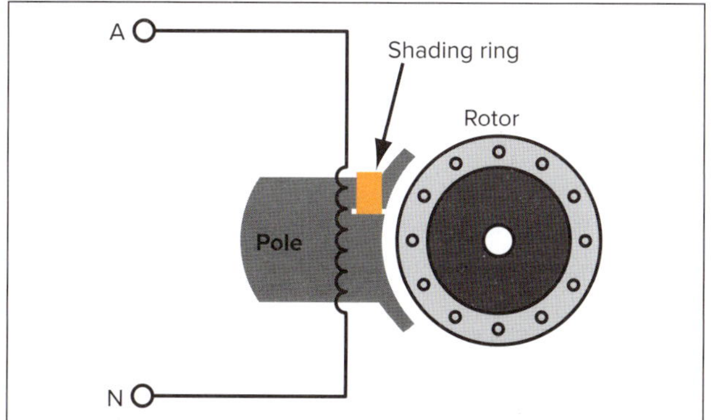

FIGURE 12.40 Shaded pole motor circuit

FIGURE 12.41 Salient poles, windings and shaded ring
Simon Dand

The supply current produces an alternating flux in each pole. This alternating flux cuts the shading ring, inducing an EMF in it. Because of the low-resistance path, the current flowing through the ring is relatively high. Also, according to Lenz's Law, the induced current in the shading ring will produce a flux that will tend to oppose the change of the main flux.

When the supply current rises rapidly from A to B (as in Figure 12.42(d)), an induced voltage is established in the shading ring. The current in the ring produces a flux that opposes the build-up of the main flux. As a result, the main flux is concentrated in the unshaded section of the pole (see Figure 12.42(a)).

When the current changes from B to C, there is little change in value of current and very little voltage is induced in the shading ring. Consequently, practically no current or flux is produced in the shaded ring. The main flux is at this time nearly always at maximum value and is uniformly distributed over the whole pole face (see Figure 12.42(b)).

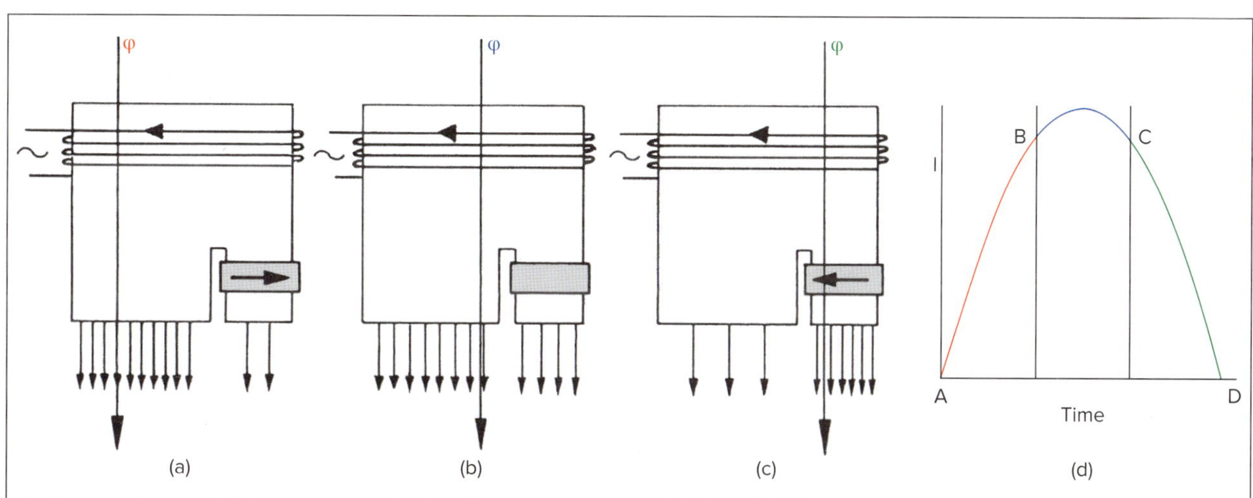

FIGURE 12.42 Movement of field in a shaded pole motor

When the supply current drops rapidly from C to D, an induced voltage is established in the shading ring. The current in the shading ring produces a flux that opposes the collapse of the main flux. The concentration of flux therefore occurs in the shaded section of the pole (see Figure 12.42(c)).

The magnetic axis shifts across the pole face from the unshaded part to the shaded part. This shifting flux is like a rotating field and produces a small torque, causing the rotor to rotate in the direction of the flux, towards the shaded section of the pole.

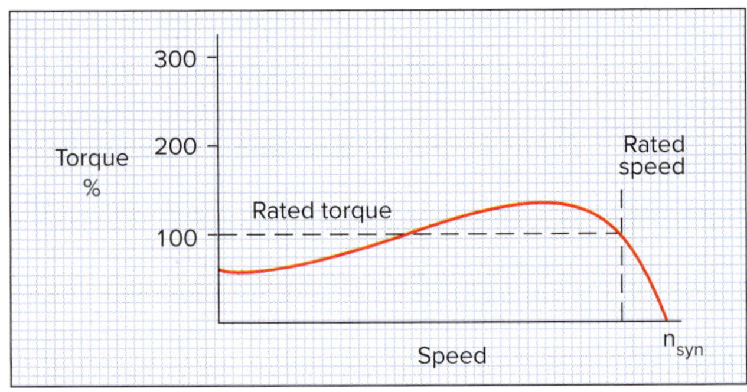

FIGURE 12.43 **Shaded pole motor speed/torque curve**

The starting torque of a shaded pole motor is extremely low, as indicated in Figure 12.43, and the motor runs with a slip speed slightly higher than the single-phase motors that were described earlier. It is simple in construction, low in cost and reliable. There are no switches, slip-rings, brushes or capacitors that might require maintenance. The motor efficiency is down, and this tends to restrict its use to low power ratings.

The direction of rotation has to be reversed by altering the direction of the rotating magnetic field across the pole face. This is done by shifting the shading ring to the other side of the pole face. Some poles are fitted with slots on both sides for this purpose, but with others the only method is to remove the stator from its housing and replace it the other way around in the frame.

Because its speed can be varied within a limited range by a series resistor or inductor, the shaded pole motor is suitable for fans and blowers, advertising signs, damper controllers, hair dryers and other applications where the starting torque requirements are minimal.

CHECK YOUR UNDERSTANDING

12.14 How does a single-phase capacitor-start motor work?

12.15 How does a single-phase capacitor-start/capacitor-run motor work?

12.16 What is a permanent split capacitor motor?

12.17 How does a shaded pole motor work?

12.6 Single-phase motors—universal motor

12.6.1 Operating principles of a series universal motor

The series motor is often called a 'universal' motor because it can operate effectively on d.c. and a.c. up to powerline frequencies. Like the normal d.c. series motor, it has a highly variable speed characteristic, with speeds up to 15 000 rpm in domestic appliances. Under some circumstances, *governors* must be used to restrict speeds to safe values.

12.6.2 Construction

The armature construction is like that of a d.c. armature, with laminations, commutator and windings. The armature windings are connected in series with the two field coils by means of carbon brushes running on the commutator (see Figures 12.44 and 12.45). Field pole construction consists of lamination stampings riveted together to form salient poles. The field coils are concentrated-type windings fitted closely around the salient poles.

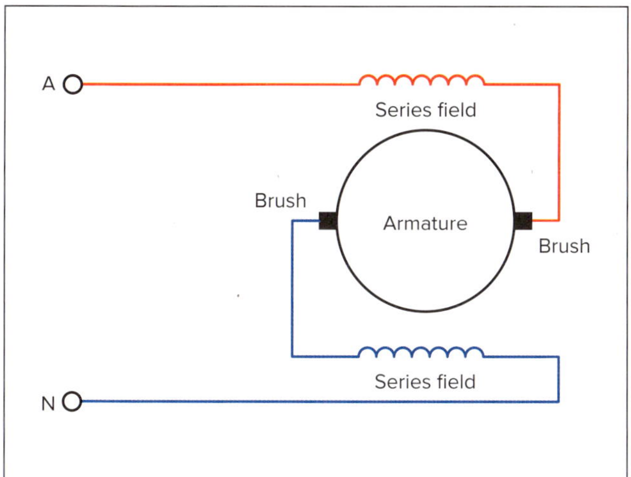

FIGURE 12.44 Universal/series motor

FIGURE 12.45 Typical series universal motor
KPixMining/Alamy Stock Photo

12.6.3 Operation

There is a common current flowing through both windings, so the two magnetic fluxes produced are in phase with each other. Interaction of the fluxes produces torque to turn the armature. As the a.c. supply alternates, the fluxes change in unison, remaining in phase. When the line current flows from A to B in Figure 12.46, north and south poles are produced as shown. Assuming that the armature current is in the direction indicated by the dots and crosses, the flux produced around the armature conductors interacts with the field flux, producing an anticlockwise rotation.

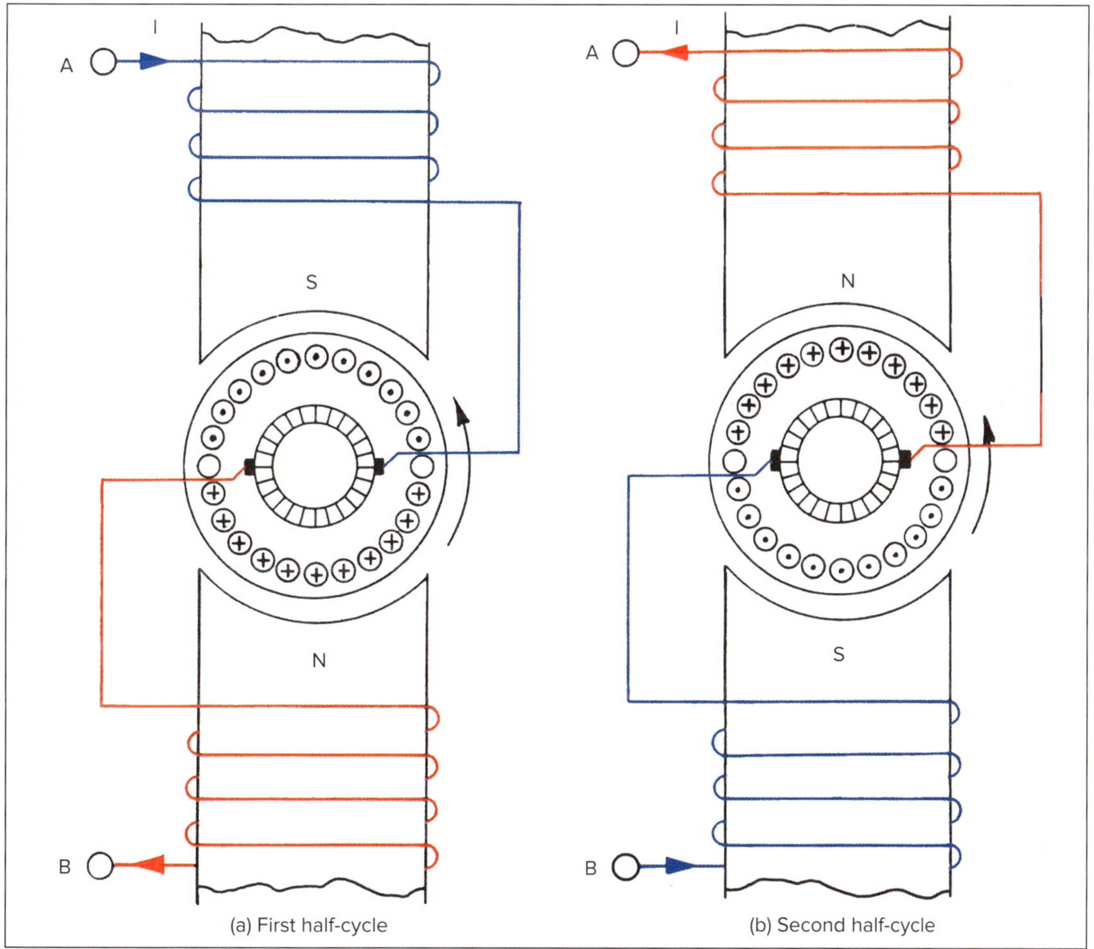

FIGURE 12.46 Torque production in a universal motor

When the line current flows from B to A on the alternate half-cycle, the polarities of the main fields are reversed (see Figure 12.46(b)). The current through the armature also reverses, reversing the armature flux. The resultant torque is still in the anticlockwise direction, so a steady rotation in one direction is maintained.

Rotation is reversed by changing the direction of current flow through the armature with respect to the field; that is, changing over the leads to the armature or the fields—but not both (see Figure 12.47). The load/speed characteristic for the series universal motor is shown in Figure 12.48. When the load is heavy, the speed is low; at light loads, the speed is extremely high. With the small series motor in domestic use, the internal losses (such as friction and windage) are large enough to limit the speed to a safe value. The motor runs at a relatively high speed and has good starting and running torque characteristics, considering its size.

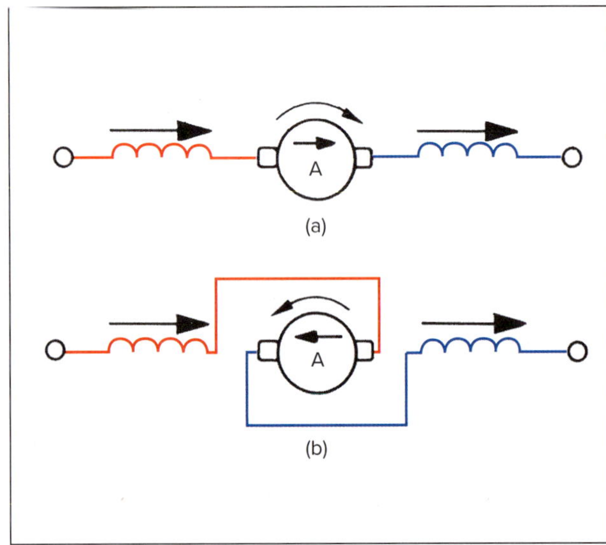

FIGURE 12.47 Reversing the direction of rotation of a universal motor

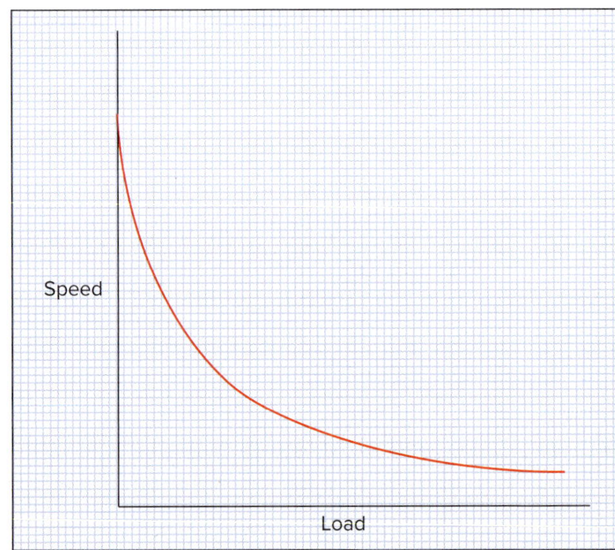

FIGURE 12.48 Speed/load characteristic of a universal motor

12.6.4 Application

The series motor is popular in portable appliances such as saws and drills, sewing machines, business machines, food mixers, small washing machines and vacuum cleaners. Table 12.2 provides a summary of single-phase motors and Table 12.3 shows the comparative advantages and disadvantages of single-phase and three-phase motors.

TABLE 12.2 A summary of single-phase motors

Motor type	Reversal	Applications	Torque
Split phase	Interchange connections of one winding	Washing machines Blowers Bench grinders	Moderate
Capacitor-start	Interchange connections of one winding	Pumps and small compressors	Moderate increase
Capacitor-start/capacitor-run	Interchange connections of one winding	Air-conditioning units	Substantial increase
Permanent split	Interchange connections of one winding	Light loads subject to regular reversals (e.g. ceiling fans)	Very low
Shaded pole	Move shading rings to opposite side of pole or reverse stator in body	Fans within appliances (e.g. fan heaters)	Low
Series	Interchange connections of either field or armature	Domestic appliances Hand tools	Low

TABLE 12.3 Comparative advantages and disadvantages of single-phase and three-phase motors

Advantages of three-phase motors	Advantages of single-phase motors
1. For the same power output, a three-phase motor is smaller and lighter. 2. More efficient use of the iron core. 3. Higher efficiency—less input power for the same output power. 4. For the same output power, line currents are smaller. 5. Suitable for powerline frequencies in excess of several hundred hertz. 6. Less mechanical vibration, owing to magnetic fields of more constant strength. 7. Inherently self-starting, owing to naturally rotating magnetic field. 8. No starting mechanism or switches required. 9. No additional wiring of extra conductors for fully sealed motors and reduced complications of difficult installations such as submersible pumps. 10. Direction can be reversed externally by interchanging supply lines. 11. Starting currents more easily controlled without great loss of starting torque.	1. Only two windings in use, one of which can be of lighter construction. 2. More amenable to automatic machine winding—reduced construction costs in many cases. 3. Can operate on a single-phase supply.
Disadvantages of three-phase motors	**Disadvantages of single-phase motors**
1. Three identical windings are needed. 2. Three-phase supply is needed. 3. Not conducive to machine winding—more labour intensive.	1. Higher line currents for the same power. 2. Energy distributors may limit starting currents of large motors. 3. Single-phase motor reversal is usually done internally.

CHECK YOUR UNDERSTANDING

12.18 What is a universal motor?

12.19 Describe the construction of a universal motor.

12.20 How does a universal motor operate?

12.21 How can a universal motor be reversed?

12.7 Three-phase synchronous machines

12.7.1 Introduction

A machine designed to be connected to the supply and run at synchronous speed is called a 'synchronous machine'. The description applies to both motors and generators. A synchronous condenser is a special application of a synchronous motor.

While the synchronous motor has only one generally used name, the synchronous generator is also referred to as an 'alternator' or an 'a.c generator'. When an alternator is driven at a constant speed, it produces an alternating voltage at a fixed frequency, depending on the number of poles in the machine.

12.7.2 Three-phase synchronous machine construction

The three-phase synchronous machine has two main windings:

1. a three-phase a.c. winding
2. another winding carrying d.c.

When designed for use as an alternator:

1. the a.c. winding in which the induced EMF is generated is known as the 'armature'
2. the d.c. winding that establishes the magnetic field is known as the 'excitation'.

From a theoretical point of view, it doesn't matter if the armature rotates between field poles or whether the field poles rotate inside a stationary armature. Therefore, there can be two distinct types of alternator construction:

- rotating armature
- rotating field.

In most cases, alternators are of the rotating field construction, where the rotor contains the d.c. winding and the stator contains the stationary a.c. winding. An alternator with a rotating a.c. winding and a stationary d.c. winding, while suitable for smaller outputs, is not satisfactory for the larger outputs required at power stations. With these machines, the output can be in megawatts, a value too large to be handled with brushes and slip-rings. Because the terminal voltages range up to 33 kV, the only satisfactory construction is to have the a.c. windings stationary and to supply the rotor with d.c.

This arrangement has the following advantages:

- extra winding space is available for the a.c. windings
- it is easier to insulate for higher voltages
- simple, strong rotor construction
- lower voltages and currents in the rotating windings
- high-current windings have solid connections to the external circuit
- it is more suited to the higher speeds (and smaller number of poles) of turbine drives.

12.7.3 Stator

The stator of the three-phase synchronous machine consists of a slotted laminated core into which the stator windings are fitted. The stator winding consists of three separate (but identical) windings symmetrically distributed around the stator and physically displaced from each other by 120°E. Each phase winding has several coils connected in series to form a definite number of magnetic poles. A four-pole machine, for example, has four groups of coils per phase or four-pole phase groups. The ends of the three-phase windings are connected in either the star or delta configuration to the external circuit.

A typical three-phase stator is shown in Figure 12.49.

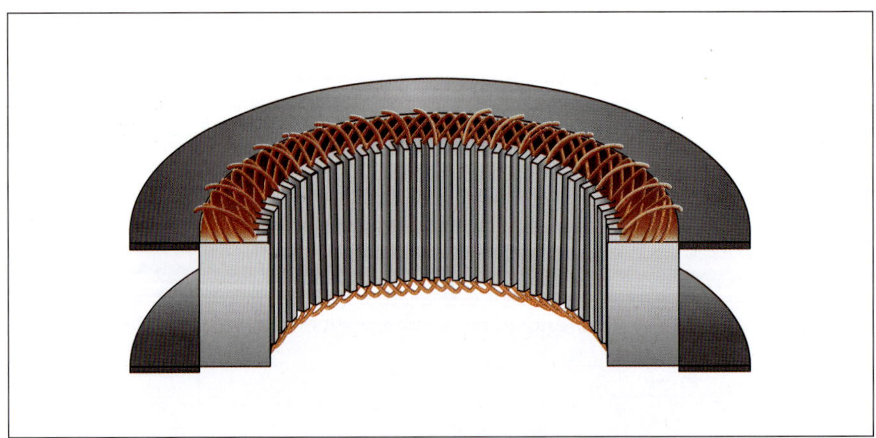

FIGURE 12.49 Stator for a three-phase synchronous machine

12.7.4 Rotor

The alternator rotor can be either low speed or high speed.

Low speed (salient pole)

This type usually consists of a 'spider' (like that used in d.c. machines) on which the field poles and the field coils are bolted (see Figure 12.50(a)). The pole faces of salient pole rotors are shaped so that their centre is closer to the stator than the leading and trailing edges (see Figure 12.51). As the peripheral forces produced on the circumference of the rotor would be excessive at high speed, physical constraints limit the use of this type of rotor to low-speed machines.

High speed (cylindrical)

The high-speed (or cylindrical) rotor was developed to meet the needs of higher-speed prime movers. To counteract centrifugal forces, its diameter must be small in comparison with its length (see Figure 12.50(b)).

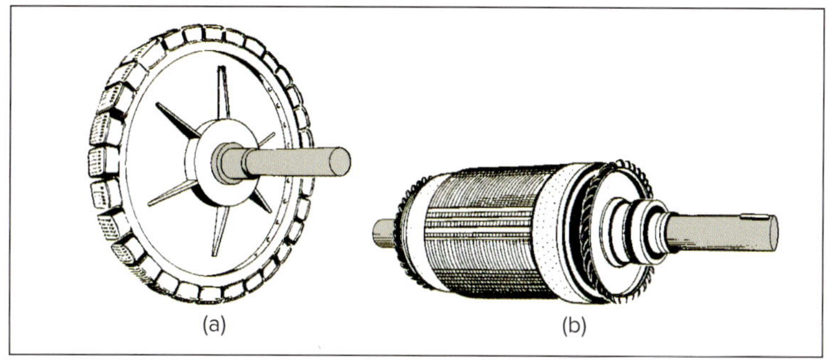

FIGURE 12.50 Main types of alternator rotors: (a) low speed; (b) high speed

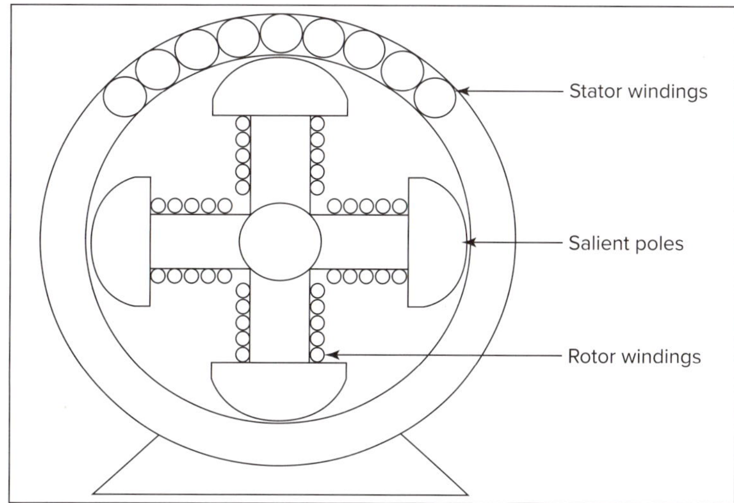

FIGURE 12.51 Salient pole faces
Simon Dand

CHECK YOUR UNDERSTANDING

12.22 What is a three-phase synchronous machine?

12.23 What basic components are part of a synchronous machine's construction?

12.24 How do the rotor and stator differ in a synchronous machine?

12.25 What are the major differences between low- and high-speed rotors?

12.8 Alternators and generators

The primary function of an alternator is to convert mechanical energy into electrical energy; Figure 12.52 shows this energy flow. The term 'alternator' has been used in previous chapters and will be used in this chapter, but it should be remembered that other terms are in use. In general, the principles of construction and operation for alternators and synchronous motors are similar.

While alternators were once seldom seen outside power houses and whole communities were supplied from a central source, it has been common for many years to use smaller sized alternators for the provision of power where a permanent electrical supply is unavailable (e.g. new building sites and remote camping). With the high level of computer usage, there is a further need for standby generating plants to ensure a continuity of supply and prevent loss of data from computer memories. So much information is now being stored in computers that even brief interruptions to their power supplies can have serious consequences for the accuracy and extent of the information stored.

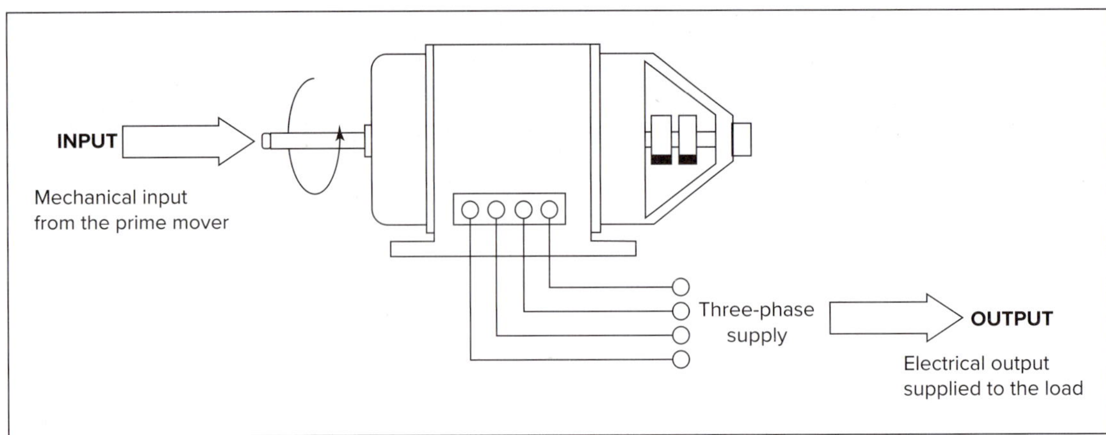

FIGURE 12.52 Energy flow through an alternator

12.8.1 Generator action

The process of generation relies upon the fact that if there is relative motion between a conductor and a magnetic field, an EMF or voltage will be induced in the conductor.

The direction in which the induced EMF acts is dependent upon the direction of motion and the magnetic field polarity in relation to the motion. Fleming's right-hand rule can be used to determine this relationship.

The magnitude of the induced EMF is governed by the:

- field strength, i.e. the amount of magnetic flux cutting the conductor
- speed at which the conductor cuts the magnetic field, i.e. the amount of relative motion
- length of conductor within the field
- angle at which the conductor cuts the magnetic field.

The instantaneous voltage can be calculated using:

$$e = Blv \sin\theta$$

where:

B = flux density
l = length of conductor
v = velocity of motion
sinθ = angle at which the conductor cuts the magnetic field.

Figure 12.53 shows a basic alternator. In the basic form, an alternator has a single loop of wire rotating in a magnetic field between two poles—one north and one south. The two ends of the wire loop are attached to slip-rings that are insulated from the driving shaft. The conductors from the external circuit connect to the slip-rings via carbon brushes that rub against them as they rotate.

A slip ring connection is a sliding contact so that no matter what position the side of the loop of wire (either ab or cd) from 0° through to 360° is positioned, either side of the loop will *always* be connected to either conductor A or conductor B.

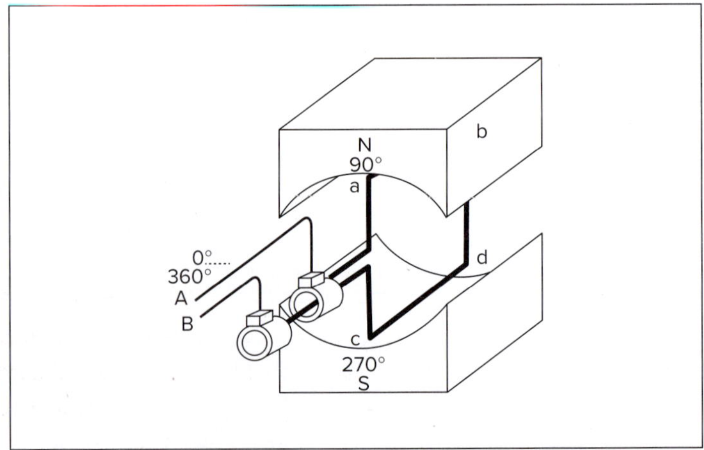

FIGURE 12.53 Basic form of an alternator

If the induced EMF in the loop is plotted for a complete 360° revolution, the waveform shown in Figure 12.54 would be produced.

If there was a load connected to the external circuit (conductors A and B from Figure 12.53), an alternating current would flow. Practical alternators use suitable construction methods to produce an output voltage that is sinusoidal in shape. Figure 12.55 shows a basic single-phase rotating field alternator.

A three-phase alternator is constructed by using three coils or phase windings spaced at 120° apart within the stator (see Figure 12.56).

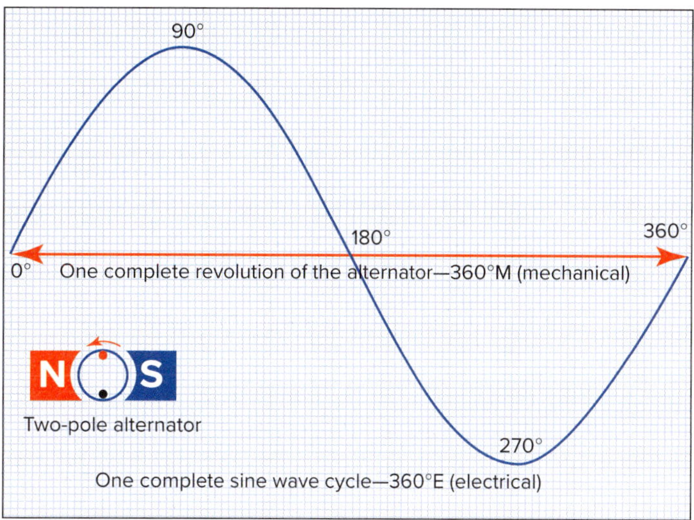

FIGURE 12.54 **Induced EMF in a basic alternator**

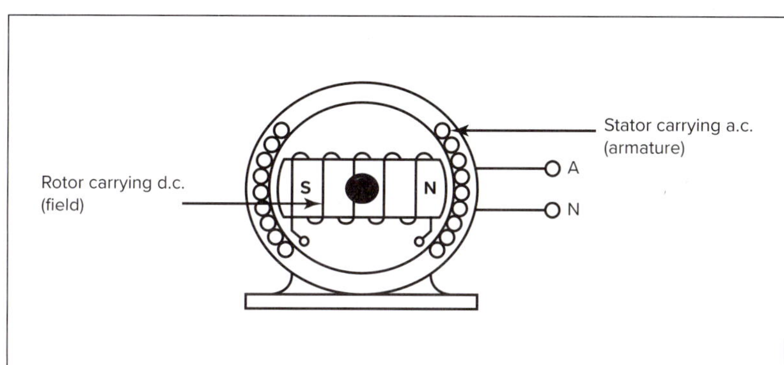

FIGURE 12.55 **Simple view of a single-phase alternator**

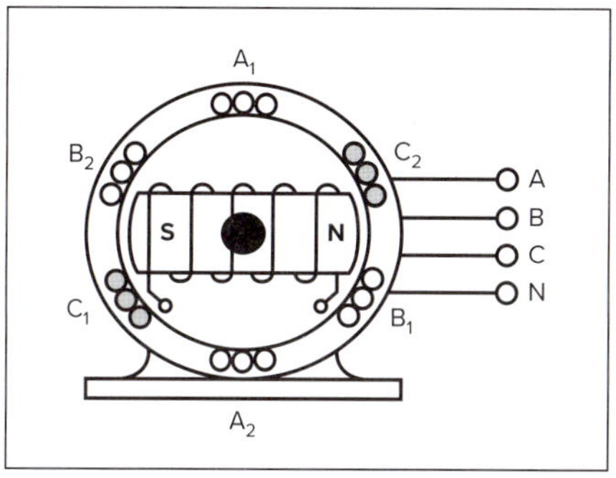

FIGURE 12.56 **Simple view of a three-phase alternator**

Whatever EMF is induced at a certain moment in the A phase winding will occur 120° later in the B phase winding and 240° later in the C phase winding; thus each of the EMFs produced in a three-phase alternator have a phase displacement of 120° and the combination produces a three-phase output (see Figure 12.57).

12.8.2 Frequency and speed relationship

Alternators are designed to generate a specific voltage at a definite frequency. A relationship exists between the speed of an alternator, its number of poles and the generated frequency. When considering one phase of a three-phase alternator, if an alternator has p number of poles on the field system, the induced EMF within the stator windings goes through $\frac{p}{2}$ cycles per revolution and the speed in revolutions per minute equals $\frac{n}{60}$ revolutions per second. Therefore, the induced EMF within the stator goes through $\frac{n \times p}{60 \times 2}$ cycles per second. As the frequency of an a.c. waveform is measured in cycles per second (hertz), the frequency of the generated EMF can be calculated by:

$$f = \frac{np}{120}$$

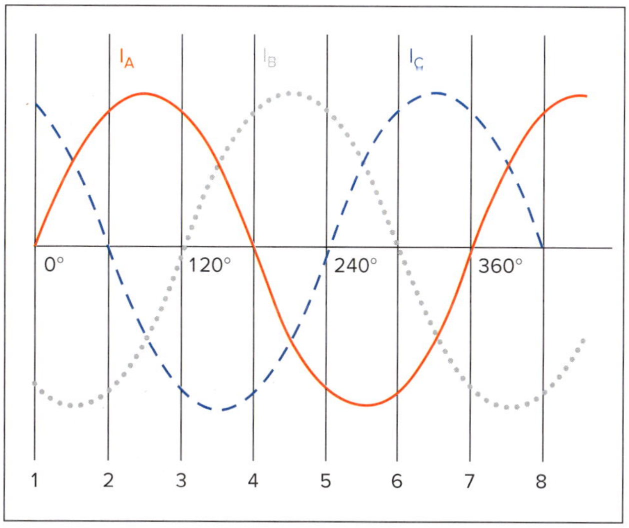

FIGURE 12.57 **Three-phase waveforms**

where:

f = frequency in hertz (Hz)

p = number of field poles

n = speed of rotation in revolutions per minute (rpm).

If the frequency of the supply is known, then the required speed of the machine can be calculated by:

$$n = \frac{120f}{p}$$

12.8.3 Excitation

The usual method for d.c. excitation of the rotor windings is for each machine to have its own d.c. generator, called an 'exciter'. The exciter can be belt driven or geared down from the synchronous machine, but the usual practice is for it to be directly coupled to the rotor shaft (see Figure 12.58).

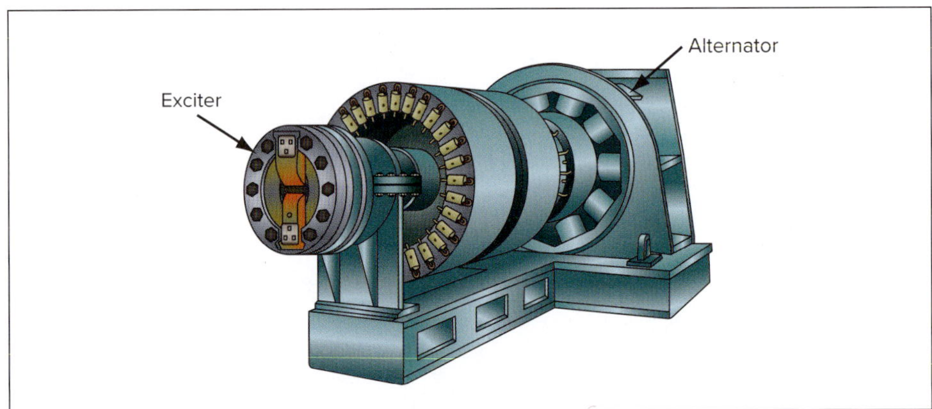

FIGURE 12.58 **Alternator and exciter**

The exciter armature rotates within the influence of the exciter field, causing a d.c. voltage to be generated in the armature. The exciter output is fed into the field windings of the synchronous machine. By adjusting a rheostat in the exciter field circuit, the strength of the magnetic field in the rotor can be varied.

The basic diagram of an alternator and its exciter is shown in Figure 12.59.

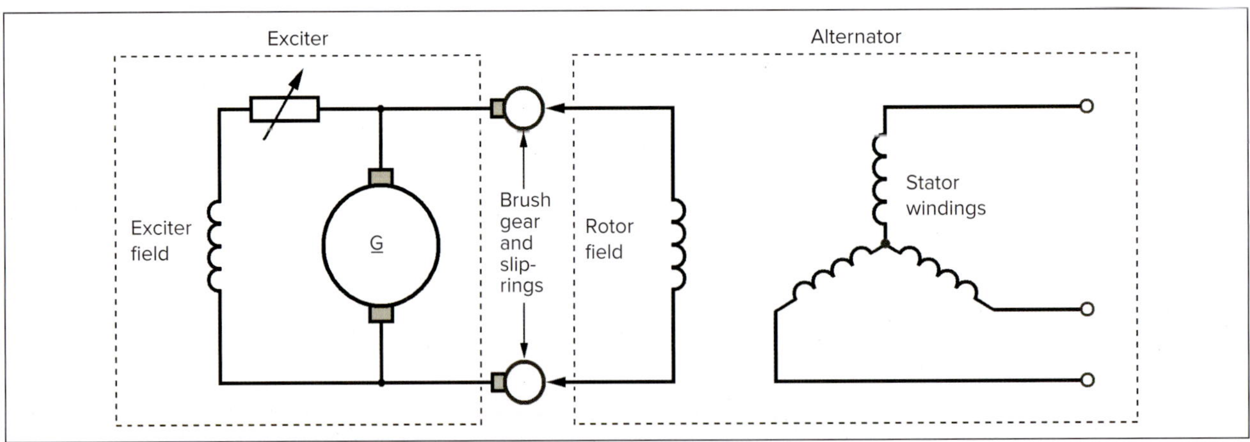

FIGURE 12.59 **Basic alternator circuit**

With very large alternators, the d.c. excitation requirements are substantial. This means that the d.c. generator must be so large that it might not be able to self-excite. Because of this, it might need an exciter of its own, one that is able to self-excite and provide power for the field of the main generator, which in turn supplies the rotor field of the alternator.

Some alternators use a brushless excitation system in which the exciter armature has been replaced by a small three-phase alternator that rotates within the influence of a small residual magnetic field. This causes a small three-phase voltage to be generated in the exciter. When converted to d.c. by an internal rectifier, it supplies the main field of the alternator, resulting in an a.c. output voltage. A sensor unit connected to the output of the machine monitors the output voltage and load current of the alternator and sends electrical signals to a controlled rectifier, which in turn controls the strength of the exciter field. The sensor unit and the controlled rectifier are, in a sense, the voltage regulator of the machine.

A basic circuit of a brushless generating system is shown in Figure 12.60.

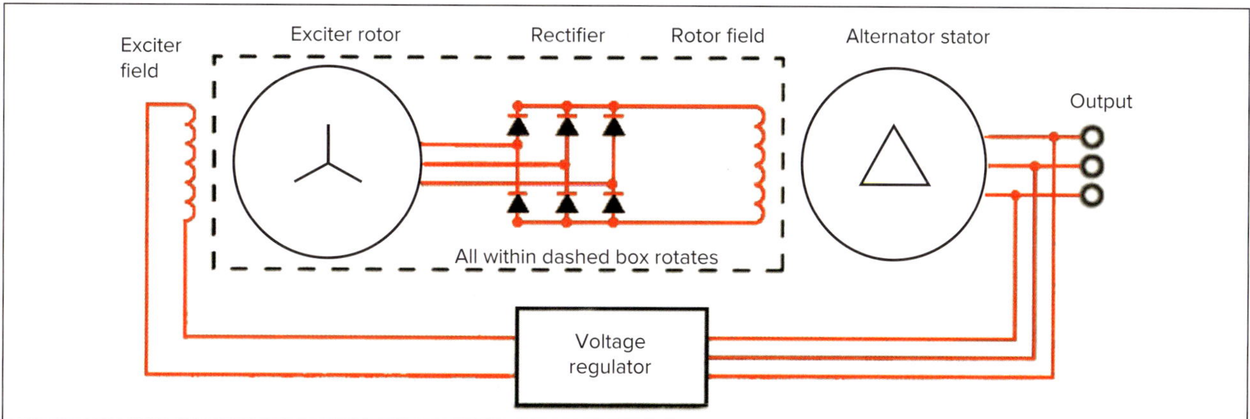

FIGURE 12.60 Brushless excitation

12.8.4 Generated voltage

The value of the generated a.c. voltage depends on the strength of the rotor flux and the speed at which it cuts the windings. Because the speed must be constant (and is linked to the frequency required), the sole remaining factor determining the value of the generated voltage is the strength of the rotor flux.

For an alternator, the generated voltage is found from:

$$V_g = 4.44\ \Phi f N k_d k_p$$

where:

V_g = generated voltage per phase (RMS)
Φ = flux per pole in weber
f = frequency in hertz
N = number of turns per phase
k_d = a constant, dependent on winding distribution
k_p = a constant, dependent on coil pitch.

EXAMPLE 12.4

Calculate the line voltage of a 50 Hz star-connected alternator, given the following details:

$$\Phi = 0.67\ \text{Wb/pole}$$
$$k_d = 0.85$$
$$k_p = 0.98$$
$$N = 36\ \text{turns per phase}$$
$$V_g = 4.44\ \Phi\ f N k_d k_p$$

Phase voltage, $V_p = 4.44 \times 0.67 \times 50 \times 36 \times 0.85 \times 0.98 = 4460\ \text{V}$

then line voltage, $V_L = \sqrt{3} \times V_p = 1.732 \times 4460 = 7725\ \text{V} = 7.73\ \text{kV}$

12.8.5 Effect of load on alternator voltage

An alternator can be considered to consist of three components in series:

- an a.c. generating source
- a resistor—representing iron and copper losses
- an inductor—representing the inductance of the windings and magnetic leakage.

Any load placed on the alternator must be assumed to be in series with these components (see Figure 12.61).

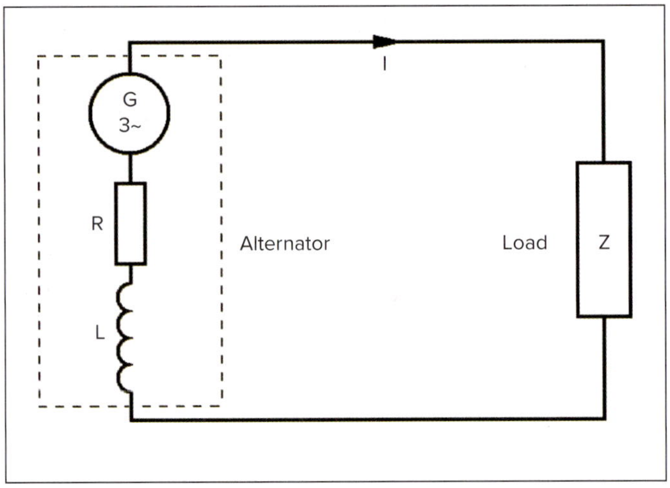

FIGURE 12.61 Equivalent circuit of an alternator

The series impedance of the resistance and inductance provides a drop in voltage before the generated voltage can reach the connected load. Additionally, the load current in the a.c. windings produces an armature reaction, which also affects the output voltage.

With a unity power factor load, the armature reaction merely distorts the main field and the effect on voltage is minimal; the voltage drop is mainly due to the series impedance. Figure 12.62(a) shows that the resistive voltage drop IR is in phase with the load current I and the voltage drop due to the reactance IX being at 90°E to the IR drop.

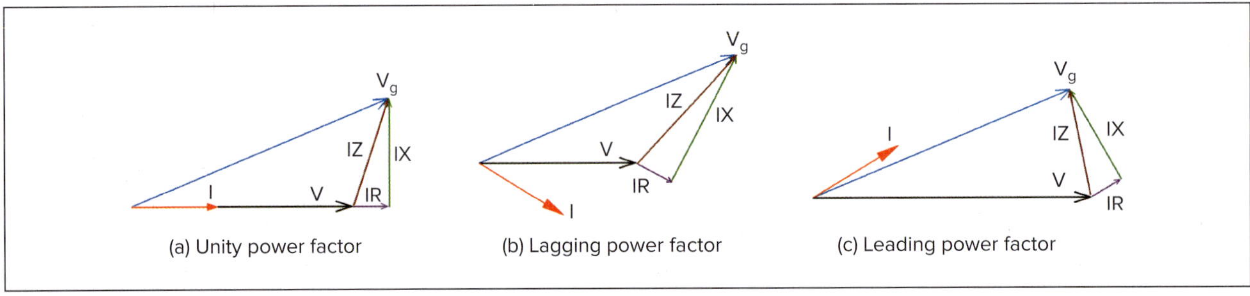

FIGURE 12.62 Phasors for various power factor loads on an alternator

These two values combine to form a voltage drop IZ due to the impedance of the alternator windings.

The phasor sum of the output voltage and IZ gives the generated voltage Vg.

For a load with a lagging power factor, however, the magnetic effect of the stator current opposes that of the rotor (see Figure 12.62(b)). This results in a weakened rotor field and reduces the output voltage further than the resistive load did. As before, IR is in phase with the load current I. IX is at 90°E to IR, so placing IZ at a different angle from the previous case. In a similar manner, Vg is equal to the phasor sum of the output voltage and IZ.

For a load with a leading power factor, the flux caused by the stator current assists that of the rotor, resulting in an increased output voltage (see Figure 12.62(c)).

The characteristics of the three types of loads are shown in Figure 12.63.

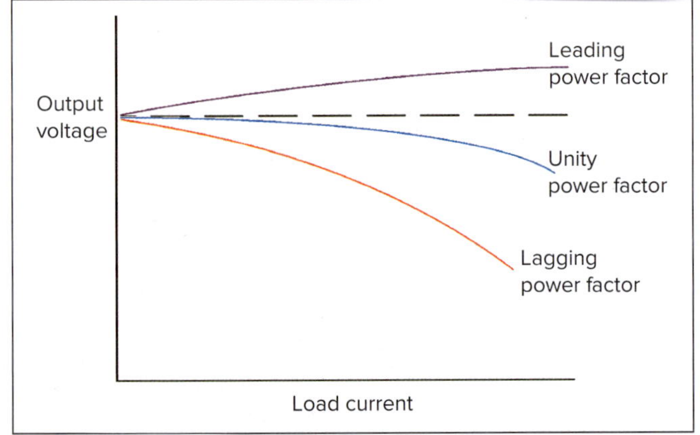

FIGURE 12.63 Effect of power factor on the output voltage of an alternator

EXAMPLE 12.5

A three-phase star-connected alternator has an output voltage of 3300 V at full load, with unity power factor. When the load is removed and the excitation is unchanged, the voltage rises to 3350 V. Find the percentage regulation.

$$\begin{aligned} V_R\% &= \frac{V_{NL} - V_{FL}}{V_{FL}} \times 100\% \\ &= \frac{3350 - 3300}{3300} \times 100\% \\ &= \frac{50}{3300} \times 100 = 1.5\% \text{ at unity power factor} \end{aligned}$$

The regulation must also be referred to the load power factor because at any other power factor these figures would be different.

12.8.6 Voltage regulation

An alternator is required to give a prescribed terminal voltage at full load. The difference in output between no load and full load is a measure of its voltage regulation. The difference is compared with the full-load value in a similar manner to that for d.c. machines.

Percent voltage regulation is calculated by:

$$V_R\% = \frac{V_{NL} - V_{FL}}{V_{FL}} \times 100\%$$

12.8.7 Alternator ratings

An alternator is rated according to three basic factors:

1. frequency
2. voltage
3. current.

The frequency fixes the speed at which the alternator must be driven, the voltage rating sets the designed output voltage and the rated current is the full-load current output. The last two factors help establish the volt-ampere rating, which is usually expressed in kVA.

The alternator rating cannot be given in kilowatts because the power factor of any load placed on the alternator is beyond the control of the manufacturer, and because its value could vary considerably during operation.

EXAMPLE 12.6

A three-phase 400 V 50 Hz alternator is rated at 150 kVA at 0.8 power factor. Calculate the:

- power loading in kilowatts when fully loaded with power factor values of 0.8 and 0.6
- full-load current of the alternator.

The machine is rated at 150 kVA at 0.8 power factor, so at this load:

$$\text{true power output} = 150 \times 0.8 = 120 \text{ kW}$$

At 0.6 power factor:

true power output = 150 × 0.6 = 90 kW

In both cases, the current flowing will be the full-load current value, which should not be exceeded because of cooling problems within the windings.

At 0.8 power factor:

$$P = \sqrt{3}\ V\ I \cos\phi$$

$$\therefore I = \frac{P}{\sqrt{3}\ V \cos\phi} = \frac{120\,000}{\sqrt{3} \times 400 \times 0.8} = 217\ A$$

This is the full-load current rating for each phase winding of this particular alternator and it applies irrespective of the load power or power factor.

12.8.8 Parallel operation of alternators

Most commercial power stations are designed to have a number of alternators operating in parallel, supplying a common load at constant voltage. As an alternator's efficiency is at its maximum near its full-load capacity, it is more economical to have each machine delivering its approximate rated output. During the early hours of the morning, for example, when there is a light load, it might be necessary to have only one machine connected to the line, delivering its rated output. As the load varies during a 24-hour period, the number of machines connected in parallel is determined.

12.8.9 Synchronising

Before a three-phase alternator can be connected in parallel with another three-phase supply, the following conditions must be fulfilled:

1. The output waveform of each supply must be identical. This is determined by the design features of the alternators. It is standard practice to generate a sinusoidal waveform supply.
2. The phase sequence or rotation of each supply must be the same. This ensures that the EMFs of each supply reach their maximum values in the same sequence—for example: A, B, C. The phase sequence is determined by the method of connection of the alternator phase windings to the terminals of the machine. This check is carried out during the commissioning process after the initial installation or following a major maintenance overhaul, and it is not necessary to do it each time the machine is connected in parallel with others.
3. The alternator and supply voltages must be the same.
4. The alternator and supply voltages must also be in phase.
5. The alternator and supply frequencies must be identical.

The last three conditions can be adjusted by the operator. The voltage of the incoming alternator is adjusted by varying the field excitation, and the frequency is determined by the speed of the prime mover.

To ensure that the alternator and supply voltages are in phase with each other before connecting them in parallel to the load, some method of indicating the phase relationship is required. Smaller-sized alternators can be synchronised with lamps, but for larger machines a more exact method is required.

12.8.10 Load sharing between alternators in parallel

To examine the operation of alternators in parallel, it is helpful to consider three typical applications of an alternator. These are an alternator operating on its own, an alternator operating in parallel with an alternator of the same size and an alternator operating in parallel with a distribution grid.

An alternator operating on its own

When an alternator is operating independently, the frequency and output voltage can be adjusted by adjusting the set points on the governor and the field current regulator (see Figure 12.64).

Adjusting the governor's set points adjusts the speed of the prime mover and, consequently, the frequency of the output. Adjusting the field current will change the output voltage.

If the load connected to the alternator increases, the governor on the prime mover will open to increase the input power at the same frequency and the automatic voltage regulator will increase the field current to maintain a constant output voltage.

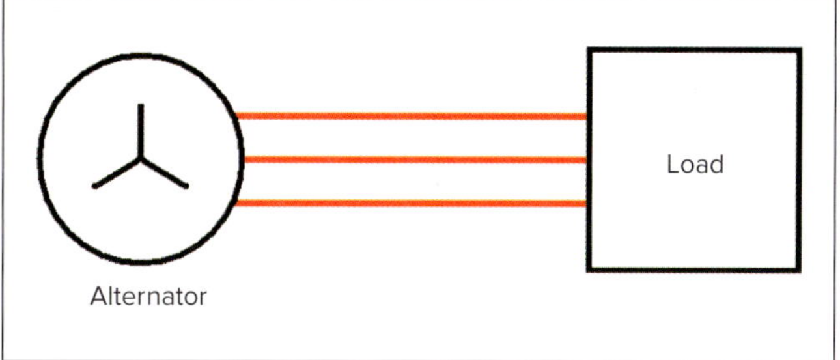

FIGURE 12.64 An alternator operating alone

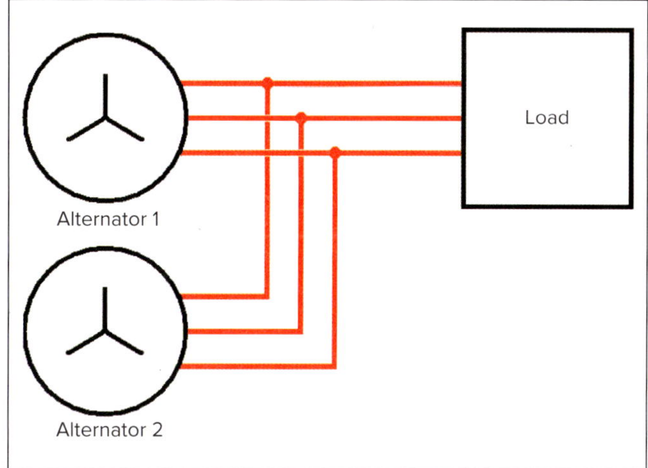

FIGURE 12.65 An alternator operating in parallel with another alternator of the same size

An alternator operating in parallel with an alternator of the same size

The main reasons for connecting two alternators in parallel are either to share any given load between them or to shift the load to the incoming machine without causing an interruption to the supply. The incoming machine, when first synchronised, should have no load on it and might even be drawing power from the supply lines. It is then necessary to adjust the controls of the incoming machine until it is delivering its appropriate share of the load to the supply lines. However, when two alternators of the same size are connected in parallel to a common system load, a change to the controls of one will affect the operation of the other (see Figure 12.65).

As they are connected together and to the system load, the alternators will always have the same voltage and frequency.

To increase the load taken by an alternator running in parallel, it is necessary to increase the set points on the governor. However, changing the governor set point on one alternator only will cause that machine to speed up and change the system frequency.

To adjust the power sharing between the two alternators, the governor set point on one alternator is increased and, at the same time, the governor set point on the other alternator is decreased. The machine whose governor's set point was increased will assume more of the load.

To change the system frequency without changing the power sharing between the two alternators, the governor set points on both must be adjusted up or down simultaneously.

Where a field-current adjustment controlled the output voltage of a single alternator, it will control the power factor of the load connected to each alternator when they are connected in parallel. If the field currents of both alternators are increased or decreased simultaneously, the system voltage will alter.

To adjust the power factor of the load supplied by each alternator, increase the field current on one alternator and simultaneously decrease the field current on the other alternator. The power factor of the alternator whose field current was increased will lag by a greater angle.

An alternator operating in parallel with a distribution grid

When an alternator is connected to a large transmission system, the operator of the alternator will have little effect on the voltage or frequency of the system. Consider the case of a single alternator connected to the Australian East Coast grid. The grid is so large that any adjustments to the controls of the alternator will not cause a change in the grid voltage or frequency.

A large grid system is sometimes referred to as an 'infinite bus'. This power system is so large that, no matter how much power is supplied to or taken from it, its voltage and frequency do not alter. So when an alternator is connected to a large grid, the frequency and terminal voltage of the alternator are controlled by that grid. This is demonstrated in Figure 12.66.

To change the power supplied by a single alternator connected to the grid, the set point on the governor of the alternator is altered. Similarly, the field current regulator in the alternator controls the power factor of the load taken by that alternator.

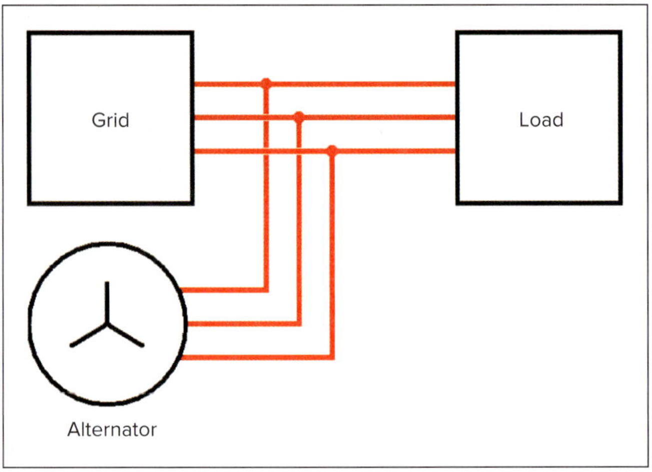

FIGURE 12.66 **An alternator operating in parallel with the grid**

12.8.11 Hunting in alternators

The driving torque and speed of a piston engine are not constant during a complete revolution but vary according to the position and speed of the pistons. This causes minute variations in the speed of the alternator shaft. The speed variations are small but cause momentary increases and decreases above and below the average rotational speed. The effect is called 'hunting' and leads to small voltage variations, which can include harmonics that will distort the waveform. It is partially neutralised by the inertia of the rotating parts.

Remedies for hunting involve the use of quite heavy flywheels and special windings in the pole faces. These are called 'amortisseur' or 'damper' windings (see Figure 12.67).

The voltage pulses created by hunting can cause circulating currents to flow between alternators connected in parallel, resulting in an increase in the mechanical oscillations of the rotating parts. Electrical losses are also increased.

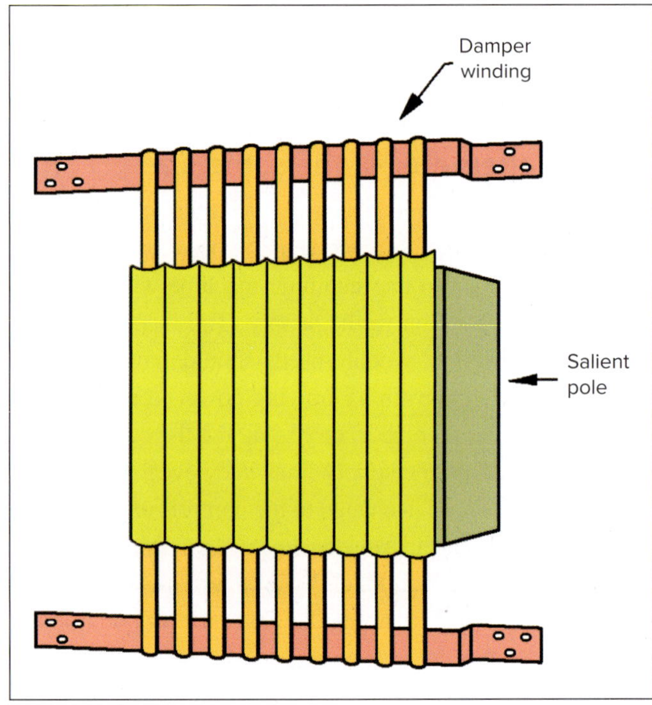

FIGURE 12.67 **Amortisseur windings for an alternator**

High-speed turbines are not affected to the same extent by hunting. The major causes of oscillation about a fixed point are the minor adjustments of the governors as load changes on the machine occur.

12.8.12 Standby power supplies

Standby power supplies are generally intended to provide mains power at a specified voltage and frequency. There are two main forms of standby power-supply units.

The first type of unit (uninterruptible) is meant for use where no interruption to a power supply can be tolerated—for example, to computer, hospital and aircraft navigation equipment. Losing power at a crucial moment in an operation could mean loss of life, or in the middle of a computer calculation could mean the loss of valuable data. There is also an increasing use of this type of power supply for portable work because it can often be run from a 12 V vehicle battery. It is quick, convenient and quiet.

The second type of standby unit is where momentary losses of power can be tolerated. Such uses would include emergency lighting, theatres and industrial uses such as fully environmental chicken-meat sheds. Delays of several seconds in restoring power can be acceptable in some circumstances. This type of standby power supply would also be suitable for lifts and high-rise buildings.

A subsection of this latter category includes portable power supplies such as small generating plants that can be carried from job to job in a vehicle.

12.8.13 Uninterruptible power supplies (UPS)

A block diagram of a UPS is shown in Figure 12.68. It can be seen that the unit runs off the mains supply with a battery permanently 'floating' on charge from an inbuilt battery charger. Effectively, the battery bank is supplying an inverter, which converts d.c. to a.c. at mains voltage and frequency. In the event of losing mains power, the unit continues to operate as long as the battery has sufficient charge.

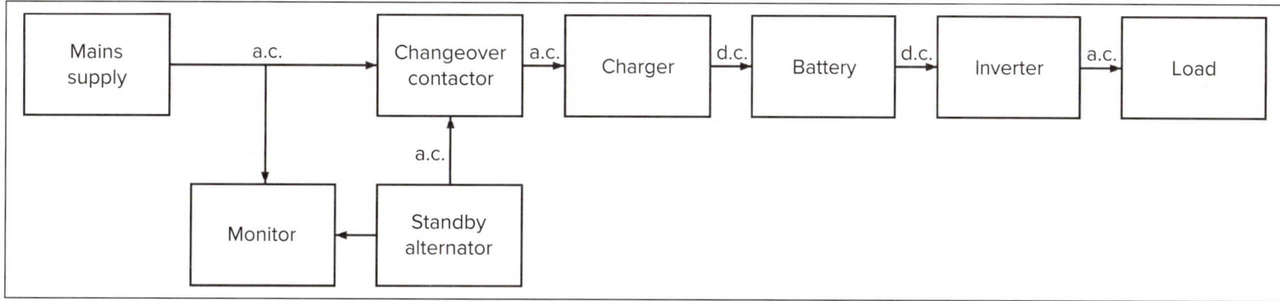

FIGURE 12.68 Block diagram of an uninterruptible power supply (UPS) with back-up

More critical loads usually have an engine-driven alternator on standby to ensure that the battery charge is maintained. The battery capacity has to be great enough to supply the circuit power while the alternator and engine are being started and run up to speed. Extra operating time allowances have to be made for non-starting incidents with the engine and also to provide greater flexibility as a back-up in an emergency.

When using a vehicle battery, the d.c. values will be high, so care must be taken to ensure that the battery does not go flat. For example, a 500 W TV draws about 20 A on 32 V d.c. The current drain would easily exceed 50 A on 12 V.

12.8.14 Engine-driven alternators

A large range of engine-driven alternators is available, so a choice has to be made on the basis of several factors. Options range from buying a small portable unit at the best-possible price through to carefully planning for the most suitable unit for a particular purpose. It is not simply a matter of selecting an alternator with respect to the load it has to supply; the choice should also take into account many other considerations. Some of these are listed below, and their order of importance is governed by the intended use for the alternator. Figure 12.69 shows the essential components of a standby power supply using an engine-driven alternator.

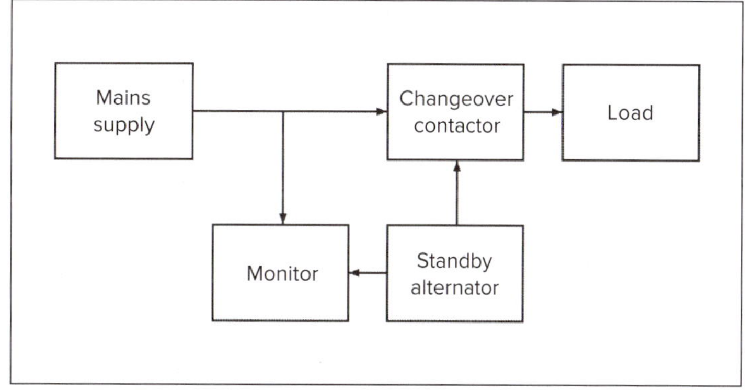

FIGURE 12.69 Block diagram for an engine-driven standby alternator

Purchase price

The overall cost of smaller units may be lower, but in terms of cost per kVA they are more expensive and operate at lower efficiencies. As the size of the unit increases, the cost per kVA reduces, while the operating efficiency increases.

Type of prime mover

The economy of the prime mover in terms of efficiency has a bearing on its selection. This in turn is affected by the type of service it will encounter. For example, a steam turbine has good economy throughout its entire load range. However, it is expensive, large and needs a long time to get the unit on load from cold. An internal combustion engine has poor efficiency at light loads but is much cheaper to buy initially. For some loads it is cheaper to buy several smaller alternators than one large unit. Problems of paralleling the units must then be considered.

The cost and availability of fuel must always be a consideration. While distillate is more expensive initially, as is the diesel engine itself, the fuel cost per hour is less, while maintenance costs are far higher than those for a petrol engine.

FIGURE 12.70 Self-contained portable power supply
maksimee/Alamy Stock Photo

The petrol engine is cheaper to buy, the fuel is readily available and the prime mover is suited to smaller units used purely for portable power supplies on intermittent duties. In the long term, the diesel engine runs better on full loads than the petrol engine. The petrol engine is more tolerant of dirty fuel than the diesel engine and does not need specialised skills for maintenance purposes. Figure 12.70 shows a portable generating unit driven by a single-cylinder petrol engine. The alternator, rated at 5.5 kVA, is driven by a petrol engine. The size and weight of the unit is such that it can be moved to any site where power is required. Two 15 A socket-outlets are available for connection of equipment.

Starting methods

Starting methods are governed by the intended use of the generating unit. The quicker the changeover to auxiliary power, the more expensive is the starting method. The cheapest method involves merely starting the unit manually when it is evident that the main power supply has failed. A more expensive method involves the use of an automatic changeover contactor that drops out when the main supply fails.

In turn, this connects a starting motor to the engine and, when the alternator is up to speed, connects it to the load.

Load sizes and alternator capacities

Smaller generating plants are usually intended for standby purposes for short periods. They usually have only one load connected to them at a time, such as a portable tool or a small lighting load. With medium- and larger-sized alternators, consideration has to be given to the possible connection of intermittent larger loads, such as the starting currents of motors. The unit then has to have the electrical capacity and engine power to maintain both the output voltage and the frequency during these current surges to avoid interruptions to other equipment connected to the same supply.

Operation of alternators

With the exception of some manually operated equipment, most standby and back-up alternator operations are now beyond the control of the operator. Where some degree of manipulation is available, there are two important factors that should always be considered—voltage and frequency. In most cases, the voltage is controlled by automatic voltage regulators while the frequency is controlled by the engine governor. The order of operation is to set the speed first, which in turn sets the frequency, and then to adjust the voltage of the unit. To do this in the reverse order is to alter the voltage each time the speed is altered.

A study of the output voltage equation will confirm the soundness of this method.

$$V_g = 4.44\, \Phi f\, n\, k_d\, k_p$$

In an operating alternator, the above equation can be reduced to the following:

$$V_g = k\, \Phi\, f$$

where k is a constant consisting of all the components of the equation that cannot be altered in an operating alternator. The flux value is altered by changing the d.c. excitation and the frequency is altered by changing the set point on the engine governor.

12.8.15 Prime movers

Low speed

Most diesel engines used as prime movers for driving alternators operate within the range 500–1000 rpm and this necessitates the use of rotors with many pairs of poles.

Hydroelectric turbines have water-driven impellers that operate at low speeds, and consequently they also drive rotors with many poles. While the diesel-driven alternator usually has its shaft in the horizontal plane, the hydroelectric unit has its shaft in the vertical plane. This method of construction means that special thrust bearings have to be fitted to take the end thrust of the rotating component.

High speed

Steam or gas turbines operate efficiently at speeds of about 3000 rpm. To produce 50Hz, an alternator driven by a steam or gas turbine at 3000 rpm must consist of only two poles.

The relationship between speed, frequency and the number of poles can be determined from:

$$f = \frac{np}{120}$$

By transposition:

$$n = \frac{120f}{p}$$

where:

n = rpm

f = frequency in hertz

p = number of poles.

For a large-diameter rotor of 24 poles at 50 Hz:

$$n = \frac{120 \times 50}{24} = 250 \text{ rpm}$$

For a small diameter rotor of two poles at 50 Hz:

$$n = \frac{120 \times 50}{2} = 3000 \text{ rpm}$$

An alternator in the low-speed range will have a large diameter and a comparatively short axial length. For high-speed turbines, the extra expense and auxiliary machinery needed restricts their use to larger sizes. Higher outputs mean that the length of the alternator must be increased, and the increase in length causes complications in cooling.

EXAMPLE 12.7

At what speed would the governor of a 12-pole diesel-driven alternator have to be set to enable a frequency of 50 Hz to be generated?

$$n = \frac{120\ f}{p}$$

$$n = \frac{120 \times 50}{12} = 500\ \text{rpm}$$

A typical standby motor-driven alternator is shown in Figure 12.71. This type of alternator is used as a back-up in the event of a power failure.

FIGURE 12.71 Skid-mounted generating unit
MH Capture/Shutterstock.com

12.8.16 Alternator cooling

Low speed

With engine-driven or hydroelectric alternators, there is no great difficulty in providing adequate ventilation because of the characteristically large diameter and short axial length. In addition to the large surface area available for direct radiation of heat, there is a fanning action due to the rotation of the fields, an action that can be increased by the addition of fan blades if necessary.

When the axial length is short, the heat developed in the embedded windings is quickly conducted to the ends where it can be dissipated by the fanning action. As the machine size becomes larger, it is often necessary to have ventilation ducts within the core to provide paths through which the cooling air can flow.

High speed

The provision of adequate cooling facilities is a problem in high-speed machines of large capacity, if the operating temperature of the windings is to be kept within safe limits. The surface area available for cooling in a high-speed machine is less than that in a low-speed machine of the same capacity.

The diameter of the rotor must be small enough to keep the surface speed down to a safe value, so for large capacities the length of the machine must be considerable. This long axial length causes difficulty in cooling the central portion of the core because the heat generated cannot be conducted away quickly enough to limit the temperature rise in the core to a value that will protect the windings and the insulation.

These considerations gave rise to the necessity of completely enclosing the alternator and allowing the use of forced ventilation to carry away the heat produced. Where cooling air is used, it must be filtered to keep it clean, and sometimes washed by passing it through a spray chamber to prevent a build-up of dust within the machine. Washing the air has the added advantage of cooling it, further reducing the temperature of the alternator, allowing the rating of the machine to be increased.

To increase alternator ratings still more, hydrogen gas is used instead of air because of its greater ability to absorb heat. The machine is completely enclosed and the hydrogen is blown through the alternator and then through a heat exchanger before being cycled through the alternator again. The total exclusion of air from the fully-sealed machine is necessary to prevent an explosive air/hydrogen mixture from forming.

Considerable care is taken to ensure the purity of the hydrogen gas. The oil pressure for the bearings is at a higher value than the pressure of the hydrogen being pumped through the machine. This ensures that the oil flow through the seal is towards the hydrogen gas so that it is retained in the machine. The oil may then be passed through a vacuum process to remove any hydrogen gas or air before being reused in the machine.

These cooling methods require considerable power and auxiliary equipment, so the output from the alternator must be increased by an appreciable amount for the method to be economically feasible. Accordingly, it is used only on very large capacity machines.

12.8.17 Induction generators

An induction generator is a type of a.c. generator that uses the same principles of operation as the induction motor, where electromagnetic induction is the means of producing electrical power. The main difference in operation is, however, where the prime mover for an induction generator has to turn the rotor *faster* than synchronous speed, and thus a normal a.c. induction motor can be used as an induction generator without any internal modifications.

For the start-up of an induction generator, an excitation field is required for induction within the rotor. This is normally supplied either by an external power supply connected initially to the stator windings or internally, via residual magnetism within the rotor itself. After induction commences in the rotor, a prime mover rotating above synchronous speed is applied to it. This produces *negative slip*, which allows the stator to continue to induce rotor currents and thus produce rotor flux. The rotor flux in turn is now cutting the stator windings, producing an active current and power that can then be supplied to the load. The active power supplied is proportional to the negative slip, i.e. slip speed above the synchronous speed.

CHECK YOUR UNDERSTANDING

12.26 How is a voltage generated in an alternator?

12.27 How is excitation achieved in an alternator?

12.28 What does the term 'voltage regulation' mean?

12.29 What factors need to be considered when connecting alternators in parallel?

12.30 List the different types of prime movers that are used with alternators.

12.9 Three-phase synchronous motors

An alternator, when supplied with electrical energy at rated voltage and frequency, will run quite well as a motor. When a synchronous machine is driven as a motor, the machine is called a 'synchronous motor'. Since the supply frequency is fixed, the motor's speed remains constant, irrespective of the load.

12.9.1 Construction

Stator

The stator has a three-phase winding and is of the same type as that in an alternator or induction motor.

When this winding is energised with a.c., it produces a magnetic flux that rotates at synchronous speed. It is the same speed at which the synchronous machine would have to be driven to generate an a.c. voltage at line frequency.

The speed of the rotating magnetic field can be derived from the same equation used for alternators in section 12.8.2:

$$n = \frac{120\,f}{p}$$

Rotor

Although of similar construction to the alternator rotor, it is usually made with salient poles. When excited with d.c., it produces alternate north and south magnetic poles, which are attracted to those produced in the stator.

12.9.2 Operating principle

A synchronous motor works on the principle of magnetic attraction between two magnetic fields of opposite polarity; one is the rotating magnetic field of the stator and the other is the magnetic field of the rotor.

A synchronous motor has torque only at synchronous speed, so special steps have to be taken to get the motor up to speed and synchronised with the supply. The two magnetic fields are then rotating at the same speed and lock in with each other.

12.9.3 Effect of load on a synchronous motor

A three-phase synchronous motor has no starting torque. It has to be brought up to speed (or as close to it as possible) by some other means so that it can pull itself into synchronism. Once up to speed, the rotor field can be excited with direct current and the rotor is, in effect, then dragged around at the same speed as the three-phase stator field. Its speed is synchronised with that of the stator field. This is markedly different in principle from the induction motor, where the rotating field of the stator is pushing against the induced rotor field. That causes the rotor to rotate, but with some slip, whereas in the synchronous motor there cannot be slip, merely a 'hanging back' due to the load imposed on the machine. This is illustrated in Figure 12.72 and shows as a torque angle. If the load becomes too great for a synchronous motor, it immediately pulls out of synchronism and stops.

When a synchronous motor runs on no load, the relative positions of stator and rotor poles coincide (see Figure 12.72(a)).

When a load is applied, the rotor must still continue to rotate at synchronous speed but, owing to the retarding action of the load, the rotor pole lags behind the stator pole. Their relative positions are displaced by what is called the 'torque angle' or 'load angle' (see Figure 12.72(b)). The greater the load applied, the greater the torque angle.

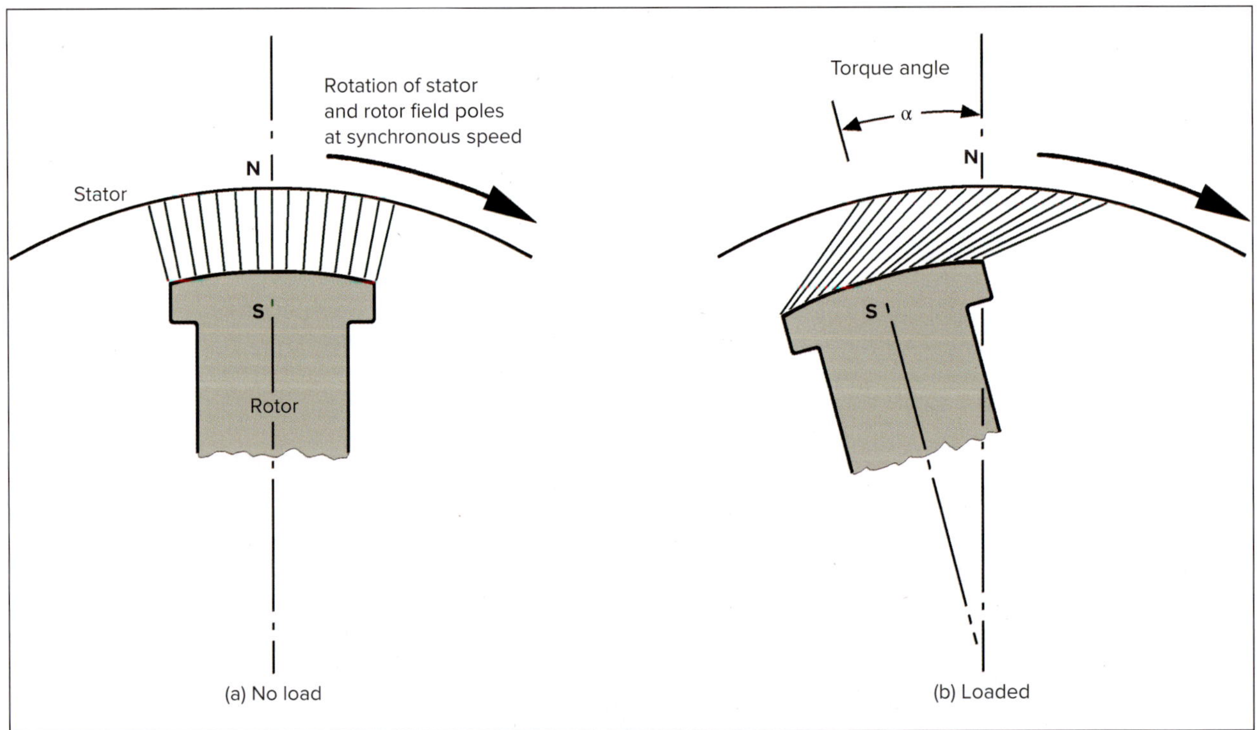

FIGURE 12.72 Relative position of stator and rotor magnetic fields

The magnetic coupling between each stator and rotor pole distorts according to the load applied. If the load on the motor becomes excessive, the magnetic coupling breaks and the rotor slows down until it stops.

When the motor is rotating at synchronous speed, with a fixed d.c. excitation in the rotor windings, the rotor flux cuts the stator windings, inducing a voltage in each phase winding. By Lenz's Law, this voltage opposes the applied voltage. The phase relationship between this induced voltage and the applied voltage depends on the relative positions of each stator and rotor pole, which in turn depend on the load applied to the motor.

Using the example of an ideal synchronous motor with no losses, the operation on no load can be examined. Neglecting motor losses, on no load the torque angle is zero, and so the induced voltage Vg and the applied voltage V are equal and opposite. The resultant voltage VR across the windings is zero, and so the current drawn from the supply is also zero. This is illustrated by the phasors in Figure 12.73(a). (Note: in a real motor, there will be a small no-load current to supply the losses.)

When a light load is applied to the motor, the torque angle increases and the induced voltage Vg in the stator windings is now (180 – α)°E out of phase with the applied voltage V (see Figure 12.73(b)).

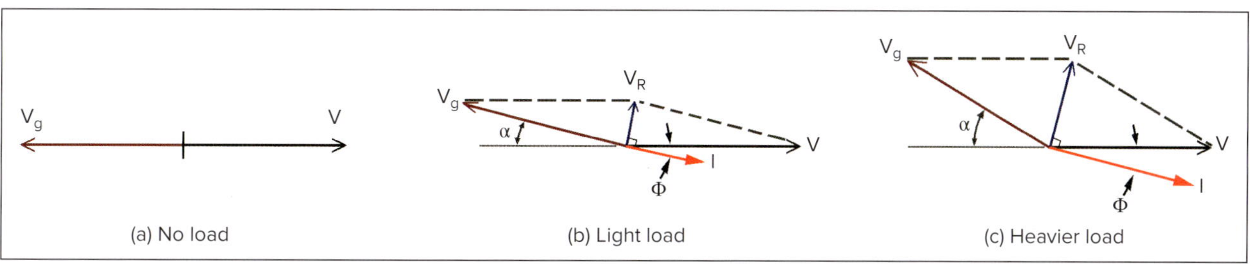

FIGURE 12.73 Effect of load on the line current with constant excitation

These two voltages combine to produce an effective voltage V_R across the stator windings, which is sufficient to draw a current I from the supply. Because of the relatively high inductance of the stator windings, the line current I in each winding lags each resultant voltage V_R by nearly 90°E. This causes the line current I to lag the applied voltage by ϕ.

As the load is increased, so does the torque angle. This causes an increase in the resultant voltage V_R across each stator winding (see Figure 12.73(c)). Due to the increase in the value of V_R, the line current I increases and the phase angle ϕ between the applied voltage V and the line current I also increases. Therefore, for fixed excitation, any increase in the load on a synchronous motor will cause an increase in the line current, at a lower power factor.

12.9.4 Effect of varying field excitation

If the load applied to a synchronous motor is constant, the power input to the motor is also constant. When the rotor field excitation is varied, the induced voltage in each stator winding is also altered.

The phasor diagram in Figure 12.74(a) represents the conditions for a given load at unity power factor. The power input per phase is VI_1. If the rotor field excitation is decreased, the induced voltage V_g decreases (see Figure 12.74(b)). This causes the line current I_2 to lag the applied voltage V by ϕ_2. Since the load—and hence the power input—is constant, the power component of I_2 must remain the same as I_1 in Figure 12.74(a). The line current I_2 must therefore increase to accommodate the lagging power factor. In short, a reduction in the d.c. field excitation causes an increase in line current and a lagging power factor.

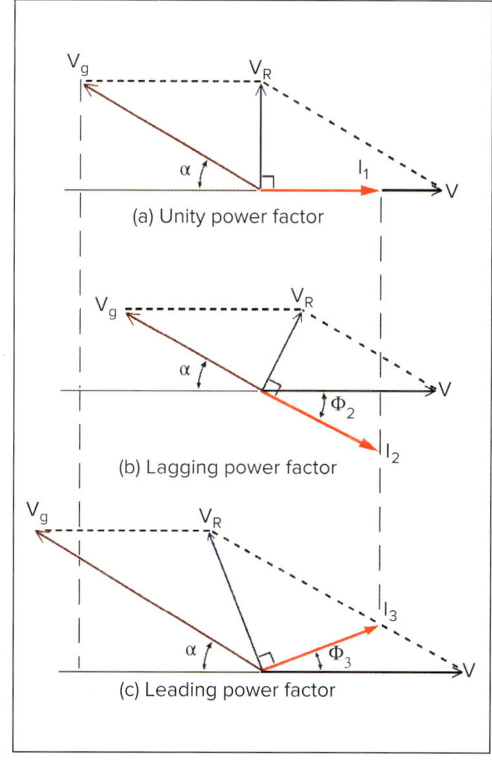

FIGURE 12.74 Effect of varying the d.c. field excitation

If the d.c. excitation is increased, the induced voltage Vg increases (see Figure 12.74(c)). The line current I_3 will therefore lead the applied voltage V by ϕ_3. It will also be greater than I_1 in Figure 12.74(a) because the power component is the same, owing to the load remaining constant. Therefore, an increase in d.c. excitation causes an increase in line current and a leading power factor.

It can be seen that if the excitation of a synchronous motor on a constant load is varied from a low to a higher value, then the:

1. stator current gradually decreases, reaches a minimum and then increases again
2. power factor, which lags at first, gradually increases, becomes unity when the stator current is a minimum and then decreases again but becomes leading.

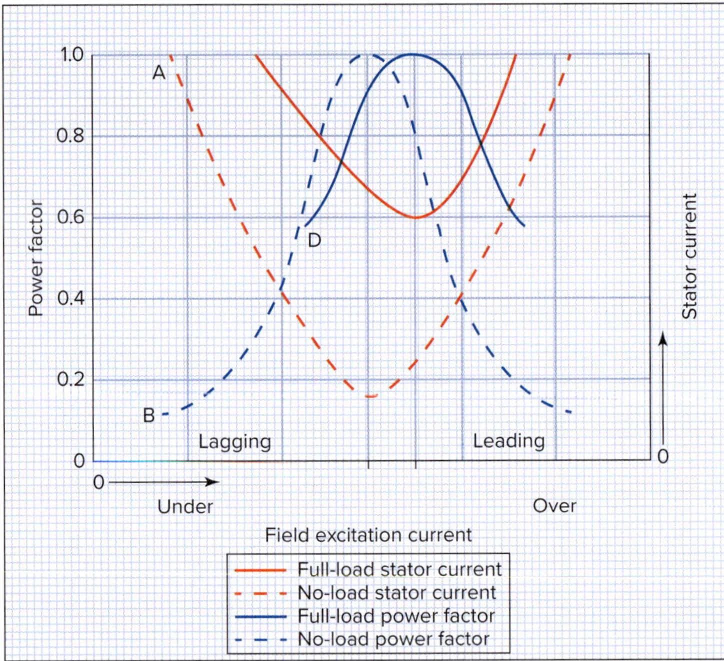

FIGURE 12.75 V curves

This can be shown graphically using 'V' curves (Figure 12.75).

Altering the power factor of a synchronous motor by adjusting the field excitation is a common method of power factor correction used in large installations (see section 12.9.6).

Care should be taken when adjusting the excitation of a synchronous motor. There are limits to which it can safely be taken. Over-excitation and under-excitation can cause a synchronous motor to become unstable. Once these limits have been exceeded, the power produced by the motor decreases and the danger of overloading becomes imminent as the machine exceeds its design limits.

The most obvious situation is one of under-excitation where the magnetic bond between the rotating field and the rotor is so weakened that the load exceeds the pull-out torque of the motor and it drops out of synchronism. Over-excitation creates a situation where the line current and mechanical load exceed the full-load rating of the machine and the magnetic bond becomes so stiff that changes in load place undue mechanical stresses on the motor shaft.

12.9.5 Hunting in synchronous motors

A change in the load on a synchronous motor causes a change in the value of the torque angle (see Figure 12.72). In general, the inertia of the rotor prevents an instant change to the new conditions, with the result that the rotor shifts past the point of equilibrium and then has to correct itself. While the rotor and the rotating field in the stator are still rotating at a synchronous average speed, the change in load on the rotor causes this periodic swing around the point of equilibrium. This hunting (or surging) causes an undesirable fluctuation in line current to the motor.

The usual method for damping these surges is to use an amortisseur winding. It consists of copper bars embedded in the pole faces of the rotor and shorted out at each end (see Figure 12.76). Any surging causes an induced voltage in the copper bars. This results in a magnetic field being created and opposing the surging effect.

Often the shorting-out bars are extended around the rotor, resulting in a squirrel cage-type rotor winding about the salient poles. While damping any tendency of the rotor to hunt, they can also assist the motor in starting by acting as sections of a squirrel cage winding. In effect, this winding enables the motor to be started as an induction motor.

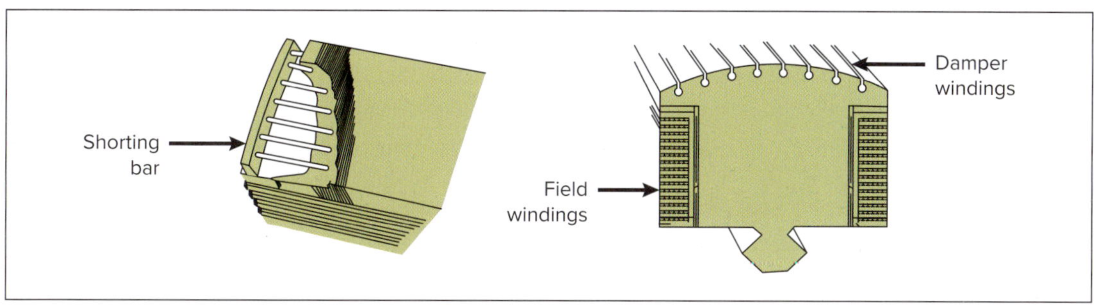

FIGURE 12.76 Salient pole with amortisseur windings

12.9.6 Applications of synchronous motors

Power factor correction

The characteristic of being able to adjust the power factor of a synchronous motor while it is running can be put to beneficial use in industry as a means of correcting the power factor of an entire installation.

The synchronous motor can be run unloaded, but more often it is used to drive some item of equipment necessary for the operation of the plant—for example, air or hydraulic compressors, high-frequency alternators, large fans and blowers or high-pressure water supplies.

An added advantage can be an economic incentive offered by distribution entities for ensuring a certain minimum-value power factor in an installation. For example, the charge per kWh may be reduced if the power factor does not drop below 0.75 (or some similar figure). Where large amounts of power are being distributed and power factor correction is needed, specially designed synchronous motors are run without any load connected. Under these circumstances, the over-excited synchronous motor is called a 'synchronous capacitor' or 'condenser'.

Voltage control

An important application of synchronous motors is in the control of voltage for transmission lines. Synchronous motors are installed at suitable positions along the line and their excitation adjusted as desired to cause them to draw lagging or leading currents in order to raise or lower the voltage. When synchronous motors are installed under these conditions, there is greater stability of the voltage on the transmission line.

Low-speed drives

A synchronous motor has good efficiency, and at low speeds its high initial cost is adequately compensated for by the comparatively lower running cost. At low speeds, the induction motor has a decreasing efficiency, whereas the synchronous motor retains its high efficiency.

Rock- and ore-crushing heads

This application requires a crushing head that moves slowly and has a very heavy rotating flywheel to provide kinetic energy as sudden shock loads are placed upon the crushing head.

12.9.7 Starting methods of synchronous motors

Auxiliary motors

Some synchronous motors are equipped with a special motor designed for use only during the starting period. The auxiliary motor (also called a 'pony motor') runs the synchronous motor up to speed, at which point it is first synchronised and then connected to the supply. It is an expensive method, particularly if high starting torques are required.

Induction motor starting

In this method, a reduced line voltage is applied to the stator windings and the d.c. winding on the rotor is short-circuited. With the aid of the amortisseur windings, the complete machine behaves as an induction motor as it accelerates up to a speed slightly below synchronism. At an appropriate time, the short is removed from the rotor winding, d.c. is applied to the rotor winding and the full line voltage is applied to the stator winding. As the speed is only slightly less than synchronous speed, the rotor field can lock in with the stator field and accelerate to synchronism.

12.9.8 Single-phase synchronous motors

The primary purpose of single-phase synchronous motors is to meet constant speed requirements, even though they have very low efficiency. Their use is usually limited to operations where speed is critical and torque requirements are low. Small synchronous motors are built in a variety of types and constructional details, but are almost invariably one of the two types described below.

Reluctance motors

The stator winding of the reluctance motor is like that of the split-phase or capacitor-start motor, and the windings are of the usual squirrel-cage type. Figure 12.77(a) shows a cross-section of a reluctance motor's stator and rotor. The rotor, however, is assembled from laminations from which several teeth are cut to form definite salient poles. This can be seen in Figure 12.77(b), where the slots in the rotor lamination will still contain the normal parts of a squirrel cage winding. The gaps or slots cut in the rotor will often be of unequal sizes to assist the starting function. *The number of poles in the stator is not necessarily equal to the number of poles on the rotor.*

Hysteresis motors

The cylindrical rotor within a hysteresis motor works on hysteresis losses induced into the rotor. The stator carries both the main and auxiliary windings. The rotor is made of magnetic material with high hysteresis-loss properties. The rotor is also produced to have high resistance to reduce eddy current losses. The motor starts like a single-phase motor, but when the rotor nears synchronous speed, the rotor speed locks with the stator's synchronous speed.

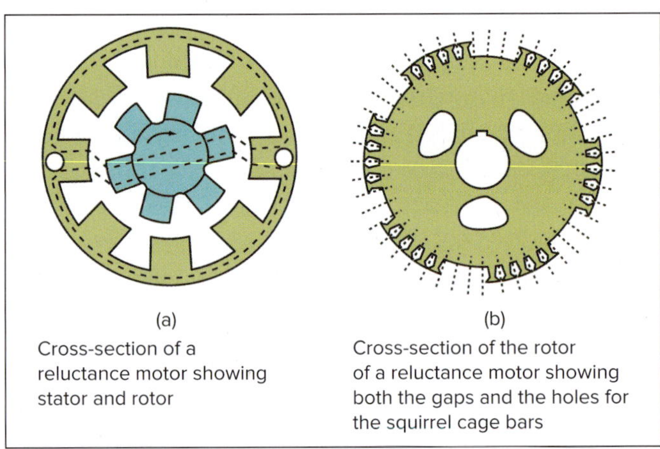

(a) Cross-section of a reluctance motor showing stator and rotor

(b) Cross-section of the rotor of a reluctance motor showing both the gaps and the holes for the squirrel cage bars

FIGURE 12.77 Internal configuration of a reluctance motor

CHECK YOUR UNDERSTANDING

12.31 Explain the construction of a synchronous motor.

12.32 What does the term 'hunting' mean in relation to a synchronous motor?

12.33 What techniques are used to start a synchronous motor?

12.10 Motor protection

12.10.1 Introduction

An electric motor will normally run unattended for many years. During its operational life, there will be occasions when an overload will be placed on it or a fault will occur in it (or in the control gear or the supply circuit). The motor and its associated circuit must have devices to protect them should such a problem occur.

Protection is normally supplied by the circuitry associated with a motor starter. In addition to monitoring an electric motor for such functions as controlled acceleration, reversing, braking or speed control, a starter should also provide the motor with some form of protection. Line voltages and motor currents need to be monitored and, when required, the motor should be isolated from the supply.

Protective devices take many forms, but most are designed to operate within the control circuit of the motor starter. In this way, a fault occurring in one phase can be used to trip a contactor to isolate all three phases and disconnect the motor from the supply.

All electrical circuits must be protected against the effects of both overload currents and fault currents. Before discussing their effects, it is important to note the difference between an overload current and a fault current.

An overload current flows when the motor is used incorrectly—that is, if it has been overloaded and draws more than its rated current. If the motor draws more that the rated circuit current, then both the motor and the circuit are overloaded. *There is nothing wrong with the motor or the circuit—the problem is how the machine is being operated.*

In Figure 12.78, if too much material were placed on the conveyor, the motor would slow down due to overload. The motor will draw excess current because of operator error; this is an overload current.

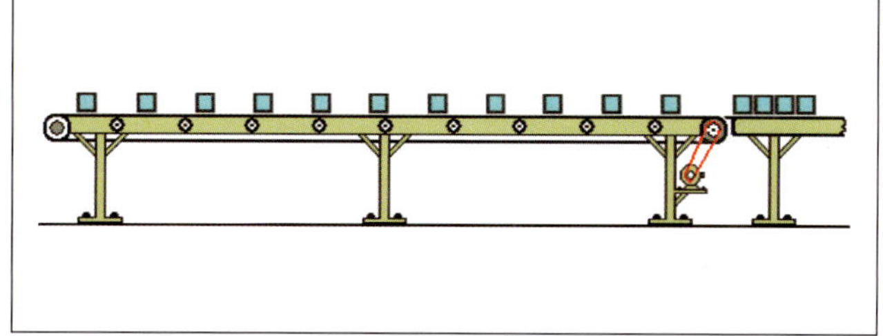

FIGURE 12.78 **Typical conveyor system**

When an overload current flows in the circuit, the circuit protection device needs to operate before damage can occur to the cable or the motor. The values of current-carrying capacity allocated to cables are generally determined so that the cable insulation does not sustain damage due to temperature rise during normal operation.

A fault current occurs when there is a short-circuit between live conductors only, or between live conductors and earth. This current can be very large and can cause considerable damage if the correct circuit protection has not been used. For a motor circuit, a short-circuit current would only be present if a serious fault developed in the motor, control gear or associated supply circuit.

One common cause of motor failure is temperatures rising above design values. This can be due to an electrical fault such as single phasing or to mechanical overloading by the driven equipment. The overload might be a relatively small one for a sustained period, or it might be a large and sudden overload such as occurs when a machine locks up because of a mechanical fault.

12.10.2 Short-duration overload

Standard motor design allows for a short-duration overload by having a normal running temperature that is well within the capability of the temperature rating of the motor insulation.

When an overload current flows in the motor windings, the heat generated by the current is proportional to the square of the current ($H \propto I^2$). If a motor rated at 10 A has an overload of 20 A, it will cause *four times* the amount of heat to be generated in the motor ($10^2 = 100$, $20^2 = 400$). This is a serious situation if the overload is sustained for more than a few seconds. To compound the problem, an overloaded induction motor will slow down with the load, and the shaft-mounted fan will supply less cooling air just when proper cooling is most required.

12.10.3 Sustained overload

If a motor has a sustained overload, there will be a high temperature rise within it. If the temperature becomes higher than the rated temperature of the insulation, the motor will most likely sustain a serious insulation failure.

12.10.4 Locked rotor

When the rotor of an induction motor is locked (meaning that the motor has stalled on an overload), the motor will draw excessive current. The value of the locked rotor current is the same as the DOL (direct online) starting current, and this is usually about seven times the rated current.

The heat generated by a 10 A motor with a locked rotor will be approximately 49 times the heat produced at full load ($10^2 = 100$, $70^2 = 4900$). At the same time, the shaft-mounted fan is not supplying any cooling air at all. This is serious, and if the correct protection has not been provided, the motor can sustain a significant insulation failure.

12.10.5 Under-voltage protection

Two kinds of under-voltage protection are available for motors controlled by contactors. The first is called 'no-volt protection' because once the voltage is reduced below the holding-in voltage level of the coils, the contactors drop out. When power is restored, the motor must be taken through the starting sequence again until it is up to speed.

The second type is an additional part of a control circuit and keeps the contactors in a holding position for momentary dips in voltage. If the power is restored before the motor can coast down appreciably in speed, full power is immediately restored to the motor.

12.10.6 Over-voltage protection

For alternating current motors connected to a distribution system, over-voltage is rare. It is more common with long-distance, high-voltage transmission lines where lightning strikes are likely to occur. It is a more prevalent problem with variable-speed mobile generating plants such as those in aeroplanes and automobiles.

As it is undesirable to have an electrical installation operating with a low power factor, consumers are encouraged to keep loads at high power factors. Sometimes, consumers adopt the uneconomical practice of connecting capacitors in parallel with individual induction motors to correct the power factor.

This can lead to over-voltage transients occurring when a motor is switched off. The size of the capacitors should be sufficient only to correct the power factor of each load, and preferably ensure that it still has a lagging power factor.

Capacitors connected across an induction motor can cause self-excitation as the motor continues to spin. In some circumstances, this generated voltage can be as high as double line voltage. In normal circumstances, this over-voltage lasts only for a short period and most equipment can cope with the voltage surge.

There are occasions when a motor application requires the fitting of surge suppression devices. One example is the submersible bore pump. In its operating location, it is a very good earthing point and, as many bores are fed from overhead lines at a distance from a source of supply, the installation is vulnerable to lightning strikes. In an attempt to protect the motor against what may well be a 500 kV over-voltage surge, suppressors are fitted. Motor protection is rather uncertain in such circumstances.

12.10.7 Repetitive starting

The starting current of an induction motor can be up to seven times the full load current. The heat generated during a normal start can be handled by the motor as the starting current is only present for a brief time. When the motor is up to speed, the shaft-mounted fan will remove any excess heat.

The total heat produced by the current can be determined using Joule's law.

$$H = I^2Rt$$

where:

H = heat in joules
I = current in amperes
R = winding resistance in ohms
t = time in seconds.

The amount of time that the high current flows for is the important factor. From the equation, it can be seen that the heat produced is directly proportional to the time in seconds. In the case of repetitive starts, the amount of heat produced will keep building up due to the extended time factor and the fan not running at full speed long enough to remove the excess heat.

12.10.8 Special requirements for motor protection in high-humidity or moist environments, high-temperature areas and corrosive atmospheres

The temperature rating of a motor indicates the maximum temperature that the windings can withstand before permanent damage is done to them.

Motors with Class 'B' insulation have a maximum operating temperature of 130°C, so they operate satisfactorily at temperatures up to that. If the winding temperature exceeds that value, some deterioration of the windings will occur. As a rule, for every 10°C of temperature rise above the rated value, the insulation life is halved.

Temperature rise

Temperature rise of a motor is the difference between the ambient temperature and the motor's winding temperature. The winding temperature is calculated by the resistance method. If the ambient temperature is 30°C and the rated maximum temperature of the motor is 105°C (Class 'A' insulation), the maximum permitted temperature rise is 65°C: 105 – 40 = 65°C.

However, if the ambient temperature is 40°C, then the maximum permitted temperature rise is only 55°C. The motor may be handling the load when the ambient temperature is low but may suffer over-temperature problems when the ambient temperature is high.

Hot-spot allowance

The above calculation assumes that the winding has the same temperature throughout. However, if the temperature has been determined using the resistance method, then it will be the average value. To offset any hot spots in the winding, an allowance is made for safety. A value of 10°C is normally used. So, for the above situation with a 40°C ambient, the maximum permitted temperature rise would be 55 – 10 = 45°C.

Calculating winding temperature

Calculations to determine the temperature of a motor winding are based on measuring the cold winding resistance and the hot winding resistance, as well as the extrapolated temperature for zero resistance. An extrapolated value can be used because the resistance of copper and aluminium windings increases in a linear manner with temperature. The extrapolated temperature for zero resistance of copper windings is –234.5°C.

Motor insulation classes

There are currently four motor insulation classes in common use.

CLASS A

- Maximum temperature rise: 60°C
- Hot-spot over-temperature allowance: 5°C
- Maximum winding temperature: 105°C

CLASS B

- Maximum temperature rise: 80°C
- Hot-spot over-temperature allowance: 10°C
- Maximum winding temperature: 130°C

CLASS F

- Maximum temperature rise: 105°C
- Hot-spot over-temperature allowance: 10°C
- Maximum winding temperature: 155°C

CLASS H

- Maximum temperature rise: 125°C
- Hot-spot over-temperature allowance: 15°C
- Maximum winding temperature: 180°C

EXAMPLE 12.8

A motor has a winding resistance of 16.5 Ω at 25°C. After running at full load for two hours, the resistance is measured as 20 Ω. What is the temperature of the windings?

$$R_2 = \frac{R_1[234.5 + t_2]}{234.5 + t_1}$$

By transposition:

$$\begin{aligned} t_2 &= \frac{R_2}{R_1} \times (234.5 + t_1) - 234.5 \\ &= \frac{20}{16.5} \times (234.5 + 25) - 234.5 \\ &= 1.21 \times 259.5 - 234.5 \\ &= 80°\text{C} \end{aligned}$$

EXAMPLE 12.9

A motor has a winding resistance of 16.5 Ω at an ambient temperature of 25°C. After running at full load for two hours, the resistance is measured as 20 Ω. Determine the temperature rise in the windings.

The resistance equation used above can be modified to give:

$$\begin{aligned} \text{Temperature rise} &= \frac{(R_2 - R_1)}{R_1} \times (234.5 + t_1) \\ &= \frac{(20 - 16.5)}{16.5} \times (234.5 + 25) \\ &= \frac{3.5}{16.5} \times 259.5 \\ &= 55°\text{C} \end{aligned}$$

An allowance of 10°C is then added to this value to give a value of 65°C. If the maximum permitted temperature of the motor is 105°C, then the motor is operating within its safe temperature range.

$$\text{Motor temperature} = \text{temperature rise} + \text{ambient} = 65 + 25 = 90°\text{C}$$

12.10.9 High humidity

Motors that operate in areas where the humidity and temperature are high are subjected to moist conditions, which can produce condensation on their inner surfaces. In humid areas, where the operating temperature of a machine has a large differential to the off state and ambient temperature, the motor is particularly susceptible to condensation. For example, a machine can sometimes go from 20°C to 40°C, with relative humidity often reaching 100%.

Condensation will occur on a motor surface when the frame temperature of the motor is lower than the dew-point temperature of the ambient air. Any condensation can cause the condition of the windings and inner surfaces to deteriorate over time. This problem is compounded if the motor is installed in an area where there is little or no natural ventilation.

To protect the motor and windings against these conditions, motors can be given an extra coating on their inner surfaces and on the windings. The stator windings are impregnated with a special electrical varnish that is designed to prevent any moisture from reaching the windings. The metal surfaces are also given a special coat of a rust-preventative paint.

12.10.10 Environmental protection

Enclosures

Induction motors are available in many different enclosures, depending on the task they are designed for.

Some of the types of enclosures are:

- open
- protected
- drip proof
- duct or force ventilated
- totally enclosed
- flame proof
- weatherproof
- submersible
- explosion proof.

AS/NZS 1359.21 lists more than a dozen methods for cooling induction motors. Adding to this a variety of different mounting types, such as horizontal or vertical, tends to make the list rather long. The type of motor enclosure chosen depends on the conditions under which it has to function. The enclosures listed below are the main general groups, but there can be many sub-groups of each.

Open motors

The ends of the machine are open, allowing free ventilation through and around the windings. Air is circulated through the motor by a fan attached to the motor shaft. The motor needs to be installed in a clean and dry atmosphere.

Protected motors

With the protected motor enclosure, the windings and live parts are protected mechanically but a free flow of air is still drawn through the motor by a fan attached to the motor shaft. The protection is often obtained by fastening wire mesh or perforated metal over enclosure openings.

Drip-proof motors

Drip proofing is an advance on the protected type of enclosure. The openings are further protected by a hood, preventing foreign materials and moisture from falling vertically onto the motor and entering it. The hood may be incorporated into the enclosure during manufacture of the motor.

Duct- or force-ventilated motors

With forced ventilation, a fan forces air through the motor at a constant rate. A motor with forced ventilation is shown in Figure 12.79. This arrangement is essential where the motor is connected to a variable speed drive and is required to run at reduced speed with a load. In these conditions, the motor's own fan will not be spinning fast enough to deliver the required amount of cooling. The motor might not be in a suitable atmosphere to allow clean air to flow through it—examples would be corrosive atmospheres, high temperatures and high humidity. For ventilation, cool and clean air must be drawn in from outside the installation.

There are two major types of ventilation in this case. In one type, air is drawn from outside through a duct and can be expelled inside the installation, or it may have to be ducted to the outside atmosphere again. In the second type, a blower is installed outside and air is forced through a duct to the motor. This is a method of cooling usually adopted for very large motors.

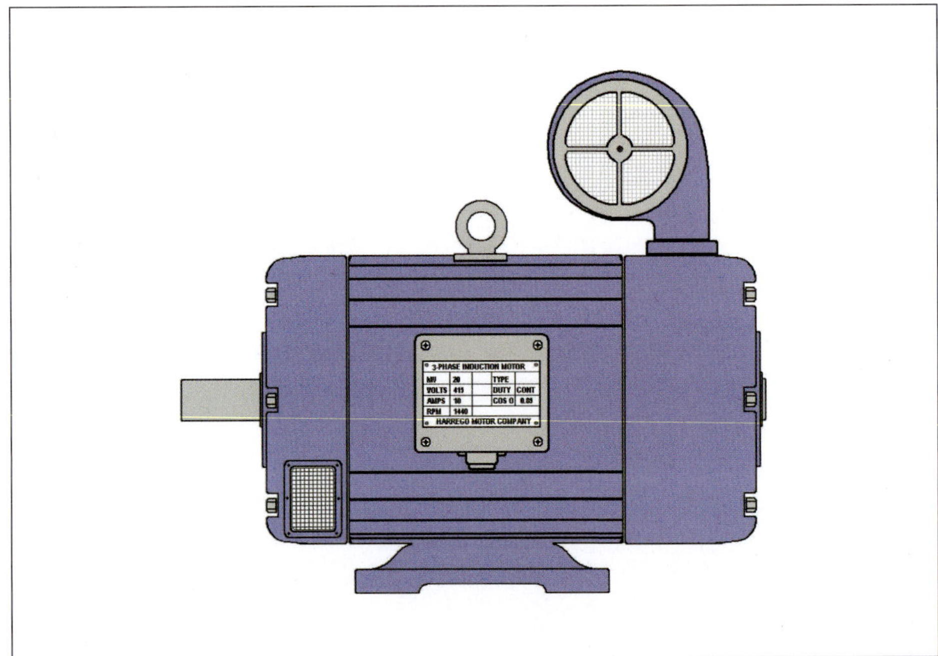

FIGURE 12.79 Three-phase motor with fan-forced cooling

Totally enclosed, fan-cooled motors

In the totally enclosed, fan-cooled (TEFC) type (see Figure 12.80), there is no contact between the air inside the machine and the air outside. The fan attached to the shaft is external to the motor proper and enclosed within its own housing. The motor housing is usually ribbed, and the air driven by the fan flows along the ribs and removes the internal heat by conduction through the housing.

The heat is removed from the motor at a slower rate than by direct cooling, but the increased surface area created by the finning process assists. Modifications to this method of enclosing a motor enable it to be classified as waterproof, weatherproof or submersible.

Flame-proof motors

The flame-proof type (see Figure 12.81) is a totally enclosed motor with additional precautions to seal the bearings and assembly contact surfaces. Electrical connections are made through a special sealed gland. The motor housing is made strong enough to withstand any internal explosion and still prevent sparks escaping to the external atmosphere.

The motor is used when there are flammable gases and the risk of explosion if sparks enter this atmosphere. The enclosure must comply with stringent regulations before being classified as flame proof.

FIGURE 12.80 **Totally enclosed, fan-cooled motor**
Jurijs Lescenko/Alamy Stock Photo

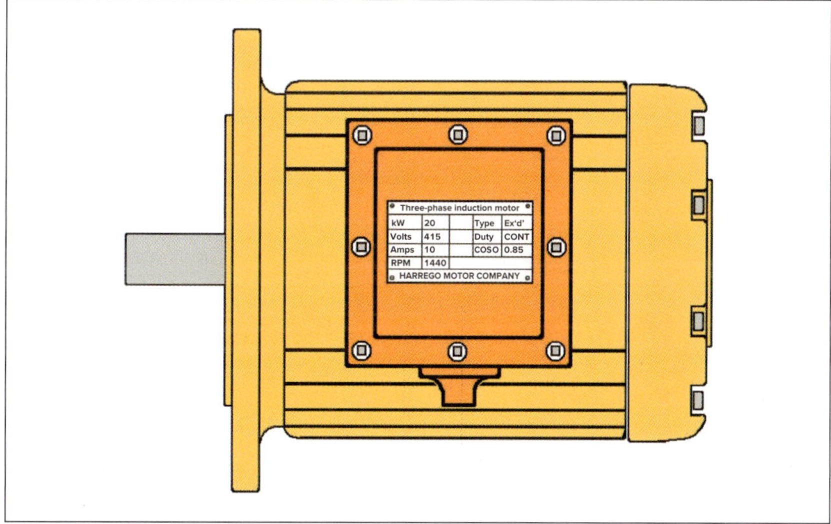

FIGURE 12.81 **Typical flange-mounted flame-proof motor**

12.10.11 Protection devices

AS/NZS 3000:2018 requires that circuits supplying motors should be protected against overload and fault currents. In addition, any unattended motor is required to be protected against:

- overloads
- fault currents
- under-voltage
- over-temperature.

There is also a requirement that all motors should have a control device for stopping and starting, as well as an isolator.

All the above requirements are usually met by installing fuses or circuit-breakers in the switchboard, a motor starter with controls at a convenient location and an isolator near the motor. Modern proprietary motor starters are available with some, or all, of the above features built into the one unit.

12.10.12 Fuses

The fuse is possibly the simplest form of circuit protection. It consists of a fuse element that is designed to melt and prevent further current flow. The major disadvantage of a blown fuse is that the active component must be replaced.

The primary purpose of a fuse is to protect a circuit rather than any load. Under short-circuit conditions, the reaction time of a high rupturing capacity (HRC) fuse in isolating a circuit is probably the fastest of all protection systems.

A fuse element must have a current rating that is high enough to allow a motor to draw starting currents yet low enough to give some protection against overloads. Because of these opposing factors, a fuse cannot provide complete protection for both the circuit and the load.

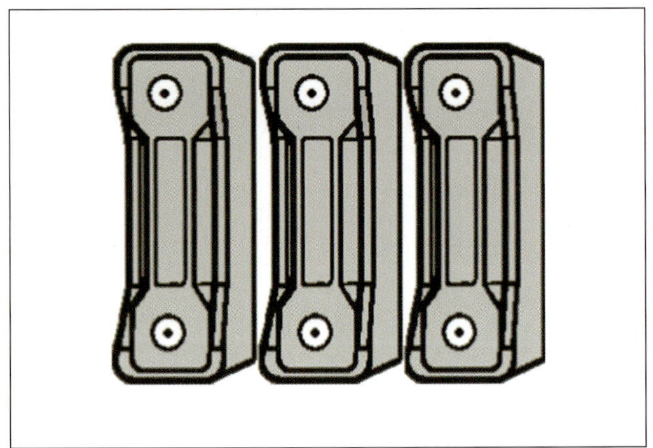

FIGURE 12.82 HRC fuses for three-phase motor circuit

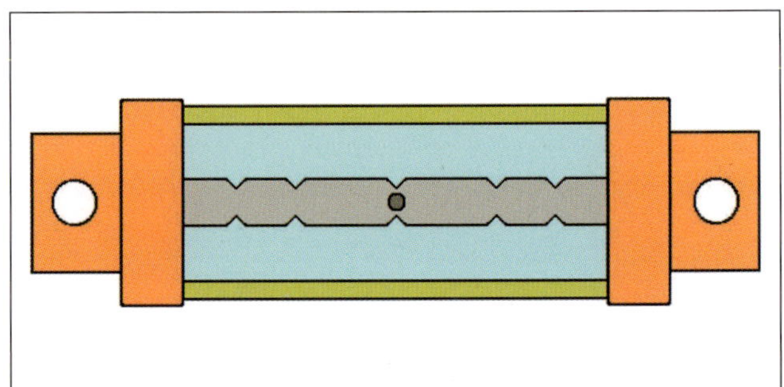

FIGURE 12.83 Cutaway of an enclosed-type fuse link

Figure 12.82 shows a set of three typical HRC fuses used for motor circuit protection. They have a plastic carrier and base and are designed to be used with enclosed fuse links. HRC fuses are extremely consistent and reliable in terms of tripping values. The element of an HRC fuse is enclosed in an insulating tube filled with powdered quartz to quench any arc that might develop.

Figure 12.83 shows a cutaway drawing of an enclosed-type fuse link. This type has lugs to allow the fuse link to be held in place with machine-threaded screws. HRC fuses up to 63 A are often manufactured with plug-in fuse links.

The silver element has restrictions cut into its section to give an extremely fast response to fault currents. If a fault current flows through the fuse, the restricted sections of the fuse element heat up very quickly by the Joule's-law effect ($H = I^2Rt$).

The fuse element of an HRC fuse has a **eutectic** bead to give a good response to prolonged overload currents. If a prolonged overload current flows, the eutectic bead will reach a point where it can no longer dissipate the heat and will melt. By careful selection of the mass of the eutectic bead, manufacturers can tailor the operation of the fuse to any time/overload-current characteristic required.

To ensure the safe operation of equipment, it is essential that only the correct size is used when replacing HRC fuses. Replacement cartridges of the correct size should be kept on hand.

When used to protect induction motor circuits, HRC fuses that are rated 'M' for 'motor starting' should be used. These can provide protection while not tripping on motor starting currents.

12.10.13 Circuit-breakers

When circuit-breakers are used for motor protection, the correct type must be selected. Figure 12.84 shows a typical time/current characteristic of type 'C' and 'D' circuit-breakers.

It can be seen from the characteristics that a standard 'C'-type circuit-breaker has the magnetic trip portion set at approximately 7.5 times the nominal current rating of the breaker. This value may not be enough to allow for the starting current of the motor, and could result in nuisance tripping. A 'D'-type circuit-breaker has the magnetic trip set to about 12.5 times the nominal circuit-breaker rating. Tripping times that lie within the thermal characteristics are identical, as both types will trip at the same time for a steady overload.

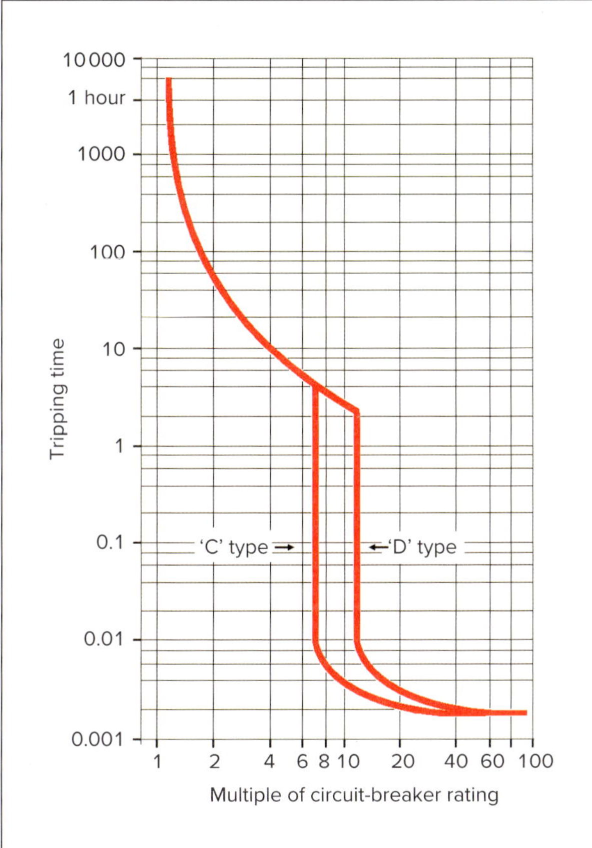

FIGURE 12.84 Typical time/current characteristic for 'C'- and 'D'-type circuit-breakers

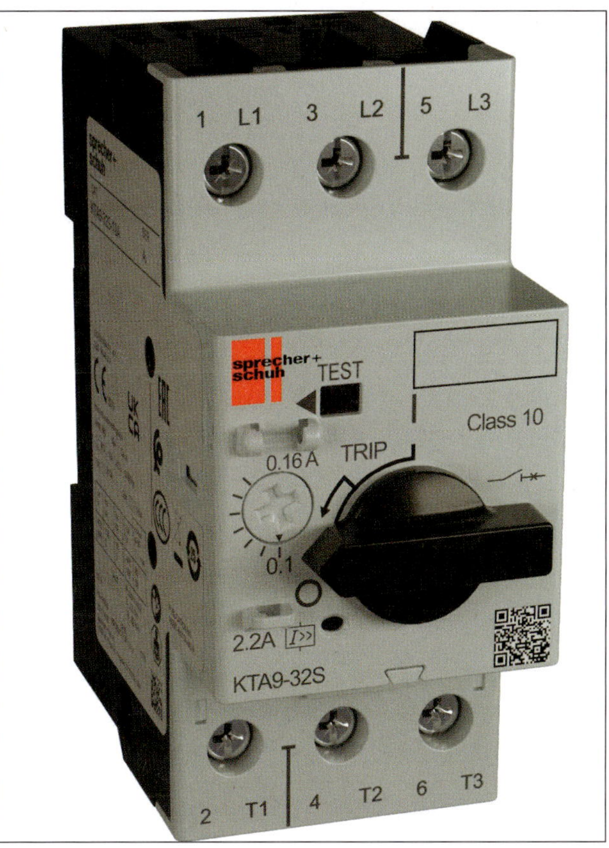

FIGURE 12.85 Combined motor circuit-breaker and isolator

Reproduced with permission of NHP Electrical Engineering Products Pty Ltd

The circuit-breaker shown in Figure 12.85 combines the features of a motor circuit-breaker and an isolator in the one unit.

12.10.14 Overcurrent relays

Magnetically activated overcurrent relays

Instant tripping relays are operated by the direct action of the motor current on an armature. The principle is illustrated in Figure 12.86. The relay consists of a series coil wound on a magnetic core. The coil is connected in one motor line and the armature is attracted to the main body of the core when the motor current exceeds a predetermined value.

The mechanical movement of the armature can be arranged to either close or open an electrical circuit as desired.

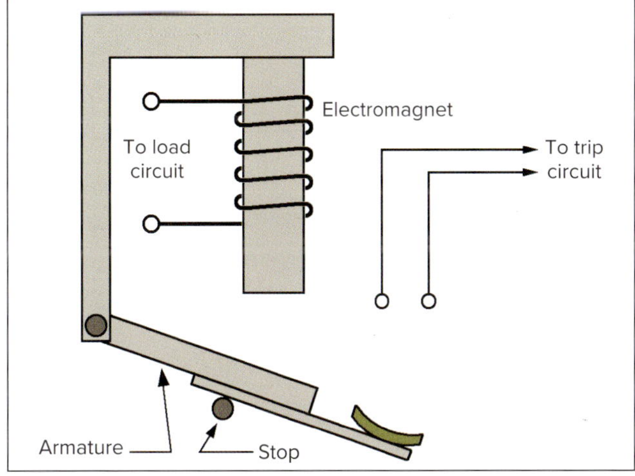

FIGURE 12.86 Simple electromagnetic overload trip

12.10.15 Thermal overloads

Many types of thermal overload relays are available for motor protection. Some operate on different principles from others, but all are designed to open a contact when a temperature-sensitive element—such as a bimetal strip—receives sufficient heat to activate it. Because the contact is usually connected in the control circuit of the starter, the opening of the contact causes the main contactors to drop out and switch off the supply to the motor.

The operating principle of most thermal overload elements relies on a bimetal strip (see Figure 12.87). Correctly designed thermal elements produce an amount of heat related to the amount of motor current. The quantity of heat stored, and hence the temperature of the bimetal strip, relates to the amount of bending of the strip. After short-duration overloads, the heat can dissipate and the temperature of the strip is reduced. If small overloads continue for any length of time, the amount of heat generated will activate the relay.

With starting currents, insufficient heat energy is developed in the strip for it to bend enough in the time taken for the motor to run up to speed and the current to reduce to the normal running value.

Ideally, there should be a thermal detecting element in each line of a three-phase motor.

Thermal overload elements are placed in the main supply lines leading to the motor, while the associated control contacts are connected in series with the control circuit. This is to ensure that if only one overload operates, it disconnects the motor from the supply.

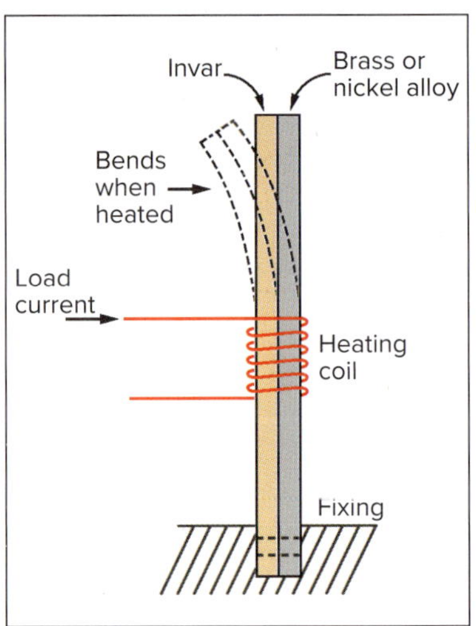

FIGURE 12.87 Operating principle of a bimetal strip

12.10.16 Combined thermal-magnetic overcurrent relays

In the thermal-magnetic version of the overcurrent relay, the advantage of the inbuilt delay of the thermal type is combined with the instantaneous tripping characteristic of the magnetic current overload relay.

The combination of the two methods is considered ideal motor protection, for the following reasons:

- For exceedingly high currents, the magnetic section of the relay acts almost instantaneously.
- For small overloads, the heat accumulated in the thermal section causes delayed tripping according to the rate of heat generation.

Depending on design and application, the combined unit might not have oil dashpots to delay the magnetic relay action. Instead, it is set at a current rating in excess of motor starting requirements, and the thermal element rating is retained at the lower value.

The combined unit shown in Figure 12.88 has three separate devices connected in series to fulfil several requirements. The components have been designed to operate together. The top part is a magnetically operated circuit-breaker and isolator, the middle part is a contactor, and the lower part is a thermal overload sensing unit.

Figure 12.89 shows a small DOL starter. The contactor and overload unit can be seen mounted inside the starter, while the stop and start button are on the cover. This type has an IP65 rating.

FIGURE 12.88 Motor circuit-breaker with contactor and overload unit

Reproduced with permission of NHP Electrical Engineering Products Pty Ltd

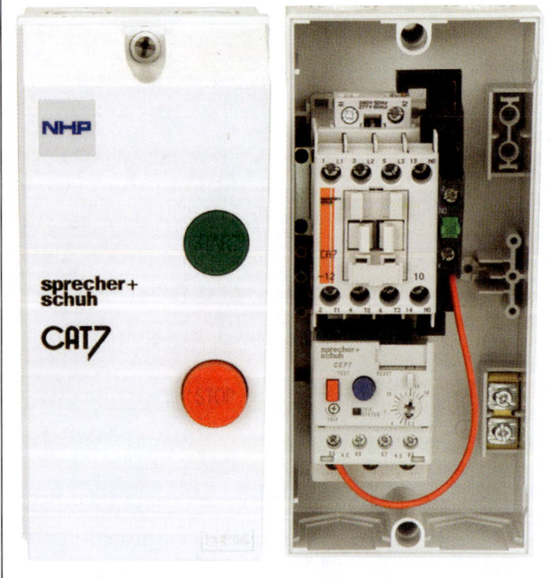

FIGURE 12.89 Small DOL starter with cover removed

Reproduced with permission of NHP Electrical Engineering Products Pty Ltd

12.10.17 Temperature-dependent resistors

Temperature-dependent resistor protection

A resistor with a positive temperature coefficient (PTC) has the characteristic of increasing its resistance only gradually until a critical temperature is reached. Above this point, its resistance increases rapidly. This type of resistor is commonly called a 'thermistor' (from 'thermal resistor').

This critical temperature can be varied by altering the composition of the material from which the thermistor is made. The determination of a critical temperature for a PTC thermistor can also determine its use (see Figure 12.90).

For example, many electric motors are designed to have a maximum operating temperature of 60°C. At this temperature, the heat generated by motor losses is approximately equal to the heat being lost by the motor. Effectively, this means that the temperature of the motor then remains constant.

PTC thermistors are made in many different shapes so they can fit the requirement of particular jobs. When suitably insulated and placed inside the windings of a motor, the internal temperature of the windings can be monitored.

If the critical temperature is, say, 65°C, then the PTC thermistor's resistance would increase rapidly above this temperature and could be an indication that there is something wrong with either the motor or its load.

Under normal conditions, thermistors are only capable of handling small values of current and must be used in conjunction with other equipment.

Figure 12.91 illustrates one method of monitoring the temperature of motor windings. During the winding process, a PTC thermistor is inserted into each phase winding of the motor and all are connected in series with the coil of a small relay. This relay controls a pair of contacts in the control circuit of the motor starter. An isolating transformer and a bridge rectifier supply this circuit with direct current.

When the start button is pressed, the transformer supplies power to the PTC thermistor circuit and, provided their collective resistance is below the critical temperature value, enough current will flow to activate the relay coil and close the contact connected in series with the main contactor coil.

Normal DOL starting procedure follows, with normal contactor action. If the temperature of any one of the three PTC thermistors rises above the critical value, the resistance of the circuit increases, the current flow through the relay coil decreases and the relay drops out. This action causes the main contactor to drop out and isolate the motor.

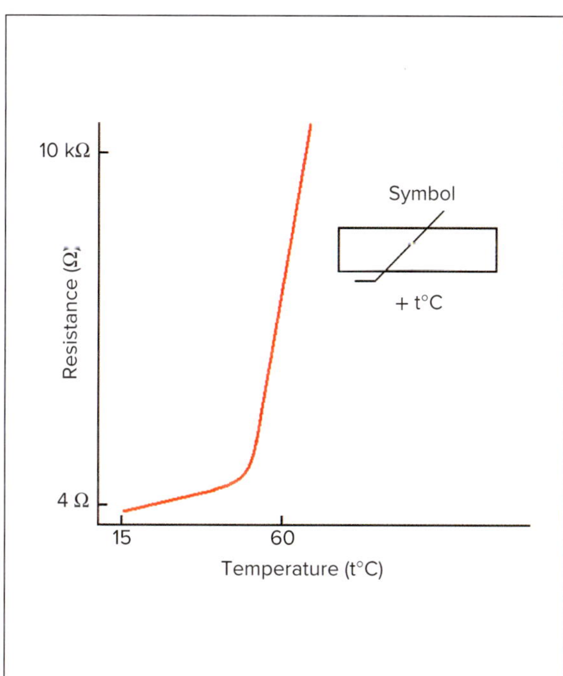

FIGURE 12.90 Typical PTC thermistor characteristic

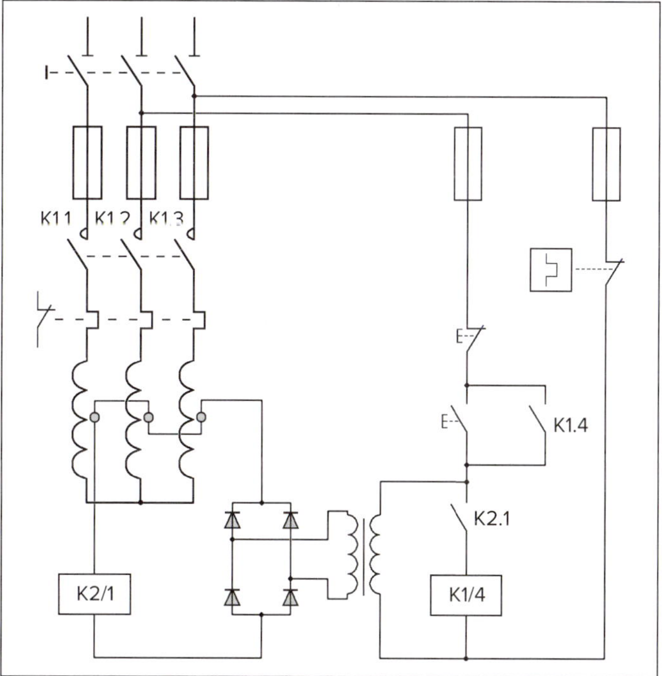

FIGURE 12.91 Using PTC resistors to protect motor windings

The complete control circuitry for a thermistor relay is usually housed in the one unit (see Figure 12.92).

As with other thermally activated devices, there is an inherent delay in a PTC thermistor cooling down and resetting itself. A thermal overload is normally made as small as possible to reduce its thermal capacity, but this has no effect on the thermistor when it is buried within the windings because they regulate the rate of cooling.

For a locked rotor situation, PTC thermistors are an inadequate form of motor protection and external thermal and magnetic overload protection should still be provided. The time taken for motor windings to heat up when the rotor is locked in a standstill position is comparatively long. Irreversible damage can be done to the motor windings before the PTC thermistor exceeds its critical temperature and disconnection occurs.

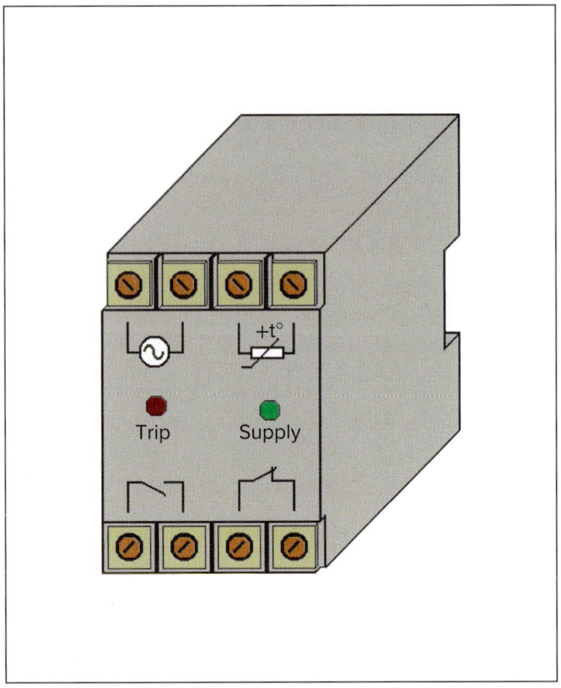

FIGURE 12.92 Typical thermistor relay

12.10.18 Single-phasing protection

A three-phase motor operating under ideal conditions will draw three equal phase currents. This suggests that the three-line voltages are also equal, a situation that rarely exists in practice. A small variation in voltage of, say, 2% can cause a current variation of around 10% to 15%.

The function of an overload relay, whether magnetically or thermally activated, is to disconnect the motor from the supply lines under specified conditions of current flow and within a set period of time.

Overload relays cannot protect a motor against internal faults—that is not their intended function; controllers are intended to handle the starting currents of induction motors. Fault currents may be many times this value, so fuses or circuit-breakers should be installed ahead of the controller.

The only protection readily available for a three-phase motor is the provision of thermal overload heating elements in each phase. An internal fault in the motor shows externally as greatly unbalanced line currents. Thermal or magnetic overloads can then disconnect the motor from the supply.

If an external fault occurs, such as a line to the motor becoming open-circuited when the motor is running, the motor is said to be 'single phasing', and the two remaining line currents increase by approximately 73% each. One of the phase windings then carries about twice as much current as the other two and motor damage can occur.

A feature that is added to some thermal overloads is 'differential tripping'. This is where three identical heating elements are used to detect a loss in one phase, such as would occur when a three-phase motor is operating under single-phasing conditions. The imbalance or difference in the heating elements causes the thermal overload to trip. Figure 12.93 shows an example of a thermal overload with differential tripping.

For smaller motors, the cost of installing phase-failure relays might be prohibitive,

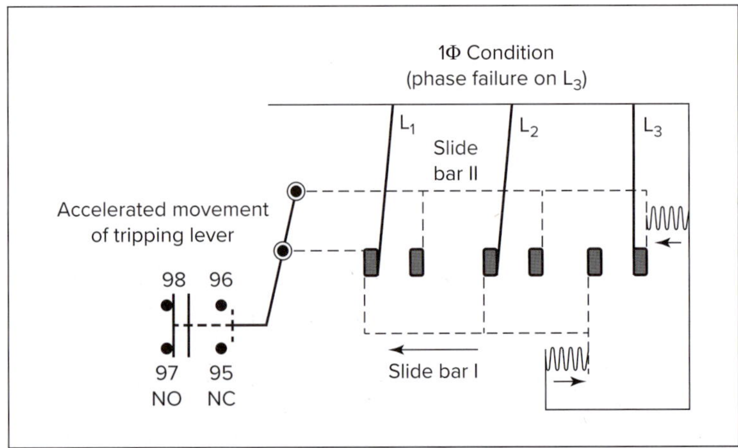

FIGURE 12.93 Thermal overload relay with differential tripping

but for larger motors it can be worthwhile as additional protection. Voltage-sensitive relays are connected across each phase, with operating contacts connected in the motor's control circuit to ensure that the motor is disconnected from the supply in the event of any phase voltage deviating beyond specified limits.

12.10.19 Reverse-phase sequence protection

Some equipment can be damaged if inadvertently driven in the wrong direction by the drive motor. This can occur when the phase sequence of the supply has been changed. A phase-sensitive relay is supplied by voltages from each phase and isolates the motor from the supply if the phase sequence is incorrect. The relay itself may be partially mechanical and operate a vane, which in turn operates contacts in the main control circuit; or it may be an electronic device.

12.10.20 Electronic overloads

The modern equivalent of the thermal overload unit is the electronic overload. These units, which are a similar size to the traditional bimetallic strip-operated units, have small current transformers inbuilt to sense the value of the current.

An electronic overload unit is shown in Figure 12.94. The advantage of the electronic unit is that, in addition to having the normal settings with trip and alarm contacts, it has added features such as single-phasing protection.

The current setting can be adjusted for different motor run-up times using DIL switches. Additional clip-on modules can be added to give further features such as remote tripping, DeviceNet, PTC thermistor relays and ethernet connections.

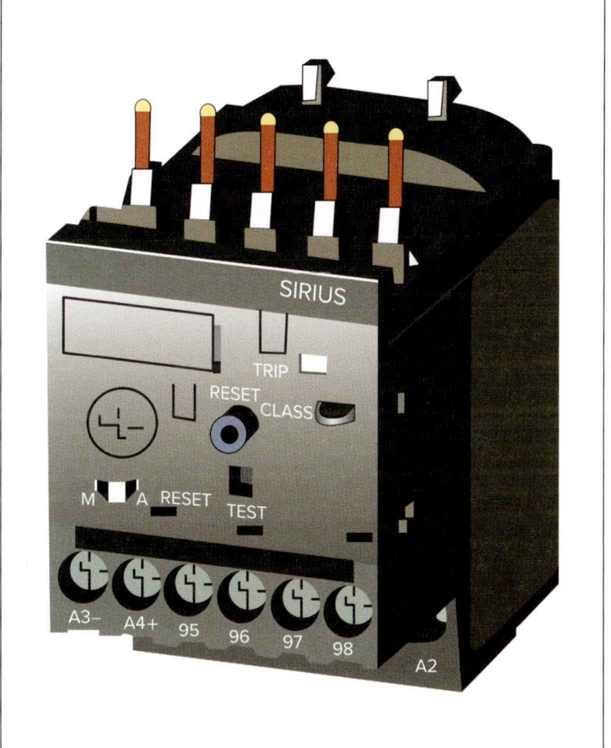

FIGURE 12.94 Electronic overload relay

12.10.21 Synchronous generator loss-of-excitation protection

Extreme weakening or complete loss of excitation on a synchronous generator is detrimental to both the generator and the power system to which it is connected. Generator loss of excitation (LoE) can be caused by:

- field winding open circuit
- field winding short-circuit
- flash over at brushes or slip-rings
- loss of supply to the main exciter
- accidental field-breaker tripping
- automatic voltage regulation (AVR) control circuit failure.

LoE accounts for more than 60% of all generator failures so protective systems need to be installed in case loss of excitation occurs.

An undetected LoE condition causes the rotor current to decrease gradually and the field voltage system slowly collapses. The result is that the synchronous generator begins to consume reactive power from the power system to which it is connected, instead of supplying it. This condition can then cause an area-wide system voltage collapse.

Partial excitation field weakening can be produced by multiple ground faults that cause sections of the field windings to be short-circuited. This condition results in uneven flux reductions within the excitation field that cause unbalancing of the rotor. When the rotor is unbalanced, damaging vibrations and overheating are created, which will damage the synchronous generator.

Several industry-accepted schemes are used to detect partial and total loss of excitation. One uses the measurement of the generator excitation field current with overcurrent and undercurrent relays.

Measurement of reactive current through the synchronous generator can also been used to detect loss-of-excitation conditions. These two schemes are used on small generators.

The most popular and reliable protection scheme for loss-of-excitation detection uses an offset mho relay, shown in Figure 12.95. The offset mho relay is connected at the generator terminals and measures terminal voltage between two phases along with the difference in current between the two phases. From the measured voltage and current the relay can calculate the impedance as viewed from the generator terminals, and operates when the impedance falls below specific operating characteristics.

The loss-of-excitation protection is normally connected to trip the main generator breaker(s), the field breaker and the transfer unit auxiliaries. The field breaker is tripped to minimise rotor field damage when a loss of field is caused by a rotor field short-circuit or a slip-ring flashover.

FIGURE 12.95 Offset mho relay

12.10.22 Over-voltage protection

Generators that are subject to a wide range of speeds when driven by a prime mover can produce high voltages. The output voltage can be three-to-four times greater than normal voltage while under the influence of full-strength magnetic fields.

Vehicles, aircraft and similar engine-driven units have generators driven either by belts or directly by solid drives. They are an accessory and are quite separate from the primary drive intention. As the engines are necessarily subject to a wide range of speeds, the generators will also have a wide speed range.

Since the generator is expected to produce a useful output at comparatively low speeds, it will need some form of voltage control when driven at high speeds. Many systems for regulating the output voltage have been used, but all rely on controlling the field current of the generator to control output voltage.

Probably the most common form of voltage control is a quick-acting relay that is sensitive to voltages above a certain level. Older-style voltage regulators often had a voltage-controlled relay that inserted resistance in the field circuit when the relay was activated.

In more modern units, voltage-sensitive semiconductor components were introduced. These conduct at specific voltages and reduce the current flowing through the field. Depending on circuit configuration, the unit can insert resistance into the circuit or divert the current around the field when voltage levels exceed a set value. The method is accurate and can also be combined with current-controlled sections.

Many mobile units use alternators with rectifier units to produce direct current. A similar control method is used for over-voltage protection as the alternator field is excited with direct current.

CHECK YOUR UNDERSTANDING

12.34 How are motors protected against short-duration overload and sustained overload?

12.35 How are motors protected against under- and over-voltage?

12.36 What does 'single phasing' mean, and how is a motor protected against it?

12.37 What are the environmental protection requirements for motors?

12.11 Safe testing methods for locating faults in low voltage (LV) a.c. machines

12.11.1 Electrical tests for three-phase motor windings

The tests applicable to an electric motor are either electrical or visual—usually both. Electrical tests are possibly the easiest to carry out (assuming that the equipment is available) because they do not usually require the motor to be dismantled; the testing can be done at the terminal box. Subject to the results of the electrical test, it may then be necessary to dismantle the motor.

Windings in either the stator or rotor (for wound-rotor motors) can have a combination of the following faults:

- open circuit or partial open circuit such as a high-resistance connection
- short-circuit of the complete winding, or partial short-circuit
- windings shorted to earth via the motor frame or some other metallic part of the motor
- motor windings' insulation partially melted, or the winding completely burnt out
- reversed connection of a winding.

12.11.2 Continuity tests

Extra-low voltage sources such as multimeters are suitable for continuity testing. The usual multimeter is a series ohmmeter type, and care must be exercised in taking any resistance readings as absolute readings, particularly at extremely low values. If a winding is supposed to read 1 Ω, then a meter reading of, say, 0.7 Ω is inconclusive: the accuracy of the meter itself might be in question. All this shows is that there is some resistance and some form of electrical continuity between the two leads being checked.

Without some knowledge of the motor circuit, it is impossible to establish whether the correct part of the circuit is being tested. For example, if a three-phase motor connected in delta configuration is being tested, an ohmmeter would give a reading across any pair of terminals, even if one phase winding was open circuit. In Figure 12.96, phase winding A is open circuit, yet a reading can be obtained between L_2 and L_3 (phase B), and between L_1 and L_3 (phase C) and between L_1 and L_2 (phases B and C in series). If the motor is large, the winding resistance is low, and it might be difficult for ohmmeters to detect the difference in readings. In such cases, it might be necessary to disconnect the delta bridges and check the phases individually.

A similar situation occurs when testing single-phase motors. It is vital to know whether the two windings are in parallel, whether there is a capacitor in the circuit, whether the motor has a start switch and whether it is operational.

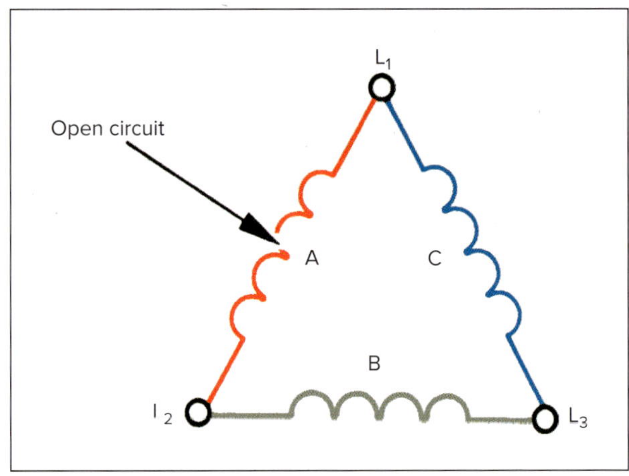

FIGURE 12.96 Delta-connected motor with one phase open-circuited

12.11.3 Insulation-to-earth test

An insulation-to-earth test must be made with instruments of the correct voltage. A 400 V circuit, for example, cannot be tested satisfactorily with a 3 V ohmmeter. Similarly, a small generator local supply with an exciter rated at 24 V should not be tested with an insulation tester of 500 V. Some prior knowledge of the circuit of the motor is necessary. Each phase or winding should be tested separately and the results compared. One phase with a reading considerably different from that of the other two could indicate a problem.

12.11.4 Insulation test between windings

For an insulation test between windings, the windings should be disconnected both from each other and from the supply source to make the test meaningful. A suitable test voltage should be used, and the relative readings compared to see if there is variation. A low reading between two phases is an indication of a problem.

12.11.5 Visual inspection

If a further check is required after electrical testing, the motor is usually dismantled for a visual inspection. With exceptionally large machines, it is possible to carry out a limited visual inspection by removing the covers and looking inside without dismantling the complete machine. In most cases, burnt-out motors have a characteristic smell that is quite easy to detect.

Probably the most obvious sign is the smell of heat within the windings. Insulation can be charred, and windings can consist of bare copper wire with all covering burnt off. The burning smell is not an infallible indication, however, as the fault can trip the supply before the burning becomes appreciable.

Non-electrical tests indicating burnt insulation include pressing the winding with the hands and listening for a crackling noise (a winding in good condition should make no noise), rubbing the bindings with the fingers to see if they crumble and, on larger motors, tapping the windings lightly with a small hammer (a faulty coil group sometimes gives a flatter sound than the rest of the windings).

Where windings have short-circuited to earth, small holes in the windings and associated copper globules can occasionally be seen. With lighter-coloured enamels and varnishes, faulty turns and coils are clearly visible as being much darker than other windings.

12.11.6 Capacitor tests

If a single-phase motor includes a capacitor for starting and/or running, the motor may not start rotating when power is applied or may run rough if the capacitors included in the motor circuit are faulty. Therefore, it may be necessary to test the capacitor(s) included in the motor circuit to ensure they are healthy and in working order. A capacitor meter or the capacitor meter function on some multimeters can indicate the capacitance of a capacitor. However, the following tests are a good indicator of good or bad capacitors.

Visual test

The first test to determine the health of a capacitor is a visual one. Capacitors that have failed usually display a bulging case and the capacitor may or may not be leaking from within. This is an obvious indicator that a capacitor should be replaced. Always ensure you replace faulty capacitors with similar capacitors in terms of, for example, voltage rating and capacitance values.

Charging test

A capacitor's ability to take a charge can be indicated by use of an ohmmeter. This type of test will only indicate whether the capacitor is functional or non-functional; it is not used to grade a functional capacitor. When a completely discharged, functional capacitor is connected to an ohmmeter, the capacitor will begin to charge towards the supply voltage of the ohmmeter. As current is flowing at a high rate, this appears to the ohmmeter to be like a closed circuit and thus it will indicate a low or near-zero ohmic value. As the capacitor continues charging and the voltage across it approaches the voltage of the ohmmeter, the current flowing will slowly decrease. This slow decrease in current flow is indicated on the ohmmeter by the reading slowly approaching an open-circuit value, i.e. high resistance or infinity. Once the capacitor is fully charged, the voltage across the capacitor will equal the ohmmeter's supply voltage. As there is no potential difference between the capacitor's voltage and the ohmmeter's supply voltage, no current will flow and the circuit will now appear as an open circuit. The ohmmeter will display this as a high-resistance reading or infinity. When performing this test, the following points must be considered:

1. Ensure the supply voltage has been removed from the capacitor before testing. Use lockout-tagout procedures to ensure no reconnection to the supply can be made during testing.
2. Discharge the capacitor before testing. For small capacitors, a simple screwdriver may be used to short the two leads of the capacitor together for discharging. However, be mindful that large-value capacitors connected to equipment operated at mains low voltage (230/400 V) can contain a substantial charge and thus enough energy to be lethal. AS/NZS 3000:2018, Clause 4.15.3 gives provision for discharging and control of capacitor equipment.
3. Select an adequate ohmmeter and use a high range between 10 kΩ and 1 MΩ. Although a digital ohmmeter can be used for this test, an analogue ohmmeter will give a better visual indication.
4. Before connecting the ohmmeter test leads, ensure the polarity is correct for polarised capacitors.

Voltage test

A fully charged capacitor should hold a voltage, and this voltage can be tested with a voltmeter. Once the supply has been removed from the capacitor, the capacitor being tested should hold a d.c. voltage across it. The voltage across the capacitor should slowly drop off as it naturally discharges. A voltmeter indicating no voltage across a capacitor or a capacitor that cannot hold a charge are other indications of malfunction.

12.11.7 Mechanical tests

As there are several mechanical and non-electrical components in rotational machines, there need to be techniques to determine mechanical faults. The following list indicates areas of rotating machines that are mechanical in nature and need to be regularly inspected (and repaired if faulty):

1. The bearings mounted on the machine shaft.
2. Mechanical load on motors, such as gear boxes, pulleys, fans and impellers etc.
3. The presence of a bent shaft.
4. Misaligned load or coupling between the motor and load, or between alternator and prime mover.
5. A locked rotor.
6. Blocked air vents preventing adequate ventilation.
7. Seized or broken centrifugal switches.

12.11.8 Specialised test equipment

There are times when the checks described above do not give a definite answer about the condition of the windings. If a motor is faulty in operation but passes the tests, further testing is necessary. One popular method involves using the windings as the secondary of a transformer with a piece of equipment called a 'growler'—a name derived from the noise made when it is in operation.

Another piece of test equipment, the Prufrex, is sometimes used with motor stators. It is plugged into an a.c. supply and moved around inside the stator. If a short-circuit exists in the stator windings, the circulating currents upset the magnetic field of the tester and a light flashes, indicating the location of the fault.

12.11.9 Dismantling three-phase induction motors

When dismantling an electric motor, the main aim is to reassemble it in its original form after any inspection or repairs. To make that easier than relying on memory or experience, it is a common practice to mark the end-shields and other pieces in some way.

Probably the most common method of doing this is to use a centre punch and make adjacent punch marks on matching surfaces. (Punch marks should be avoided on machined surfaces.) Some technicians prefer to use a cold chisel to make one mark across two surfaces so they can reassemble the motor more accurately. In either case, only a light mark is needed; all that matters is that it can be used on reassembly. Large, deep marks are not recommended because of the possibility of damaging the casing. Making what are sometimes called 'witness marks' is good practice and is widely used by experienced technicians. For example, it is common practice to use a centre punch to place two marks on the drive end and three marks on the non-drive end of the motor.

Whenever possible, an experienced operator will keep components of subassemblies separate from other subassemblies as they are removed. Taking photographs before and during disassembly can also provide an invaluable reference for reassembly. Putting all components in one container and then having to try to sort them out when they are needed for reassembly is bad practice.

Withdrawing the rotor from the stator requires care, to avoid damaging either the rotor or the stator and its windings. A mechanical and electrical examination of the motor can then take place.

Matching punch marks when rebuilding is a quick and accurate method of ensuring that the motor is assembled in its prescribed condition and manner. It also ensures that the motor shaft is protruding from the right end and the terminal block and housing is in the correct position.

Bearings should be checked for wear and re-greased on assembly with just the right amount of correct-grade grease.

CHECK YOUR UNDERSTANDING

12.38 What mandatory tests are required before a motor can be placed in service?

12.39 How are capacitors tested for single-phase motors?

12.40 What mechanical tests need to be considered for motors?

12.41 What needs to be considered when dismantling a three-phase induction motor?

SUMMARY

- a.c. machines are divided into two groups: a.c. motors and a.c. generators.
- a.c. machines operate at either asynchronous or synchronous speed.
- a.c. machines are also designed for either single-phase supply or three-phase supply.
- An a.c. motor converts electrical energy into mechanical or rotary energy.
- a.c. machines require an excitation field and armature to produce motor action in a machine and armature action in a generator.
- Motor action is the interaction between a conductor and a magnetic field that produces torque.
- Torque is used in motors to create rotation and thus produce mechanical power to turn a load.
- Armature action is the interaction between a moving conductor and a magnetic field to generate a voltage and thus produce electrical power to power a load.
- A typical a.c. machine has a stationary armature called the stator. The stator is the component that either consumes electrical power, as in a motor, or supplies electrical power, as in a generator.
- The various types of single-phase motors derive their name from the motor's starting circuitry. They are:
  - split phase
  - capacitor-start
  - capacitor-start/capacitor-run
  - permanent split capacitor (PSC)
  - shaded pole.
- A synchronous machine requires d.c. excitation within the rotor.
- Motor protection includes devices to protect the motor against adverse conditions such as:
  - short-duration overload
  - sustained and locked rotor overloads
  - short-circuits
  - overtemperature
  - under-voltage
  - over-voltage
  - phase reversal
  - single-phasing.
- Motor protection uses devices such as:
  - circuit-breakers
  - HRC fuses
  - thermistors.

END-OF-CHAPTER QUESTIONS

12.1 Describe how the rotating magnetic field is produced in a three-phase motor.

12.2 Explain why an induction motor always runs at less than synchronous speed.

12.3 What is meant by the:

(a) synchronous speed of an induction motor?
(b) actual speed of an induction motor?
(c) slip speed of an induction motor?

12.4 What is the relationship between synchronous speed, actual speed and slip speed in an induction motor?

12.5 How can the direction of rotation be reversed in an a.c. series universal motor?

12.6 Describe a method for finding which phase has an earth fault in a three-phase delta-connected motor.

12.7 Explain why the power factor of an induction motor increases towards unity as the load increases.

12.8 Briefly describe the construction of:
(a) squirrel cage rotors
(b) wound rotors.

12.9 What is meant by the term 'locked rotor torque'?

12.10 What is meant by the term 'breakdown torque'?

12.11 What is meant by the term 'split phase'?

12.12 Briefly describe the split-phase method of starting single-phase induction motors.

12.13 Why is a centrifugal switch used in most single-phase induction motors?

12.14 Briefly describe the principle of operation of the capacitor-start, induction-run motor.

12.15 How is the direction of rotation reversed in a split-phase motor?

12.16 Explain the principle of operation of the shaded pole motor.

12.17 Draw a simple circuit diagram for a permanent split capacitor motor. Give the typical operating characteristics and list an application for this type of motor.

12.18 List three types of single-phase induction motor and briefly describe the characteristics of each one.

12.19 Give one reason why an overload current could occur in a motor circuit.

12.20 Give one reason why a fault current could occur in a motor circuit.

12.21 Explain why the temperature of a motor increases when it is loaded.

12.22 What is the function of an HRC fuse in a motor circuit?

12.23 Why must a magnetically operated overload have a delaying device fitted when starting an electric motor?

12.24 Explain the difference between a 20 A type 'C' circuit-breaker and a 20 A type 'D' circuit-breaker.

12.25 What is meant by a thermal overload device? Where is it connected into a circuit?

12.26 Why are PTC resistors in a motor control circuit considered to be automatically resetting?

12.27 Explain how electronic overloads sense the value of current.

12.28 State two features that are available on an electronic overload unit.

12.29 What are the constructional differences between low-speed and high-speed alternators?

12.30 Explain why a low-speed synchronous machine has a large salient pole-type rotor.

12.31 What are the types of losses that reduce the efficiency of an electric motor? List all the types of losses and give a means to reduce each.

12.32 Determine the percentage slip for the following three-phase motors:
(a) 50 Hz four-pole motor running at 1420 rpm
(b) 50 Hz six-pole motor running at 960 rpm
(c) 50 Hz eight-pole motor running at 720 rpm.

12.33 Determine the slip speed of the following three-phase motors:
(a) 50 Hz four-pole motor at 4% slip
(b) 50 Hz two-pole motor at 6% slip
(c) 50 Hz eight-pole motor at 5% slip.

12.34 The rotor frequency of an unloaded two-pole motor is 0.5 Hz when running on a 50 Hz supply. Calculate the:

(a) synchronous speed
(b) actual speed.

12.35 A six-pole motor runs at 875 rpm on a 50 Hz supply obtained from a diesel-powered alternator.

(a) Calculate the synchronous speed of the motor
(b) Calculate the slip speed, rotor slip frequency and percentage slip
(c) What would be the synchronous speed of the motor if the frequency was allowed to drift to 55 Hz?

12.36 A two-pole motor is operating at 50 Hz. Determine the synchronous speed of the rotating magnetic field.

12.37 The synchronous speed of a four-pole capacitor-start, induction-run motor is 1500 rpm. Determine the frequency of supply.

12.38 The rotor speed of a 50 Hz capacitor-start/capacitor-run motor is 2800 rpm. How many poles would this motor have?

12.39 Complete the diagram of the capacitor-start motor in Figure 12.97.

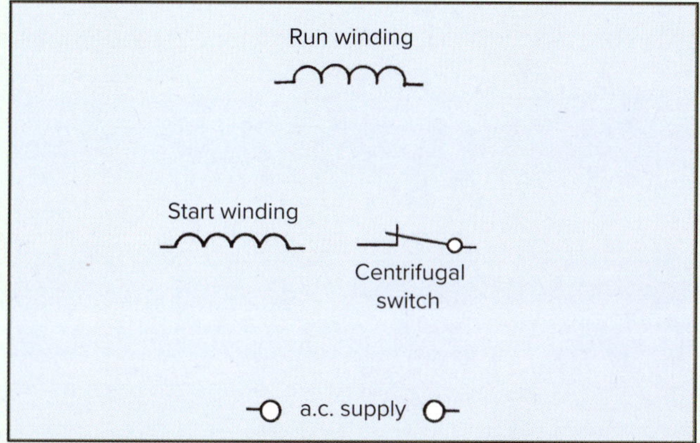

FIGURE 12.97 Wiring diagram for a split-phase motor

12.40 A motor has a winding resistance of 15 Ω at 22°C. After running, the resistance is measured as 18 Ω. What is the temperature of the windings after running?

12.41 How many poles must a synchronous machine have to operate at 250 rpm and a frequency of 50 Hz?

12.42 What supply frequency would be required to run a four-pole synchronous motor at 3300 rpm?

12.43 A diesel-driven alternator is governed to 600 rpm. If the alternator has ten poles, what is the frequency of the alternator output?

12.44 Find the power rating of a three-phase 3300 V 100 kVA alternator at power factors of 1.0, 0.8 and 0.65. What is the maximum current of the alternator in each instance?

12.45 A three-phase 415 V 125 kVA alternator supplies its rated load at a power factor of 0.8 lagging. Given that the efficiency of the alternator is 89%, what power is the prime mover required to deliver?

CHAPTER 13

Develop and connect electrical control circuits

LEARNING OBJECTIVES

- Understand basic relay circuits
- Identify relay circuits and drawing conventions shown in AS/NZS 1102
- Identify remote stop/start control and electrical interlocks
- Examine time-delay relays
- Design circuits using contactors
- Design circuits that include jogging and electrical interlocks
- Examine the operation of various control devices
- Examine the operation and applications of programmable relays
- Examine three-phase induction motor starters
- Explain the need for reduced-voltage starting on three-phase induction motors
- Examine three-phase induction motor reversal and braking
- Examine three-phase induction motor speed control

PREREQUISITE KNOWLEDGE

- Work health and safety in the workplace
- Control and protection for electrical installations
- Direct current (d.c.) machines
- Magnetic and electromagnetic devices
- Single-path and multiple-path circuits

INTRODUCTION

There are many contexts in which graphical symbols are used in everyday life, as can be seen in Figure 13.1. For example, they are used on road signs to indicate slippery road conditions, while symbols in signage show first-aid locations. Their general use in signage is a form of workplace communication, an example of that being how stylised male and female symbols show the locations of washrooms and conveniences. What makes a graphical symbol effective is its ability to be interpreted correctly regardless of language, and its capacity to convey a message quickly, clearly and accurately.

Graphical symbols are used in electrical drawing practices, including the development and connection of electrical control circuits.

People first started to notionalise electrical circuits in the 19th century by drawing simple sketches of what they saw. Usually, a symbol key was provided.

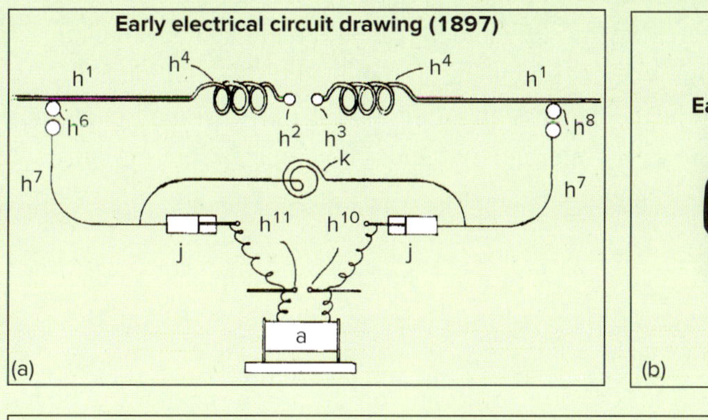

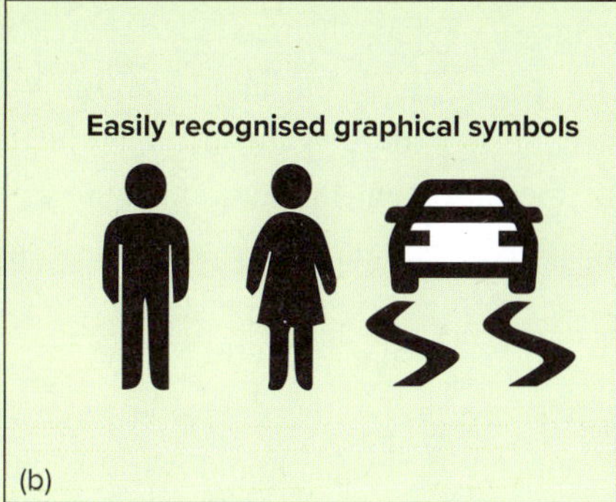

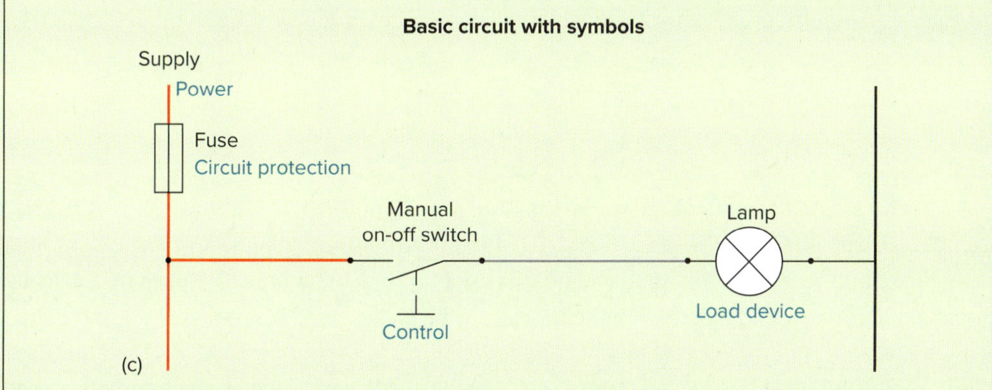

FIGURE 13.1 **Graphical symbolic representations**

Over time, improvements were made. Drawings were simplified and stylised, and components were shown by symbols. Wires and components were now deliberately separated and drawn distinct from one another, even though the drawing no longer looked like the real layout.

13.1 Developing circuits

Developing circuits requires an understanding of how a circuit operates safely and correctly as well as how to design, develop and draw *circuit diagrams* (known as 'schematics') and *wiring diagrams*.

This chapter covers symbols, types of circuit representations, how to convert from one diagrammatic format to another and drawing conventions. It goes on to cover circuit components such as *relays, contactors* and *timers,* and examines the operation of common electrical control circuits before discussing basic circuit fault finding and testing techniques. The chapter also discusses common *control devices, programmable relays, programmable logic controllers (PLCs), induction motor starters, motor reversal, braking methods and speed control.*

13.1.1 Circuit diagrams

Circuit diagrams are stylised graphical representations of components and the electrical connections between them. Their purpose is to illustrate the power flow (or signal flow) through the circuit.

Control circuits are electrical circuits made up of five parts that are identified via symbols to show the power source (also known as the 'supply'), circuit protection, on-off and timed control devices, load devices performing work and the wires connecting these components. Figure 13.1(c) shows a basic representation of the concept of a control circuit with a fuse, a switch and a lamp.

As the graphical representation of circuits evolved, local and national styles of drawing and standard symbols were adopted by groups of people who needed to share drawings. Eventually, an international body, the International Electrotechnical Committee (IEC), formulated an international standard of graphical representation.

Although some countries and organisations kept their own symbols, most of those who adopted the SI (or metric) standard of measurement also adopted the IEC (International Electrotechnical Commission) symbols. IEC 60617:2012 (Part 2 of 13) has become the international standard on the subject of graphical symbols for diagrams.

Even today, however, some countries that manufacture and supply machinery and equipment produce electrical wiring and schematic diagrams with symbols that are different from those mandated by the IEC. (One reason for this is the development of new technologies for which the IEC has yet to create symbols.) This is relevant as Australia utilises equipment and machinery from countries and regions around the world, including China, Japan, the UK, the US and Europe (sometimes certain European countries will even use circuit symbols that differ from those used by other European countries). Usually, a key or diagram legend will be provided to label and identify symbols.

Reading and interpreting electrical schematic and wiring diagrams is a necessary industry skill for electrotechnology installation, maintenance and repair workers.

AS/NZS 1102 is very close to the IEC standard, with most Australian organisations following its symbol conventions. In December 2016, AS/NZS 1102, sections 101–113 were withdrawn and have yet to be replaced. If the responsible Technical Committee considers that a standard is not up to date technically, does not reflect current practice and/or is not suitable for new and existing applications (products, systems or processes), it may request that the standard is given withdrawn status.

It is still possible for a withdrawn standard to be used within an industry or referenced by a government if they choose to do so. One reason for this may be because there are no replacement technical publications readily available.

13.1.2 Electrical symbols

Symbols are used to identify circuit components in electrical circuit drawings and architectural floor plans. Some of the more common AS/NZS 1102 symbols are shown in Figure 13.2. Although there is no fixed size for symbols, all those in the same drawing should be the same relative size and proportion.

In a drawing that shows switch contacts being operated, either manually or from their associated coils, the line representing the operating part of the switch rotates clockwise.

13.1.3 Symbols as physical components

Electrical circuits are controlled by devices that fall into two main categories: the devices that physically either open or close a circuit and the devices that cause them to operate. *Make-and-break contacts* are in the first category. Referred to as 'contacts', they perform a switching function. They can be thought of as electrical contacts that switch or change state, to open or close a circuit path.

The second category refers to the device components that cause electrical contacts to change state. These include electrically operated contactor, relay and timer coils and pushbutton mechanisms. They cause electrical contacts to change from being normally:

- open (NO, see section 13.4.2) to being closed; or
- closed (NC) to being open.

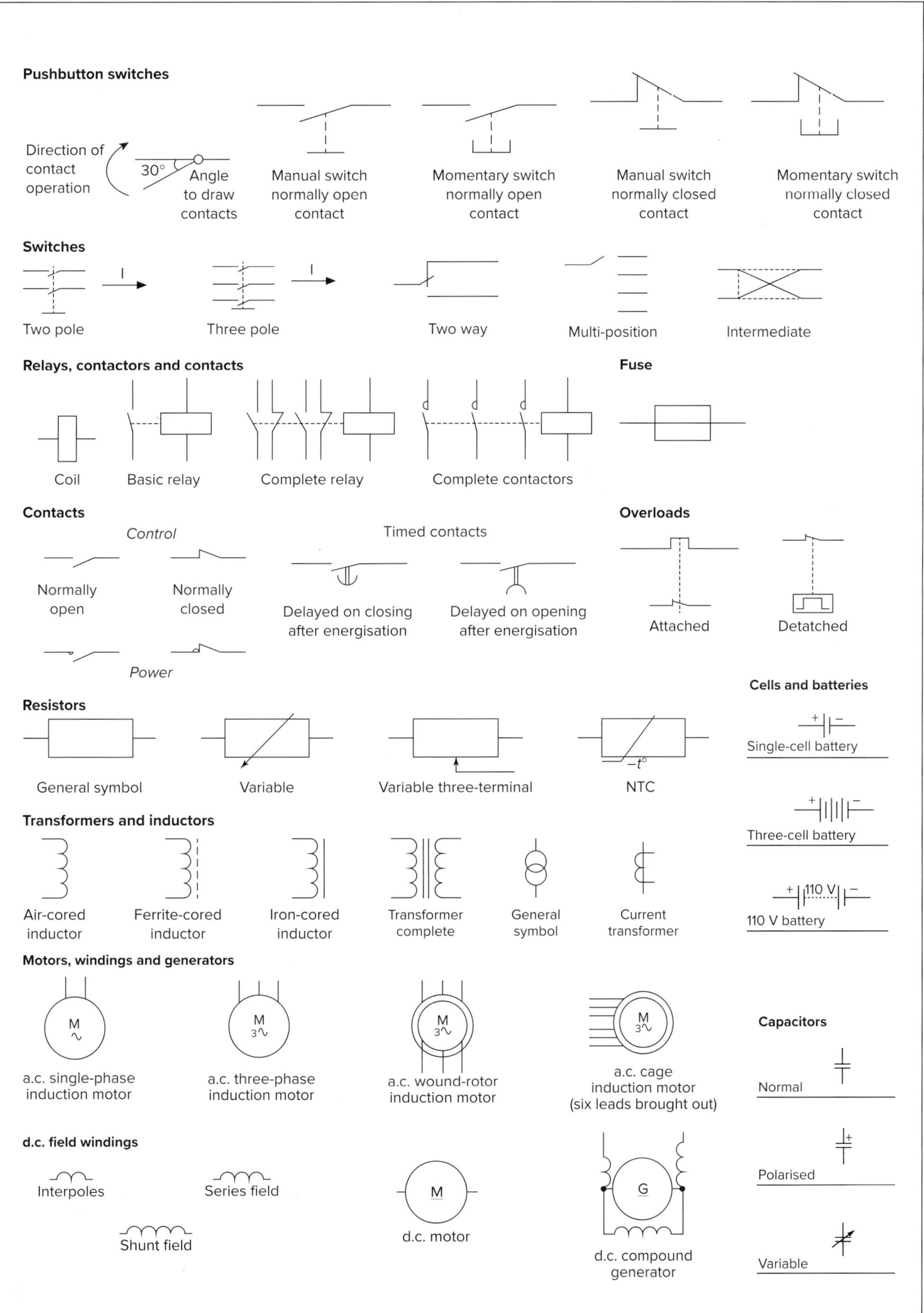

FIGURE 13.2 **Commonly used electrical circuit symbols**

Pushbutton control

Pushbutton switches have two parts to their construction, the electrical contacts and the mechanical operator. The contacts may be normally open, normally closed or a combination of both. Figure 13.3 shows examples of three pushbutton devices that are manually operated and the electrical circuit symbols for normally open and normally closed switches.

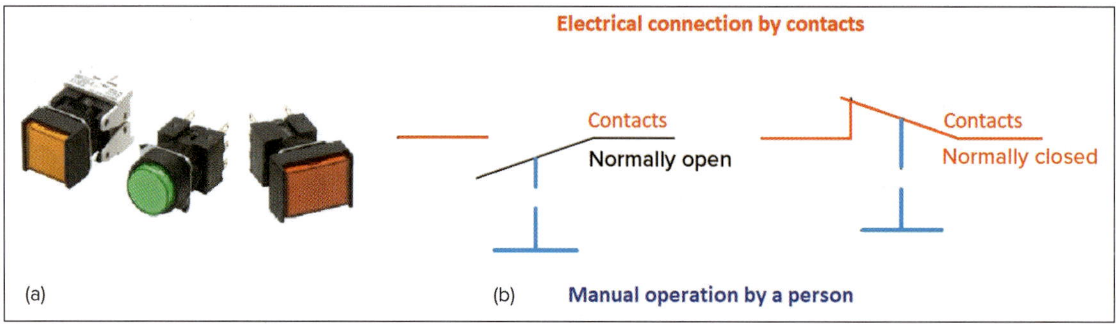

FIGURE 13.3 (a) Examples of manual pushbuttons and (b) the electrical circuit symbols for normally open and normally closed switches

Pushbuttons are used extensively in the electrical industry in applications such as start pushbuttons and simple light switch control. Depending on the task they are required to perform, pushbutton switches may be selected to have multiple inputs and outputs—double or triple pole, for example, they may also allow for multiple throws.

Relay control

Contacts can be manually operated by a person (as seen in Figure 13.3) or electrically using coil-operated devices known as 'relays' and 'contactors' (both are shown in Figure 13.4). When the coil is energised, the associated contacts simultaneously change state via an internal mechanical linkage, as illustrated with symbols in Figure 13.4(c).

Relays and contactors have contacts (often make-and-break contacts) as part of their design. The 'coil' is an electromagnetic device that physically causes the relay such as that shown in Figure 13.4(a) and the contactor such as that in Figure 13.4(b) to operate. This occurs when current is allowed to flow through the coil. The resulting effect causes the normally closed contacts to open and the normally open contacts to close. Although the contactor or relay and the contacts and coil are usually in the same component housing, they are normally shown separately in schematic diagrams.

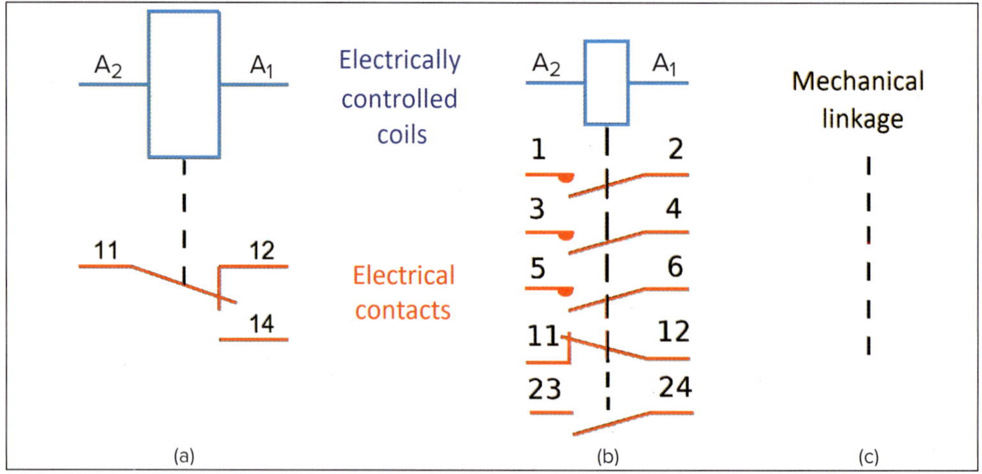

FIGURE 13.4 Coil-operated contacts

CHECK YOUR UNDERSTANDING

13.1 Draw the symbols for a momentary pushbutton and a manual pushbutton.

13.2 Draw two contacts, one normally open (NO) the other normally closed (NC).

13.2 Circuit representations

Circuits are drawn for the benefit of electrotechnology industry workers, for the commissioning and testing of new equipment and general electrical installations, for maintenance, repairs and fault finding and for building plant and equipment modifications.

There are different types of drawings, each of which has specific benefits and applications. The most common circuit representations include:

- circuit diagrams (also called schematics)
- ladder diagrams
- block diagrams
- single-line diagrams
- wiring diagrams.

13.2.1 Circuit diagrams

A circuit diagram is a visual representation of a circuit that communicates information quickly, clearly and effectively. Circuit diagrams (or schematics) have been refined over the years, and many of the component symbols have been changed to make them easier to recognise and draw.

A circuit diagram shows how an electrical circuit *operates*. It is important to remember that a circuit diagram might have no resemblance to the actual physical components of the circuit—it represents the components and the connections between them.

Figure 13.5 is a circuit diagram of an electronic circuit for a radio receiver. It would be suitable for fault finding and repair by an electronics technician.

Circuit diagrams are the most useful type of diagram for understanding the operation of an electric circuit, but they can be represented in other ways such as block, single-line and wiring diagrams.

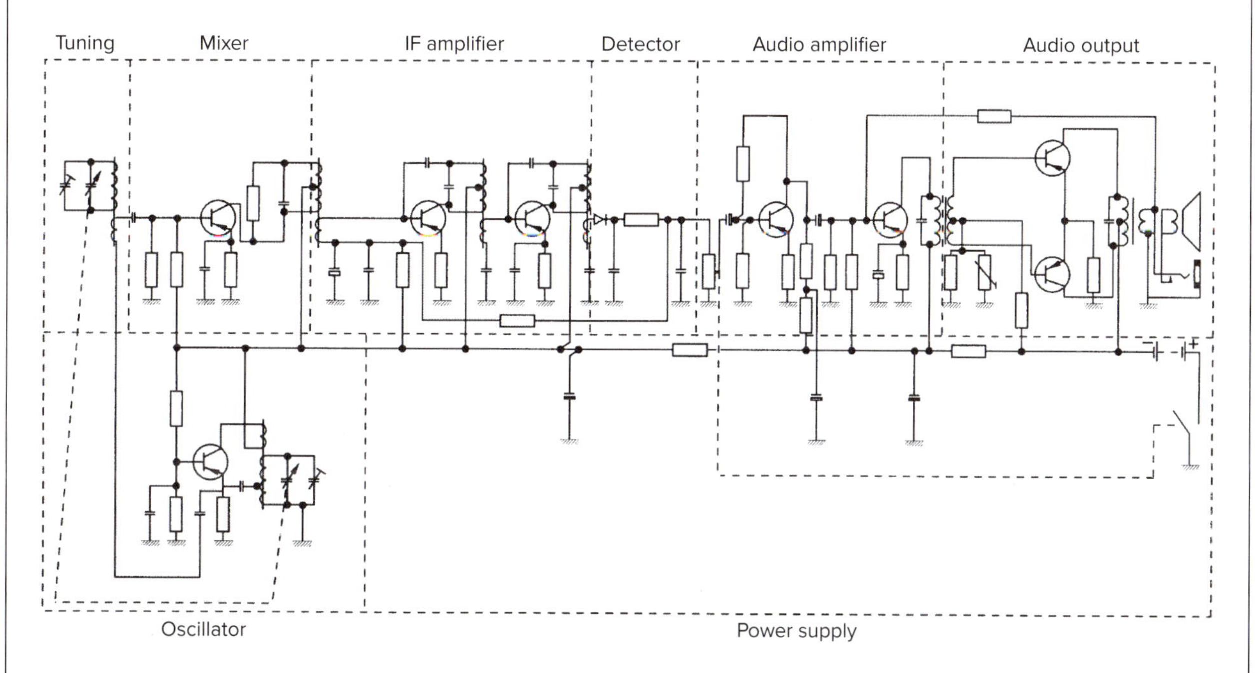

FIGURE 13.5 Circuit diagram of a small AM radio receiver

13.2.2 Ladder diagrams

Control circuits in particular can be drawn as 'ladder' diagrams, which are so called because the supply lines are drawn on each side like the stiles of a ladder and the circuit component parts are drawn across them like rungs.

These diagrams can be seen as a step towards programming logic controllers and are mostly used with programming in mind.

13.2.3 Block diagrams

A block diagram shows what a circuit *does*, not how it works. It is an overall picture of why a particular circuit is used and the function of groups of components within it. Block diagrams are useful when a circuit is relatively complex. In many cases, an electrical worker might first look at a block diagram of a circuit before reading the circuit diagram.

Block diagrams usually do not show any power supply. There is no need to include these details as block diagrams do not represent actual circuit connections—they merely explain what the circuit does.

Figure 13.6 is a block diagram of a small AM transistor radio receiver. Although the actual circuit might appear complex, it is made from seven separate sections, which the block diagram shows. Without knowing exactly how the radio circuit operates, the function of the circuit can easily be explained from the block diagram. It is instructive to compare this with the actual circuit diagram shown in Figure 13.5. This electronic circuit is an illustration of how a complicated circuit can be represented by a simple block diagram.

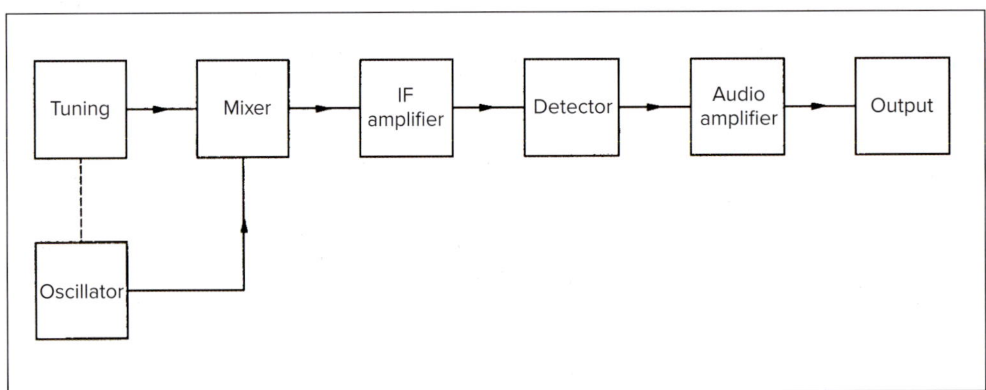

FIGURE 13.6 Block diagram of an AM radio receiver

Figure 13.7 is a simpler block diagram. It represents a regulated supply connected to an electrical load. The input from the mains passes to a block on the left called the 'power supply'. This could be an a.c.-to-d.c. rectifier unit. Its output then passes through another block called a 'filter' to a block labelled 'regulator'. This regulates the voltage to a predetermined value. The regulated voltage is then passed to the load.

Figure 13.8 is a block diagram of a motor starting circuit. The arrows on the lines indicate the sequence and purpose of each circuit block.

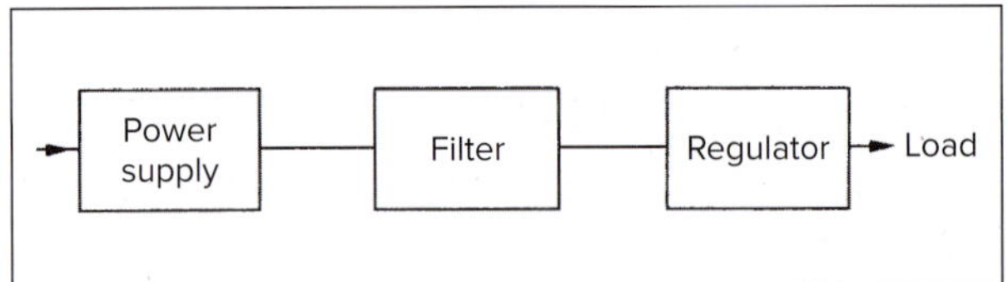

FIGURE 13.7 Block diagram of a regulated power supply

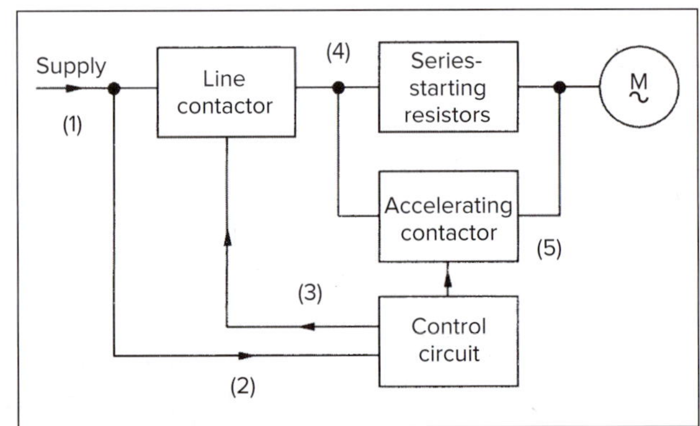

FIGURE 13.8 Block diagram of a motor starting circuit
Note: see text for explanation of numbered points.

The a.c. supply from the mains is connected to the line contactor (1). However, it cannot operate until allowed to do so by the control and timing circuit (2). When the control circuit is activated (usually by a pushbutton switch), the line contactor closes (3) and supplies energy to the motor through the current-limiting starting resistors (4). The motor starts up and, after a predetermined time, when it is nearly up to full speed, the accelerating contactor closes and bypasses the starting resistors (5). Full supply is now provided to the motor, enabling it to develop full power. By operating a stop pushbutton switch, the control circuit opens all the contactors and the motor stops.

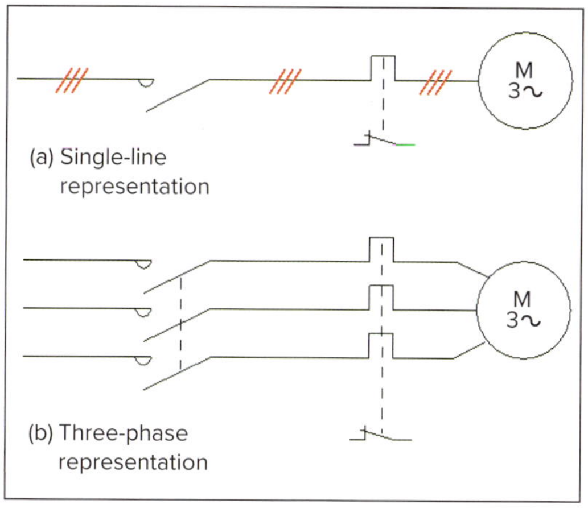

(a) Single-line representation

(b) Three-phase representation

FIGURE 13.9 (a) Single-line diagram; (b) three-phase diagram

13.2.4 Single-line diagrams

In power and transmission lines, a single-line diagram is a simplified notation for representing a three-phase power system. The single-line diagram has its largest application in power-flow studies. Figure 13.9 shows this application with a three-phase motor circuit. The three parallel line marks shown in the single-line representation in (a) indicates the three-phase representation shown in (b).

13.2.5 Wiring diagrams

A wiring diagram is much closer to the real thing than a circuit diagram; it is a stylised true representation of the components and wiring of an electrical circuit. Taking a photograph of the circuit wiring would be of little use if you needed to know how the connections are made (the problem being that the wires could be bunched together). Also, connections may be made in an area below or behind another object. This would make a photograph extremely difficult to follow.

In a wiring diagram, lines representing the conductors are drawn straight and separate. They are usually evenly spaced and are all connected to circles representing the terminals on the circuit components. Where more than one conductor line terminates at a terminal, lines are angled to the terminal circles. There is a clear indication where each line (representing a conductor) starts and finishes.

A representation of the wiring of two lights can be seen in Figure 13.10. Here, the actual circuit components (switches and lampholders in this case) are drawn in the same relative position as they would be in an actual circuit. The terminals of the lamp holders and switches have been emphasised to show exactly where each conductor connects. For clarity, the lamps have been separated from the lamp holders.

A wiring diagram is seldom drawn to scale, although in some cases it may be in direct proportion to the shape of the object represented. In Figure 13.10, both the lamps and switches could be some distance apart. If drawn to scale, or even in proportion, the component parts would be too small. All lines representing wires have been shortened so that their actual connections to the components are clear.

The wiring diagram in Figure 13.11 relates to the motor starter block diagram in Figure 13.8. It is a true representation of the motor starter but is not drawn to

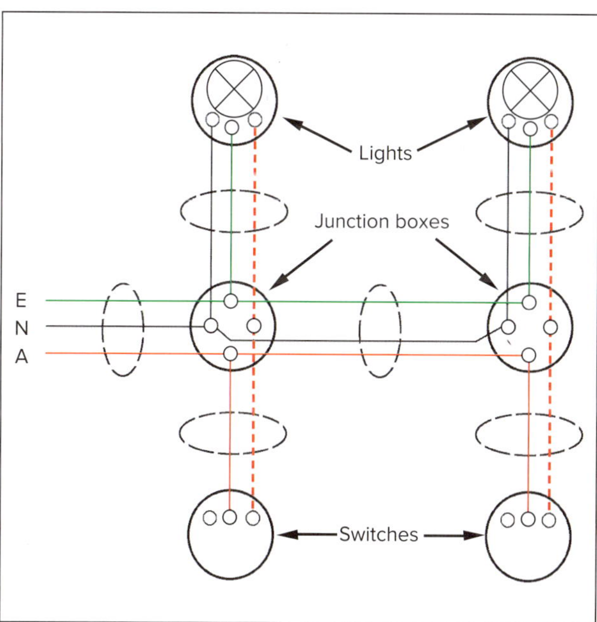

FIGURE 13.10 Wiring diagram of a simple lighting circuit

any particular scale. The components are in proportion both in outline and position. Note that the outlines of all components are drawn as dashed lines. Circuit diagram symbols of the pushbutton switches, contactors and motor have been inserted to show actual operation of the circuit components. Greatest prominence is given to the lines representing the wiring conductors and the circles representing the terminals.

The main point to note about wiring diagrams is that the conductors are all separated and, as far as possible, equally spaced. Note also the manner in which many conductors have been angled as they connect to the terminals. In a wiring diagram, connections are only made at terminals. In a circuit diagram, connections between conductors are made away from the components and may be shown at any point on a given conductor. Circuit diagrams do not represent the actual physical circuit layout; wiring diagrams do.

Drawing a wiring diagram is, in effect, wiring up the circuit or apparatus on paper. Each conductor line drawn between terminals represents an actual insulated cable, cut to length, bared of insulation at the ends and connected properly to each terminal. Because of this, a wiring diagram is invaluable for actually wiring up a piece of equipment or checking equipment when looking for a fault.

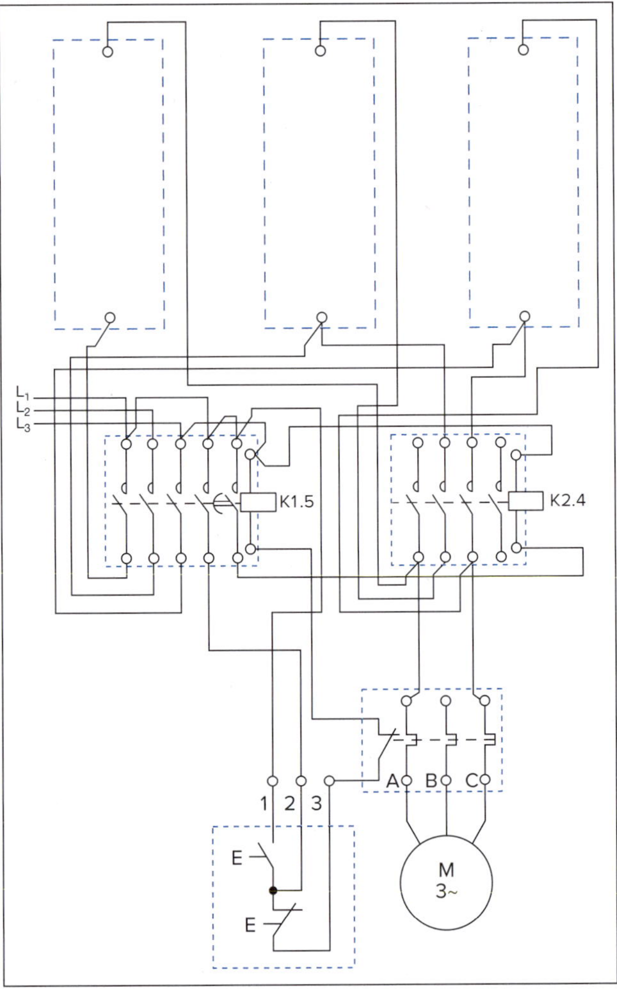

FIGURE 13.11 Wiring diagram of an automatic motor starter

In many cases, the designer of a piece of equipment may first produce a block diagram, then design a circuit diagram and finally produce a wiring diagram. A circuit diagram shows how an electrical circuit *operates* whereas wiring diagrams show how a circuit *is actually constructed.*

Figure 13.11 has been drawn with errors to allow us to consider the benefit of using circuit diagrams for fault finding in conjunction with a wiring diagram. The Figure 13.8 motor starter block diagram also gives an overview of the automatic motor starter operation. The Figure 13.12(a) power circuit schematic representation has been correctly drawn for the wiring diagram. Figure 13.12(b) shows the missing wires.

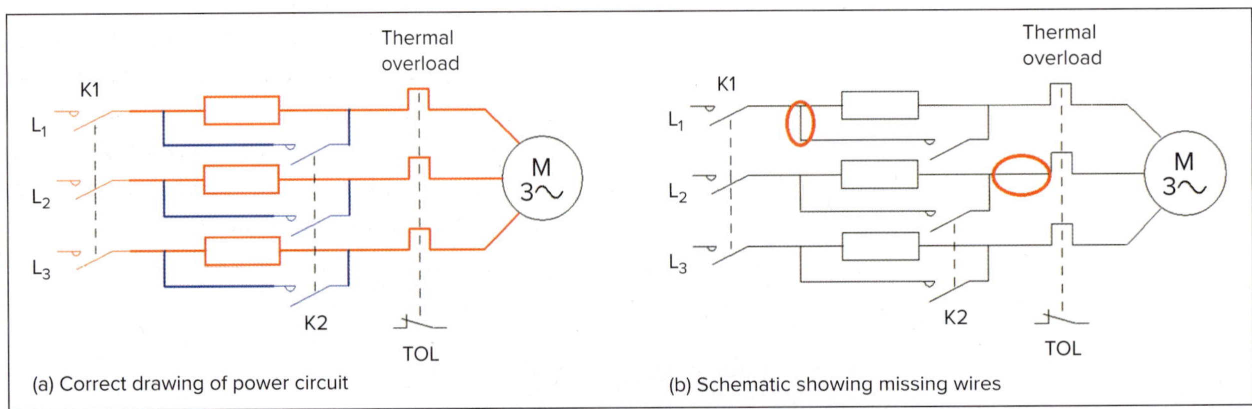

FIGURE 13.12 Schematic of power circuit from wiring diagram

A quick comparison between the wiring diagram in Figure 13.11 and the power circuit schematics shown in Figure 13.12 will reveal the problem: the wiring diagram has two missing connections.

The problem identified in the wiring diagram in Figure 13.11 can be avoided using a variety of methods. If redrawn from a schematic, a way of identifying the component connections needs to be followed. Drawing schematic diagrams and converting from one type of electrical drawing to another will often require line or rung numbering, cable numbering and component labels.

13.3 Drawing circuit diagrams

Schematic drawings can be drawn in either vertical or horizontal orientation. Figure 13.13 shows the power and control schematics in both layouts. The control circuit in this example is a basic two-line design.

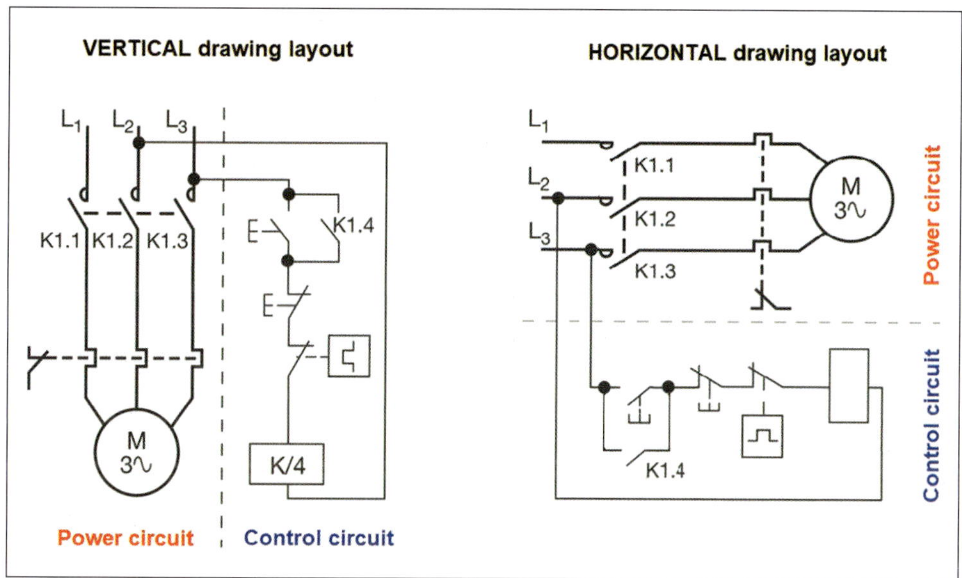

FIGURE 13.13 Vertical and horizontal schematic drawing layouts

To make circuit diagrams easier to read, certain layout conventions are used. One is the direction flow of energy and the other is the sequence, or events of operation.

Figure 13.14 shows a ladder diagram in horizontal layout with power flow (red arrows) from left to right and the control sequence (blue arrows) rung-by-rung, from top to bottom. We can see from this drawing that circuit components have identifying labels, rung or line numbers and cable numbering.

These drawing conventions assist electrical installation, repair and maintenance staff to read and interpret the circuit operation.

Figure 13.15 shows a simple schematic diagram in a vertical layout. The control sequence is from left to right (blue arrows) and the power flow (red arrows) is from the top supply rail to the bottom rail. It is not always possible to adhere exactly to this, particularly with the control sequence in complicated schematic diagrams, but it is followed as closely as possible.

CHECK YOUR UNDERSTANDING

13.3 Refer to Figure 13.14 and then list the components in the correct order of operation to permit power flow in Rung 4.

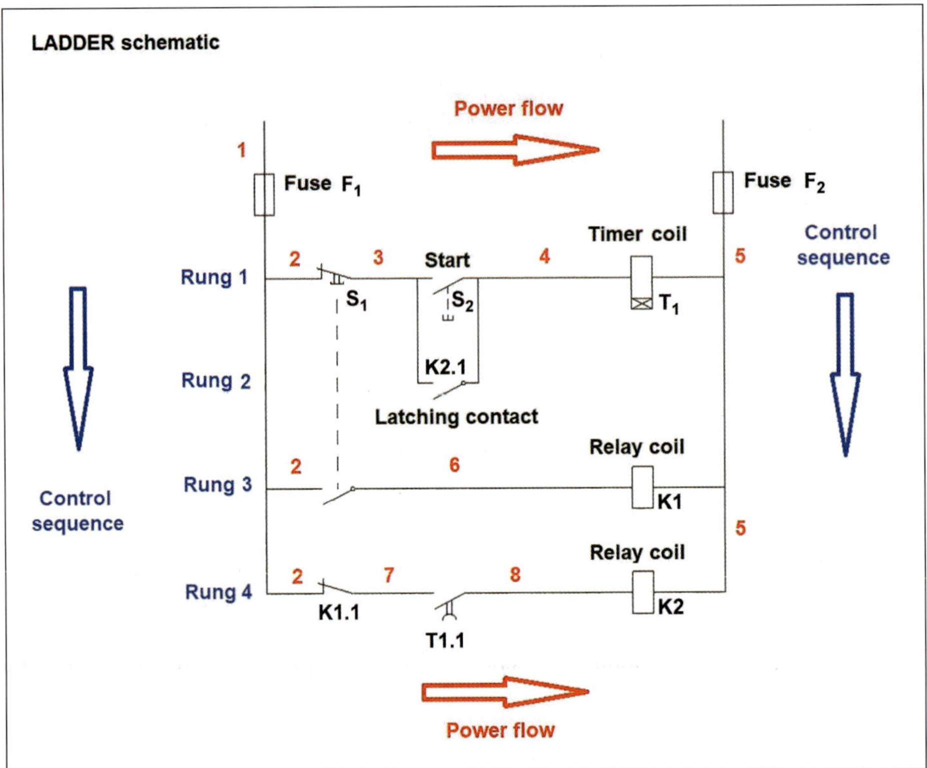

FIGURE 13.14 Ladder schematic drawn in a horizontal layout

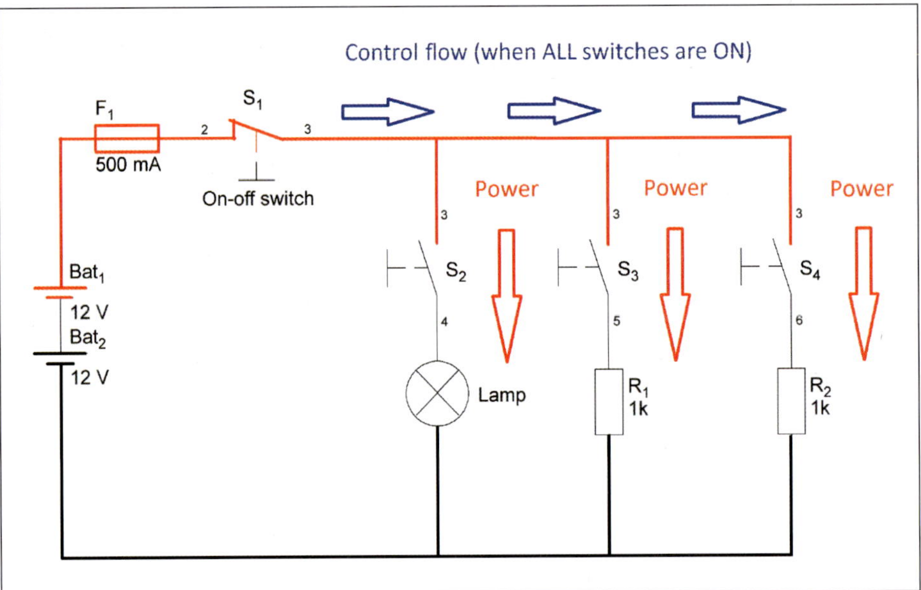

FIGURE 13.15 Vertical layout schematic

Figure 13.15 shows the power supply on the left. Power flows to the right through the main switch S_1 or, more correctly, provides a supply voltage for current flow through the three loads. When switches S_1, S_2, S_3 and S_4 are operated in sequence, load 1 (lamp) operates first, followed by load R_1 and then load R_2.

When each switch is operated, current flows from top to bottom through the respective load as shown by the red arrows. The circuit components are presented in line symmetrically, to make the diagram easier to read and understand. Many schematic drawings use the vertical layout, grouping together lamps, solenoids, relay and contactor coils. This allows electrotechnology staff to locate devices more easily and use the schematic as a tool for fault finding and determining correct operation of equipment.

13.3.1 Conventions in line work

The components in real circuits are connected by conductors, which may be actual wires or copper tracks on a printed circuit board. In most circuits, the conductors are considered to have negligible resistance (none, to all intents and purposes). The line used to represent a conductor is therefore nothing more than a statement that the two components are connected. The length of the conductor is not important to the circuit diagram. This is illustrated in Figure 13.16, where a battery is connected to a load.

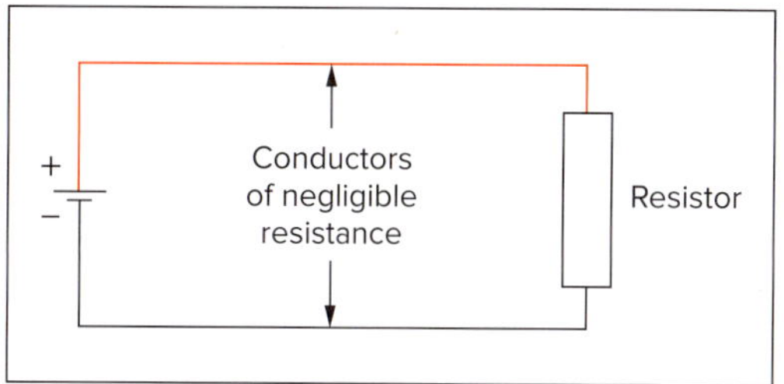

FIGURE 13.16 Conductors between components in a circuit

Straight lines

Conductors are always represented by straight lines with right-angle turns, regardless of the path the actual conductor takes. These lines may join the circuit components or other conductors.

Line types

Ordinary conductors are a simple single-weight line as shown in Figures 13.14 to 13.16; in many circuit diagrams, this is the only type necessary. In some, however, certain conductors carry more or less power than ordinary conductors. Figure 13.13 shows power conductors as bold or double-width lines, while signal conductors are shown as thin or half-weight lines. Temporary connections may be shown as dashed or dotted lines.

Joins in conductors

When conductors connect to other conductors, the joint is usually indicated by a distinctive dot, with the more common usage shown in Figures 13.17(a), (d) and (e). Drawings with crossing wires as shown in Figure 13.17(b) do not join (although they did in early drawing conventions). The no-longer-used symbol 13.17(c) indicated that wires crossing were not connected.

When lines representing conductors in a schematic circuit diagram have to cross, they should do so at right angles or as close to right angles as possible. There is no need to indicate that crossing conductors are not joined and no dot is shown at the point of crossing. Figure 13.17(b) shows conductors crossing but not connected to one another.

In some diagrams, conductors cross while others join. Figure 13.17(f) illustrates a convention that was popular in some older electronics drawings. While not as common in schematic drawings, it is used frequently in many electrical wiring diagrams for terminal connections (see Figure 13.11). To avoid any confusion, conductor joins are drawn offset. The ends of the lines representing the conductor joins are displaced sideways at 45° for a distance of 2 or 3 mm. A dot is then placed at the point of each join. Figure 13.17(a) shows a more commonly used method, the *double offset*, which is popular with some versions of *computer-aided design (CAD)* software programs. Figure 13.17(d) shows a method that is *not* recommended for hand-drawn circuits. This is because a spot of ink or some other mark on the page could inadvertently turn crossing conductors like those shown in Figure 13.17(b) into 'joined' conductors.

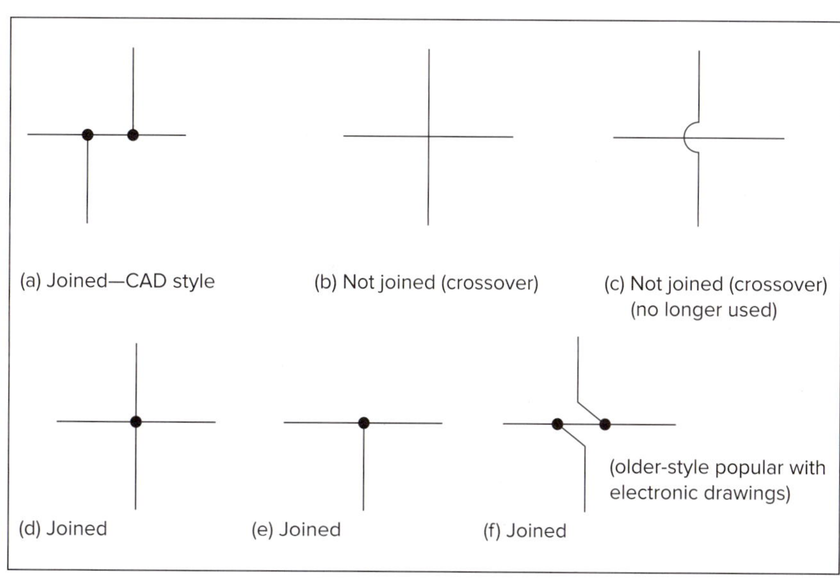

FIGURE 13.17 Representation of conductors crossing but not connected

13.3.2 Line numbering and cable identification

The circuit diagram in Figure 13.18 shows cable numbering, component labels and an alpha-numeric drawing grid for reference. It becomes hard to identify circuit components and cable connections without using some form of reference when drawing a schematic.

Drawing and circuit interpretation can be made easier by applying the following:

- symmetrical placement of circuit components
- using labels for components
- showing relay contact locations
- a drawing cross-reference
- a system for terminal and cable numbering.

Symmetrical component placement on the drawing shows that the start pushbutton S_1 on the first line of the schematic can be found at grid reference C2. In this example drawing, the lines and circuit components are evenly spaced and symmetrically aligned for easy identification, circuit interpretation and analysis.

The particular convention used is often up to the customer or design engineer but will usually be based on labels, page reference, line numbering or terminal device input/output. Using for example an alpha-numeric drawing grid for referencing as in Figure 13.18, the relay coil labelled K1 can be found at grid reference G2. We can also see additional information provided adjacent to the coil—that K1 has four contacts, two normally open and two normally closed. More complicated drawings may include grid locations for each component. In this example, relay contact K1.1 is located at grid reference C4.

Complicated control circuit drawings may not fit on one page, requiring component referencing to other pages. K1 relay contacts K1.3 and K1.4 do not appear on the page shown in Figure 13.18, which means there must be at least a second page. Printing press circuits can have a drawing set of 2, 3, 4 or more pages. Page and grid referencing

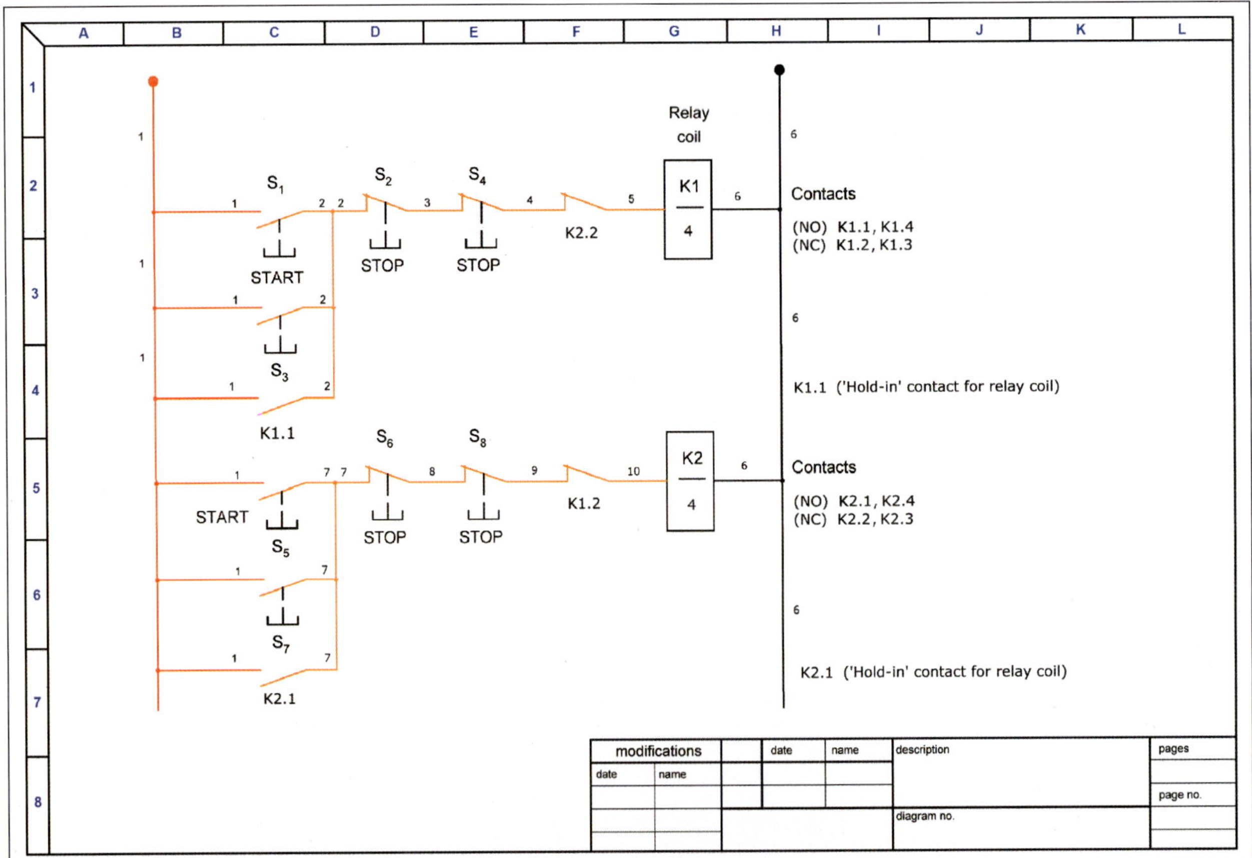

FIGURE 13.18 Ladder schematic with grid reference

makes it easier to locate components in a circuit, particularly for multiple pages in a set of drawings. A page and grid reference assists with exact component placement.

A wire path can also appear on multiple pages. In standard electrical installations, wires have only one identifying label which is the same at both ends.

Fault finding circuit wiring in complicated machinery and equipment will require cable tracing across more than one drawing page. Unlike lighting and power installations, equipment wiring can be labeled differently at either end. One end may connect to a terminal strip while the other end might connect to a printed circuit card socket. When tracing the path of equipment wiring from one drawing page to another, a grid reference is given to identify where to find the wire.

Figure 13.18 shows consecutive wiring identification numbering.

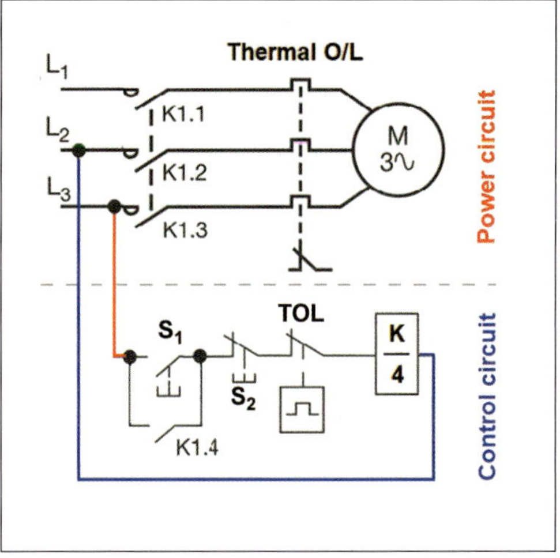

FIGURE 13.19 Symbols placed symmetrically for ease of interpretation

13.3.3 Placement of circuit components

Figure 13.19 shows an identifying number (L_1, L_2 and L_3) for the supply line (line voltage) connections, wire connections to symbols and arrangement of circuit components found in schematic diagram representations. Note also the line-thickness convention for power and control circuits. The control circuit is drawn in thinner lines because it plays no part in transferring energy to the motor and is usually wired with a smaller size cable.

13.3.4 Connections to symbols

Connections to circuit symbols in circuit diagrams should be made at some distance from the circuit symbols (around 5 mm, depending on the drawing scale). They should not be made on the symbols themselves. This is done so that the outlines of the symbols are not confused with closely drawn conductor lines. Computer-drawn symbols usually have *connection tails* to connect to rather than the symbol itself. CAD and schematic editing programs adopt this practice.

13.3.5 Arrangement of components

Energy consuming devices such as indicator lamps, relay, timer and contactor coils are generally placed at the ends of a circuit path or rung. This can be seen in Figure 13.18, with coils K1 and K2. Lamps and coils in control circuits can be connected in parallel but not in series.

Symbols should be placed in line (aligned) or in the same relative position if they are similar (see Figures 13.18 and 13.19). They should be spaced evenly so they are easy to see and interpret. We can see in the Figure 13.19 power circuit that the contacts K1.1, K1.2 and K1.3 are symmetrically aligned, as are the motor contactor thermal overloads. Repetitive parts of the circuit should be identical to avoid confusion.

13.3.6 Parallel components

In electrical circuits, two components are often placed in parallel. If one symbol is more important, it is placed in the same line as the conductor and the parallel component symbol is offset. If they both have the same importance, then they are placed evenly each side of the conductor line. These features can be seen in Figure 13.20. Temporary connections should be shown with a dashed line.

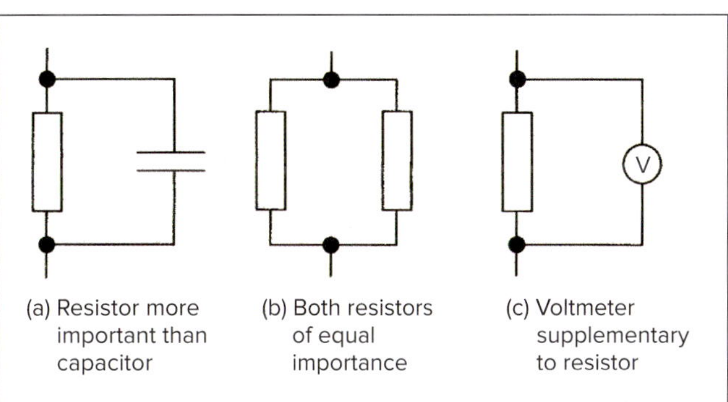

FIGURE 13.20 Parallel connection of circuit components

CHECK YOUR UNDERSTANDING

13.4 What do the dashed lines in Figure 13.19 for K1.1, K1.2, K1.3 and the thermal overload contacts represent?

13.3.7 Drawing conventions for circuit diagrams

Figure 13.21 illustrates the control and power circuits for a three-phase motor. The three main contacts in the power circuit (K1.1, K1.2 and K1.3) are normally open so that when the relay coil (K1) is energised, line voltage is applied to the motor.

Drawing conventions allow easy identification of the power and control circuits through line thickness and type of contact. Lines L_1, L_2 and L_3 represent the power conductors and have been drawn thicker than the control conductor lines. This method differentiates between power and control circuits. A broken line is shown separating the two parts of the circuit, but this is not regular practice. The K1 contactor and thermal overload each have an auxiliary contact in the control circuit.

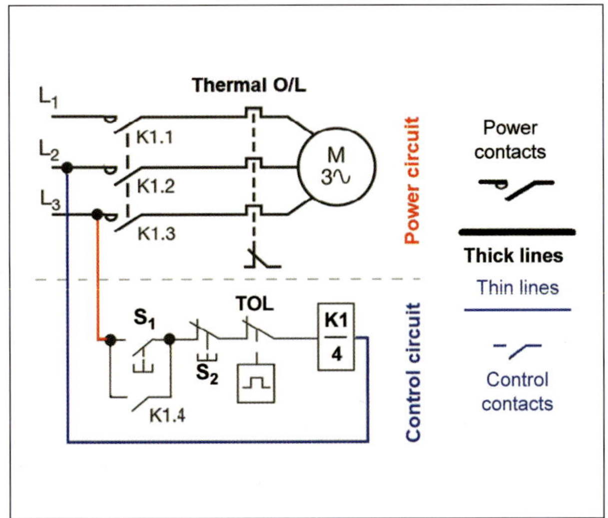

FIGURE 13.21 Power and control circuits for a three-phase motor

13.3.8 Control circuit operation

Using standard circuit component symbols helps in reading the schematic and determining how the circuit operates. The following explanation of a control and power circuit operation relates to Figure 13.21.

The control circuit is powered by two supply lines connected to L_2 and L_3. This circuit uses line voltage for the control circuit.

The stop and start pushbutton switches provide a convenient way of operating motor and other control circuits. It is easier to press a button than operate a toggle or turn a handle. The start-stop pushbuttons have contacts which change state when operated. The control circuit switches and contacts are not rated to carry full current. The on or start pushbutton has normally open contacts and a *momentary effect* when pressed. The symbol indicates that when the button is released, the switch contact will open again.

To start the motor, the start button S_1 is pressed. Current passes through the normally closed stop pushbutton S_2, energising the K1 contactor coil. The moment K1 is energised, all K1 contacts change state. In this case, they are all normally open contacts so they all close.

The momentary pushbutton does not have to be held in manually by an operator for the circuit to remain energised. A normally open auxiliary contact (K1.4) is connected in parallel across the start pushbutton in the control circuit. Contact K1.4 takes the place of the start pushbutton switch when the latter is released.

The purpose of this normally-open auxiliary contact is to act as a latching or hold-in contact, to maintain supply for the contactor coil. When it closes, a current path is created for maintaining coil operation and so the coil remains energised.

The stop pushbutton has normally closed contacts and is connected in series with the coil. Pushing the *momentary stop* pushbutton switch breaks the current path, de-energising the K1 contactor coil and causing all the associated contacts to open. The motor then stops and cannot be re-started until the start pushbutton switch is pressed again.

13.3.9 Power circuit operation

When the start pushbutton (S_1) is operated, contactor coil K1 energises. All the K1 contacts change state and the three power contacts K1.1, K1.2 and K1.3 in the power circuit close, allowing line-voltage supply to the motor. The broken line linking these contacts indicates that all three contacts close simultaneously.

The motor will operate continuously until it is halted by the stop pushbutton (S_2) being pressed in the control circuit, or by the thermal overload operating due to excessive current. If an overload condition occurs, the thermal overload unit senses the increase in motor current and mechanically operates the thermal overload (TOL) contact in the control circuit. This de-energises the contactor coil K1, causing the contactor contacts in the power circuit to open, disconnecting line voltage and again the motor will stop. The motor will not restart automatically and cannot be restarted until the overload (if tripped) is reset.

13.4 Relays, contactors and timers

The basic concept of operation for these devices is a coil being operated electrically, causing sets of contacts to change state for control of electrical circuits. They can be classified as 'pneumatic', 'general purpose', 'solid state' and 'programmable'.

A general-purpose relay is simply an electromagnetically operated switch. A contactor is also an electromagnetically operated switch that controls the electrical supply for energising high-current load devices like heating elements and motors. A contactor is in some ways just a large relay; it is sometimes difficult to say whether a device should be called a relay or contactor in any given circuit application. Standard convention is that relays are typically low powered and contactors are high powered.

Solid state relays (SSRs) do not have moving contacts. In terms of operation, they employ semiconductor switching elements such as **thyristors**, **TRIACs**, **diodes** and transistors.

13.4.1 Drawing relays and contactors in circuits

Mechanical or electromechanical relays and contactors are devices that consist of three basic parts: the operating coil, the associated magnetic circuit and the contacts that are actuated by the coil. Coils and contacts with particular operational characteristics can be represented in schematic drawings by specific circuit symbols.

When drawing contactors and relays in circuits, the coil component is part of the control circuit. The same symbol is used for relay and contactor coils, but the coils for relay timers are drawn to illustrate different types and function.

Contactors may be drawn in circuits in a number of ways, one of which is illustrated in Figure 13.22(a). This shows a contactor with its operating coil and all its associated contacts. The broken line enclosing the components indicates that they are all part of the one assembly. This type of attached representation is seldom used in schematic circuits because it does not lend itself to a logical circuit arrangement.

A second method is to indicate by a dashed line that the contactor coil operates the contacts joined by the line. This method of **semi-detached representation** is sometimes seen in diagrams, especially when the components on the contactor are close together or in line. It is illustrated in Figure 13.22(b).

The most common method is termed 'detached representation'. The contactor coil is labelled

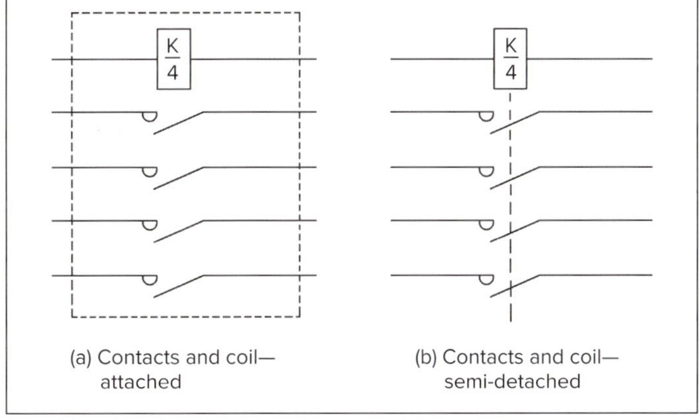

FIGURE 13.22 Symbolic representation of coils and contactors

with the contactor designation and with the number of contacts it operates (see Figure 13.21). The contactor is designated as 'K1', and the number 4 in the coil symbol indicates that K1 has four associated contacts. Sometimes the coil designation is placed beside the coil symbol. In this case there are four normally open contacts (three for power and one for the control circuit). Contactors can have more or fewer contacts; some could be normally closed and others timed.

13.4.2 Construction and operation of relays and contactors

The principle of operation for relays and contactors is similar to that represented in Figure 13.23. A very old type of contactor representation clearly shows the coil, the operating part of the magnetic circuit and a single set of contacts. Modern contactors look like the one shown in Figure 13.24.

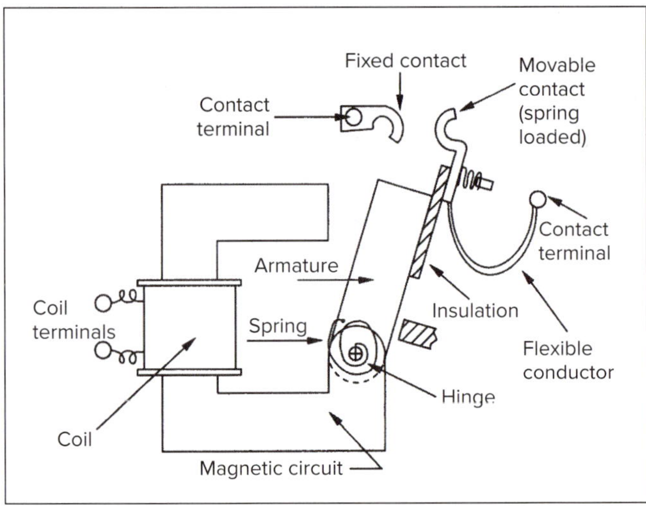

FIGURE 13.23 Operating components of a contactor

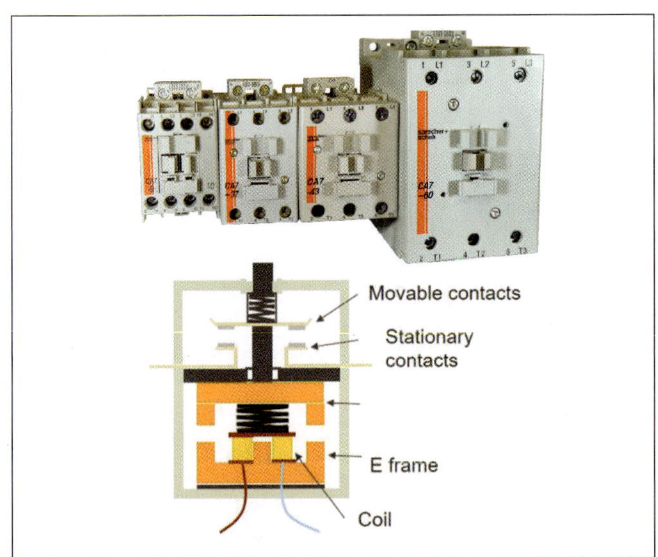

FIGURE 13.24 Typical contactor

Reproduced with permission of NHP Electrical Engineering Products Pty Ltd

Electromagnetic operation

When the coil is energised, a magnetic field is produced in the magnetic circuit. This attracts the hinged armature against the tension of the spring to complete the magnetic circuit. The movable contact, attached to—but insulated from—the armature, closes against the fixed contact. When the coil is de-energised, the armature spring pulls the armature open, overcoming any residual magnetism.

Electrical contacts

Contactors consist of normally open or normally closed contacts which are indicated by labelling. This labelling may be the schematic symbol or the letters NO (standing for 'normally open') for a make contact or NC ('normally closed') for a break contact. Typical contactors, especially larger ones, will have a combination of NO or NC contacts activated by the same coil.

AS/NZS 1102 makes no real distinction between power contacts and control contacts. It merely provides two symbols labelled 'Form 1' and 'Form 2'.

Figure 13.25 shows the different types of contact symbols for power and control circuit drawings. For clarity, Form 1 contact symbols are used extensively in this textbook for control contacts and the contactor symbol is used for power circuit contacts.

Most contactors have at least three power contacts. When used for motor-starting duty, they are designed to carry five times their rated current for a short time. Other contacts on the contactor are often termed 'auxiliary contacts', or sometimes 'control contacts'.

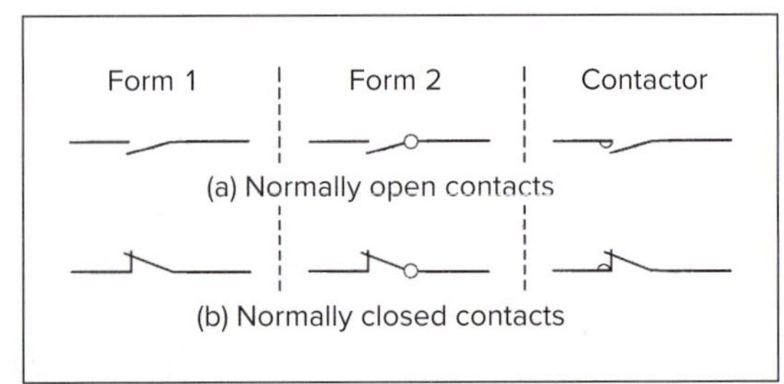

FIGURE 13.25 Symbols for electrical contacts

Power and control contacts may be either 'normally open' (NO) or 'normally closed' (NC). The word 'normally' in this context means that the contactor itself is not energised (i.e. power is not applied to the coil). When the contactor coil is energised, the contacts change their state. Normally open contacts close and normally closed contacts open. In drawings they are *always* shown in the de-energised state. The operation of contact symbols (whether opening or closing) is always considered to be in a *clockwise* direction.

13.4.3 Timers

Switching based on certain timings is often required—for example, for lights that are turned on and off for a specific period of time. The timing device may be pneumatic (like those used in switches for stairwell lighting), electro-mechanical, electronic (solid state) or programmable. Some contactors, such as K1 in the wiring diagram in Figure 13.11, have additional pneumatically 'timed' contacts. Older and some customised 'timers' can be pneumatic delay contacts as clip-on or add-on contactor accessories.

A mechanical wheel or dial can be set in older-style analogue systems. When the dial reaches set points (determined by the installer or an operator), the switch will operate to either close or open, depending on the settings. This type of mechanical timer was common for some industrial, consumer and commercial equipment such as glasswashers in clubs and bars, dishwashers and washing machines. These timers worked by having cams on a rotary shaft operating micro-limit switches to control machine sequence and cycles. On electronic systems, the installer programs on and off times. With more advanced devices such as dedicated controllers or programmable relays, multiple set points per day and per week can be programmed.

While more accurate and reliable technology has replaced mechanical timers, mention of their operating concept and principle is relevant as it shows that physical positioning of moving parts, machine sequence, cycle and timing are descriptions sometimes used interchangeably as control terms. Timing requires a reference point. It is worth noting too that count registers are used for 'clocking' in electronic, programmable and logic-control systems rather than relying on a time-based reference. Start and stop control sequences based on seconds and minutes in some systems use counts or pulses as the time reference.

Relay timers in general have one set of contacts controlled by the timing mechanism, with both a normally closed and a normally open contact sharing a common contact. Along with the contacts are 'coil terminals', which are designated as 'A_1' and 'A_2' or 'L' and 'N', depending on manufacturer labelling. The installer has to read the manufacturer's specific wiring diagram to identify the terminals correctly.

Timed contacts will close or open after an adjustable designated period. Timed contacts are designated by a special symbol—a semicircle or 'parachute'—to indicate the contact is slowed down in operation. Timed contact symbols of various forms are shown in Figure 13.26. Care needs to be taken when interpreting the operation represented by these symbols. Electronic and programmable relays provide advanced features with more functions while using these symbols.

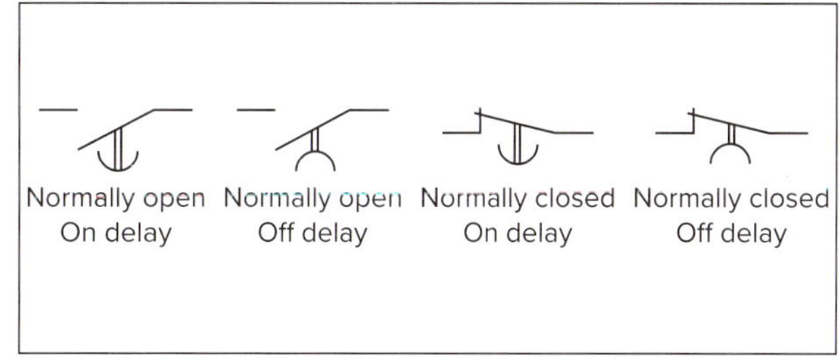

FIGURE 13.26 Timed contact symbols

On-delay timing relays are drawn with two diagonals in a rectangle on one end of their coil symbols. They have a delay before their contacts change state, after energising the coil. Off-delay timer relay contacts are slow to change state after de-energisation. They are designated with a filled-in rectangle on one end of their coil symbols. The symbols for the three types of relay coil are shown in Figure 13.27. The contacts of slow-closing coils are commonly referred to as 'on-delay' contacts, and for slow-releasing coils as 'off-delay' contacts.

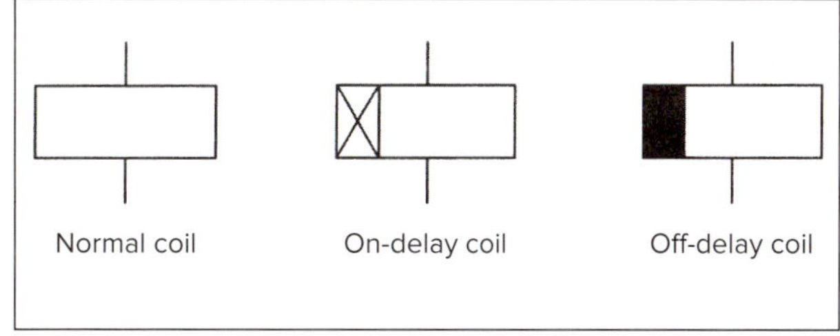

FIGURE 13.27 Normal and timed relay coils

Programmable timer relays (discussed in section 13.4.3) are available which incorporate combinations of instantaneous operating contacts, on-delay and off-delay contact operations.

Timer applications are common in equipment for operation sequence and cycles. Processing, production and manufacturing equipment, for example, can incorporate timers as part of their start-up sequence. This can be thought of as using two start buttons that are pressed one after the other, each incorporating separate timer circuits. The pushbutton labels may be START and RUN. Pressing the first or START pushbutton initiates a start-up sequence with flashing lights and warning buzzer alerting workers and warning operators that the equipment is about to start. This START initiates a short time-delay window which may be between three and five seconds. During this period, the second (or RUN) pushbutton will not function. The second time-delay window commences with the first and may be, for example, up to 10 seconds. The start sequence for safety is to have the warning period for about 5 seconds, followed by a time window during which the RUN pushbutton needs to be pressed for the machinery to operate. If the RUN is not initiated during this time, the control circuit resets and a new start-up sequence is required. The start-up sequence described is similar in principle to those for printing presses used in the newspaper industry.

Unplanned stoppages during production and manufacturing cycles can result in moving parts such as mechanical arms and levers being in the wrong physical position. Using timers for the start cycle can afford the time needed for these mechanical parts to 'position reset' (or return to their home position), preventing damage and equipment breakdown on starting up.

Timers are also used for motor starters. Primary and secondary resistance starters use timers to switch out resistors in controlled stages. Another example is a star–delta control timer. These are adjusted so that after a set time, when the required motor speed is reached, their contacts operate for coil control of the interlocked motor circuit contactors. A changeover from the star-start contactor (low start current and torque) to the delta-run contactor (full load speed and maximum torque) changes the motor wiring circuit connections.

13.4.4 Thermal overloads

Fuses and circuit-breakers provide circuit protection while thermal overloads provide protection for the motor windings. They are current-protection devices with adjustable settings, electrically connected on the contactor load side, in series with the motor windings. Thermal overloads need to be matched to the rating of the contactor. The thermal overload operates like a switch. Heat is generated by current flow to the motor windings.

Thermal overloads work by having a bimetallic strip bending on over-temperature (due to the heating effect of current), which causes normally open contacts to close and normally closed contacts to open (see Figure 13.28).

The contactor coil in the control circuit is wired in series with the normally closed auxiliary contacts of the thermal overload—if the overload is activated, the **thermal overload (TOL)** contacts open, de-energising the contactor coil. The contactor main contacts in the power circuit then open, disconnecting the line voltage and protecting the motor windings from damage due to over-temperature conditions.

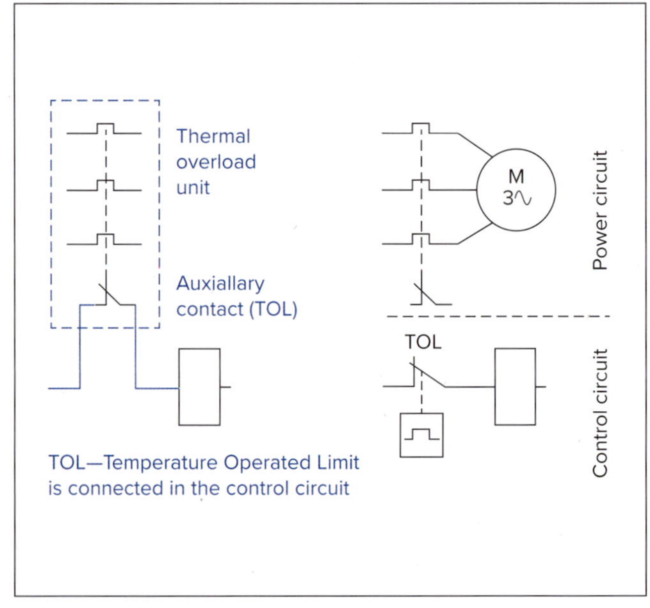

FIGURE 13.28 **Thermal overload unit**

Thermal overloads are attached to the output or load side of a motor contactor. They often look like a plug with straight pins on one side connected to the contactor, with terminals for cable connection to the motors on the other side. The auxiliary contacts are normally labelled with either 'NO' or 'NC' lettering or symbols. Although overloads act as a form of protection for any device where over-temperature due to higher-than-normal current is a problem, they are mainly used for electric motors.

CHECK YOUR UNDERSTANDING

13.5 What do contactors and relays have in common in terms of how they are drawn in schematic circuits?

13.6 What is the name of the most common method of drawing contactors in circuits?

13.7 Timers can have a combination of which contacts?

13.8 Where are the auxiliary contacts in thermal overload units?

13.5 Fault finding control circuits

Discovering and diagnosing faults successfully requires common sense, awareness and knowledge of how the equipment operates, as well as the application of a logical approach to analysing faults and checking and testing techniques. Effective communication skills and good workplace relationships are also useful as they foster information sharing and co-operation.

The main cause or causes of faults will usually be one or a combination of:

- equipment design faults—production tasks may be outside the equipment's normal operation specifications, for example, if it was purchased from overseas or has been modified. Certain functions can result in unexpected production stops.
- wiring and equipment faults—electrical and/or mechanical damage due to work environment, wear and tear, sensor or detector adjustments, predictive-preventative programmed maintenance and repair 'windows of opportunity' and schedules.
- operator error—lack of training or support, set-up mistakes, fatigue, stress, emergency stop operation, blocked and misaligned sensors or detectors.

Fault finding is a major part of electrical work and is a skill that requires knowledge of the circuit operation through:

- familiarity with equipment operation and control under normal operation
- observation of equipment operation and the sequence that led to the fault condition
- careful study of the circuit schematics and wiring diagrams.

Plant maintenance and breakdown staff will develop this knowledge through experience over time. The number of recurring faults in production environments can be high. Staff may have to deal with anything from straightforward machine stoppages or process-sequence resets to attending to minor adjustments of sensors and detectors to more significant fault repairs. After between three and six months on the job and under the guidance of experienced mentors, technical specialists (including plant electricians) can diagnose and carry out arguably 80% of the required fault rectification, repair and maintenance tasks.

Knowing how to operate, understand and repair equipment enables electricians and technicians to become the technical 'gun' operator.

Symptoms, faults and causes

WORKPLACE SCENARIO

As the in-house electrician, you are advised that one of the equipment plant conveyor systems has suddenly stopped and will not restart. This is a symptom of a fault condition, and the cause could be mechanical, electrical or operator related. The state of the equipment illustrates the importance of an electrician/technician being able to quickly identify and discern between symptoms, fault condition and causes.

The cause of equipment stopping suddenly and then not being able to be restarted could be loss of supply, circuit-breaker or overload trip, an operator forgetting to reset the emergency stop pushbutton or a mechanical jam—there are multiple possibilities. When determining what the fault is, its cause also needs to be found. For example, was an overload trip due to the motor drawing excessive current because of a jam or a dropped phase? And what was the root cause of the jam or the dropped phase?

If you are not familiar with the equipment and its operation, or you do not have enough information about the fault that has been reported, listen carefully to the operator's description of what happened and then ask questions. (In cases where the equipment is still functioning, observe its status closely during operation or when the fault recurs.)

The type of questions that you could ask the operator are:

- What happened?
- What was the equipment/machine doing when the fault occurred?
- How does it work normally?

The operator's responses can provide useful information. You can also increase your awareness of the situation by doing the following:

- *Looking*—for something obvious, such as smoke, sparks, blown fuses, overloads, broken mechanical parts or brackets, misaligned sensors, mechanical jamming of moving parts, broken cables or exposed, overheating or burnt wiring, coils or components.
- *Listening*—for mechanical noises such as clunking, grinding or scraping metal; noisy bearings, labouring motors, humming, buzzing and chattering of relays or contactors.
- *Smelling*—a smell of plastic or rubber may indicate overheating cables; a smell of varnish could indicate burnt out or overheated coils, contactors, transformer windings or motors.
- *Touching*—check motors for excessive vibration or overheating (while of course avoiding burnt hands and electric shocks).

Taking these steps can rule out many simple faults and help determine whether there is power present. If you find something at fault, damaged or out of adjustment, follow the necessary equipment shutdown, isolate, lock-out and tag-out procedures.

Applying a logical process of elimination and checking allows faults to be more readily identified and rectified. Following adjustment, replacement or repair, check the rest of the circuit. There may be other fault causes to identify and correct. Damaged electrical components—particularly circuit cards—could be damaged further (or even fail) when the circuit is re-energised.

13.5.1 Checking and testing techniques and procedures

If the procedures outlined in the Workplace Scenario do not establish the fault and its cause, two steps may need to be taken. One is to follow a fault escalation procedure (many production and processing environments have them). This procedure should state which staff should be notified and when, as well as fault-finding processes and time frames.

The second step is to follow fault-finding processes for electrical equipment. Depending on the fault condition and the equipment, this can involve one of the following approaches:

- *Systemic—live testing* for voltage (see Figure 13.29) or *dead testing* for continuity (Figure 13.30), using a multimeter or similar function-testing device. This approach involves one probe lead of the meter

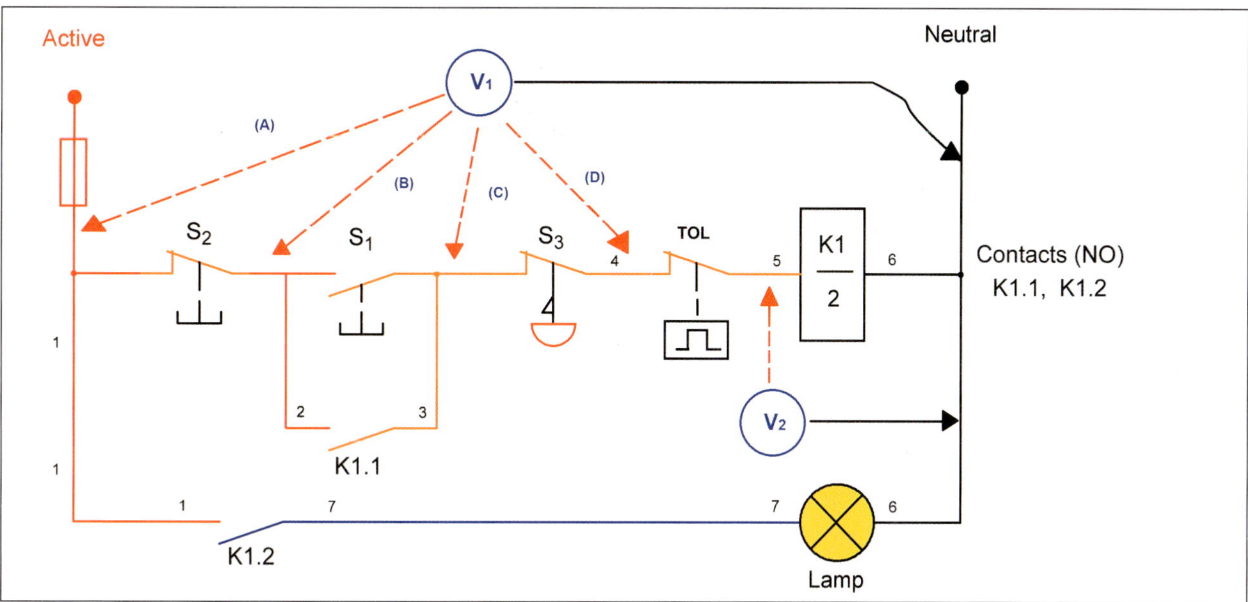

FIGURE 13.29 Live testing using a multimeter

being attached to the circuit neutral. The other test probe is moved along the circuit path, checking each component or device in series to determine the fault.

- *Circuit division*—dead testing (for continuity) or live testing (for voltage) by dividing the circuit in half. This is particularly common in emergency stop loops to determine an emergency stop operation and its location.

A multimeter can be used to determine circuit path continuity, as shown in Figure 13.30. As the circuit is de-energised (or 'dead'), it is possible to check from the contactor coil through to the active supply rail. By pressing the S_1 start pushbutton, the circuit path is established (continuity) without needing supply.

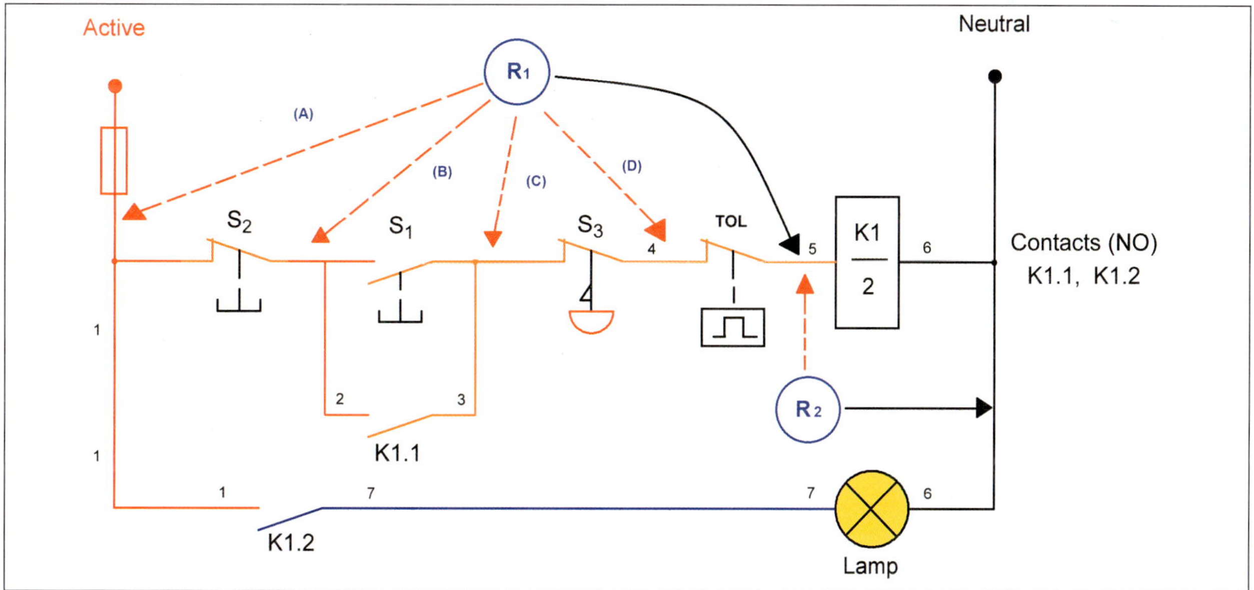

FIGURE 13.30 Dead testing a de-energised circuit

13.5.2 Applying circuit-checking and testing techniques to an electrical control circuit

It is a requirement of electrical work that the safe and correct operation of powered plant and equipment is ensured. The correct type of test equipment can, if used properly, prevent unwanted and 'false' equipment sequence start-up.

(Circuit and voltage testers with inherent capacitance as part of their design should not be used for this.) Common test equipment includes:

- Voltage testers—usually high-impedance multimeters on the voltage range, although test lamps (low impedance) that are fused and insulated may be used in some situations.
- Ohmmeter—for checking circuit continuity and resistance of circuit components. They are used for open circuits, zero resistance for switch operation, open and closed contacts, conductors and terminal connections. They provide relay coil and element resistance readings.
- Clamp meter—'tong tester' for non-contact testing of current, unbalanced three-phase loads and current-leakage situations. Modern clamp meters include voltage and resistance capability using test probes.
- Insulation resistance (IR) meters—for testing insulation integrity of conductors, and testing between phase windings of motors. Some IR testers incorporate resistance and continuity functions.

It is advantageous to use dedicated test equipment. While a general-use multimeter can measure voltage, current, resistance and continuity, selecting the wrong setting can damage the equipment under test.

Live testing by inexperienced electrotechnology workers can be dangerous and can lead to an electric shock situation and/or equipment damage. Before attempting any testing, assess the situation, use common sense and follow safe work practices and procedures. It is useful to keep the following in mind:

1. Are you trained, qualified and authorised to carry out testing?
2. Does the workplace have organisational policies and procedures for working on machinery and equipment/live equipment? Do you know what they are?
3. Do you have the appropriate personal protective equipment, tools and test equipment to perform the work?
4. Have you checked and tested your tools and equipment for damage, safe use and correct operation?
5. Are you familiar with the equipment, do you have a basic understanding or working knowledge of it and can you perform work following manufacturer recommendations?
6. Do you have the required wiring and schematic drawings and manuals?
7. Do you need to deploy a rescue kit and have an observer?
8. Have you conducted a risk assessment?

Appropriate in-house and manufacturer equipment training is usually provided to qualified, authorised on-site staff. However, if you are engaged as an outside contractor, the questions above need to be asked and addressed. Making a mistake can result in equipment damage, lost production and downtime while sourcing replacement components and spare parts.

13.6 Fault finding for single- and three-phase motor circuits

Sometimes motors need to be checked under operating conditions during production periods. This may be due to a reported fault condition or symptom. For example, the motor exterior casing may be hot to touch, there may be a grinding or high-pitched squeal coming from the motor itself or the motor may be tripping the overloads.

Figure 13.31 shows some of the common test points for using a multimeter to measure voltage and a clamp meter to measure current for single- and three-phase motors. If the motor is running hot or tripping out on an overload condition, the fault cause may be either electrical or mechanical. If the motor is making strange noises, the fault is more likely to be mechanical. Table 13.1 lists the test point locations and the type of test or measurement that is carried out at each.

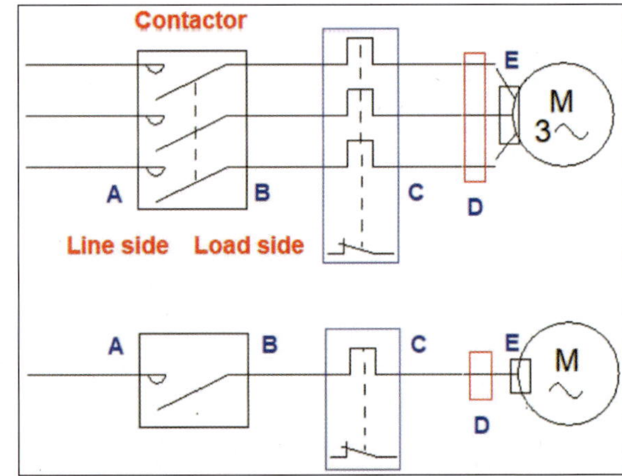

FIGURE 13.31 Single- and three-phase motor testing points

TABLE 13.1 Motor test points

Label	Test point location	Type of device and test/measurement
A	Supply or line side of contactor	Multimeter—terminal voltage
B	Load side of contactor	Multimeter—terminal voltage Clamp meter—line current
C	Output side of thermal overloads	Multimeter—terminal voltage Clamp meter—load current
D	Motor isolator (adjacent motor)	Multimeter—terminal voltage Clamp meter—load current
E	Motor terminal/junction box	Multimeter—terminal voltage Clamp meter—load current
Test point A	Check the supply at the contactor line side terminals. It should be the same on each phase. Are there three phases?	
Test point B	Check the contactor load side terminals for voltage on each phase. Do the load side and supply side voltage measurements line up? Are the contacts sticking, stuck or not making good contact?	
Test point C	Check voltage at the output side of the thermal overloads. Again, are the voltage levels the same as those at the contactor load side terminals? Is the thermal overload unit suspect or faulty? Check the motor current using a clamp meter. Are the phases each drawing a similar current? (This is a check that the load is balanced.)	
Test point D	Check line and load side of isolator contacts. Do the line and load side voltage measurements line up? Is there a problem with the isolator? Check the motor current using a clamp meter. Are the phases each drawing a similar current? (This is a check that the load is balanced.)	
Test point E	Check phase currents and phase voltages at motor terminals.	

The decision whether to take measurements and readings from the motor isolator or motor terminal box is based on safety and ease of access. Sometimes it may be easier to access the motor isolator to take voltage and current measurements, but ultimately, if there appears to be a problem, checks will probably need to confirm whether it is at the motor itself.

Single-phase voltage checks (three tests):

- active—neutral
- active—earth
- neutral—earth.

Three-phase voltage checks (ten tests, involving phase actives, neutral and earth):

- phase—phase (three tests)
- phase—neutral (three tests)
- phase—earth (three tests)
- neutral—earth.

13.7 Control circuit variations

Control circuits for conveyors, industrial processes and machines are designed to be operated in many different ways. Equipment and machinery may need to be operated and stopped from more than one location and may involve two, three or more people as operators. The following are some common application examples of control circuit variations:

- multiple stop controls
- two-position (3-wire) control

- local or remote operation
- 2-wire control
- 2-wire and pushbutton control
- jogging control
- reversing circuits
- interlocks.

13.7.1 Multiple stop controls

Figure 13.21 showed a basic start-stop control circuit. A variation on that control circuit is the provision of only one start position with multiple stop positions (see Figure 13.32). This can arise when a number of emergency stop positions might be required for a piece of equipment. Only one pushbutton switch will start the operation, but if there is a malfunction the equipment can be stopped by operators at any number of positions (this requires that all the stop pushbutton switches are connected in series).

FIGURE 13.32 Multiple stop control positions

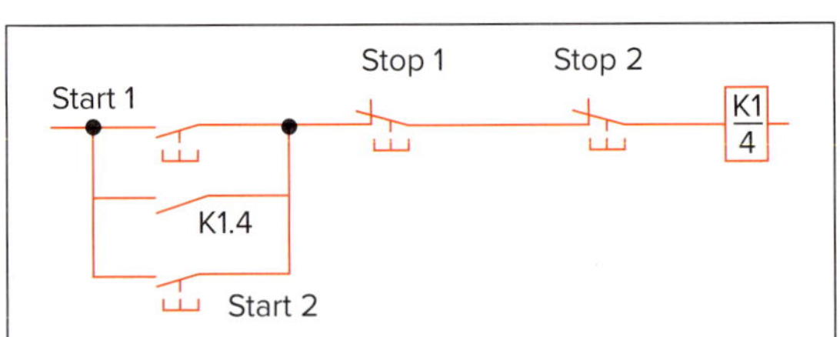

FIGURE 13.33 Two start and stop control positions

13.7.2 Two-position (3-wire) control

So far, only circuits with one start pushbutton have been discussed. To provide extra start positions, all that is needed is to connect extra start pushbuttons in parallel with the first. To stop or de-energise the circuit, extra stop pushbutton switches are placed in series with the first. In Figure 13.33 there are two start and two stop pushbutton switches. As is usual, the start pushbuttons have normally open contacts and the stop pushbutton contacts are normally closed. These are labelled 'Start 1', 'Start 2', 'Stop 1' and 'Stop 2' for clarity.

This type of control circuit is also referred to as '3-wire control'. The schematic representation in Figure 13.34 is shown for local and remote operation.

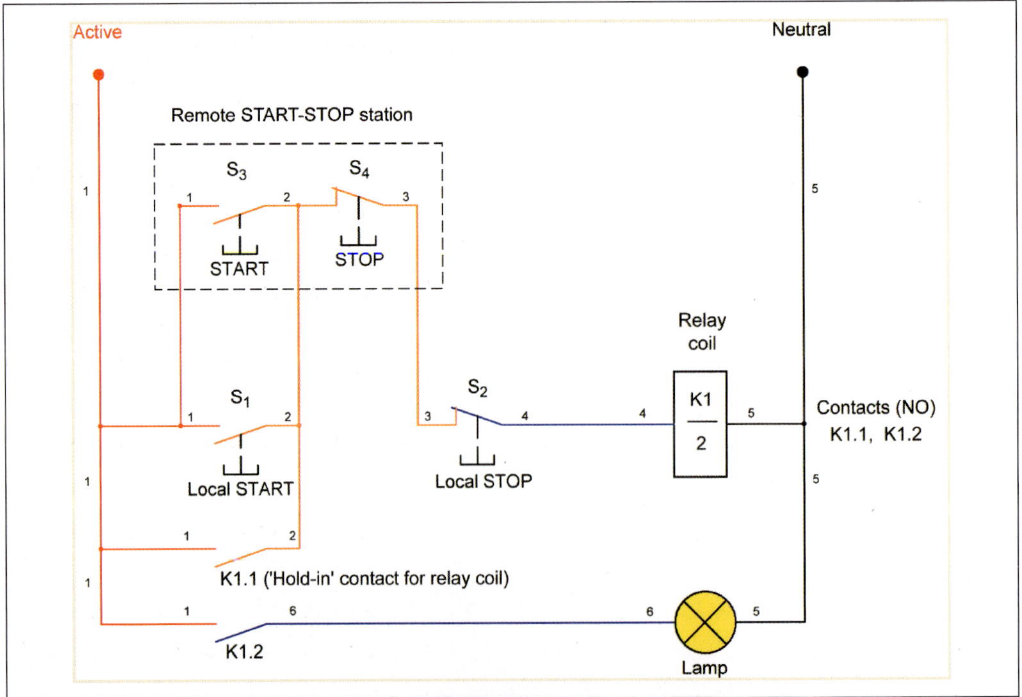

FIGURE 13.34 Local-remote 3-wire control circuit

The remote start-stop station is shown as a separate piece of equipment by a dotted line box on the diagram. Additional start-stop stations can be added using the 3-wire method. Note the component spacing, layout and sequential wire numbering provided before and after circuit components. The two normally open K1 contacts used in the circuit are identified adjacent to the coil.

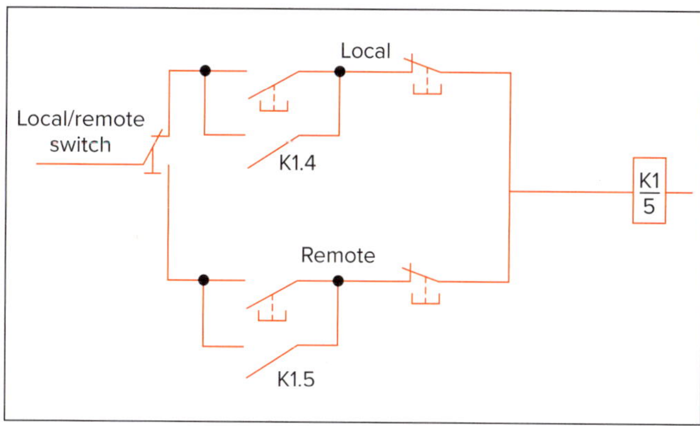

FIGURE 13.35 Remote (or local) operation control

13.7.3 Local or remote operation

In some operations it may be necessary to shift the operating position of the stop/start pushbutton switches. To avoid the possibility of someone operating the machine at the incorrect position, the circuit of Figure 13.33 is amended so that only one position at a time can be used. This is often referred to as 'local' or 'remote' operation. Figure 13.35 is a circuit representation of this type. On the left side of the diagram is a hand-operated changeover switch, which switches into circuit either the upper-local or lower-remote pushbutton switches. This type of circuit requires an extra contactor contact and only a simple changeover switch.

Processing and production equipment requires the local or remote option of control for maintenance. There may also be a need to switch between the normal run mode and set-up modes. Once selected in run or production mode, operators may require multiple start-stop control stations. It should be noted for this example that if the equipment is switched to 'local' mode, only the local stop pushbuttons are in circuit to stop the equipment.

13.7.4 2-wire control

Many motor-control circuits use an automatic start control. This could be, for example, from a thermostat on a refrigerator, a float switch on a water tank or a pressure switch on an air compressor. Because there is no other stop/start control and only two wires need to be run to the actuating device, is it usually referred to as '2-wire control'. Figure 13.36 is a simple 2-wire control circuit. In this case, the controlling device is a pressure switch, which is represented by the lower-case 'p' in the square.

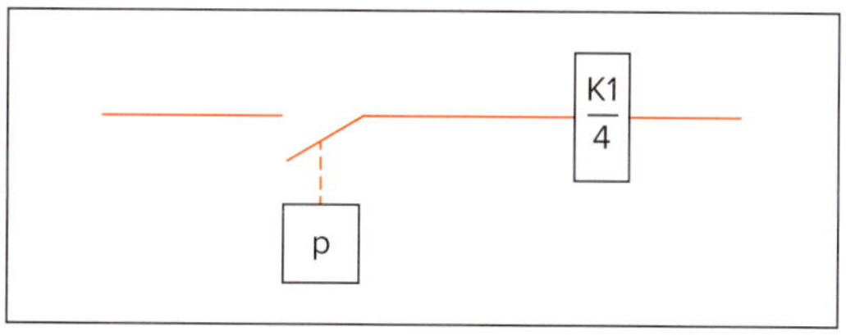

K1
4
p

FIGURE 13.36 2-wire control by a pressure switch

13.7.5 2-wire and pushbutton control

In some cases, it might be necessary to start a motor-controlled device using a pushbutton switch but allow another control to turn it off. This type of circuit is shown in Figure 13.37. It has normal stop/start control with a float switch connected in series with the stop pushbutton. This would enable a pump to be started and then switch off automatically when a tank is full. The motor could be stopped at any time while running but could not be restarted after an automatic stop until the water level fell and the float switch closed again.

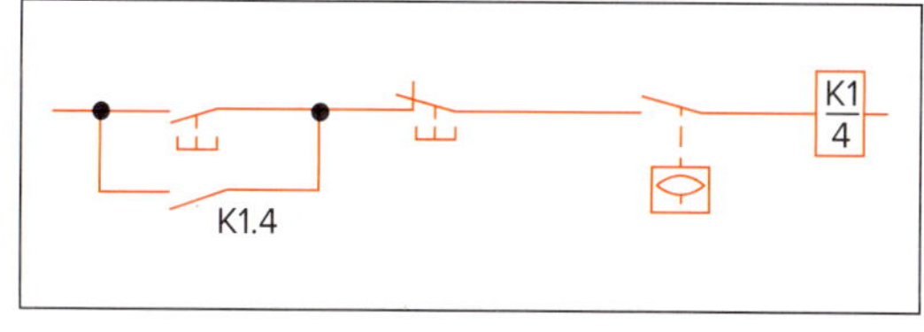

FIGURE 13.37 Combined stop/start and automatic control

13.7.6 Jogging control

Jogging control is where a pushbutton switch is connected for a circuit to operate only while the switch is held depressed. It is sometimes necessary to *jog* or *inch* a machine to a certain position so that setting up can be done and adjustments made. Attempting to jog using start and stop pushbuttons is neither good practice nor reliable—and it could be dangerous. It is more efficient to install a special circuit with a jogging pushbutton for this function. These circuits require special motor starters (AC4 type) that are rated for jogging duty. Contactors are designed and rated for different applications including lighting, heating, motor starting and jogging control.

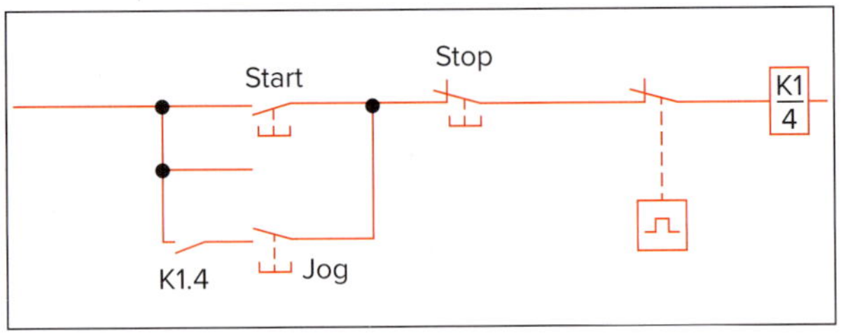

FIGURE 13.38 Jogging pushbutton switch in a control circuit

Such a circuit requires a changeover switch between a RUN condition and SET-UP mode. In RUN mode, the machinery operates with high motor speeds. SET-UP modes allow plant and equipment staff to carry out pre-production tasks on rotating equipment set to run at very low 'crawl' speeds.

In the schematic circuit diagram in Figure 13.38, the jogging pushbutton switch is a pushbutton changeover switch with a normally closed and a normally open position. In its normal position, the circuit appears to operate as a simple stop/start control. This type of motor-stopping control would not be viable in production environments.

This circuit shows the relationship between incremental movements or jogging control and **crawl speed** for the slow-roll operation of motor-driven equipment and machinery. When the jogging button is pressed, the hold-in contact K1.4 is isolated and the start pushbutton switch is bypassed. While the jogging button is pressed, coil K1.4 is energised. When the 'break before make' jogging button is released, the circuit to the coil is opened. For this circuit, the start pushbutton would need to be pressed so that normal crawl-speed operation is possible.

The jogging pushbutton sometimes includes a small delay on reclosing to give the contactor time to open contact K1.4. Its contacts would be described as 'break before make'. The simple circuit shown is used for very slow speeds. Working on rotating equipment is dangerous and, even with machine speeds set under 10 rpm to mitigate risk of injury, rotating machinery hazards are still present and have in the past resulted in the loss of fingers and limbs in industrial accidents.

Newspaper printing presses are an example of where inching- and jogging-control functions are used for feeding paper leads through rollers, attaching to (or removing printing plates from) cylinders and equipment set up in a production environment. Other industrial examples include small controlled movement to accurately position or align tools and machine parts and to thread sheet steel in mills and cloth in the textile industry.

13.7.7 Reversing circuits

To reverse a three-phase motor, all that is necessary is to interchange two supply lines to the motor. This can be accomplished by using two contactors, as in the circuit diagram in Figure 13.39. When K1.1, K1.2 and K1.3 contacts close, the motor will operate in a forward direction. When K2.1, K2.2 and K2.3 contacts close, the motor will operate in the reverse direction. Examination of the power circuit will show that two supply lines would be short-circuited if both contactors closed at the same time. Following the description for circuit operation, we will look closely at the methods used to prevent operation of both contactors at the same time. These methods include mechanical and electrical interlocking.

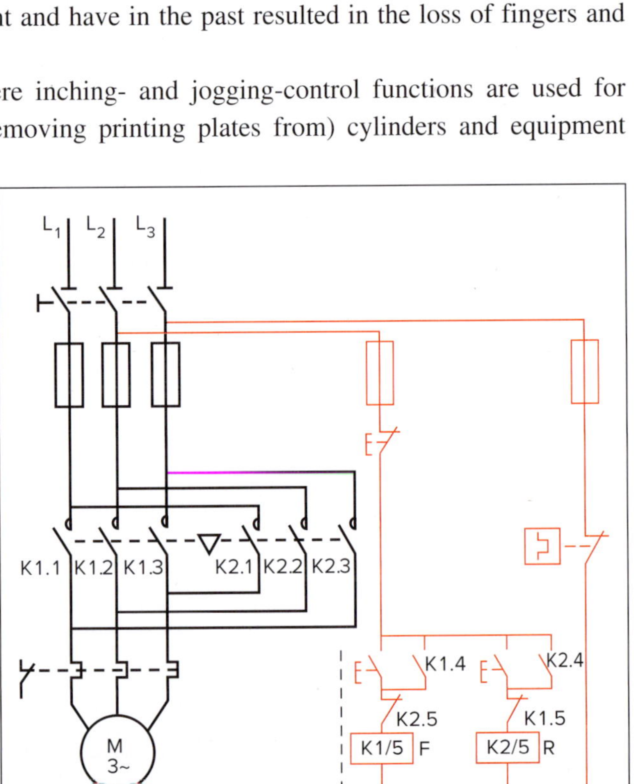

FIGURE 13.39 Schematic circuit diagram of a reversing contactor

Although not labelled on the diagram, the forward pushbutton will energise the K1 contactor and the reverse pushbutton will energise the K2 contactor. The figure 5 in the contactor coils K1/5 and K2/5 signifies that each contactor has five associated contacts. When either of these coils are energised, all five contacts change state.

Reversing contactor circuit description

K1 and K2 are the coil parts of separate contactors that allow line voltage supply to the motor. One contactor allows the motor to rotate in one direction while the other allows reverse rotation by reversing the connections to two of the motor phase windings. When either coil is energised, all of their associated switch contacts operate for motor power and control.

When the forward start pushbutton switch is pressed, current flows through the normally closed stop pushbutton, then through the normally closed interlock K2.5, energising contactor coil K1. Current flows through the K1 coil and the normally closed overload contact, back to the line supply (L_3) to complete the circuit path.

All of the associated switch contacts for contactor K1 (K1.1, K1.2, K1.3, K1.4 and K1.5) will change state. The first three switch contacts (K1.1, K1.2 and K1.3) close, allowing full line voltage supply to the motor,

Switch contacts K1.4 and K1.5 are for circuit control. Contact K1.4 is in parallel with the start pushbutton switch and operates as the 'latch' or hold-in contact maintaining supply to the contactor coil K1. The start switch can be released and K1 remains energised. These latching contacts are commonly referred to as 'auxiliary hold-in contacts'.

When K1 is energised, interlock contact K1.5 opens to prevent the reversing contactor K2 from being energised. If the reverse pushbutton switch is pressed, coil K2 cannot be energised. The same applies to the reverse operation.

To reverse the motor direction, the stop pushbutton switch is pressed and all coils and contacts return to their 'off' or de-energised state. The motor can then be started in the opposite direction.

To stop the motor, the stop pushbutton switch is pressed and all coils and contacts return to their 'off' or de-energised state.

13.7.8 Interlocks

The circuit diagram of a reversing contactor in Figure 13.39 shows the start-stop control for a three-phase motor with two interlock methods, mechanical and electrical interlocking. Both interlocking methods should be incorporated in the design of control and power circuits.

Mechanical interlocks

Short-circuits due to reversing of the supply lines by activation of the forward and reverse contactors at the same time can be prevented in two ways—mechanical and electrical interlocking. The small equilateral triangle and the broken-lines symbol between the two sets of power contacts are shown in Figure 13.40 to signify that the two contactors are mechanically interlocked. This means that if one contactor is closed, it is mechanically impossible for the other to close or make connection.

Figure 13.40 (a) shows the de-energised state for both contactors. In Figure 13.40(b) we can see that contactor K1 is energised and the mechanical interlock prevents the line contacts of K2 from closing. We see in Figure 13.40(c) that if contactor K2 is energised first, the mechanical interlock prevents line contacts for K1 closing.

Using an interlock device normally positioned between the two contactors provides the physical connection. Two pegs or bars from the interlock device are attached to the contactors and are moved when one contactor is actuated. The pegs are attached inside the interlock to two flaps that physically block the other from moving. As the contactor moves, the pegs also move. It is then positioned such that, even if the second contactor's coil was to be energised, the contactor armature would not be able to close.

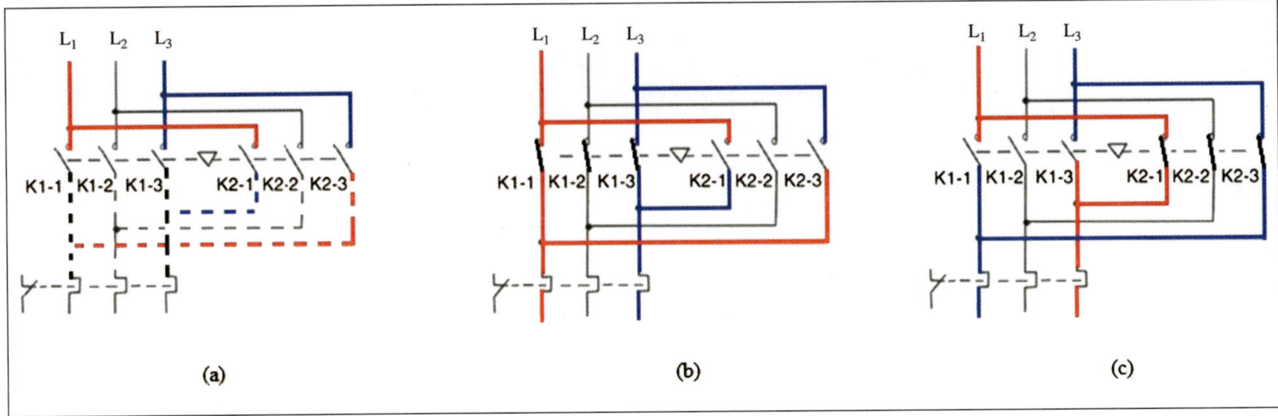

FIGURE 13.40 Mechanical interlocking

The use of a mechanical interlock allows the device to be physically isolated. This is generally the preferred method and is the only method recognised by the Australian Standards as providing complete assurance of interlocking should a contact fail.

Electrical interlocks

The second method is to use electrical interlocks in contactor control circuits. Electrical interlocking can be used in a variety of ways for different purposes, but the most common is the forward-reverse control, where the two contactors are used to operate the motor in either direction.

Figure 13.41 shows the three conditions for an electrically interlocked circuit. In Figure 13.41(a), the circuit path to coils K1 and K2 show a normally closed contact in series with each coil. When neither coil is energised as shown, the associated (NC) contacts remain closed. However, when K1 is energised as seen in Figure 13.41(b), the normally closed contact in series with coil K2 opens, preventing K2 from being energised.

The reverse is also true. If coil K2 is energised first, as seen in Figure 13.41(c), then the normally closed contact K2 opens, preventing coil K1 from being energised.

This can be seen in Figure 13.39, where the normally closed contact K2.5 is in the forward coil circuit and the normally closed contact K1.5 is in the reverse coil circuit. This means that when contactor coil K1 is energised, contact K1.5 will open, preventing contactor coil K2 from being energised.

Electrical interlocks should only be used as a back-up for mechanical interlocks or on systems where the risk of harm is minimal. Without mechanical interlocks, failure of the normally-closed contacts on the energised contactor to open when energised will allow the second contactor to energise and its contacts on closing will cause a short-circuit and *flashover* between phases.

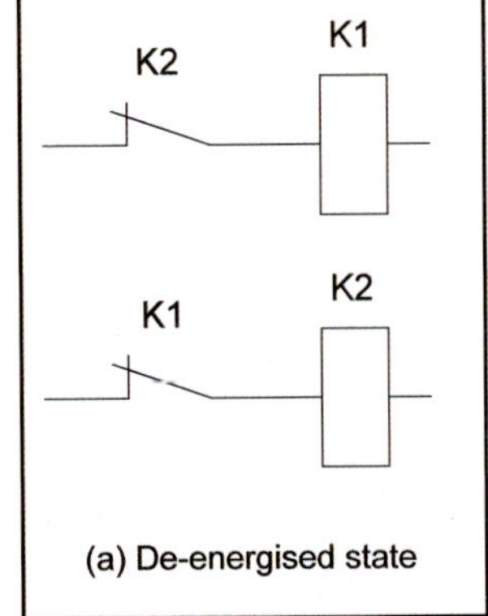

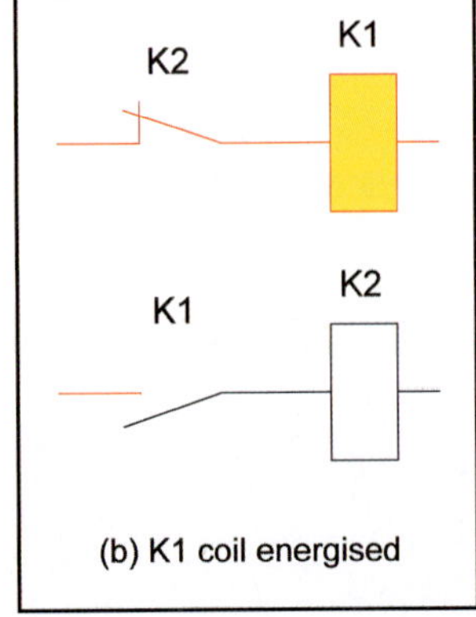

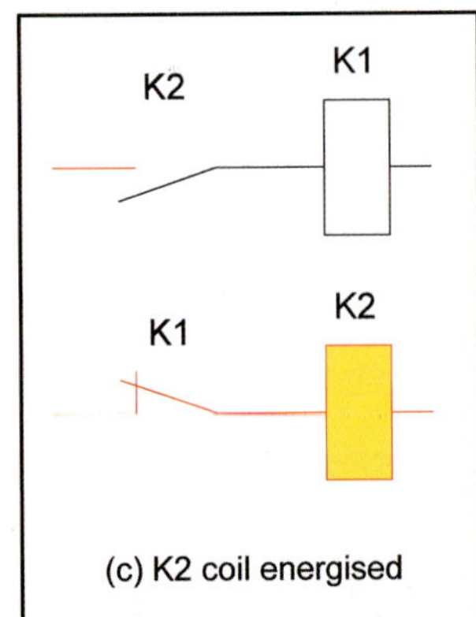

FIGURE 13.41 Electrical interlocking

WORKPLACE SCENARIO: APPLICATION OF REVERSING CONTACTORS

This type of circuit is common in dispatch docks and loading bays, where conveyors are used to move packages, parcels and other products for distribution by trucks and contractor vehicles. Apart from the start-stop operation for belt conveyors, additional safety devices, emergency stop switches and multiple start-stop control stations are installed for dispatch operators and contractors loading their vehicles.

Most dispatch conveyors can extend and retract, and raise and lower for the convenience of contractors with vehicle trays at different heights and loading depths from front to back of the tray. When the conveyor extends into the truck or tray area, it needs to also raise or lower for safe manual handling by operators. The conveyor extend-retract and raise-lower functions have separate motors which operate on a similar principle to the 'forward-reverse' contactor. Position limit switches are also installed for safety and to de-energise the motor contactors. Safety limit switches are necessary for conveyor extend-retract and raise-lower maximum travel limits. These limits act as equipment and operator safeguards.

Factories and warehouses have an upper and lower limit (switch) to cut supply to roller shutter door motors. The roller door can raise and lower by using mechanically interlocked contactors to interchange two supply lines to the motor.

Example applications for these loading bay dispatch conveyor systems include mail processing facilities, parcel centres and newspaper production plants.

NOTE: The switching system where the contacts for one contactor must de-energise and open before the second contactor can energise is called a 'break-before-make' system. Often a timer is also required to ensure a 'momentary state' where both contactors are in the de-energised state.

CHECK YOUR UNDERSTANDING

13.9 What does mechanical interlocking prevent?

13.10 What does electrical interlocking prevent?

13.11 Provide two examples of the use of mechanical and electrical interlocking in motor forward-reverse applications.

The control circuit of Figure 13.39 has been redrawn in a horizontal orientation and is shown in Figure 13.42. Symbols as recommended in AS/NZS 1102 are used. The circuit, however, is identical to that of the control circuit in Figure 13.39.

In other countries, different standard symbols are often used. In the USA, more than one set of standards is used, but a popular one is the NEMA (National Electrical Manufacturers Association) standard. Figure 13.43 shows the control circuit of Figure 13.42 redrawn to NEMA standards. The circuits are identical but in the NEMA version each overload is shown separately as three normally closed contacts.

Many of the programmable logic controllers both in Australia and overseas use this standard, and the programmers are oriented towards it. Circuit diagrams supplied with the controllers are, more often than not, drawn to NEMA standards.

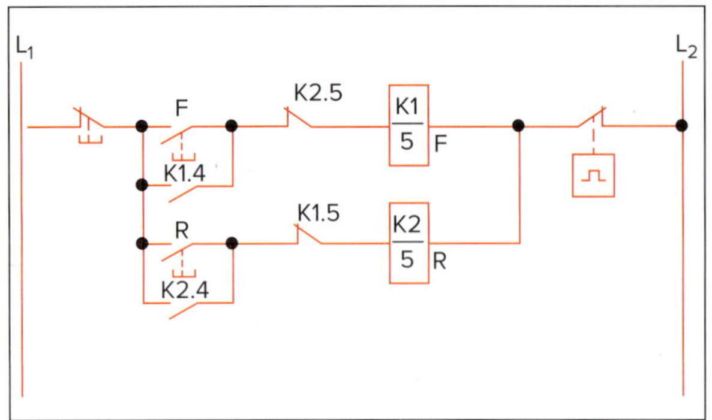

FIGURE 13.42 Control circuit of Figure 13.39 drawn with a horizontal layout

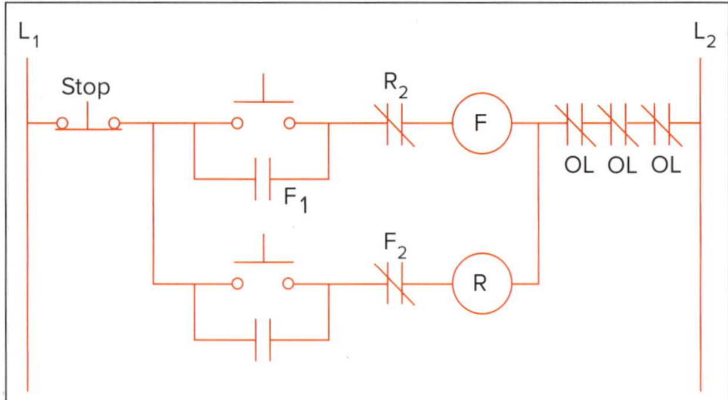

FIGURE 13.43 Reversing control circuit drawn to NEMA standards

13.8 Developing a control schematic from a description or list of conditions

There are a few steps in developing schematic control circuits from a description. The description is itself a list of necessary operational conditions for the equipment to function correctly and safely. The steps are:

1. Identify and list the circuit components.
2. Establish circuit relationships.
3. Lay out the circuit components, observing proper drawing conventions.

13.9 Control devices

Control devices allow for control of some function on the circuit. Although they are often mentioned alongside circuit protection, it is important to understand that control and protection are two distinct aspects of circuit design and circuit components. Confusion often arises on this point as the actions of the two aspects seem similar—both provide some form of manual and automatic disconnection. Protection devices respond to overcurrent, overloading, overvoltage or earth leakage and automatically disconnect via blowing a fuse or tripping a circuit-breaker or residual current device. Control devices respond to nearly any other circumstance or input. Although switches or sensors are the most common types of control devices, variable resistors and *resistor banks* can also be used as control devices.

13.9.1 Types of control

The two general types of control are control that limits or affects the:

1. voltage
2. current.

Basic electrical theory states the relationship between voltage, current and resistance, and how either the voltage or current can be varied by varying the resistance. Some types of resistor devices can be manually adjusted or switched into circuits using timing and contactor control. Two common types of variable-resistor control devices are the potentiometer and the rheostat.

This use of timing and contactor control is applied as a method for both reduced-voltage starters and speed control. By increasing the resistance, the voltage drop across the resistor is increased, reducing the amount of voltage available for the remainder of the circuit. The concept also allows voltage control for motor starting by the use of a variable resistor or resistor load banks. Types of three-phase induction motor reduced-line voltage starters include:

- star–delta starters
- primary resistance starters
- autotransformer starters.

Secondary resistance starters with a wound rotor use resistors or resistor load banks as a method for speed regulation: the reduction in voltage causes a reduction in speed.

Current control is the more common type of control, and is often simpler to understand. Since there is only one current path in a series circuit, it should be clear that open-circuiting the series path means that current will no longer flow and the circuit will stop operating. This is usually done via switches or the use of contacts and relays.

13.9.2 Types of control devices

Control devices can be manually operated switches such as pushbuttons or micro or limit switches that are operated by product position or machine sequence. There are also non-contact switches and sensors that operate automatically. Common types of control devices are listed in Table 13.3.

TABLE 13.3 Common control devices

Manually operated (contact) switches	Mechanically operated switches	Relays	Sensors (contact and non-contact)
▶ Pushbutton switches (latching and momentary) ▶ Toggle switches ▶ Selector switches ▶ Thumbwheel switches ▶ DIP switches	▶ Limit switches ▶ Proximity switches ▶ Pressure switches ▶ Temperature control ▶ Float and flow switches	▶ General-purpose relays ▶ Solid state ▶ Controllers (timer, counter and temperature)	▶ Photoelectric (fibre optic and laser) ▶ Capacitive ▶ Inductive ▶ Temperature ▶ Ultrasonic ▶ Pressure/flow ▶ Code readers/ OCR

Control devices include:

- Switches that respond to an input by changing state.
- Bimetallic switches used in thermostats or overcurrent devices that respond to heat, causing the metal to bend.
- Contactors and relays responding to current flowing in their coil, causing them to operate.
- Float switches like those used in bore water pumps to stop the pump from pumping dry, responding to water level by movement of a ball operating contacts or a magnetic reed switch.
- Light sensors used for street lighting that operate by detecting light falling on either phototransistors or on photoresistors. The amount of light falling on the sensor (or light level) during daytime will vary the amount of resistance, limiting current.
- Limit switches and proximity switches that operate based on the position of objects. Limit switches are operated by an object moving to a position where it stops, for example by an automatic door opening. The limit switch is actuated by the door reaching its limit of travel.
- Proximity switches that operate by producing either electromagnetic or infrared radiation as either a beam or electromagnetic field and detecting the return signal. Proximity switches can be seen in use detecting items on supermarket conveyer belts.

The problem with manual control switches is that their contacts inevitably wear out and cause failure. Control devices that use resistors while not requiring manual input are more susceptible to over-power situations, which cause them to burn out. Limit and proximity switches require correct placement and alignment; if they are misaligned, they may detect the wrong object or not detect an object at all. The advantage that manual control devices have over automatic devices is that they are robust and comparatively cheap (depending on the application). Some devices do not need manual input as they receive an automatic input (such as light) rather than human intervention.

Switches operate in two ways: by interrupting the flow of current when opening the circuit or by allowing current flow by closing the circuit. As simple and limited as these may seem, it is possible to create complex responses to inputs. An example of this is the float switch that will not allow a pump to turn on if the water level is too low, even if it is set via a timer to do so. This is an example of what is called a 'logic circuit'. Logic circuits are the foundation for automation and are commonly used in large plants and factories.

The selection of control devices for a circuit depends on the following factors:

- the power the device is expected to handle
- the type and amount of voltage seen across the device
- the type and amount of current flowing in the device, and how long it will do so
- the type and number of circuits to be controlled
- how the circuit is to be affected (meaning whether it is to be opened or closed).

These factors are often called 'duty ratings' and must be considered when selecting all manner of components—from pushbuttons and pilot lamps to relays and timers. The selection of devices and components may have been specified in a bill of materials or specification from the client or engineer. Equally, though, the installer may be required to source components themselves from the manufacturer's catalogue or website.

When installing the control components into a circuit, the general rule is to place them at the start but after the circuit protection. The multiple devices that perform control allow for parallel configurations, but care should be taken when using limit and proximity switches: while they are control devices, they perform functions not dissimilar to protection devices and therefore should be installed in series with emergency stops and other control devices.

13.9.3 Sensors, actuators and transducers

While sensors generally provide an input to a control circuit, the control outcome is an action or response. These types of devices that result in some form of action or movement are referred to as 'actuators'. A third classification of devices is the **transducer**. Depending on their operation, transducers can be classified as either sensors or actuators.

Sensors

Control sensors are, in a way, very similar to our own senses of sight, hearing, taste, smell and touch. Human biological sensors detect sensory signals such as light and sound, and these are converted into electrical signals as messages to the brain, which then outputs a response.

Sensors perform an 'input' function to the control process. They usually detect either a physical change or some type of energy such as heat, light, chemical or motion and then convert these signals into either an analogue or digital representation of the input signal. Sensors are commonly used to detect levels, presence, proximity, temperature and pressure. They can be categorised as follows.

Temperature sensors

- Thermocouple—a thermoelectromotive force resulting from two different types of metal joined together at one end. When the joined end is heated, a potential difference occurs between the wires at the open end. This is known as the 'Seebeck effect'.
- Thermistor—produces a decrease in resistance when heated.
- Resistance temperature detector—when they are heated, material resistance increases.
- Semiconductor temperature sensor—silicon (positive temperature coefficient) and germanium (negative temperature coefficient) can both be used for temperature sensing.

Pressure/flow sensors

These devices are equipped with a pressure-sensitive element. This measures the pressure of a gas or liquid against a diaphragm which converts the measured value into an electrical output signal. Diaphragms are typically stainless steel or silicon. Different sensors are used for liquids, gases, flammable substances and corrosive substances.

Pressure sensors

- Strain gauge transducer—measure strain or tension. They are accurate and reliable.
- Piezo-resistive transducer—used to sense ranges from 10 to 5000 psi (the engineering unit relating to pressure). They are used in lubrication, pneumatic and hydraulic systems.

Code readers/optical character recognition (OCR)

These devices read barcodes. A barcode is a display of information in the form of bars (black portions) and spaces (areas between the bars) of varying widths. Barcodes are used in a variety of different industries and applications, notably in distribution and logistics. Outwards remittance (billing forms and statements) uses code-reader technology for scanning, collating and mail-out to customers.

Photoelectric sensors

Two common types of photoelectric sensors are those that operate by line of sight (also referred to as 'through beam' sensors) and those that use a reflector. Both systems detect presence. They comprise a transmitted light source known as the 'emitter' and a receiver to detect the light. Both systems detect objects when a light beam is broken. The line of sight type has an emitter and a receiver lined up directly opposite each other. The reflector type has an emitter and receiver housed together in the one unit. Light from the emitter is directed towards a reflector, which reflects it back to the receiver.

Reflectors can be in various sizes, and can be round or rectangular or can be reflective tape. These systems can be affected by dust, dirt, smoke, moisture, airborne contaminants, direct and reflected sunlight. They require regular emitter, receiver and reflector wiping and cleaning to operate effectively.

Fibre optic and laser sensors

The use of optical fibre is another method of transmitting light. It can be ideal for detecting small objects or focusing on very small sensing areas at close range. Lasers are also sometimes used as high-intensity visible light sources for detection of extremely small objects at a distance. Industrial, commercial and control applications for photoelectric, fibre optic and laser sensors include:

- people, object and product detection
- product counting—bottles, cans, cartons, boxes and packages.

Measurement and displacement sensors

These are non-contact devices for presence detection, proximity, displacement and measurement. They function on optical, inductive, capacitive and ultrasonic technology and principles. Displacement sensors are devices that measure distance between the sensor and an object and any changes (displacement) in position or some form of movement by part of the object. A measurement sensor measures the position and dimensions and the height, width and thickness of an object.

Inductive proximity sensors

Inductive sensors are non-contact proximity sensors that produce an electromagnetic field to detect ferrous targets. They consist of four major components: a ferrite core with coils, an oscillator, a Schmitt trigger and an output amplifier. When a ferrous target enters this magnetic field, eddy currents are induced on the metal's surface, changing the reluctance of the magnetic circuit.

Eddy current sensors are different from proximity and displacement sensors in that they use an air-core coil instead of a ferromagnetic core. Due to the air core, the sensor measurement distance needs to be closer to the product, so there is a narrower air gap.

Capacitive sensors

Capacitive proximity sensors produce an electrostatic field. They are passive transducers that require an external force for operation, converting changes in capacitance into an electrical signal. They can detect through some containers, sense metals and non-metallic materials such as paper, glass, cloth and liquids.

When an object nears the sensing surface, the electrostatic field of the electrodes changes the capacitance in the oscillator circuit. Changes in capacitance will result in changes in the oscillator's amplitude. Product sensing and detection causes an increase in amplitude, triggering an output signal.

Ultrasonic sensors

Ultrasonic proximity sensors are transducer devices that are capable of presence detection, displacement and product dimension measurement of all materials. They can transmit and receive high-frequency sound signals as sound waves to detect a target or object. When these sound waves reflect off an object, an echo is created. This echo is the reflected sound wave bouncing from the detected object back to the sensor. The time taken for this echo is directly proportional to the distance between the object and sensor. This feature makes ultrasonic sensors useful for:

- people, object and product detection
- level measurement in small containers
- height and stack height sensing
- contour recognition
- product counting.

Actuators

An actuator is a specific type of transducer. Actuators perform an 'output' function, usually to carry out a physical process or task, or to control an external device such as a motor. An actuator converts energy into motion.

A bimetallic strip is an example of a thermal actuator device, directly converting thermal energy into motion as thermal expansion. When heat is applied to two dissimilar strips of metal joined along their entire lengths, the bimetallic strip bends in the direction of the metal with the smaller coefficient of thermal expansion.

Relays, contactors and solenoids are examples of actuators. Although these are considered as 'outputs', relays and their contacts form part of the internal logic control for programmable relays and programmable logic controllers. In those applications only the contactors, solenoids and indicating lamps are considered as outputs. An electric motor acts as both a transducer and an actuator. Motors convert electrical energy to magnetic energy and then to mechanical energy or motion.

Transducers

By definition, a transducer is a device where any variation in energy magnitude of any form is able to reproduce that variation in another measurable form—generally these days as electrical voltage, even though it may only be in millivolts.

Rotary encoders are sensors that are sometimes referred to as 'shaft encoders'. They are electromechanical transducers which can provide an output signal or pulse, due to shaft rotation, shaft position and speed. The output pulse from multiple encoders can be used for speed synchronisation and control of multiple conveyors as part of an industrial production or processing environment.

A thermocouple is a type of transducer where two dissimilar metals produce a voltage on being heated at their junction. For example, a copper/constantan thermocouple with the cold end kept at a constant temperature produces 4.3 mV at 100°C and 14.8 mV at 300°C. A graph of these values is approximately linear over a restricted range, and can be used to indicate the temperature on a voltmeter that has been suitably calibrated to read temperature.

Alternatively, the thermocouple can be connected to a dedicated controller, programmable relay or an input for a PLC controlling a process.

Measurement and displacement sensors are examples of transducers. Table 13.4 gives further examples of transducers and their application.

TABLE 13.4 Examples of transducers and their applications

Transducer type	Application(s)
Electroacoustic transducer	Loudspeaker—converts electrical signals into sound Microphone—converts sound waves into analogue electrical signals
Electromagnetic transducer	Generator—converts motion in a magnetic field into electrical energy
Electromechanical transducer	Strain gauge—converts the deformation (strain) of an object into electrical resistance Galvanometer—converts the electric current of a coil in a magnetic field into movement Generator—converts mechanical energy (motion) into electrical energy Motor—converts electrical energy into mechanical energy
Electrochemical transducer	Battery—converts chemical energy directly into electrical energy
Thermoelectric transducer	Thermocouple—converts heat energy into electrical energy Temperature sensitive resistor (a thermistor)—changes heat energy to electrical energy

13.10 Programmable relays

Relays and contactors were the origins of automatic control. The basic relay provided adequate control for automatic control and industrial processing for many years. But as industry demand, need for improvement and greater automation grew, the control limits of the relay became apparent.

Mass-production industries and original equipment manufacturers (OEMs) required improvements that the relay could not provide in terms of high speeds, reliability of service and minimum maintenance.

The programmable relay was originally introduced as a relay replacer bringing multiple timers, relays and counters into one unit. They did little more than manage the on/off sequencing of motors and solenoids, providing basic functions that could previously only be implemented with individually installed and wired devices.

The rise of home automation is in part due to a programmable relay's ability to control the domestic environment.

13.10.1 Programmable relays—basic

Dedicated controllers with more features than basic relays and timers evolved through automation needs, innovation, the rise of cheap electronics and technological developments. Many industrial automated applications still rely on dedicated controllers for specific temperature, timing or counting control tasks.

Solid-state circuitry was introduced initially because of its ability to operate as an on/off device and to repeat the operation millions of times—more than a relay could and with minimal maintenance. The control function being in either an 'On' or an 'Off' state depended on an input signal from common control devices such as those listed in Table 13.3.

Sensors and switches provide input signals to the solid-state device or dedicated controller and this in turn operates the required device or an actuator. For example, a thermostat can be manufactured to close its contacts at a specific temperature. Closure of the contacts acts on an electronic component, which can then start an air-conditioning system. The thermostat contacts open when the temperature drops and the air-conditioner stops. The system has only two states—on or off.

In this example, the thermostat replaced the start pushbutton of the relay circuit and a solid-state device such as a TRIAC replaced the contacts of the relay (TRIACs are control devices for switching circuits). For industrial process control, this on/off method does not provide the accuracy needed for many products. Devices such as the thermostat mentioned above cannot operate at the one temperature for both on and off states. This introduces a differential—if the contacts open at 22°C, it is extremely unlikely that they will also close at this temperature. Depending on the circuit, the closing temperature might be 18°C, giving a differential of 4°C. A more advanced programmable relay or programmable logic controller (PLC) is required to meet the on-off and differential temperature control conditions required.

13.10.2 Programmable relays—advanced

The programmable relay is a relay that can have multiple sensors (such as photovoltaic, temperature and flow-level sensors) or control inputs (such as start-stop pushbuttons and timers) and be able to perform decisions based on the logic that has been programmed into it via its on-board software and human-machine interface (HMI).

Programmable relays are more suited to small, stand-alone, low-complexity applications. Many varieties also provide an LCD and operating buttons to allow setting, monitoring of inputs, outputs and ladder programming. Many industrial automated applications still rely on dedicated controllers for specific temperature, timing or counting control tasks. Table 13.5 lists a number of programmable relay applications for building automation systems and control in the commercial and building sectors.

'Programmable relay' is a broad term that now encompasses electronic control devices that range from dedicated controllers to those with full programmable logic controller (PLC) functionality (see section 13.10.3). Some brands of programmable relays are promoted in the market as mini and micro PLCs; modern PLCs can be regarded as little computers made up of solid-state circuitry.

TABLE 13.5 Programmable relays for building automation systems and control

OEM building automation applications	
Door/gate control	Entertainment/theatre controls
Security controls	Compressor control
Automated window blind control	Conveyor control
Roller shutter control	Pump control
Access control	Awning control
Weather condition controls	Air quality monitoring
Lighting control and automation • daylight dependent • timer dependent • energy management	Heating, ventilation and air-conditioning (HVAC) • temperature control • time/temperature dependent • energy management

Advantages of programmable relays over relay controls

- Separate control devices are no longer needed. The built-in features on a single programmable relay can take the place of many control devices (timers, relays and counters), combining sensing, 2-wire and start-stop controls.
- Reliable circuit control provided electronically with no moving contacts, reducing wear, tear and the frequency for relay repair or replacement.
- Industrial process functions and applications become more efficient, and easier to manage and troubleshoot.
- Installation of a programmable relay control system reduces labour costs and time required for wiring, system commissioning and installation testing.
- Integrated display shows alarms and input-output status for better troubleshooting. This often eliminates or minimises the need for a technician to use a multimeter or logic probe for trouble shooting and fault finding.

13.10.3 Programmable logic controllers (PLCs)

The modern programmable logic controller (PLC) has a cost that is relatively low compared with the installation, wiring and hardware costs of relay logic systems. PLC programs can be reprogrammed easily to monitor something completely new when manufacturing or production requirements change. The micro-processor block shown in Figure 13.44 symbolically contains the central processing unit (CPU), random access memory (RAM), program logic and an internal memory storage device. A stored logic control program is accessed in the working memory by the central processing unit. The PLC reads the program and examines the input device status as part of its scan cycle operation. The CPU uses the program logic to determine when certain events should take place and the sequence in which they should occur. These events will be internal processes such as timer, count and internal relay functions, as well as external events that involve control of outputs. The CPU makes these decisions based on information it receives from sensors connected to its input terminals combined with the user-programmed instructions.

The PLC, input and output modules all require a power source or supply. Line voltage is transformed to extra-low voltage (24/12/5 V), usually 5 V for the CPU and memory functions of the microprocessor. In Figure 13.44, block modules (4 and 5) represent the communication and programming interface options for the family of PLCs that are currently on the market. Smaller PLCs include a built-in human-machine user interface (HMI) for checking and programming. Larger industrial PLCs may require a personal computer (PC) with capability for data access (keyboard) and a display monitor (or a laptop) to enter and modify the program logic.

Some human–machine interface units are detachable and may include storage, upload and/or download of program logic via external storage and back-up options (USB or through ethernet). Older models, traditional and legacy equipment may still use a serial cable as an option. Wireless and Bluetooth technologies are common.

It is worth noting the different use of name labels and terminology within industry. Computer technicians and IT workers refer to keyboards and mice as both 'peripherals' and 'human interface devices' (HIDs). Hardware devices used for programmable relays and a programmable logic controllers are referred to as 'human-machine interfaces' (HMIs).

Programmable relays and PLCs are becoming so common that there are now very few situations where they are not used. As PLCs are also replacement relays, they can be installed wherever a

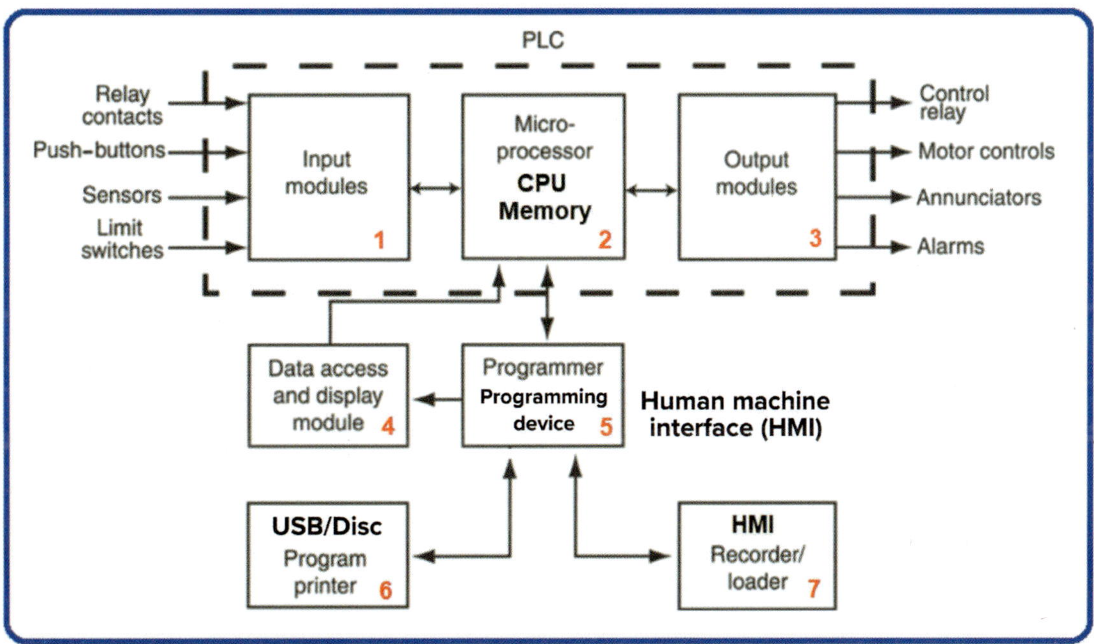

FIGURE 13.44 Block diagram for a programmable logic controller

relay would have been installed, for example instead of a timer and a relay. Instead of a stop/start control circuit with relays, a PLC can handle the inputs and control the motor. Other sensors and inputs include:

- Mechanical devices that can be used as electrical limit switches to indicate to a PLC that a machine has reached the limit of its position.
- Pressure switches that can be fitted with a pair of contacts to indicate an on/off relationship, or, if fitted with certain types of resistive material, can provide an infinite number of pressure readings to a PLC.
- Positive and negative temperature coefficient resistors (thermistors) can also be used as temperature indicators, provided they are suitably calibrated.

Table 13.6 lists the respective features of programmable relays and PLCs.

TABLE 13.6 Comparison of programmable relays and programmable logic controllers

Programmable relays	Programmable logic controllers (PLCs)
Often include a built-in LCD or human–machine interface (HMI) for programming.	Require a PC to program or an additional-cost human–machine interface (HMI).
Include function keys and buttons to navigate, enter and edit the program, as well as to start and stop the configuration.	Need software to accomplish the same tasks as programmable relays.
More suited to stand-alone, low-complexity applications, for example home automation. (See Table 13.5 for more examples.)	More suited to managing, monitoring and controlling complex industrial automation systems.
(Usually) non-expandable system; set number of input and output connections.	Expandable system; input and output modules can be added as needed.
Lower-cost investment.	Higher-cost investment.

13.10.4 Programming

A group of instructions that enables a microprocessor to perform a specific task is called a 'program'. A program is a step-by-step procedure to solve a problem, initiate certain actions or manipulate data that it stores within its memory.

Prior to the adoption of graphical user interfaces (GUIs), the use of PLCs required a technician skilled in the use of one of the textual or graphical PLC programming languages shown in Table 13.7.

The most widely used of these languages are ladder diagrams. The concept of PLC ladder logic is similar to that of the relay ladder logic and gate logic used for the integrated circuits of solid-state electronics, and is underpinned by Boolean algebra.

TABLE 13.7 The most popular PLC programming languages

Programming language type	
Textual language	Graphical language
▶ Instructions list (IL) ▶ Structured text (ST)	▶ Ladder diagrams (LD) ▶ Function block diagram (FBD) ▶ Sequential function chart (SFC)

The arrival of the GUI has, in part, removed the need for technicians, electricians and other industry workers, and the specialist programming skill sets needed in these languages. Programming can be done without being able to code in PLC programming languages as some manufacturers now provide products using a drag-and-drop graphical user interface.

Each brand of PLC has its own proprietary programming software, and while it is designed to be easy to use, it is recommended that appropriate training is undertaken. In general terms, the user selects what circuit components they need, then on the interface page drags down function blocks and connects the inputs that trigger the functions. The user specifies what connecting output is to be controlled and the software compiles the machine code ladder logic. The completed program can be loaded onto the PLC.

Programming the PLC requires some knowledge of what it needs to operate. For the most part, this is in relation to the input and output modules of the PLC. The inputs to a PLC are from what is known as a 'peripheral' device (meaning in this case from outside the PLC). These inputs can either be digital (i.e. a switch with binary states of ON or OFF, where ON may be the presence of a voltage and OFF the absence) or analogue (e.g. a thermostat). When it receives the inputs, the PLC processor and program writes the information into a list or table to be processed.

Programming rules

The program scan sequence or flow through the program by the PLC follows the ladder diagram, examining the 'inputs' on each line and adjusting the 'outputs' as necessary. 'Scan time' can be defined as the time it takes for the PLC to complete one scan cycle.

The main programming rules are as follows:

- The scan is from left to right, rung by rung, top to bottom until the end of the program is reached.
- Electricity can flow in any direction in a circuit but not program scan cycles.
- When the end of the program is reached, the computer restarts at the top and cycles through again.
- Each line starting from the left bus must end in a coil, timer or counter.
- Contacts are not to be placed to the right of (after) a coil, timer or counter.
- Coil outputs can only be used once in the program but contacts with the same coil can be used as many times as necessary.
- All program components are individually addressed.
- Contacts are programmed 'Normally Open', including normally closed input devices.
- PLC programs should have an 'END STATEMENT' or program command 'END'.

13.10.5 Drawing PLC ladder diagrams and symbols

Figure 13.45 shows the start-stop circuit for a direct online (DOL) starter drawn in relay ladder logic and the equivalent PLC ladder logic.

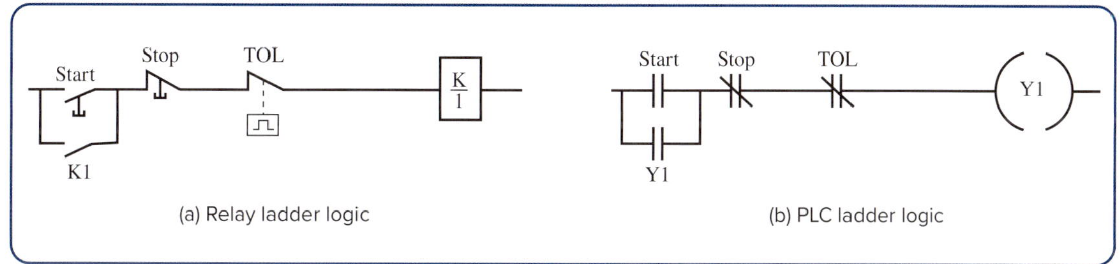

FIGURE 13.45 DOL starter circuit

Ladder contacts

PLC programs have slightly different ways of displaying normally open and normally closed contacts in their de-energised state. Figure 13.46 shows versions of the de-energised contacts and output coil on the top line. The energised condition can be the symbol itself shown in a different colour or the symbol highlighted to show the change.

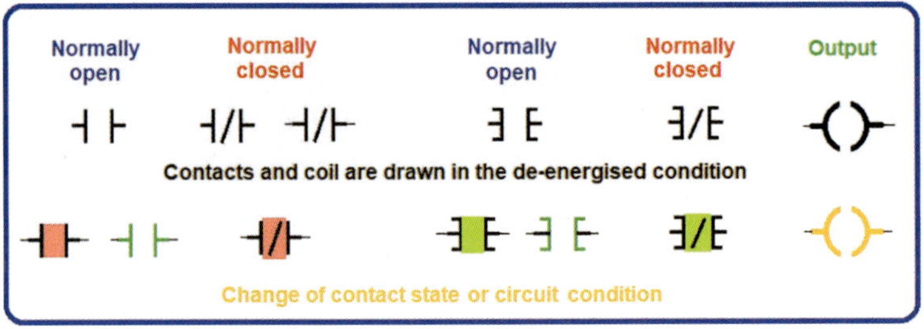

FIGURE 13.46 PLC contact symbols

Figure 13.47 shows a block diagram for a programmable logic controller. The inputs to the PLC receive information from external sources while the PLC outputs send information to the controlled process, which could be the starting of an electric motor or some other appropriate task.

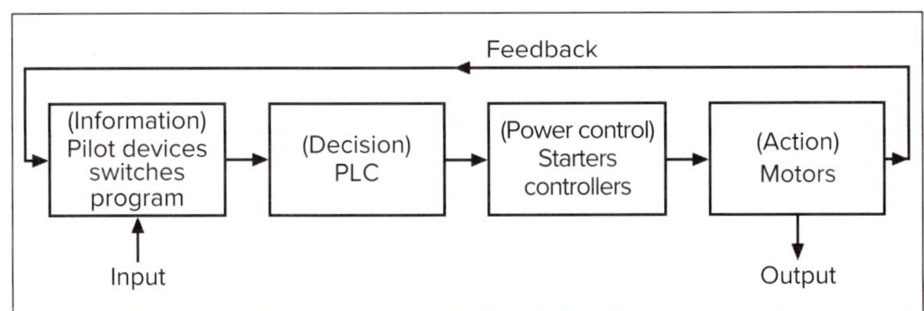

FIGURE 13.47 Block diagram for an automatic starter

By introducing a PLC into the starting process for the motor controller shown in Figure 13.47, there is no longer any need for an attendant to be on hand to start or stop the motor.

At the heart of the PLC is a decision-making unit called a 'microprocessor'. The basis of this is shown in the block diagram of Figure 13.48. The microprocessor is part of the PLC.

Technically, there is a difference between a microcomputer and a microprocessor. A microcomputer contains a microprocessor and various other circuits to store information, provisions to connect it to the process being automated and a clock. The clock provides specific electric pulses at specific times to control the internal processes of the unit. This is known as the 'scan cycle'. The PLC processes this information and sends signals to the outputs. The outputs are either actuators or contactors connected to the various output modules using relay, transistor SCR or TRIAC driver circuits (see Figure 13.49).

Photo-transistor technology can be used for switching control for d.c. and a.c. input and output modules, while photo-triac drivers are for a.c. only. These modules in most cases are optically coupled for electrical isolation, separating and protecting the sensitive internal PLC microprocessor electronics from external load conditions and faults. Figure 13.50 illustrates the concept of optical isolation.

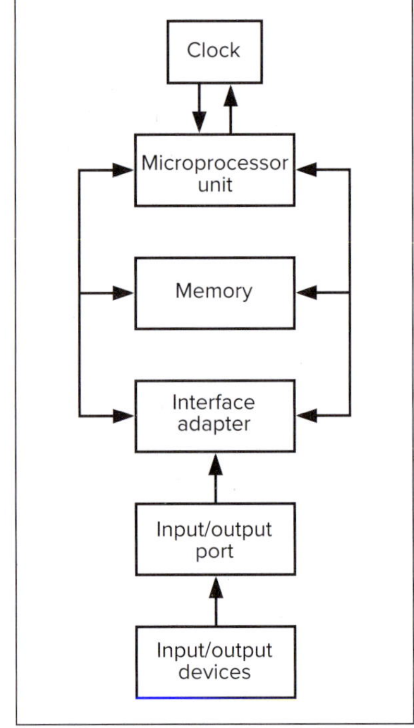

FIGURE 13.48 Block diagram of a microcomputer

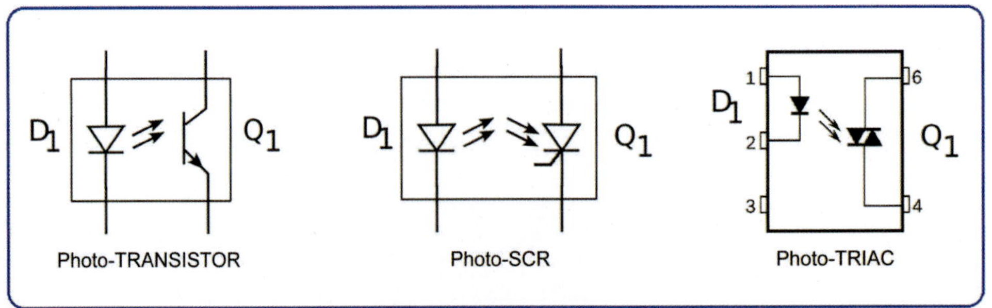

FIGURE 13.49 PLC input-output module driver circuits

Program scans perform checks on changes to inputs, counts, timers and change of state for internal contacts. Closed contacts are examined if open ('XIO') and open contacts are examined if closed ('XIC'). Relay logic requires all stop pushbuttons and emergency stops to be normally closed when not actuated and contact open when the stop is operated to de-energise the electrical circuit. This is not always the case for stops and e-stops (emergency stops) in a PLC installation. The examine-if-open instruction for a stop button is not considered good practice.

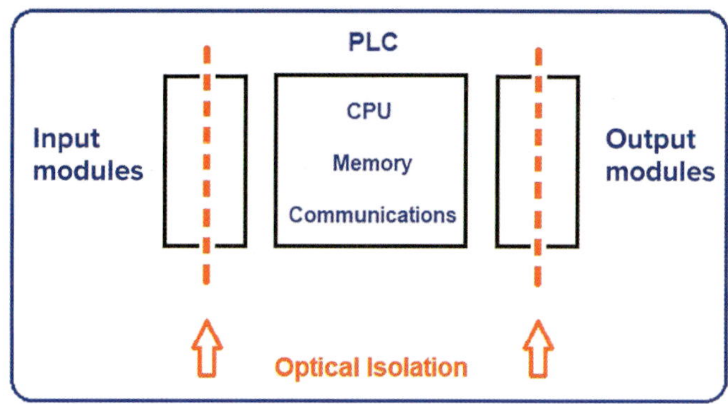

FIGURE 13.50 PLC module optical isolation

Figure 13.51 shows the result of relay logic directly converted to PLC ladder logic. While both versions are drawn in the de-energised state, it is important to note that microprocessor controls and PLC programs are often programmed as normally open (NO) contacts for stop pushbuttons as part of a fail-safe design approach. This is to ensure that when the program executes an examine-if-closed (XIC) instruction during the scan cycle, the PLC will detect that the stop pushbutton has been operated (the instruction is true). An advantage of this for PLC program design is that the system can also stop a machine when a wire to the button breaks or open circuits.

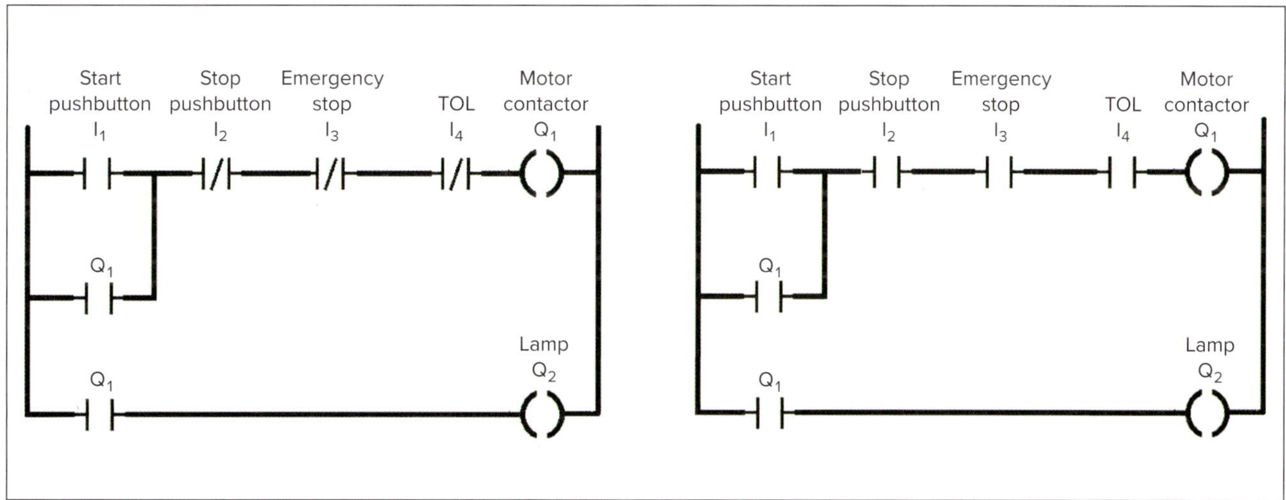

FIGURE 13.51 Relay logic schematic converted to PLC ladder diagram

The various sections of a typical modular PLC can be seen in Figure 13.52. From the left end are the power supply, the processor, then various input and output modules.

FIGURE 13.52 A typical PLC
Reproduced with permission of NHP Electrical Engineering Products Pty Ltd

13.10.6 Connecting, checking and troubleshooting

Most PLC and programmable relays are DIN mounted and installed in the same way as a circuit-breaker or RCD. Connecting peripheral devices like proximity switches or thermostats to the PLC is made easier by using cable and terminal numbers corresponding to a diagram and an input-output (IO) listing. Common methods of identifying the inputs and outputs include using straight numbering (1, 2, 3 etc.) or letters and numbers (I_1, I_2, O_1, O_2 etc., where the use of I indicates input and O output). When developing PLC control circuits, the use of an IO or input-output listing is commonplace. Figure 13.53 illustrates that as each PLC has a number of inputs, the diagrams showing connections will invariably also show the numbers so they can be connected.

Any control device can be connected as an input; most can also be connected as an output. Switches like proximity, limit or light sensors cannot be used as outputs, for obvious reasons. If the device connected to the PLC is connected as a normally closed device, care should be taken so that the desired operation is obtained when programmed into PLC logic. This method can be problematic as a PLC may interpret the lack of input from a pressed NC switch as being not operated when in fact the wire to the switch is damaged. Rather than opening the circuit as expected, the PLC does nothing. Figure 13.54 shows the external input listing and symbols, and the internal program symbols also used for relay ladder logic.

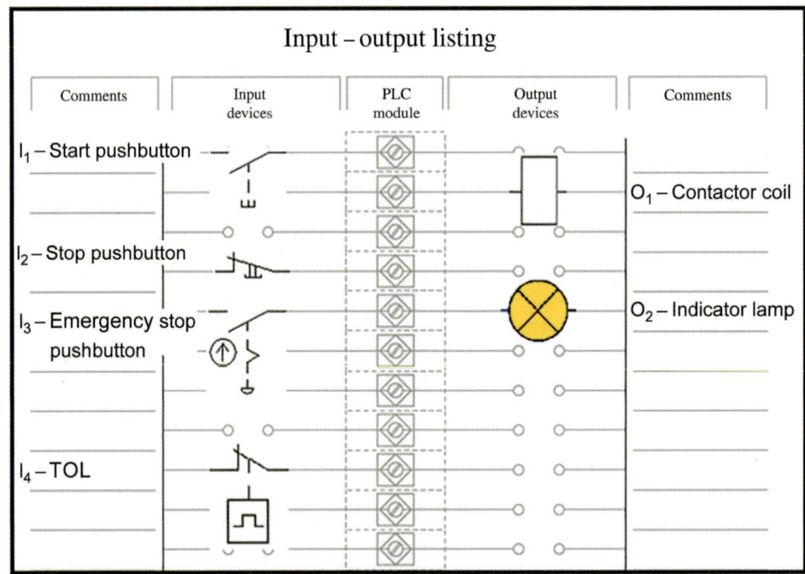

FIGURE 13.53 Input-output device listing for PLC

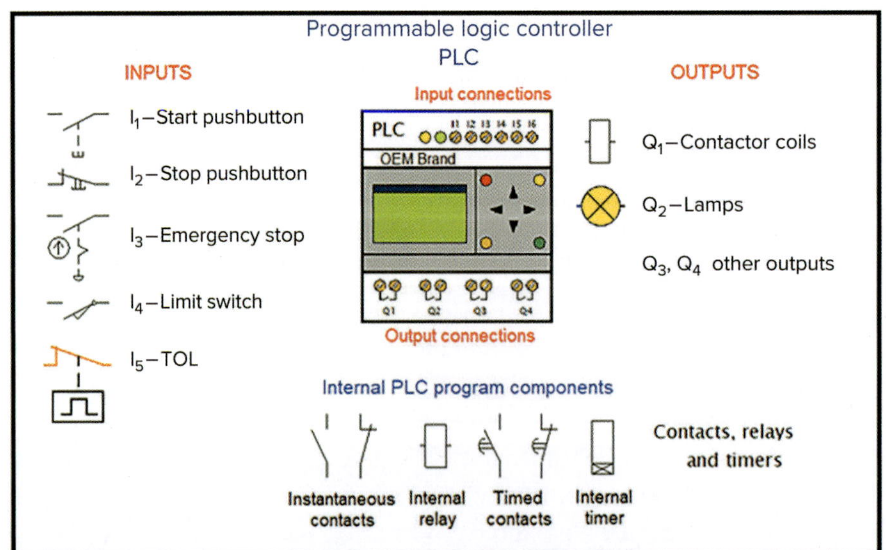

FIGURE 13.54 PLC input-output connections

The software that comes with the PLC will include a checker that will verify that it has been correctly set up. This software will also allow for the simulation of the PLC operation, with the user able to operate the switches and check its operation. This includes the ability to check the set value of timers and time remaining. The error checking of the PLC software is invaluable for fault finding. The most common faults found as part of a PLC installation are the other components failing, i.e. the switches used as inputs or relays used as outputs. The first step towards diagnosing the fault is to operate the switches manually and see if the PLC operates correctly. If the other components are working but the PLC is not controlling them properly by either turning the wrong outputs on or failing to react to an input, download the program from the PLC to a laptop and run the error checking. If no errors are found and the circuit works correctly when simulated, re-upload the software to the PLC. If the PLC program has either been altered or become corrupted by a failing battery or memory chip, re-uploading the software will fix this type of error.

13.10.7 Features, advantages and disadvantages

Features and advantages

Programmable logic controllers offer numerous advantages and almost limitless features, thanks to the software handling the logic. For example, as a PLC has its own clock it is possible to set up multiple timers for multiple events. PLCs can accept almost any sensor input provided to it via various types of transducers. Also, having fewer moving parts makes them more reliable than the older mechanical relays; and having fewer components makes fault finding easier.

Disadvantages

Programmable logic controllers have significant shortcomings: a computer with the proprietary software installed is required to work on the ladder program and make changes; they need a battery to ensure the PLC works; they are costly.

13.11 PLCs and safety

Safety in relation to programmable controllers needs to be looked at from three angles: PLC safety and protection, equipment safety and safety of personnel.

13.11.1 PLC safety and protection

PLCs in an industrial environment require a certain amount of mechanical protection as well as basic electrical protection. The transducer connections radiate out from the controller and are subject to spurious voltages being induced in the connecting cables. This is generally referred to as 'electrical noise' and can be misinterpreted by the microprocessor as a form of input. It can produce erratic equipment operation and incorrect data entry—and can prove difficult to trace as the source of trouble.

In the industrial context, there are usually many machines and other equipment being switched on and off continually. This causes voltage spikes and surges in the main supply. Since a PLC typically operates at extra-low voltage, a transformer is employed to reduce the mains voltage to a suitable value, and consequently these surges in voltages are transferred through the transformer to the processor.

Suppression of this electrical noise becomes an essential requirement for satisfactory operation of a PLC. This might mean applying special runs with shielded wiring to the transducers as well as noise suppression at the transducer.

PLC input-output modules use *photocouplers*, also known as 'opto-isolators' or 'optocouplers'. The main function of an opto-isolator is to block high voltages, voltage surges and transients. These are designed to prevent the system and other sensitive parts from damage and disruption of the PLC operation. Photo-transistor optocouplers can be used for input and output modules, while photo-SCR (silicon-controlled rectifier) and photo-TRIAC types are generally used for output modules.

13.11.2 Controlled equipment safety

A PLC cannot be relied upon and used as the only control for the safety-related parts of the machine. Safety switches, safeguarding and emergency stops are incorporated into the equipment and circuit design. These are only briefly mentioned here to make the reader aware of them, but their purpose and function are referenced in AS/NZS 3000:2018.

Where machinery is controlled by a PLC, precautions have to be taken to ensure that erratic operation by a PLC does not lead to the destruction of that machinery. This usually means that extra equipment has to be installed. For example, a second limit switch might have to be installed as a back-up to one that is actually controlled by the PLC. In the case of heating baths, a second temperature sensor might have to be installed as a back-up to the main one that sends information to the processor and overrides any function of the PLC.

Safety relays can be installed for the monitoring and control of door safety switches, guards and emergency stop switches. They are installed to provide additional control equipment safety by detecting circuit fault conditions including:

- wire breaks
- faulty contactors
- faulty safety actuators
- timing faults.

Figure 13.55 illustrates the concept of safety relays, door safety and emergency stop as an equipment safety system separate to—and not solely relying on—PLC control. You can see that the emergency stop pushbutton has two sets of contacts. The normally open contacts are connected as the PLC Input 3, more for monitoring than stop control. If the PLC hardware fails or the logic program software does not respond to this input, there is also a hard-wired fail-safe redundancy measure. Figure 13.55 shows that when the safety relay operates, all power is removed from the PLC output module and actuators.

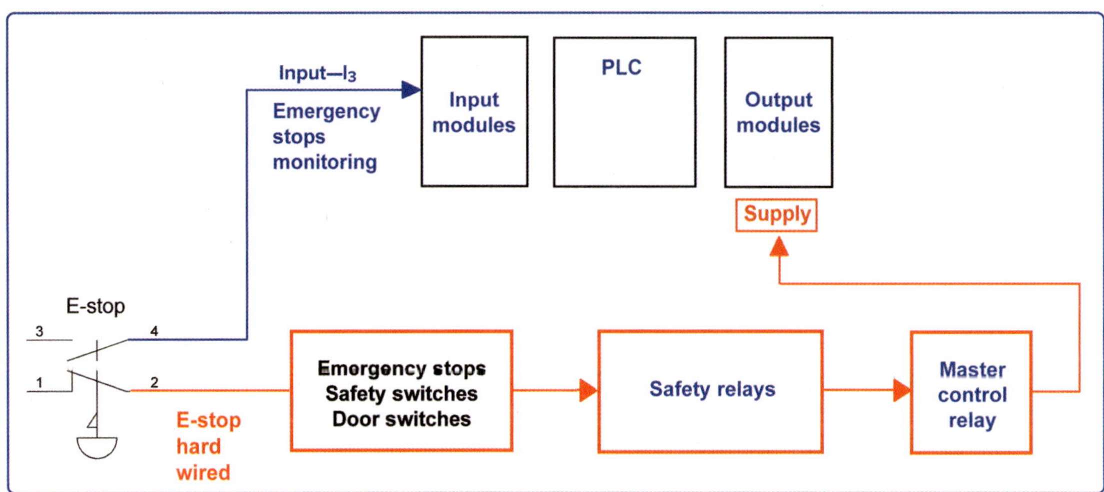

FIGURE 13.55 Block diagram of PLC stop control

The door safety switches, emergency stop loop and other safeguarding measures are series connected with one or more safety relay circuits. These relays are wired to safely and appropriately isolate power, including PLC output module actuators.

Safeguarding prevents a person from coming into contact with hazards, or even aims to eliminate hazards before that becomes a risk. Examples of equipment safeguarding include:

- equipment and machinery guards—fixed, movable and powered guards
- light curtains—presence detection for hands, arms, head and upper body

- safety mats—(pressure-sensitive mats) ensuring safe positioning of workers operating equipment and preventing them from leaning into a machine process or operation
- two-hand control—for example, the safe operation of guillotines for the prevention of hand, arm and body parts against amputation and crush injuries.

Two-hand control is having two separate, hand-operated switches connected in series with a stop control circuit. These are commonly combined with light curtains, guard covers and operator pressure-sensitive mats to ensure no part of the body can be caught, cut, crushed or severed. Equipment design allows for a number of different machine stop situations. These include:

- Category 0—an uncontrolled stop, equivalent to pulling the plug; immediate removal of power to the machine actuators.
- Category 1—a controlled stop, equivalent to a graceful stop then pulling the plug; a controlled stop with power available to the machine actuators to achieve the stop and then removal of power.
- Category 2—a controlled stop with power still available to the machine actuators.

13.11.3 Safety of personnel

Operators or maintenance personnel must be protected against any erratic behaviour of the controlled process or task. Unexpected starting or stopping of equipment due to spurious signals getting into the system wiring can cause accidents. An emergency stop button should be included as part of the installation to override all other input signals.

Properly designed emergency stop systems ensure that operating the emergency stop button while the machine is in mid-cycle should result in equipment and machinery coming to a stop quickly and safely. Operation of the emergency stop will also prevent machine start-up. This would in most situations be equivalent to a Category 0 stop condition.

Emergency stops need to be hard wired to cut power to PLC outputs through safety relay or master control relay circuits as a fail-safe safety measure. E-stops may have additional contacts used as PLC inputs for monitoring, indication and control circuit function, as can be seen in Figure 13.55.

AS/NZS 3000 has made it mandatory for isolating switches to be provided in equipment not directly under the control of a worker. These isolation switches have to be capable of preventing the PLC from bypassing them. The isolation must be a physical barrier and not just the switching off of electronic equipment.

CHECK YOUR UNDERSTANDING

13.12 What are the three key requirements of the operation of a programmable controller?

13.13 Explain the purpose of emergency stop systems.

13.14 Explain what safety relays are.

13.12 Converting diagrams

Section 13.2 lists a number of circuit representations. Converting diagrams also involves converting:

- a schematic diagram from vertical to horizontal orientation, or vice versa
- a wiring diagram to a schematic diagram
- a schematic diagram (relay logic) to a ladder diagram (PLC logic).

The secret to successfully converting from one orientation to another is to observe three things: the direction of power flow, the sequence of operation and correct orientation of circuit symbols.

13.13 Three-phase induction motor starters

13.13.1 Introduction

This section gives a brief overview of the operating principles, applications and starter circuit schematics for the control of three-phase induction motors. The full advantages of an electric motor drive can be realised only when there is compatibility between the following three major components:

- drive motor
- driven machine
- control equipment.

There are many forms of motor-starting methods, and at least as many forms of speed control. Motors, particularly large ones, might have to meet starting current requirements as well as running a machine up to operating speed without imposing severe mechanical shocks on the system.

Three main methods of achieving speed regulation in d.c. motors are flux control, voltage control and armature resistance control. The use of resistors and rheostats in series with windings creates voltage drop across these resistors. This technique of reducing voltage by adding resistance is also used as a method of starting control for three-phase motors.

Control equipment for a.c. and d.c. motors includes:

- solid state reduced-voltage starters
- electronic (soft) starters
- variable speed drives (VSDs) and variable frequency drives (VFDs).

13.13.2 Requirements of motor-control equipment

The prime function of a motor starter is to connect the motor and its coupled machine to the supply without disturbance to other machines and users. It must do this with due consideration to the mechanical inertia of the driven machine, its permitted acceleration and the allowable time taken to get it up to operating speed. It must do this repeatedly and with minimal maintenance problems. Under these conditions, the selection of a motor starter must take into consideration the following factors:

- the limitation of starting current to values acceptable to distribution entities, thus causing minimal disturbance to the line voltages of other local users
- control of starting and accelerating torque from the viewpoint of mechanical shocks to the machine system and the motor driving shaft
- protection of the motor against overloads and overheating
- isolation of the motor in the event of a fault
- provision for interlocking the motor's operation with that of other motors and machines
- motor reversal
- speed control
- motor braking.

13.13.3 Limitation of starting currents

All Australian distribution entities require the starting currents of motors to be contained within certain limits, depending on the power of the motor. The distribution entities in turn have varying requirements, and enquiries might have to be made to them in specific cases.

Since there are so many different circumstances and so many types of installations, it is almost impossible for a distribution entity to lay down firm rules and hope to cover all cases for all installations. It is worth noting that their information booklets often use the phrase: 'Notwithstanding any of the above, the entity's engineer may decide …'

The regulations governing the starting and running currents drawn by electric motors and the maximum demands of an installation are long and involved. There is the further complication that the local rules might vary from one distribution entity to another. The intent of these regulations is to prevent the creation of large transient currents that could cause voltage surges on consumer mains, to the detriment of other users.

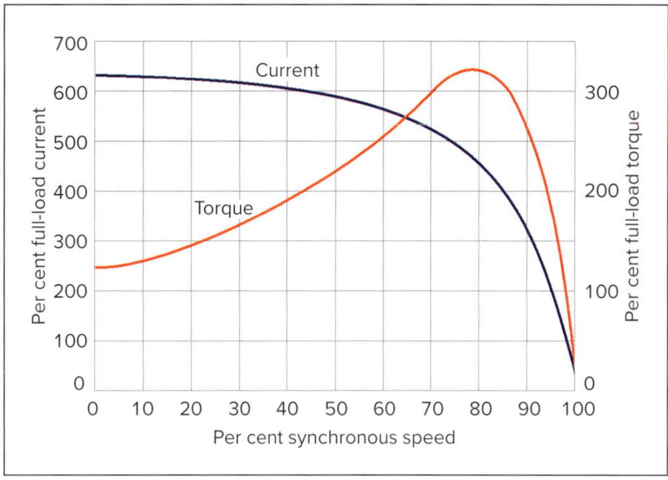

FIGURE 13.56 **Speed vs current and torque for a standard squirrel cage motor**

13.13.4 Speed–torque relationships

When a three-phase squirrel cage motor is connected directly to the supply, it will draw a large starting current. This current, which usually lasts only a second or two, can be up to seven times the motor's full-load current. Figure 13.56 shows the characteristics of the current and torque for a standard squirrel cage motor. It can be seen that the current is almost proportional to the rotor slip; as the motor builds up to speed, the slip decreases and the current decreases.

A principal aim of a starter on an induction motor is to reduce the starting current as much as possible while at the same time providing adequate torque to meet the requirements of the load.

This will be a compromise because, in order to reduce the starting current, it is necessary to reduce the voltage supplied to the motor. However, a reduction in voltage will also result in a reduction in torque, as the motor torque is proportional to the fraction of the voltage squared. That is:

$$T \propto V^2$$

This means that with 100% of the rated voltage, 100% of the rated torque will be available. However, if 90% of the rated voltage is applied to the motor, the torque will be reduced to $(90\%)^2 = 81\%$ of the available torque. If the motor is being started in a no-load condition, this will be adequate. However, if the motor is loaded, there may not be enough torque to overcome the load and the motor will stall.

13.14 Motor-control equipment

Soft-starter systems, variable speed drives and variable frequency drives provide a gentle ramping up of electrical motors to full speed. Motor speed can also be adjusted for full control during the run cycle by varying the output from the VSD or VFD. However, this is achieved differently for a.c. and d.c. motors.

Innovation and technological developments introduce changes in terminology. Some of the biggest changes in recent years have related to what certain technologies used to be called and what they could do in the past and what they can do now. Seemingly familiar but sometimes interchanged terms include 'solid state', 'electronic' and 'soft' starters, 'inverters', 'variable speed drive' and 'variable frequency drive' technologies.

13.14.1 Solid state reduced-voltage starter

Figure 13.57 shows a block diagram for a solid state reduced-voltage starter for a three-phase motor. A typical three-phase soft starter uses six thyristors or silicon-controlled rectifiers (SCRs).

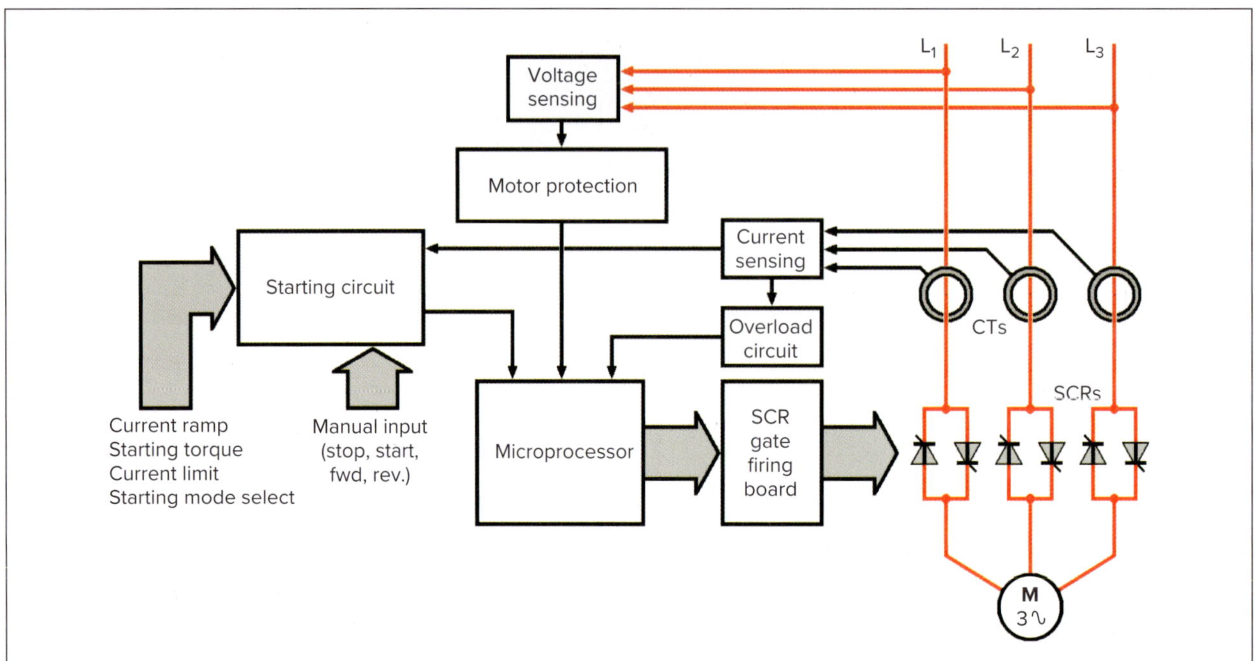

FIGURE 13.57 Block diagram of an electronic reduced-voltage motor-starter

The central processor unit has three inputs to receive information from the starting circuit, the line voltage and the line current. This enables it to start and operate the motor efficiently. It monitors the line current at all times, particularly during the start-up sequence to limit the amount of current that can flow into the motor at that time. This is often set at around 300% of full-load current. If at any time any one phase voltage deviates from set limits or disappears altogether, the motor protection circuit conveys this information to the processor and the processor takes the required action to protect the motor. Similarly, the motor is also protected against overloading and overheating. Note that there has to be an operator on hand to push the necessary buttons for the motor to start and stop.

13.14.2 Soft starter

A soft starter is an electronic or solid-state device which provides electronic control to electro-mechanical starters and regulates the voltage for a.c. electric motors at start-up. The motor starts 'softly' due to an initially lower current, ramping up to full speed without pulling excess current. Soft starters operate on a fixed frequency and have full torque only at full voltage.

These starters are used in applications that require speed and torque control during start-up and stopping. Some soft starters have limited slow speed control between starting and stopping. They provide a controlled acceleration at start-up, ramping to full speed. They can also be used for braking through a controlled ramp-down when a stop sequence is initiated. This principle is applied to printing presses in the newspaper industry as an 'immediate-slow-to-stop' function that is used to prevent product damage.

Conveyor systems used in product movement operations benefit from soft-starter control, accelerating from rest and stopping without jerking. Soft starters are popular upgrades for older reduced-voltage starters because they are more controllable, programmable and efficient. However, care needs to be exercised in control gear retrofit situations that the existing installation cable insulation resistance is adequate for the spike signals generated by the controller switching.

Three-phase induction motors are designed to operate at standard line voltages and frequencies. In Australia, low voltage induction motors are intended to run on a three-phase line-to-line voltage of (usually) 400 V and a frequency of 50 Hz. This results in the flux density in the air gap of the motor being within defined limits.

Any motor starter designed to reduce the line voltage to a motor also has a tendency to reduce the current flowing to the motor, thereby reducing starting torque.

The autotransformer method of starting goes part of the way to solving the problem of maintaining starting torque and current at a reduced voltage, but the method has disadvantages.

It was reasoned that if the motor-starting current could be manipulated to satisfy the above requirements, then the voltage and frequency might also be controlled. This gave rise to the electronically controlled starter, which was developed to include such refinements as control over the following:

- starting currents
- overload currents
- over- and under-voltage monitoring
- motor protection
- frequency
- motor isolation
- sequencing other motors
- dynamic braking
- low-speed operation
- slip compensation (better speed regulation)
- remote computer control.

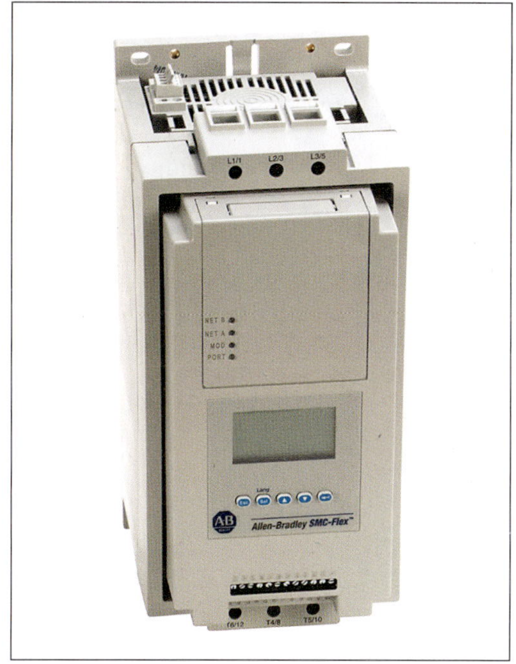

FIGURE 13.58 Electronic soft starter

Reproduced with permission of NHP Electrical Engineering Products Pty Ltd

In an electronically controlled starter (see Figure 13.58), the controlling circuits are more involved than in a normal relay-operated starter. The general principle is to convert the alternating current supply to direct current, which is then fed through a filter to remove most of the transients. The d.c. is then supplied to an inverter to convert it back to a.c. using a high-speed switching sequence. This is shown in block diagram form in Figure 13.59.

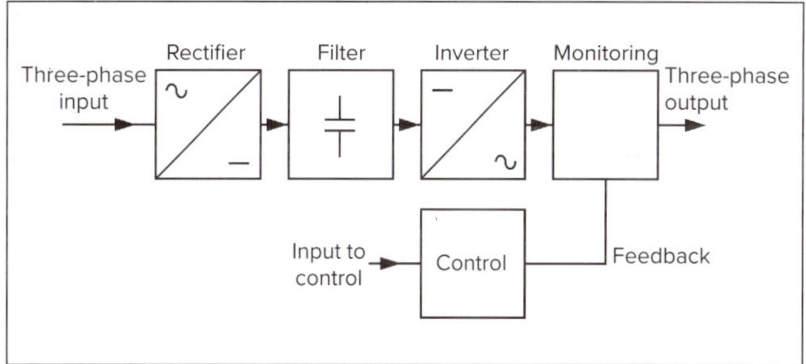

FIGURE 13.59 Block diagram for electronic motor control

The units can be programmed to monitor and control motor currents during starting and stopping sequences. This gave rise to the term 'ramping', which describes the control of a motor current during the starting and stopping sequences. If a graph is drawn for motor current against time, the characteristic is a straight line. In those terms, the motor can be ramped up or down, as shown in Figure 13.60. These linear characteristics are highly variable and depend on the information programmed into the unit.

The motor current is controlled by the starter programming to remain within two values:

1. a predetermined maximum value
2. a sufficient minimum value to ensure that the motor has enough torque to enable starting.

During the starting sequence, as the motor accelerates up to speed and the motor current tends to decrease, the starter increases the available current,

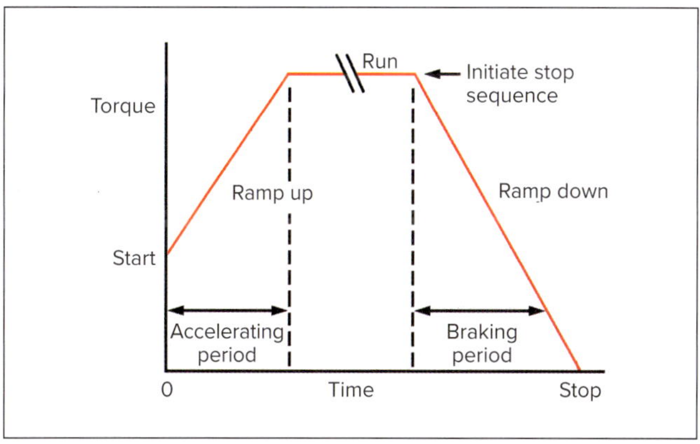

FIGURE 13.60 Ramping control

keeping the available torque constant to keep the motor accelerating up to a speed governed by the frequency of the supply.

During the stopping sequence, the motor is disconnected from the supply and brought to a rapid and controlled stop by the starter—that is, it is ramped down. (See section 13.17 for various methods of braking a.c. motors.) On a production line, time spent waiting for a machine to coast down to a stop is time wasted, and is usually not tolerated.

In a starter of this type, there is considerable interaction between the applied voltage, the motor current and the starter-generated frequency. All three are monitored electronically within the starter, and the result is a voltage and a frequency supplied to the motor that meet the prevailing motor conditions.

For example, an increased voltage will increase the starting current (say, 10%), while an increased frequency will decrease starting current (approximately 5%). These are two opposing effects. Similarly, increasing the applied voltage will increase the starting torque, but an increase in frequency will decrease the starting torque. The resulting alternating voltage wave is only vaguely similar to a sine wave, and as a result generates harmonics at a ratio many times higher than the fundamental frequency. The current waveform is filtered to some extent by the inductance of the motor windings and is closer to a sinusoidal shape.

Applications

This method can be a more expensive way of starting squirrel cage motors, although initial costs are being reduced constantly and for smaller motor starters the initial cost is now approaching a comparable value. Care should be taken in selecting motors to be used in conjunction with these types of starters. They are capable of running squirrel cage motors at speeds well above and below their designed levels. In circumstances where there is a need to start them regularly or run them at slow speeds for extended periods, consideration must be given to cooling requirements.

The units can be made at ratings in excess of 300 kW. Since costs for the units are high, careful consideration must be given to their possible use. Centrifugal fans are one possible use, and this is more than justified in printing workshops where a printing press must be run up to speed slowly and smoothly because of the rolls of paper being fed through. Another typical use is on production lines, where several motors must be controlled and their speeds integrated to ensure smooth and coordinated processing.

13.14.3 Inverters, variable speed drives and variable frequency drives

Inverters are actually frequency converters and are also called 'a.c. drives' or 'variable frequency drives'. They can also be used to convert from a single-phase supply to three phase for motor speed and torque control. Supply frequency and voltage are adjusted to cause changes in motor speed and torque based on load requirements.

Inverters are electronic devices that can initially convert a fixed a.c. supply frequency and voltage into d.c. which is then conditioned, smoothed and converted back to a.c. as pulses (pulse-width modulation), to control speed and torque for a.c. electric motors.

13.14.4 Variable speed drives—a.c. and d.c. motor control

A variable speed drive changes the speed of a motor by changing the input voltage and can be used with both a.c. and d.c. motors. VSDs can often be retrofitted to existing motors and used in conjunction with soft starters. VSDs use a rectifier circuit to convert a.c. to d.c. at a specified voltage and current for the drive to adjust. Changing the voltage of the d.c. changes the speed of the motor.

A VSD controls the speed and torque of an a.c. motor by converting supply frequency and voltage input to a variable frequency and voltage output. Modern electronic VSDs are also known as 'variable frequency drives' as they work by varying the a.c. electrical input frequency to control drive speed.

System performance can be improved by controlling speed to precisely match the load. Matching a VSD with the load can significantly reduce energy use.

13.14.5 Variable frequency drives—a.c. motor control

A variable frequency drive (VFD) is applicable to a.c. drives only while a variable speed drive can refer to either a.c. drives or d.c. drives. VFDs vary motor speed by varying frequency to the motor. VSDs specifically for d.c. motors vary motor speed by varying the motor voltage.

Variable-voltage/variable-frequency (VVVF) drives are used in the mining industry. The term refers to an a.c. drive automatically controlling voltage to suit the frequency the motor is running at.

VFD applications are usually a constant torque, a variable torque or constant kW. Constant and variable torque loads are the most common applications. Constant torque loads occur where torque is independent of speed. A 50% reduction in speed will deliver a 50% reduction in power. Examples of constant torque loads include cranes, hoists, conveyors, positive displacement pumps, reciprocating air compressors and rotary screw air compressors, punch presses, wire drawing machines, paper machines and printing presses.

Variable torque loads are where torque is proportional to speed. Loads have low torque at high speeds and high torque at low speeds. Pumps and fans typically have variable torque loads. Reducing the motor speed by 20% can reduce the power required by 50%.

13.14.6 Electronic frequency control

The rapid advances made in semiconductor technology have led to much more efficient and effective means of altering mains frequencies. Figure 13.61 shows block diagrams of inverter drives. Figure 13.62 shows a typical control unit for these drives.

The three-phase supply is first converted to a d.c. supply. There are at least two possible conversion methods in general use. One is to use a fixed or uncontrolled rectifier circuit and the other is to use a controlled rectifier circuit. Each type of circuit has its advantages and disadvantages.

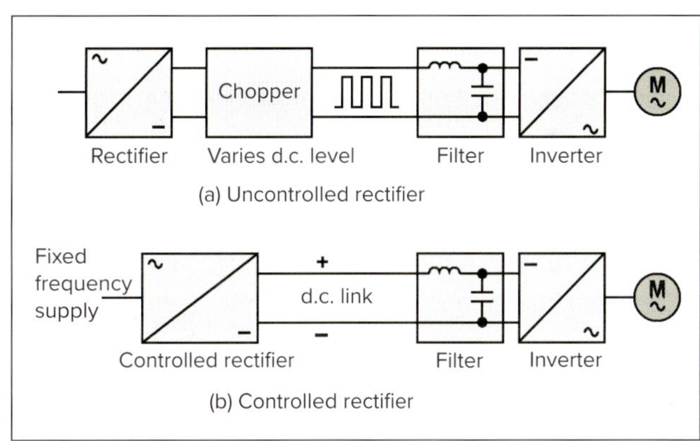

FIGURE 13.61 Block diagrams of variable frequency motor drives

FIGURE 13.62 Typical electronic frequency control unit for an induction motor

Reproduced with permission of NHP Electrical Engineering Products Pty Ltd

13.14.7 Uncontrolled rectifiers

The output voltage from an uncontrolled rectifier circuit (for example, a bridge rectifier) is governed by the a.c. input voltage. The d.c. output is then fed to a 'chopper' circuit whose function is to switch the d.c. on and off at a rate higher than the fundamental frequency of the supply. The on/off action adjusts the average d.c. voltage level supplied to the inverter.

Figure 13.61(a) shows the approximate square-wave shape of the chopper output. This is then fed through a filter circuit to remove as many transients and spikes as economically reasonable, and it then becomes the input to the inverter circuit.

13.14.8 Controlled rectifiers

A controlled rectifier circuit such as one using silicon-controlled rectifiers with phase control has the advantages of a fast response, relative cheapness and the ability to maintain a regenerative action of feeding power back into the mains. It has the disadvantage of operating at a lagging power factor.

Because the d.c. from the controlled rectifier does not have to be fed through a chopper circuit, it tends to be less complicated and is possibly a more efficient circuit. The output is generally filtered before being supplied to the inverter section (see Figure 13.61(b)).

To maintain optimum performance of an a.c. motor with a varying frequency supply, it is necessary to maintain the designed magnetic flux conditions in the magnetic circuit of the motor. This is generally achieved by ensuring that the ratio of supply voltage to frequency is kept constant. Any change in frequency must then be accompanied by a change in voltage. That is:

$$\frac{V}{f} = k$$

The actual type of circuit and the control method depend largely on the motor application. A balance has to be found between the desired factors of speed, torque and power. Other factors are the type of circuits in the unit and whether the final output sends a signal of its own performance to the unit's input for monitoring and self-adjustment of the unit.

13.14.9 Direct online starting

DOL starting is the simplest way to start an induction motor. Full voltage is applied directly to the stator windings of the stationary motor. The effect is often severe. At the instant of connection to the supply, the stator is completely demagnetised and, as the winding resistance is low, there is a high inrush of current from the mains. At switch-on, the rotor bars behave like the short-circuited secondary winding of a transformer and aggravate the effect. It is quite usual for the starting current to reach a value seven times that of normal full-load current (see Figure 13.63).

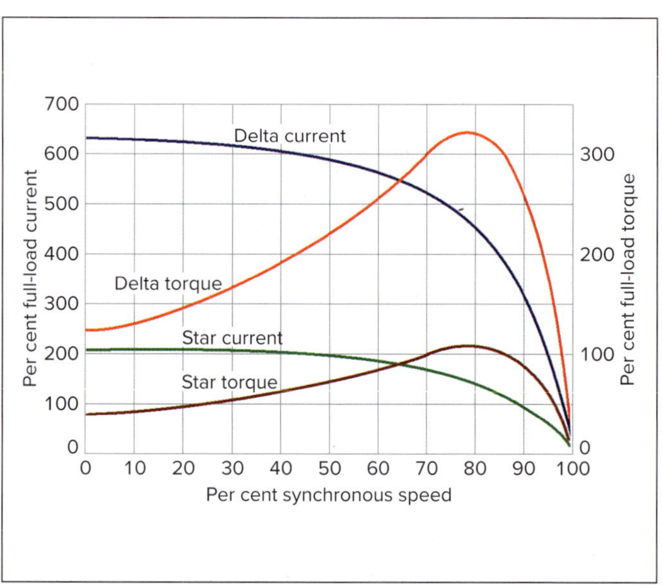

FIGURE 13.63 Comparison of current and torque characteristics for star and delta connections

The starting torque produced is two-to-three times that of the full-load torque. This is transmitted to shafts, bearings, belts and the driven machine so quickly that a considerable mechanical shock is transmitted to all connected parts. If the motor is capable of starting the connected load, acceleration is swift. For very large motors (which implies a heavy load), the mechanical shock from DOL starting can shear shafts or cause severe belt-slip, which results in accelerated wear.

Applications

DOL starting is generally restricted to comparatively small-size motors. Motors up to 4 kW are normally started DOL, but in special circumstances those up to 25 kW have been started this way.

DOL starting is also usually restricted to situations where there is little or no load imposed on the motor when it is being started. DOL starting a centrifugal pump is one instance where the major part of the load appears after the motor has been started when the load is gradually imposed by an increase in pressure on the liquid being forced through the pump.

13.14.10 Star–delta starting

For star–delta starting, the six ends of the three windings have to be brought out to the motor terminals (see Figure 13.64). A reduced voltage to the windings is achieved by first connecting the windings in the star configuration, then switching to delta to apply rated voltage when the motor is up to speed. For a 400 V winding, the applied voltage for each phase is reduced to 230 V or 58% of the rated voltage. Effectively, this reduces the starting torque to 33% $(T \propto V^2)$ of that obtained with DOL starting. Note that $(0.58)^2 = 0.33 = 33\%$.

If the star connection has sufficient torque to run the motor up to around 75% or 80% of full-load speed, the motor can be reconnected in delta configuration to enable it to accelerate to full speed for normal operation. It also means that there has to be a short time-lag between the starting and running periods while the motor windings are isolated from the power supply and reconnected in delta configuration.

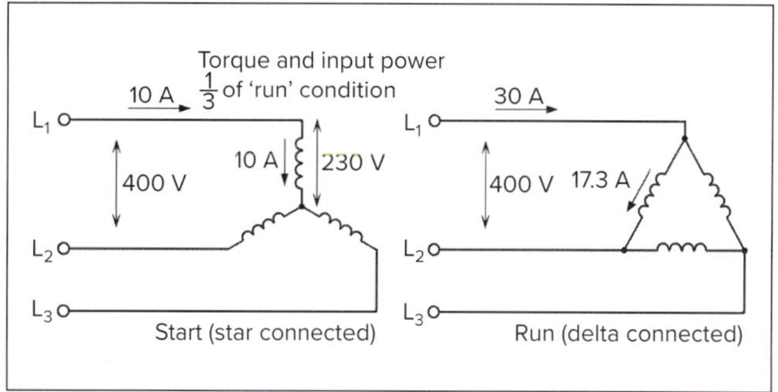

FIGURE 13.64 **Comparison of star and delta starting**

In the star connection, the line and phase currents are equal, but when the windings are reconnected in delta configuration this condition no longer applies. The phase voltage increases by a ratio of $\sqrt{3}$ or an increase of 173%. Consequently, the phase currents increase by the same ratio. This results in line current increasing to three times the value of the star connection. These values are illustrated in Figure 13.64 for a winding impedance of 23 Ω.

During the transition period, the inertia of the motor and connected machine must allow free running with little deceleration. It also means that while 'coasting' it might generate a voltage of its own and, on reconnection to the supply, this voltage can randomly add to, or subtract from, the applied line voltage. In most cases, it causes a transient current that can be up to twenty times that of normal running current. While it only lasts for a few milliseconds, it still causes voltage surges or spikes on the supply lines. These are referred to as 'changeover transients'.

In an attempt to eliminate any changeover transients caused during the open-circuit interval, a closed-circuit transition star–delta circuit was developed. It has resistors in parallel with each phase winding. The resistors must be substantial enough to withstand starting currents.

In this method, when the star point of the windings is opened for reconnection to delta, the resistor star point remains closed and current flows through the resistors and windings in series. This reduces the tendency for the motor to slow down and eliminates many of the transients. After connection to delta, the resistors are short-circuited to eliminate any current flow through them.

This starter is much more expensive than a conventional star–delta starter. It needs at least one extra contactor as well as the extra resistors. Because of the expense and its relatively few advantages, the starter is seldom used. It is only mentioned here to illustrate the point that transients can be troublesome in an installation.

Applications

Star–delta starting is a relatively simple way to get a motor up to speed without causing excessive starting currents. It can be used in any application where the motor can be started 'unloaded'. Other applications include driving centrifugal pumps where little starting torque is required until the pump is up to speed and working. Other uses are on farm dam pumps, and at the end of long runs where DOL starting would cause an appreciable voltage drop.

Another use is in lathes where a clutch is incorporated in the mechanism. The motor is started and later the clutch is engaged to turn the lathe. Many large fans or blowers can be started with a star–delta starter.

13.14.11 Primary resistance

For primary resistance starting, resistors are connected in series between the supply lines and the motor terminals, although inductors are occasionally used in place of resistors. The intention is to reduce the actual voltage at the motor terminals and so limit starting current. Starting torque is reduced at the same time.

Metallic resistor starters

The traditional method of primary resistance starting is to insert metal resistors in series with the supply for starting (see Figure 13.65). Once the motor has accelerated to around 75% to 80% of its rated speed, full voltage is then applied to the motor. This is usually done by short-circuiting the resistors in each phase. While this type of starter is more expensive than a DOL-type starter, the method does have advantages.

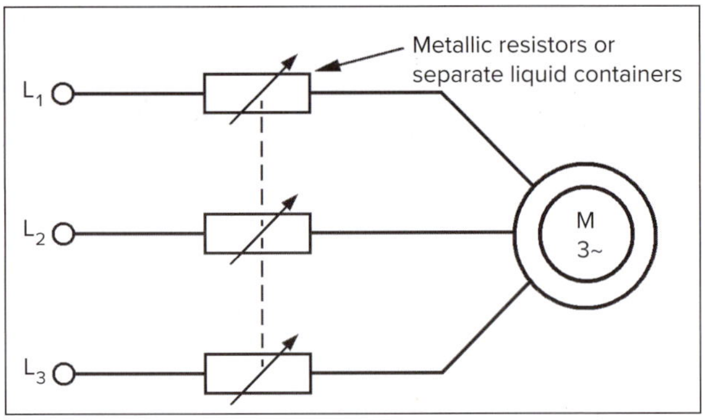

As the motor accelerates, its starting current decreases. This in turn causes a decreased voltage drop across the resistors and an increased voltage at the motor terminals. The overall effect is to give an increased torque as the motor accelerates. Opposing this is the effect of the resistors getting hot during the starting process and slightly increasing their resistance.

Because the motor is not disconnected from the supply when the resistors are shorted out at the changeover from the start to run connections, there is no transient current.

Metallic resistors or separate liquid containers

L_1

L_2

L_3

M 3~

FIGURE 13.65 Basic power circuit for primary resistance motor starting

Liquid resistor starters

A similar technique uses liquids instead of metallic resistors. Liquid containers are far bulkier and less robust than metallic resistors, but modern liquid electrolytes have the ability to reduce their resistance as they are heated up by starting currents flowing in the liquid.

Compared with older types, where metal electrodes were manually wound down into a liquid to accelerate the motor, today's starters use a liquid that can be adjusted in content to suit particular applications. Often non-corrosive, it is placed in sealed and insulated containers to reduce contact with the atmosphere. Occasionally a non-reactive oil may be placed on the surface to further isolate the electrolyte from the atmosphere and prevent evaporation.

The electrodes are permanently immersed in the liquid (see Figure 13.66). Once in position, they need no further adjustment. Depending on the manufacturer, the plates may be flat and parallel or consist of a series of short concentric cylinders placed one inside another. Each phase winding has its own separate container and set of electrodes.

When the motor is first started, the liquid has a resistance modified to suit the application. Starting current and torque are limited to predetermined values. The voltage appearing at the motor windings is reduced and the starting torque is also reduced. When properly adjusted, the motor is able to start and accelerate smoothly without mechanical shocks.

As the starting current flows, it heats the liquid and its internal resistance is reduced. Consequently, the voltage to the motor increases and this action is a progressive one up to around 80% to 90% of full-load speed, when the resistance of the liquid stabilises. Then a timer operates a contactor, which bridges out the electrodes in the liquid and the motor can accelerate to full speed.

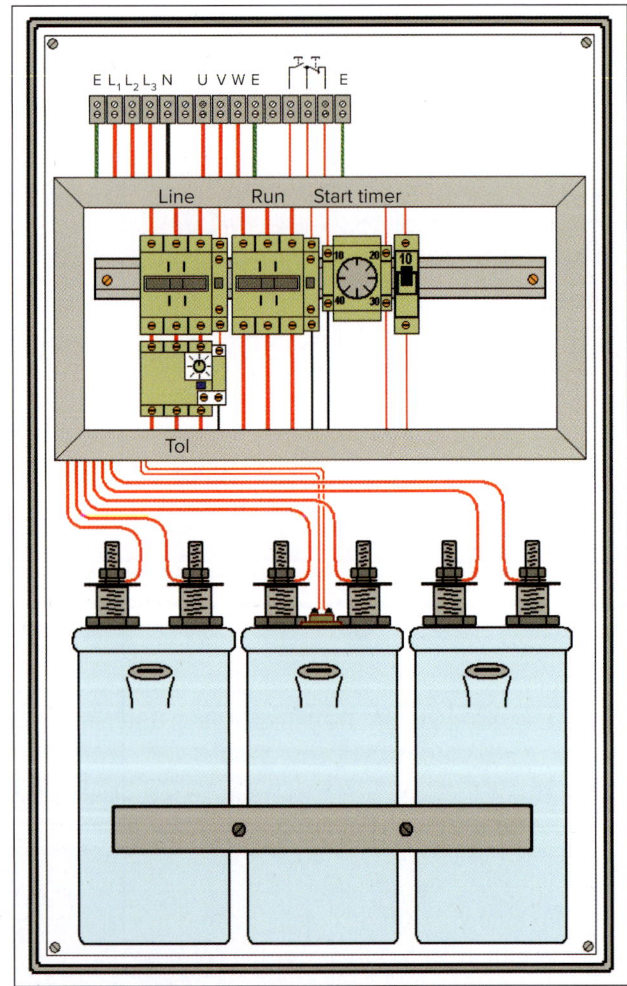

FIGURE 13.66 Typical liquid resistance starter

The primary resistance starter takes more power from the line than some starters but does so at a higher power factor. The liquid starter possibly increases the acceleration rate of the motor so that it reaches full-load speed more quickly. Like the metallic resistor starter, there is no open-circuit transition stage, so it eliminates transients in both starting current and torque. The first method uses a metallic positive temperature coefficient resistor while the second method uses a liquid negative temperature coefficient element. It is this fact that enables the liquid-type primary resistance starter to have an advantage over the metallic resistor type.

Applications

Because of the lowered starting torque, the primary resistance method of starting induction motors is limited to loads where initial starting torque requirements are comparatively low—that is, loads where the running torque only comes into play at full speed. Fans, blowers and water pumps are common applications.

EXAMPLE 13.1

A 400 V three-phase induction motor draws 160 A when connected DOL. If a primary resistance starter is connected to the motor so that the voltage to the motor is reduced to 280 V for starting, determine the:

- percentage of rated voltage applied to the motor during starting
- starting current taken by the motor
- percentage of DOL starting torque produced.

Percentage rated voltage:

$$\text{Percentage voltage} = \frac{280 \times 100}{400} = 70\%$$

Starting current:

$$\begin{aligned} I_S &= 70\% \text{ of DOL starting current} \\ &= 70\% \text{ of } 160 \text{ A} \\ &= 0.7 \times 160 \\ &= 112 \text{ A} \end{aligned}$$

Percentage starting torque:

$$\begin{aligned} T_S\% &= \left(\frac{70\%}{100}\right)^2 = 0.7^2 = 0.49 \\ &= 0.49 \times 100 = 49\% \text{ DOL torque} \end{aligned}$$

13.14.12 Autotransformer starters

Autotransformer starters generally use two autotransformers connected in an open-delta configuration to provide reduced-voltage starting. Taps are usually provided on the transformers to enable selection of the required starting torque. The principle is illustrated in Figure 13.67.

Points to note are:

1. The motor-starting current varies directly with the applied motor voltage.
2. The line current varies as the square of the motor voltage.
3. The torque varies as the square of the motor voltage.

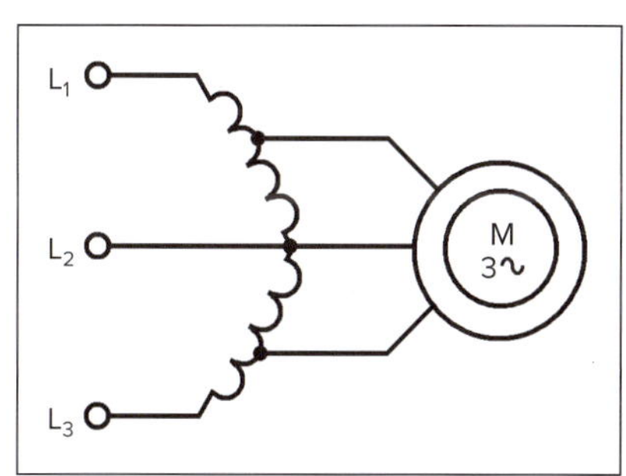

FIGURE 13.67 Starting connections for an induction motor using an open-delta transformer

Point 2 means that the starting current with an autotransformer starter is less than that obtained with a primary resistance starter. This is its major advantage—for the same starting current, the motor torque is increased.

Point 3 means that if a 50% voltage tapping is chosen, the resultant starting torque is 25% of full-load torque. A tapping has to be selected that will be the minimum that enables the motor to start against its load.

The major characteristics of this type of starter are:

1. low line current
2. low line power
3. low power factor
4. open-circuit transition periods
5. acceleration is in a series of steps; it is not continuous or smooth.

The provision of tappings on the transformers makes it possible for a choice of voltages (and hence currents) to be available for starting purposes.

The use of a transformer makes it possible to reduce the line input current at a greater rate than that at which the torque is reduced. With a transformer:

$$\text{input voltage} \times \text{input current} = \text{output voltage} \times \text{output current}$$

$$\text{That is, } V_1 \times I_1 = V_2 \times I_2 \text{ (neglecting all losses)}$$

During starting, a reduced voltage V_2 is applied to the motor, thereby reducing the starting current I_2. Because of transformer action, however, the input current is reduced still further. It can be illustrated by Example 13.2.

EXAMPLE 13.2

A 400 V three-phase induction motor draws 160 A when connected DOL (see Figure 13.68). If an autotransformer starter with the motor connected to the 70% tapping is used to start the motor, determine the:

- voltage applied to the motor during starting
- starting current taken by the motor
- starting current drawn from the supply.

For 70% tapping:

$$\begin{aligned} \text{motor voltage} &= 70\% \text{ of input voltage} \\ &= 70\% \text{ of } 400 \text{ V} \\ &= 0.7 \times 400 \\ &= 280 \text{ V} \end{aligned}$$

For 70% tapping:

$$\begin{aligned} \text{motor current} &= 70\% \text{ of DOL starting current} \\ &= 70\% \text{ of } 160 \text{ A} \\ &= 0.7 \times 160 \\ &= 112 \text{ A} \end{aligned}$$

For a transformer:

$$\begin{aligned} V_1 I_1 &= V_2 I_2 \\ \therefore I_1 &= \frac{V_2 I_2}{V_2} = \frac{280 \times 112}{400} \\ &= 78.4 \text{ A} \end{aligned}$$

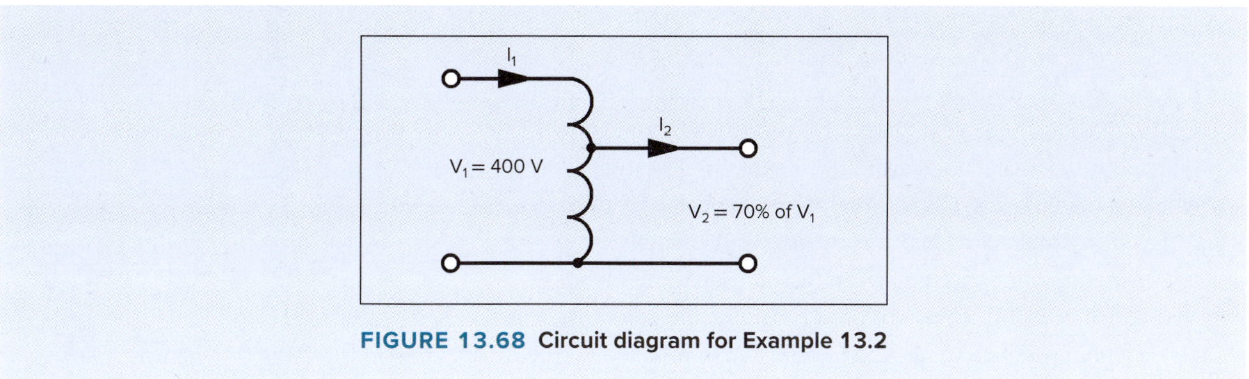

FIGURE 13.68 Circuit diagram for Example 13.2

Note that the motor voltage has been reduced by 70% to 280 V. Because I ∝ V^2, the torque will have been reduced to $(70\%)^2 = 49\%$ of the DOL value. While the motor current has been reduced to 70% of the DOL value, the line input current has been reduced to 49% of the DOL value by transformer action.

For a primary resistance starter to give the same reduction in line input current, the voltage applied to the motor would need to be reduced to 49% of the DOL value, further reducing the starting torque.

For comparison purposes, see Table 13.8.

TABLE 13.8 Comparisons between primary resistance and autotransformer starting

Starter type	Voltage applied to motor	Starting current	Starting torque
Primary resistance	49%	49%	$(49\%)^2 = 24\%$
Autotransformer	70%	49%	$(70\%)^2 = 49\%$

A more expensive form of autotransformer starter that was designed to alleviate the problem of open-circuit transients is the *Korndorfer starter*. This uses three autotransformers and an extra contactor. Its use is limited to applications where there is no reasonable alternative. It maintains a continuous torque during the transition periods of reconnection (see Figure 13.69).

The diagram in Figure 13.69 shows that, during the transition period, the motor is connected to the supply by part of the transformer windings, which act as series inductors during the transition time.

Applications

Owing to the high cost of autotransformer starters, their use is restricted to heavy loads that have to be started from rest. Such applications are larger-type refrigeration units and air-compressors where the motor might have to start against a substantial head pressure. In some cases, electrically operated relief valves might also have to be fitted to release the head pressure to enable the motors to start.

13.14.13 Secondary resistance

Despite the predominant use of squirrel cage motors and their improved starting methods, there are still applications where the wound-rotor motor has advantages that more than compensate for its initial cost. Apart from the high starting torque characteristics, there are still current-surge limitations that cannot be met by a squirrel cage motor combined with any of the starters discussed above.

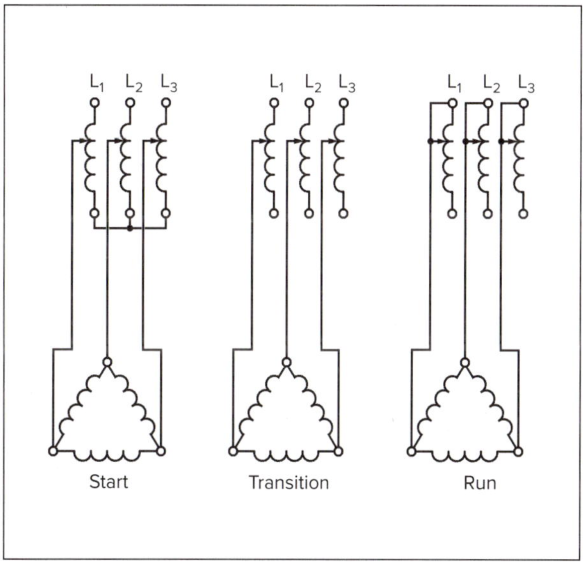

FIGURE 13.69 Autotransformer Korndorfer switching

The speed–torque characteristic of any induction motor depends largely on the relative proportions of resistance and reactance in the rotor circuit. When the rotor resistance is equal to the rotor inductive reactance, maximum torque is produced. At the instant of starting, while the rotor is still stationary, the frequency of the currents in the rotor will equal the line frequency. This causes increased inductive reactance in the rotor circuit. With the introduction of external resistance into the wound-rotor circuit, the rotor resistance and impedance values can be adjusted to produce maximum torque at starting, and at the same time minimise starting currents (see Figure 13.70).

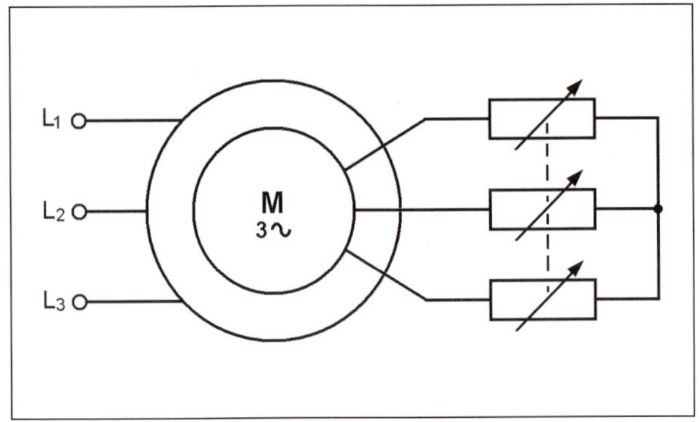

FIGURE 13.70 Basic power circuit for secondary resistance motor starting

Metallic resistor starters

For a wound-rotor motor, the rotor windings are connected internally in either star or delta configuration and the connecting leads are brought out to slip-rings. With secondary resistance starting, it is possible to control motor performance and still satisfy the local requirements for starting currents.

If the speed–torque characteristics of a wound-rotor motor are examined, it will be found that the incidence of maximum torque can be made to occur at different speeds by adjusting the value of external resistance placed in the rotor circuit.

While external resistance is connected in the rotor circuit, the starting torque is almost proportional to the starting current. It cannot be truly proportional because of the inductive reactance of the rotor windings.

Full line voltage is applied to the stator windings for starting and the external rotor resistance is progressively reduced as the motor accelerates. An equal value of resistance in the circuit of each phase is the norm, but occasionally these resistances can be made unequal, resulting in unbalanced rotor currents. As the resistance is reduced either manually or by contactors, the motor accelerates in a series of stages. The greater the number of resistance stages, the smoother the acceleration.

Liquid resistor starters

The introduction of liquids in chambers was an attempt to replace metallic resistors and eliminate the steps or stages in the starting process. This method makes starting smooth, with a gradual acceleration in speed until the final step where the liquid resistor is finally shorted out. Consequently, transient effects are minimal.

The passage of current through the liquid causes partial vaporisation of the electrolyte in the chamber. This affects the value of the resistance between electrodes. The greater the current flow, the greater the vaporisation. As the liquid heats up, the resistance between the electrodes is progressively reduced and the motor can accelerate smoothly up to working speed since the current flow controls the relative proportions of liquid and vapour in each of the liquid chambers. The final stage of the starting sequence occurs when a timed contactor shorts out the liquid chambers.

Applications

The wound-rotor motor is ideal for driving machines designed to handle impact loads. Examples of such machines are presses, drop forging hammers and guillotines. Heavy loads are applied suddenly and reliance is placed on the inertia of the heavy rotating parts to complete the action of the machine. It means also that the drive motors can be of lesser power since, apart from starting the machine, their sole purpose is to make up the losses of the rotating parts.

It follows that heavy rotating parts of machines, including large flywheels, might take considerable energy to get them rotating at the required speed. For an electric motor to get the machine rotating at this speed, a lengthy starting sequence is usually required.

Overhead cranes sometimes use wound-rotor motors since the resistance in the rotor circuit forms part of a variable speed process. Variable speed characteristics are discussed in section 13.18.

13.15 Three-phase motor-starter circuits

This section outlines the operation of pushbutton motor-starter circuits for the more common types of starters. It must be emphasised that the circuits presented are only representative. There are many variations between each type and further variations from one manufacturer to another.

These circuits present the principles of operation only and it is to be expected that other starters might have different circuits.

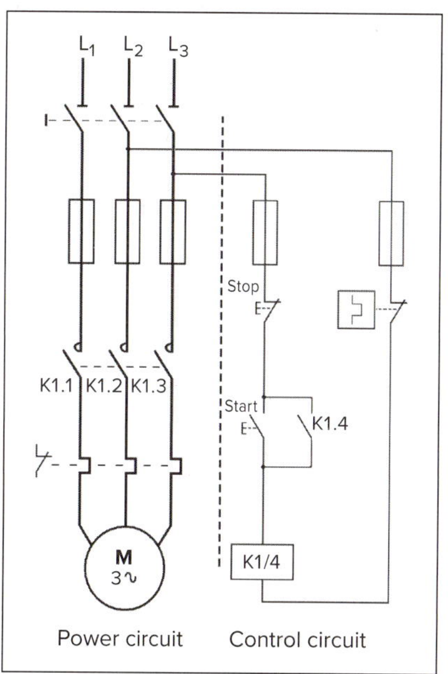

FIGURE 13.71 **Contactor circuit for DOL starting**

13.15.1 Direct-on-line contactor starter circuit

Circuit operation

1. In Figure 13.71, pressing the start button completes a circuit from L_3 through the normally closed stop button to coil K1 and the overload to L_2.
2. Main contactor coil K1 then energises and applies full line voltage directly to the motor via contacts K1.1, K1.2 and K1.3.
3. Contact K1.4 bridges out the start button contacts so that, on the release of the start button, the contactor remains in the energised state. That is, the control circuit is latched in the 'on' position. Pressing the stop button disables the latching circuit and allows the main contactor to revert to the 'off' state.

13.15.2 Star–delta contactor starter circuit

Circuit operation

1. In Figure 13.72, pressing the start button completes a circuit from L_3 through the normally closed stop button and two normally closed contactor contacts (K4.1 and K3.4) to coil K2 and the overload contact to L_2.
2. When K2 energises, it causes the 'ends' of the three windings to be joined in star configuration via contacts K2.1, K2.2 and K2.3.
3. Simultaneously, coil K3 is open-circuited by K2.5. This is the delta-connecting coil and it must be isolated when the star connection is in operation. Similarly, when the delta connection is in operation, the star connection must be isolated, using a method called 'electrical interlocking'. As a precaution, the star and delta connecting contactors are often mechanically interlocked, in addition to the electrical interlocking provided by contacts K2.5 and K3.4.
4. When K2.4 closes, a voltage is applied to the timer K4 and to coil K1. This allows K1.4 to close and bridge the start button.
5. Contacts K1.1, K1.2 and K1.3 close and apply a voltage to the 'starts' of the motor windings.
6. The voltage applied across the windings at this time is only a proportion of full line voltage (58%), and the starting current is reduced accordingly.

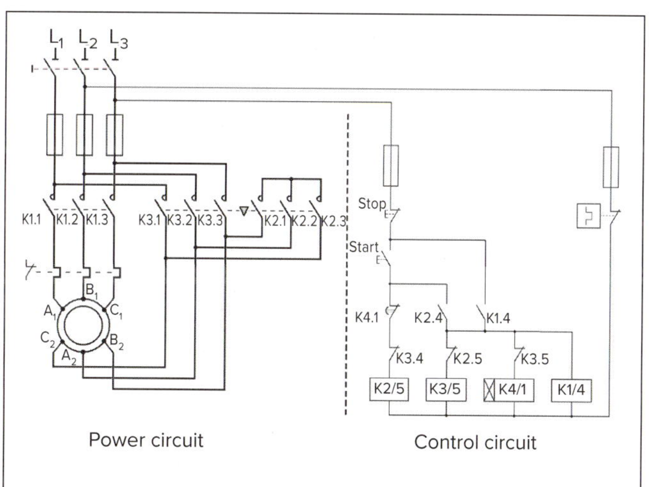

FIGURE 13.72 **Contactor circuit for star–delta starting**

7. When the time-delay period has elapsed, contact K4.1 opens and forces contactor K2 to de-energise and disconnect the star connection. K2 de-energising also causes contact K2.5 to close and energise coil K3. This action open-circuits the interlocking contact K3.4 and also switches off the timer K4 via K3.5.
8. Contacts K3.1, K3.2 and K3.3 close and complete the delta connection, allowing full line voltage to be applied to the motor.
9. Pressing the stop button de-energises all coils and allows the starter to revert to the 'off' state.

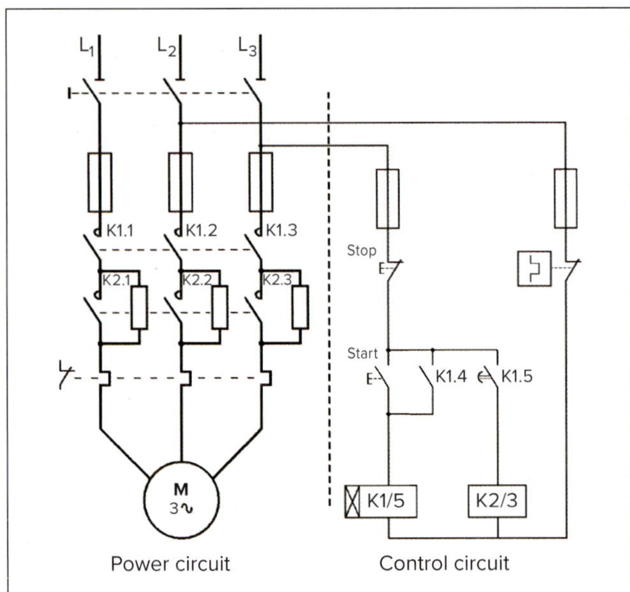

FIGURE 13.73 Contactor circuit for primary resistance starting

13.15.3 Primary resistance contactor starter circuit

Circuit operation

1. In Figure 13.73, pressing the start button completes a circuit from L_3 through the normally closed stop button to coil K1 and the overload to L_2.
2. The main contactor K1 energises, contact K1.4 closes and bridges out the start button contacts, so that on release of the start button the K1 contactor remains energised.
3. Contacts K1.1, K1.2 and K1.3 close, and a reduced line voltage is applied to the motor through the resistors in series with each line to the motor. The starting current is limited by the resistors to a value below that of DOL starting.
4. Delayed-action contact K1.5 operates after a predetermined delay and completes the circuit for coil K2. Its operation causes contacts K2.1, K2.2 and K2.3 to close, shorting out the resistors and allowing full line voltage to be applied to the motor.
5. Pressing the stop button de-energises all coils and allows the starter to revert to the 'off' state.

13.15.4 Autotransformer contactor starter circuit

Circuit operation

1. In Figure 13.74, pressing the start button completes a circuit from L_3 through the normally closed stop button, a normally closed delay contact K1.5, electrical interlock K3.3, coil K2 and the normally closed thermal overload contact to L_2.
2. When K2 is energised, it closes the contacts K2.1 and K2.2, connecting the ends of the autotransformers to line L_2 in an open-delta configuration.
3. The operation of K2 simultaneously closes contact K2.3 and opens contact K2.4, the electrical interlock, to prevent K3 operating while K2 is energised.

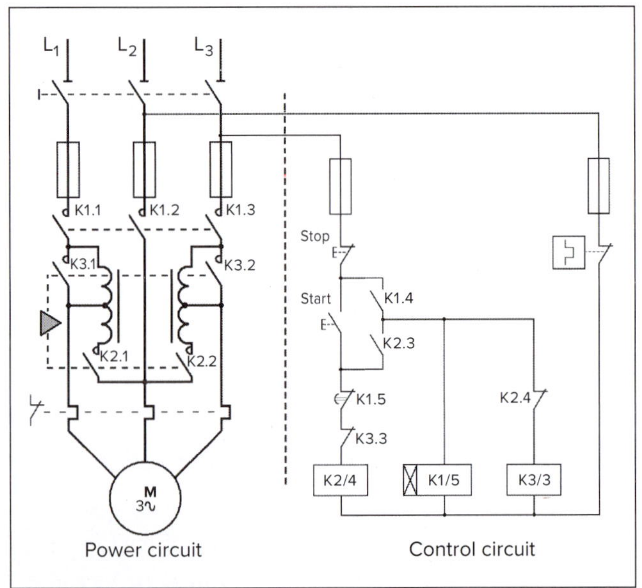

FIGURE 13.74 Contactor circuit for autotransformer starting

4. K2.3 supplies power to coil K1, which energises. Contacts K1.1, K1.2, K1.3 and K1.4 close. Full line voltage is connected to the autotransformers and a reduced line voltage is supplied to the motor via the transformer tapping. Contact K1.4 ensures that a voltage is available to the control circuit when the start button is released.
5. The delayed opening contact K1.5 opens after a predetermined time lapse, de-energising K2 and open-circuiting the delta connection. Contact K2.4 then closes again and coil K3 is energised.
6. Contacts K3.1 and K3.2 close, and full line voltage is applied to the motor through the K1 contacts in series with two lines. The electrical interlock contact K3.3 opens and isolates coil K2.
7. Pressing the stop button de-energises all coils and allows the starter to revert to the 'off' state.

13.15.5 Secondary resistance contactor starter circuit

Circuit operation

1. In Figure 13.75, pressing the start button completes a circuit from L_3 through the normally closed stop button, coil K1, and the thermal overload contact to L_2. Coil K2, which is in parallel with coil K1, is energised at the same time as K1, but its contact K2.1 only operates after a predetermined time delay.
2. Contact K1.4 bridges out the start button contacts so that, on the release of the start button, the contactor remains in the operational state—that is, the control circuit is latched in the 'on' position.
3. Contacts K1.1, K1.2 and K1.3 close and apply full line voltage to the stator terminals of the motor. The rotor has two resistors in series with each winding and, because the ends are connected in star configuration, current flows in the rotor windings and the motor is able to generate torque and commence turning. After the delay time, contact K2.1 closes. This causes K4 to be energised, along with K3, a second time-delay relay.
4. Contacts K4.1 and K4.2 then close and reduce the amount of resistance connected across the slip-rings. This action enables the motor to attain a higher speed.
5. After a further time delay, contact K3.1 closes. Coil K5 is then energised and closes contacts K5.1 and K5.2. This action removes the remainder of the resistance in the rotor circuit and the motor is in its normal running mode.
6. Pressing the stop button de-energises all coils and allows the starter to revert to the 'off' state.

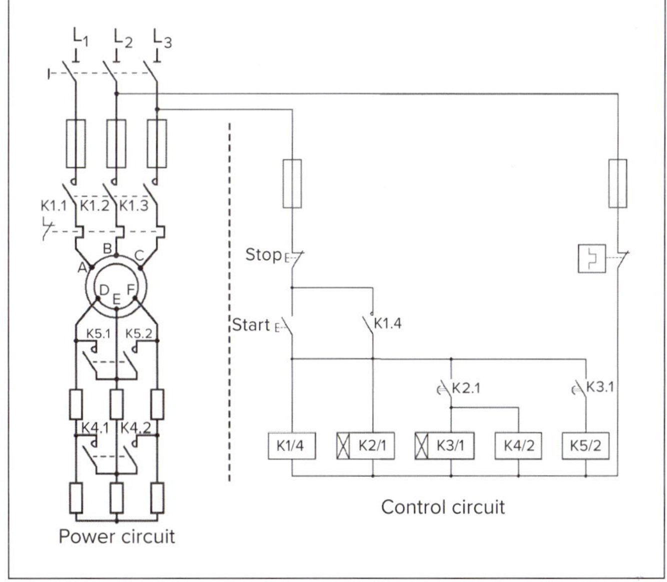

FIGURE 13.75 Contactor circuit for secondary resistor starting

13.16 Three-phase motor reversal

13.16.1 Introduction

The rotor of a three-phase motor always tries to rotate in the same direction as the rotating magnetic field. Since the direction of the rotating magnetic field depends on the phase sequence of the applied voltages, the direction of rotation can be reversed by interchanging two stator leads.

In practice, this is generally achieved by using two mechanically interlocked contactors to change the stator connections when required. In addition, the two contactors are usually interlocked electrically to prevent energising the contactor not in use. If both contactors were to be allowed to energise at the same time, two phases would be connected together as a dead short.

Where the contactor panel is located on moving machinery such as a crane or hoist, it is vital to have mechanical as well as electrical interlocks as the two contactors could be closed due to inertia if the crane runs into the end stops.

Figure 13.76 shows typical circuit connections for a DOL forward-reverse starter circuit. The same principle applies to all forms of three-phase motor starting.

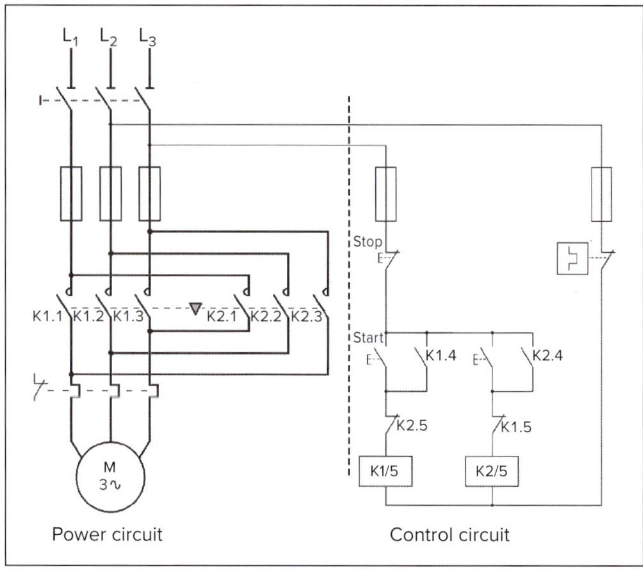

FIGURE 13.76 Two-contactor DOL reversing circuit

13.17 Three-phase motor braking

13.17.1 Introduction

In many installations, it is quite satisfactory to allow a machine to coast to a halt as its inertia is dissipated in friction losses within the machine. This inertia, which can be considerable in larger machines, can be dissipated more quickly by some form of braking. The braking system used must be of a type to suit the machine and its requirements.

The major types of braking in use are:

- mechanical braking
- eddy-current disc braking
- dynamic braking
- regenerative braking
- plug braking.

13.17.2 Mechanical braking

The principle of mechanical braking is to bring equipment to a complete halt and act as a parking mechanism. Generally, mechanical braking consists of creating deliberate friction between rotating and stationary components. Machine braking systems might use more than one braking method. For example, an overhead crane might use the dynamic method for slowing down a load, and a solenoid-operated mechanical brake for holding the load stationary. In the interests of safe working, the mechanical brake has to be made fail-safe by being applied automatically when power is removed. This is a protection in the event of a power failure. Travelling cranes commonly use this form of solenoid braking on all directional movements. In Figure 13.77, the brake shoes are held in the 'on'

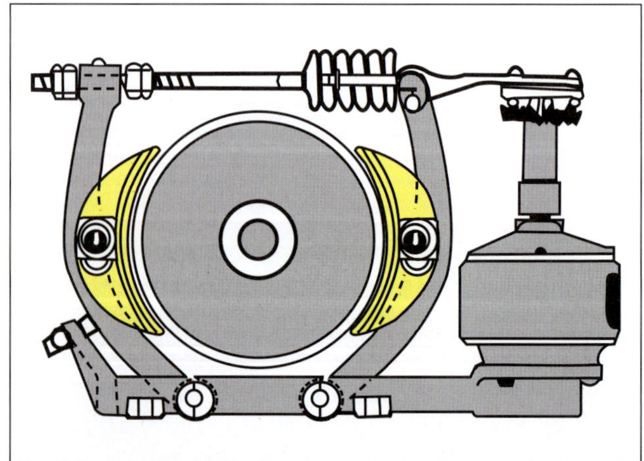

FIGURE 13.77 Mechanical braking with brake shoes and solenoid

position against a flat pulley by a substantial spring. An application of power to the motor also energises the solenoid, which releases the brake and allows the pulley to rotate.

Three-phase motors can be supplied by the manufacturer with a mechanically operated disc brake as an integral part of the motor.

In Figure 13.78, the brake pressure is applied by means of a spring. The spring tension can be adjusted so that the braking pressure suits the particular application. The brake is released by a magnetic coil, which is supplied with a d.c. voltage by a rectifier in the motor terminal box.

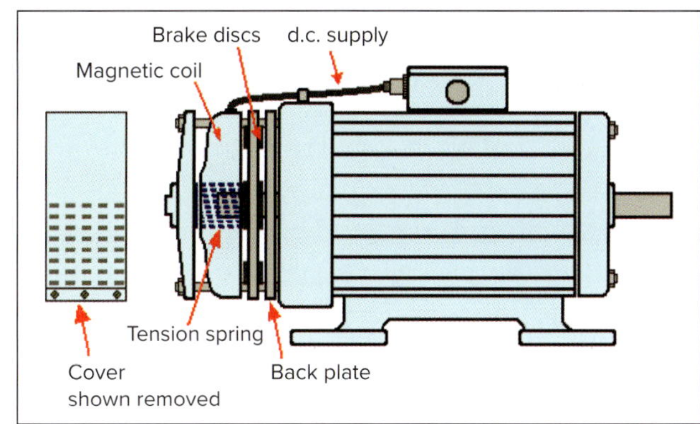

FIGURE 13.78 Typical three-phase motor with integral disc brake

When supply is applied to the motor terminals, it is also connected to the rectifier and hence the brake is released. This type of brake is inherently fail-safe because, if the supply to the motor is interrupted at any time, the brake is automatically applied.

Another system of mechanical braking uses a flat disc rotating in a mixture of finely powdered iron dust. The dust may be in a dry form or a paste immersed in a liquid. A coil surrounds the container and, on the application of a d.c. voltage, the iron powder grips the disc and holds it firmly against the container. If two discs that are free to rotate are used, the device can be used as a clutch mechanism.

It is worth noting that the mechanical system of braking usually brings the machine to a complete halt and can be used as a holding brake as well.

13.17.3 Eddy-current disc braking

An eddy-current disc consists of a sturdy disc that is connected to the machine shaft and is free to rotate with the machine between a set of coils held firmly in a stationary position. When a voltage is applied to the stationary coils, eddy currents are set up in the rotating disc and form a load on the machine. As the machine slows down, the induced voltages and currents decrease, and so does the rate of deceleration.

Eddy-current discs are not capable of bringing a machine to a complete stop, nor are they capable of being used as a holding brake. They simply increase the rate of slowing down of the machine.

13.17.4 Dynamic braking

Dynamic braking works on the principle of using the motor as a generator and dissipating the machine's inertia as electrical energy.

In a.c. motors, this is often achieved by disconnecting the rotating motor from the power supply and applying d.c. to the windings. Because the rotor is still moving, circulating currents are generated within the rotor. These form a load on the machine and slow the motor rather more quickly than just coasting to a halt. Like the eddy-current disc method, it only hastens the slowing process and cannot bring the motor to a complete stop. A mechanical braking system is still needed as a holding brake.

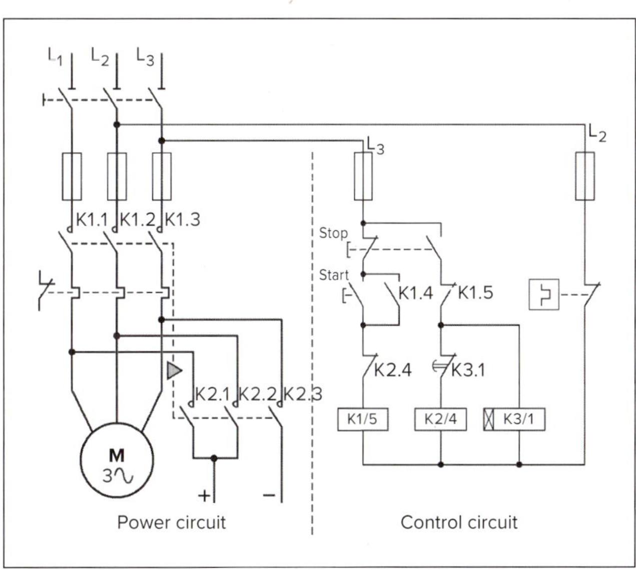

FIGURE 13.79 Braking an a.c. motor by d.c. injection (dynamic braking)

A typical circuit is shown in Figure 13.79. The main contactor and the contactor applying direct current to the stator windings are electrically interlocked and, on pressing the start button, the main contactor K1 is energised, K1.1, K1.2 and K1.3 connect the supply to the motor and contact K1.5 isolates contactors K2 and K3.

When the stop button is pressed, K1 drops out and the normally open section of the switch completes the circuit to contactor K2. When it is activated, it isolates the main contactor and simultaneously applies direct current to the stator windings. At the same time as K2/4 is energised, the time delay contactor K3 is also energised. After a preset time lapse, it operates and switches off the direct current.

During the stopping process, the stop button must be held in the stop position for a short period to activate the braking system. The direct current is usually obtained from a rectified a.c. supply.

A typical use for dynamic braking is in electric trains, where the driving motors are used as generators and the energy generated is dissipated in banks of resistors. Some large cranes also use this system but, as with other applications, a system of mechanical braking is also required for bringing all movement completely to rest.

Another form of dynamic braking is the use of what is known as an 'induction generator'. Capacitors are connected across the input terminals of a three-phase motor. When it is necessary to stop the motor, the power supply source is removed and resistors are connected across the motor terminals.

The motor, while it still rotates, generates alternating currents, which are dissipated within the resistors. The system involves extra contactors, resistors and large capacitors, but is an excellent way of controlling the speed of motors and equipment connected to overhauling loads.

13.17.5 Regenerative braking

Regenerative braking uses the inertia of a moving load to convert mechanical energy into electrical energy and feed it back into the power supply source. This method is not used very often with a.c. sources; it is a more involved method than those described above and it uses extra equipment.

The electrical energy fed back into the supply source has to be considerable to justify the additional expense. It follows that the mechanical energy available to supply it also has to be considerable.

A typical application of regenerative braking is its use in electric traction systems such as trains or trams. There are often thousands of tonnes on the move, and this constitutes considerable inertia. If this energy can be transformed into electrical energy and slow down the train or tram, a large saving in electricity costs can be achieved. There will also be a saving in wear and tear on the brake shoes used in a mechanical system.

The system becomes less effective as the vehicle slows, and mechanical braking is also required. At some point, the electrical energy being generated will be insufficient to be fed back into the supply line and the system has to be disconnected from the supply. Occasionally, the system may then use dynamic braking as a further slowing process.

Because of the expense of fitting extra equipment to machines, plus the fact that the method cannot completely stop and hold machinery in a stationary position, the applications of regenerative braking are limited. For a.c. working, its main use is in controlling overhauling loads, such as cranes lowering heavy loads.

13.17.6 Plug braking

Plug braking with three-phase motors is the system of reconnecting a motor to rotate in the reverse direction while still rotating in the forward direction. It is a sudden, and almost violent, method of bringing a motor to a complete stop. The actual time taken depends on the amount of inertia in the accompanying machine.

In order to use plugging as a stopping mechanism, some means must be provided to remove all power from the motor at the instant of change in direction. This can be done with a friction-operated single-pole changeover switch mounted on the motor driving shaft. Another method uses an eddy-current disc rotating between magnets to activate contacts, which in turn control the main contactors.

The starter circuit has pushbuttons to activate rotation in the required direction and the movement of the shaft closes the appropriate contact, allowing the main contactor for that direction to energise. A stop button allows the contactor in use to drop out and also activates the contactor for the opposite direction. This is latched in until the first amount of reverse movement occurs. This movement opens the holding-in contact of the starter and removes all power from the motor.

A circuit for a three-phase motor using the plugging method of braking is shown in Figure 13.80.

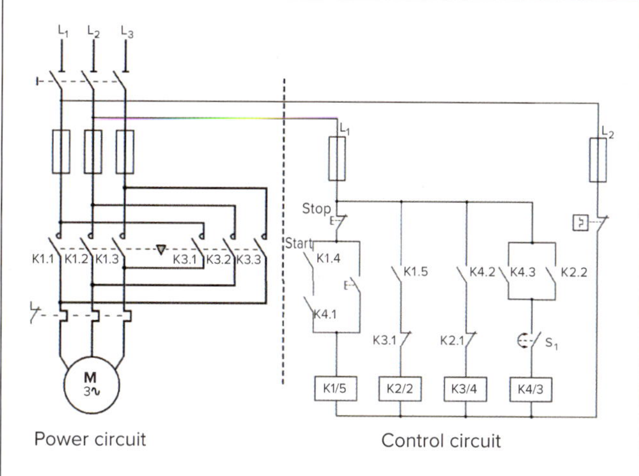

FIGURE 13.80 Typical diagram for plug braking of an a.c. motor

Circuit operation

- Pressing the start button energises K1 via the stop button and the TOL contact. K1.1, K1.2 and K1.3 close and connect the supply to the motor so it runs in the normal direction. K1.5 energises K2.
- K2.1 opens and prevents K3 from energising. K2.2 closes and energises K4 via the now-closed shaft rotation switch S_1. Note that S_1 opens whenever the shaft is stationary.
- K4.1 and K1.4 close and bridge out the start button so that, on release of the button, K1 remains energised.
- When the stop button is pressed, K1 drops out and K1.5 opens and de-energises K2. Contact K2.1 now closes and as K4 is still energised via S_1, K3 energises and connects the supply to the motor in the reverse direction. When the shaft stops rotating, switch S_1 opens and K4 and K3 both drop out.

Motors generally have to be specially designed for this application by having stronger driving shafts. The drive shaft has to withstand the forces created by the driven machine's inertia in bringing the machine to a halt. The rotor bars also have extra mechanical forces exerted on them. Again, if it is necessary, some mechanical brake may have to be applied to hold the machine in position. One plugging stop is generally recognised as being equivalent to around three repetitive normal starts.

The motor windings might also have to be specially designed if the application calls for repeated starting and stopping. It is characteristic of a three-phase motor that the amount of current flowing when plugging is applied is almost equal to normal starting current. Also, the starting current flow is applied for almost the same length of time as normal starting.

Probably the most common application is in larger production lathes doing repetitive process work. In such an application, a mechanical holding brake might not be needed.

Electronic starters equipped with current-limiting or ramping circuits can use this system of braking with some success. The inverter circuits can sometimes be reversed in action and the excess generated energy converted to direct current and dissipated by dynamic braking. Normally, it cannot be fed back into the supply source as can be done with a standard regenerative braking system.

13.18 Speed control of a.c. induction motors

13.18.1 Introduction

Torque is produced in an induction motor through the interaction of two magnetic fields. The first rotating magnetic field is created by currents flowing in the stator windings. This rotating field cuts the conductors in the rotor and induces a voltage in them. The rotor voltage causes currents to flow in the rotor and produce a second magnetic field.

The two magnetic fields interact and cause the rotor to rotate in the direction of the rotating stator field. It accelerates to approximately 96% of the speed of the rotating stator field. This 4% difference in speed on full load is the slip speed. Without slip, an induction motor cannot develop torque.

The speed of an induction motor is always governed by the rotating magnetic field in the stator. This stator field always rotates at a synchronous speed that is governed by two factors:

1. the number of pairs of poles
2. the frequency of the applied voltage.

The synchronous speed can be found from:

$$n = \frac{120f}{p}$$

where:

n = synchronous speed
f = line frequency
p = number of poles.

The number 120 is derived from the product of the number of seconds in a minute and the fact that magnetic poles always come in pairs. The other two quantities are called 'variables'.

It is important to note that there are only two variables that determine speed.

If the frequency increases, the speed increases:

$$n \propto f$$

If the number of poles increases, the speed decreases:

$$n \propto \frac{1}{p}$$

These two basic principles are the only factors that can affect the change in speed of an induction motor, although the methods adopted to achieve this are many.

13.18.2 Speed control by changing the number of poles

Changing the number of poles in a stator winding always involves an abrupt step change from one speed to another. On a 50 Hz supply, a two-pole motor will rotate at 3000 rpm (ignoring slip speed). If a change is made to a four-pole stator, the speed will quickly change to 1500 rpm. The change in speed can transmit minor transients into the supply lines. With larger motors, a short time delay should be introduced when changing from one winding to the other.

The most common method is to design windings that can be interconnected to change the number of poles. It is invariably a 1:2 ratio—that is, a two-pole winding converts to a four-pole winding, or a four-pole to an eight-pole winding, and so on.

Figure 13.81(a) illustrates the principle and connections involved for a four-pole to eight-pole conversion. Only one phase is drawn; small rectangles are used to represent pole-phase groups and the arrows indicate the sense of winding direction. It is necessary for the pole windings to be connected in pairs opposite each other as shown.

Note that the centre-tap (T_A) of the winding has been brought out so that it can be accessed externally. In Figure 13.81(b), the four-pole phase windings have

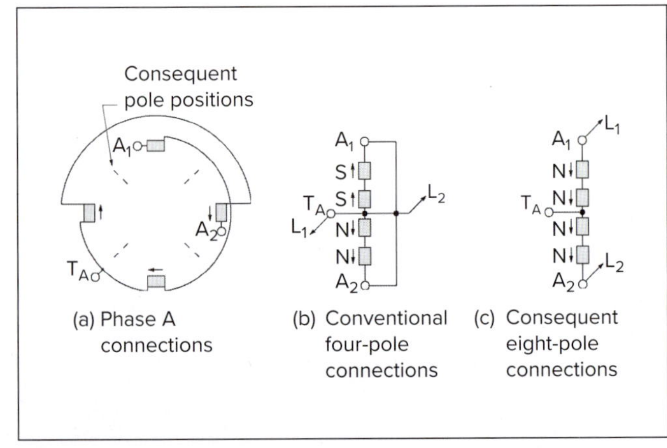

FIGURE 13.81 Series-parallel connection for a two-speed motor

been redrawn in a vertical line. If A_1 and A_2 are bridged and connected to line L_2 and the centre-tap T_A connected to line L_1, the motor will have four conventional poles, as indicated by the arrows.

In Figure 13.81(c) the bridge has been removed. A_1 is reconnected to line L_1 and A_2 left connected to line L_2. As indicated by the arrows, the current flows through all four pole-groups in series and all coils give the same north-to-south direction polarity—for example, as shown there would be four north poles (coils aligned south to north) from A_1 to A_2.

The magnetic flux is diverted in the stator and exits between the north poles, as indicated by the broken lines. The resulting magnetic circuit of the motor is that of an eight-pole machine.

To get around this 1:2 ratio limitation, some stators have been designed to accommodate two electrically separate windings. Only one winding is used at a time. These windings need not be in the ratio of 1:2 but can be in any reasonable relationship. For example, one winding could be a two-pole winding, the other a six-pole winding.

Activating the two-pole winding would cause the motor to rotate at 3000 rpm. During this period, the other winding, if not connected in star configuration, must have the delta bridges open-circuited to prevent induced currents flowing in the unused winding. When provided with a suitable switch, the windings could be exchanged without stopping the motor or its coupled machine. The motor would then change speed to 1000 rpm.

Step speed control with *pole amplitude modulation (PAM)* is a rather lesser-known system. It was developed for close speed ratios, such as changing a four-pole to a six-pole and an eight-pole to a ten-pole. It relies on the principle of unequal coil groupings within the motor when manufactured. Connections are made with special contactors to control speed steps. PAM windings are covered by copyright but can be manufactured under licence. PAM motors are made as small as 0.5 kW but have been made in sizes up to 7 MW.

13.18.3 Speed control by changing frequency

One important aspect of this method of speed control is that, at higher frequencies, the standard induction motor runs at speeds well above the base design value. At increased speeds, air circulation is improved, resulting in improved cooling.

Better cooling permits higher current densities to be used, even though there is increased friction and windage losses due to higher speeds. There are also increased iron losses due to the higher frequencies. At higher frequencies, the impedance of the windings is also increased and, to ensure a constant flux density in the air gap, a higher supply voltage is required.

At constant flux density in the air gap, torque is proportional to current flow. Since power is dependent on both torque and speed, it can be seen that the power output of the motor increases at a faster rate than the speed increase.

Increasing the frequency of a complete plant would ensure that all motors ran at a faster speed, and this might not always be desirable. Decreasing the frequency would ensure that all motors ran more slowly, and would produce the same undesirable result. In general, frequency changing as a method of speed control is limited to specific machines or groups of machines, as in a series of transport rollers in a steel mill.

There are two main methods for frequency changing. The first uses rotating machinery to achieve the desired result. It is expensive, although less efficient than some other methods; but it is extremely reliable, with very few maintenance problems. Therefore, it is still in use.

The second method uses electronic switching to synthesise an irregular-shaped alternating current wave from d.c.

13.18.4 Rotating machinery for frequency control

One common motor-control system is the *Schrage motor*, where the speed is altered by adjusting the brush positions for each phase.

Figure 13.82 illustrates another method for speed control—the *Kramer method* for controlling the speed of a wound-rotor motor. Four rotating machines are used to control the speed of one motor.

The system can only be justified economically for extremely large motors or integrated groups of motors in heavy industry. Where Schrage motors have practical limitations, the Kramer system can be built in megawatt ranges. Reliability and minimal maintenance have been well established in both systems.

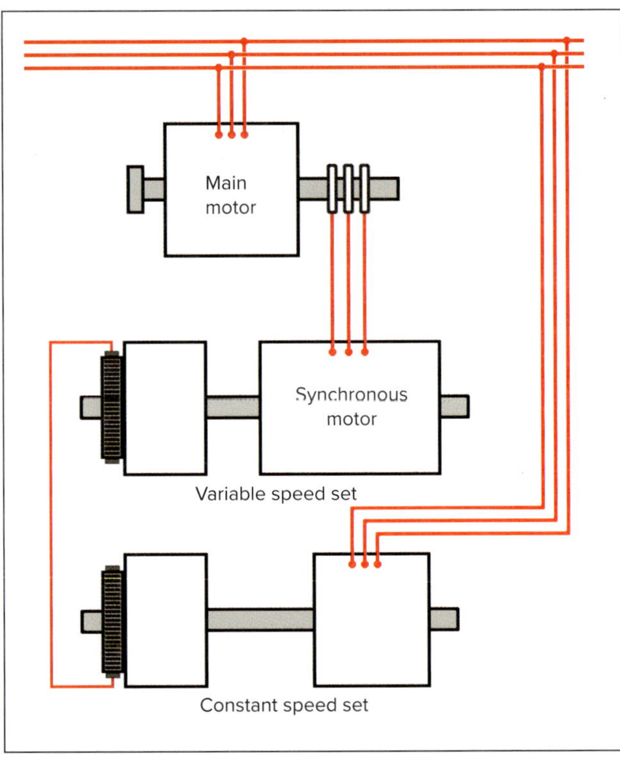

FIGURE 13.82 Kramer drive using rotating machines

13.18.5 Wound-rotor motors

The slip in any induction motor is proportional to the rotor copper losses. In a wound-rotor motor, the rotor resistance can be varied with the addition of external resistance, so rotor copper losses and speed can be adjusted with a controller.

Because rotor current is proportional to the developed torque, it follows that rotor losses vary with the applied load, thereby affecting the speed. This method of speed control has the characteristic of a variation in speed for a variation in load; that is, increasing the load causes the speed to decrease, while decreasing the load results in a speed increase.

For this reason, speed control of a wound-rotor motor by varying the rotor resistance is satisfactory only for a steady load. Speeds lower than half full-load speed are not practical and increased losses at lower speeds lead to high operating temperatures, which might exceed the ratings of the motor. Motor efficiency is poor, speed regulation is poor and the external resistances consume power wastefully.

Various methods for speed control have been tried. The two most common are unequal:

- voltages applied to stator windings
- resistances inserted in rotor windings.

Unlike the rotating machinery systems described above, these speed changes are step changes. Probably the most common application is in the hoist and lowering mechanism of overhead cranes.

Types of motor speed control are listed in Table 13.9.

TABLE 13.9 Types of motor speed control

Type of motor	Speed characteristic, no load to full load	Type of speed control
a.c., squirrel cage, multi-speed	Speed drop up to 5% from two or more initial speeds.	Pole changing. Windings of different pole numbers, or reconnect one winding to change the pole number.
a.c., squirrel cage, single-speed	Speed drop up to 15%, depending on design.	Primary voltage control. Stator frequency control at constant volts per cycle.
a.c., slip-ring	Speed drop up to 50%, depending on rotor resistance.	Secondary resistor connected to slip-rings. Machine and solid-state conversion feedback of rotor power.
a.c., synchronous	No speed drop. Speed set by stator frequency.	Adjustable frequency from motor generator set or solid-state frequency converter.

SUMMARY

- Schematics (circuit diagrams) show how a circuit operates.
- Block diagrams are useful because they provide an overview of what a circuit does.
- Relays and contactors are both electromagnetically operated switches.
- A contactor is used in a power circuit.
- A contactor can have both power and control contacts.
- Contactors and relays consist of an operating coil, a magnetic circuit and associated contacts.
- There are various types of contacts controlled by contactors and relays. Examples include normally open (NO), normally closed (NC), changeover and timed.
- In motor-control circuit diagrams, the control circuit is drawn in lighter lines than the power circuit.
- In detached representation in circuit diagrams, the contactor coil has both the contactor designation and the number of contacts it operates. The coil designation may be placed on or beside the coil symbol. Each contact has the designation of the contactor and a number usually representing its importance or its order of operation, for example K1.1.
- In pushbutton switch control additional start pushbutton switches are all placed in parallel. Additional stop pushbutton switches are all placed in series.
- Control by automatically operated switches is termed '2-wire control'.
- When reversing contactors are used, the contactors are mechanically interlocked while their contacts are electrically interlocked in the control circuit.
- When drawing circuit diagrams, the top-to-bottom and left-to-right rules should be followed when possible.
- Control circuits are usually drawn as ladder diagrams.
- A motor starter limits starting currents, monitors any overloads and stops or isolates the motor if necessary.
- There are several methods for starting motors and each has its own advantages.
- Starting torque varies with each motor-starting method.
- Alternating current starters can be manual, pushbutton or automatic.
- Three-phase motors are always reversed by reversing the phase rotation, i.e. reversing any two lines.
- Motor braking methods include:
  - mechanical—friction with brake shoes; the only holding method
  - dynamic—electrical energy dissipated in resistance; includes eddy-current discs
  - regenerative—converts mechanical energy to electrical energy and returns it to the supply source
  - plugging—full reversing power applied; power to be removed when motor stops.
- Direct current motor braking is usually by dynamic braking backed up by mechanical braking.
- The speed of a three-phase motor can be controlled by changing the number of poles in the windings or changing the line frequency. The number of poles can be changed by altering winding connections or by having more than one winding.
- The a.c. motor inverter principle is to convert the alternating-current supply to direct current, which is then fed through a filter to remove most of the transients. The d.c. is then supplied to an inverter to convert it back to a.c. using a high-speed switching sequence.
- Direct current motor starters can be manually or pushbutton operated. Full automatic starting and stopping is available with special starters.
- Acceleration, torque and starting currents can be controlled with timed relays, voltage-sensitive relays and electronic control.
- Automatic control of electrical equipment is attained with programmable logic controllers that are pre-programmed to make decisions based on signals conducted to them by transducers.

END-OF-CHAPTER QUESTIONS

13.1 When switching a large current or voltage, what is the advantage of using a relay or contactor rather than an ordinary switch?

13.2 Name the four parts of a basic relay circuit.

13.3 Name the international standard that relates to electrical drawing convention.

13.4 Draw the circuit diagram symbols for the following:

(a) fuse
(b) relay coil
(c) timer contact, NO, delay on
(d) timer contact, NC, delay off.
(e) manually operate switch, NO
(f) pushbutton switch, NC.

13.5 What is the convention for representing the flow of energy, or flow of 'signal' and sequence of operation or events that was adopted to make circuit diagrams easier to read?

13.6 What is the main difference between a circuit diagram and a wiring diagram?

13.7 Briefly state what a block diagram of an electrical circuit represents.

13.8 Name the three basic parts of a contactor.

13.9 Explain the operating principle of a contactor.

13.10 To provide two-position control in a control circuit, how are the two stop pushbuttons connected? How are the two start pushbuttons connected?

13.11 Draw a detached control circuit showing two start and two stop pushbutton switches, plus the maintaining/holding contact.

13.12 Explain what is meant by 'jogging control'.

13.13 How is a three-phase motor's direction of rotation reversed by electrical means?

13.14 Name two types of interlocks.

13.15 State an application for ladder diagrams.

13.16 State two general types of control allowed by control devices.

13.17 Name five common types of switches that are used as control devices.

13.18 Name five factors that must be taken into consideration when selecting control devices.

13.19 Define a transducer.

13.20 Why is a direct online starter (DOL) not suitable to start large induction motors?

13.21 Why is a star–delta starter referred to as a 'reduced-voltage' starter?

13.22 Explain why the starting torque in a star–delta starter is 33% of full-load torque.

13.23 Briefly describe the main advantage of using a primary resistance starter rather than a DOL starter.

13.24 List the five major characteristics of an autotransformer starter.

13.25 Name a major advantage of using a soft starter rather than reduced-voltage starters.

13.26 Name five refinements that a soft starter can control, but a reduced voltage starter cannot.

13.27 Name a major advantage of a secondary resistance starter (wound-rotor motor).

13.28 Explain the operation of a DOL starter.

13.29 Name the five major types of three-phase motor braking.

13.30 Explain how mechanical braking is accomplished.

13.31 Explain the operation of eddy-current disc braking.

13.32 Explain the principle of dynamic braking.

13.33 Explain the principle of regenerative braking.

13.34 Explain the principle of plug braking.

13.35 Explain how torque is produced in an a.c. induction motor.

13.36 The stator field rotates at synchronous speed. State the two factors that govern the synchronous speed.

CHAPTER 14

Trade calculations

LEARNING OBJECTIVES

- Understand the SI system of units
- Use SI base units in calculations and in deriving other units
- Use derived units and recognise that they are built from base units
- Understand the purpose of multipliers and sub-multipliers
- Use multipliers and sub-multipliers in calculations
- Use engineering notation in calculations
- Use and transpose typical electrical formulas and laws as equations in calculations
- Understand the concepts of work, energy and power as applied to electrical and mechanical calculations
- Recognise and use scalar and vector quantities
- Use Pythagoras' theorem and trigonometry in solving basic right-angle triangle problems
- Understand the purpose of typical types of graphs and charts, and be able to read and interpret simple graphs

PREREQUISITE KNOWLEDGE

Year 10-level English and Mathematics

INTRODUCTION

This chapter presents the units of measurement, mathematical processes and simple mechanics that are used by electrotechnology workers. It also covers the application of fundamental mathematical processes and techniques to solve routine problems.

To understand electrical principles, it is necessary to understand mechanical principles. That in turn requires a working knowledge of a system of units—in the case of Australia and New Zealand (and many other countries), the SI system (see section 14.2.4).

Electrotechnology workers also require what the Australian Core Skills Framework (ACSF) refers to as 'core skills' in learning, reading, writing, oral communication and numeracy. Industry workers need not only be able to learn but also apply their knowledge practically. Electricians in particular often need to manipulate mechanical and electrical values when planning electrical work, fault finding and repairing electrical equipment. An understanding of graphs, methods of graphical solution and trigonometry is required, as is knowledge of the relevant mathematical processes.

Reading allows electrotechnology workers to make sense of information in the form of electrical and mechanical concepts, terminology, measurements and values. Effective writing skills are needed to impart information such as the solutions to problems and to report outcomes and solutions. (Oral communication skills are vital here, too.) Written communication involves clearly conveying information through words, mathematical calculations, graphs, drawings and diagrams.

The ACSF states that numeracy has three interrelated indicators:

- Identification (through reading, observing or listening) of the mathematics required in a given situation
- Understanding mathematical procedures and processes
- Representing and communicating the relevant mathematics.

Electrical problem solving requires an understanding of electrical concepts, terminology and naming conventions, symbols and mathematical formulas. Electrical concepts give us knowledge about how circuits and devices function, and formulas provide tools for problem solving. Electrical formulas are more than mathematical symbols and values—they serve as meaningful explanations for how electrical phenomena work. Behind every symbol and term is an electrical concept or principle of operation.

A formula is a mathematical rule that is expressed symbolically, and a formula populated with given values is an equation. An equation basically says, 'this equals that'. We use the term 'equation' for any formula or law (such as Ohm's Law) that is used in problem-solving and calculation processes.

Calculations enable electrotechnology workers to estimate such necessary values as the voltage drop across circuit components, the current flowing through a circuit and the power consumed by electrical installations, machinery and equipment.

14.1 Solving problems—trade calculations

When you need to do a written trade calculation, it is useful to ask yourself these three questions:

1. What is the question asking me to provide? (Key words could be 'find', 'calculate' or 'determine'.)
2. What information have I been given? (This could be, for example, electrical terms or values for V, I, R and P.)
3. What problem-solving tools will I need? (This relates to the selection of relevant formulas and problem-solving steps.)

There are some basic steps for written problem solving, particularly in situations with more complicated circuits, installations and electrical calculations. The following eight steps constitute a useful framework that can be customised to the particular problem:

1. Read the question and determine what it is asking (the unknown value).
2. Identify the individual pieces of information provided (the 'givens' or known values).
3. List the knowns and the unknowns.
4. Draw a diagram and mark on it the known values given in the question.
5. Determine what you need to do to solve the problem, including what formula(s) to use.
6. Rearrange the equation (by transposition) to make the unknown(s) the subject.
7. Populate the transposed expression with the known values.
8. Calculate the answer and express it in the format required. This may be, for example, engineering notation.

14.2 Mathematics, numbers and units

Many of the mathematical concepts and principles presented in this chapter can be used as tools for solving trade calculation problems. Having a good grasp of school-level maths will be useful in getting the most from the content, as will the use of an electronic calculator (one that is approved for use in your training course). Calculator features include functions for:

- scientific and engineering notation: EXP or ($\times 10^x$)
- exponentials: 10^x and e^x
- powers and roots: x^2 and $\sqrt{x}$; x^3 and $\sqrt[3]{x}$
- reciprocals: x^{-1} or $\frac{1}{x}$
- trigonometric and inverse trigonometric ratios: sin, cos, tan, $\sin^{-1}$, $\cos^{-1}$, $\tan^{-1}$
- percentages: %.

Using a calculator is a skill that requires learning, practice, an understanding of mathematical concepts and, in the case of the electrotechnology industry, SI units (see section 14.2.5).

14.2.1 Types of mathematics

Arithmetic and other types of mathematics such as algebra, geometry and trigonometry relate to the science of numbers. This includes their operations, interrelationships, combinations and structure, as well as logical reasoning and quantitative calculation.

Arithmetic deals with the everyday calculations for working with numbers—addition (+), subtraction (–), multiplication (×) and division (÷). These are sometimes referred to as the 'four branches of arithmetic', and they involve using fractions, decimals, whole numbers, percentages relating to division and 'exponents' relating to multiplication.

- Examples of fractions are $\frac{1}{4}$, $\frac{2}{3}$ and $\frac{17}{64}$.
- Examples of decimal fractions (usually referred to simply as 'decimals') are 0.6, 0.06 and 0.006.
- Examples of whole numbers or integers are 2, 5, 9, 22, 87 and 135.

14.2.2 Algebraic conventions for (+), (–) and (x)

When two signs follow each other, for example + – b, brackets ( ) are used to show which sign belongs to the number. The other sign shows the function being performed.

For example, 'a + – b' or 'a + (–b)' is simply written as 'a – b'.

The multiplication sign (×) is often replaced with a dot or is removed altogether.

For example, 'a × b' can be written as 'a.b' or simply 'ab'.

14.2.3 BODMAS

The acronym BODMAS represents a sequence of use of functions in simple maths. It stands for:

- **B**rackets
- **O**rders (exponents and roots)
- **D**ivision/**M**ultiplication
- **A**ddition/**S**ubtraction.

In this sequence, the functions look like:

$$B = \text{Brackets } (X), \{X\}, [X], \langle X \rangle$$

$$O = \text{Orders } \sqrt{X}, X^5$$

$$D = \text{Division } X \div Y, \frac{X}{Y}$$

$$M = \text{Multiplication } X \times Y$$

$$A = \text{Addition } X + Y, X - (-Y), \text{sum}(X), \Sigma X$$

$$S = \text{Subtraction } X - Y, X + (-Y)$$

Sometimes the alternative 'BOMDAS' is used, where the priority of calculation is Brackets, Orders, Multiplication/Division and Addition/Subtraction. Another variation is 'BIDMAS', which stands for 'Brackets, Indices, Division/Multiplication and Addition/Subtraction'.

These orders of functions give subtraction the same importance as addition, and division the same importance as multiplication.

Consider the effect that BODMAS has in the following expressions:

$$2 + 4 \times 3 - 1$$
$$2 - 1 + 4 \times 3$$

Working from left to right is the normal process, but BODMAS must be applied first. The multiplication must be done before the addition and subtraction, so the correct answer for both expressions is 13.

BODMAS: the importance of brackets

The function of brackets is very important. Consider the effect of brackets in the following calculations:

$$4 \times 3 - 2 + 1 \quad = 11$$
$$4 \times 3 - (2 + 1) = 9$$
$$4 \times (3 - 2) + 1 = 5$$
$$4 \times (3 - 2 + 1) = 8$$

Consider the result of performing addition before subtraction in the calculation $4 \times 3 - 2 + 1$. The correct answer, working from left to right as per BODMAS, is 11.

BODMAS: the importance of orders

Orders are numbers that involve powers, (indices, exponents) or roots (square roots, cube roots). Examples are shown in Table 14.1.

TABLE 14.1 Powers and roots

Number (3, 4, 5) with power, index or exponent (2)			Square root of number		
$3^2 = 9$	$4^2 = 16$	$5^2 = 25$	$\sqrt{9} = 3$	$\sqrt{16} = 4$	$\sqrt{25} = 5$

Proficient use of BODMAS brings the ability to deal with all of the equations in electrical trade studies.

EXAMPLE 14.1

Use BODMAS to solve the following expression: $2 + 3^2 (25 \div 5)$.

Brackets first: $(25 \div 5) = 5$
Our expression now reads: $2 + 3^2 \times 5$
Followed by Orders: $3^2 = 3 \times 3 = 9$
Our expression now reads: $2 + 9 \times 5$
Followed by Multiplication: $9 \times 5 = 45$
Followed by Addition: $45 + 2 = 47$

CHECK YOUR UNDERSTANDING

14.1 Use BODMAS to solve 3 – 2 + 1.

14.2 Use BODMAS to solve 1 – 2 + 3.

14.3 Use BODMAS to solve $A = \pi r^2$ where r = 25 mm.

14.4 Use BODMAS to solve $W = \frac{1}{2}LI^2$.

14.2.4 SI units

In the early 1970s, Australia legislated to use of the metric system or, to give it its proper title, the 'Système international (d'unités)'. The Australian Standard is defined in AS ISO 1000-1998. The system consists of seven SI base units and many derived units, and are involved in much of the maths used by electrotechnology workers and electricians.

Written electrical problems feature the SI unit quantity term (such as 'voltage', 'current', 'resistance' and 'power'), the quantity value and quantity symbols (V, I, R, P etc.), which are also referred to as the 'dimension symbols'. These, like the unit symbols (V, A, Ω, W etc.), should be memorised.

14.2.5 Problem solving using SI units

EXAMPLE 14.2

Customise and apply the written problem-solving steps suggested at the beginning of this chapter to this example, which involves the use of Ohm's Law.

1. Read the description.
   Connecting a component to a 12 V supply voltage results in a current of 500 mA flowing in the circuit. Determine the value of the component resistance and power consumed.
2. Identify the SI quantities given (see Table 14.2).

TABLE 14.2 Identifying known and unknown values

Quantity symbol	SI unit (term) quantity	Value/quantity	(SI) Unit name	Unit symbol
V	**Voltage**	12	volt	V
I	Electric **current**	0.5	ampere	A
R	Electric **resistance**	**FIND**	ohm	Ω
P	**Power**	**FIND**	watt	W

3. List the givens as known and unknown values to calculate.
   V = 12 V
   I = 500 mA (or 0.5 A)
   R = ?
   P = ?

4. Draw a diagram (see Figure 14.1).

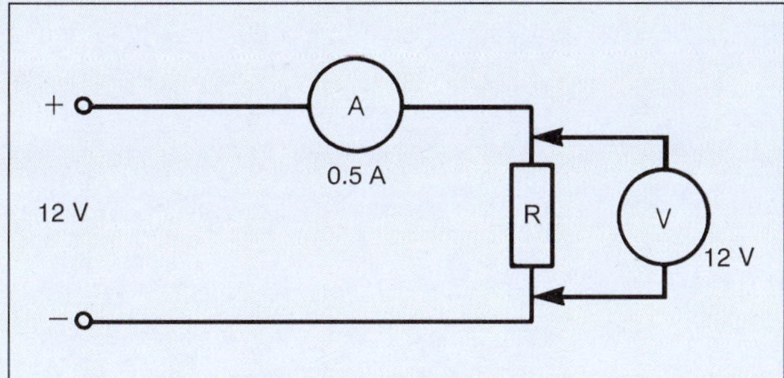

FIGURE 14.1 Basic series circuit

5. Determine what formula(s) to use.

$$I = \frac{V}{R}$$

$$P = V \times I$$

6. Rearrange (transpose) if necessary.

$$I = \frac{V}{R}$$

$$\text{becomes } R = \frac{V}{I}$$

7. Solve.

$$R = \frac{V}{I} = \frac{12}{0.5} \text{ or } \frac{12}{500 \times 10^{-3}} = \frac{12}{0.5} = 24 \text{ ohms}$$

$$P = V \times I = 12 \times 0.5 \text{ (or } 12 \times 500 \times 10^{-3}) = 6 \text{ W}$$

14.3 Fundamental skills in mathematics

14.3.1 Using decimals

A decimal expresses whole numbers and fractional parts of a whole number, separating them with a decimal point. Or, put another way, a decimal number can be defined as 'a number, the whole number part and the fractional part, separated by a decimal point'.

Take for example the number 334.56 (see Figure 14.2). The whole part (to the LEFT of the decimal point) represents integers as Units, Tens, Hundreds and Thousands, which can also be represented as powers of ten. The whole-number part—334—consists of three hundred plus thirty plus four.

The fractional part (to the RIGHT of the decimal point) represents one- to nine-tenths of a whole number $\left(\frac{1}{10} \ldots \frac{9}{10}\right)$, one- to nine-hundredths $\left(\frac{\times}{100}\right)$, one- to nine-thousandths $\left(\frac{\times}{1000}\right)$. The fractional number part (to the *right* of the decimal point)—0.56—consists of five tenths $\left(\frac{5}{10}\right)$ and six hundredths $\left(\frac{6}{100}\right)$.

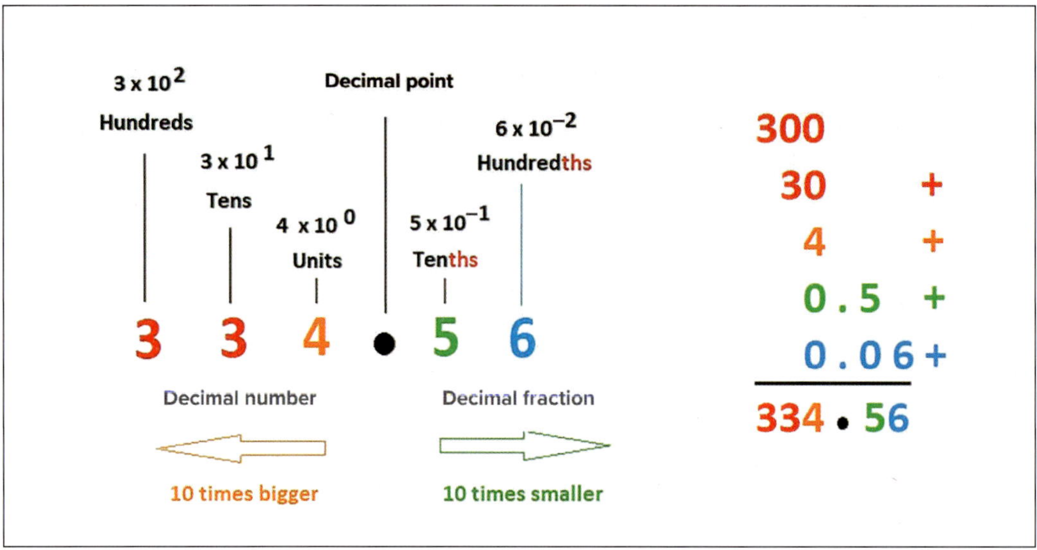

FIGURE 14.2 Decimals and decimal fractions

14.3.2 Decimals and powers of 10

The example number 334.56 in Figure 14.2 shows the relationship of Units, Tens and Hundreds to powers of 10. The Hundreds numeral three (3) value is ten times bigger than the Tens numeral three (3).

The powers of 10 in the number 334.56 show these powers for the whole numbers as 3×10^2 for 300, 3×10^1 for 30 and 4×10^0 for the units. Note that $10^0 = 1$ so 4×10^0 is the same as four times one. The number ten is said to be raised by the powers 0, 1 and 2. Powers are also known in maths as 'exponents' or 'indices'. The fractional part of the number 334.56 shows these powers as 10^{-1} and 10^{-2} (the index or exponent values are –1 and –2). Indices are the number exponent or power.

So, the fractional decimal 0.707, moving from the right of the decimal point, represents:

$$0.707 = \frac{7}{10\text{s}} + \frac{0}{100\text{s}} + \frac{7}{1\,000\text{s}}$$

Consider the following number: 765.432

We have:

$$\begin{aligned} 7 \text{ Hundreds } (7 \times 10^2) &= 700.000 \\ 6 \text{ Tens } (6 \times 10^1) &= 060.000 \\ 5 \text{ Units } (5 \times 10^0) &= 005.000 \\ 4 \text{ Tenths } (4 \times 10^{-1}) &= 000.400 \\ 3 \text{ Hundredths } (3 \times 10^{-2}) &= 000.030 \\ 2 \text{ Thousandths } (2 \times 10^{-3}) &= 000.002 \end{aligned}$$

14.3.3 Multiplication and division by powers of 10

When a whole number is multiplied by 10, a zero is added to the right-hand side of the number. When multiplying by 10, 100, 1000 or a power of 10, the decimal point needs to move to the right by the same number of zeros of the multiplier. For example:

- $10 \times 1 = 10$
- $10 \times 0.707 = 7.07$
- $100 \times 0.707 = 70.7$
- $1000 \times 1.414 = 1414$

When multiplying decimal fractions, count the number of digits to the right of the decimal point for each number so that this number of digits is to the right of the decimal point in your answer.

The following are examples of decimal fractions:

$$15 \times 24 \text{ (0 decimal places)} = 360 \text{ (0 decimal places)}$$
$$15 \times 2.4 \text{ (1 decimal place)} = 36.0 \text{ (1 decimal place)}$$
$$1.5 \times 2.4 \text{ (2 decimal places)} = 3.60 \text{ (2 decimal places)}$$
$$0.15 \times 2.4 \text{ (3 decimal places)} = 0.360 \text{ (3 decimal places)}$$

When dividing decimals with fractions (particularly without the aid of a calculator), you can multiply both numbers by multiples of 10. This helps avoid putting the decimal point in the wrong place in your answer.

For example, 2.4 divided by 1.5 would be: $2.4 \div 1.5 = 1.6$ $(2.4 \times 10) \div (1.5 \times 10)$.

This becomes: $24 \div 15 = 1.6$.

14.3.4 Scientific notation

When dealing with very large or very small numbers, the standard number system becomes too hard to work with—numbers like 0.0 000 000 301 or 6 060 842 000 000 are difficult to calculate, either manually or with a calculator.

Scientific notation is a system that represents these numbers differently. The magnitude of the number is converted into an exponent, and the value of the number (sometimes referred to as the 'mantissa') is expressed with one digit before a decimal place.

For example, the number 6 060 842 000 000 expressed in scientific notation would be:

6.060842×10^{12}—the exponent 10^{12} means that the number is 1 000 000 000 000 multiplied by the mantissa, or the decimal point is moved 12 places to the right.

A similar rule applies to small numbers like 0.0 000 000 301, which would be:

3.01×10^{-9}—the exponent 10^{-9} means the mantissa (3.01) is multiplied by 0.00 000 001, or the decimal point is moved nine places to the left.

Variations in calculators make it important to understand how individual models manage scientific notation. For example, the exponent key may have 'EXP' on it, or it may have '10x'—it depends on the model.

The advantages of using scientific notation include representing very large or very small numbers more easily as well as its direct use in calculations. The use of significant figures generally means scientific notation uses no more than three or four digits in the mantissa.

14.3.5 Engineering notation

Engineers and tradespeople often need to talk about the numbers they use, perhaps when buying components and making installation decisions. Scientific notation is cumbersome for use in language, so *engineering notation* is used instead. In engineering notation, the mantissa is not 1 to 9 as in scientific notation, but 1 to 999, and the exponent is given in multiples of three—10^3, 10^6, 10^9 etc. These are also given names that are easy to remember and in fairly common usage, for example, 10^3 = 'kilo', 10^6 = 'mega' and 10^9 = 'giga'.

14.3.6 Using multiples and sub-multiples

The words used before a unit are known as 'multiples' and 'sub-multiples'. To get the most from this textbook, try to memorise the multipliers in Table 14.3. It is helpful to start by using familiar values; thinking about, for example, how many millimetres are in a metre or how many metres are in a kilometre makes the leap to thinking about milliamps less daunting.

TABLE 14.3 Engineering notation

Type of notation	Multipliers								
	Tera (T)	Giga (G)	Mega (M)	kilo (k)	Base	milli (m)	micro (μ)	nano (n)	pico (p)
Engineering	10^{12}	10^{9}	10^{6}	10^{3}	10^{0}	10^{-3}	10^{-6}	10^{-9}	10^{-12}
Exponential	E12	E9	E6	E3	E0	E–3	E–6	E–9	E–12

EXAMPLE 14.3

How many metres would a person have to walk to cover 2.37 kilometres?

The term 'kilo' means 'thousand', so 2.37 kilometres is the same as 2.37 thousand metres or 2370 metres.

EXAMPLE 14.4

How many farads are there in 125 picofarads (125 pF)? From Table 14.3, it can be seen that there are 1 000 000 000 000 picofarads in one farad.

Therefore: $\frac{125}{1\,000\,000\,000\,000} = 0.000\,000\,000\,125$ farads.

The convenience of the former figure of 125 pF is evident. '125E−12' (125×10^{-12}) would be entered into a calculator.

CHECK YOUR UNDERSTANDING

14.5 When the prefix 'milli' is used, how much is a unit multiplied by?

14.6 When the prefix 'mega' is used, how much is a unit multiplied by?

14.7 What is the prefix for a unit that is multiplied by 10^{-6}?

14.8 What is the prefix for a unit that is multiplied by 10^{3}?

14.9 How many microhms does 0.0002 ohms equal?

14.3.7 Significant and non-significant figures

Measurement depends on the sensitivity of the measuring instrument and the skill deployed in taking the measurement. The number of significant figures in a measurement indicates the precision of the value (see Table 14.4). The rules for non-significant figures are shown in Table 14.5.

TABLE 14.4 Rules for significant figures

General rules for significant figures	Examples
All non-zero numbers are significant.	6.25 mm, 0.514 cm, 25.3 cm and 0.000432 cm all contain three significant figures.
Zeros between two non-zero digits are significant.	401 cm, 20.5 cm and 1.03 mm all contain three significant figures.
Trailing zeros to the right of the decimal are significant.	0.400 cm, 0.330 cm and 2.00 mm all contain three significant figures.
Trailing zeros in a whole number with the decimal shown are significant.	250.0 mm, 510.0 mm and 200.0 cm all contain four significant figures.

TABLE 14.5 Rules for non-significant figures

General rules for non-significant figures	Examples
Trailing zeros in a whole number without the decimal shown are not significant (but see the note following this table).	Of the numbers 250, 510 and 200, only 25, 51 and 2 are significant.
Leading zeros are not significant (zeros preceding the first non-zero digit are not significant).	0.0 101 cm, 0.00 100 cm and 0.000 101 cm only contain three significant figures.

Note: Trailing zeros in a whole number without the decimal shown may in some situations be considered as significant. While trailing zeros for the values 250 mm, 510 mm and 200 cm may not be considered significant in terms of measurement accuracy, the significance of measurement figures is *resolution dependent.*

14.3.8 Rounding off

Rounding can be either to significant figures or decimal places. When problem solving, it is important to read the question to determine which rounding method is required. Trade maths calculations will ask you to round off a number to either one, two or three decimal places or to a certain number of significant figures. Except in specific cases, it is generally not necessary to calculate past three decimal places. It should be noted also that trade maths calculation answers are expressed in engineering notation.

$$\frac{3}{16} = 0.1875 \text{ correct to one place} = 0.2$$

$$\text{correct to two places} = 0.19$$

$$\text{correct to three places} = 0.188$$

Rounding off decimals depends on three considerations:

1. How many decimal places are we rounding off to?
2. What is the value of the digit that follows the number of digits that we are rounding off?
3. Are we rounding up or rounding down?

For example, if 1.47 needs to be rounded to one decimal place, the next or following digit is a 7. If the following number value is between 5 and 9 inclusive, we round up. As the value 7 falls within this range, we round the number 4 up so that 1.47 rounds up to a 1.5 value.

Take 1.414 rounded to two decimal places. The next or following digit is a 4. If the following number value is between 0 and 4 inclusive, we round down. That is, the number value remains unchanged—1.414 to two decimal places remains a 1.41 value.

CHECK YOUR UNDERSTANDING

14.10 What is 1.9792 rounded to two decimal places?

14.11 What is 18.738 rounded to two decimal places?

14.12 What is 1.9792 rounded to two significant figures?

14.13 What is 1.738 rounded to two significant figures?

14.3.9 Fractions

Learning the concepts that govern the use of fractions is important for transposing equations. Electronic calculators have either the $\frac{1}{x}$ or x^{-1} keys to simplify the use of fractions and reciprocals in electrotechnology problem solving.

Proper fractions are parts of a whole number less than 1 (<1) and can, for example, be represented as eight parts (eighths) or sixteen parts (sixteenths). Improper fractions are greater than 1 (>1). For example, pi (π) can be expressed as a decimal 3.14, as $3\frac{1}{7}$ or as an improper fraction $\frac{22}{7}$.

A fraction is expressed as a *numerator* divided by a *denominator*. The fraction $\frac{7}{16}$ represents the number 7 as the numerator and the number 16 as the denominator. The denominator here represents the fraction in sixteenths. This fraction can be converted to a decimal fraction by dividing 16 into 7.

Seven sixteenths as a decimal fraction is $\frac{7}{16} = 0.4375$

In section 14.3.2, the number 0.707 is represented as a decimal fraction. This can be written in the expanded form of $\left(7 \times \frac{1}{10}\right) + 0 + \left(\frac{7}{1000}\right)$.

CHECK YOUR UNDERSTANDING

14.14 Write $\frac{7}{10} + \frac{5}{100}$ as a decimal fraction.

14.15 Write $\frac{2}{10} + \frac{4}{100} + \frac{8}{1000}$ as a decimal fraction.

14.16 Convert $\frac{1}{8}$ as a decimal fraction.

14.17 Convert $\frac{5}{16}$ as a decimal fraction.

Working with fractions

When the numerator and denominator of a fraction are both multiplied by the same number, a different form results—but not a different value.

For example, $\frac{1}{8} = \frac{2}{16} = \frac{4}{32} = \frac{8}{64}$

and $\frac{1}{2} = \frac{2}{4} = \frac{4}{8} = \frac{8}{16} = \frac{16}{32}$

Adding and subtracting fractions—establish a common denominator

The most common examples in trade maths of adding fractions are in calculating capacitors in series or the equivalent resistance for two, three, four or more parallel resistances. One of the key points to remember when estimating what this value could be is that the total equivalent resistance value is always less than the lowest-value resistor in parallel. It is worth repeating that the $\frac{1}{x}$ or x^{-1} calculator keys can simplify working with fractions and reciprocals.

EXAMPLE 14.5

Calculate the following parallel resistance values without a calculator.

$R_T = R_1 // R_2 // R_3$

The lowest common denominator between 2, 4 and 8 will be 8. How many times do 2 and 4 divide into 8?

For the reciprocal values of R_1: $\frac{1}{2} = \frac{4}{8}$; and for R_2: $\frac{1}{4} = \frac{2}{8}$

See the diagram of three parallel resistors (Figure 14.3).

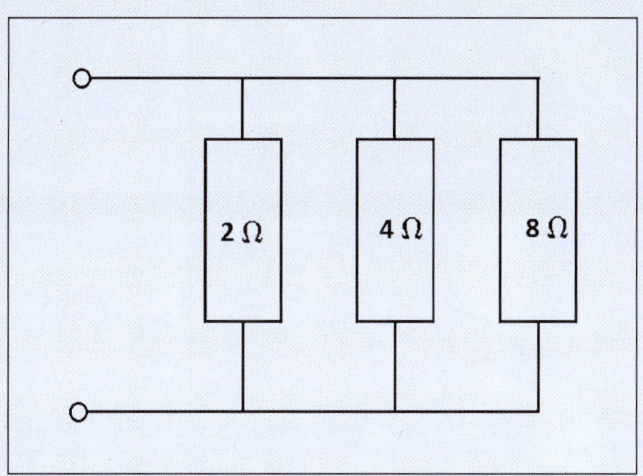

FIGURE 14.3 **Three parallel resistors**

$$\frac{1}{R_T} = \frac{1}{R_1} + \frac{1}{R_2} + \frac{1}{R_3}$$

$$\frac{1}{R_T} = \frac{4}{8} + \frac{2}{8} + \frac{1}{8}$$

$$\frac{1}{R_T} = \frac{7}{8}$$

$$\frac{R_T}{1} = \frac{8}{7} = 1.14\ \Omega$$

CHECK YOUR UNDERSTANDING

14.18 Add $\frac{1}{5}$ and $\frac{1}{8}$, giving your answer as a fraction.

14.19 Add $\frac{1}{4}$ and $\frac{1}{5}$, giving your answer as a fraction.

14.20 Calculate $\frac{5}{8} - \frac{5}{16}$, giving your answer as a fraction.

14.21 Calculate $\frac{5}{16} - \frac{7}{32}$, giving your answer as a fraction.

14.3.10 Percentages

'Percentage' means 'per 100' or 'out of one hundred', and percentages are fractions with 100 on the bottom (the denominator). When writing fractions and decimals as percentages, multiply by 100 and put the % sign in place. Figure 14.4 shows fractions converted to decimals, which, when multiplied by 100, gives the value as a percentage.

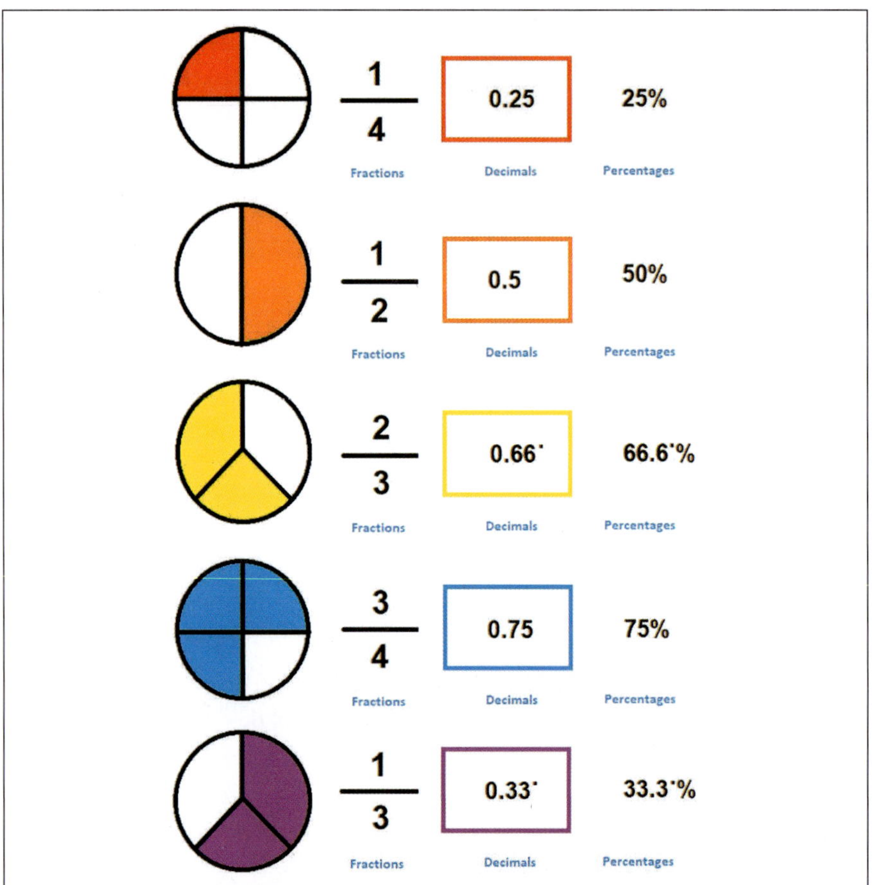

FIGURE 14.4 Converting fractions and decimals to percentages

14.3.11 Transposition

The ability to work with formulas, laws and equations using *transposition* is vital in performing electrical calculations. Transposition requires knowledge of fractions (and it also requires not inadvertently swapping over numerators and denominators).

A formula is like a recipe for how to solve a particular problem. To avoid writing a new equation for each variation of the same problem, a formula can be transposed when a different variable needs to be found. The most common formula in the electrical trade is Ohm's Law, $V = IR$ or $E = IR$, which, when transposed, can be written as $I = \frac{V}{R}$

Although V is typically used to express voltage, E is sometimes used to express a source of voltage.

EXAMPLE 14.6

The formula associated with Ohm's Law is:

$$V = I \times R \qquad (1)$$

Consider the calculation using the values:

$$V = 12 \qquad (2)$$
$$I = 4 \qquad (3)$$
$$R = 3 \qquad (4)$$
$$\therefore 12 = 4 \times 3 \qquad (5)$$

Transpose the equation to find I:

$$I = \frac{V}{R} \quad (6)$$

$$\therefore 4 = \frac{12}{3} \quad (7)$$

Transpose the equation to find R:

$$R = \frac{V}{I} \quad (8)$$

$$\therefore 3 = \frac{12}{4} \quad (9)$$

Note that the two sides are equal in each case.

Transposition works by the principle that the two sides of an equation are equal and must always remain so. Two common methods of ensuring this are:

1. Do the same operation to both sides—if a number is added to one side, it must be added to the other side. The same goes for multiplication, division, subtraction, squaring or even using Sine, Cosine and Tangent.
2. Change sides, change signs—this means that a number can be moved from one side of an equation to the other by taking the inverse of it on both sides. For example, to move +8 from the right- to the left-hand side, you would subtract 8 from both sides.

Remember, however, to follow the rules of BODMAS. For example, do not take a number out of a bracket, and do the calculation between the brackets first whenever possible.

Transposition techniques include:

- using the cross-multiplication method
- dividing both sides of the equation by the same quantity
- adding or subtracting the same quantity to both sides of the equation
- applying a function to both sides of the equation.

The following examples demonstrate how these techniques can be applied.

Transposition using the cross-multiplication method

When transposing an equation, working with fractions can sometimes be avoided by a process called 'cross-multiplication'. The following are three examples of how it works.

(a) Find ϕ:	(b) Find B:	(c) Find mmf:
$B = \frac{\phi}{A}$	$\mu = \frac{B}{H}$	$R_m = \frac{mmf}{\phi}$
$\frac{B}{1} = \frac{\phi}{A}$	$\frac{\mu}{1} = \frac{B}{H}$	$\frac{Rm}{1} = \frac{mmf}{\phi}$
$B \times A = \phi \times 1$	$\mu \times H = B \times 1$	$Rm \times \phi = mmf \times 1$
$\therefore \phi = BA$	$\therefore B = \mu H$	$\therefore mmf = Rm \cdot \phi$

Dividing both sides of the equation by the same quantity

Using the equation Power (P) = Voltage (V) × Current (I):

(a) Find V:	(b) Find I:
$P = V \times I$	$P = V \times I$
Divide both sides by I: $\frac{P}{I} = V \times \frac{I}{I}$	Divide both sides by V: $\frac{P}{V} = \frac{V}{V} \times I$
$I \div I$ cancels out, therefore: $V = \frac{P}{I}$	$V \div V$ cancels out, therefore: $I = \frac{P}{V}$

Adding or subtracting the same quantity to/from both sides of the equation

(a) Add the same quantity to both sides of the equation.	(b) Subtract the same quantity from both sides of the equation.
Find the power input: $P_{out} = P_{in} - \text{Losses}$ $P_{out} + \textbf{Losses} = P_{in} - \text{Losses} + \textbf{Losses}$ $P_{in} = P_{out} + \text{Losses}$	Find the power output: $P_{in} = P_{out} + \text{Losses}$ $P_{in} - \textbf{Losses} = P_{out} + \text{Losses} - \textbf{Losses}$ $P_{out} = P_{in} - \textbf{Losses}$

Applying a function to both sides of the equation

Examples of how this is used can be shown by flipping the numerator and denominator on both sides (the *reciprocal*) and obtaining the square root of both sides of the equation.

(a) Finding the reciprocal of both sides:	(b) Obtaining the square root of both sides of the equation:
$\frac{1}{R_T} = \frac{7}{8}$ $\frac{R_T}{1} = \frac{8}{7} = 1.14\ \Omega$	If $P = \frac{V^2}{R}$ then find V $\sqrt{V^2} = \sqrt{(P.R)}$ $V = \sqrt{(P.R)}$

In Example 14.7, in (10) a squared number on the right can be balanced by taking the square root of the left side. Lines (12)–(15) show the same process for a more complex equation. The result is that x is left on its own.

EXAMPLE 14.7

The following are examples to demonstrate transposition. In each, find x.

Treat each side the same:

$$z = x + y \quad (1)$$
$$z - y = x + y - y \quad (2)$$
$$\therefore z - y = \underline{x} \quad (3)$$

Or change sides, change signs:

$$z = x + y \quad (4)$$
$$\therefore z - y = \underline{x} \quad (5)$$

As above:

$$z = x \times y \quad (6)$$
$$\therefore z \div y = x \times y \div y \quad (7)$$
$$\text{(or) } \frac{z}{y} = \frac{xy}{y} \quad (8)$$
$$\therefore \frac{z}{y} = \underline{x} \quad (9)$$

Note:

$$\text{if } z = x^2 \qquad (10)$$

$$\therefore \sqrt{z} = x \qquad (11)$$

$$\therefore \text{if } z = \sqrt{x^2 + y^2} \qquad (12)$$

$$\therefore z^2 = x^2 + y^2 \qquad (13)$$

$$\therefore z^2 - y^2 = x^2 \qquad (14)$$

$$\therefore \sqrt{z^2 - y^2} = x \qquad (15)$$

14.3.12 Ratios

One of the most common applications of mathematical ratios is in electrical transformers. An example would be a downlight, with the supply voltage (nominally 240 V) stepping down to supply a 12 V lamp. Dividing $\frac{240}{12}\left(\frac{V_1}{V_2}\right)$ gives a voltage ratio of 20:1. The ratio relationship between voltage and coil winding turns is shown in Figure 14.5.

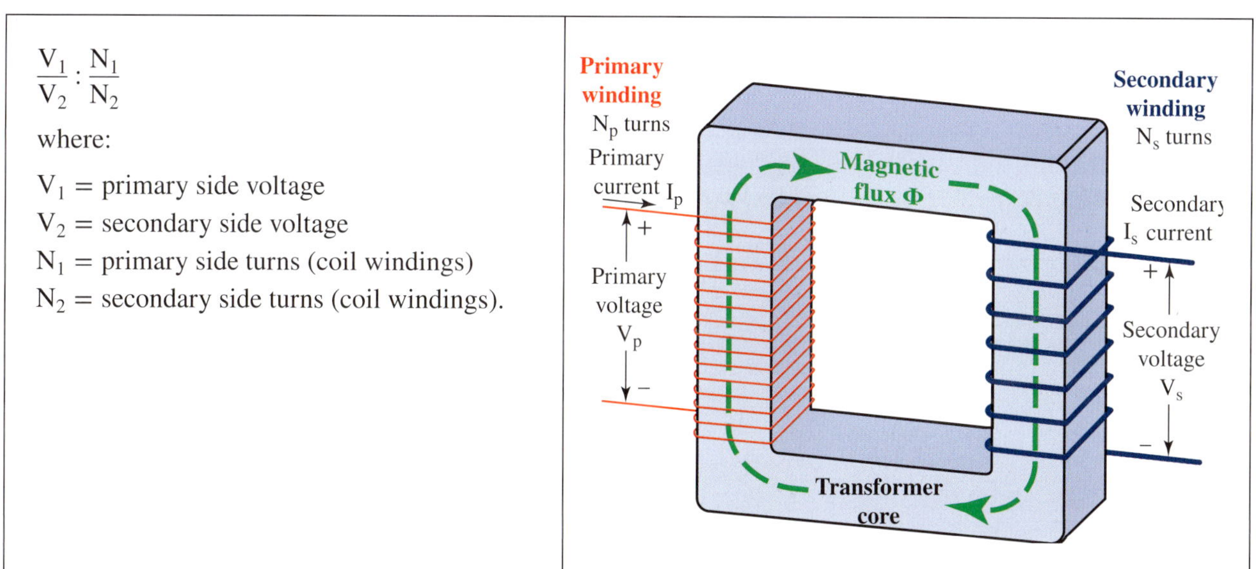

FIGURE 14.5 Transformer voltage and turns ratio

Architectural, electrical and mechanical drawing use a measurement to scale. This 'drawing to scale' is another application of using ratios (e.g. 1:100, 1:500, 1:1000). Floor plans are drawn to a scale for ease of representation. Actual dimensions are scaled from larger to smaller metric units.

WORKPLACE SCENARIO

You have been called out to a hotel kitchen to attend to a bain-marie. A typical bain-marie consists of either a fixed-shape or a flexible-length heating element (the flexible type can be in different lengths for the same wattage). When you arrive at the job, you find that you need to replace a flexible element. Having been a diligent student, you recall from your studies of direct current that doubling the length of a conductor also doubles the resistance. This provides a ratio of resistance based on length.

Frustratingly, you find that the existing element is not in one piece; but you've got a good grasp of your trade maths and you remember that knowing the resistance without the correct element length requires determining that length using resistance measurements and a ratio expression. This takes the following form.

$$\frac{R_1}{R_2} : \frac{l_1}{l_2}$$

Cross-multiply so:

$$R_1 \times l_2 = R_2 \times l_1$$

Transpose for l_2:

$$l_2 = \frac{R_2 \times l_1}{R_1}$$

where:

R_1 = resistance for section of element
R_2 = element resistance based on wattage
l_1 = length for section of broken element
l_2 = element length to source as replacement.

Knowing the wattage and supply voltage for the element allows you to determine the total resistance. Measuring the resistance for the broken section of element, the resistance length ratio equation tells you the total element length.

CHECK YOUR UNDERSTANDING

14.22 State the ratio of circles to squares in the following pattern:

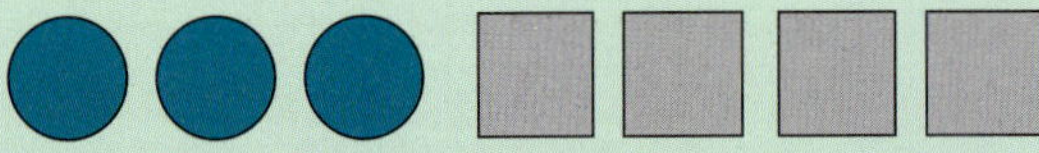

14.23 State the ratio of squares to circles in the following pattern:

14.3.13 Shapes, area and angles

Types of geometric shapes found in length measurements and area calculations for electrical problems include circles, squares, rectangles, rhombuses and parallelograms. Triangles, various types of angles and trigonometry are used in calculations for more advanced vector and phasor calculations.

Figure 14.6 shows circle dimensions and calculations. These involve radius, diameter, circumference and area measurements of cable and conduit for electrical installations.

Squares and rectangles are common in site plans, floor plans and items such as electrical switchboards. Measurements and area calculations involve length and width, and there are other three-dimensional measurements involving height or depth as well, for area and volume calculations.

The measurements on floor plans allow workers to determine where socket-outlets, switches, ceiling fans and light fittings need to be installed. Orthogonal drawings have separate floor-plan, side-section and elevation views with length, width and height measurements for electricians to refer to. Room area calculations are used for determining a variety of installation requirements for lighting, heating and cooling.

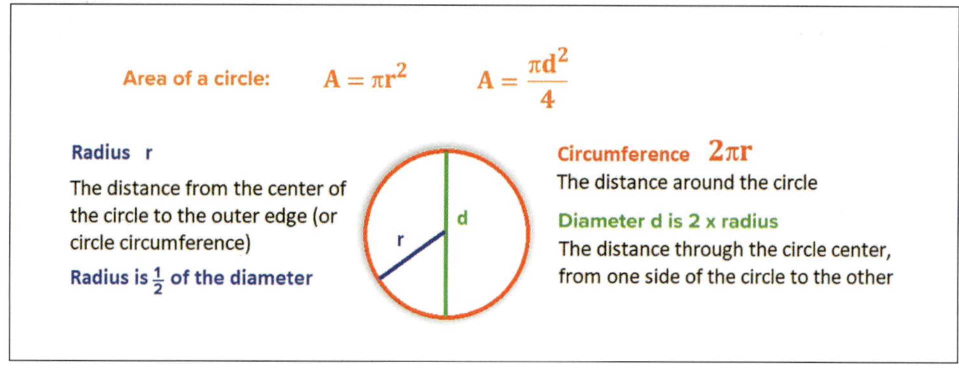

FIGURE 14.6 Circle dimensions and calculations

In addition to circles, squares and rectangles, area calculations may need to be based on a combination of different shapes. These could include triangles, rhombuses and trapeziums, as seen in Figure 14.7.

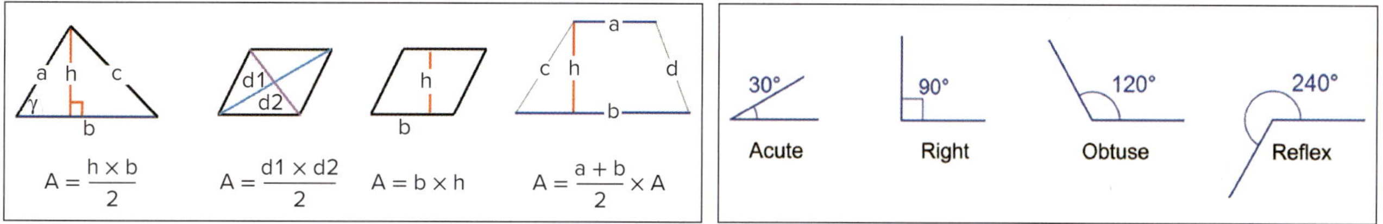

FIGURE 14.7 Calculating area

FIGURE 14.8 Types of geometric angles

Geometric angles

Figure 14.8 shows the names of the various types of angles found in geometric shapes and used in electrical problem solving.

14.3.14 Pythagoras' theorem

Table 14.6 shows three sets of squares. Counting the individual squares shows that 'a' is 3 × 3 square, 'b' is 4 × 4 and 'h' is 5 × 5.

TABLE 14.6 Pythagoras' theorem

Squares as area	Pythagoras' theorem
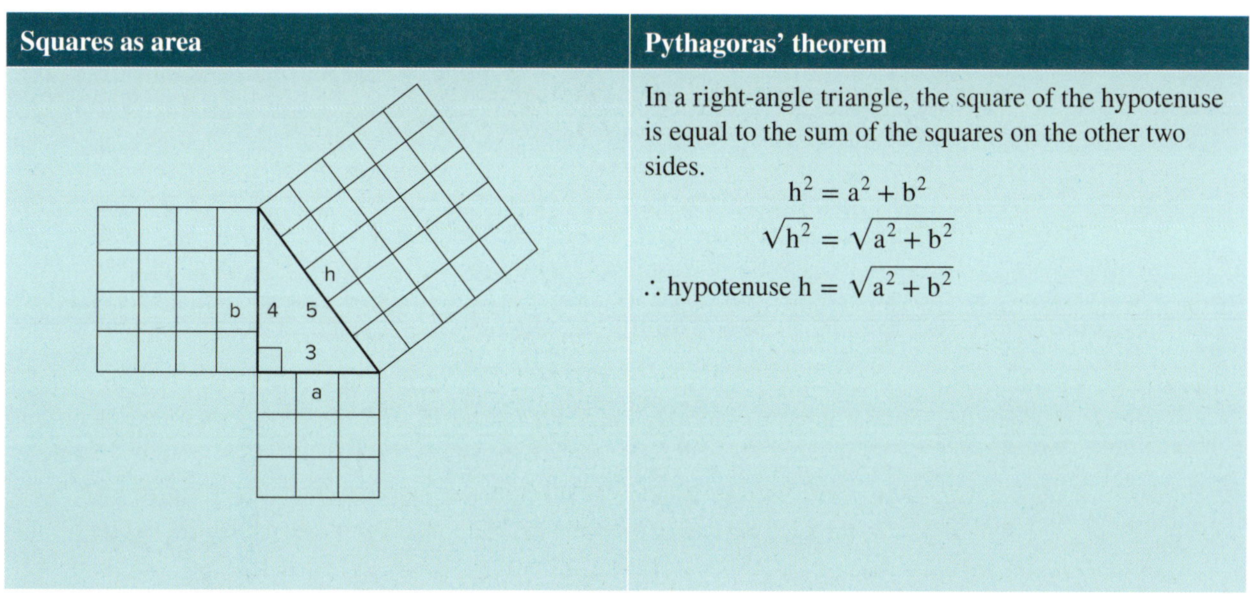	In a right-angle triangle, the square of the hypotenuse is equal to the sum of the squares on the other two sides. $h^2 = a^2 + b^2$ $\sqrt{h^2} = \sqrt{a^2 + b^2}$ $\therefore$ hypotenuse $h = \sqrt{a^2 + b^2}$

Figure 14.9 presents electrical problem-solving applications for Pythagoras' theorem. The impedance and power triangles shown are used in a.c. calculations.

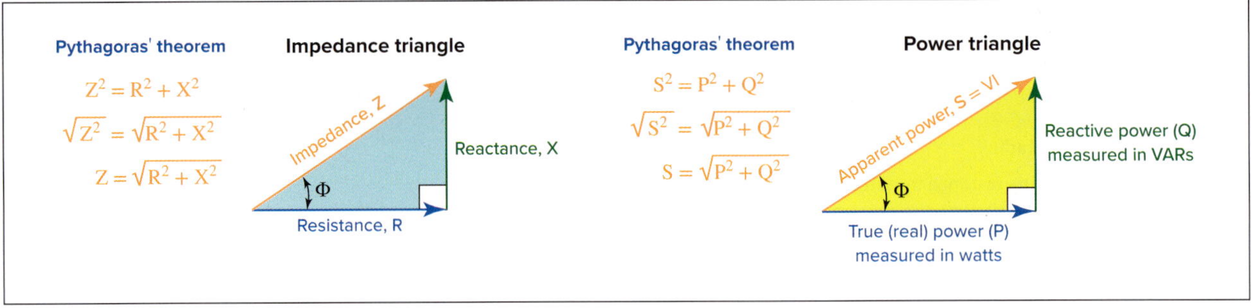

FIGURE 14.9 Impedance and power triangles

14.3.15 Graphs and charts

Information can be presented in the form of charts, tables, graphs and diagrams. Four of the more common types of chart are pie charts, histograms, bar charts and line charts. They are visual methods of showing comparisons between different values of quantities.

Pie charts are like actual pies in that their slices show the percentage relationship between different quantities. Graphs are a type of chart. They show the mathematical relationship between values of different quantities (see Figure 14.10).

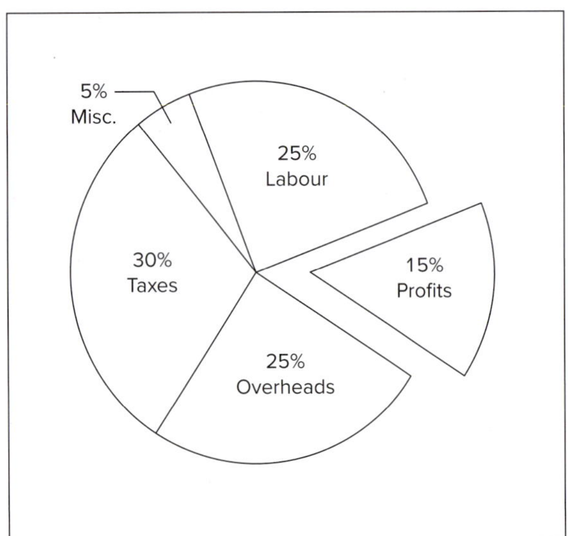

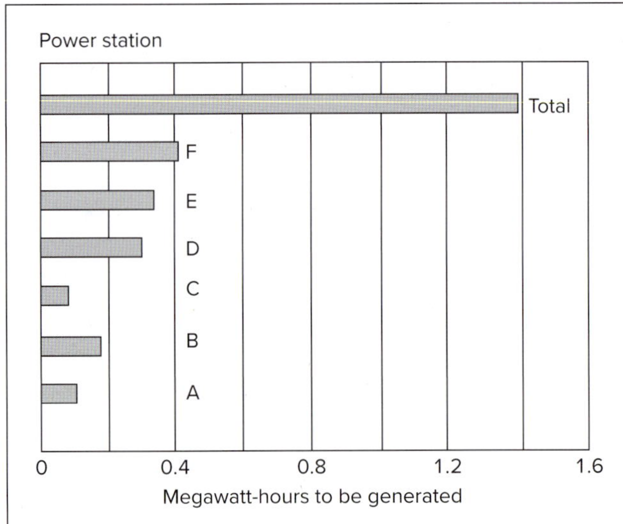

FIGURE 14.10 (a) Pie chart with 'slices' (b) horizontal bar graph

Line charts are also referred to as 'line graphs'. Information values, referred to as 'data', are plotted against two axes. The x-axis is shown as a horizontal line across the graph while the y-axis is drawn perpendicular (at 90 degrees) to the x-axis. The example in Figure 14.11(a) shows the x- and y-axes joining at a common zero point. Figure 14.11(b) shows engine rpm on a common x-axis with two separate vertical axes—one for power output and one for torque.

Line graphs and *Cartesian graphs* are used extensively in electrical and electrotechnological problem solving. Both have an x-axis and a y-axis for plotting data co-ordinates. (Graphs do not always start at 0, and quite often 0,0 is the mid-point of the graph.)

Cartesian graphs compare two sets of numbers or co-ordinates, each set being represented as (x, y). The x-axis is shown as a horizontal line across the graph while the y-axis is drawn vertical to (and intersects) the x-axis perpendicularly. These (x, y) data points are referred to as 'Cartesian co-ordinates'.

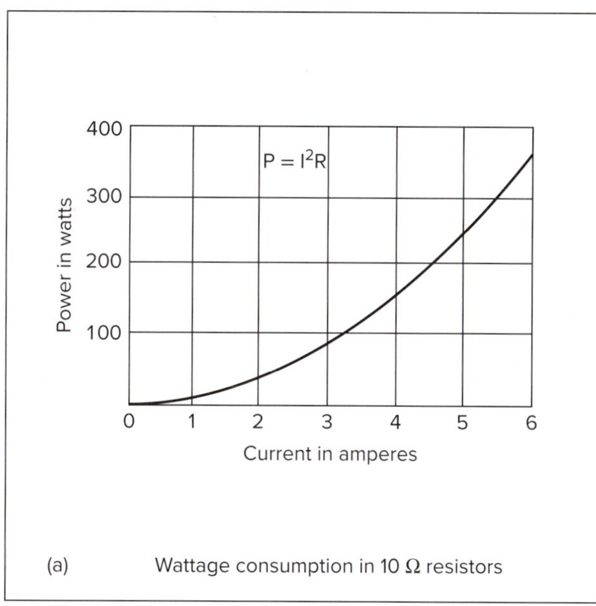

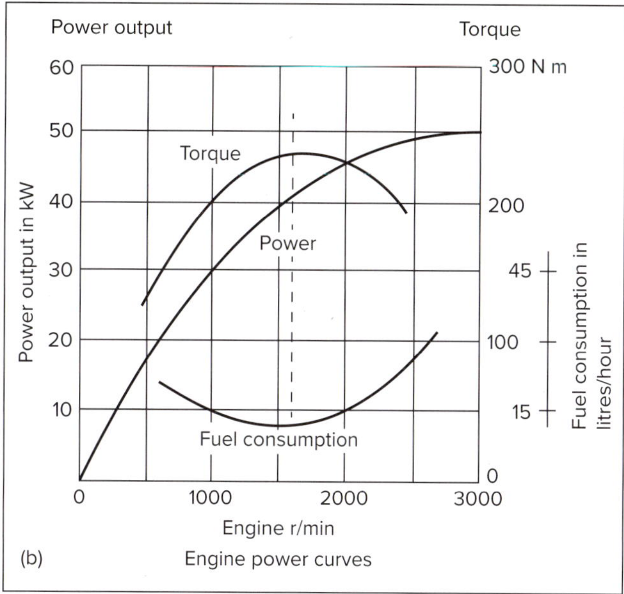

FIGURE 14.11 **Graphs: (a) single-curve graph (b) multiple curves graphed on a common axis**

Electrical problem solving employs a range of mathematical skills, including the use of geometry, trigonometry, *polar* and Cartesian co-ordinates for plotting vectors and two- and three-dimensional values. The ability to read graph axes and scales and identify and plot co-ordinates for specific values is a necessary trade maths skill for interpreting data and conveying information for problem solving.

CHECK YOUR UNDERSTANDING

14.24 In Figure 14.12, what is the voltage (Line A) when the current is 0.4 A?

14.25 In Figure 14.12, what is the voltage (Line B) when the current is 0.8 A?

14.26 In Figure 14.12, what is the current (Line A) when the voltage is 40 V?

14.27 In Figure 14.12, what is the current (Line B) when the voltage is 30 V?

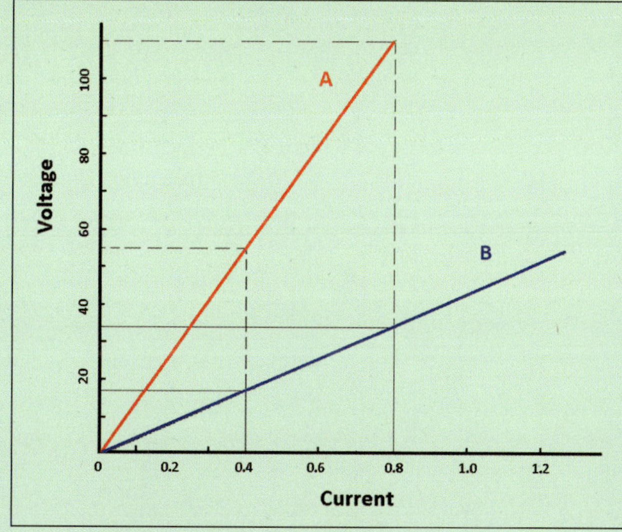

FIGURE 14.12 **Line graph**

14.4 Trade maths examples and solutions

14.4.1 d.c. circuit calculations

Ohm's Law and VIRP

The acronym 'VIRP' represents voltage (V, volts), current (I, amperes), resistance (R, ohms) and power (P, watts).

$I = \frac{V}{R}$ $V = R \times I$ $R = \frac{V}{I}$ $P = V \times I$ $P = \frac{V^2}{R}$	$P = I^2R$

Series-parallel circuit calculations

The majority of electrical problems will be with series and series-parallel resistance circuits. The mechanics for performing resistance and capacitance calculations are similar, while the mathematic calculation difference is to do with applying the appropriate formula:

Series resistance $R_T = R_1 + R_2 + R_3$	Series capacitance $\frac{1}{C_T} = \frac{1}{C_1} + \frac{1}{C_2} + \frac{1}{C_3}$

Parallel resistance $\frac{1}{R_T} = \frac{1}{R_1} + \frac{1}{R_2} + \frac{1}{R_3}$	Parallel capacitance $C_T = C_1 + C_2 + C_3$

The adding of capacitor values to determine total capacitance (CT) is due to adding capacitor plate areas together.

EXAMPLE 14.8

A coil takes a current of 10 mA and has a resistance of 20 Ω. Calculate the value of resistance to be connected in series to allow the coil to be connected across a 24 V supply.

This is an Ohm's Law question $I = \left(\frac{V}{R}\right)$, where R is the adding of the coil resistance to an unknown value. We know that to solve this problem, the maximum circuit current has to be the value of coil current.

List the givens:

$V = 24$ V

$I = 10$ mA

R_1 (coil) = 20 Ω

$R_{Total} = R_1$ (coil) + R_2 (unknown)

Find R_2 = **2380 Ω**

Step 1. From $I = \frac{V}{R}$

transpose the equation to $R = \frac{V}{I}$

Step 2. Substitute the known values:

$R = \frac{24}{10}$ mA = 2400 Ω or 2.4 kΩ

Step 3. If $R_1 + R_2 = 24$ kΩ, then solve for R_2

2400 Ω (R_{Total}) − 20 Ω = **2380 Ω**

EXAMPLE 14.9

Two resistors with respective values of 90 Ω and 70 Ω are connected in parallel. The total current taken from the supply is 2 A. Determine the supply voltage and the branch currents.

This is another Ohm's Law question $I = \left(\frac{V}{R}\right)$, where R_1 and R_2 are in parallel.

List the givens in Figure 14.13:

$R_1 = 90\ \Omega$

$R_2 = 70\ \Omega$

$I_{Total} = 2$ A

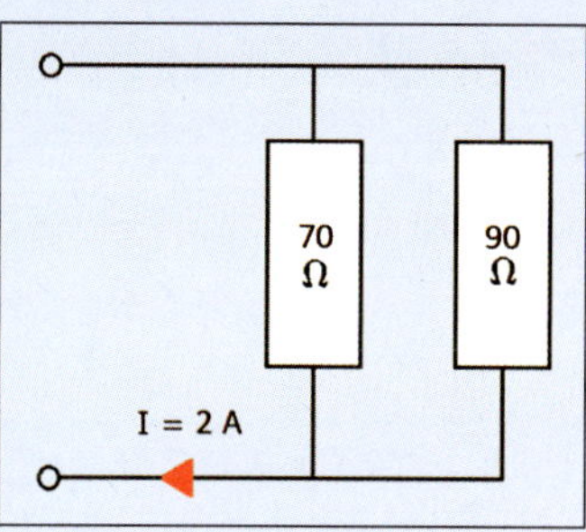

FIGURE 14.13 Diagram for Example 14.9

Find $R_{Total} = R_1 // R_2 =$ **39.375 Ω**

Find V (supply voltage) = **78.75 V**

Find $I_1 =$ **0.875 A**

Find $I_2 =$ **1.125 A**

Step 1. Determine the total circuit resistance ($R_1 // R_2$) (ANS: **39.375 Ω**)

Note: do not round at this point as doing so and then multiplying or dividing the rounded number would increase the margin for error.

Step 2. From $I = \frac{V}{R}$ transpose the equation to read $V = R \times I$

Step 3. Substitute the known values: $V = R \times I = (39.375 \times 2) = \mathbf{78.75\ V}$

Step 4. From $I = \frac{V}{R}$ current $I_1 = \mathbf{0.875\ A}$

Step 5. From $I = \frac{V}{R}$ current $I_2 = \mathbf{1.125\ A}$

EXAMPLE 14.10

A 25 Ω resistor and an 80 Ω resistor are connected in parallel. The current through the 25 Ω resistor is 3.4 A. What is the value of a third resistor that has to be added in parallel to make the supply current 5.4 A?

It may be slightly easier to solve this problem with the aid of a drawing. Reading the question carefully is the key in determining what you are actually being asked to solve. What do all three of these resistors have in common? Firstly, they are connected in parallel; secondly, the voltage across each resistor will be the same.

List the givens in Figure 14.14:

$I_{(Total)} = 5.4$ A

$I\ (R_1) = 3.4$ A

$R_1 = 25\ \Omega$

$R_2 = 80\ \Omega$

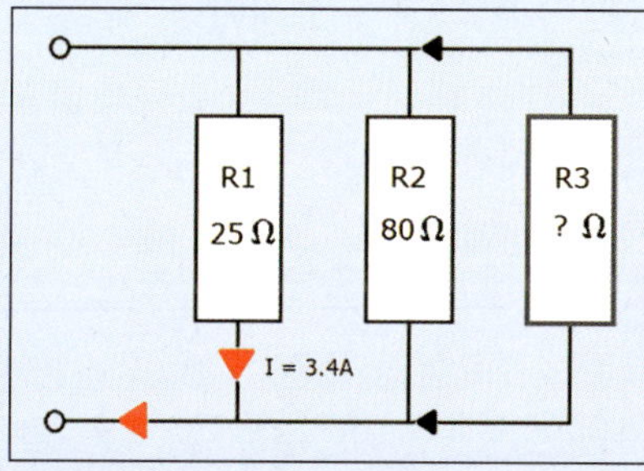

FIGURE 14.14 Diagram for Example 14.10

Step 1. **V = 85 V**

Step 2. **I = 1.0625 A**

Step 3. **I (R_3) = 0.9375 A**

Find $R_3 =$ **90.67 Ω**

Step 1. We can determine the voltage across the 25 Ω resistor, knowing the current flow through it. So, from $V = R \times I$, $V = 25 \times 3.4 = \mathbf{85\ V}$

Step 2. Determine the current through the 80 Ω resistor. For $I = \frac{V}{R}$ populate the known values so that

$I = \frac{85}{80} = \mathbf{1.0625\ A}$

Step 3. Determine the current through the unknown resistor.

If $I_{Total} = 5.4$ A, $(I_1 + I_2 + I_3)$

Then $5.4 = 3.4 + 1.0625 + I3$

$\therefore I3 = 5.4 - (3.4 + 1.0625) = \mathbf{0.9375\ A}$

Step 4. Transpose from $I = \frac{V}{R}$ so $R = \frac{V}{I}$

Step 5. Calculate the value for the third resistor.

For $R = \frac{V}{I}$, populate the known values so that

$R = \frac{85}{0.9375} = \mathbf{90.66}$ or **90.67 Ω** to two decimal places.

EXAMPLE 14.11

Two resistors R_1 and R_2 are connected in series across a 200 V supply. The resistance of R_1 is 140 Ω and the voltage drop across R_2 is 88 V. Calculate the:

(a) circuit current
(b) power dissipated by each resistor
(c) total power taken by the circuit.

List the givens in Figure 14.15:

$V = 200\ V$

$V_2 = 88\ V$

$R_1 = 140\ \Omega$

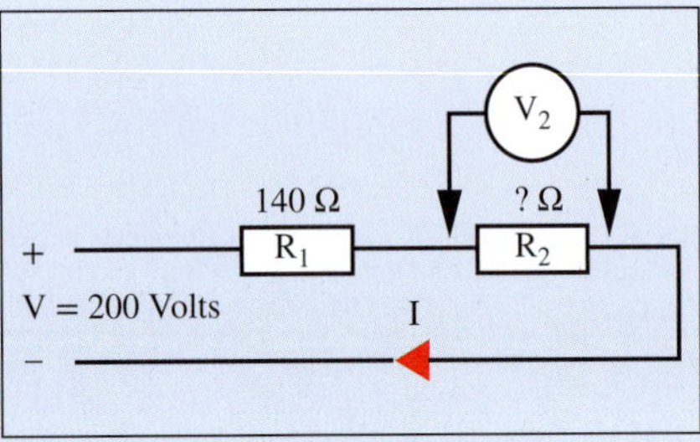

FIGURE 14.15 Diagram for Example 14.11

(a) Circuit current = **0.8 A**
(b) Power dissipated by each resistor.
Power dissipated in R_1 = **89.6 W**
Power dissipated in R_2 = **70.4 W**
(c) Total power taken by the circuit = **160 W**

Find V_1 (voltage drop across R_1)
Find I (circuit current)

Step 1. Determine V_1
If $V = V_1 + V_2$, then $V_1 = V - V_2$
Populate values so $V_1 = 200 - 88 = 112\ V$

Step 2. Calculate I from $I = \frac{V}{R}$
where:
$V = V_1 = 112\ V$
$R = R_1 = 140\ \Omega$
Populate $I = \frac{V}{R} = \frac{112}{140} = \mathbf{0.8\ A}$

Step 3. Calculate power dissipated in R_1
From $P = I \times V = 0.8 \times 112 = \mathbf{89.6\ W}$

Step 4. Calculate power dissipated in R_2
From $P = I \times V = 0.8 \times 88 = \mathbf{70.4\ W}$

Step 5. Determine total power taken by circuit
From $P = I \times V = 0.8 \times 200 = \mathbf{160\ W}$

Confirmation for Step 5:
Total power = 89.6 + 70.4 = 160 W

EXAMPLE 14.12

A circuit consists of two parallel resistors of 100 Ω and 150 Ω respectively. They are connected in series with a third resistor of 75 Ω. If the current through the 75 Ω resistor is 0.45 A, find the current through the 100 Ω resistor.

List the givens in Figure 14.16:

$R_1 = 100\ \Omega$
$R_2 = 150\ \Omega$
$R_3 = 75\ \Omega$
$I_{Total} = 0.45\ A$

Step 1. Calculate for $R_1 // R_2 = \mathbf{60\ \Omega}$

Step 2. Rearrange (transpose) $I = \frac{V}{R}$ for V
So $V = R \times I$

Step 3. Where $R = 60 + 75\ \Omega\ (135\ \Omega)$
Populate $V = R \times I = 135 \times 0.45 = \mathbf{60.75\ V}$

Step 4. Where $R = 75\ \Omega$
Populate $V = R \times I = 75 \times 0.45 = \mathbf{33.75\ V}$

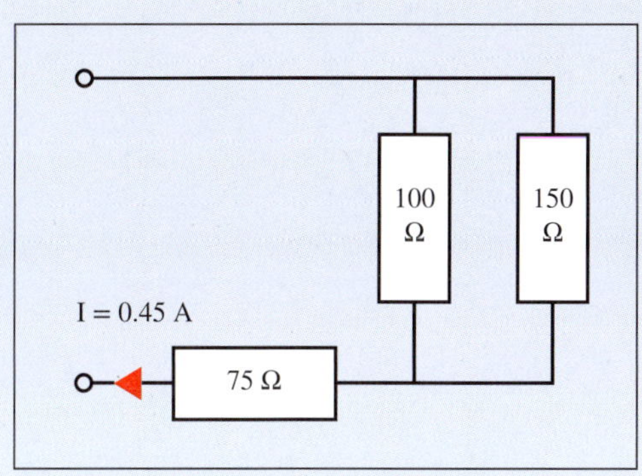

FIGURE 14.16 Diagram for Example 14.12

Step 1. R_1 // R_2 = **60 Ω**

Step 2. **V = R × I**

Step 3. Supply voltage = **60.75 V**

Step 4. V = R × I = **33.75 V**

Step 5. Voltage across R_1 // R_2 = **27 V**

Step 6. Current through 100 Ω resistor = **270 mA**

Step 5. Determine voltage drop across R_1 // R_2

V_{Total} (60.75 V) = V_3 (33.75 V) + V (R_1 // R_2)

V (R_1 // R_2) = V_{Total} (60.75 V) – V_3 (33.75 V)

∴ V (R_1 // R_2) = **27 V**

Step 6. Find current through 100 Ω resistor

V = 27 V

R = 100 Ω

From $I = \frac{V}{R} = \frac{27}{100} = \mathbf{0.27\ A}$

EXAMPLE 14.13

When three resistors are connected in parallel across a supply, the supply current is 8 A. The current measured in branches 1 and 3 is 1.5 A and 4 A respectively. The power measured in branch 2 is 600 W. Calculate the:

(a) supply voltage
(b) resistances of each of the resistors.

List the givens in Figure 14.17:

$I_1 = 1.5$ A

$I_3 = 4$ A

$I_{Total} = 8$ A

P (R_2) = 600 W

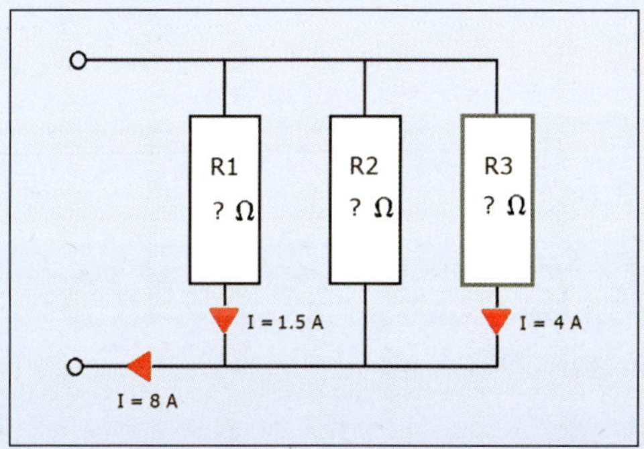

FIGURE 14.17 Diagram for Example 14.13

Step 1. Determine I_2

If $I_{Total} = I_1 + I_2 + I_3$ then $I_2 = I_{Total} - (I_1 + I_3)$

Populate values so $I_2 = 8 - (1.5 + 4) = 2.5$ A

Step 2. Determine supply voltage (V)

If P = V × I then $V = \frac{P}{I}$

Populate values so $V = \frac{600}{2.5} = \mathbf{240\ V}$

Step 3. Calculate R_1

From $I = \frac{V}{R}$

Transpose the equation to read $R = \frac{V}{I}$

Populate values so $R_1 = \frac{240}{1.5} = \mathbf{160\ \Omega}$

Step 4. Calculate R_2

From $I = \frac{V}{R}$

Transpose the equation to read $R = \frac{V}{I}$

Populate values so $R_2 = \frac{240}{2.5} = \mathbf{96\ \Omega}$

(a) Supply voltage = **240 V**
(b) Resistance
R_1 = **160 Ω**
R_2 = **96 Ω**
R_3 = **60 Ω**

Step 5. Calculate R_3

From $I = \frac{V}{R}$

Transpose the equation to read $R = \frac{V}{I}$

Populate values so $R_3 = \frac{240}{4} = \mathbf{60\ \Omega}$

EXAMPLE 14.14

Two resistors with respective values of 1 k2 and 1 k8 are connected in series. If they are connected to a 300 V supply, determine the power consumed by each resistor and the total power of the circuit.

For this problem we need to determine the total series circuit current using the Ohm's Law equation and then use it to determine the power using the equation $P = V \times I$.

Step 1. Determine total circuit current from $I = \frac{V}{R} = \frac{300}{(1200 + 1800\ \Omega)}$

Step 2. Using I = 0.1 A, from $P = V \times I$ determine power consumed by 1 k2

From $P = I^2R = 0.1 \times 0.1 \times 1200 = 12$ W

Step 3. Using I = 0.1 A, from $P = I^2R$ determine power consumed by 1 k8

From $P = I^2R = 0.1 \times 0.1 \times 1800 = 18$ W

From $P = V \times I = 300 \times 0.1 = 30$ W

Confirmation check: 18 + 12 = 30 W

EXAMPLE 14.15

Two resistors with respective values of 25 Ω and 35 Ω when connected in series with a d.c. supply have a circuit current of 2 A. What value of resistance must be added in series to limit current to 1 A?

Part 1
Find the givens in Figure 4.18:

R_1 = 25 Ω
R_2 = 35 Ω
I = 2 A
Find V = **120 V**

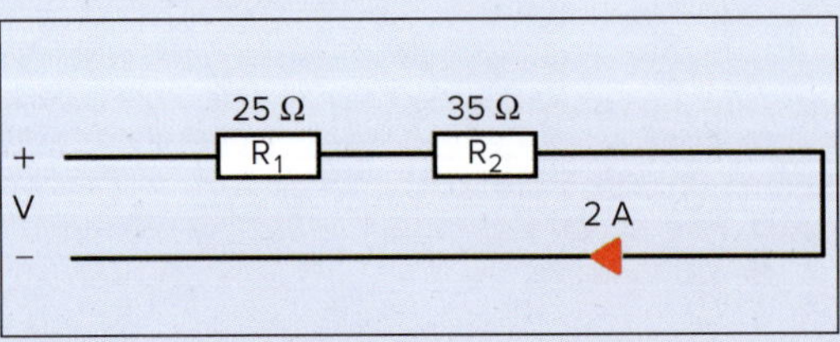

FIGURE 14.18 **Diagram for Example 14.15, Part 1**

Part 1: From $I = \frac{V}{R}$
where:

I = 2 A
R = 25 + 35 Ω (60 Ω)

Find V

Step 1. Rearrange (transpose) $I = \frac{V}{R}$ for V
So $V = R \times I$

Step 2. Populate $V = R \times I = 60 \times 2 = \mathbf{120\ V}$

Part 2

Find the givens in Figure 14.19:

$R_1 = 25\ \Omega$

$R_2 = 35\ \Omega$

$I = 1\ A$

Find $R_3 =$ **60 Ω**

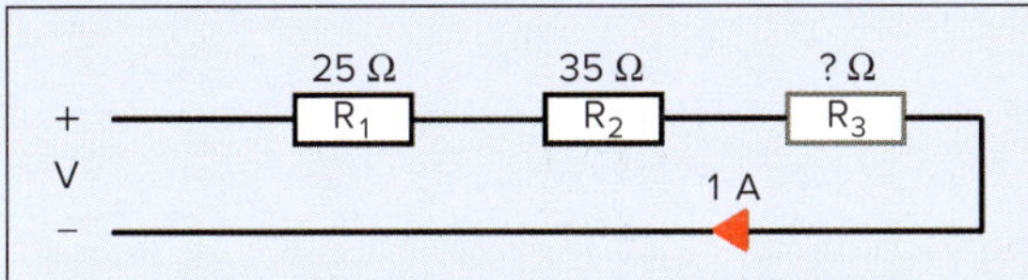

FIGURE 14.19 **Diagram for Example 14.15, Part 2**

Part 2: From $I = \frac{V}{R}$

where:

$I = 1\ A$

$V = 120\ V$

Step 3. Rearrange (transpose) $I = \frac{V}{R}$ for R

So $R = \frac{V}{I}$

Find new R (new $R = 25 + 35 + R_3$)

Step 4. Populate new $R = \frac{V}{I} = \frac{120}{1} = 120\ \Omega$

Step 5. Determine value of added resistance

$R_3 = 120 - 60 =$ **60 Ω**

A simpler way to approach this problem is to recognise the relationship between resistance and current in the equation $I = \frac{V}{R}$: doubling the resistance halves the current. The problem-solving steps can be used with other, more difficult, problems.

EXAMPLE 14.16

A magnetic circuit has a reluctance of 20 000 ampere turns per weber. If the total flux required in the circuit is 0.3 weber and the coil can carry a current of 5 amperes, calculate the number of turns required.

Step 1. Read the question—what is it really asking? What is being asked is often in the first line of the problem description or in the final sentence. In this case, you are being asked to calculate the number of turns required.

Reading the question includes identifying the device. Knowing what the device is and having some general knowledge about how it operates can assist in problem solving. Is the device a coil, a contactor, a generator or a motor? In this example, the magnetic circuit is a coil.

Step 2. Locate and identify the individual pieces of information provided (the givens). The question has been rewritten and colour coded to show the different SI unit quantities, values and unit names.

A magnetic circuit has a **reluctance** of 20 000 ***ampere turns per weber.*** If the total **flux** required in the circuit is 0.3 *weber* and the coil can carry a **current** of 5 ***amperes,*** calculate the number of **turns** required.

Step 3. List the givens. This step helps to identify what information has been provided and what information is missing. In Table 14.7, the quantity symbols have been added to give you an idea of which formulas to use to solve this problem. Listing the givens also provides a procedure to neatly set out work for checking and review.

TABLE 14.7 **Givens for Example 14.16**

Quantity symbol	SI unit (term) quantity	Value/ quantity	SI unit name	Unit symbol
Rm	**Reluctance**	20000	*ampere turns per weber*	At/Wb
ϕ	Magnetic **flux**	0.3	*weber*	Wb
I	Electric **current**	5	*amperes*	A
N	**turns**	**FIND**	*(-) turns**	(-) t

Note: *refer to turns

List the givens:

Reluctance (R_m) = 20 000 At/Wb

Magnetic flux (ϕ) = 0.3 Wb

Electric current (I) = 5 A

With practice, the givens can be abbreviated as:

R_m = 20 000 At/Wb

ϕ = 0.3 Wb

I = 5 A

Step 4. Draw a diagram and mark on it the information given. This can assist in identifying missing information or indicate a hidden step in solving the problem. Although a diagram is not needed in this particular example, the use of diagrams is an essential skill in solving most calculation problems.

Step 5. Determine the steps necessary to solve the problem and what formulas will be needed (see Table 14.8). Being able to identify and contextualise the givens from concepts and theory knowledge is important in selecting and applying the correct formula(s). The formula to select in solving this problem is given as:

$$R_m = \frac{IN}{\Phi}$$

Step 6. Rearrange the equation to make the unknown the subject via transposition (see Table 14.8).

TABLE 14.8 Electromagnetism problem solutions for Example 14.16

Step 5. Select the formula	Step 6. Transposition steps	Step 7. Populate the equation
$R_m = \frac{IN}{\phi}$	Cross multiply: $R_m \times \phi = I.N$ Divide both sides by I: $N = \frac{Rm.\phi}{I}$	$N = \frac{R_m.\phi}{I}$ $N = \frac{20000.0.3}{5}$ $\therefore N = 1200$ turns

Step 7. Populate the transposed equation with the givens (see Table 14.8).

14.4.2 d.c. motors and generators

Types of d.c. motors and generators are classified by their windings, the more common types for electrical calculation problems being the series field, shunt field and compound (either cumulative or differential). Applying knowledge of the type of device—how it is constructed, electrically connected and wired—is often combined with problem solving and performing mathematical calculations.

Drawings and diagrams can be very helpful in solving problems. They can help identify missing information and assist in determining the steps required for problem solving.

Two equations—one for current the other for voltage—are used for solving d.c. machine problems. Both apply Kirchoff circuit voltage and current laws. The current equation is illustrated on the following generator and motor diagrams (Figures 14.20 and 14.21) as armature (I_a), field (I_{sh}) and load current (I_L) relationships. The equation follows the rule that the sum of the current flowing into a junction equals the sum of the current flowing out of the junction.

For a generator such as that in Figure 14.20, this is the armature current (I_a) equals the sum of field (I_{sh}) and load current (I_L). Voltage values are based on terminal voltage (V), generated voltage (E_g) and armature circuit voltage drop ($V_d = I_aR_a$).

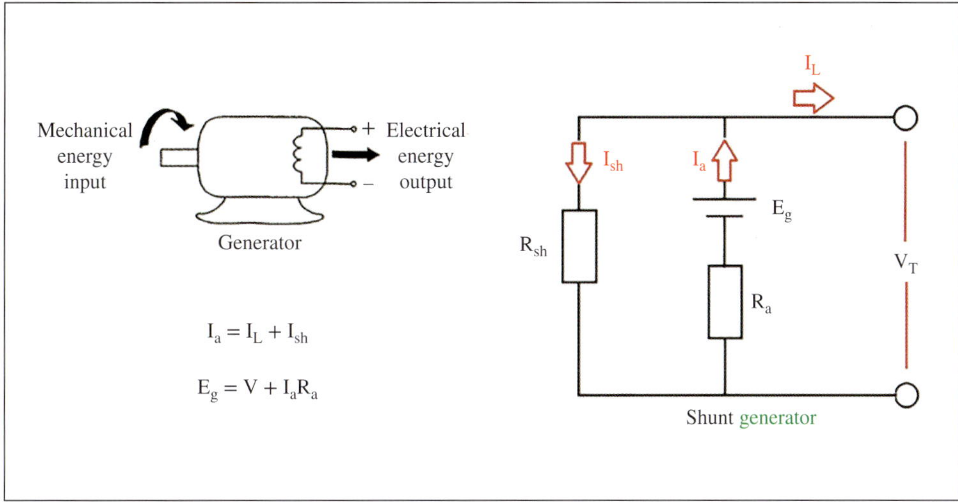

FIGURE 14.20 Shunt generator

Something that the example motor and generator have in common is that the shunt field winding has an applied voltage at full load that is equal to the terminal voltage. This visual information assists in calculating the shunt field current using Ohm's Law, where:

$$I_{sh} = \frac{V_T}{R_{sh}}$$

You will notice that the generator and motor equations are different. This is because they relate to the machines' respective principles of operation. The generator in Figure 14.20 supplies current and voltage for a connected load device.

The motor in Figure 14.21 is connected to a supply source so the total motor load current (I_L) equals the sum of armature (I_a) and field current (I_{sh}). The motor Eg value is referred to as 'back EMF', and under normal operating conditions is always lower than the terminal voltage.

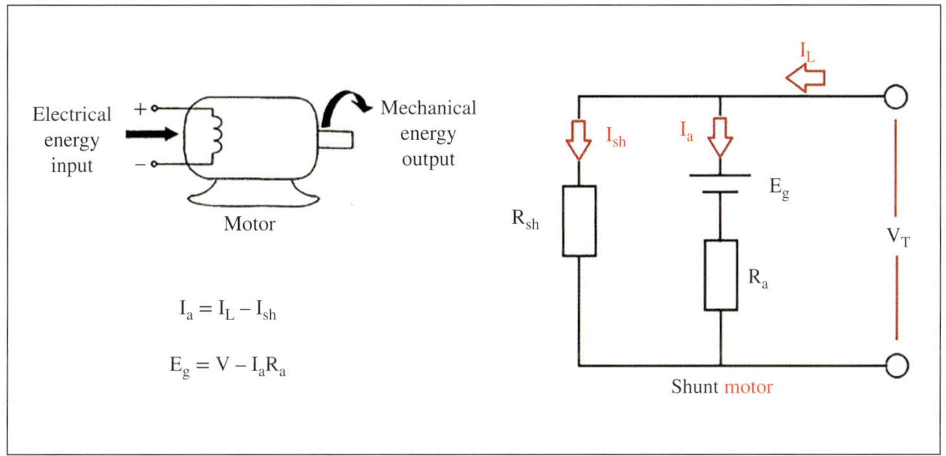

FIGURE 14.21 Shunt motor

14.4.3 Losses in a machine

An electric motor rated at 3 kW is said to have a nominal output of 3 kW; but that is not a real indication of the energy delivered to it. In both mechanical and electrical systems, there are losses that can have an important bearing on the operation of the system. The major losses are friction and windage, while other losses such as copper and iron losses must also be accounted for.

EXAMPLE 14.17

If a device has a power input of 160 W and a power output of 120 W, find the efficiency and the loss (W).

$$\text{efficiency} = \frac{\text{power output}}{\text{power input}} \times 100$$

$$\therefore \text{efficieny} = \frac{120}{160} \times 100 = 75\%$$

The loss is the difference between the power output and input, and in this example is 160 – 120 = 40 W loss.

Extra power has to be supplied to a system to compensate for these losses. Therefore, the power into a system is the sum of the power out plus losses:

$$P_{in} = P_{out} + \text{losses}$$

or:

$$P_{out} = P_{in} - \text{losses}$$

This can be expressed as the ratio of the power output to the power input as a percentage, which is called the 'efficiency' of the system.

As efficiency is a ratio, it has no units:

$$\eta = \frac{(P_{out} \times 100)}{P_{in}} \%$$

EXAMPLE 14.18

Find the efficiency of a 3 kW electric motor if the losses are found to be 357 W.

$$\text{input power} = 3000 + \text{ losses} = 3357 \text{ W}$$

$$\text{output power} = 3000 \text{ W(motor rating)}$$

$$\eta = \frac{3000}{3357} \times 100 = 89.4\%$$

The usual symbol for efficiency is η (eta, pronounced 'eeta') and is expressed as a number followed by the per cent symbol, e.g. 'efficiency = 89%' or 'η = 89%'.

14.5 Energy, work and power

Energy, work and power are not the same thing. Energy is the ability or capacity to do work, work is the energy that is expended and power is the rate of doing work or expending energy.

Work can also be described as the process of converting energy from one form into another. For example, converting the chemical energy of fuel into the heat energy required to drive a steam turbine is a form of work, and the speed with which the heat is generated defines the power.

$$\text{energy} = \text{work} = \text{power} \times \text{time}$$

There are many equations relating to energy, not just in the electrical and mechanical contexts but also in the chemical and nuclear contexts—and in many others.

One common mechanical equation calculates the work involved in moving an object a known distance by using a known force. When a body is moved through a distance by a force acting on it, work is done. That is, if a force of F newtons acts through a given distance, then:

$$W = Fd$$

where:

W = work in joules (J)
F = force in newtons (N)
d = distance in metres (m).

EXAMPLE 14.19

A force of 100 N is required to move a box 5 m along a horizontal surface. Find the value of work done.

$$\text{work} = F \times d = 100 \times 5 = 500\ \text{J}$$

Power is the rate of doing work. It can be found by dividing the work value by the time in seconds, and is expressed in units as J/s = W (watts):

$$\text{power} = \frac{\text{work}}{\text{time}}$$

EXAMPLE 14.20

If the box in Example 14.19 was moved first in 10 s and later in 5 s, calculate the power used in both cases.

$$\text{power} = \frac{\text{work}}{\text{time}} = \frac{500}{10} = 50\ \text{W}$$
$$\text{power} = \frac{\text{work}}{\text{time}} = \frac{500}{5} = 100\ \text{W}$$

The potential to do work can be found from power multiplied by its time of application, that is:

$$\text{energy} = \text{power} \times \text{time}$$
$$\text{but power} = \frac{\text{work}}{\text{time}}$$
$$\therefore \text{energy} = \left(\frac{\text{work}}{\text{time}}\right) \times \text{time}$$
$$\text{but work} = F \times d$$
$$\therefore \text{energy} = \left(\frac{F \times d}{\text{time}}\right) \times \text{time}$$
$$\therefore \text{energy} = F \times d\ [\text{joules}]$$

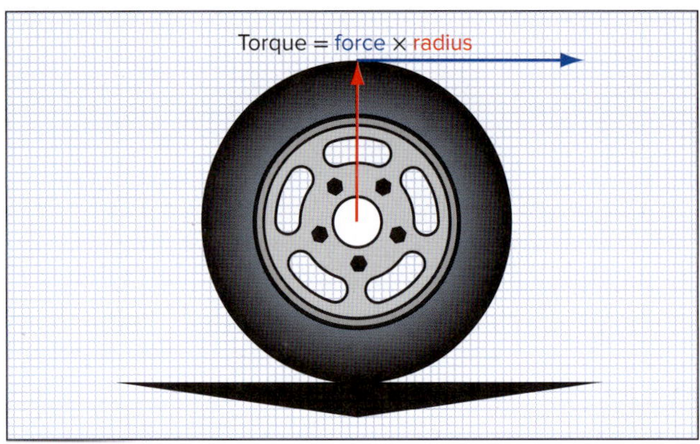

FIGURE 14.22 Torque

14.5.1 Torque

Torque is a force applied at some distance from the centre of a rotating mass with the effect that the mass will rotate unless an equal and opposite torque is simultaneously applied. Common examples of torque are tightening a nut with a spanner or turning the steering wheel of a car. If insufficient force is applied, these items will not turn, yet a torque has been applied. The actual value of torque is due to two factors—the applied force (newtons) and the lever arm length or radius (metres) from the axis of rotation (see Figure 14.22).

$$T = Fr \text{ (Nm—newton metres)}$$

where:

T = Torque

F = Force in newtons

r = distance from the axis of rotation to the point where force is being applied, in metres

$P = \omega T$.

EXAMPLE 14.21

A force of 150 N is applied to the end of a spanner 0.4 m long to tighten a nut. Calculate the torque applied to the nut.

$$\text{Torque} = Fr = 150 \times 0.4 = 60 \text{ Nm}$$

EXAMPLE 14.22

Find the torque exerted by a 3 kW electric motor operating at 1440 rpm.

$$P = \omega T = 2\pi nT\ 60$$

$$3 \text{ kW} = 3000 \text{ W}$$

$$\text{That is, } 3000 = 2 \times \pi \times 1440 \times \frac{T}{60}$$

$$\therefore T = \frac{3000 \times 60}{2\pi \times 1440} = 19.89 \text{ Nm}$$

14.6 Scalar and vector quantities

All quantities can be classified as being either 'scalar' or 'vector', according to these two criteria:

1. Scalar quantities are those that have magnitude but no direction (e.g. mass, volume, energy and time).
2. Vector quantities are those that must be expressed in both magnitude and direction (e.g. velocity, acceleration and force).

It is easier to understand the terms 'scalar' and 'vector' when thinking of mechanical quantities, but they also occur in electrical theory. When applied to electrical systems, vectors are called 'phasors', but the basic principles are the same as for mechanical vectors.

14.6.1 Scalar quantities

A number and a unit are sufficient to specify many physical quantities. These quantities can be added by ordinary arithmetic. For example, 5 seconds + 3 seconds = 8 seconds, or 1 km + 2.6 km = 3.6 km. Vector quantities acting in the same direction can also be treated as scalar quantities and ordinary arithmetic applies. Scalar quantities may be drawn as a straight line proportional to their magnitude, and negative values may be drawn on the same line but in the opposite direction. For example, a person is given $110, but owes $60 for a speeding fine. They can draw a line 110 mm to the right, and then from the end of that line draw another line 60 mm to the left—they will have 50 mm left, telling them they have $50 available for cheering themselves up after getting the fine. This is a graphical means of calculating the addition of two scalar quantities.

14.6.2 Vector quantities

Unlike scalar quantities, vector quantities cannot be satisfactorily described without giving a direction as well as a quantity and a unit. A vector quantity can be represented by a straight line which, when drawn to a scale, is able to represent both magnitude and direction. In mechanics, the vector quantity is often a force. Weight is a special case of force. The unit 'kilogram' is used to measure mass, while the unit 'newton' is used for force and weight. It is convenient for the purpose of this work to use force as a means of examining methods for solving vector problems.

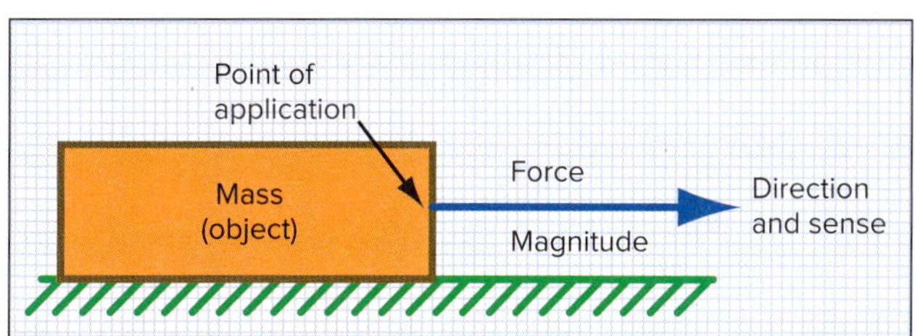

FIGURE 14.23 Characteristics of a force

The characteristics of a force are shown in Figure 14.23. The force has both a magnitude and a direction, and also a point of application and a sense of direction.

14.6.3 Forces acting at a point

Where more than one force acts on a body simultaneously, they can assist or oppose one another. Cyclists are familiar with tail winds and head winds, but staying balanced becomes more complicated with a wind blowing from neither of these directions. The cyclist then has to lean 'against' the wind to continue on a desired path, a situation usually associated with turning. A similar situation exists with a car in a crosswind. The steering wheel has to be held at an angle to counteract the side forces. In effect, the car is actually being steered across the road to counteract the side force, so that the resultant motion is straight along the road.

The resultant value of two forces acting on a body depends on the angle between the directions of the forces, as well as their respective magnitudes.

In Figure 14.24, F_1 is added directly to F_2 as a straight arithmetical addition and the resultant force is $F_1 + F_2$ (i.e. riding with a tail wind).

In Figure 14.24(a), two forces are acting together on a body. In Figure 14.24(b), the forces are in opposition and the resultant force is $F_1 - F_2$, which will act in the direction of the larger force (i.e. riding into a head wind). In Figure 14.24(c), the simple arithmetical process cannot be used. The combination of F_1 and F_2 gives a resultant force FR acting in a direction between F_1 and F_2. The strength of the force is a value somewhere between the larger of the two forces and the sum of the two.

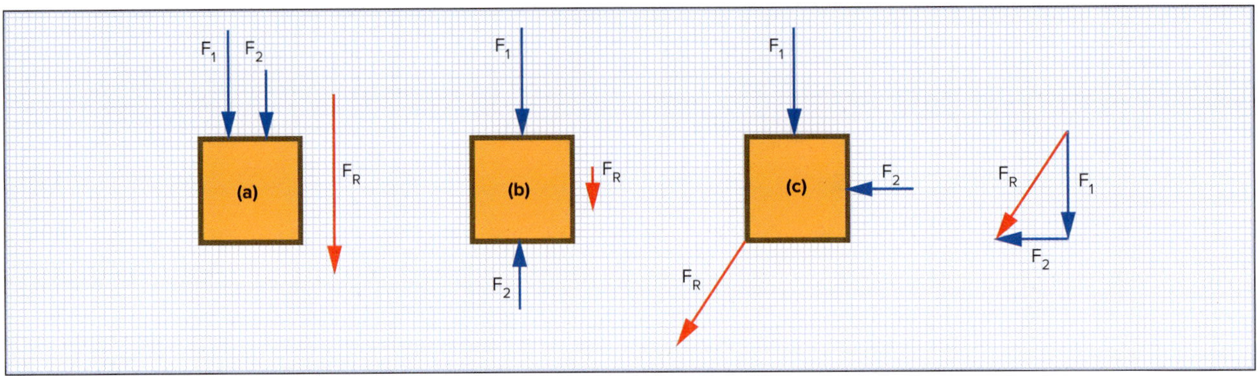

FIGURE 14.24 Two forces acting on a body

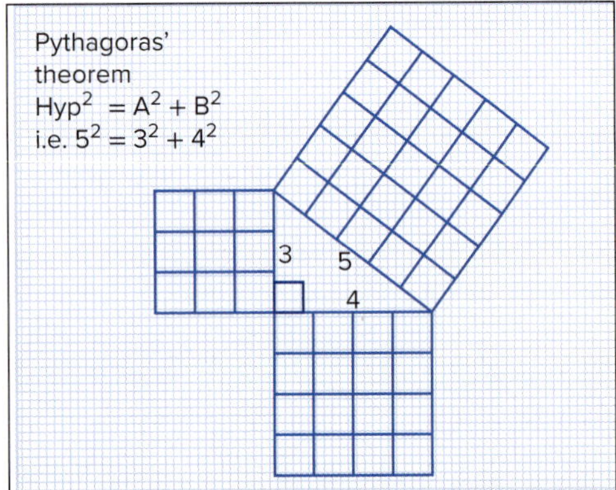

FIGURE 14.25 Pythagoras' theorem as it relates to a 3:4:5 ratio triangle

14.6.4 Forces acting at 90 degrees

In a special case where F_1 and F_2 are acting at right angles to each other, the value of the resultant force can be derived mathematically by Pythagoras' theorem.

The right-angle triangle method can be used to analyse the effects of two forces acting at 90° to each other. One of the best-known examples is the triangle with sides in the ratio of 3:4:5. It can be seen that the square on the Hypotenuse ($5^2 = 25$) is equal to the sum of the squares on the other two sides:

$$3^2 + 4^2 = 9 + 16 = 25$$

That is:

$$5^2 = 3^2 + 4^2$$
$$\text{or } h = \sqrt{a^2 + b^2}$$

EXAMPLE 14.23

Two forces (F_1 and F_2), each of 25 N, act at right angles to each other on a body (see Figure 14.26). Determine the value of the resultant force (F_R) acting on the body.

$$\begin{aligned} F_R &= \sqrt{F_1^2 + F_2^2} \\ &= \sqrt{25^2 + 25^2} \\ &= \sqrt{625 + 625} \\ &= \sqrt{1250} \\ &= 33.35 \text{ N} \end{aligned}$$

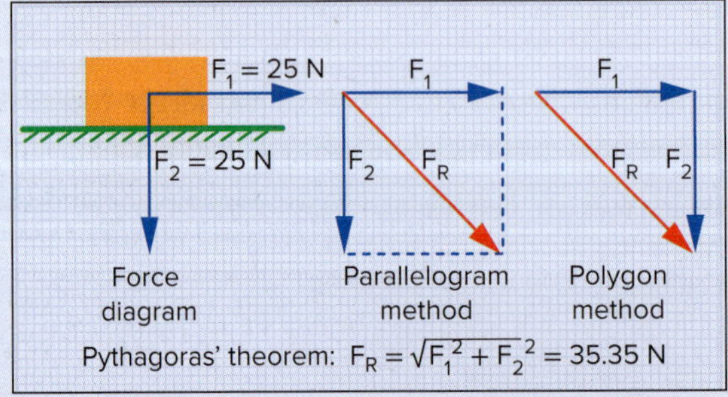

FIGURE 14.26 Force diagrams for Example 14.23

This method can only be used when the forces act at an angle of 90° to each other.

EXAMPLE 14.24

With the aid of trigonometry and a calculator, solve the unknown values of the sides in the triangle in Figure 14.27.

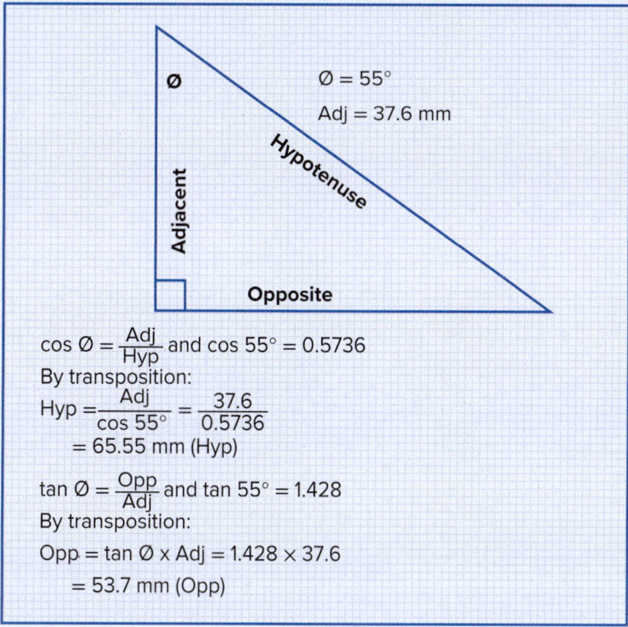

FIGURE 14.27 Triangle for Example 14.24

14.6.5 Parallelogram method

Figure 14.28 shows the parallelogram method, where two forces F_1 and F_2 are drawn to scale from the point of application O. Lines are drawn from the ends of each force, (A–C and B–C) parallel to the other forces, completing the parallelogram. The resultant force is then drawn from the point of application (O) to the intersection of the parallel lines. This is also known as the 'graphical' or 'vector parallelogram' method of solution. It can be used to solve both the magnitude and direction of resultant forces, irrespective of the direction of the separate forces. The only limitation is the draftsperson's accuracy of scale and angle.

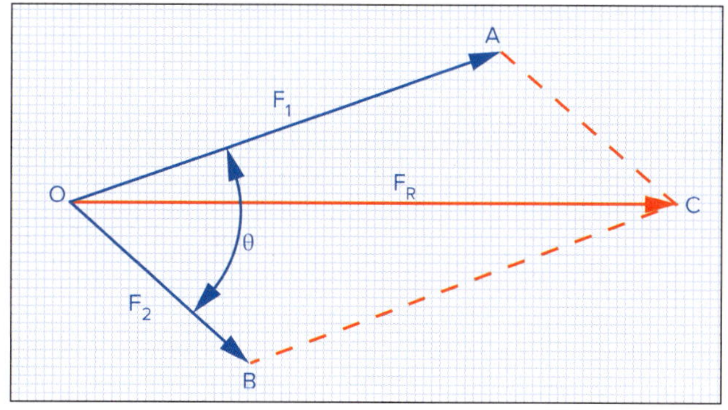

FIGURE 14.28 Parallelogram method of adding vectors

EXAMPLE 14.25

Two forces of 8 N and 5 N respectively act outwards from a point with an angle of 60° between them. Find the resultant force being exerted at the point and the angle at which it acts with respect to the 8 N force.

Resultant = 11.4 N
Angle to OB = 22°

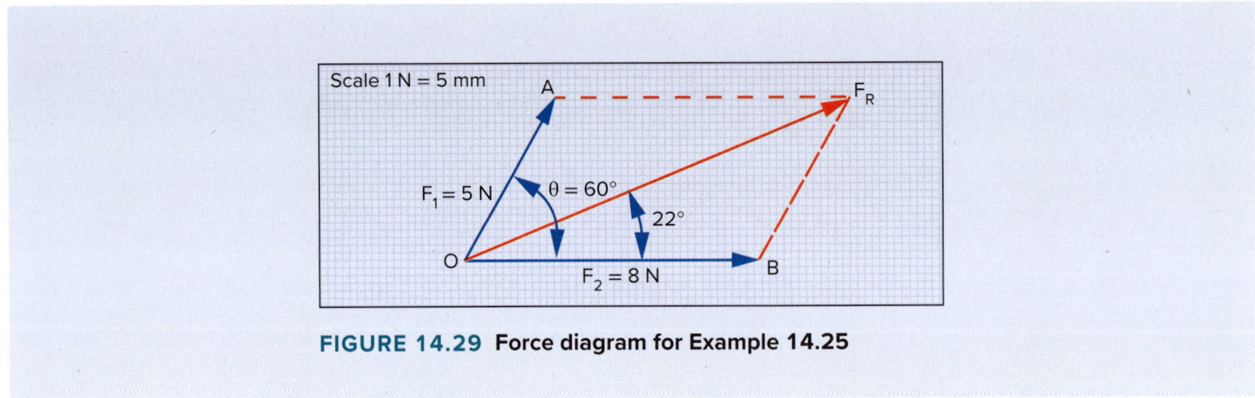

FIGURE 14.29 Force diagram for Example 14.25

In Figure 14.29 two forces F_1 and F_2 act on the point O as indicated by their arrowheads. OA and OB are drawn to scale at the appropriate angles to each other. The value of FR is taken from the scale to which the figure is drawn. Probably the most common method of construction is that of intersecting arcs with the aid of compasses. The compass is set to length OA and an arc is drawn from the tip of arrow OB. Another arc is drawn with the compass at the length of OB and the centre at the tip of OA. To get the most out of this explanation, access to a scale rule, a set of compasses and a protractor is useful.

14.6.6 Vector polygon method

The parallelogram method becomes more cumbersome when three or more forces are involved.

The polygon method (sometimes referred to as the 'tip-to-tail' method), requires each vector to be drawn to scale and angle with its origin at the end of the previous vector. The vectors can be drawn in any order, provided the requirements of magnitude and direction are met. The resultant is the distance between the origin of the vector diagram and the end of the last vector. The method is best illustrated in Figure 14.30.

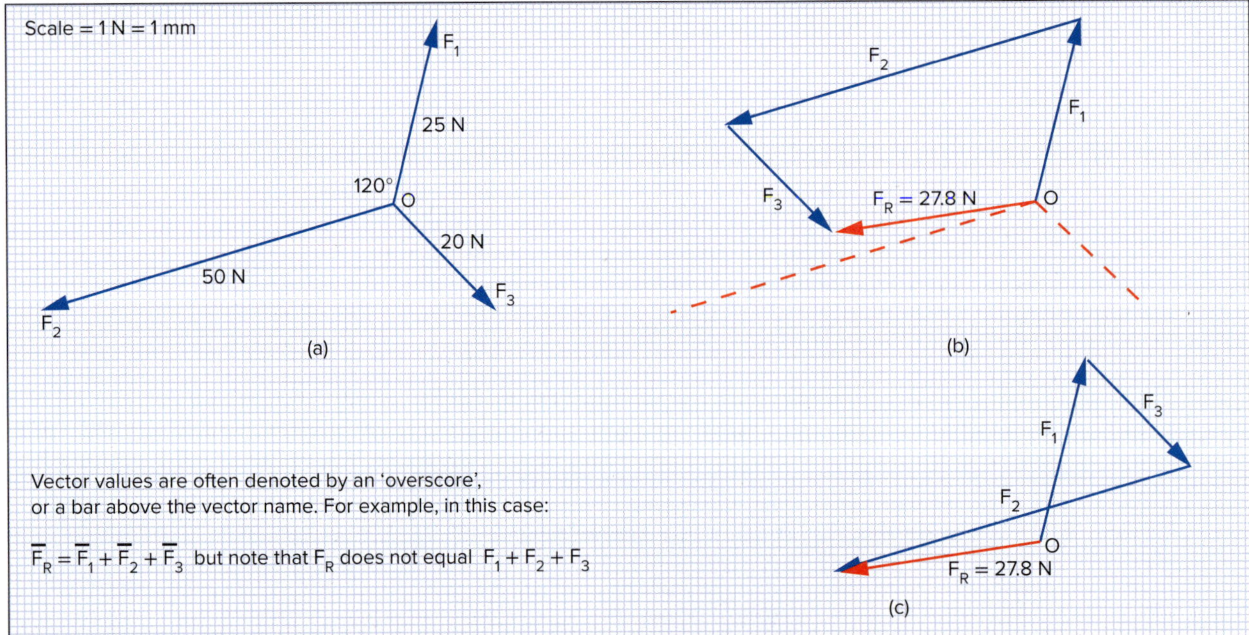

FIGURE 14.30 Polygon method of adding vectors

In Figure 14.30 three vectors are drawn as F_1, F_2 and F_3. Vector addition is shown in (b) but the order is altered in (c), giving a different figure. It should be noted that the two resultants have the same magnitudes and directions. FR is the value and direction of a single force that could replace all three original forces and produce the same effect. In vector diagrams, the resultant can replace all of the vectors. The resultant starts at the origin and finishes at the arrowhead end of the last vector.

14.6.7 Vector components

Vectors can be drawn on graph paper with vertical and horizontal axes and added as vector components in the same way as scalar quantities.

The forces F_1, F_2 and F_3 shown in Figure 14.30(a) can be drawn to scale as right-angle triangles, and their horizontal and vertical components measured—or trigonometry used—to calculate the horizontal and vertical values of each force.

The angle of the resultant can be determined using a ruler, graph paper, protractor and compasses. The angle can also be calculated using trigonometry. Once the sum of the horizontal and vertical forces is known, they can be combined using Pythagoras' theorem to have one single vector resultant.

Using Pythagoras' theorem, $\sqrt{((-27.32)^2 + (-5.33)^2)} = 27.83$ N. The direction of the resultant is downwards to the left, which would also be the same as that found by the polygon method. Mathematical values of vectors and vector components can be obtained by trigonometry.

14.6.8 Rectangular vs polar

A vector has magnitude and direction (angle) and is usually written in polar form—for example, a 40 N force at an angle of 60° is written as 40 N/60°. If it is converted into horizontal and vertical components, it is in a rectangular form and would be written [20,34.6]N. Modern scientific calculators can easily convert between polar and rectangular forms, but without knowing how to do this conversion, it is hard to fully appreciate how and why the methods work.

14.7 Trigonometry

The component values for forces could be obtained by graphical means or by trigonometry, which is based on the measurements of triangles (and the trigonometry of right-angle triangles is relatively simple). Trigonometry can be used to calculate the magnitude and direction of vectors with an accuracy greater than that of graphical means.

14.7.1 Ratios of lengths of sides

The right-angle triangles shown in Figure 14.31 are different sizes and have different lengths of sides, but the angles remain constant. Therefore, the relationships between the sides remain the same and they have the same trigonometry.

As all the corresponding angles are equal, the various figures are called 'similar triangles'. Also, the ratios between the sides remain constant, irrespective of the sizes of the triangles. A check of Figure 14.32 will show that the ratio of the horizontal side to the Hypotenuse for each triangle is 1:2 or h (0.5).

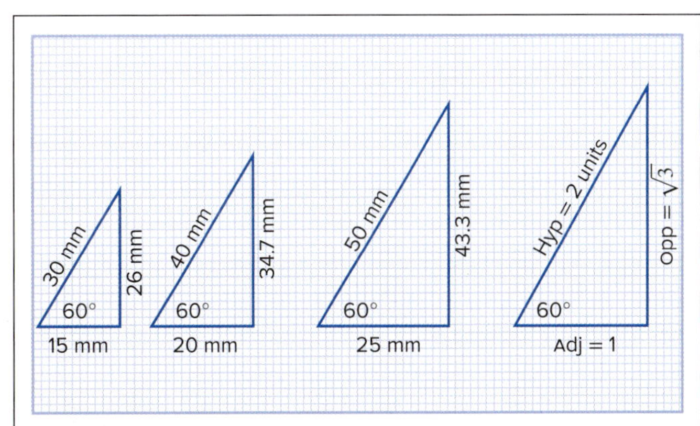

FIGURE 14.31 Similar triangles

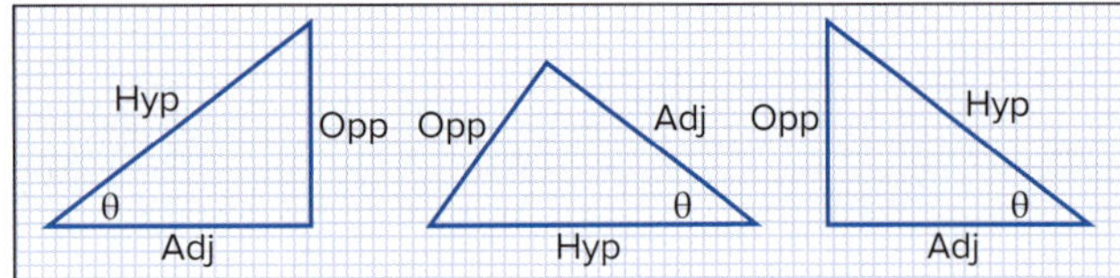

FIGURE 14.32 **Hypotenuse, Opposite and Adjacent**

Reference to trigonometrical tables shows that the Cosine for 60° is 0.5. What this really means is that, in the right-angle triangle, the ratio of the two sides adjacent to a 60° angle will always be 1:2 or $\frac{1}{2}$ or 0.5 (as it is usually expressed).

In any right-angle triangle, the longest side is called the 'Hypotenuse', the side nearest a given angle is called the 'Adjacent' and the side of the triangle opposite the angle is called the 'Opposite' side.

There are three ratios of sides for a right-angle triangle:

- Sine of angle = Opposite side/Hypotenuse, $\sin\theta$ = Opposite/Hypotenuse
- Cosine of angle = Adjacent side/Hypotenuse, $\cos\theta$ = Adjacent/Hypotenuse
- Tangent of angle = Opposite side/Adjacent side, $\tan\theta$ = Opposite/Adjacent

These sides are illustrated in Figure 14.32 and apply when the triangle is rotated into any position.

SUMMARY

- The SI system of units is a metric system and is, with very few exceptions, a world-wide standard.
- There are seven base units in the system.
- There are many types of units derived from the base units, for example mechanical, electrical and magnetic.
- The engineering notation system has standard multiples and sub-multiples with specific names.
- Scientific notation is a method for expressing quantities using a mantissa and an exponent value.
- Work, power, energy and torque are components of the electrical circuit or mechanical system.
- All machines have losses. Efficiency is the ratio of the input and output values, usually expressed as a percentage.
- Scalar values have magnitude only.
- Vector (phasor) values have magnitude and direction.
- Mechanical forces are expressed as vectors.
- Electrical quantities are graphically expressed as phasors.
- Combinations of forces give rise to a resultant force.
- A resultant force can be evaluated graphically by drawing it to scale, with due regard to length and direction.
- Polar form is in the form of force and angle.
- Rectangular form is in the form of horizontal and vertical components.
- Vectors can be converted from polar into rectangular form and then added.
- Resultants can be found as the sum of vectors in rectangular form converted to polar form.
- Trigonometry is a mathematical system where angles are expressed in terms of the ratios of the sides of a triangle.
- A graph is a pictorial representation of a series of quantities.
- Graphs have axes and scales for those axes.

GLOSSARY

A

absolute energy level The energy level in a material that is governed by its absolute temperature (K), mass (kg) and specific energy value.

absolute permittivity In a capacitor, the ratio of the electric flux density to the electric field strength.

1. The absolute permittivity of a capacitor is equal to the product of the permittivity of free space and the relative permittivity of the dielectric used in the capacitor.
2. Absolute permittivity = $\sum_0 \sum_r$

active component An active component is a device that requires an external energy source to operate, along with the circuit current that it is connected to.

actuator That part of a relay or solenoid that converts electrical energy to mechanical movement.

adjustable resistor A resistor whose resistance can change or be changed e.g. with a contact point that can be moved so that the resistance between one end and the contact point can be varied.

air-core inductor An inductor wound in the form of a coil with air or vacuum as the core.

alloy A combination of two or more metals into one material.

Alnico magnet A magnet made from an alloy of aluminium (Al), nickel (Ni) and cobalt (Co): Al-ni-co.

alternating current (a.c.) A current flow which increases to maximum in one direction, decreases to zero, then reverses direction and increases to maximum in the opposite direction; this process is repeated continuously. This is opposed to direct current (d.c.) which flows in one direction only.

alternator A rotating machine for producing an alternating voltage and hence an alternating current (a.c.).

ambient temperature The temperature of the environment of a given location.

- Indoors, ambient temperature can be taken as room temperature.
- In instrument work, Standard Temperature and Pressure (STP) is taken as 25ºC.
- For electrical circuit wiring, the Australian Standard AS/NZS 3000 gives a value of 40°C as normal ambient temperature in Australia and 30° C in New Zealand.

ammeter An instrument for measuring the current (in amperes) flowing in a circuit.

amortisseur or damper windings Windings in the pole face of synchronous machines used to reduce hunting.

ampere

1. The SI unit of electric current that results in a force of 2ΠE–7 Nm force between two parallel conductors.
2. The rate of flow of electrons in a circuit; 1 amp is equal to 1 coulomb per second or 6.24E18 electrons per second.
3. The current flowing in a circuit when 1 volt is applied to a resistance of 1 ohm.

analogue voltmeter A voltmeter in which the voltage is indicated by a physical pointer moving over a calibrated scale.

anodise To deposit an oxide coating on a metal by means of an electrolytic process.

annunciator A panel that indicates the condition of a labelled device in an electrical or mechanical system.

apparent power (S) In an a.c. circuit, the product of the supply voltage and the current. Given the unit VA.

appliances Devices or machines that perform a task by converting electrical energy into mechanical or heat energy etc.

arc, arcing A luminous electric current between two electrodes; the presence of a luminous electrical discharge between two electrodes.

armature

1. The winding of an electrical machine where an EMF is induced.
2. The moving part of an electrical machine, relay or solenoid.

atom An atom is the smallest unit of matter that represents a chemical element.

atomic number The number of protons in the nucleus of an atom; all atomic numbers are whole numbers.

atomic weight The weight in grams of Avogadro's number of atoms; Avogadro's number (or constant) = 6.02E23.

autotransformer A transformer with only one winding, in which part of the winding is common to both the primary and secondary.

auxiliary An extra part of a process or piece of equipment. Auxiliary contacts in a relay are contacts in addition to the main contacts.

B

ballast A resistor, inductor or capacitor used to limit the current in an a.c. circuit e.g. found in the circuit for gas-discharge lighting.

battery A voltage source consisting of a number of cells connected in a series and/or parallel arrangement.

bearings The supports for a rotating shaft.

bleed resistor (*see* discharge resistor) A resistor placed in a circuit so as to draw a current to drain or discharge a capacitor.

block diagram A diagram in which the principal elements of a system are shown in the form of blocks, with their relationship shown by connecting lines.

breakdown torque The maximum torque produced by a motor.

brown-out A condition of low voltage in an electrical installation, experienced during a supply fault, where incandescent lights are dim due to the low voltage and electric motors may stop or overheat.

bypass circuit A circuit connected in parallel to a main circuit so that a portion of the current passes through the bypass circuit rather than the main circuit.

C

capacitance

1. A measure of the energy storage capacity of a capacitor.
2. Capacitance (measured in farads) is the charge in coulombs that can be stored for a given applied voltage.
3. $C = \frac{Q}{V}$

capacitance bridge A bridge network circuit consisting of an a.c. supply, a galvanometer, two resistors and two capacitors, used for determining the value of an unknown capacitor.

capacitive circuit An a.c. circuit in which the capacitive reactance is larger than the inductive reactance. That is, the current leads the voltage.

capacitive effect The ability of some circuits to produce a leading current even if a capacitor is not present in the circuit.

capacitive reactance The reactance offered to an a.c. circuit by a capacitor: $X_c = \frac{1}{(2\Pi fC)}$

capacitor A device for storing electrical energy consisting of two conducting materials separated by a dielectric material (insulator).

capacity (electrical) The capacitance of a capacitor measured in farads.

carbon-compound resistor A resistor that consists of a mixture of carbon particles and a binder, moulded into a particular shape and then baked at high temperature.

catalyst A substance that facilitates a chemical reaction but remains unchanged at the end of the reaction.

cell (electrolytic) An electrochemical single unit d.c. voltage source as used to assemble a battery.

cell (photovoltaic) A semiconductor device that produces energy when exposed to sunlight.

Celsius The unit of temperature measurement, formerly called the centigrade system, named after the Swedish astronomer Anders Celsius.

Centigrade A temperature scale that uses two fixed points: the lower fixed point 0°C is the temperature at which ice melts, while the upper fixed point 100°C is the temperature at which pure water boils. The scale is divided into 100 equal parts between the two points.

charge The electrical property possessed by protons and electrons that produces a force of attraction or repulsion between them.

chemical bonding The force that bonds elemental atoms into molecules.

choke An inductor used to limit the current in an a.c. circuit; usually found in the circuits for gas-discharge lighting. The term 'choke' arose from observations that the inductor appears to reduce (choke) the current flow.

circuit A current path that forms a complete loop including a source, path and load. Circuits may also include control elements such as switches and fuses.

circuit diagram A diagram in which lines are used to represent the circuit conductors, and standard symbols are used to represent the circuit components.

closed circuit A circuit in which all the connections between supply and the load are joined, enabling a current to flow. In a normal circuit, a closed circuit is formed when the control switch is closed.

coercive force The magnetising force required to be applied to a magnetic material, in the opposite direction to the residual magnetism, to reduce the residual magnetism to zero.

commutation

1. In rotating d.c. machines, the process whereby the connections of an armature coil are reversed to convert generated a.c. into d.c., or d.c. supply into a.c. in the armature.
2. In electronics, the switching of currents back and forth between various sections of the circuit, as needed for correct operation.

commutator The part of a d.c. machine that performs the task of commutation.

complex circuit A circuit that cannot be reduced to simple KVL or KCL terms i.e. series or parallel circuit elements.

complex impedance (Z) An impedance consisting of resistance and reactance.

1. The general form of impedance for electrical calculations is: $Z = \sqrt{(R^2 + (XL\text{–}XC)^2)}$
2. In complex maths form, resistance (R) is a real number and reactance (X) is an imaginary number: $Z = R \pm jX$

compound A material that consists of two or more elements combined together chemically, in definite proportions.

compound circuit A circuit consisting of both series and parallel elements.

conduction The transmission of energy through a conductor: electric current through an electrical conductor or heat energy through a thermal conductor.

conductor A material that will pass energy through it from one point to another. From an electrical point of view, conductors are materials that conduct electricity easily.

convection The transmission of heat through liquid, air or a gas by molecular movement of the material due to density differences.

contactor An electrically operated switch by which one circuit can be opened or closed by the opening or closing of an independent circuit, the current in the controlling circuit usually being much smaller than the current in the controlled circuit. Contactors are relays designed to carry larger currents.

cooling The action of removing excess heat from a machine or process.

copper loss The power losses (as heat) in copper conductors due to current flow through their resistance. The copper loss is equal to I^2R. Copper losses are also called I^2R losses.

corrosion The gradual conversion of a metal by a chemical action such as oxidation.

coulomb (C) The quantity of electric charge transferred per second by a current of 1 ampere, nominally equal to 6.24E18 electrons.

coupling

1. The connection of two or more circuits so that energy may be transferred from one to another, e.g. magnetic or capacitive coupling.
2. The connection between the shaft of a motor and the input shaft of a machine.

crest factor The ratio of the maximum value to the rms value of an a.c. waveform; for a sinusoidal waveform, a value of $\sqrt{2}$ or 1.414.

crystalline Made of crystal or formed by the crystallisation of a material.

current The movement of electrons through a conductor.

current paths The different paths that current can take through a circuit.

current-carrying capacity The maximum current which an electric device or cable is rated to carry without excessive overheating and failure.

D

dead short A short circuit with negligible impedance.

decoupling Separating signals from other circuits.

deflection value (f.s.d.) The amount of deflection in the needle of an analogue meter; full scale deflection (f.s.d) occurs when the needle deflects to the maximum reading on the scale of the meter.

deteriorate To become worse in quality.

diac A two-terminal electronic device that will conduct in either direction when a certain 'break-over' voltage is reached.

diamagnetic The property of a material to produce a magnetic field that opposes a magnetic force. When freely suspended in a magnetic field, diamagnetic materials align themselves at right angles to the magnetic field.

dielectric The insulating material that separates the plates of a capacitor and so permits the storage of an electrostatic field; also the property of permittivity.

dielectric constant The ratio by which a dielectric material will increase the charge (on a capacitor) compared to a vacuum.

dielectric heating The heating of an insulating material by placing it between two plates to form a capacitor with the material to be heated acting as the dielectric; a high voltage at high frequency is then applied to the plates, causing the insulating material to heat up due to internal losses.

dielectric strength

1. The property of a material to withstand the forces of an electrostatic field resulting from high voltages; once this threshold is exceeded, the material allows current to pass.
2. The volts per unit thickness necessary to cause breakdown of a dielectric e.g. glass has a dielectric strength of 30 kV/mm.

diesel alternator An alternator that has a diesel engine as the prime mover.

diode A two-terminal (anode and cathode) semiconductor device that will pass current in one direction only.

dipole (magnet)

1. A magnetic field with two poles, north and south, as in the Earth's magnetic field.
2. A pair of electrons forming a magnetic dipole pair.

direct current (d.c.) A current that flows in one direction only; it never changes its polarity.

discharge resistor A resistor connected across a capacitor to discharge the capacitor safely when the circuit is de-energised.

distribution transformer A transformer for stepping down high transmission voltages to voltages suitable for use with household electrical appliances and equipment e.g. a transformer for stepping down from 11 kV to 400/230 V.

drift or electron drift The random movement of electrons through a material.

E

earth fault A fault current that flows to earth as a result of a low impedance path from a live conductor to the general mass of earth.

eddy current A circulating current induced within the core of a conducting material by a varying magnetic field.

effective area (plates) The effective area of the plates of a capacitor, taking into account that the outside plates in a stack are not a part of the capacitor.

Effective Series Resistance (ESR) meter An instrument used to measure the very low internal series resistance of batteries and electrolytic capacitors, in order to establish their health.

efficiency The efficiency of a process or machine is the ratio of the output to the input: $\frac{\text{output}}{\text{input}} \times 100\%$.

electric current The flow of electrons between two points in a circuit due to the presence of an electromotive force (EMF).

electric field The electrostatic field caused by a difference in potential (voltage) between two points.

electrical current flow The movement of electrical charge, also known as 'drift'.

electrical energy Energy stored in an electrostatic or electromagnetic field, also known as an electric charge.

electrical intensity (EI) The intensity of an electric current (measured in amperes) flowing in a circuit.

electrochemical process A process where electrical energy is used to create a chemical effect or where a chemical effect is used to create electrical energy.

electrode The conducting element by which current passes into or out of a device or process.

electrolyte An electrically conductive liquid or paste that allows the passage of current by the dissociation of ions within the material.

electrolytic capacitor A capacitor in which a film of oxide, deposited on a plate, acts as the dielectric.

electromagnet A magnet consisting of a coiled conductor through which a current is passed.

electromagnetic chuck A chuck that uses the holding power of electromagnets to retain magnetic materials firmly in position during a machining process.

electromagnetic clutch A clutch that uses the attractive power of electromagnets to provide the coupling between two rotating shafts.

electromagnetic force (EMF) The force that moves electrons: an electrostatic force resulting from a difference in potential that causes a flow of electrons from the negative potential to the positive potential.

electromotive force (EMF) The force that moves electrons: an electrostatic force resulting from a difference in potential that causes a flow of electrons from the negative potential to the positive potential.

electron The negatively charged particles that form the outer shells of an atom; an electron has a negative charge equal and opposite to that of a proton.

electron current The flow of electrons through a conductor from a negative potential to a positive potential.

electron drift The random movement of electrons through a material.

electron flow The movement of electrons through a conductor from a negative potential to a positive potential.

electronic amplification The increase in amplitude of an electrical signal by an electronic circuit.

electronic ruler An ultrasonic transmitter and receiver that measures the time taken for an ultrasonic sound to travel to a distant surface and return; it then calculates the distance from the known speed of sound.

electroplating The electro-deposition of a metal on the surface of another metal by the process of electrolysis.

electrostatic charge An electric charge on an object due to stationary electrical particles.

electrostatic field A physical field that surrounds electrical charged particles and exerts a force on all other charged particles within the field.

element An element is a pure substance consisting only of atoms that all have the same numbers of protons.

encapsulated (e.g. capacitor) An encapsulated item is completely surrounded by a protective coating of a water-resistant material such as plastic, epoxy or wax.

end cover The covers of a motor, alternator or generator that support the bearings, between which the rotating part is suspended.

energy consumption The amount of energy that a machine or process consumes.

engine-driven alternator An alternator that has an engine as the prime mover.

equilibrium A state in which opposing forces are exactly balanced.

equivalent resistance The amount of resistance that, when connected to a supply, will draw the same total current as a combination of other resistors.

excess The amount by which one quantity exceeds another.

excitation The process of producing a magnetic field by passing a current through the windings of an electric machine.

expand To make larger.

F

farad

1. The unit of measurement of capacitance, named after Michael Faraday.
2. A farad is the capacity that exists between two plates of a capacitor if the transfer of 1 coulomb from one plate to the other creates a potential difference of 1 volt.

fault A failing in a circuit due to either:

- an open circuit in a conductor
- a short circuit between conductors of different potentials; or
- a short circuit between a conductor and earthed equipment.

ferrite A material with ferromagnetic properties made from a compound of different metallic oxides and ceramic materials that has been powdered, compressed and sintered at high temperature.

ferromagnetic Iron and alloys of iron that can be magnetised.

field An area under the influence of electricity or magnetism such as the electric field of a capacitor or the magnetic field of a magnet.

field coil Insulated windings around the field poles of d.c. machines for the purpose of generating an electromagnetic field.

field poles The magnetic poles of d.c. machines, usually fabricated from sheet-steel stampings or pressure-cast from ferro-ceramic materials. Their high permeability provides a high concentration of magnetic flux at a particular position in the machine.

filtering The act of removing unwanted frequencies from a waveform of a current or voltage.

flexible (metal) A metal that has the property of flexibility, such as a spring.

flux density The flux per unit area of an electric, luminous or magnetic field.

flux leakage The flux set up in a magnetic circuit which does not follow the desired path and returns to the source via the surrounding air.

form factor (waveform) The ratio of the rms value to the average value in an a.c. waveform.

force To move an object from one position to another requires an amount of 'force'. Force is defined by the mass of the object multiplied by the change in velocity, or acceleration, of the object.

fossil fuel Any fuel that results from the fossilisation of animal or plant matter under high pressure for long periods; a non-renewable carbon energy resource.

free electron An electron that has escaped the influence of an atom and is free to move away.

frequency The rate at which an instance is repeated e.g. the number of cycles per second of an alternating waveform; 1 cycle per second is called 1 hertz (Hz).

frequency (audio) Any frequency that can be heard by the human ear; usually considered to be between 15 hertz and 20 kilohertz.

friction The mechanical or electrostatic opposition to motion of objects in close contact.

fuel cell A source of EMF; a cell that converts the chemical energy stored in a fuel into electrical energy.

Full Scale Deflection (F.S.D) The maximum capacity of a meter is termed its 'full-scale deflection' (FSD) value.

fuse A protective control device that breaks a circuit when the circuit current exceeds the value for which it is designed.

fusion The joining of two of more materials by melting them together.

G

galvanometer An instrument for measuring small values of electricity calibrated either as a current or as a voltage instrument.

gas turbine A rotating engine that extracts energy from the expansion of a combustible gas impinging on turbine blades.

gaseous In the form of a gas.

gauss

1. The older cgs unit of measurement of magnetic flux density, now replaced by the tesla.
2. The gauss was named after the German mathematician and physicist Carl Friedrich Gauss.
3. 1 tesla is equal to 1E4 gauss.

generator A machine or device that converts one form of energy to another e.g. an electrical generator converts mechanical energy into electrical energy.

global warming is the gradual increase in the overall temperature of the earth's atmosphere and long-term heating of Earth's climate system, as an aspect of climate change, attributed to the greenhouse effect causing increased concentrations of greenhouse gases in the atmosphere.

gravity The force that exists between any two masses e.g. between an object and the general mass of the Earth.

greenhouse gases (GHGs) A gas or gaseous compound (such as carbon dioxide or methane) that absorbs and emits radiant energy (within the thermal infrared range) in Earth's atmosphere, that traps heat causing the greenhouse effect.

H

henry

1. The unit of inductance named after the American scientist Joseph Henry.
2. The inductance of a closed circuit in which an EMF of 1 volt is produced when an electric current flowing in the circuit varies uniformly at the rate of 1 ampere per second.

hunting The periodic swinging of the rotor around the point of equilibrium due to a sudden change in the load on a synchronous motor.

hysteresis Lagging behind; the time delay between a change and the effect that it causes.

1. The delay between the change to a system and the resulting output.
2. The amount by which the magnetic flux density lags behind the magnetising force in a magnetic circuit or material.

hysteresis curve The S-shaped curve, also known as the BH curve, of a magnetic material which results from the residual flux remaining after the magnetising force has been removed.

I

impedance The opposition to current flow in an a.c. circuit.

in phase (phasor) The relationship between two waveforms of the same frequency such that they appear in the same angular position in a phasor diagram.

in series The linear connection of circuit components (one after the other) so that there is only a single path for the current through the circuit.

induced field A magnetic field set up by current flowing through a conductor.

induced voltage A voltage set up due to the movement of a magnetic field relative to a conductor.

inductance The property of a circuit that enables an EMF to be induced in it.

induction The setting up of an electric or magnetic field by proximity to another electric or magnetic field.

inductive reactance The impedance offered to an a.c. circuit by an inductor: $X_L = 2\pi fL$

inductor A conductor, wound in the shape of a coil, that possesses the property of inductance.

inferred zero method A method of determining the temperature or resistance of a copper conductor based on the inferred temperature at which the resistance of copper is zero ohms.

ion An atom or molecule that has gained or loss an electron, i.e. has gained a negative or positive charge.

insulation Materials of high dielectric strength that oppose the conduction of current.

insulation resistance The resistance between live conductors, or between live conductors and earth, of an electrical circuit or electrical equipment.

insulation resistance (IR) meter A device that measures the total resistance between any two points separated by electrical insulation and therefore determines how effective the insulation is in resisting current flow.

insulator

1. A non-conducting material in which the electrons are tightly bound to the nucleus of the atom, or within chemical bonds, so that they are not free to move.
2. In high-voltage circuits, the device that insulates the live conductors from the supporting structure.

integer A whole number, without decimal or fractional parts.

International Protection (IP) rating The rating given to the degree of protection of electrical equipment, in accordance with *AS 60529*.

inversely proportional A relationship between two values where one value is proportional to the inverse of the other as one increases the other decreases and visa-versa.

ionisation A process whereby a previously neutral atom or molecule has lost or gained an electron.

iron armature The moving part of an electric machine, relay or solenoid that is manufactured from iron.

iron losses The losses in the iron core of an electrical machine due to magnetic hysteresis and eddy currents.

J

joule (J)

1. The unit of energy and work in the SI system of measurement, named after James Prescott Joule.
2. The work required to move an electric charge of 1 coulomb through an electrical potential difference of 1 volt.
3. The work required to produce 1 watt of power for 1 second.
4. The amount of work done or energy expended when 1 newton is applied over a distance of 1 metre.

Joule's law The law discovered by James Prescott Joule. It states that the heat energy produced by an electric current is equal to the current squared, multiplied by the resistance of the circuit, multiplied by the time in seconds for which the current flows: $H = I^2Rt$.

K

kilowatt hour (kWh) The everyday unit of energy, traditionally used to measure electric energy when large rates of energy use are extended over hours or days; 1 kilowatt hour = 3,600,000 joules or 3.6 MJ.

Kirchhoff's Current Law (KCL) Kirchhoff's Current Law states that the algebraic sum of the current entering a junction is equal to the algebraic sum of the current leaving a junction: $\sum I_{in} = \sum I_{out}$

Kirchhoff's Voltage Law (KVL) Kirchhoff's Voltage Law states that the algebraic sum of the voltage drops around a series circuit equals the applied voltage: $\sum_E = \sum_V$ or $V_{Total} = V_1 + V_2 + V_3 + \ldots$

L

lagging

1. The displacement, in electrical degrees, between two waveforms, where the peak of one waveform occurs after (lags) the other.
2. On a phasor diagram, a phasor that lags behind another when using conventional anticlockwise rotation.

laminated core The core of an electromagnetic circuit such as an electric motor, alternator or transformer which has been laminated from thin sheets of a ferrous material.

lamination Using a thin sheet of a ferrous material such as silicon steel to make the core of an electromagnetic circuit for an electric motor, alternator or transformer.

LCR bridge An instrument based on a bridge circuit for measuring inductance, capacitance and resistance against known standard components.

leading

1. When the peak of one waveform occurs before the peak of another waveform, that phasor is said to be leading the other.
2. On a phasor diagram, a phasor that is leading another when using conventional anticlockwise rotation.

leakage flux The magnetic flux set up in a magnetic circuit that does not follow the desired path and returns to the source via the surrounding air.

Lenz's Law

1. Named after the Russian physicist Heinrich Lenz, this law relates to electromagnetic induction and is the electrical counterpart of Newton's third law which states that action and reaction are equal and opposite.
2. Lenz's Law states that the direction of an induced EMF is such that the resulting current flow produces a magnetic field which tends to oppose the original motion causing the induced EMF.

lethal voltage A voltage high enough to cause death by electric shock due to the passage of a lethal value of current flow.

light emitting diode (LED) A semiconductor device that emits light when forward biased.

light energy Radiant energy within the visible spectrum; wavelengths between 380 and 760 nm.

limit of travel The maximum distance that the moving part of an instrument or machine can travel before its movement is prevented or stopped.

load (circuit)

1. The power used by a circuit or machine when used for the intended purpose.
2. The machine or device connected to a circuit specifically to convert energy from one form into another.

local equilibrium A balance of forces or chemicals in one place but not necessarily throughout a system.

loops Any current path which forms a complete loop that conforms to the requirements of KVL.

low-volt relay (see brown-out) A relay which de-energises at a predetermined low value of voltage, to ensure that damage to electrical equipment due to low voltage cannot occur.

M

magnetic chuck A chuck that uses the holding power of magnets to retain magnetic materials firmly in position during a machining process.

magnetic circuit The path of the magnetic flux as it progresses from the north pole to the south pole of a magnetic source.

magnetic core A core of magnetic material in an electromagnetic device such as a transformer, electric motor, relay or solenoid intended to reduce the reluctance of the magnetic circuit.

magnetic field The extent of the magnetic flux around a magnet or magnetic circuit.

magnetic flux The magnetic flux established by a magnetic circuit or magnet.

magnetic fringing The tendency of magnetic flux to spread out when crossing an air gap, causing the flux density in the air gap to be less than in the magnetic material on either side of the air gap.

magnetic induction The process whereby the molecules of a magnetic material align themselves with the magnetic field of another magnet, thus becoming a temporary magnet.

magnetic leakage The flux set up in a magnetic circuit which does not follow the desired path and returns to the source via the surrounding air.

magnetic losses The magnetic flux set up in a magnetic circuit which does not link with the desired elements in the circuit.

magnetic relay A switch controlled by a solenoid with a fixed iron core and a movable armature, which is attracted to the iron core when the solenoid is energised and held away by spring action when the solenoid is de-energised.

magnetic reluctance The opposition of a material to being magnetised.

magnetic saturation A stage of magnetisation in a magnetic circuit where an increase in the magnetising force (H) will have a negligible effect on the flux density (B) of the magnetic circuit.

magnetically hard Ferromagnetic materials whose magnetic properties are difficult to alter.

magnetically soft Ferromagnetic materials whose magnetic properties are easily altered.

magnetism A property of certain materials by which they can exert a mechanical force on other magnetic materials.

magneto-motive force (MMF)

1. The force causing magnetic flux to circulate around a magnetic circuit.
2. The MMF of a current-carrying coil is equal to the product of the current and the number of turns in the coil: $F_m = IN$

mass The reluctance of a body to have its state of motion changed; in SI units, force = mass X acceleration (F = ma).

matter Any physical substance which is affected by gravity.

megger The Megger brand insulation resistance tester was originally manufactured by Evershed & Vignoles Limited in the UK. Because of its widespread use in the Australian electrical industry, the term 'megger' is taken to mean both:

1. an insulation resistance tester
2. to test the insulation resistance of an electrical circuit or appliance.

megohm One million ohms (1E6 Ω).

metal-film resistor A resistor made by placing a thin layer of metal-film compound, in spiral form, on a ceramic former.

meter (measuring device) An instrument for measuring a value of voltage, current or resistance etc.

microtherm device A small semi-conductor device for detecting temperature changes in electrical equipment.

molecule The smallest part of matter that can be identified by the properties of that material.

multimeter An instrument designed to measure electric current, voltage, and usually resistance, typically over several ranges of value

Mu-metal A high-permeability alloy consisting of 77% nickel and 15% iron with copper and molybdenum; Mu-metal is used for shielding magnetic fields.

mutual inductance The inductance between two current-carrying conductors where the magnetic field of each one links with the magnetic field of the other.

N

negative (-ve) charge

1. The charge on an electron.
2. Any object which has an excess of electrons has a negative charge.

negative temperature coefficient (of resistance) A material whose resistance decreases with a rise in temperature has a negative temperature coefficient.

neutral Neither positive nor negative.

neutron An elementary part of the nucleus of an atom with no electric charge.

newton metre (Nm) One newton metre is the torque developed by a shaft when a force of 1 newton is applied at a radius of 1 metre.

node A junction point in a circuit where two or more components are joined.

non-inductive resistance (see pure resistance) A wire-wound resistor that has been wound in such a way so as to have negligible inductance.

non-renewable Energy sources that will eventually be depleted on Earth.

no-volt relay

1. A safety relay designed to de-energise when the supply voltage is removed from a circuit, so as to prevent the circuit from functioning when the supply voltage is restored.
2. No-volt relays are used with unattended electrical machines to prevent their sudden starting after a power-supply failure.

noxious (fumes) Harmful fumes.

nuclear energy The energy released from a nuclear reaction.

nuclear reactor An energy source that can sustain nuclear fission in a continuous manner.

nucleus The core of an atom consisting of protons and neutrons; hydrogen has only one proton and no neutrons.

O

Ohm's Law (V = IR) Named after George Ohm who first proposed the theory, Ohm's Law states:

- The current flowing between any two points in an electric circuit is directly proportional to the potential difference between these points, and also inversely proportional to the resistance of the circuit between these points.

ohmic materials When connected in an electric circuit, ohmic materials will pass a current, the value of which can be determined by Ohm's Law.

ohmmeter A meter for measuring the resistance of a circuit (in ohms).

open circuit An electric circuit in which the current is prevented from flowing by a break in the circuit.

open circuit characteristic (OCC) shows the relationship between generated EMF at no load and the field current at a given fixed speed.

orbit (atomic) The path of an electron around the nucleus of an atom.

out of phase (phasor) The relationship between two waveforms that are occurring simultaneously and are at different angular positions in a phasor diagram.

overload relay A relay designed to operate at a predetermined value of overload current and so cause the disconnection of an overloaded device from the supply.

P

parallel circuit A circuit in which the same voltage appears across each of two or more alternate current paths for the current.

parallax error The apparent change in the position of an object, resulting from a change in the position from which it is viewed. In an analogue instrument, parallax error occurs when the viewer is not directly in line with the pointer on the dial.

particle (atomic) Parts of an atom that make up the atom.

path (circuit) The path taken by an electric current in completing one full circuit.

passive component A passive component is a device that does not require energy to operate, except for the available circuit current that it is connected to.

Periodic Table The periodic table is a table containing all the chemical elements, arranged by atomic number, electron configuration, and chemical properties.

period (waveform) The time taken for one cycle of an alternating waveform.

periodic function A function that repeats itself periodically, at fixed points in time.

periodic time (waveform) The time taken for one cycle of an alternating waveform; equal to the inverse of the frequency: $T_p = \frac{1}{f}$

permanent magnet A magnet made of a material that enables it to maintain its magnetic properties over an extended period of time.

permeability The ease with which a magnetic material can be magnetised.

permittivity The ease with which an electrostatic field can be set up in a material.

phase angle

1. The angular displacement, in electrical degrees, between two waveforms of the same frequency.
2. The angle, in electrical degrees, between two phasors on a phasor diagram.

phasor A line that, by its length and direction, represents the magnitude and phase relationships of a.c. quantities on a Cartesian plane; phasors represent rotating vectors.

photoelectric The generation of an EMF by exposure to light.

photons A quantity of electromagnetic energy (packet) that exhibits both particle and wave behaviour.

photoresistor

1. A resistor made from semiconductor material which changes its resistance with a change in its level of exposure to light.
2. An LDR (light dependent resistor) whose resistance reduces considerably when exposed to sunlight.

physiological Having an effect on the body, human or animal.

piezoelectric effect

1. The production of an EMF across the faces of a material (usually crystalline) by the application of a mechanical force.
2. The production of a mechanical movement (force) across the faces of a material (usually quartz) by the application of a voltage.

plasma

1. The fourth state of matter: highly energised with freely moving ions and electrons.
2. Plasma is present in lightning, welding arcs and fluorescent lamps.

plunger solenoid A coil with a moveable core (plunger) that moves when current flows through the coil, providing a linear mechanical force.

polarised relay A relay that is designed to operate only when the correct polarity is applied to its terminals; it will only operate when the current flows in the correct direction between its terminals.

polarity A condition of having two opposing poles, such as north and south for a magnetic circuit or positive and negative for an electric circuit.

porous An apparently solid material that has many small holes, allowing the passage of a liquid or gas.

positive (+ve) charge

1. A condition where a charge exists that has a deficiency of electrons.
2. A charge that attracts electrons (negative charges).

positive temperature coefficient (of resistance) A material whose resistance increases with a rise in temperature

potential The difference in the EMF, measured in volts, between any two points.

potential difference (PD) The difference in the EMF, measured in volts, between any two points.

potentiometer A variable resistor set up so that the voltage at its variable pointer is some fraction of the voltage between its ends; a voltage divider circuit.

power factor

1. The factor by which the apparent power in an a.c. circuit must be multiplied to obtain the true power consumed by the circuit.
2. The ratio of the true power to the apparent power in an a.c. circuit.

powered coil (*see* primary coil) The winding of a transformer connected to the supply voltage.

practical inductor An inductor with characteristics that make it possible to manufacture.

precipitation

1. Small particles of smoke, dust or moisture etc falling in the atmosphere.
2. The separation of substances in a solution.

pressure

1. The force per unit area.
2. The SI unit of pressure is the pascal; 1 pascal is the force exerted when a force of 1 newton is applied to an area of 1 square metre.

primary coil (primary winding) The winding of a transformer that is connected to the supply voltage.

prime mover A machine that transforms energy from one form of energy into mechanical form; typically the driving force for a generator or an alternator.

prime mover The driving force for an alternator or generator. Examples include petrol and diesel engines, and steam or water driven turbines.

printed circuit motor The printed circuit motor has no iron in the core. The rotor windings are printed or stamped onto a thin fibreglass disc that rotates in the air gap between pairs of permanent magnets. Mainly used for small control applications.

programmable unijunction transistor (PUT) A low-current, four-layer semi-conductor device similar to an SCR except that it has an anode gate rather than a cathode gate.

proton A positively charged particle in the nucleus of an atom that has a charge equal but opposite to that of an electron.

pure capacitance A theoretical capacitance that has no resistance or inductance.

pure inductance A theoretical inductance that has no resistance or capacitance.

pure resistance A theoretical resistance that has no inductance or capacitance.

R

radiation The propagation of energy radial from the source e.g. in the form of electromagnetic waves.

rare earth materials Ferromagnetic materials such as samarium-cobalt and neodymium that are used to make magnets with special high-strength properties.

reactance The opposition, measured in ohms, to the flow of current through the capacitive or inductive elements of an a.c. circuit; the symbol for reactance in an a.c. circuit is X.

reactive power

1. The product of the line voltage and that portion of the line current that does not consume power.
2. Reactive power Q = VI sinϕ, measured in VAR

reduction (current) The removal of oxidation in order to return an oxidised metal to its elemental form such as iron, copper or aluminium.

reference phasor In a phasor diagram, the phasor which all angular measurements are referenced to.

relative permeability

1. The ratio of the permeability of a magnetic material to the permeability of free space.
2. The symbol for relative permeability is μr.

relay An electrically operated switch by which one circuit can be opened or closed by the opening or closing of an independent circuit, the current in the controlling circuit usually being much smaller than the current in the controlled circuit.

renewable Energy sources that can all be replenished or come from a source that won't run out in our life-time.

resistivity A measure of the resisting power of a specified material to the flow of an electric current.

residual magnetism That portion of magnetic flux that remains in a ferromagnetic material when the magnetising force is removed.

resistance (resistivity) The opposition to the current flow, measured in ohms, in a d.c. circuit or a purely resistive a.c. circuit.

resistive circuit

1. A circuit consisting only of resistive elements.
2. A circuit that has no capacitance or inductance.

resistor A circuit element that is designed to provide opposition to current flow.

resonance A condition in an a.c. circuit in which the inductive reactance exactly equals the capacitive reactance.

rheostat A variable resistor with only two terminals.

rotor The rotating part of an electrical machine.

S

saturate To saturate the core of a magnetic circuit is to reach a state where an increase in the magnetising force (H) will have a negligible effect on the flux density (B).

scalar Is a physical quantity that can be described by its magnitude, e.g. volume, energy, speed, and mass.

secondary coil (winding) Windings other than the primary winding.

self-inductance The term 'self-inductance' is used when a conductor or a coil has a voltage induced in it by its own magnetic field which, according to Lenz's Law, limits the rate of change of the current.

semiconductor A material with four valence electrons in the outermost layer or shell; semiconductor materials, in their pure intrinsic state, are neither good conductors nor good insulators.

series circuit A circuit in which there is only one path for the current flow when it is moving from the higher potential terminal to the lower potential terminal.

servo motor A motor used in a servo system that is controlled by an amplified control signal.

shaft The axial spindle on which torque is applied in a motor or generator.

shell (atom) The three-dimensional area in which the electrons orbit around the nucleus of an atom.

shells The electrons in an atom are positioned in shells that surround the nucleus. The shells are like rings the orbit the nucleus, with each additional shell being a greater distance from the nucleus.

short circuit A fault of negligible impedance between two points in a circuit, resulting in excessive current flow between the two points; the excessive current usually causes damage to the circuit.

shorted A path of negligible impedance between two points in a circuit.

shunt A parallel circuit.

simmerstat A device for controlling the current applied to a circuit by varying the 'mark-space-ratio' of the current.

simple circuit A circuit with only one supply voltage, one path and one load.

sine wave Abbreviation for sinusoidal waveform.

single-phase motor An electric motor designed to operate from a single-phase supply.

single-phase transformer A transformer designed to operate from a single-phase supply.

sinusoidal voltage An a.c. voltage with a sinusoidal waveform.

sinusoidal waveform A waveform whose amplitude is proportional to the sine of an angle from 0–360°.

slip-rings Rings attached to the shaft which are designed to carry current from the rotating section to the stationary section of an alternator, or from the stationary section to the rotating section of a motor.

slip speed The difference between the synchronous speed of the rotating field and the actual speed of the rotor.

slug A ferrite core that can be moved into and out of the central part of a coil, for tuning purposes.

snubber A circuit that allows the energy in an inductive circuit to pass safely around the inductor.

solar cell A photovoltaic cell that converts light or other radiant energy directly into electrical energy.

solder The joining of two surfaces or conductors by melting a lower melting point alloy, usually lead/tin, between the two.

solenoid An electromagnet, usually in coil form, with an armature or plunger that moves when the electromagnet is energised.

solenoid coil The coil of wire that forms the electromagnet of a solenoid.

solid core The core of an electromagnetic device that consists of a solid metal form; that is, the core is not laminated.

solid-state relay (SSR) A relay, the operational components of which consist of semi-conductor elements.

source (circuit) The supply voltage element of a circuit.

spindle The axial shaft on which torque is applied in a motor or generator.

split-phase motor A single-phase squirrel cage motor having two separate windings (start and run) with the currents displaced electrically, resulting in a phase displacement between the two fluxes to produce a rotating field.

squirrel cage motor An induction motor with a rotor whose basic construction resembles a cage.

specific heat capacity The quantity of energy required to raise the temperature of a mass of 1 kilogram of that material through 1 kelvin.

state (of matter) Matter can exist as a solid, a liquid, a gas or plasma, these being known as the four states of matter.

static Staying still.

1. Forces in equilibrium.
2. Electrical discharges in the atmosphere resulting from the release of static electricity which interfere with good reception of AM radio.

static electricity The charge set up by stationary particles of ions and electrons, produced by friction, heat or pressure.

stator

1. The windings of an induction motor that generate the rotating field.
2. The windings in an alternator where the EMF is generated.

steam turbine A turbine whose rotational force is produced by the pressure of steam on a set of blades.

stepper motor A small motor in which digital pulses are converted to rotational movement of the rotor.

stress

1. Force per unit area; stress $= \frac{\text{force}}{\text{area}} = \frac{\text{newtons}}{\text{square metre}}$.
2. Stress is measured in newtons per metre^2.

substation A group of electrical apparatus, usually at high voltages, designed to change the level of voltage and control its distribution to specific areas.

superconductor A conductor that exhibits zero opposition to current flow at some very low temperature.

switch A device for controlling the flow of current in a circuit by opening or closing the circuit.

synchronous capacitor An a.c. generator run at synchronous speed with a light load, with its excitation adjusted so that it draws current at a leading power factor. Used for power factor correction.

synchronous motor An a.c. motor designed to run at the synchronous speed of the supply; a two-pole a.c. motor will run at 3000 rpm when connected to a 50 Hz supply.

synchronous speed The synchronous speed is the speed of the revolution of the magnetic field in the stator winding of the motor. It is determined by the frequency and the number of magnetic poles of the motor.

T

tariff A table or rate of charges for electrical energy.

tesla (T)

1. The unit of magnetic flux density.
2. 1 tesla is equal to 1 weber per square metre.

thermal conductivity The rate of flow of heat through a material.

thermionic valve An electronic device with a heated cathode that relies on the emission of electrons by heat.

thermistor A resistor made from a semiconductor material that changes its resistance with a change in temperature.

thermocouple A junction of two dissimilar metals placed at the point where the temperature is to be measured; the junction generates millivolts of potential difference proportional to temperature over a limited range at the open end

thermoelectric The production of electricity by the action of heat on a thermocouple.

thermonuclear Relating to the heat released in a nuclear reaction.

thermopile A number of thermocouples connected in series, to give a larger output voltage, when measuring temperature.

three-phase circuit A circuit supplied by three voltages of the same value but different in phase by 120 electrical degrees.

three-phase motor An electric motor designed to be connected to a three-phase a.c. supply.

three-phase transformer A transformer designed to be connected to three-phase a.c. supply.

thyristor A bistable semi-conductor device, with three or more P-N junctions that can be switched to the on-state or the off-state. The thyristor family includes SCRs, triacs, GTOs and PUTs.

time constant In an inductive circuit, the time in seconds that the current takes to reach 63.2% of its final value; the time constant for an inductive circuit $\tau = \frac{L}{R}$. In a capacitive circuit, the time constant is equal to CR.

tolerance (resistor) single-phase a.c. The maximum percentage deviation from the stated value.

torque A force that tends to produce rotational motion. Torque is measured in newton metres (Nm)

toroid A coil wound around a ring-type core.

toroidal core A ring-shaped core for a coil.

tractive The type of force exerted by a magnet on ferromagnetic material.

transducer

1. A sensor/detector device that converts one type of energy into another.
2. In electric circuits, the conversion is usually from mechanical, magnetic, pressure, photovoltaic, etc., to electrical.

transformer

1. An electrical device for changing the value of an alternating voltage.
2. A step-up transformer has a larger output voltage than the input voltage, while a step-down transformer has a lower output voltage than the input voltage.

transformer effect The effect of changing an alternating voltage to a different value.

transformer losses The iron and copper losses in a transformer. The iron losses are due to eddy currents and magnetic hysteresis while the copper losses are due to the heating effect of an electric current ($H = I^2Rt$).

transistor A three-terminal semiconductor device that can perform the actions of switching and amplification.

transmission The action of sending electrical energy over distances, usually by means of high-voltage lines.

TRIAC A three-terminal solid-state bidirectional rectifier that functions in the same manner as two SCRs connected in inverse parallel. A triac can be triggered in either forward or reverse conduction by a pulse applied to its gate electrode.

trimmer or trimpot resistor Small resistors that are adjusted by using a tool such as a screwdriver

true power

1. The power consumed by the purely resistive components of an a.c. circuit.
2. In an a.c. circuit, the square of the current through the resistive component multiplied by the value of the resistive component.
3. In an a.c. circuit, the product of the supply voltage, the current and the cosine of the phase angle between the voltage and the current. Measured in Watts (W)

turbine An engine, the rotational torque of which is supplied by the expansion of steam, gas or air or the pressure and velocity of water against a set of blades.

U

ultrasonic Above the frequency of sound; a frequency that is higher than the audible frequencies.

unijunction transistor (UJT) A three-terminal semi-conductor device (Base 1, Base 2 and emitter) which has a negative-resistance characteristic between its emitter and base 1 terminals.

universal motor A series connected motor that can run on both d.c and single phase a.c.

V

valency ring (orbit) The outermost shell of orbiting electrons in an atom.

valence electrons Electrons contained within the valence shell. The number of valence electrons determines an atom's ability to gain or lose an electron.

valence shells The atomic shell furthest from an atom's nucleus. The number of electrons it contains determines an atom's ability to gain or lose electrons.

Van de Graaff generator A machine that produces high-voltage static electricity by the action of a motor-driven rubber belt rubbing lightly against a comb.

vaporisation The action of changing from a liquid state to a gaseous state.

variable capacitor A capacitor with a means of adjusting the value of capacitance.

variable resistor A resistor with a means of adjusting the value of resistance.

Variac An autotransformer that has a toroidal winding and an adjustable carbon-brush so that the value of the output voltage can be adjusted to any value between zero and the input voltage.

varistor (voltage-dependent resistor (VDR)) A resistor made from semiconductor material that changes its resistance with a change in the value of its applied voltage.

vector A measurement value that has both magnitude and direction.

velocity A vector quantity that has both magnitude (speed) and direction (against a frame of reference).

ventricular fibrillation This is a state of rapid, uncoordinated muscle spasms. Blood stops flowing around the body and in around four minutes the brain starts dying

voltage

1. The electrical force that causes current to flow in a circuit.
2. The general term for electromotive force.

voltage comparator A device that compares two input voltages and produces different output voltages depending on the relative values of the inputs.

voltage drop (V_d) The potential difference, measured in volts, between two points in a circuit.

voltmeter An instrument for measuring the value of electromotive force in volts.

W

Watt The SI units for power and is equal to one joule of work performed per second.

wattless power (see apparent power)

1. The product of the line voltage and that portion of the line current that does not consume power.
2. Reactive power Q = VI sin φ
3. Wattless power is measured in volt-amperes reactive (VARs).

wattmeter An instrument for measuring the true power consumed by a circuit.

weber (Wb) The SI unit of magnetic flux; 1 weber equals 1E8 magnetic lines of force.

Wheatstone bridge A null-type measuring circuit in which a resistance is measured by comparison with a standard resistance value.

white goods Large heavy appliances that usually connect to a power point but do not lend themselves to being moved about.

Wimshurst machine A machine that produces high-voltage electricity by the action of sliding contacts connected to two parallel rotating plates rotating in opposite directions.

windage The losses in an electrical machine due to the energy consumed by the rotation of the cooling fans and other rotating components.

winding A number of turns of a conductor, wound around a core or former for the purpose of producing a magnetic field in an electrical machine such as a transformer, motor or alternator.

wind generator An electric generator that obtains its rotational torque from the action of the wind on a set of blades.

wire-wound resistor

1. A resistor consisting of a ceramic former over which resistance wire has been wound.
2. Wire-wound resistors are used for higher wattage values where the dissipation of heat is important.

working voltage (of capacitor) The recommended maximum voltage that should be continuously applied to a capacitor.

Z

zero voltage switching (ZVS) The technique of switching a supply on or off by only switching at the instant that the a.c. supply passes through zero.

INDEX

A

E

F

G

H

I

J

K

L

M

N

S

T

U

V

W